CARBON DIOXIDE CAPTURE AND STORAGE

This Intergovernmental Panel on Climate Change (IPCC) Special Report provides information for policymakers, scientists and engineers in the field of climate change and reduction of CO_2 emissions. It describes sources, capture, transport, and storage of CO_2. It also discusses the costs, economic potential, and societal issues of the technology, including public perception and regulatory aspects. Storage options evaluated include geological storage, ocean storage, and mineral carbonation. Notably, the report places CO_2 capture and storage in the context of other climate change mitigation options, such as fuel switch, energy efficiency, renewables and nuclear energy.

This report shows that the potential of CO_2 capture and storage is considerable, and the costs for mitigating climate change can be decreased compared to strategies where only other climate change mitigation options are considered. The importance of future capture and storage of CO_2 for mitigating climate change will depend on a number of factors, including financial incentives provided for deployment, and whether the risks of storage can be successfully managed. The volume includes a *Summary for Policymakers* approved by governments represented in the IPCC, and a *Technical Summary*.

The IPCC Special Report on *Carbon Dioxide Capture and Storage* provides invaluable information for researchers in environmental science, geology, engineering and the oil and gas sector, policymakers in governments and environmental organizations, and scientists and engineers in industry.

IPCC Special Report on
Carbon Dioxide Capture and Storage

Edited by

Bert Metz **Ogunlade Davidson** **Heleen de Coninck**

Manuela Loos **Leo Meyer**

Prepared by Working Group III of the
Intergovernmental Panel on Climate Change

Published for the Intergovernmental Panel on Climate Change

CAMBRIDGE
UNIVERSITY PRESS

CAMBRIDGE UNIVERSITY PRESS
Cambridge, New York, Melbourne, Madrid, Cape Town, Singapore, São Paulo

Cambridge University Press
40 West 20th Street, New York, NY 10011–4211, USA

Published in the United States of America by Cambridge University Press, New York

www.cambridge.org
Information on this title:www.cambridge.org/9780521863360

First published 2005

Printed in Canada

A catalogue record for this publication is available from the British Library

ISBN-13 978-0-521-86643-9 hardback
ISBN-10 0-521-86643-X hardback

ISBN-13 978-0-521-68551-1 paperback
ISBN-10 0-521-68551-6 paperback

Please use the following reference to the whole report:
IPCC, 2005: IPCC Special Report on Carbon Dioxide Capture and Storage. Prepared by
Working Group III of the Intergovernmental Panel on Climate Change [Metz, B.,
O. Davidson, H. C. de Coninck, M. Loos, and L. A. Meyer (eds.)]. Cambridge University
Press, Cambridge, United Kingdom and New York, NY, USA, 442 pp.

Cover image: Schematic of geological storage options (Courtesy CO2CRC).

Contents

Foreword

The Intergovernmental Panel on Climate Change (IPCC) was jointly established by the World Meteorological Organization (WMO) and the United Nations Environment Programme (UNEP) in 1988. Its terms of reference include: (i) to assess available scientific and socio-economic information on climate change and its impacts and on the options for mitigating climate change and adapting to it and (ii) to provide, on request, scientific/technical/socio-economic advice to the Conference of the Parties (COP) to the United Nations Framework Convention on Climate Change (UNFCCC). From 1990, the IPCC has produced a series of Assessment Reports, Special Reports, Technical Papers, methodologies and other products that have become standard works of reference, widely used by policymakers, scientists and other experts.

At COP7, a draft decision was taken to invite the IPCC to write a technical paper on geological storage of carbon dioxide[a]. In response to that, at its 20th Session in 2003 in Paris, France, the IPCC agreed on the development of the Special Report on Carbon dioxide Capture and Storage.

This volume, the Special Report on Carbon dioxide Capture and Storage, has been produced by Working Group III of the IPCC and focuses on carbon dioxide capture and storage (CCS) as an option for mitigation of climate change. It consists of 9 chapters covering sources of CO_2, the technical specifics of capturing, transporting and storing it in geological formations, the ocean, or minerals, or utilizing it in industrial processes. It also assesses the costs and potential of CCS, the environmental impacts, risks and safety, its implications for greenhouse gas inventories and accounting, public perception, and legal issues.

As is usual in the IPCC, success in producing this report has depended first and foremost on the knowledge, enthusiasm and cooperation of many hundreds of experts worldwide, in many related but different disciplines. We would like to express our gratitude to all the Coordinating Lead Authors, Lead Authors, Contributing Authors, Review Editors and Expert Reviewers. These individuals have devoted enormous time and effort to produce this report and we are extremely grateful for their commitment to the IPCC process. We would like to thank the staff of the Working Group III Technical Support Unit and the IPCC Secretariat for their dedication in coordinating the production of another successful IPCC report. We are also grateful to the governments, who have supported their scientists' participation in the IPCC process and who have contributed to the IPCC Trust Fund to provide for the essential participation of experts from developing countries and countries with economies in transition. We would like to express our appreciation to the governments of Norway, Australia, Brazil and Spain, who hosted drafting sessions in their countries, and especially the government of Canada, that hosted a workshop on this subject as well as the 8th session of Working Group III for official consideration and acceptance of the report in Montreal, and to the government of The Netherlands, who funds the Working Group III Technical Support Unit.

We would particularly like to thank Dr. Rajendra Pachauri, Chairman of the IPCC, for his direction and guidance of the IPCC, Dr. Renate Christ, the Secretary of the IPCC and her staff for the support provided, and Professor Ogunlade Davidson and Dr. Bert Metz, the Co-Chairmen of Working Group III, for their leadership of Working Group III through the production of this report.

Michel Jarraud
Secretary-General,
World Meteorological Organization

Klaus Töpfer
Executive Director,
United Nations Environment Programme and
Director-General,
United Nations Office in Nairobi

---

[a] See http://unfccc.int, Report of COP7, document FCCC/CP/2001/13/Add.1, Decision 9/CP.7 (Art. 3.14 of the Kyoto Protocol), Draft decision -/CMP.1, para 7, page 50: "*Invites* the Intergovernmental Panel on Climate Change, in cooperation with other relevant organisations, to prepare a technical paper on geological carbon storage technologies, covering current information, and report on it for the consideration of the Conference of the Parties serving as the meeting of the Parties to the Kyoto Protocol at its second session".

Preface

This Special Report on Carbon dioxide Capture and Storage (SRCCS) has been prepared under the auspices of Working Group III (Mitigation of Climate Change) of the Intergovernmental Panel on Climate Change (IPCC). The report has been developed in response to an invitation of the United Nations Framework Convention on Climate Change (UNFCCC) at its seventh Conference of Parties (COP7) in 2001. In April 2002, at its 19th Session in Geneva, the IPCC decided to hold a workshop, which took place in November 2002 in Regina, Canada. The results of this workshop were a first assessment of literature on CO_2 capture and storage, and a proposal for a Special Report. At its 20th Session in 2003 in Paris, France, the IPCC endorsed this proposal and agreed on the outline and timetable[b]. Working Group III was charged to assess the scientific, technical, environmental, economic, and social aspects of capture and storage of CO_2. The mandate of the report therefore included the assessment of the technological maturity, the technical and economic potential to contribute to mitigation of climate change, and the costs. It also included legal and regulatory issues, public perception, environmental impacts and safety as well as issues related to inventories and accounting of greenhouse gas emission reductions.

This report primarily assesses literature published after the Third Assessment Report (2001) on CO_2 sources, capture systems, transport and various storage mechanisms. It does not cover biological carbon sequestration by land use, land use change and forestry, or by fertilization of oceans. The report builds upon the contribution of Working Group III to the Third Assessment Report Climate Change 2001 (Mitigation), and on the Special Report on Emission Scenarios of 2000, with respect to CO_2 capture and storage in a portfolio of mitigation options. It identifies those gaps in knowledge that would need to be addressed in order to facilitate large-scale deployment.

The structure of the report follows the components of a CO_2 capture and storage system. An introductory chapter outlines the general framework for the assessment and provides a brief overview of CCS systems. Chapter 2 characterizes the major sources of CO_2 that are technically and economically suitable for capture, in order to assess the feasibility of CCS on a global scale. Technological options for CO_2 capture are discussed extensively in Chapter 3, while Chapter 4 focuses on

methods of CO_2 transport. In the next three chapters, each of the major storage options is then addressed: geological storage (chapter 5), ocean storage (chapter 6), and mineral carbonation and industrial uses (chapter 7). The overall costs and economic potential of CCS are discussed in Chapter 8, followed by an examination of the implications of CCS for greenhouse gas inventories and emissions accounting (chapter 9).

The report has been written by almost 100 Lead and Coordinating Lead Authors and 25 Contributing Authors, all of whom have expended a great deal of time and effort. They came from industrialized countries, developing countries, countries with economies in transition and international organizations. The report has been reviewed by more than 200 people (both individual experts and representatives of governments) from around the world. The review process was overseen by 19 Review Editors, who ensured that all comments received the proper attention.
In accordance with IPCC Procedures, the Summary for Policymakers of this report has been approved line-by-line by governments at the IPCC Working Group III Session in Montreal, Canada, from September 22-24, 2005. During the approval process the Lead Authors confirmed that the agreed text of the Summary for Policymakers is fully consistent with the underlying full report and technical summary, both of which have been accepted by governments, but remain the full responsibility of the authors.

We wish to express our gratitude to the governments that provided financial and in-kind support for the hosting of the various meetings that were essential to complete this report. We are particularly are grateful to the Canadian Government for hosting both the Workshop in Regina, November 18-22, 2002, as well as the Working Group III approval session in Montreal, September 22-24, 2005. The writing team of this report met four times to draft the report and discuss the results of the two consecutive formal IPCC review rounds. The meetings were kindly hosted by the government of Norway (Oslo, July 2003), Australia (Canberra, December 2003), Brazil (Salvador, August 2004) and Spain (Oviedo, April 2005), respectively. In addition, many individual meetings, teleconferences and interactions with governments have contributed to the successful completion of this report.

---

[b] See: http://www.ipcc.ch/meet/session20/finalreport20.pdf

We endorse the words of gratitude expressed in the Foreword by the Secretary–General of the WMO and the Executive Director of UNEP to the writing team, Review Editors and Expert Reviewers.

We would like to thank the staff of the Technical Support Unit of Working Group III for their work in preparing this report, in particular Heleen de Coninck for her outstanding and efficient coordination of the report, Manuela Loos and Cora Blankendaal for their technical, logistical and secretarial support, and Leo Meyer (head of TSU) for his leadership. We also express our gratitude to Anita Meier for her general support, to Dave Thomas, Pete Thomas, Tony Cunningham, Fran Aitkens, Ann Jenks, and Ruth de Wijs for the copy-editing of the document and to Wout Niezen, Martin Middelburg, Henk Stakelbeek, Albert van Staa, Eva Stam and Tim Huliselan for preparing the final layout and the graphics of the report. A special word of thanks goes to Lee-Anne Shepherd of CO2CRC for skillfully preparing the figures in the Summary for Policymakers. Last but not least, we would like to express our appreciation to Renate Christ and her staff and to Francis Hayes of WMO for their hard work in support of the process.

We, as co-chairs of Working Group III, together with the other members of the Bureau of Working Group III, the Lead Authors and the Technical Support Unit, hope that this report will assist decision-makers in governments and the private sector as well as other interested readers in the academic community and the general public in becoming better informed about CO_2 capture and storage as a climate change mitigation option.

Ogunlade Davidson and Bert Metz
Co-Chairs IPCC Working Group III on Mitigation of Climate Change

IPCC Special Report

Carbon Dioxide Capture and Storage

Summary for Policymakers

A Special Report of Working Group III
of the Intergovernmental Panel on Climate Change

This summary, approved in detail at the Eighth Session of IPCC Working Group III (Montreal, Canada, 22-24 September 2005), represents the formally agreed statement of the IPCC concerning current understanding of carbon dioxide capture and storage.

Based on a draft by:
Juan Carlos Abanades (Spain), Makoto Akai (Japan), Sally Benson (United States), Ken Caldeira (United States), Heleen de Coninck (Netherlands), Peter Cook (Australia), Ogunlade Davidson (Sierra Leone), Richard Doctor (United States), James Dooley (United States), Paul Freund (United Kingdom), John Gale (United Kingdom), Wolfgang Heidug (Germany), Howard Herzog (United States), David Keith (Canada), Marco Mazzotti (Italy and Switzerland), Bert Metz (Netherlands), Leo Meyer (Netherlands), Balgis Osman-Elasha (Sudan), Andrew Palmer (United Kingdom), Riitta Pipatti (Finland), Edward Rubin (United States), Koen Smekens (Belgium), Mohammad Soltanieh (Iran), Kelly (Kailai) Thambimuthu (Australia and Canada)

Contents

What is CO₂ capture and storage and how could it contribute to mitigating climate change?

1. Carbon dioxide (CO₂) capture and storage (CCS) is a process consisting of the separation of CO₂ from industrial and energy-related sources, transport to a storage location and long-term isolation from the atmosphere. This report considers CCS as an option in the portfolio of mitigation actions for stabilization of atmospheric greenhouse gas concentrations.

Other mitigation options include energy efficiency improvements, the switch to less carbon-intensive fuels, nuclear power, renewable energy sources, enhancement of biological sinks, and reduction of non-CO₂ greenhouse gas emissions. CCS has the potential to reduce overall mitigation costs and increase flexibility in achieving greenhouse gas emission reductions. The widespread application of CCS would depend on technical maturity, costs, overall potential, diffusion and transfer of the technology to developing countries and their capacity to apply the technology, regulatory aspects, environmental issues and public perception (Sections 1.1.1, 1.3, 1.7, 8.3.3.4).

2. The Third Assessment Report (TAR) indicates that no single technology option will provide all of the emission reductions needed to achieve stabilization, but a portfolio of mitigation measures will be needed.

Most scenarios project that the supply of primary energy will continue to be dominated by fossil fuels until at least the middle of the century. As discussed in the TAR, most models also indicate that known technological options[1] could achieve a broad range of atmospheric stabilization levels but that implementation would require socio-economic and institutional changes. In this context, the availability of CCS in the portfolio of options could facilitate achieving stabilization goals (Sections 1.1, 1.3).

What are the characteristics of CCS?

3. Capture of CO₂ can be applied to large point sources. The CO₂ would then be compressed and transported for storage in geological formations, in the ocean, in mineral carbonates[2], or for use in industrial processes.

Large point sources of CO₂ include large fossil fuel or biomass energy facilities, major CO₂-emitting industries, natural gas production, synthetic fuel plants and fossil fuel-based hydrogen production plants (see Table SPM.1). Potential technical storage methods are: geological storage (in geological formations, such as oil and gas fields, unminable coal beds and deep saline formations[3]), ocean storage (direct release into the ocean water column or onto the deep seafloor) and industrial fixation of CO₂ into inorganic carbonates. This report also discusses industrial uses of CO₂, but this is not expected to contribute much to the reduction of CO₂

Table SPM.1. Profile by process or industrial activity of worldwide large stationary CO₂ sources with emissions of more than 0.1 million tonnes of CO₂ (MtCO₂) per year.

Process	Number of sources	Emissions (MtCO₂ yr⁻¹)
Fossil fuels		
Power	4,942	10,539
Cement production	1,175	932
Refineries	638	798
Iron and steel industry	269	646
Petrochemical industry	470	379
Oil and gas processing	Not available	50
Other sources	90	33
Biomass		
Bioethanol and bioenergy	303	91
Total	**7,887**	**13,466**

[1] "Known technological options" refer to technologies that exist in operation or in the pilot plant stage at the present time, as referenced in the mitigation scenarios discussed in the TAR. It does not include any new technologies that will require profound technological breakthroughs. Known technological options are explained in the TAR and several mitigation scenarios include CCS

[2] Storage of CO₂ as mineral carbonates does not cover deep geological carbonation or ocean storage with enhanced carbonate neutralization as discussed in Chapter 6 (Section 7.2).

[3] Saline formations are sedimentary rocks saturated with formation waters containing high concentrations of dissolved salts. They are widespread and contain enormous quantities of water that are unsuitable for agriculture or human consumption. Because the use of geothermal energy is likely to increase, potential geothermal areas may not be suitable for CO₂ storage (see Section 5.3.3).

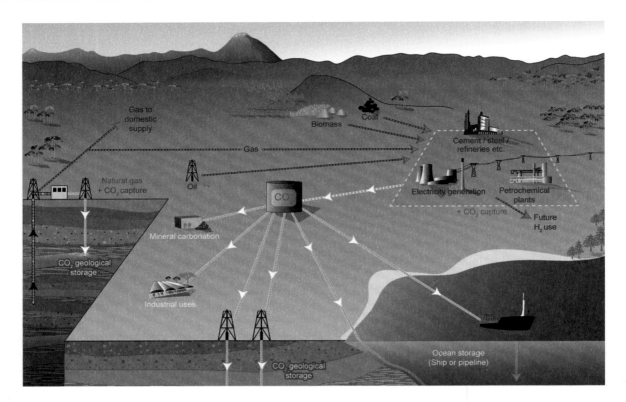

Figure SPM.1. Schematic diagram of possible CCS systems showing the sources for which CCS might be relevant, transport of CO_2 and storage options (Courtesy of CO2CRC).

emissions (see Figure SPM.1) (Sections 1.2, 1.4, 2.2, Table 2.3).

4. *The net reduction of emissions to the atmosphere through CCS depends on the fraction of CO_2 captured, the increased CO_2 production resulting from loss in overall efficiency of power plants or industrial processes due to the additional energy required for capture, transport and storage, any leakage from transport and the fraction of CO_2 retained in storage over the long term.*

Available technology captures about 85–95% of the CO_2 processed in a capture plant. A power plant equipped with a CCS system (with access to geological or ocean storage) would need roughly 10–40%[4] more energy than a plant of equivalent output without CCS, of which most is for capture and compression. For secure storage, the net result is that a power plant with CCS could reduce CO_2 emissions to the atmosphere by approximately 80–90% compared to a plant without CCS (see Figure SPM.2). To the extent that leakage might occur from a storage reservoir, the fraction retained is defined as the fraction of the cumulative amount of injected CO_2 that is retained over a specified period of time. CCS systems with storage as mineral carbonates would need 60–

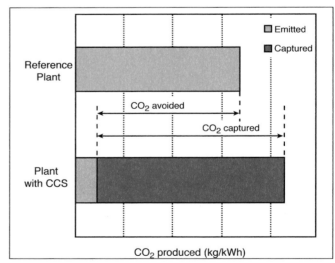

Figure SPM.2. CO_2 capture and storage from power plants. The increased CO_2 production resulting from the loss in overall efficiency of power plants due to the additional energy required for capture, transport and storage and any leakage from transport result in a larger amount of "CO_2 produced per unit of product" (lower bar) relative to the reference plant (upper bar) without capture (Figure 8.2).

[4] The range reflects three types of power plants: for Natural Gas Combined Cycle plants, the range is 11–22%, for Pulverized Coal plants, 24–40% and for Integrated Gasification Combined Cycle plants, 14–25%.

180% more energy than a plant of equivalent output without CCS. (Sections 1.5.1, 1.6.3, 3.6.1.3, 7.2.7).

What is the current status of CCS technology?

5. *There are different types of CO_2 capture systems: post-combustion, pre-combustion and oxyfuel combustion (Figure SPM.3). The concentration of CO_2 in the gas stream, the pressure of the gas stream and the fuel type (solid or gas) are important factors in selecting the capture system.*

Post-combustion capture of CO_2 in power plants is economically feasible under specific conditions[5]. It is used to capture CO_2 from part of the flue gases from a number of existing power plants. Separation of CO_2 in the natural gas processing industry, which uses similar technology, operates in a mature market[6]. The technology required for pre-combustion capture is widely applied in fertilizer manufacturing and in hydrogen production. Although the initial fuel conversion steps of pre-combustion are more elaborate and costly, the higher concentrations of CO_2 in the gas stream and the higher pressure make the separation easier. Oxyfuel combustion is in the demonstration phase[7] and uses high purity oxygen. This results in high CO_2 concentrations in the gas stream and, hence, in easier separation of CO_2 and in increased energy requirements in the separation of oxygen from air (Sections 3.3, 3.4, 3.5).

6. *Pipelines are preferred for transporting large amounts of CO_2 for distances up to around 1,000 km. For amounts smaller than a few million tonnes of CO_2 per year or for larger distances overseas, the use of ships, where applicable, could be economically more attractive.*

Pipeline transport of CO_2 operates as a mature market technology (in the USA, over 2,500 km of pipelines transport more than 40 $MtCO_2$ per year). In most gas pipelines, compressors at the upstream end drive the flow, but some pipelines need intermediate compressor stations. Dry CO_2 is not corrosive to pipelines, even if the CO_2 contains contaminants. Where the CO_2 contains moisture, it is removed from the CO_2 stream to prevent corrosion and to avoid the costs of constructing pipelines of corrosion-

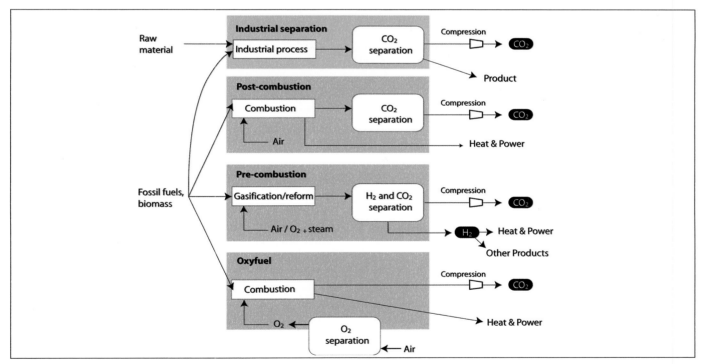

Figure SPM.3. Schematic representation of capture systems. Fuels and products are indicated for oxyfuel combustion, pre-combustion (including hydrogen and fertilizer production), post-combustion and industrial sources of CO_2 (including natural gas processing facilities and steel and cement production) (based on Figure 3.1) (Courtesy CO2CRC).

---

[5] "Economically feasible under specific conditions" means that the technology is well understood and used in selected commercial applications, such as in a favourable tax regime or a niche market, processing at least 0.1 $MtCO_2$ yr[-1], with few (less than 5) replications of the technology.

[6] "Mature market" means that the technology is now in operation with multiple replications of the commercial-scale technology worldwide.

[7] "Demonstration phase" means that the technology has been built and operated at the scale of a pilot plant but that further development is required before the technology is ready for the design and construction of a full-scale system.

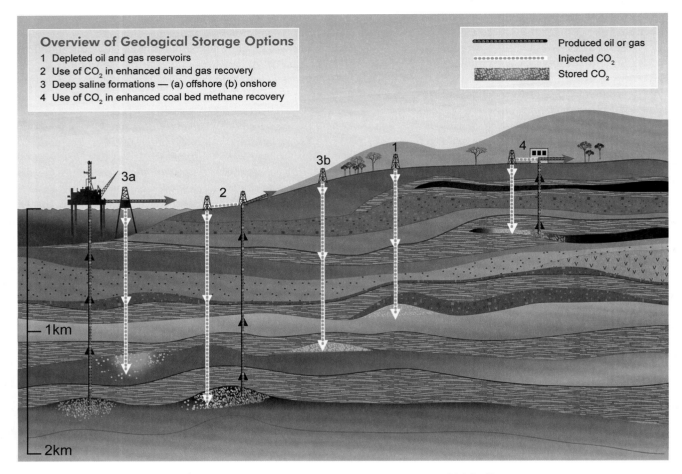

Figure SPM.4. Overview of geological storage options (based on Figure 5.3) (Courtesy CO2CRC).

resistant material. Shipping of CO_2, analogous to shipping of liquefied petroleum gases, is economically feasible under specific conditions but is currently carried out on a small scale due to limited demand. CO_2 can also be carried by rail and road tankers, but it is unlikely that these could be attractive options for large-scale CO_2 transportation (Sections 4.2.1, 4.2.2, 4.3.2, Figure 4.5, 4.6).

7. *Storage of CO_2 in deep, onshore or offshore geological formations uses many of the same technologies that have been developed by the oil and gas industry and has been proven to be economically feasible under specific conditions for oil and gas fields and saline formations, but not yet for storage in unminable coal beds[8] (see Figure SPM.4).*

If CO_2 is injected into suitable saline formations or oil or gas fields, at depths below 800 m[9], various physical and geochemical trapping mechanisms would prevent it from migrating to the surface. In general, an essential physical trapping mechanism is the presence of a caprock[10]. Coal bed storage may take place at shallower depths and relies on the adsorption of CO_2 on the coal, but the technical feasibility largely depends on the permeability of the coal bed. The combination of CO_2 storage with Enhanced Oil Recovery (EOR[11]) or, potentially, Enhanced Coal Bed Methane recovery (ECBM) could lead to additional revenues from the oil or gas recovery. Well-drilling technology, injection technology, computer simulation of storage reservoir performance and monitoring methods from existing applications are being

---

[8] A coal bed that is unlikely to ever be mined – because it is too deep or too thin – may be potentially used for CO_2 storage. If subsequently mined, the stored CO_2 would be released. Enhanced Coal Bed Methane (ECBM) recovery could potentially increase methane production from coals while simultaneously storing CO_2. The produced methane would be used and not released to the atmosphere (Section 5.3.4).

[9] At depths below 800–1,000 m, CO_2 becomes supercritical and has a liquid-like density (about 500–800 kg m^{-3}) that provides the potential for efficient utilization of underground storage space and improves storage security (Section 5.1.1).

[10] Rock of very low permeability that acts as an upper seal to prevent fluid flow out of a reservoir.

[11] For the purposes of this report, EOR means CO_2-driven Enhanced Oil Recovery.

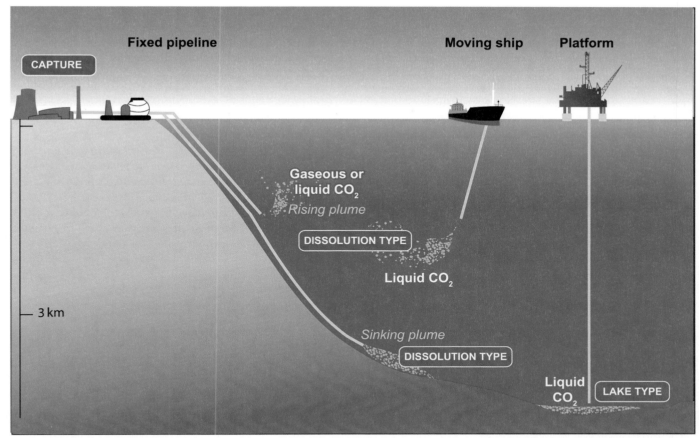

Figure SPM.5. Overview of ocean storage concepts. In "dissolution type" ocean storage, the CO_2 rapidly dissolves in the ocean water, whereas in "lake type" ocean storage, the CO_2 is initially a liquid on the sea floor (Courtesy CO2CRC).

developed further for utilization in the design and operation of geological storage projects.

Three industrial-scale[12] storage projects are in operation: the Sleipner project in an offshore saline formation in Norway, the Weyburn EOR project in Canada, and the In Salah project in a gas field in Algeria. Others are planned (Sections 5.1.1, 5.2.2, 5.3, 5.6, 5.9.4, Boxes 5.1, 5.2, 5.3).

8. *Ocean storage potentially could be done in two ways: by injecting and dissolving CO_2 into the water column (typically below 1,000 meters) via a fixed pipeline or a moving ship, or by depositing it via a fixed pipeline or an offshore platform onto the sea floor at depths below 3,000 m, where CO_2 is denser than water and is expected to form a "lake" that would delay dissolution of CO_2 into the surrounding environment (see Figure SPM.5). Ocean storage and its ecological impacts are still in the research phase[13].*

The dissolved and dispersed CO_2 would become part of the global carbon cycle and eventually equilibrate with the CO_2 in the atmosphere. In laboratory experiments, small-scale ocean experiments and model simulations, the technologies and associated physical and chemical phenomena, which include, notably, increases in acidity (lower pH) and their effect on marine ecosystems, have been studied for a range of ocean storage options (Sections 6.1.2, 6.2.1, 6.5, 6.7).

9. *The reaction of CO_2 with metal oxides, which are abundant in silicate minerals and available in small quantities in waste streams, produces stable carbonates. The technology is currently in the research stage, but certain applications in using waste streams are in the demonstration phase.*

The natural reaction is very slow and has to be enhanced by pre-treatment of the minerals, which at present is very energy intensive (Sections 7.2.1, 7.2.3, 7.2.4, Box 7.1).

---

[12] "Industrial-scale" here means on the order of 1 $MtCO_2$ per year.

[13] "Research phase" means that while the basic science is understood, the technology is currently in the stage of conceptual design or testing at the laboratory or bench scale and has not been demonstrated in a pilot plant.

10. *Industrial uses[14] of captured CO_2 as a gas or liquid or as a feedstock in chemical processes that produce valuable carbon-containing products are possible, but are not expected to contribute to significant abatement of CO_2 emissions.*

The potential for industrial uses of CO_2 is small, while the CO_2 is generally retained for short periods (usually months or years). Processes using captured CO_2 as feedstock instead of fossil hydrocarbons do not always achieve net lifecycle emission reductions (Sections 7.3.1, 7.3.4).

11. *Components of CCS are in various stages of development (see Table SPM.2). Complete CCS systems can be assembled from existing technologies that are mature or economically feasible under specific conditions, although the state of development of the overall system may be less than some of its separate components.*

There is relatively little experience in combining CO_2 capture, transport and storage into a fully integrated CCS system. The utilization of CCS for large-scale power plants (the potential application of major interest) still remains to be implemented (Sections 1.4.4, 3.8, 5.1).

What is the geographical relationship between the sources and storage opportunities for CO_2?

12. *Large point sources of CO_2 are concentrated in proximity to major industrial and urban areas. Many such sources are within 300 km of areas that potentially hold formations suitable for geological storage (see Figure SPM.6). Preliminary research suggests that, globally, a small proportion of large point sources is close to potential ocean storage locations.*

Table SPM.2. Current maturity of CCS system components. The X's indicate the highest level of maturity for each component. For most components, less mature technologies also exist.

CCS component	CCS technology	Research phase [13]	Demonstration phase [7]	Economically feasible under specific conditions [5]	Mature market [6]
Capture	Post-combustion			X	
	Pre-combustion			X	
	Oxyfuel combustion		X		
	Industrial separation (natural gas processing, ammonia production)				X
Transportation	Pipeline				X
	Shipping			X	
Geological storage	Enhanced Oil Recovery (EOR)				X[a]
	Gas or oil fields			X	
	Saline formations			X	
	Enhanced Coal Bed Methane recovery (ECBM)		X		
Ocean storage	Direct injection (dissolution type)	X			
	Direct injection (lake type)	X			
Mineral carbonation	Natural silicate minerals	X			
	Waste materials		X		
Industrial uses of CO_2					X

[a] CO_2 injection for EOR is a mature market technology, but when this technology is used for CO_2 storage, it is only 'economically feasible under specific conditions'

[14] Industrial uses of CO_2 refer to those uses that do not include EOR, which is discussed in paragraph 7.

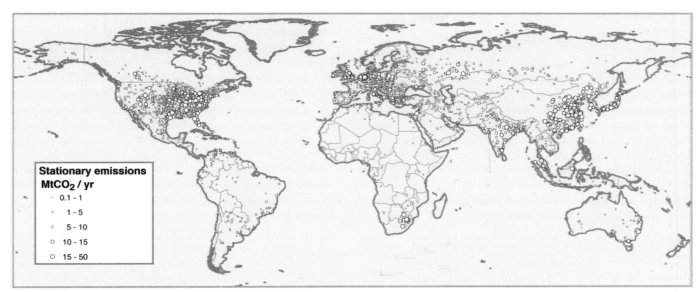

Figure SPM.6a. Global distribution of large stationary sources of CO_2 (Figure 2.3) (based on a compilation of publicly available information on global emission sources; IEA GHG 2002)

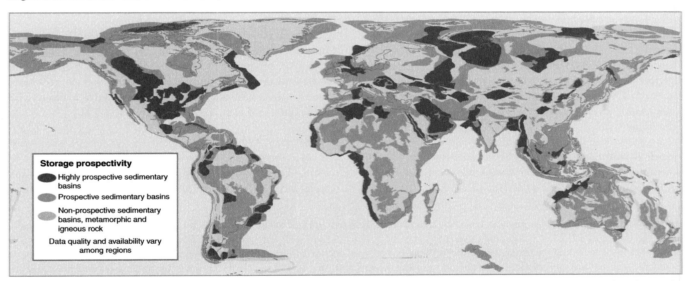

Figure SPM.6b. Prospective areas in sedimentary basins where suitable saline formations, oil or gas fields or coal beds may be found. Locations for storage in coal beds are only partly included. Prospectivity is a qualitative assessment of the likelihood that a suitable storage location is present in a given area based on the available information. This figure should be taken as a guide only because it is based on partial data, the quality of which may vary from region to region and which may change over time and with new information (Figure 2.4) (Courtesy of Geoscience Australia).

Currently available literature regarding the matches between large CO_2 point sources with suitable geological storage formations is limited. Detailed regional assessments may be necessary to improve information (see Figure SPM.6b).

Scenario studies indicate that the number of large point sources is projected to increase in the future, and that, by 2050, given expected technical limitations, around 20–40% of global fossil fuel CO_2 emissions could be technically suitable for capture, including 30–60% of the CO_2 emissions from electricity generation and 30–40% of those from industry. Emissions from large-scale biomass conversion facilities could also be technically suitable for capture. The proximity of future large point sources to potential storage sites has not been studied (Sections 2.3, 2.4.3).

13. *CCS enables the control of the CO_2 emissions from fossil fuel-based production of electricity or hydrogen, which in the longer term could reduce part of the dispersed CO_2*

emissions from transport and distributed energy supply systems.

Electricity could be used in vehicles, and hydrogen could be used in fuel cells, including in the transport sector. Gas and coal conversion with integrated CO_2 separation (without storage) is currently the dominant option for the production of hydrogen. More fossil fuel or biomass-based hydrogen or electricity production would result in an increased number of large CO_2 sources that are technically suitable for capture and storage. At present, it is difficult to project the likely number, location and size of such sources (Sections 2.5.1).

What are the costs[15] for CCS and what is the technical and economic potential?

14. Application of CCS to electricity production, under 2002 conditions, is estimated to increase electricity generation costs by about 0.01–0.05 US dollars[16] per kilowatt hour (US$/kWh), depending on the fuel, the specific technology, the location and the national circumstances. Inclusion of the benefits of EOR would reduce additional electricity production costs due to CCS by around 0.01– 0.02 US$/kWh[17] (see Table SPM.3 for absolute electricity production costs and Table SPM.4 for costs in US$/tCO$_2$ avoided). Increases in market prices of fuels used for power generation would generally tend to increase the cost of CCS. The quantitative impact of oil price on CCS is uncertain. However, revenue from EOR would generally be higher with higher oil prices. While applying CCS to biomass-based power production at the current small scale would add substantially to the electricity costs, co-firing of biomass in a larger coal-fired power plant with CCS would be more cost-effective.

Costs vary considerably in both absolute and relative terms from country to country. Since neither Natural Gas Combined Cycle, Pulverized Coal nor Integrated Gasification Combined Cycle systems have yet been built at a full scale with CCS, the costs of these systems cannot be stated with a high degree of confidence at this time. In the future, the costs of CCS could be reduced by research and technological development and economies of scale. Economies of scale could also considerably bring down the cost of biomass-based CCS systems over time. The application of CCS to biomass-fuelled or co-fired conversion facilities would lead to lower or negative[18] CO_2 emissions, which could reduce the costs for this option, depending on the market value of CO_2 emission reductions (Sections 2.5.3, 3.7.1, 3.7.13, 8.2.4).

15. Retrofitting existing plants with CO_2 capture is expected to lead to higher costs and significantly reduced overall efficiencies than for newly built power plants with capture. The cost disadvantages of retrofitting may be reduced in the case of some relatively new and highly efficient existing plants or where a plant is substantially upgraded or rebuilt.

The costs of retrofitting CCS to existing installations vary. Industrial sources of CO_2 can more easily be retrofitted with CO_2 separation, while integrated power plant systems would need more profound adjustment. In order to reduce future retrofit costs, new plant designs could take future CCS application into account (Sections 3.1.4, 3.7.5).

16. In most CCS systems, the cost of capture (including compression) is the largest cost component.

Costs for the various components of a CCS system vary widely, depending on the reference plant and the wide range

Table SPM.3. Costs of CCS: production costs of electricity for different types of generation, without capture and for the CCS system as a whole. The cost of a full CCS system for electricity generation from a newly built, large-scale fossil fuel-based power plant depends on a number of factors, including the characteristics of both the power plant and the capture system, the specifics of the storage site, the amount of CO_2 and the required transport distance. The numbers assume experience with a large-scale plant. Gas prices are assumed to be 2.8-4.4 US$ per gigajoule (GJ), and coal prices 1-1.5 US$ GJ^{-1} (based on Tables 8.3 and 8.4).

Power plant system	Natural Gas Combined Cycle (US$/kWh)	Pulverized Coal (US$/kWh)	Integrated Gasification Combined Cycle (US$/kWh)
Without capture (reference plant)	0.03 - 0.05	0.04 - 0.05	0.04 - 0.06
With capture and geological storage	0.04 - 0.08	0.06 - 0.10	0.05 - 0.09
With capture and EOR[17]	0.04 - 0.07	0.05 - 0.08	0.04 - 0.07

[15] As used in this report, "costs" refer only to market prices but do not include external costs such as environmental damages and broader societal costs that may be associated with the use of CCS. To date, little has been done to assess and quantify such external costs.

[16] All costs in this report are expressed in 2002 US$.

[17] Based on oil prices of 15–20 US$ per barrel, as used in the available literature.

[18] If, for example, the biomass is harvested at an unsustainable rate (that is, faster than the annual re-growth), the net CO_2 emissions of the activity might not be negative.

Table SPM.4. CO_2 avoidance costs for the complete CCS system for electricity generation, for different combinations of reference power plants without CCS and power plants with CCS (geological and EOR). The amount of CO_2 avoided is the difference between the emissions of the reference plant and the emissions of the power plant with CCS. Gas prices are assumed to be 2.8-4.4 US\$ GJ^{-1}, and coal prices 1-1.5 US\$ GJ^{-1} (based on Tables 8.3a and 8.4).

Type of power plant with CCS	Natural Gas Combined Cycle reference plant US\$/tCO_2 avoided	Pulverized Coal reference plant US\$/tCO_2 avoided
Power plant with capture and geological storage		
Natural Gas Combined Cycle	40 - 90	20 - 60
Pulverized Coal	70 - 270	30 - 70
Integrated Gasification Combined Cycle	40 - 220	20 - 70
Power plant with capture and EOR[17]		
Natural Gas Combined Cycle	20 - 70	0 - 30
Pulverized Coal	50 - 240	10 - 40
Integrated Gasification Combined Cycle	20 - 190	0 - 40

Table SPM.5. 2002 Cost ranges for the components of a CCS system as applied to a given type of power plant or industrial source. The costs of the separate components cannot simply be summed to calculate the costs of the whole CCS system in US\$/CO_2 avoided. All numbers are representative of the costs for large-scale, new installations, with natural gas prices assumed to be 2.8-4.4 US\$ GJ^{-1} and coal prices 1-1.5 US\$ GJ^{-1} (Sections 5.9.5, 8.2.1, 8.2.2, 8.2.3, Tables 8.1 and 8.2).

CCS system components	Cost range	Remarks
Capture from a coal- or gas-fired power plant	15-75 US\$/tCO_2 net captured	Net costs of captured CO_2, compared to the same plant without capture.
Capture from hydrogen and ammonia production or gas processing	5-55 US\$/tCO_2 net captured	Applies to high-purity sources requiring simple drying and compression.
Capture from other industrial sources	25-115 US\$/tCO_2 net captured	Range reflects use of a number of different technologies and fuels.
Transportation	1-8 US\$/tCO_2 transported	Per 250 km pipeline or shipping for mass flow rates of 5 (high end) to 40 (low end) MtCO_2 yr^{-1}.
Geological storage[a]	0.5-8 US\$/tCO_2 net injected	Excluding potential revenues from EOR or ECBM.
Geological storage: monitoring and verification	0.1-0.3 US\$/tCO_2 injected	This covers pre-injection, injection, and post-injection monitoring, and depends on the regulatory requirements.
Ocean storage	5-30 US\$/tCO_2 net injected	Including offshore transportation of 100-500 km, excluding monitoring and verification.
Mineral carbonation	50-100 US\$/tCO_2 net mineralized	Range for the best case studied. Includes additional energy use for carbonation.

[a] Over the long term, there may be additional costs for remediation and liabilities.

in CO_2 source, transport and storage situations (see Table SPM.5). Over the next decade, the cost of capture could be reduced by 20–30%, and more should be achievable by new technologies that are still in the research or demonstration phase. The costs of transport and storage of CO_2 could decrease slowly as the technology matures further and the scale increases (Sections 1.5.3, 3.7.13, 8.2).

17. *Energy and economic models indicate that the CCS system's major contribution to climate change mitigation would come from deployment in the electricity sector. Most*

modelling as assessed in this report suggests that CCS systems begin to deploy at a significant level when CO_2 prices begin to reach approximately 25–30 US\$/tCO_2.
Low-cost capture possibilities (in gas processing and in hydrogen and ammonia manufacture, where separation of CO_2 is already done) in combination with short (<50 km) transport distances and storage options that generate revenues (such as EOR) can lead to the limited storage of CO_2 (up to 360 MtCO_2 yr^{-1}) under circumstances of low or no incentives (Sections 2.2.1.3, 2.3, 2.4, 8.3.2.1)

18. *Available evidence suggests that, worldwide, it is likely[19] that there is a technical potential[20] of at least about 2,000 GtCO₂ (545 GtC) of storage capacity in geological formations[21].*

There could be a much larger potential for geological storage in saline formations, but the upper limit estimates are uncertain due to lack of information and an agreed methodology. The capacity of oil and gas reservoirs is better known. Technical storage capacity in coal beds is much smaller and less well known.

Model calculations for the capacity to store CO_2 in the oceans indicate that this capacity could be on the order of thousands of $GtCO_2$, depending on the assumed stabilization level in the atmosphere[22] and on environmental constraints such as ocean *p*H change. The extent to which mineral carbonation may be used can currently not be determined, since it depends on the unknown amount of silicate reserves that can be technically exploited and on environmental issues such as the volume of product disposal (Sections 5.3, 6.3.1, 7.2.3, Table 5.2).

19. *In most scenarios for stabilization of atmospheric greenhouse gas concentrations between 450 and 750 ppmv CO_2 and in a least-cost portfolio of mitigation options, the economic potential[23] of CCS would amount to 220– 2,200 GtCO₂ (60–600 GtC) cumulatively, which would mean that CCS contributes 15–55% to the cumulative mitigation effort worldwide until 2100, averaged over a range of baseline scenarios. It is likely[20] that the technical potential[21] for geological storage is sufficient to cover the high end of the economic potential range, but for specific regions, this may not be true.*

Uncertainties in these economic potential estimates are significant. For CCS to achieve such an economic potential, several hundreds to thousands of CO_2 capture systems would need to be installed over the coming century, each capturing some 1–5 $MtCO_2$ per year. The actual implementation of CCS, as for other mitigation options, is likely to be lower than the economic potential due to factors such as environmental impacts, risks of leakage and the lack of a clear legal framework or public acceptance (Sections 1.4.4, 5.3.7, 8.3.1, 8.3.3, 8.3.3.4).

20. *In most scenario studies, the role of CCS in mitigation portfolios increases over the course of the century, and the inclusion of CCS in a mitigation portfolio is found to reduce the costs of stabilizing CO_2 concentrations by 30% or more.*

One aspect of the cost competitiveness of CCS systems is that CCS technologies are compatible with most current energy infrastructures.

The global potential contribution of CCS as part of a mitigation portfolio is illustrated by the examples given in Figure SPM.7. The present extent of analyses in this field is limited, and further assessments may be necessary to improve information (Sections 1.5, 8.3.3, 8.3.3.4, Box 8.3).

What are the local health, safety and environment risks of CCS?

21. *The local risks[24] associated with CO_2 pipeline transport could be similar to or lower than those posed by hydrocarbon pipelines already in operation.*

For existing CO_2 pipelines, mostly in areas of low population density, accident numbers reported per kilometre pipeline are very low and are comparable to those for hydrocarbon pipelines. A sudden and large release of CO_2 would pose immediate dangers to human life and health, if there were exposure to concentrations of CO_2 greater than 7–10% by volume in air. Pipeline transport of CO_2 through populated areas requires attention to route selection, overpressure protection, leak detection and other design factors. No major obstacles to pipeline design for CCS are foreseen (Sections 4.4.2, AI.2.3.1).

22. *With appropriate site selection based on available subsurface information, a monitoring programme to detect problems, a regulatory system and the appropriate use of remediation methods to stop or control CO_2 releases if they arise, the local health, safety and environment risks of geological storage would be comparable to the risks of current activities such as natural gas storage, EOR and deep underground disposal of acid gas.*

Natural CO_2 reservoirs contribute to the understanding of the behaviour of CO_2 underground. Features of storage sites with a low probability of leakage include highly impermeable caprocks, geological stability, absence of leakage paths

---

[19] "Likely" is a probability between 66 and 90%.

[20] "Technical potential" as defined in the TAR is the amount by which it is possible to reduce greenhouse gas emissions by implementing a technology or practice that already has been demonstrated

[21] This statement is based on the expert judgment of the authors of the available literature. It reflects the uncertainty about the storage capacity estimates (Section 5.3.7)

[22] This approach takes into account that the CO_2 injected in the ocean will after some time reach equilibrium with the atmosphere.

[23] Economic potential is the amount of greenhouse gas emissions reductions from a specific option that could be achieved cost-effectively, given prevailing circumstances (i.e. a market value of CO_2 reductions and costs of other options).

[24] In discussing the risks, we assume that risk is the product of the probability that an event will occur and the consequences of the event if it does occur.

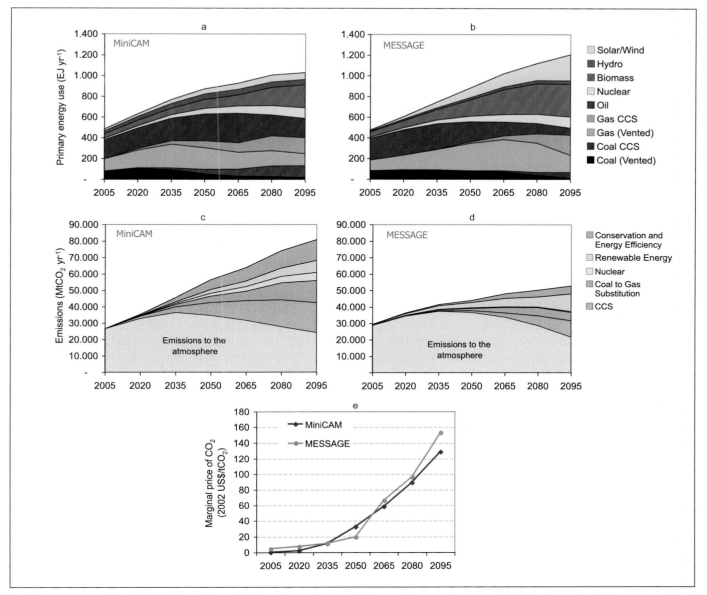

Figure SPM.7. These figures are an illustrative example of the global potential contribution of CCS as part of a mitigation portfolio. They are based on two alternative integrated assessment models (MESSAGE and MiniCAM) while adopt the same assumptions for the main emissions drivers. The results would vary considerably on regional scales. This example is based on a single scenario and, therefore, does not convey the full range of uncertainties. Panels a and b show global primary energy use, including the deployment of CCS. Panels c and d show the global CO_2 emissions in grey and corresponding contributions of main emissions reduction measures in colour. Panel e shows the calculated marginal price of CO_2 reductions (Section 8.3.3, Box 8.3).

and effective trapping mechanisms. There are two different types of leakage scenarios: (1) abrupt leakage, through injection well failure or leakage up an abandoned well, and (2) gradual leakage, through undetected faults, fractures or wells. Impacts of elevated CO_2 concentrations in the shallow subsurface could include lethal effects on plants and subsoil animals and the contamination of groundwater. High fluxes in conjunction with stable atmospheric conditions could lead to local high CO_2 concentrations in the air that could harm animals or people. Pressure build-up caused by CO_2 injection could trigger small seismic events.

While there is limited experience with geological storage, closely related industrial experience and scientific knowledge could serve as a basis for appropriate risk management, including remediation. The effectiveness of the available risk management methods still needs to be demonstrated for use with CO_2 storage. If leakage occurs at a storage site, remediation to stop the leakage could involve standard well repair techniques or the interception and extraction of the CO_2 before it would leak into a shallow groundwater aquifer. Given the long timeframes associated with geological storage of CO_2, site monitoring may be required for very long periods (Sections 5.6, 5.7, Tables 5.4, 5.7, Figure 5.25).

23. *Adding CO_2 to the ocean or forming pools of liquid CO_2 on the ocean floor at industrial scales will alter the local chemical environment. Experiments have shown that sustained high concentrations of CO_2 would cause mortality of ocean organisms. CO_2 effects on marine organisms will have ecosystem consequences. The chronic effects of direct CO_2 injection into the ocean on ecosystems over large ocean areas and long time scales have not yet been studied.*

Model simulations, assuming a release from seven locations at an ocean depth of 3,000 m, where ocean storage provides 10% of the mitigation effort for stabilization at 550 ppmv CO_2, resulted in acidity increases (*p*H decrease >0.4) over approximately 1% of the ocean volume. For comparison purposes: in such a stabilization case without ocean storage, a *p*H decrease >0.25 relative to pre-industrial levels at the entire ocean surface can be expected. A 0.2 to 0.4 *p*H decrease is significantly greater than pre-industrial variations in average ocean acidity. At these levels of pH change, some effects have been found in organisms that live near the ocean's surface, but chronic effects have not yet been studied. A better understanding of these impacts is required before a comprehensive risk assessment can be accomplished. There is no known mechanism for the sudden or catastrophic release of stored CO_2 from the ocean to the atmosphere. Gradual release is discussed in SPM paragraph 26. Conversion of molecular CO_2 to bicarbonates or hydrates before or during CO_2 release would reduce the *p*H effects and enhance the retention of CO_2 in the ocean, but this would also increase the costs and other environmental impacts (Section 6.7).

24. *Environmental impacts of large-scale mineral carbonation would be a consequence of the required mining and disposal of resulting products that have no practical use.*

Industrial fixation of one tonne of CO_2 requires between 1.6 and 3.7 tonnes of silicate rock. The impacts of mineral carbonation are similar to those of large-scale surface mines. They include land-clearing, decreased local air quality and affected water and vegetation as a result of drilling, moving of earth and the grading and leaching of metals from mining residues, all of which indirectly may also result in habitat degradation. Most products of mineral carbonation need to

be disposed of, which would require landfills and additional transport (Sections 7.2.4, 7.2.6).

Will physical leakage of stored CO_2 compromise CCS as a climate change mitigation option?

25. *Observations from engineered and natural analogues as well as models suggest that the fraction retained in appropriately selected and managed geological reservoirs is very likely[25] to exceed 99% over 100 years and is likely[20] to exceed 99% over 1,000 years.*

For well-selected, designed and managed geological storage sites, the vast majority of the CO_2 will gradually be immobilized by various trapping mechanisms and, in that case, could be retained for up to millions of years. Because of these mechanisms, storage could become more secure over longer timeframes (Sections 1.6.3, 5.2.2, 5.7.3.4, Table 5.5).

26. *Release of CO_2 from ocean storage would be gradual over hundreds of years.*

Ocean tracer data and model calculations indicate that, in the case of ocean storage, depending on the depth of injection and the location, the fraction retained is 65–100% after 100 years and 30–85% after 500 years (a lower percentage for injection at a depth of 1,000 m, a higher percentage at 3,000 m) (Sections 1.6.3, 6.3.3, 6.3.4, Table 6.2)

27. *In the case of mineral carbonation, the CO_2 stored would not be released to the atmosphere (Sections 1.6.3, 7.2.7).*

28. *If continuous leakage of CO_2 occurs, it could, at least in part, offset the benefits of CCS for mitigating climate change. Assessments of the implications of leakage for climate change mitigation depend on the framework chosen for decision-making and on the information available on the fractions retained for geological or ocean storage as presented in paragraphs 25 and 26.*

Studies conducted to address the question of how to deal with non-permanent storage are based on different approaches: the value of delaying emissions, cost minimization of a specified mitigation scenario or allowable future emissions in the context of an assumed stabilization of atmospheric greenhouse gas concentrations. Some of these studies allow future leakage to be compensated by additional reductions in emissions; the results depend on assumptions regarding the future cost of reductions, discount rates, the amount of CO_2 stored and the atmospheric concentration stabilization level assumed. In other studies, compensation is not seen as an option because of political and institutional uncertainties, and the analysis focuses on limitations set by the assumed

---

[25] "Very likely" is a probability between 90 and 99%.

stabilization level and the amount stored. While specific results of the range of studies vary with the methods and assumptions made, all studies imply that, if CCS is to be acceptable as a mitigation measure, there must be an upper limit to the amount of leakage that can take place (Sections 1.6.4, 8.4).

What are the legal and regulatory issues for implementing CO_2 storage?

29. Some regulations for operations in the subsurface do exist that may be relevant or, in some cases, directly applicable to geological storage, but few countries have specifically developed legal or regulatory frameworks for long-term CO_2 storage.

Existing laws and regulations regarding inter alia mining, oil and gas operations, pollution control, waste disposal, drinking water, treatment of high-pressure gases and subsurface property rights may be relevant to geological CO_2 storage. Long-term liability issues associated with the leakage of CO_2 to the atmosphere and local environmental impacts are generally unresolved. Some States take on long-term responsibility in situations comparable to CO_2 storage, such as underground mining operations (Sections 5.8.2, 5.8.3, 5.8.4).

30. No formal interpretations so far have been agreed upon with respect to whether or under what conditions CO_2 injection into the geological sub-seabed or the ocean is compatible.

There are currently several treaties (notably the London[26] and OSPAR[27] Conventions) that potentially apply to the injection of CO_2 into the geological sub-seabed or the ocean. All of these treaties have been drafted without specific consideration of CO_2 storage (Sections 5.8.1, 6.8.1).

What are the implications of CCS for emission inventories and accounting?

31. The current IPCC Guidelines[28] do not include methods specific to estimating emissions associated with CCS.

The general guidance provided by the IPCC can be applied to CCS. A few countries currently do so, in combination with their national methods for estimating emissions. The IPCC guidelines themselves do not yet provide specific methods for estimating emissions associated with CCS. These are expected to be provided in the 2006 IPCC Guidelines for National Greenhouse Gas Inventories. Specific methods may be required for the net capture and storage of CO_2, physical leakage, fugitive emissions and negative emissions associated with biomass applications of CCS systems (Sections 9.2.1, 9.2.2).

32. The few current CCS projects all involve geological storage, and there is therefore limited experience with the monitoring, verification and reporting of actual physical leakage rates and associated uncertainties.

Several techniques are available or under development for monitoring and verification of CO_2 emissions from CCS, but these vary in applicability, site specificity, detection limits and uncertainties (Sections 9.2.3, 5.6, 6.6.2).

33. CO_2 might be captured in one country and stored in another with different commitments. Issues associated with accounting for cross-border storage are not unique to CCS.

Rules and methods for accounting may have to be adjusted accordingly. Possible physical leakage from a storage site in the future would have to be accounted for (Section 9.3).

What are the gaps in knowledge?

34. There are gaps in currently available knowledge regarding some aspects of CCS. Increasing knowledge and experience would reduce uncertainties and thus facilitate decision-making with respect to the deployment of CCS for climate change mitigation (Section TS.10).

---

[26] Convention on the Prevention of Marine Pollution by Dumping of Wastes and Other Matter (1972), and its London Protocol (1996), which has not yet entered into force.

[27] Convention for the Protection of the Marine Environment of the North-East Atlantic, which was adopted in Paris (1992). OSPAR is an abbreviation of Oslo-Paris.

[28] Revised 1996 IPCC Guidelines for National Greenhouse Gas Inventories, and Good Practice Guidance Reports; Good Practice Guidance and Uncertainty Management in National Greenhouse Gas Inventories, and Good Practice Guidance for Land Use, Land-Use Change and Forestry

IPCC Special Report

Carbon Dioxide Capture and Storage

Technical Summary

Coordinating Lead Authors
Edward Rubin (United States), Leo Meyer (Netherlands), Heleen de Coninck (Netherlands)

Lead Authors
Juan Carlos Abanades (Spain), Makoto Akai (Japan), Sally Benson (United States), Ken Caldeira (United States), Peter Cook (Australia), Ogunlade Davidson (Sierra Leone), Richard Doctor (United States), James Dooley (United States), Paul Freund (United Kingdom), John Gale (United Kingdom), Wolfgang Heidug (Germany), Howard Herzog (United States), David Keith (Canada), Marco Mazzotti (Italy and Switzerland), Bert Metz (Netherlands), Balgis Osman-Elasha (Sudan), Andrew Palmer (United Kingdom), Riitta Pipatti (Finland), Koen Smekens (Belgium), Mohammad Soltanieh (Iran), Kelly (Kailai) Thambimuthu (Australia and Canada), Bob van der Zwaan (Netherlands)

Review Editor
Ismail El Gizouli (Sudan)

Contents

1. Introduction and framework of this report

Carbon dioxide capture and storage (CCS), the subject of this Special Report, is considered as one of the options for reducing atmospheric emissions of CO_2 from human activities. The purpose of this Special Report is to assess the current state of knowledge regarding the technical, scientific, environmental, economic and societal dimensions of CCS and to place CCS in the context of other options in the portfolio of potential climate change mitigation measures.

The structure of this Technical Summary follows that of the Special Report. This introductory section presents the general framework for the assessment together with a brief overview of CCS systems. Section 2 then describes the major sources of CO_2, a step needed to assess the feasibility of CCS on a global scale. Technological options for CO_2 capture are then discussed in Section 3, while Section 4 focuses on methods of CO_2 transport. Following this, each of the storage options is addressed. Section 5 focuses on geological storage, Section 6 on ocean storage, and Section 7 on mineral carbonation and industrial uses of CO_2. The overall costs and economic potential of CCS are then discussed in Section 8, followed by an examination in Section 9 of the implications of CCS for greenhouse gas emissions inventories and accounting. The Technical Summary concludes with a discussion of gaps in knowledge, especially those critical for policy considerations.

Overview of CO_2 capture and storage

CO_2 is emitted principally from the burning of fossil fuels, both in large combustion units such as those used for electric power generation and in smaller, distributed sources such as automobile engines and furnaces used in residential and commercial buildings. CO_2 emissions also result from some industrial and resource extraction processes, as well as from the burning of forests during land clearance. CCS would most likely be applied to large point sources of CO_2, such as power plants or large industrial processes. Some of these sources could supply decarbonized fuel such as hydrogen to the transportation, industrial and building sectors, and thus reduce emissions from those distributed sources.

CCS involves the use of technology, first to collect and concentrate the CO_2 produced in industrial and energy-related sources, transport it to a suitable storage location, and then store it away from the atmosphere for a long period of time. CCS would thus allow fossil fuels to be used with low emissions of greenhouse gases. Application of CCS to biomass energy sources could result in the net removal of CO_2 from the atmosphere (often referred to as 'negative emissions') by capturing and storing the atmospheric CO_2 taken up by the biomass, provided the biomass is not harvested at an unsustainable rate.

Figure TS.1 illustrates the three main components of the CCS process: capture, transport and storage. All three components are found in industrial operations today, although mostly not for the purpose of CO_2 storage. The capture step involves separating CO_2 from other gaseous products. For fuel-burning processes such as those in power plants, separation technologies can be used to capture CO_2 after combustion or to decarbonize the fuel before combustion. The transport step may be required to carry captured CO_2 to a suitable storage site located at a distance from the CO_2 source. To facilitate both transport and storage, the captured CO_2 gas is typically compressed to a high density at the capture facility. Potential storage methods include injection into underground geological formations, injection into the deep ocean, or industrial fixation in inorganic carbonates. Some industrial processes also might utilize and store small amounts of captured CO_2 in manufactured products.

The technical maturity of specific CCS system components varies greatly. Some technologies are extensively deployed in mature markets, primarily in the oil and gas industry, while others are still in the research, development or demonstration phase. Table TS.1 provides an overview of the current status of all CCS components. As of mid-2005, there have been three commercial projects linking CO_2 capture and geological storage: the offshore Sleipner natural gas processing project in Norway, the Weyburn Enhanced Oil Recovery (EOR)[1] project in Canada (which stores CO_2 captured in the United States) and the In Salah natural gas project in Algeria. Each captures and stores 1–2 $MtCO_2$ per year. It should be noted, however, that CCS has not yet been applied at a large (e.g., 500 MW) fossil-fuel power plant, and that the overall system may not be as mature as some of its components.

---

[1] In this report, EOR means enhanced oil recovery using CO_2

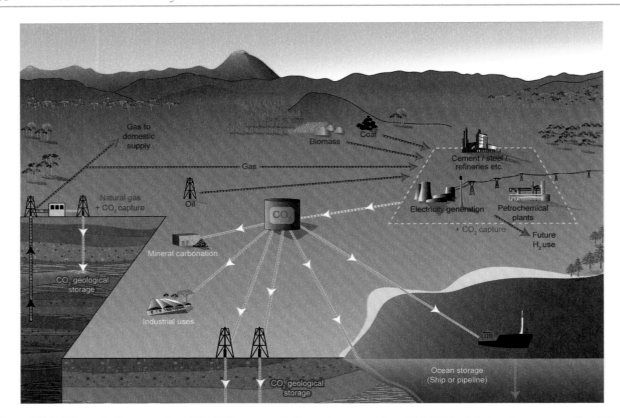

Figure TS.1. Schematic diagram of possible CCS systems. It shows the sources for which CCS might be relevant, as well as CO_2 transport and storage options (Courtesy CO2CRC).

Why the interest in CO_2 capture and storage?

In 1992, international concern about climate change led to the United Nations Framework Convention on Climate Change (UNFCCC). The ultimate objective of that Convention is the "stabilization of greenhouse gas concentrations in the atmosphere at a level that prevents dangerous anthropogenic interference with the climate system". From this perspective, the context for considering CCS (and other mitigation options) is that of a world constrained in CO_2 emissions, consistent with the international goal of stabilizing atmospheric greenhouse gas concentrations. Most scenarios for global energy use project a substantial increase of CO_2 emissions throughout this century in the absence of specific actions to mitigate climate change. They also suggest that the supply of primary energy will continue to be dominated by fossil fuels until at least the middle of the century (see Section 8). The magnitude of the emissions reduction needed to stabilize the atmospheric concentration of CO_2 will depend on both the level of future emissions (the baseline) and the

desired target for long-term CO_2 concentration: the lower the stabilization target and the higher the baseline emissions, the larger the required reduction in CO_2 emissions. IPCC's Third Assessment Report (TAR) states that, depending on the scenario considered, cumulative emissions of hundreds or even thousands of gigatonnes of CO_2 would need to be prevented during this century to stabilize the CO_2 concentration at 450 to 750 ppmv[2]. The TAR also finds that, "most model results indicate that known technological options[3] could achieve a broad range of atmospheric CO_2 stabilization levels", but that "no single technology option will provide all of the emissions reductions needed". Rather, a combination of mitigation measures will be needed to achieve stabilization. These known technological options are available for stabilization, although the TAR cautions that, "implementation would require associated socio-economic and institutional changes".

---

[2] ppmv is parts per million by volume.

[3] "Known technological options" refer to technologies that are currently at the operation or pilot-plant stages, as referred to in the mitigation scenarios discussed in IPCC's Third Assessment Report. The term does not include any new technologies that will require drastic technological breakthroughs. It can be considered to represent a conservative estimate given the length of the scenario period.

Table TS.1. Current maturity of CCS system components. An X indicates the highest level of maturity for each component. There are also less mature technologies for most components.

CCS component	CCS technology	Research phase [a]	Demonstration phase [b]	Economically feasible under specific conditions [c]	Mature market [d]
Capture	Post-combustion			X	
	Pre-combustion			X	
	Oxyfuel combustion		X		
	Industrial separation (natural gas processing, ammonia production)				X
Transportation	Pipeline				X
	Shipping			X	
Geological storage	Enhanced Oil Recovery (EOR)				X[e]
	Gas or oil fields			X	
	Saline formations			X	
	Enhanced Coal Bed Methane recovery (ECBM)[f]		X		
Ocean storage	Direct injection (dissolution type)	X			
	Direct injection (lake type)	X			
Mineral carbonation	Natural silicate minerals	X			
	Waste materials		X		
Industrial uses of CO_2					X

[a] Research phase means that the basic science is understood, but the technology is currently in the stage of conceptual design or testing at the laboratory or bench scale, and has not been demonstrated in a pilot plant.

[b] Demonstration phase means that the technology has been built and operated at the scale of a pilot plant, but further development is required before the technology is required before the technology is ready for the design and construction of a full-scale system.

[c] Economically feasible under specific conditions means that the technology is well understood and used in selected commercial applications, for instance if there is a favourable tax regime or a niche market, or processing on in the order of 0.1 $MtCO_2$ yr^{-1}, with few (less than 5) replications of the technology.

[d] Mature market means that the technology is now in operation with multiple replications of the technology worldwide.

[e] CO_2 injection for EOR is a mature market technology, but when used for CO_2 storage, it is only economically feasible under specific conditions.

[f] ECBM is the use of CO_2 to enhance the recovery of the methane present in unminable coal beds through the preferential adsorption of CO_2 on coal. Unminable coal beds are unlikely to ever be mined, because they are too deep or too thin. If subsequently mined, the stored CO_2 would be released.

In this context, the availability of CCS in the portfolio of options for reducing greenhouse gas emissions could facilitate the achievement of stabilization goals. Other technological options, which have been examined more extensively in previous IPCC assessments, include: (1) reducing energy demand by increasing the efficiency of energy conversion and/or utilization devices; (2) decarbonizing energy supplies (either by switching to less carbon-intensive fuels (coal to natural gas, for example), and/or by increasing the use of renewable energy sources and/or nuclear energy (each of which, on balance, emit little or no CO_2); (3) sequestering CO_2 through the enhancement of natural sinks by biological fixation; and (4) reducing non-CO_2 greenhouse gases.

Model results presented later in this report suggest that use of CCS in conjunction with other measures could significantly reduce the cost of achieving stabilization and would increase flexibility in achieving these reductions. The heavy worldwide reliance on fossil fuels today (approximately 80% of global energy use), the potential for CCS to reduce CO_2 emissions over the next century, and the compatibility of CCS systems with current energy infrastructures explain the interest in this technology.

Major issues for this assessment

There are a number of issues that need to be addressed in trying to understand the role that CCS could play in mitigating climate change. Questions that arise, and that are addressed in different sections of this Technical Summary, include the following:

- What is the current status of CCS technology?
- What is the potential for capturing and storing CO_2?
- What are the costs of implementation?
- How long should CO_2 be stored in order to achieve significant climate change mitigation?
- What are the health, safety and environment risks of CCS?
- What can be said about the public perception of CCS?
- What are the legal issues for implementing CO_2 storage?
- What are the implications for emission inventories and accounting?
- What is the potential for the diffusion and transfer of CCS technology?

When analyzing CCS as an option for climate change mitigation, it is of central importance that all resulting emissions from the system, especially emissions of CO_2, be identified and assessed in a transparent way. The importance of taking a "systems" view of CCS is therefore stressed, as the selection of an appropriate system boundary is essential for proper analysis. Given the energy requirements associated with capture and some storage and utilization options, and the possibility of leaking storage reservoirs, it is vital to assess the CCS chain as a whole.

From the perspectives of both atmospheric stabilization and long-term sustainable development, CO_2 storage must extend over time scales that are long enough to contribute significantly to climate change mitigation. This report expresses the duration of CO_2 storage in terms of the 'fraction retained', defined as the fraction of the cumulative mass of CO_2 injected that is retained in a storage reservoir over a specified period of time. Estimates of such fractions for different time periods and storage options are presented later. Questions arise not only about how long CO_2 will remain stored, but also what constitutes acceptable amounts of slow, continuous leakage[4] from storage. Different approaches to this question are discussed in Section 8.

CCS would be an option for countries that have significant sources of CO_2 suitable for capture, that have access to storage sites and experience with oil or gas operations, and that need to satisfy their development aspirations in a carbon-constrained environment. Literature assessed in the IPCC Special Report 'Methodological and Technological Issues and Technology

Transfer' indicates that there are many potential barriers that could inhibit deployment in developing countries, even of technologies that are mature in industrialized countries. Addressing these barriers and creating conditions that would facilitate diffusion of the technology to developing countries would be a major issue for the adoption of CCS worldwide.

2.   Sources of CO_2

This section describes the major current anthropogenic sources of CO_2 emissions and their relation to potential storage sites. As noted earlier, CO_2 emissions from human activity arise from a number of different sources, mainly from the combustion of fossil fuels used in power generation, transportation, industrial processes, and residential and commercial buildings. CO_2 is also emitted during certain industrial processes like cement manufacture or hydrogen production and during the combustion of biomass. Future emissions are also discussed in this section.

Current CO_2 sources and characteristics

To assess the potential of CCS as an option for reducing global CO_2 emissions, the current global geographical relationship between large stationary CO_2 emission sources and their proximity to potential storage sites has been examined. CO_2 emissions in the residential, commerical and transportation sectors have not been considered in this analysis because these emission sources are individually small and often mobile, and therefore unsuitable for capture and storage. The discussion here also includes an analysis of potential future sources of CO_2 based on several scenarios of future global energy use and emissions over the next century.

Globally, emissions of CO_2 from fossil-fuel use in the year 2000 totalled about 23.5 $GtCO_2$ yr^{-1} (6 GtC yr^{-1}). Of this, close to 60% was attributed to large (>0.1 $MtCO_2$ yr^{-1}) stationary emission sources (see Table TS.2). However, not all of these sources are amenable to CO_2 capture. Although the sources evaluated are distributed throughout the world, the database reveals four particular clusters of emissions: North America (midwest and eastern USA), Europe (northwest region), East Asia (eastern coast of China) and South Asia (Indian subcontinent). By contrast, large-scale biomass sources are much smaller in number and less globally distributed.

Currently, the vast majority of large emission sources have CO_2 concentrations of less than 15% (in some cases, substantially less). However, a small portion (less than 2%) of the fossil fuel-based industrial sources have CO_2 concentrations in excess of 95%. The high-concentration sources are potential candidates for the early implementation

---

[4] With respect to CO_2 storage, leakage is defined as the escape of injected fluid from storage. This is the most common meaning used in this Summary. If used in the context of trading of carbon dioxide emission reductions, it may signify the change in anthropogenic emissions by sources or removals by sinks which occurs outside the project boundary.

Table TS.2. Profile by process or industrial activity of worldwide large stationary CO_2 sources with emissions of more than 0.1 $MtCO_2$ per year.

Process	Number of sources	Emissions ($MtCO_2$ yr^{-1})
Fossil fuels		
Power	4,942	10,539
Cement production	1,175	932
Refineries	638	798
Iron and steel industry	269	646
Petrochemical industry	470	379
Oil and gas processing	N/A	50
Other sources	90	33
Biomass		
Bioethanol and bioenergy	303	91
Total	**7,887**	**13,466**

of CCS because only dehydration and compression would be required at the capture stage (see Section 3). An analysis of these high-purity sources that are within 50 km of storage formations and that have the potential to generate revenues (via the use of CO_2 for enhanced hydrocarbon production through ECBM or EOR) indicates that such sources currently emit approximately 360 $MtCO_2$ per year. Some biomass sources like bioethanol production also generate high-concentration CO_2 sources which could also be used in similar applications.

The distance between an emission location and a storage site can have a significant bearing on whether or not CCS can play a significant role in reducing CO_2 emissions. Figure TS.2a depicts the major CO_2 emission sources (indicated by dots), and Figure TS.2b shows the sedimentary basins with geological storage prospectivity (shown in different shades of grey). In broad terms, these figures indicate that there is potentially good correlation between major sources and prospective sedimentary basins, with many sources lying either directly above, or within reasonable distances (less than 300 km) from areas with potential for geological storage. The basins shown in Figure TS.2b have not been identified or evaluated as suitable storage reservoirs; more detailed geological analysis on a regional level is required to confirm the suitability of these potential storage sites.

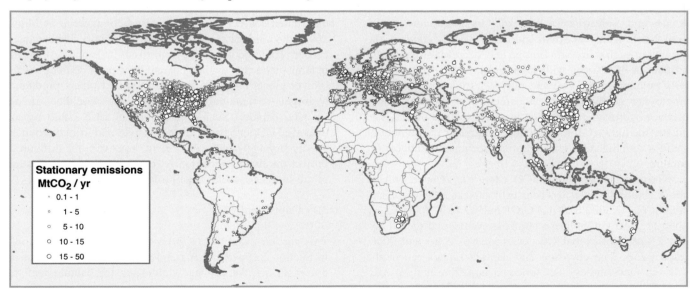

Stationary emissions $MtCO_2$ / yr
- 0.1 - 1
- 1 - 5
- 5 - 10
- 10 - 15
- 15 - 50

Figure TS.2a. Global distribution of large stationary sources of CO_2 (based on a compilation of publicly available information on global emission sources, IEA GHG 2002)

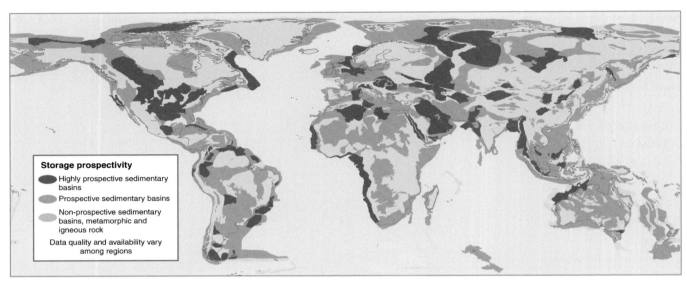

Figure TS.2b. Prospective areas in sedimentary basins where suitable saline formations, oil or gas fields, or coal beds may be found. Locations for storage in coal beds are only partly included. Prospectivity is a qualitative assessment of the likelihood that a suitable storage location is present in a given area based on the available information. This figure should be taken as a guide only, because it is based on partial data, the quality of which may vary from region to region, and which may change over time and with new information (Courtesy of Geoscience Australia).

Future emission sources

In the IPCC Special Report on Emission Scenarios (SRES), the future emissions of CO_2 are projected on the basis of six illustrative scenarios in which global CO_2 emissions range from 29 to 44 $GtCO_2$ (8–12 GtC) per year in 2020, and from 23 to 84 $GtCO_2$ (6–23 GtC) per year in 2050. It is projected that the number of CO_2 emission sources from the electric power and industrial sectors will increase significantly until 2050, mainly in South and East Asia. By contrast, the number of such sources in Europe may decrease slightly. The proportion of sources with high and low CO_2 content will be a function of the size and rate of introduction of plants employing gasification or liquefaction of fossil fuels to produce hydrogen, or other liquid and gaseous products. The greater the number of these plants, the greater the number of sources with high CO_2 concentrations technically suitable for capture.

The projected potential of CO_2 capture associated with the above emission ranges has been estimated at an annual 2.6 to 4.9 $GtCO_2$ by 2020 (0.7–1.3 GtC) and 4.7 to 37.5 $GtCO_2$ by 2050 (1.3–10 GtC). These numbers correspond to 9–12%, and 21–45% of global CO_2 emissions in 2020 and 2050, respectively. The emission and capture ranges reflect the inherent uncertainties of scenario and modelling analyses, and the technical limitations of applying CCS. These scenarios only take into account CO_2 capture from fossil fuels, and not from biomass sources. However, emissions from large-

scale biomass conversion facilities could also be technically suitable for capture.

The potential development of low-carbon energy carriers is relevant to the future number and size of large, stationary CO_2 sources with high concentrations. Scenarios also suggest that large-scale production of low-carbon energy carriers such as electricity or hydrogen could, within several decades, begin displacing the fossil fuels currently used by small, distributed sources in residential and commercial buildings and in the transportation sector (see Section 8). These energy carriers could be produced from fossil fuels and/or biomass in large plants that would generate large point sources of CO_2 (power plants or plants similar to current plants producing hydrogen from natural gas). These sources would be suitable for CO_2 capture. Such applications of CCS could reduce dispersed CO_2 emissions from transport and from distributed energy supply systems. At present, however, it is difficult to project the likely number, size, or geographical distribution of the sources associated with such developments.

3. Capture of CO_2

This section examines CCS capture technology. As shown in Section 2, power plants and other large-scale industrial processes are the primary candidates for capture and the main focus of this section.

Capture technology options and applications

The purpose of CO_2 capture is to produce a concentrated stream of CO_2 at high pressure that can readily be transported to a storage site. Although, in principle, the entire gas stream containing low concentrations of CO_2 could be transported and injected underground, energy costs and other associated costs generally make this approach impractical. It is therefore necessary to produce a nearly pure CO_2 stream for transport and storage. Applications separating CO_2 in large industrial plants, including natural gas treatment plants and ammonia production facilities, are already in operation today. Currently, CO_2 is typically removed to purify other industrial gas streams. Removal has been used for storage purposes in only a few cases; in most cases, the CO_2 is emitted to the atmosphere. Capture processes also have been used to obtain commercially useful amounts of CO_2 from flue gas streams generated by the combustion of coal or natural gas. To date, however, there have been no applications of CO_2 capture at large (e.g., 500 MW) power plants.

Depending on the process or power plant application in question, there are three main approaches to capturing the CO_2 generated from a primary fossil fuel (coal, natural gas or oil), biomass, or mixtures of these fuels:

Post-combustion systems separate CO_2 from the flue gases produced by the combustion of the primary fuel in air. These systems normally use a liquid solvent to capture the small fraction of CO_2 (typically 3–15% by volume) present in a flue gas stream in which the main constituent is nitrogen (from air). For a modern pulverized coal (PC) power plant or a natural gas combined cycle (NGCC) power plant, current post-combustion capture systems would typically employ an organic solvent such as monoethanolamine (MEA).

Pre-combustion systems process the primary fuel in a reactor with steam and air or oxygen to produce a mixture consisting mainly of carbon monoxide and hydrogen ("synthesis gas"). Additional hydrogen, together with CO_2, is produced by reacting the carbon monoxide with steam in a second reactor (a "shift reactor"). The resulting mixture of hydrogen and CO_2 can then be separated into a CO_2 gas stream, and a stream of hydrogen. If the CO_2 is stored, the hydrogen is a carbon-free energy carrier that can be combusted to generate power and/or heat. Although the initial fuel conversion steps are more elaborate and costly than in post-combustion systems, the high concentrations of CO_2 produced by the shift reactor (typically 15 to 60% by volume on a dry basis) and the high pressures often encountered in these applications are more favourable for CO_2 separation. Pre-combustion would be used at power plants that employ integrated gasification combined cycle (IGCC) technology.

Oxyfuel combustion systems use oxygen instead of air for combustion of the primary fuel to produce a flue gas that is mainly water vapour and CO_2. This results in a flue gas with high CO_2 concentrations (greater than 80% by volume). The water vapour is then removed by cooling and compressing the gas stream. Oxyfuel combustion requires the upstream separation of oxygen from air, with a purity of 95–99% oxygen assumed in most current designs. Further treatment of the flue gas may be needed to remove air pollutants and non-condensed gases (such as nitrogen) from the flue gas before the CO_2 is sent to storage. As a method of CO_2 capture in boilers, oxyfuel combustion systems are in the demonstration phase (see Table TS.1). Oxyfuel systems are also being studied in gas turbine systems, but conceptual designs for such applications are still in the research phase.

Figure TS.3 shows a schematic diagram of the main capture processes and systems. All require a step involving the separation of CO_2, H_2 or O_2 from a bulk gas stream (such as flue gas, synthesis gas, air or raw natural gas). These separation steps can be accomplished by means of physical or chemical solvents, membranes, solid sorbents, or by cryogenic separation. The choice of a specific capture technology is determined largely by the process conditions under which it must operate. Current post-combustion and pre-combustion systems for power plants could capture 85–95% of the CO_2 that is produced. Higher capture efficiencies are possible, although separation devices become considerably larger, more energy intensive and more costly. Capture and compression need roughly 10–40% more energy than the equivalent plant without capture, depending on the type of system. Due to the associated CO_2 emissions, the net amount of CO_2 captured is approximately 80–90%. Oxyfuel combustion systems are, in principle, able to capture nearly all of the CO_2 produced. However, the need for additional gas treatment systems to remove pollutants such as sulphur and nitrogen oxides lowers the level of CO_2 captured to slightly more than 90%.

As noted in Section 1, CO_2 capture is already used in several industrial applications (see Figure TS.4). The same technologies as would be used for pre-combustion capture are employed for the large-scale production of hydrogen (which is used mainly for ammonia and fertilizer manufacture, and for petroleum refinery operations). The separation of CO_2 from raw natural gas (which typically contains significant amounts of CO_2) is also practised on a large scale, using technologies similar to those used for post-combustion capture. Although commercial systems are also available for large-scale oxygen separation, oxyfuel combustion for CO_2 capture is currently in the demonstration phase. In addition, research is being conducted to achieve higher levels of system integration, increased efficiency and reduced cost for all types of capture systems.

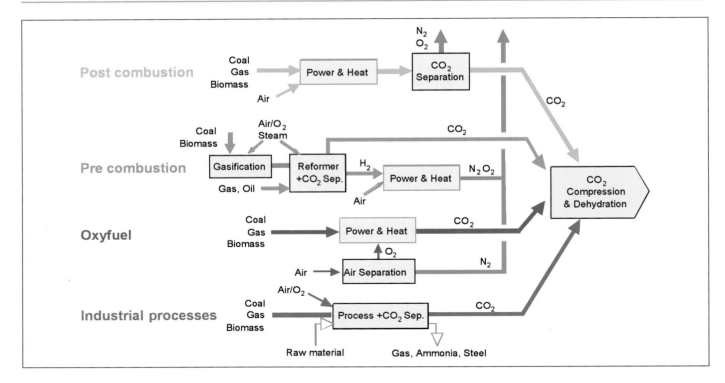

Figure TS.3. Overview of CO_2 capture processes and systems.

(a)

(b)

Figure TS.4. (a) CO_2 post-combustion capture at a plant in Malaysia. This plant employs a chemical absorption process to separate 0.2 $MtCO_2$ per year from the flue gas stream of a gas-fired power plant for urea production (Courtesy of Mitsubishi Heavy Industries). (b) CO_2 pre-combustion capture at a coal gasification plant in North Dakota, USA. This plant employs a physical solvent process to separate 3.3 $MtCO_2$ per year from a gas stream to produce synthetic natural gas. Part of the captured CO_2 is used for an EOR project in Canada.

CO₂ capture: risks, energy and the environment

The monitoring, risk and legal implications of CO_2 capture systems do not appear to present fundamentally new challenges, as they are all elements of regular health, safety and environmental control practices in industry. However, CO_2 capture systems require significant amounts of energy for their operation. This reduces net plant efficiency, so power plants require more fuel to generate each kilowatt-hour of electricity produced. Based on a review of the literature, the increase in fuel consumption per kWh for plants capturing 90% CO_2 using best current technology ranges from 24–40% for new supercritical PC plants, 11–22% for NGCC plants, and 14–25% for coal-based IGCC systems compared to similar plants without CCS. The increased fuel requirement results in an increase in most other environmental emissions per kWh generated relative to new state-of-the-art plants without CO_2 capture and, in the case of coal, proportionally larger amounts of solid wastes. In addition, there is an increase in the consumption of chemicals such as ammonia and limestone used by PC plants for nitrogen oxide and sulphur dioxide emissions control. Advanced plant designs that further reduce CCS energy requirements will also reduce overall environmental impacts as well as cost. Compared to many older existing plants, more efficient new or rebuilt plants with CCS may actually yield net reductions in plant-level environmental emissions.

Costs of CO₂ capture

The estimated costs of CO_2 capture at large power plants are based on engineering design studies of technologies in commercial use today (though often in different applications and/or at smaller scales than those assumed in the literature), as well as on design studies for concepts currently in the research and development (R&D) stage. Table TS.3 summarizes the results for new supercritical PC, NGCC and IGCC plants based on current technology with and without CO_2 capture. Capture systems for all three designs reduce CO_2 emissions per kWh by approximately 80–90%, taking into account the energy requirements for capture. All data for PC and IGCC plants in Table TS.3 are for bituminous coals only. The capture costs include the cost of compressing CO_2 (typically to about 11–14 MPa) but do not include the additional costs of CO_2 transport and storage (see Sections 4–7).

The cost ranges for each of the three systems reflect differences in the technical, economic and operating assumptions employed in different studies. While some differences in reported costs can be attributed to differences in the design of CO_2 capture systems, the major sources of variability are differences in the assumed design, operation and financing of the reference plant to which the capture technology is applied (factors such as plant size, location, efficiency, fuel type, fuel cost, capacity factor and cost of capital). No single set of assumptions applies to all situations or all parts of the world, so a range of costs is given.

For the studies listed in Table TS.3, CO_2 capture increases the cost of electricity production[5] by 35–70% (0.01 to 0.02 US$/kWh) for an NGCC plant, 40–85% (0.02 to 0.03 US$/kWh) for a supercritical PC plant, and 20–55% (0.01 to 0.02 US$/kWh) for an IGCC plant. Overall, the electricity production costs for fossil fuel plants with capture (excluding CO_2 transport and storage costs) ranges from 0.04–0.09 US$/kWh, as compared to 0.03–0.06 US$/kWh for similar plants without capture. In most studies to date, NGCC systems have typically been found to have lower electricity production costs than new PC and IGCC plants (with or without capture) in the case of large base-load plants with high capacity factors (75% or more) and natural gas prices between 2.6 and 4.4 US$ GJ⁻¹ over the life of the plant. However, in the case of higher gas prices and/or lower capacity factors, NGCC plants often have higher electricity production costs than coal-based plants, with or without capture. Recent studies also found that IGCC plants were on average slightly more costly without capture and slightly less costly with capture than similarly-sized PC plants. However, the difference in cost between PC and IGCC plants with or without CO_2 capture can vary significantly according to coal type and other local factors, such as the cost of capital for each plant type. Since full-scale NGCC, PC and IGCC systems have not yet been built with CCS, the absolute or relative costs of these systems cannot be stated with a high degree of confidence at this time.

The costs of retrofitting existing power plants with CO_2 capture have not been extensively studied. A limited number of reports indicate that retrofitting an amine scrubber to an existing plant results in greater efficiency loss and higher costs than those shown in Table TS.3. Limited studies also indicate that a more cost-effective option is to combine a capture system retrofit with rebuilding the boiler and turbine to increase plant efficiency and output. For some existing plants, studies indicate that similar benefits could be achieved by repowering with an IGCC system that includes CO_2 capture technology. The feasibility and cost of all these options is highly dependent on site-specific factors, including the size, age and efficiency of the plant, and the availability of additional space.

---

[5] The cost of electricity production should not be confused with the price of electricity to customers.

Table TS.3. Summary of CO_2 capture costs for new power plants based on current technology. Because these costs do not include the costs (or credits) for CO_2 transport and storage, this table should not be used to assess or compare total plant costs for different systems with capture. The full costs of CCS plants are reported in Section 8.

Performance and cost measures	New NGCC plant			New PC plant			New IGCC plant		
	Range		Rep.	Range		Rep.	Range		Rep.
	Low	High	value	Low	High	value	Low	High	value
Emission rate without capture (kgCO₂/kWh)	0.344 - 0.379		0.367	0.736 - 0.811		0.762	0.682 - 0.846		0.773
Emission rate with capture (kgCO₂/kWh)	0.040 - 0.066		0.052	0.092 - 0.145		0.112	0.065 - 0.152		0.108
Percentage CO_2 reduction per kWh (%)	83 - 88		86	81 - 88		85	81 - 91		86
Plant efficiency with capture, LHV basis (%)	47 - 50		48	30 - 35		33	31 - 40		35
Capture energy requirement (% increase input/kWh)	11 - 22		16	24 - 40		31	14 - 25		19
Total capital requirement without capture (US$/kW)	515 - 724		568	1161 - 1486		1286	1169 - 1565		1326
Total capital requirement with capture (US$/kW)	909 - 1261		998	1894 - 2578		2096	1414 - 2270		1825
Percent increase in capital cost with capture (%)	64 - 100		76	44 - 74		63	19 - 66		37
COE without capture (US$/kWh)	0.031 - 0.050		0.037	0.043 - 0.052		0.046	0.041 - 0.061		0.047
COE with capture only (US$/kWh)	0.043 - 0.072		0.054	0.062 - 0.086		0.073	0.054 - 0.079		0.062
Increase in COE with capture (US$/kWh)	0.012 - 0.024		0.017	0.018 - 0.034		0.027	0.009 - 0.022		0.016
Percent increase in COE with capture (%)	37 - 69		46	42 - 66		57	20 - 55		33
Cost of net CO_2 captured (US$/tCO₂)	37 - 74		53	29 - 51		41	13 - 37		23
Capture cost confidence level (see Table 3.6)	moderate			moderate			moderate		

Abbreviations: Representative value is based on the average of the values in the different studies. COE=cost of electricity production; LHV=lower heating value. See Section 3.6.1 for calculation of energy requirement for capture plants.

Notes: Ranges and representative values are based on data from Special Report Tables 3.7, 3.9 and 3.10. All PC and IGCC data are for bituminous coals only at costs of 1.0-1.5 US$ GJ⁻¹ (LHV); all PC plants are supercritical units. NGCC data based on natural gas prices of 2.8-4.4 US$ GJ⁻¹ (LHV basis). Cost are stated in constant US$2002. Power plant sizes range from approximately 400-800 MW without capture and 300-700 MW with capture. Capacity factors vary from 65-85% for coal plants and 50-95% for gas plants (average for each=80%). Fixed charge factors vary from 11-16%. All costs include CO_2 compression but not additional CO_2 transport and storage costs.

Table TS.4 illustrates the cost of CO_2 capture in the production of hydrogen. Here, the cost of CO_2 capture is mainly due to the cost of CO_2 drying and compression, since CO_2 separation is already carried out as part of the hydrogen production process. The cost of CO_2 capture adds approximately 5% to 30% to the cost of the hydrogen produced.

CCS also can be applied to systems that use biomass fuels or feedstock, either alone or in combination with fossil fuels. A limited number of studies have looked at the costs of such systems combining capture, transport and storage. The capturing of 0.19 $MtCO_2$ yr⁻¹ in a 24 MWe biomass IGCC plant is estimated to be about 80 US$/tCO₂ net captured (300

US$/tC), which corresponds to an increase in electricity production costs of about 0.08 US$/kWh. There are relatively few studies of CO_2 capture for other industrial processes using fossil fuels and they are typically limited to capture costs reported only as a cost per tonne of CO_2 captured or avoided. In general, the CO_2 produced in different processes varies widely in pressure and concentration (see Section 2). As a result, the cost of capture in different processes (cement and steel plants, refineries), ranges widely from about 25–115 US$/tCO₂ net captured. The unit cost of capture is generally lower for processes where a relatively pure CO_2 stream is produced (e.g. natural gas processing, hydrogen production and ammonia production), as seen for the hydrogen plants

Table TS.4. Summary of CO_2 capture costs for new hydrogen plants based on current technology

Performance and cost measures	New hydrogen plant			
	Range		Representative value	
	Low	High		
Emission rate without capture (kgCO$_2$ GJ^{-1})	78	-	174	137
Emission rate with capture (kgCO$_2$ GJ^{-1})	7	-	28	17
Percent CO$_2$ reduction per GJ (%)	72	-	96	86
Plant efficiency with capture, LHV basis (%)	52	-	68	60
Capture energy requirement (% more input GJ^{-1})	4	-	22	8
Cost of hydrogen without capture (US\$ GJ^{-1})	6.5	-	10.0	7.8
Cost of hydrogen with capture (US\$ GJ^{-1})	7.5	-	13.3	9.1
Increase in H$_2$ cost with capture (US\$ GJ^{-1})	0.3	-	3.3	1.3
Percent increase in H$_2$ cost with capture (%)	5	-	33	15
Cost of net CO$_2$ captured (US\$/tCO$_2$)	2	-	56	15
Capture cost confidence level	moderate to high			

Notes: Ranges and representative values are based on data from Table 3.11. All costs in this table are for capture only and do not include the costs of CO$_2$ transport and storage. Costs are in constant US\$2002. Hydrogen plant feedstocks are natural gas (4.7-5.3 US\$ GJ^{-1}) or coal (0.9-1.3 US\$ GJ^{-1}); some plants in dataset produce electricity in addition to hydrogen. Fixed charge factors vary from 13-20%. All costs include CO$_2$ compression but not additional CO$_2$ transport and storage costs (see Section 8 for full CCS costs).

in Table TS.4, where costs vary from 2–56 US\$/tCO$_2$ net captured.

New or improved methods of CO$_2$ capture, combined with advanced power systems and industrial process designs, could reduce CO$_2$ capture costs and energy requirements. While costs for first-of-a-kind commercial plants often exceed initial cost estimates, the cost of subsequent plants typically declines as a result of learning-by-doing and other factors. Although there is considerable uncertainty about the magnitude and timing of future cost reductions, the literature suggests that, provided R&D efforts are sustained, improvements to commercial technologies can reduce current CO$_2$ capture costs by at least 20–30% over approximately the next ten years, while new technologies under development could achieve more substantial cost reductions. Future cost reductions will depend on the deployment and adoption of commercial technologies in the marketplace as well as sustained R&D.

4. Transport of CO$_2$

Except when plants are located directly above a geological storage site, captured CO$_2$ must be transported from the point of capture to a storage site. This section reviews the principal methods of CO$_2$ transport and assesses the health, safety and environment aspects, and costs.

Methods of CO$_2$ transport

Pipelines today operate as a mature market technology and are the most common method for transporting CO$_2$. Gaseous CO$_2$ is typically compressed to a pressure above 8 MPa in order to avoid two-phase flow regimes and increase the density of the CO$_2$, thereby making it easier and less costly to transport. CO$_2$ also can be transported as a liquid in ships, road or rail tankers that carry CO$_2$ in insulated tanks at a temperature well below ambient, and at much lower pressures.

The first long-distance CO$_2$ pipeline came into operation in the early 1970s. In the United States, over 2,500 km of pipeline transports more than 40 MtCO$_2$ per year from natural and anthropogenic sources, mainly to sites in Texas, where the CO$_2$ is used for EOR. These pipelines operate in the 'dense phase' mode (in which there is a continuous progression from gas to liquid, without a distinct phase change), and at ambient temperature and high pressure. In most of these pipelines, the flow is driven by compressors at the upstream end, although some pipelines have intermediate (booster) compressor stations.

In some situations or locations, transport of CO_2 by ship may be economically more attractive, particularly when the CO_2 has to be moved over large distances or overseas. Liquefied petroleum gases (LPG, principally propane and butane) are transported on a large commercial scale by marine tankers. CO_2 can be transported by ship in much the same way (typically at 0.7 MPa pressure), but this currently takes place on a small scale because of limited demand. The properties of liquefied CO_2 are similar to those of LPG, and the technology could be scaled up to large CO_2 carriers if a demand for such systems were to materialize.

Road and rail tankers also are technically feasible options. These systems transport CO_2 at a temperature of -20°C and at 2 MPa pressure. However, they are uneconomical compared to pipelines and ships, except on a very small scale, and are unlikely to be relevant to large-scale CCS.

Environment, safety and risk aspects

Just as there are standards for natural gas admitted to pipelines, so minimum standards for 'pipeline quality' CO_2 should emerge as the CO_2 pipeline infrastructure develops further. Current standards, developed largely in the context of EOR applications, are not necessarily identical to what would be required for CCS. A low-nitrogen content is important for EOR, but would not be so significant for CCS. However, a CO_2 pipeline through populated areas might need a lower specified maximum H_2S content. Pipeline transport of CO_2 through populated areas also requires detailed route selection, over-pressure protection, leak detection and other design factors. However, no major obstacles to pipeline design for CCS are foreseen.

CO_2 could leak to the atmosphere during transport, although leakage losses from pipelines are very small. Dry (moisture-free) CO_2 is not corrosive to the carbon-manganese steels customarily used for pipelines, even if the CO_2 contains contaminants such as oxygen, hydrogen sulphide, and sulphur or nitrogen oxides. Moisture-laden CO_2, on the other hand, is highly corrosive, so a CO_2 pipeline in this case would have to be made from a corrosion-resistant alloy, or be internally clad with an alloy or a continuous polymer coating. Some pipelines are made from corrosion-resistant alloys, although the cost of materials is several times larger than carbon-manganese steels. For ships, the total loss to the atmosphere is between 3 and 4% per 1000 km, counting both boil-off and the exhaust from ship engines. Boil-off could be reduced by capture and liquefaction, and recapture would reduce the loss to 1 to 2% per 1000 km.

Accidents can also occur. In the case of existing CO_2 pipelines, which are mostly in areas of low population density, there have been fewer than one reported incident per year (0.0003 per km-year) and no injuries or fatalities. This is consistent with experience with hydrocarbon pipelines,

and the impact would probably not be more severe than for natural gas accidents. In marine transportation, hydrocarbon gas tankers are potentially dangerous, but the recognized hazard has led to standards for design, construction and operation, and serious incidents are rare.

Cost of CO_2 transport

Costs have been estimated for both pipeline and marine transportation of CO_2. In every case the costs depend strongly on the distance and the quantity transported. In the case of pipelines, the costs depend on whether the pipeline is onshore or offshore, whether the area is heavily congested, and whether there are mountains, large rivers, or frozen ground on the route. All these factors could double the cost per unit length, with even larger increases for pipelines in populated areas. Any additional costs for recompression (booster pump stations) that may be needed for longer pipelines would be counted as part of transport costs. Such costs are relatively small and not included in the estimates presented here.

Figure TS.5 shows the cost of pipeline transport for a nominal distance of 250 km. This is typically 1–8 US$/$tCO_2$ (4–30 US$/tC). The figure also shows how pipeline cost depends on the CO_2 mass flow rate. Steel cost accounts for a significant fraction of the cost of a pipeline, so fluctuations in such cost (such as the doubling in the years from 2003 to 2005) could affect overall pipeline economics.

In ship transport, the tanker volume and the characteristics of the loading and unloading systems are some of the key factors determining the overall transport cost.

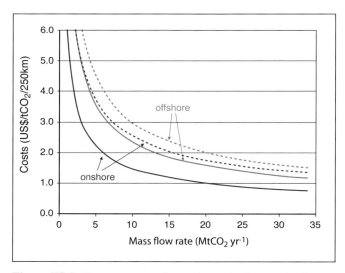

Figure TS.5. Transport costs for onshore pipelines and offshore pipelines, in US$ per tCO_2 per 250 km as a function of the CO_2 mass flow rate. The graph shows high estimates (dotted lines) and low estimates (solid lines).

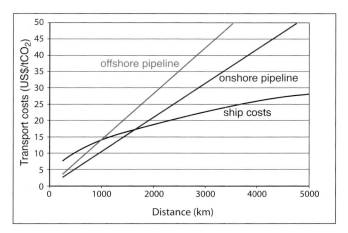

Figure TS.6. Costs, plotted as US$/tCO$_2$ transported against distance, for onshore pipelines, offshore pipelines and ship transport. Pipeline costs are given for a mass flow of 6 MtCO$_2$ yr^{-1}. Ship costs include intermediate storage facilities, harbour fees, fuel costs, and loading and unloading activities. Costs include also additional costs for liquefaction compared to compression.

The costs associated with CO$_2$ compression and liquefaction are accounted for in the capture costs presented earlier. Figure TS.6 compares pipeline and marine transportation costs, and shows the break-even distance. If the marine option is available, it is typically cheaper than pipelines for distances greater than approximately 1000 km and for amounts smaller than a few million tonnes of CO$_2$ per year. In ocean storage the most suitable transport system depends on the injection method: from a stationary floating vessel, a moving ship, or a pipeline from shore.

5. Geological storage

This section examines three types of geological formations that have received extensive consideration for the geological storage of CO$_2$: oil and gas reservoirs, deep saline formations and unminable coal beds (Figure TS.7). In each case, geological storage of CO$_2$ is accomplished by injecting it in dense form into a rock formation below the earth's surface. Porous rock formations that hold or (as in the case of depleted oil and gas reservoirs) have previously held fluids, such as natural gas, oil or brines, are potential candidates for CO$_2$ storage. Suitable storage formations can occur in both onshore and offshore sedimentary basins (natural large-scale depressions in the earth's crust that are filled with sediments). Coal beds also may be used for storage of CO$_2$ (see Figure TS.7) where it is unlikely that the coal will later be mined and provided that permeability is sufficient. The option of storing CO$_2$ in coal beds and enhancing methane production is still in the demonstration phase (see Table TS.1).

Existing CO$_2$ storage projects

Geological storage of CO$_2$ is ongoing in three industrial-scale projects (projects in the order of 1 MtCO$_2$ yr^{-1} or more): the Sleipner project in the North Sea, the Weyburn project in Canada and the In Salah project in Algeria. About 3–4 MtCO$_2$ that would otherwise be released to the atmosphere is captured and stored annually in geological formations. Additional projects are listed in Table TS.5.

In addition to the CCS projects currently in place, 30 MtCO$_2$ is injected annually for EOR, mostly in Texas, USA, where EOR commenced in the early 1970s. Most of this CO$_2$ is obtained from natural CO$_2$ reservoirs found in western regions of the US, with some coming from anthropogenic sources such as natural gas processing. Much of the CO$_2$ injected for EOR is produced with the oil, from which it is separated and then reinjected. At the end of the oil recovery, the CO$_2$ can be retained for the purpose of climate change mitigation, rather than vented to the atmosphere. This is planned for the Weyburn project.

Storage technology and mechanisms

The injection of CO$_2$ in deep geological formations involves many of the same technologies that have been developed in the oil and gas exploration and production industry. Well-drilling technology, injection technology, computer simulation of storage reservoir dynamics and monitoring methods from existing applications are being developed further for design and operation of geological storage. Other underground injection practices also provide relevant operational experience. In particular, natural gas storage, the deep injection of liquid wastes, and acid gas disposal (mixtures of CO$_2$ and H$_2$S) have been conducted in Canada and the U.S. since 1990, also at the megatonne scale.

CO$_2$ storage in hydrocarbon reservoirs or deep saline formations is generally expected to take place at depths below 800 m, where the ambient pressures and temperatures will usually result in CO$_2$ being in a liquid or supercritical state. Under these conditions, the density of CO$_2$ will range from 50 to 80% of the density of water. This is close to the density of some crude oils, resulting in buoyant forces that tend to drive CO$_2$ upwards. Consequently, a well-sealed cap rock over the selected storage reservoir is important to ensure that CO$_2$ remains trapped underground. When injected underground, the CO$_2$ compresses and fills the pore space by partially displacing the fluids that are already present (the 'in situ fluids'). In oil and gas reservoirs, the displacement of in situ fluids by injected CO$_2$ can result in most of the pore volume being available for CO$_2$ storage. In saline formations, estimates of potential storage volume are lower, ranging from as low as a few percent to over 30% of the total rock volume.

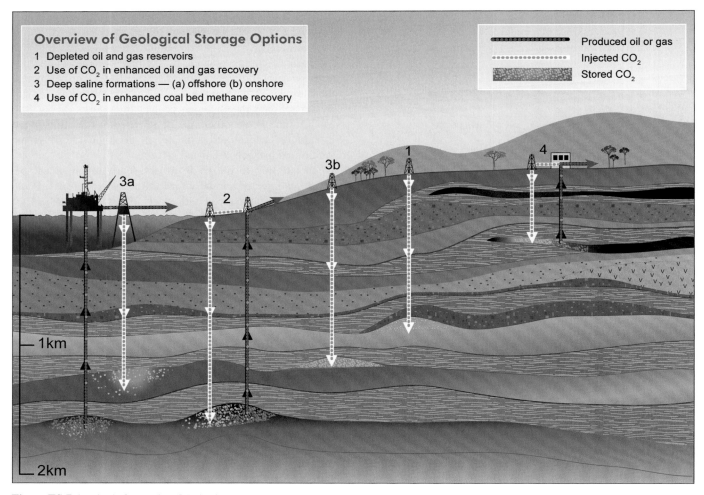

Figure TS.7. Methods for storing CO_2 in deep underground geological formations. Two methods may be combined with the recovery of hydrocarbons: EOR (2) and ECBM (4). See text for explanation of these methods (Courtesy CO2CRC).

Once injected into the storage formation, the fraction retained depends on a combination of physical and geochemical trapping mechanisms. Physical trapping to block upward migration of CO_2 is provided by a layer of shale and clay rock above the storage formation. This impermeable layer is known as the "cap rock". Additional physical trapping can be provided by capillary forces that retain CO_2 in the pore spaces of the formation. In many cases, however, one or more sides of the formation remain open, allowing for lateral migration of CO_2 beneath the cap rock. In these cases, additional mechanisms are important for the long-term entrapment of the injected CO_2.

The mechanism known as geochemical trapping occurs as the CO_2 reacts with the in situ fluids and host rock. First, CO_2 dissolves in the in situ water. Once this occurs (over time scales of hundreds of years to thousands of years), the CO_2-laden water becomes more dense and therefore sinks down into the formation (rather than rising toward the surface).

Next, chemical reactions between the dissolved CO_2 and rock minerals form ionic species, so that a fraction of the injected CO_2 will be converted to solid carbonate minerals over millions of years.

Yet another type of trapping occurs when CO_2 is preferentially adsorbed onto coal or organic-rich shales replacing gases such as methane. In these cases, CO_2 will remain trapped as long as pressures and temperatures remain stable. These processes would normally take place at shallower depths than CO_2 storage in hydrocarbon reservoirs and saline formations.

Geographical distribution and capacity of storage sites

As shown earlier in Section 2 (Figure TS.2b), regions with sedimentary basins that are potentially suitable for CO_2 storage exist around the globe, both onshore and offshore. This report focuses on oil and gas reservoirs, deep saline

Table TS.5. Sites where CO_2 storage has been done, is currently in progress or is planned, varying from small pilots to large-scale commercial applications.

Project name	Country	Injection start (year)	Approximate average daily injection rate (tCO$_2$ day^{-1})	Total (planned) storage (tCO$_2$)	Storage reservoir type
Weyburn	Canada	2000	3,000-5,000	20,000,000	EOR
In Salah	Algeria	2004	3,000-4,000	17,000,000	Gas field
Sleipner	Norway	1996	3,000	20,000,000	Saline formation
K12B	Netherlands	2004	100 (1,000 planned for 2006+)	8,000,000	Enhanced gas recovery
Frio	U.S.A	2004	177	1600	Saline formation
Fenn Big Valley	Canada	1998	50	200	ECBM
Qinshui Basin	China	2003	30	150	ECBM
Yubari	Japan	2004	10	200	ECBM
Recopol	Poland	2003	1	10	ECBM
Gorgon (planned)	Australia	~2009	10,000	unknown	Saline formation
Snøhvit (planned)	Norway	2006	2,000	unknown	Saline formation

formations and unminable coal beds. Other possible geological formations or structures (such as basalts, oil or gas shales, salt caverns and abandoned mines) represent niche opportunities, or have been insufficiently studied at this time to assess their potential.

The estimates of the technical potential[6] for different geological storage options are summarized in Table TS.6. The estimates and levels of confidence are based on an assessment of the literature, both of regional bottom-up, and global top-down estimates. No probabilistic approach to assessing capacity estimates exists in the literature, and this would be required to quantify levels of uncertainty reliably. Overall estimates, particularly of the upper limit of the potential, vary widely and involve a high degree of uncertainty, reflecting conflicting methodologies in the literature and the fact that our knowledge of saline formations is quite limited in most parts of the world. For oil and gas reservoirs, better estimates are available which are based on the replacement of hydrocarbon volumes with CO_2 volumes. It should be noted that, with the exception of EOR, these reservoirs will not be available for CO_2 storage until the hydrocarbons are depleted, and that pressure changes and geomechanical effects due to hydrocarbon production in the reservoir may reduce actual capacity.

Another way of looking at storage potential, however, is to ask whether it is likely to be adequate for the amounts of CO_2 that would need to be avoided using CCS under different greenhouse gas stabilization scenarios and assumptions about the deployment of other mitigation options. As discussed later in Section 8, the estimated range of economic potential[7] for CCS over the next century is roughly 200 to 2,000 GtCO$_2$. The lower limits in Table TS.6 suggest that, worldwide, it is virtually certain[8] that there is 200 GtCO$_2$ of geological storage capacity, and likely[9] that there is at least about 2,000 GtCO$_2$.

Site selection criteria and methods

Site characterization, selection and performance prediction are crucial for successful geological storage. Before selecting a site, the geological setting must be characterized to determine if the overlying cap rock will provide an effective seal, if there is a sufficiently voluminous and permeable storage formation, and whether any abandoned or active wells will compromise the integrity of the seal.

Techniques developed for the exploration of oil and gas reservoirs, natural gas storage sites and liquid waste disposal sites are suitable for characterizing geological storage sites for CO_2. Examples include seismic imaging, pumping tests for evaluating storage formations and seals, and cement integrity logs. Computer programmes that model underground CO_2 movement are used to support site characterization and selection activities. These programmes were initially developed for applications such as oil and

[6] Technical potential is the amount by which it is possible to reduce greenhouse gas emissions by implementing a technology or practice that already has been demonstrated.
[7] Economic potential is the amount of greenhouse gas emissions reductions from a specific option that could be achieved cost-effectively, given prevailing circumstances (the price of CO_2 reductions and costs of other options).
[8] "Virtually certain" is a probability of 99% or more.
[9] "Likely" is a probability of 66 to 90%.

Table TS.6. Storage capacity for several geological storage options. The storage capacity includes storage options that are not economical.

Reservoir type	Lower estimate of storage capacity (GtCO$_2$)	Upper estimate of storage capacity (GtCO$_2$)
Oil and gas fields	675[a]	900[a]
Unminable coal seams (ECBM)	3-15	200
Deep saline formations	1,000	Uncertain, but possibly 10^4

[a] These numbers would increase by 25% if 'undiscovered' oil and gas fields were included in this assessment.

gas reservoir engineering and groundwater resources investigations. Although they include many of the physical, chemical and geomechanical processes needed to predict both short-term and long-term performance of CO$_2$ storage, more experience is needed to establish confidence in their effectiveness in predicting long-term performance when adapted for CO$_2$ storage. Moreover, the availability of good site characterization data is critical for the reliability of models.

Risk assessment and environmental impact

The risks due to leakage from storage of CO$_2$ in geological reservoirs fall into two broad categories: global risks and local risks. Global risks involve the release of CO$_2$ that may contribute significantly to climate change if some fraction leaks from the storage formation to the atmosphere. In addition, if CO$_2$ leaks out of a storage formation, local hazards may exist for humans, ecosystems and groundwater. These are the local risks.

With regard to global risks, based on observations and analysis of current CO$_2$ storage sites, natural systems, engineering systems and models, the fraction retained in appropriately selected and managed reservoirs is very likely[10] to exceed 99% over 100 years, and is likely to exceed 99% over 1000 years. Similar fractions retained are likely for even longer periods of time, as the risk of leakage is expected to decrease over time as other mechanisms provide additional trapping. The question of whether these fractions retained would be sufficient to make impermanent storage valuable for climate change mitigation is discussed in Section 8.

With regard to local risks, there are two types of scenarios in which leakage may occur. In the first case, injection well failures or leakage up abandoned wells could create a sudden and rapid release of CO$_2$. This type of release is likely to be detected quickly and stopped using techniques that are available today for containing well blow-outs. Hazards associated with this type of release primarily affect workers in the vicinity of the release at the time it occurs, or those called in to control the blow-out. A concentration of CO$_2$ greater than 7–10% in air would cause immediate dangers to human life and health. Containing these kinds of releases may take hours to days and the overall amount of CO$_2$ released is likely to be very small compared to the total amount injected. These types of hazards are managed effectively on a regular basis in the oil and gas industry using engineering and administrative controls.

In the second scenario, leakage could occur through undetected faults, fractures or through leaking wells where the release to the surface is more gradual and diffuse. In this case, hazards primarily affect drinking-water aquifers and ecosystems where CO$_2$ accumulates in the zone between the surface and the top of the water table. Groundwater can be affected both by CO$_2$ leaking directly into an aquifer and by brines that enter the aquifer as a result of being displaced by CO$_2$ during the injection process. There may also be acidification of soils and displacement of oxygen in soils in this scenario. Additionally, if leakage to the atmosphere were to occur in low-lying areas with little wind, or in sumps and basements overlying these diffuse leaks, humans and animals would be harmed if a leak were to go undetected. Humans would be less affected by leakage from offshore storage locations than from onshore storage locations. Leakage routes can be identified by several techniques and by characterization of the reservoir. Figure TS.8 shows some of the potential leakage paths for a saline formation. When the potential leakage routes are known, the monitoring and remediation strategy can be adapted to address the potential leakage.

Careful storage system design and siting, together with methods for early detection of leakage (preferably long before CO$_2$ reaches the land surface), are effective ways of reducing hazards associated with diffuse leakage. The available monitoring methods are promising, but more experience is needed to establish detection levels and resolution. Once leakages are detected, some remediation techniques are available to stop or control them. Depending on the type of leakage, these techniques could involve standard well repair techniques, or the extraction of CO$_2$ by intercepting its leak into a shallow groundwater aquifer (see Figure TS.8).

[10] "Very likely" is a probability of 90 to 99%.

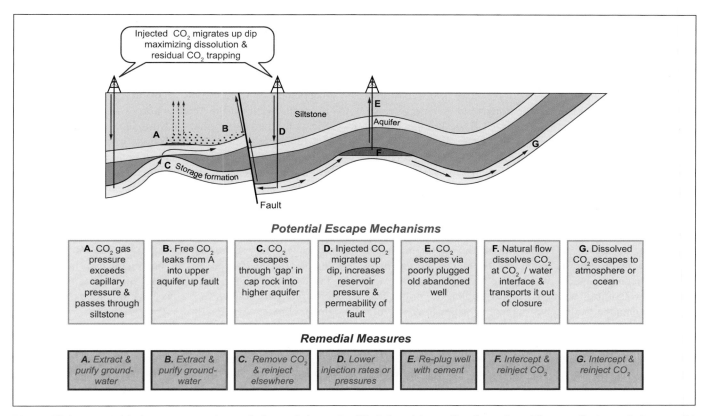

Figure TS.8. Potential leakage routes and remediation techniques for CO_2 injected into saline formations. The remediation technique would depend on the potential leakage routes identified in a reservoir (Courtesy CO2CRC).

Techniques to remove CO_2 from soils and groundwater are also available, but they are likely to be costly. Experience will be needed to demonstrate the effectiveness, and ascertain the costs, of these techniques for use in CO_2 storage.

Monitoring and verification

Monitoring is a very important part of the overall risk management strategy for geological storage projects. Standard procedures or protocols have not been developed yet but they are expected to evolve as technology improves, depending on local risks and regulations. However, it is expected that some parameters such as injection rate and injection well pressure will be measured routinely. Repeated seismic surveys have been shown to be useful for tracking the underground migration of CO_2. Newer techniques such as gravity and electrical measurements may also be useful. The sampling of groundwater and the soil between the surface and water table may be useful for directly detecting CO_2 leakage. CO_2 sensors with alarms can be located at the injection wells for ensuring worker safety and to detect leakage. Surface-based techniques may also be used for detecting and quantifying surface releases. High-quality baseline data improve the

reliability and resolution of all measurements and will be essential for detecting small rates of leakage.

Since all of these monitoring techniques have been adapted from other applications, they need to be tested and assessed with regard to reliability, resolution and sensitivity in the context of geological storage. All of the existing industrial-scale projects and pilot projects have programmes to develop and test these and other monitoring techniques. Methods also may be necessary or desirable to monitor the amount of CO_2 stored underground in the context of emission reporting and monitoring requirements in the UNFCCC (see Section 9). Given the long-term nature of CO_2 storage, site monitoring may be required for very long periods.

Legal issues

At present, few countries have specifically developed legal and regulatory frameworks for onshore CO_2 storage. Relevant legislation include petroleum-related legislation, drinking-water legislation and mining regulations. In many cases, there are laws applying to some, if not most, of the issues related to CO_2 storage. Specifically, long-term liability issues, such as global issues associated with the leakage of CO_2 to the atmosphere, as well as local concerns about environmental impact, have not yet been addressed. Monitoring and verification regimes and risks of leakage may play an important role in determining liability, and vice-versa. There are also considerations such as the longevity of institutions, ongoing monitoring and transferability of institutional knowledge. The long-term perspective is essential to a legal framework for CCS as storage times extend over many generations as does the climate change problem. In some countries, notably the US, the property rights of all those affected must be considered in legal terms as pore space is owned by surface property owners.

According to the general principles of customary international law, States can exercise their sovereignty in their territories and could therefore engage in activities such as the storage of CO_2 (both geological and ocean) in those areas under their jurisdiction. However, if storage has a transboundary impact, States have the responsibility to ensure that activities within their jurisdiction or control do not cause damage to the environment of other States or of areas beyond the limits of national jurisdiction.

Currently, there are several treaties (notably the UN Convention on the Law of the Sea, and the London[11] and OSPAR[12] Conventions) that could apply to the offshore injection of CO_2 into marine environments (both into the ocean and the geological sub-seabed). All these treaties have been drafted without specific consideration of CO_2 storage. An assessment undertaken by the Jurists and Linguists Group to the OSPAR Convention (relating to the northeast Atlantic region), for example, found that, depending on the method and purpose of injection, CO_2 injection into the geological sub-seabed and the ocean could be compatible with the treaty in some cases, such as when the CO_2 is transported via a pipeline from land. A similar assessment is now being conducted by Parties to the London Convention. Furthermore, papers by legal commentators have concluded that CO_2 captured from an oil or natural gas extraction operation and stored offshore in a geological formation (like the Sleipner operation) would not be considered 'dumping' under, and would not therefore be prohibited by, the London Convention.

Public perception

Assessing public perception of CCS is challenging because of the relatively technical and "remote" nature of this issue at the present time. Results of the very few studies conducted to date about the public perception of CCS indicate that the public is generally not well informed about CCS. If information is given alongside information about other climate change mitigation options, the handful of studies carried out so far indicate that CCS is generally regarded as less favourable than other options, such as improvements in energy efficiency and the use of non-fossil energy sources. Acceptance of CCS, where it occurs, is characterized as "reluctant" rather than "enthusiastic". In some cases, this reflects the perception that CCS might be required because of a failure to reduce CO_2 emissions in other ways. There are indications that geological storage could be viewed favourably if it is adopted in conjunction with more desirable measures. Although public perception is likely to change in the future, the limited research to date indicates that at least two conditions may have to be met before CO_2 capture and storage is considered by the public as a credible technology, alongside other better known options: (1) anthropogenic global climate change has to be regarded as a relatively serious problem; (2) there must be acceptance of the need for large reductions in CO_2 emissions to reduce the threat of global climate change.

Cost of geological storage

The technologies and equipment used for geological storage are widely used in the oil and gas industries so cost estimates for this option have a relatively high degree of confidence for storage capacity in the lower range of technical potential. However, there is a significant range and variability of costs due to site-specific factors such as onshore versus offshore, reservoir depth and geological characteristics of the storage formation (e.g., permeability and formation thickness).

Representative estimates of the cost for storage in saline formations and depleted oil and gas fields are typically between 0.5–8 US$/t$CO_2$ injected. Monitoring costs of 0.1–0.3 US$/t$CO_2$ are additional. The lowest storage costs are for onshore, shallow, high permeability reservoirs, and/or storage sites where wells and infrastructure from existing oil and gas fields may be re-used.

When storage is combined with EOR, ECBM or (potentially) Enhanced Gas Recovery (EGR), the economic value of CO_2 can reduce the total cost of CCS. Based on data and oil prices prior to 2003, enhanced oil production for onshore EOR with CO_2 storage could yield net benefits of 10–16 US$/t$CO_2$ (37–59 US$/tC) (including the costs of geological storage). For EGR and ECBM, which are still under development, there is no reliable cost information based on actual experience. In all cases, however, the economic benefit of enhanced production

---

[11] Convention on the Prevention of Marine Pollution by Dumping of Wastes and Other Matter (1972), and its London Protocol (1996), which has not yet entered into force.

[12] Convention for the Protection of the Marine Environment of the North-East Atlantic, which was adopted in Paris (1992). OSPAR is an abbreviation of Oslo-Paris.

depends strongly on oil and gas prices. In this regard, the literature basis for this report does not take into account the rise in world oil and gas prices since 2003 and assumes oil prices of 15–20 US$ per barrel. Should higher prices be sustained over the life of a CCS project, the economic value of CO_2 could be higher than that reported here.

6.   Ocean storage

A potential CO_2 storage option is to inject captured CO_2 directly into the deep ocean (at depths greater than 1,000 m), where most of it would be isolated from the atmosphere for centuries. This can be achieved by transporting CO_2 via pipelines or ships to an ocean storage site, where it is injected into the water column of the ocean or at the sea floor. The dissolved and dispersed CO_2 would subsequently become part of the global carbon cycle. Figure TS.9 shows some of the main methods that could be employed. Ocean storage has not yet been deployed or demonstrated at a pilot scale, and is still in the research phase. However, there have been small-scale field experiments and 25 years of theoretical, laboratory and modelling studies of intentional ocean storage of CO_2.

Storage mechanisms and technology

Oceans cover over 70% of the earth's surface and their average depth is 3,800 m. Because carbon dioxide is soluble in water, there are natural exchanges of CO_2 between the atmosphere and waters at the ocean surface that occur until equilibrium is reached. If the atmospheric concentration of CO_2 increases, the ocean gradually takes up additional CO_2. In this way, the oceans have taken up about 500 $GtCO_2$ (140 GtC) of the total 1,300 $GtCO_2$ (350 GtC) of anthropogenic emissions released to the atmosphere over the past 200 years. As a result of the increased atmospheric CO_2 concentrations from human activities relative to pre-industrial levels, the oceans are currently taking up CO_2 at a rate of about 7 $GtCO_2$ yr^{-1} (2 GtC yr^{-1}).

Most of this carbon dioxide now resides in the upper ocean and thus far has resulted in a decrease in pH of about 0.1 at the ocean surface because of the acidic nature of CO_2 in water. To date, however, there has been virtually no change in pH in the deep ocean. Models predict that over the next several centuries the oceans will eventually take up most of the CO_2 released to the atmosphere as CO_2 is dissolved at the ocean surface and subsequently mixed with deep ocean waters.

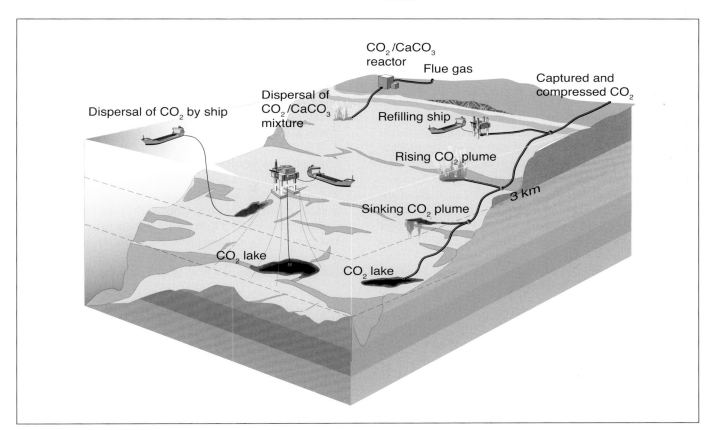

Figure TS.9. Methods of ocean storage.

There is no practical physical limit to the amount of anthropogenic CO_2 that could be stored in the ocean. However, on a millennial time scale, the amount stored will depend on oceanic equilibration with the atmosphere. Stabilizing atmospheric CO_2 concentrations between 350 ppmv and 1000 ppmv would imply that between 2,000 and 12,000 $GtCO_2$ would eventually reside in the ocean if there is no intentional CO_2 injection. This range therefore represents the upper limit for the capacity of the ocean to store CO_2 through active injection. The capacity would also be affected by environmental factors, such as a maximum allowable pH change.

Analysis of ocean observations and models both indicate that injected CO_2 will be isolated from the atmosphere for at least several hundreds of years, and that the fraction retained tends to be higher with deeper injection (see Table TS.7). Ideas for increasing the fraction retained include forming solid CO_2 hydrates and/or liquid CO_2 lakes on the sea floor, and dissolving alkaline minerals such as limestone to neutralize the acidic CO_2. Dissolving mineral carbonates, if practical, could extend the storage time scale to roughly 10,000 years, while minimizing changes in ocean pH and CO_2 partial pressure. However, large amounts of limestone and energy for materials handling would be required for this approach (roughly the same order of magnitude as the amounts per tonne of CO_2 injected that are needed for mineral carbonation; see Section 7).

Ecological and environmental impacts and risks

The injection of a few $GtCO_2$ would produce a measurable change in ocean chemistry in the region of injection, whereas the injection of hundreds of $GtCO_2$ would produce larger changes in the region of injection and eventually produce measurable changes over the entire ocean volume. Model simulations that assume a release from seven locations at 3,000 m depth and ocean storage providing 10% of the mitigation effort for stabilization at 550 ppmv CO_2 projected acidity changes (pH changes) of more than 0.4 over approximately 1% of the ocean volume. By comparison, in

a 550 ppmv stabilization case without ocean storage, a pH change of more than 0.25 at the ocean surface was estimated due to equilibration with the elevated CO_2 concentrations in the atmosphere. In either case, a pH change of 0.2 to 0.4 is significantly greater than pre-industrial variations in ocean acidity. Over centuries, ocean mixing will result in the loss of isolation of injected CO_2. As more CO_2 reaches the ocean surface waters, releases into the atmosphere would occur gradually from large regions of the ocean. There are no known mechanisms for sudden or catastrophic release of injected CO_2 from the ocean into the atmosphere.

Experiments show that adding CO_2 can harm marine organisms. Effects of elevated CO_2 levels have mostly been studied on time scales up to several months in individual organisms that live near the ocean surface. Observed phenomena include reduced rates of calcification, reproduction, growth, circulatory oxygen supply and mobility, as well as increased mortality over time. In some organisms these effects are seen in response to small additions of CO_2. Immediate mortality is expected close to injection points or CO_2 lakes. The chronic effects of direct CO_2 injection into the ocean on ocean organisms or ecosystems over large ocean areas and long time scales have not yet been studied.

No controlled ecosystem experiments have been performed in the deep ocean, so only a preliminary assessment of potential ecosystem effects can be given. It is expected that ecosystem consequences will increase with increasing CO_2 concentrations and decreasing pH, but the nature of such consequences is currently not understood, and no environmental criteria have as yet been identified to avoid adverse effects. At present, it is also unclear how or whether species and ecosystems would adapt to the sustained chemical changes.

Costs of ocean storage

Although there is no experience with ocean storage, some attempts have been made to estimate the costs of CO_2 storage projects that release CO_2 on the sea floor or in the deep ocean. The costs of CO_2 capture and transport to the shoreline (e.g

Table TS.7. Fraction of CO_2 retained for ocean storage as simulated by seven ocean models for 100 years of continuous injection at three different depths starting in the year 2000.

Year	Injection depth		
	800 m	**1500 m**	**3000 m**
2100	0.78 ± 0.06	0.91 ± 0.05	0.99 ± 0.01
2200	0.50 ± 0.06	0.74 ± 0.07	0.94 ± 0.06
2300	0.36 ± 0.06	0.60 ± 0.08	0.87 ± 0.10
2400	0.28 ± 0.07	0.49 ± 0.09	0.79 ± 0.12
2500	0.23 ± 0.07	0.42 ± 0.09	0.71 ± 0.14

Table TS.8. Costs for ocean storage at depths deeper than 3,000 m.

Ocean storage method	Costs (US$/tCO$_2$ net injected)	
	100 km offshore	**500 km offshore**
Fixed pipeline	6	31
Moving ship/platform[a]	12-14	13-16

[a] The costs for the moving ship option are for injection depths of 2,000-2,500 m.

via pipelines) are not included in the cost of ocean storage. However, the costs of offshore pipelines or ships, plus any additional energy costs, are included in the ocean storage cost. The costs of ocean storage are summarized in Table TS.8. These numbers indicate that, for short distances, the fixed pipeline option would be cheaper. For larger distances, either the moving ship or the transport by ship to a platform with subsequent injection would be more attractive.

Legal aspects and public perception

The global and regional treaties on the law of the sea and marine environment, such as the OSPAR and the London Convention discussed earlier in Section 5 for geological storage sites, also affect ocean storage, as they concern the 'maritime area'. Both Conventions distinguish between the storage method employed and the purpose of storage to determine the legal status of ocean storage of CO$_2$. As yet, however, no decision has been made about the legal status of intentional ocean storage.

The very small number of public perception studies that have looked at the ocean storage of CO$_2$ indicate that there is very little public awareness or knowledge of this subject. In the few studies conducted thus far, however, the public has expressed greater reservations about ocean storage than geological storage. These studies also indicate that the perception of ocean storage changed when more information was provided; in one study this led to increased acceptance of ocean storage, while in another study it led to less acceptance. The literature also notes that 'significant opposition' developed around a proposed CO$_2$ release experiment in the Pacific Ocean.

7. Mineral carbonation and industrial uses

This section deals with two rather different options for CO$_2$ storage. The first is mineral carbonation, which involves converting CO$_2$ to solid inorganic carbonates using chemical reactions. The second option is the industrial use of CO$_2$, either directly or as feedstock for production of various carbon-containing chemicals.

Mineral carbonation: technology, impacts and costs

Mineral carbonation refers to the fixation of CO$_2$ using alkaline and alkaline-earth oxides, such as magnesium oxide (MgO) and calcium oxide (CaO), which are present in naturally occurring silicate rocks such as serpentine and olivine. Chemical reactions between these materials and CO$_2$ produces compounds such as magnesium carbonate (MgCO$_3$) and calcium carbonate (CaCO$_3$, commonly known as limestone). The quantity of metal oxides in the silicate rocks that can be found in the earth's crust exceeds the amounts needed to fix all the CO$_2$ that would be produced by the combustion of all available fossil fuel reserves. These oxides are also present in small quantities in some industrial wastes, such as stainless steel slags and ashes. Mineral carbonation produces silica and carbonates that are stable over long time scales and can therefore be disposed of in areas such as silicate mines, or re-used for construction purposes (see Figure TS.10), although such re-use is likely to be small relative to the amounts produced. After carbonation, CO$_2$ would not be released to the atmosphere. As a consequence, there would be little need to monitor the disposal sites and the associated risks would be very low. The storage potential is difficult to estimate at this early phase of development. It would be limited by the fraction of silicate reserves that can be technically exploited, by environmental issues such as the volume of product disposal, and by legal and societal constraints at the storage location.

The process of mineral carbonation occurs naturally, where it is known as 'weathering'. In nature, the process occurs very slowly; it must therefore be accelerated considerably to be a viable storage method for CO$_2$ captured from anthropogenic sources. Research in the field of mineral carbonation therefore focuses on finding process routes that can achieve reaction rates viable for industrial purposes and make the reaction more energy-efficient. Mineral carbonation technology using natural silicates is in the research phase but some processes using industrial wastes are in the demonstration phase.

A commercial process would require mining, crushing and milling of the mineral-bearing ores and their transport to a processing plant receiving a concentrated CO$_2$ stream from a capture plant (see Figure TS.10). The carbonation process

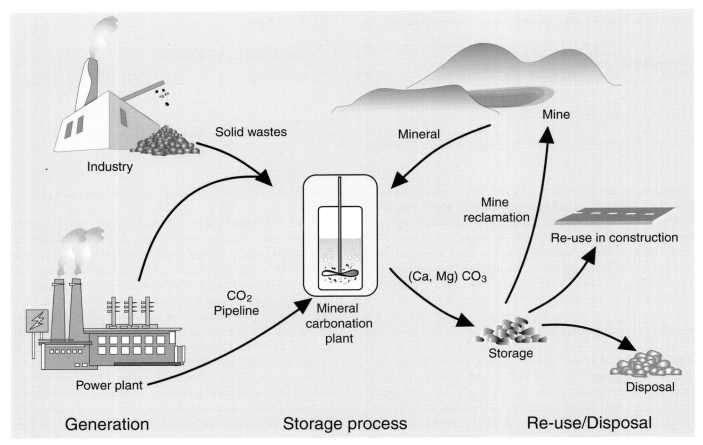

Figure TS.10. Material fluxes and process steps associated with the mineral carbonation of silicate rocks or industrial residues (Courtesy ECN).

energy required would be 30 to 50% of the capture plant output. Considering the additional energy requirements for the capture of CO_2, a CCS system with mineral carbonation would require 60 to 180% more energy input per kilowatt-hour than a reference electricity plant without capture or mineral carbonation. These energy requirements raise the cost per tonne of CO_2 avoided for the overall system significantly (see Section 8). The best case studied so far is the wet carbonation of natural silicate olivine. The estimated cost of this process is approximately 50–100 US$/t$CO_2$ net mineralized (in addition to CO_2 capture and transport costs, but taking into account the additional energy requirements). The mineral carbonation process would require 1.6 to 3.7 tonnes of silicates per tonne of CO_2 to be mined, and produce 2.6 to 4.7 tonnes of materials to be disposed per tonne of CO_2 stored as carbonates. This would therefore be a large operation, with an environmental impact similar to that of current large-scale surface mining operations. Serpentine also often contains chrysotile, a natural form of asbestos. Its presence therefore demands monitoring and mitigation measures of the kind available in the mining industry. On the other hand, the products of mineral carbonation are chrysotile-free, since this is the most reactive component of the rock and therefore the first substance converted to carbonates.

A number of issues still need to be clarified before any estimates of the storage potential of mineral carbonation can be given. The issues include assessments of the technical feasibility and corresponding energy requirements at large scales, but also the fraction of silicate reserves that can be technically and economically exploited for CO_2 storage. The environmental impact of mining, waste disposal and product storage could also limit potential. The extent to which mineral carbonation may be used cannot be determined at this time, since it depends on the unknown amount of silicate reserves that can be technically exploited, and environmental issuessuch as those noted above.

Industrial uses

Industrial uses of CO_2 include chemical and biological processes where CO_2 is a reactant, such as those used in urea and methanol production, as well as various technological applications that use CO_2 directly, for example in the horticulture industry, refrigeration, food packaging, welding, beverages and fire extinguishers. Currently, CO_2 is used at a rate of approximately 120 $MtCO_2$ per year (30 MtC yr^{-1}) worldwide, excluding use for EOR (discussed in Section 5). Most (two thirds of the total) is used to produce urea, which is used in the manufacture of fertilizers and other products. Some of the CO_2 is extracted from natural wells, and some originates from industrial sources – mainly high-concentration sources such as ammonia and hydrogen production plants – that capture CO_2 as part of the production process.

Industrial uses of CO_2 can, in principle, contribute to keeping CO_2 out of the atmosphere by storing it in the "carbon chemical pool" (i.e., the stock of carbon-bearing manufactured products). However, as a measure for mitigating climate change, this option is meaningful only if the quantity and duration of CO_2 stored are significant, and if there is a real net reduction of CO_2 emissions. The typical lifetime of most of the CO_2 currently used by industrial processes has storage times of only days to months. The stored carbon is then degraded to CO_2 and again emitted to the atmosphere. Such short time scales do not contribute meaningfully to climate change mitigation. In addition, the total industrial use figure of 120 $MtCO_2$ yr^{-1} is small compared to emissions from major anthropogenic sources (see Table TS.2). While some industrial processes store a small proportion of CO_2 (totalling roughly 20 $MtCO_2$ yr^{-1}) for up to several decades, the total amount of long-term (century-scale) storage is presently in the order of 1 $MtCO_2$ yr^{-1} or less, with no prospects for major increases.

Another important question is whether industrial uses of CO_2 can result in an overall net reduction of CO_2 emissions by substitution for other industrial processes or products. This can be evaluated correctly only by considering proper system boundaries for the energy and material balances of the CO_2 utilization processes, and by carrying out a detailed life-cycle analysis of the proposed use of CO_2. The literature in this area is limited but it shows that precise figures are difficult to estimate and that in many cases industrial uses could lead to an increase in overall emissions rather than a net reduction. In view of the low fraction of CO_2 retained, the small volumes used and the possibility that substitution may lead to increases in CO_2 emissions, it can be concluded that the contribution of industrial uses of captured CO_2 to climate change mitigation is expected to be small.

8. Costs and economic potential

The stringency of future requirements for the control of greenhouse gas emissions and the expected costs of CCS systems will determine, to a large extent, the future deployment of CCS technologies relative to other greenhouse gas mitigation options. This section first summarizes the overall cost of CCS for the main options and process applications considered in previous sections. As used in this summary and the report, "costs" refer only to market prices but do not include external costs such as environmental damages and broader societal costs that may be associated with the use of CCS. To date, little has been done to assess and quantify such external costs. Finally CCS is examined in the context of alternative options for global greenhouse gas reductions.

Cost of CCS systems

As noted earlier, there is still relatively little experience with the combination of CO_2 capture, transport and storage in a fully integrated CCS system. And while some CCS components are already deployed in mature markets for certain industrial applications, CCS has still not been used in large-scale power plants (the application with most potential).

The literature reports a fairly wide range of costs for CCS components (see Sections 3–7). The range is due primarily to the variability of site-specific factors, especially the design, operating and financing characteristics of the power plants or industrial facilities in which CCS is used; the type and costs of fuel used; the required distances, terrains and quantities involved in CO_2 transport; and the type and characteristics of the CO_2 storage. In addition, uncertainty still remains about the performance and cost of current and future CCS technology components and integrated systems. The literature reflects a widely-held belief, however, that the cost of building and operating CO_2 capture systems will decline over time as a result of learning-by-doing (from technology deployment) and sustained R&D. Historical evidence also suggests that costs for first-of-a-kind capture plants could exceed current estimates before costs subsequently decline. In most CCS systems, the cost of capture (including compression) is the largest cost component. Costs of electricity and fuel vary considerably from country to country, and these factors also influence the economic viability of CCS options.

Table TS.9 summarizes the costs of CO_2 capture, transport and storage reported in Sections 3 to 7. Monitoring costs are also reflected. In Table TS.10, the component costs are combined to show the total costs of CCS and electricity generation for three power systems with pipeline transport and two geological storage options.

For the plants with geological storage and no EOR credit, the cost of CCS ranges from 0.02–0.05 US$/kWh for PC plants and 0.01–0.03 US$/kWh for NGCC plants (both employing post-combustion capture). For IGCC plants (using pre-combustion capture), the CCS cost ranges from 0.01–0.03 US$/kWh relative to a similar plant without CCS. For all electricity systems, the cost of CCS can be reduced by about 0.01–0.02 US$/kWh when using EOR with CO_2 storage because the EOR revenues partly compensate for the CCS costs. The largest cost reductions are seen for coal-based plants, which capture the largest amounts of CO_2. In a few cases, the low end of the CCS cost range can be negative,

Table TS.9. 2002 Cost ranges for the components of a CCS system as applied to a given type of power plant or industrial source. The costs of the separate components cannot simply be summed to calculate the costs of the whole CCS system in US$/CO$_2$ avoided. All numbers are representative of the costs for large-scale, new installations, with natural gas prices assumed to be 2.8-4.4 US$ GJ^{-1} and coal prices 1-1.5 US$ GJ^{-1}.

CCS system components	Cost range	Remarks
Capture from a coal- or gas-fired power plant	15-75 US$/tCO$_2$ net captured	Net costs of captured CO$_2$, compared to the same plant without capture.
Capture from hydrogen and ammonia production or gas processing	5-55 US$/tCO$_2$ net captured	Applies to high-purity sources requiring simple drying and compression.
Capture from other industrial sources	25-115 US$/tCO$_2$ net captured	Range reflects use of a number of different technologies and fuels.
Transportation	1-8 US$/tCO$_2$ transported	Per 250 km pipeline or shipping for mass flow rates of 5 (high end) to 40 (low end) MtCO$_2$ yr^{-1}.
Geological storage[a]	0.5-8 US$/tCO$_2$ net injected	Excluding potential revenues from EOR or ECBM.
Geological storage: monitoring and verification	0.1-0.3 US$/tCO$_2$ injected	This covers pre-injection, injection, and post-injection monitoring, and depends on the regulatory requirements.
Ocean storage	5-30 US$/tCO$_2$ net injected	Including offshore transportation of 100-500 km, excluding monitoring and verification.
Mineral carbonation	50-100 US$/tCO$_2$ net mineralized	Range for the best case studied. Includes additional energy use for carbonation.

[a] Over the long term, there may be additional costs for remediation and liabilities.

indicating that the assumed credit for EOR over the life of the plant is greater than the lowest reported cost of CO$_2$ capture for that system. This might also apply in a few instances of low-cost capture from industrial processes.

In addition to fossil fuel-based energy conversion processes, CO$_2$ could also be captured in power plants fueled with biomass, or fossil-fuel plants with biomass co-firing. At present, biomass plants are small in scale (less than 100 MW$_e$). This means that the resulting costs of production with and without CCS are relatively high compared to fossil alternatives. Full CCS costs for biomass could amount to 110 US$/tCO$_2$ avoided. Applying CCS to biomass-fuelled or co-fired conversion facilities would lead to lower or negative[13] CO$_2$ emissions, which could reduce the costs for this option, depending on the market value of CO$_2$ emission reductions. Similarly, CO$_2$ could be captured in biomass-fueled H$_2$ plants. The cost is reported to be 22–25 US$/tCO$_2$ (80–92 US$/tC) avoided in a plant producing 1 million Nm^3 day^{-1} of H$_2$, and corresponds to an increase in the H$_2$ product costs of about 2.7 US$ GJ^{-1}. Significantly larger biomass plants could potentially benefit from economies of scale, bringing down costs of the CCS systems to levels broadly similar to coal plants. However, to date, there has been little experience with large-scale biomass plants, so their feasibility has not been proven yet, and costs and potential are difficult to estimate.

The cost of CCS has not been studied in the same depth for non-power applications. Because these sources are very diverse in terms of CO$_2$ concentration and gas stream pressure, the available cost studies show a very broad range. The lowest costs were found for processes that already separate CO$_2$ as part of the production process, such as hydrogen production (the cost of capture for hydrogen production was reported earlier in Table TS.4). The full CCS cost, including transport and storage, raises the cost of hydrogen production by 0.4 to 4.4 US$ GJ^{-1} in the case of geological storage, and by -2.0 to 2.8 US$ GJ^{-1} in the case of EOR, based on the same cost assumptions as for Table TS.10.

Cost of CO$_2$ avoided

Table TS.10 also shows the ranges of costs for 'CO$_2$ avoided'. CCS energy requirements push up the amount of fuel input (and therefore CO$_2$ emissions) per unit of net power output. As a result, the amount of CO$_2$ produced per unit of product (a kWh of electricity) is greater for the power plant with CCS than the reference plant, as shown in Figure TS.11. To determine the CO$_2$ reductions one can attribute to CCS, one needs to compare CO$_2$ emissions per kWh of the plant with capture to that of a reference plant without capture. The difference is referred to as the 'avoided emissions'.

[13] If for example the biomass is harvested at an unsustainable rate (that is, faster than the annual re-growth), the net CO$_2$ emissions of the activity might not be negative.

Table TS.10. Range of total costs for CO_2 capture, transport and geological storage based on current technology for new power plants using bituminous coal or natural gas

Power plant performance and cost parameters[a]	Pulverized coal power plant	Natural gas combined cycle power plant	Integrated coal gasification combined cycle power plant
Reference plant without CCS			
Cost of electricity (US$/kWh)	0.043-0.052	0.031-0.050	0.041-0.061
Power plant with capture			
Increased fuel requirement (%)	24-40	11-22	14-25
CO_2 captured (kg/kWh)	0.82-0.97	0.36-0.41	0.67-0.94
CO_2 avoided (kg/kWh)	0.62-0.70	0.30-0.32	0.59-0.73
% CO_2 avoided	81-88	83-88	81-91
Power plant with capture and geological storage[b]			
Cost of electricity (US$/kWh)	0.063-0.099	0.043-0.077	0.055-0.091
Cost of CCS (US$/kWh)	0.019-0.047	0.012-0.029	0.010-0.032
% increase in cost of electricity	43-91	37-85	21-78
Mitigation cost (US$/t$CO_2$ avoided)	30-71	38-91	14-53
(US$/tC avoided)	110-260	140-330	51-200
Power plant with capture and enhanced oil recovery[c]			
Cost of electricity (US$/kWh)	0.049-0.081	0.037-0.070	0.040-0.075
Cost of CCS (US$/kWh)	0.005-0.029	0.006-0.022	(-0.005)-0.019
% increase in cost of electricity	12-57	19-63	(-10)-46
Mitigation cost (US$/t$CO_2$ avoided)	9-44	19-68	(-7)-31
(US$/tC avoided)	31-160	71-250	(-25)-120

[a] All changes are relative to a similar (reference) plant without CCS. See Table TS.3 for details of assumptions underlying reported cost ranges.

[b] Capture costs based on ranges from Table TS.3; transport costs range from 0-5 US$/t$CO_2$; geological storage cost ranges from 0.6-8.3 US$/t$CO_2$.

[c] Same capture and transport costs as above; Net storage costs for EOR range from -10 to -16 US$/t$CO_2$ (based on pre-2003 oil prices of 15-20 US$ per barrel).

Introducing CCS to power plants may influence the decision about which type of plant to install and which fuel to use. In some situations therefore, it can be useful to calculate a cost per tonne of CO_2 avoided based on a reference plant different from the CCS plant. Table TS.10 displays the cost and emission factors for the three reference plants and the corresponding CCS plants for the case of geological storage. Table TS.11 summarizes the range of estimated costs for different combinations of CCS plants and the lowest-cost reference plants of potential interest. It shows, for instance, that where a PC plant is planned initially, using CCS in that plant may lead to a higher CO_2 avoidance cost than if an NGCC plant with CCS is selected, provided natural gas is available. Another option with lower avoidance cost could be to build an IGCC plant with capture instead of equipping a PC plant with capture.

Economic potential of CCS for climate change mitigation

Assessments of the economic potential of CCS are based on energy and economic models that study future CCS deployment and costs in the context of scenarios that achieve economically efficient, least-cost paths to the stabilization of atmospheric CO_2 concentrations.

While there are significant uncertainties in the quantitative results from these models (see discussion below), all models indicate that CCS systems are unlikely to be deployed on a large scale in the absence of an explicit policy that substantially limits greenhouse gas emissions to the atmosphere. With greenhouse gas emission limits imposed, many integrated assessments foresee the deployment of CCS systems on a large scale within a few decades from the start of any significant climate change mitigation regime. Energy and economic models indicate that CCS systems

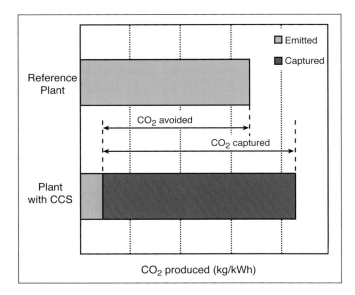

Figure TS.11. CO_2 capture and storage from power plants. The increased CO_2 production resulting from loss in overall efficiency of power plants due to the additional energy required for capture, transport and storage, and any leakage from transport result in a larger amount of "CO_2 produced per unit of product" (lower bar) relative to the reference plant (upper bar) without capture.

are unlikely to contribute significantly to the mitigation of climate change unless deployed in the power sector. For this

to happen, the price of carbon dioxide reductions would have to exceed 25–30 US$/t$CO_2$, or an equivalent limit on CO_2 emissions would have to be mandated. The literature and current industrial experience indicate that, in the absence of measures for limiting CO_2 emissions, there are only small, niche opportunities for CCS technologies to deploy. These early opportunities involve CO_2 captured from a high-purity, low-cost source, the transport of CO_2 over distances of less than 50 km, coupled with CO_2 storage in a value-added application such as EOR. The potential of such niche options is about 360 MtCO_2 per year (see Section 2).

Models also indicate that CCS systems will be competitive with other large-scale mitigation options such as nuclear power and renewable energy technologies. These studies show that including CCS in a mitigation portfolio could reduce the cost of stabilizing CO_2 concentrations by 30% or more. One aspect of the cost competitiveness of CCS technologies is that they are compatible with most current energy infrastructures.

In most scenarios, emissions abatement becomes progressively more constraining over time. Most analyses indicate that notwithstanding significant penetration of CCS systems by 2050, the majority of CCS deployment will occur in the second half of this century. The earliest CCS deployments are typically foreseen in the industrialized nations, with deployment eventually spreading worldwide. While results for different scenarios and models differ (often

Table TS.11. Mitigation cost ranges for different combinations of reference and CCS plants based on current technology for new power plants. Currently, in many regions, common practice would be either a PC plant or an NGCC plant[14]. EOR benefits are based on oil prices of 15 - 20 US$ per barrel. Gas prices are assumed to be 2.8 -4.4 US$/GJ[-1], coal prices 1-1.5 US$/GJ[-1] (based on Table 8.3a).

CCS plant type	NGCC reference plant US$/t$CO_2$ avoided (US$/tC avoided)	PC reference plant US$/t$CO_2$ avoided (US$/tC avoided)
Power plant with capture and geological storage		
NGCC	40 - 90 (140 - 330)	20 - 60 (80 - 220)
PC	70 - 270 (260 - 980)	30 - 70 (110 - 260)
IGCC	40 - 220 (150 - 790)	20 - 70 (80 - 260)
Power plant with capture and EOR		
NGCC	20 - 70 (70 - 250)	0 - 30 (0 - 120)
PC	50 - 240 (180 - 890)	10 - 40 (30 - 160)
IGCC	20 - 190 (80 - 710)	0 - 40 (0 - 160)

[14] IGCC is not included as a reference power plant that would be built today since this technology is not yet widely deployed in the electricity sector and is usually slightly more costly than a PC plant.

significantly) in the specific mix and quantities of different measures needed to achieve a particular emissions constraint (see Figure TS.12), the consensus of the literature shows that CCS could be an important component of the broad portfolio of energy technologies and emission reduction approaches.

The actual use of CCS is likely to be lower than the estimates of economic potential indicated by these energy and economic models. As noted earlier, the results are typically based on an optimized least-cost analysis that does not adequately account for real-world barriers to technology development and deployment, such as environmental impact, lack of a clear legal or regulatory framework, the perceived investment risks of different technologies, and uncertainty as to how quickly the cost of CCS will be reduced through R&D and learning-by-doing. Models typically employ simplified assumptions regarding the costs of CCS for different applications and the rates at which future costs will be reduced.

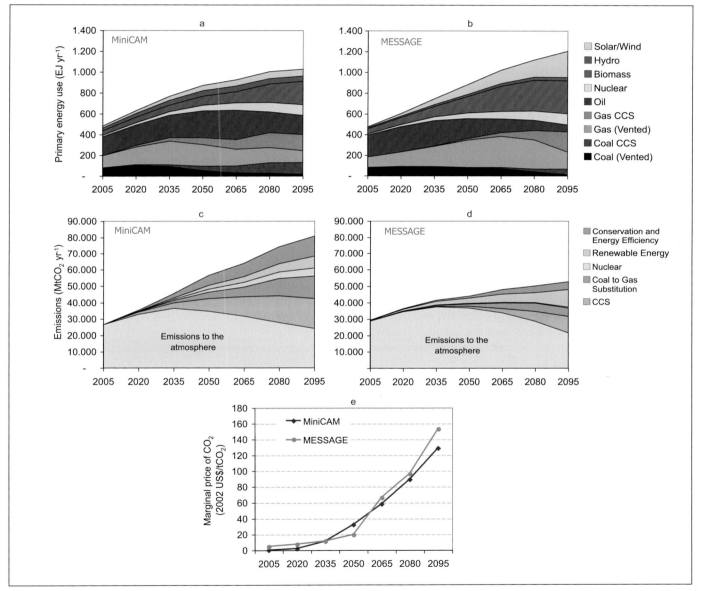

Figure TS.12. These figures are an illustrative example of the global potential contribution of CCS as part of a mitigation portfolio. They are based on two alternative integrated assessment models (MESSAGE and MiniCAM) adopting the same assumptions for the main emissions drivers. The results would vary considerably on regional scales. This example is based on a single scenario and therefore does not convey the full range of uncertainties. Panels a) and b) show global primary energy use, including the deployment of CCS. Panels c) and d) show the global CO_2 emissions in grey and corresponding contributions of main emissions reduction measures in colour. Panel e) shows the calculated marginal price of CO_2 reductions.

For CO_2 stabilization scenarios between 450 and 750 ppmv, published estimates of the cumulative amount of CO_2 potentially stored globally over the course of this century (in geological formations and/or the oceans) span a wide range, from very small contributions to thousands of gigatonnes of CO_2. To a large extent, this wide range is due to the uncertainty of long-term socio-economic, demographic and, in particular, technological changes, which are the main drivers of future CO_2 emissions. However, it is important to note that the majority of results for stabilization scenarios of 450–750 ppmv CO_2 tend to cluster in a range of 220–2,200 $GtCO_2$ (60–600 GtC) for the cumulative deployment of CCS. For CCS to achieve this economic potential, several hundreds or thousands of CCS systems would be required worldwide over the next century, each capturing some 1–5 $MtCO_2$ per year. As indicated in Section 5, it is likely that the technical potential for geological storage alone is sufficient to cover the high end of the economic potential range for CCS.

Perspectives on CO_2 leakage from storage

The policy implications of slow leakage from storage depend on assumptions in the analysis. Studies conducted to address the question of how to deal with impermanent storage are based on different approaches: the value of delaying emissions, cost minimization of a specified mitigation scenario, or allowable future emissions in the context of an assumed stabilization of atmospheric greenhouse gas concentrations. Some of these studies allow future releases to be compensated by additional reductions in emissions; the results depend on assumptions regarding the future cost of reductions, discount rates, the amount of CO_2 stored, and the assumed level of stabilization for atmospheric concentrations. In other studies, compensation is not seen as an option because of political and institutional uncertainties and the analysis focuses on limitations set by the assumed stabilization level and the amount stored.

While specific results of the range of studies vary with the methods and assumptions made, the outcomes suggest that a fraction retained on the order of 90–99% for 100 years or 60–95% for 500 years could still make such impermanent storage valuable for the mitigation of climate change. All studies imply that, if CCS is to be acceptable as a mitigation measure, there must be an upper limit to the amount of leakage that can take place.

9. Emission inventories and accounting

An important aspect of CO_2 capture and storage is the development and application of methods to estimate and report the quantities in which emissions of CO_2 (and associated emissions of methane or nitrous oxides) are reduced, avoided, or removed from the atmosphere. The two elements involved here are (1) the actual estimation and reporting of emissions for national greenhouse gas inventories, and (2) accounting for CCS under international agreements to limit net emissions.[15]

Current framework

Under the UNFCCC, national greenhouse gas emission inventories have traditionally reported emissions for a specific year, and have been prepared on an annual basis or another periodic basis. The IPCC Guidelines (IPCC 1996) and Good Practice Guidance Reports (IPCC 2000; 2003) describe detailed approaches for preparing national inventories that are complete, transparent, documented, assessed for uncertainties, consistent over time, and comparable across countries. The IPCC documents now in use do not specifically include CO_2 capture and storage options. However, the IPCC Guidelines are currently undergoing revisions that should provide some guidance when the revisions are published in 2006. The framework that already has been accepted could be applied to CCS systems, although some issues might need revision or expansion.

Issues relevant to accounting and reporting

In the absence of prevailing international agreements, it is not clear whether the various forms of CO_2 capture and storage will be treated as reductions in emissions or as removals from the atmosphere. In either case, CCS results in new pools of CO_2 that may be subject to physical leakage at some time in the future. Currently, there are no methods available within the UNFCCC framework for monitoring, measuring or accounting for physical leakage from storage sites. However, leakage from well-managed geological storage sites is likely to be small in magnitude and distant in time.

Consideration may be given to the creation of a specific category for CCS in the emissions reporting framework but this is not strictly necessary since the quantities of CO_2 captured and stored could be reflected in the sector in which the CO_2 was produced. CO_2 storage in a given location could include CO_2 from many different source categories, and even from sources in many different countries. Fugitive

---

[15] In this context, "estimation" is the process of calculating greenhouse gas emissions and "reporting" is the process of providing the estimates to the UNFCCC. "Accounting" refers to the rules for comparing emissions and removals as reported with commitments (IPCC 2003).

emissions from the capture, transport and injection of CO_2 to storage can largely be estimated within the existing reporting methods, and emissions associated with the added energy required to operate the CCS systems can be measured and reported within the existing inventory frameworks. Specific consideration may also be required for CCS applied to biomass systems as that application would result in reporting negative emissions, for which there is currently no provision in the reporting framework.

Issues relevant to international agreements

Quantified commitments to limit greenhouse gas emissions and the use of emissions trading, Joint Implementation (JI) or the Clean Development Mechanism (CDM) require clear rules and methods to account for emissions and removals. Because CCS has the potential to move CO_2 across traditional accounting boundaries (e.g. CO_2 might be captured in one country and stored in another, or captured in one year and partly released from storage in a later year), the rules and methods for accounting may be different than those used in traditional emissions inventories.

To date, most of the scientific, technical and political discussions on accounting for stored CO_2 have focused on sequestration in the terrestrial biosphere. The history of these negotiations may provide some guidance for the development of accounting methods for CCS. Recognizing the potential impermanence of CO_2 stored in the terrestrial biosphere, the UNFCCC accepted the idea that net emissions can be reduced through biological sinks, but has imposed complex rules for such accounting. CCS is markedly different in many ways from CO_2 sequestration in the terrestrial biosphere (see Table TS.12), and the different forms of CCS are markedly different from one another. However, the main goal of accounting is to ensure that CCS activities produce real and quantifiable reductions in net emissions. One tonne of CO_2 permanently stored has the same benefit in terms of atmospheric CO_2 concentrations as one tonne of CO_2 not emitted, but one tonne of CO_2 temporarily stored has less benefit. It is generally accepted that this difference should be reflected in any system of accounting for reductions in net greenhouse gas emissions.

The IPCC Guidelines (IPCC 1996) and Good Practice Guidance Reports (IPCC 2000; 2003) also contain guidelines for monitoring greenhouse gas emissions. It is not known whether the revised guidelines of the IPCC for CCS can be satisfied by using monitoring techniques, particularly for geological and ocean storage. Several techniques are available for the monitoring and verification of CO_2 emissions from geological storage, but they vary in applicability, detection limits and uncertainties. Currently, monitoring for geological storage can take place quantitatively at injection and qualitatively in the reservoir and by measuring surface fluxes of CO_2. Ocean storage monitoring can take place by

Table TS.12. Differences in the forms of CCS and biological sinks that might influence the way accounting is conducted.

Property	Terrestrial biosphere	Deep ocean	Geological reservoirs
CO_2 sequestered or stored	Stock changes can be monitored over time.	Injected carbon can be measured.	Injected carbon can be measured.
Ownership	Stocks will have a discrete location and can be associated with an identifiable owner.	Stocks will be mobile and may reside in international waters.	Stocks may reside in reservoirs that cross national or property boundaries and differ from surface boundaries.
Management decisions	Storage will be subject to continuing decisions about land-use priorities.	Once injected there are no further human decisions about maintenance once injection has taken place.	Once injection has taken place, human decisions about continued storage involve minimal maintenance, unless storage interferes with resource recovery.
Monitoring	Changes in stocks can be monitored.	Changes in stocks will be modelled.	Release of CO_2 can be detected by physical monitoring.
Expected retention time	Decades, depending on management decisions.	Centuries, depending on depth and location of injection.	Essentially permanent, barring physical disruption of the reservoir.
Physical leakage	Losses might occur due to disturbance, climate change, or land-use decisions.	Losses will assuredly occur as an eventual consequence of marine circulation and equilibration with the atmosphere.	Losses are unlikely except in the case of disruption of the reservoir or the existence of initially undetected leakage pathways.
Liability	A discrete land-owner can be identified with the stock of sequestered carbon.	Multiple parties may contribute to the same stock of stored CO_2 and the CO_2 may reside in international waters.	Multiple parties may contribute to the same stock of stored CO_2 that may lie under multiple countries.

detecting the CO_2 plume, but not by measuring ocean surface release to the atmosphere. Experiences from monitoring existing CCS projects are still too limited to serve as a basis for conclusions about the physical leakage rates and associated uncertainties.

The Kyoto Protocol creates different units of accounting for greenhouse gas emissions, emissions reductions, and emissions sequestered under different compliance mechanisms. 'Assigned amount units' (AAUs) describe emissions commitments and apply to emissions trading, 'certified emission reductions' (CERs) are used under the CDM, and 'emission reduction units' (ERUs) are employed under JI. To date, international negotiations have provided little guidance about methods for calculating and accounting for project-related CO_2 reductions from CCS systems (only CERs or ERUs), and it is therefore uncertain how such reductions will be accommodated under the Kyoto Protocol. Some guidance may be given by the methodologies for biological-sink rules. Moreover, current agreements do not deal with cross-border CCS projects. This is particularly important when dealing with cross-border projects involving CO_2 capture in an 'Annex B' country that is party to the Kyoto Protocol but stored in a country that is not in Annex B or is not bound by the Protocol.

Although methods currently available for national emissions inventories can either accommodate CCS systems or be revised to do so, accounting for stored CO_2 raises questions about the acceptance and transfer of responsibility for stored emissions. Such issues may be addressed through national and international political processes.

10.   Gaps in knowledge

This summary of the gaps in knowledge covers aspects of CCS where increasing knowledge, experience and reducing uncertainty would be important to facilitate decision-making about the large-scale deployment of CCS.

Technologies for capture and storage

Technologies for the capture of CO_2 are relatively well understood today based on industrial experience in a variety of applications. Similarly, there are no major technical or knowledge barriers to the adoption of pipeline transport, or to the adoption of geological storage of captured CO_2. However, the integration of capture, transport and storage in full-scale projects is needed to gain the knowledge and experience required for a more widespread deployment of CCS technologies. R&D is also needed to improve knowledge of emerging concepts and enabling technologies for CO_2 capture that have the potential to significantly reduce the costs of capture for new and existing facilities. More specifically, there are knowledge gaps relating to large coal-

based and natural gas-based power plants with CO_2 capture on the order of several hundred megawatts (or several $MtCO_2$). Demonstration of CO_2 capture on this scale is needed to establish the reliability and environmental performance of different types of power systems with capture, to reduce the costs of CCS, and to improve confidence in the cost estimates. In addition, large-scale implementation is needed to obtain better estimates of the costs and performance of CCS in industrial processes, such as the cement and steel industries, that are significant sources of CO_2 but have little or no experience with CO_2 capture.

With regard to mineral carbonation technology, a major question is how to exploit the reaction heat in practical designs that can reduce costs and net energy requirements. Experimental facilities at pilot scales are needed to address these gaps.

With regard to industrial uses of captured CO_2, further study of the net energy and CO_2 balance of industrial processes that use the captured CO_2 could help to establish a more complete picture of the potential of this option.

Geographical relationship between the sources and storage opportunities of CO_2

An improved picture of the proximity of major CO_2 sources to suitable storage sites (of all types), and the establishment of cost curves for the capture, transport and storage of CO_2, would facilitate decision-making about large-scale deployment of CCS. In this context, detailed regional assessments are required to evaluate how well large CO_2 emission sources (both current and future) match suitable storage options that can store the volumes required.

Geological storage capacity and effectiveness

There is a need for improved storage capacity estimates at the global, regional and local levels, and for a better understanding of long-term storage, migration and leakage processes. Addressing the latter issue will require an enhanced ability to monitor and verify the behaviour of geologically stored CO_2. The implementation of more pilot and demonstration storage projects in a range of geological, geographical and economic settings would be important to improve our understanding of these issues.

Impacts of ocean storage

Major knowledge gaps that should be filled before the risks and potential for ocean storage can be assessed concern the ecological impact of CO_2 in the deep ocean. Studies are needed of the response of biological systems in the deep sea to added CO_2, including studies that are longer in duration and larger in scale than those that have been performed until now. Coupled with this is a need to develop techniques and sensors to detect and monitor CO_2 plumes and their biological and geochemical consequences.

Legal and regulatory issues

Current knowledge about the legal and regulatory requirements for implementing CCS on a larger scale is still inadequate. There is no appropriate framework to facilitate the implementation of geological storage and take into account the associated long-term liabilities. Clarification is needed regarding potential legal constraints on storage in the marine environment (ocean or sub-seabed geological storage). Other key knowledge gaps are related to the methodologies for emissions inventories and accounting.

Global contribution of CCS to mitigating climate change

There are several other issues that would help future decision-making about CCS by further improving our understanding of the potential contribution of CCS to the long-term global mitigation and stabilization of greenhouse gas concentrations. These include the potential for transfer and diffusion of CCS technologies, including opportunities for developing countries to exploit CCS, its application to biomass sources of CO_2, and the potential interaction between investment in CCS and other mitigation options. Further investigation is warranted into the question of how long CO_2 would need to be stored. This issue is related to stabilization pathways and intergenerational aspects.

1

Introduction

Coordinating Lead Author
Paul Freund (United Kingdom)

Lead Authors
Anthony Adegbulugbe (Nigeria), Øyvind Christophersen (Norway), Hisashi Ishitani (Japan),
William Moomaw (United States), Jose Moreira (Brazil)

Review Editors
Eduardo Calvo (Peru), Eberhard Jochem (Germany)

Contents

EXECUTIVE SUMMARY

According to IPCC's Third Assessment Report:
- 'There is new and stronger evidence that most of the warming observed over the past 50 years is attributable to human activities.
- Human influences are expected to continue to change atmospheric composition throughout the 21st century.'

The greenhouse gas making the largest contribution from human activities is carbon dioxide (CO_2). It is released by burning fossil fuels and biomass as a fuel; from the burning, for example, of forests during land clearance; and by certain industrial and resource extraction processes.
- 'Emissions of CO_2 due to fossil fuel burning are virtually certain to be the dominant influence on the trends in atmospheric CO_2 concentration during the 21st century.
- Global average temperatures and sea level are projected to rise under all (…) scenarios.'

The ultimate objective of the UN Framework Convention on Climate Change, which has been accepted by 189 nations, is to achieve '(…) stabilization of greenhouse gas concentrations in the atmosphere at a level that would prevent dangerous anthropogenic interference with the climate system', although a specific level has yet to be agreed.

Technological options for reducing net CO_2 emissions to the atmosphere include:
- reducing energy consumption, for example by increasing the efficiency of energy conversion and/or utilization (including enhancing less energy-intensive economic activities);
- switching to less carbon intensive fuels, for example natural gas instead of coal;
- increasing the use of renewable energy sources or nuclear energy, each of which emits little or no net CO_2;
- sequestering CO_2 by enhancing biological absorption capacity in forests and soils;
- capturing and storing CO_2 chemically or physically.

The first four technological options were covered in earlier IPCC reports; the fifth option, the subject of this report, is Carbon dioxide Capture and Storage (CCS). In this approach, CO_2 arising from the combustion of fossil and/or renewable fuels and from processing industries would be captured and stored away from the atmosphere for a very long period of time. This report analyzes the current state of knowledge about the scientific and technical, economic and policy dimensions of this option, in order to allow it to be considered in relation to other options for mitigating climate change.

At present, the global concentration of CO_2 in the atmosphere is increasing. If recent trends in global CO_2 emissions continue, the world will not be on a path towards stabilization of greenhouse gas concentrations. Between 1995 and 2001, average global CO_2 emissions grew at a rate of 1.4% per year, which is slower than the growth in use of primary energy but higher than the growth in CO_2 emissions in the previous 5 years. Electric-power generation remains the single largest source of CO_2 emissions, emitting as much CO_2 as the rest of the industrial sector combined, while the transport sector is the fastest-growing source of CO_2 emissions. So meeting the ultimate goal of the UNFCCC will require measures to reduce emissions, including the further deployment of existing and new technologies.

The extent of emissions reduction required will depend on the rate of emissions and the atmospheric concentration target. The lower the chosen stabilization concentration and the higher the rate of emissions expected in the absence of mitigation measures, the larger must be the reduction in emissions and the earlier that it must occur. In many of the models that IPCC has considered, stabilization at a level of 550 ppmv of CO_2 in the atmosphere would require a reduction in global emissions by 2100 of 7–70% compared with current rates. Lower concentrations would require even greater reductions. Achieving this cost-effectively will be easier if we can choose flexibly from a broad portfolio of technology options of the kind described above.

The purpose of this report is to assess the characteristics of CO_2 capture and storage as part of a portfolio of this kind. There are three main components of the process: capturing CO_2, for example by separating it from the flue gas stream of a fuel combustion system and compressing it to a high pressure; transporting it to the storage site; and storing it. CO_2 storage will need to be done in quantities of gigatonnes of CO_2 per year to make a significant contribution to the mitigation of climate change, although the capture and storage of smaller amounts, at costs similar to or lower than alternatives, would make a useful contribution to lowering emissions. Several types of storage reservoir may provide storage capacities of this magnitude. In some cases, the injection of CO_2 into oil and gas fields could lead to the enhanced production of hydrocarbons, which would help to offset the cost. CO_2 capture technology could be applied to electric-power generation facilities and other large industrial sources of emissions; it could also be applied in the manufacture of hydrogen as an energy carrier. Most stages of the process build on known technology developed for other purposes.

There are many factors that must be considered when deciding what role CO_2 capture and storage could play in mitigating climate change. These include the cost and capacity of emission reduction relative to, or in combination with, other options, the resulting increase in demand for primary energy sources, the range of applicability, and the technical risk. Other important factors are the social and environmental consequences, the safety of the technology, the security of storage and ease of monitoring and verification, and the extent of opportunities to transfer the technology to developing countries. Many of these features are interlinked. Some aspects are more amenable to rigorous evaluation than others. For example, the literature about the societal aspects of this new mitigation option is limited. Public attitudes, which are influenced by many factors, including how judgements are made about the technology, will also exert an important influence on its application. All of these aspects are discussed in this report.

1.1   Background to the report

IPCC's Third Assessment Report stated 'there is new and stronger evidence that most of the warming observed over the past 50 years is attributable to human activities'. It went on to point out that 'human influences will continue to change atmospheric composition throughout the 21st century' (IPCC, 2001c). Carbon dioxide (CO_2) is the greenhouse gas that makes the largest contribution from human activities. It is released into the atmosphere by: the combustion of fossil fuels such as coal, oil or natural gas, and renewable fuels like biomass; by the burning of, for example, forests during land clearance; and from certain industrial and resource extraction processes. As a result 'emissions of CO_2 due to fossil fuel burning are virtually certain to be the dominant influence on the trends in atmospheric CO_2 concentration during the 21st century' and 'global average temperatures and sea level are projected to rise under all … scenarios' (IPCC, 2001c).

The UN Framework Convention on Climate Change (UNFCCC), which has been ratified by 189 nations and has now gone into force, asserts that the world should achieve an atmospheric concentration of greenhouse gases (GHGs) that would prevent 'dangerous anthropogenic interference with the climate system' (UNFCCC, 1992), although the specific level of atmospheric concentrations has not yet been quantified. Technological options for reducing anthropogenic emissions[1] of CO_2 include (1) reducing the use of fossil fuels (2) substituting less carbon-intensive fossil fuels for more carbon-intensive fuels (3) replacing fossil fuel technologies with near-zero-carbon alternatives and (4) enhancing the absorption of atmospheric CO_2 by natural systems. In this report, the Intergovernmental Panel on Climate Change (IPCC) explores an additional option: Carbon dioxide Capture and Storage (CCS)[2]. This report will analyze the current state of knowledge in order to understand the technical, economic and policy dimensions of this climate change mitigation option and make it possible to consider it in context with other options.

1.1.1   What is CO_2 capture and storage?

CO_2 capture and storage involves capturing the CO_2 arising from the combustion of fossil fuels, as in power generation, or from the preparation of fossil fuels, as in natural-gas processing.

It can also be applied to the combustion of biomass-based fuels and in certain industrial processes, such as the production of hydrogen, ammonia, iron and steel, or cement. Capturing CO_2 involves separating the CO_2 from some other gases[3]. The CO_2 must then be transported to a storage site where it will be stored away from the atmosphere for a very long time (IPCC, 2001a). In order to have a significant effect on atmospheric concentrations of CO_2, storage reservoirs would have to be large relative to annual emissions.

1.1.2   Why a special report on CO_2 capture and storage?

The capture and storage of carbon dioxide is a technically feasible method of making deep reductions in CO_2 emissions from sources such as those mentioned above. Although it can be implemented mainly by applying known technology developed for other purposes, its potential role in tackling climate change was not recognized as early as some other mitigation options. Indeed, the topic received little attention in IPCC's Second and Third Assessment Reports (IPCC 1996a, 2001b) – the latter contained a three-page review of technological progress, and an overview of costs and the environmental risks of applying such technology. In recent years, the technical literature on this field has expanded rapidly. Recognizing the need for a broad approach to assessing mitigation options, the potential importance of issues relating to CO_2 capture and storage and the extensive literature on other options (due to their longer history), IPCC decided to undertake a thorough assessment of CO_2 capture and storage. For these reasons it was thought appropriate to prepare a Special Report on the subject. This would constitute a source of information of comparable nature to the information available on other, more established mitigation options. In response to the invitation from the 7th Conference of the Parties to the UNFCCC in Marrakech[4], the IPCC plenary meeting in April 2002 decided to launch work on CO_2 capture and storage.

1.1.3   Preparations for this report

In preparation for this work, the 2002 Plenary decided that IPCC should arrange a Workshop under the auspices of Working Group III, with inputs from Working Groups I and II, to recommend how to proceed. This workshop took place in Regina, Canada, in November 2002 (IPCC, 2002). Three options were considered at the workshop: the production of a Technical Report, a Special Report, or the postponement of any action until the Fourth Assessment Report. After extensive discussion, the Workshop decided to advise IPCC to produce a Special Report on CO_2 capture and storage. At IPCC's Plenary Meeting in February 2003, the Panel acknowledged the importance of issues relating to CO_2 capture and storage and decided that a Special Report would be the most appropriate way of assessing the technical, scientific and socio-economic implications of capturing anthropogenic CO_2 and storing it in natural reservoirs. The Panel duly gave approval for work to begin on such a report with 2005 as the target date for publication.

The decision of the 2002 Plenary Meeting required the report to cover the following issues:

---

[1] In this report, the term 'emissions' is taken to refer to emissions from anthropogenic, rather than natural, sources.

[2] CO_2 capture and storage is sometimes referred to as carbon sequestration. In this report, the term 'sequestration' is reserved for the enhancement of natural sinks of CO_2, a mitigation option which is not examined in this report but in IPCC 2000b.

[3] For example, in the flue gas stream of a power plant, the other gases are mainly nitrogen and water vapour.

[4] This draft decision called on IPCC to prepare a 'technical paper on geological carbon storage technologies'.

- sources of CO_2 and technologies for capturing CO_2;
- transport of CO_2 from capture to storage;
- CO_2 storage options;
- geographical potential of the technology;
- possibility of re-using captured CO_2 in industrial applications;
- costs and energy efficiency of capturing and storing CO_2 in comparison with other large-scale mitigation options;
- implications of large-scale introduction, the environmental impact, as well as risks and risk management during capture, transport and storage;
- permanence and safety of CO_2 storage, including methods of monitoring CO_2 storage;
- barriers to the implementation of storage, and the modelling of CO_2 capture and storage in energy and climate models;
- implications for national and international emission inventories, legal aspects and technology transfer.

This report assesses information on all these topics in order to facilitate discussion of the relative merits of this option and to assist decision-making about whether and how the technology should be used.

1.1.4   *Purpose of this introduction*

This chapter provides an introduction in three distinct ways: it provides the background and context for the report; it provides an introduction to CCS technology; and it provides a framework for the CCS assessment methods used in later chapters.

Because this report is concerned with the physical capture, transport and storage of CO_2, the convention is adopted of using physical quantities (i.e. tonnes) of CO_2 rather than quantities of C, as is normal in the general literature on climate change. In order to make possible comparison of the results with other literature, quantities in tonnes of C are given in parenthesis.

1.2   Context for considering CO_2 Capture and Storage

1.2.1   *Energy consumption and CO_2 emissions*

CO_2 continued an upward trend in the early years of the 21st century (Figures 1.1, 1.2). Fossil fuels are the dominant form of energy utilized in the world (86%), and account for about 75% of current anthropogenic CO_2 emissions (IPCC, 2001c). In 2002, 149 Exajoules (EJ) of oil, 91 EJ of natural gas, and 101 EJ of coal were consumed by the world's economies (IEA, 2004). Global primary energy consumption grew at an average rate of 1.4% annually between 1990 and 1995 (1.6% per year between 1995 and 2001); the growth rates were 0.3% per year (0.9%) in the industrial sector, 2.1% per year (2.2%) in the transportation sector, 2.7% per year (2.1%) in the buildings sector, and –2.4% per year (–0.8%) in the agricultural/other sector (IEA, 2003).

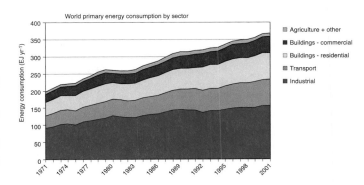

Figure 1.1 World primary energy use by sector from 1971 to 2001 (IEA, 2003).

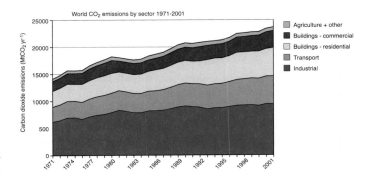

Figure 1.2 World CO_2 emissions from fossil fuel use by sector, 1971 to 2001 (IEA, 2003).

Average global CO_2 emissions[5] increased by 1.0% per year between 1990 and 1995 (1.4% between 1995 and 2001), a rate slightly below that of energy consumption in both periods. In individual sectors, there was no increase in emissions from industry between 1990 and 1995 (0.9% per year from 1995 to 2001); there was an increase of 1.7% per year (2.0%) in the transport sector, 2.3% per year (2.0%) in the buildings sector, and a fall of 2.8% per year (1.0%) in the agricultural/other sector (IEA, 2003).

Total emissions from fossil fuel consumption and flaring of natural gas were 24 $GtCO_2$ per year (6.6 GtC per year) in 2001 – industrialized countries were responsible for 47% of energy-related CO_2 emissions (not including international bunkers[6]). The Economies in Transition accounted for 13% of 2001 emissions; emissions from those countries have been declining at an annual rate of 3.3% per year since 1990. Developing countries in the Asia-Pacific region emitted 25% of the global total of CO_2; the rest of the developing countries accounted for 13% of the total (IEA, 2003).

---

[5] There are differences in published estimates of CO_2 emissions for many countries, as Marland et al. (1999) have shown using two ostensibly similar sources of energy statistics.

[6] Emissions from international bunkers amounted to 780 Mt CO_2 (213 MtC) in 2001 (IEA, 2003).

Table 1.1 Sources of CO_2 emissions from fossil fuel combustion (2001).

	Emissions	
	(MtCO$_2$ yr^{-1})	**(MtC yr^{-1})**
Public electricity and heat production	8,236	2,250
Autoproducers	963	263
Other energy industries	1,228	336
Manufacturing & construction	4,294	1,173
Transport	5,656	1,545
of which: Road	4,208	1,150
Other sectors	3,307	903
of which: Residential	1,902	520
TOTAL	23,684	6,470

Source: IEA, 2003.

1.2.2   Sectoral CO$_2$ emissions

The CO_2 emissions from various sources worldwide have been estimated by the IEA (2003). These are shown in Table 1.1, which shows that power generation is the single largest source of emissions. Other sectors where emissions arise from a few large point sources are Other Energy Industries[7] and parts of the Manufacturing and Construction sector.

Emissions from transport, which is the second largest sector (Table 1.1), have been growing faster than those from energy and industry in the last few decades (IPCC, 2001a); a key difference is that transport emissions are mainly from a multiplicity of small, distributed sources. These differences have implications for possible uses of CO_2 capture and storage, as will be seen later in this chapter.

1.2.3   Other greenhouse gas emissions

Anthropogenic climate change is mainly driven by emissions of CO_2 but other greenhouse gases (GHGs) also play a part[8]. Since some of the anthropogenic CO_2 comes from industrial processes and some from land use changes (mainly deforestation), the contribution from fossil fuel combustion alone is about half of the total from all GHGs.

In terms of impact on radiative forcing, methane is the next most important anthropogenic greenhouse gas after CO_2 (currently accounting for 20% of the total impact) (IPCC, 2001b). The energy sector is an important source of methane but agriculture and domestic waste disposal contribute more to the global total (IPCC, 2001c). Nitrous oxide contributes directly to climate change (currently 6% of the total impact of all GHGs); the main source is agriculture but another is

the industrial production of some chemicals; other oxides of nitrogen have an indirect effect. A number of other gases make significant contributions (IPCC, 2001c).

1.2.4   Scenarios of future emissions

Future emissions may be simulated using scenarios which are: 'alternative images of how the future might unfold and are (…) tools (…) to analyse how driving forces may influence future emissions (….) and to assess the associated uncertainties.' 'The possibility that any single emissions path will occur as described in scenarios is highly uncertain' (IPCC, 2000a). In advance of the Third Assessment Report, IPCC made an effort to identify future GHG emission pathways. Using several assumptions, IPCC built a set of scenarios of what might happen to emissions up to the year 2100. Six groups of scenarios were published (IPCC, 2000a): the 'SRES scenarios'. None of these assume any specific climate policy initiatives; in other words, they are base cases which can be used for considering the effects of mitigation options. An illustrative scenario was chosen for each of the groups. The six groups were organized into four 'families' covering a wide range of key 'future' characteristics such as demographic change, economic development, and technological change (IPCC, 2000a). Scenario families A1 and A2 emphasize economic development, whilst B1 and B2 emphasize global and local solutions for, respectively, economic, social and environmental sustainability. In addition, two scenarios, A1F1 and A1T, illustrate alternative developments in energy technology in the A1 world (see Figure TS.1 in IPCC, 2001a).

Given the major role played by fossil fuels in supplying energy to modern society, and the long periods of time involved in changing energy systems (Marchetti and Nakicenovic, 1979), the continued use of fossil fuels is arguably a good base-case scenario. Further discussion of how CCS may affect scenarios can be found in Chapter 8.

Most of these scenarios yield future emissions which are significantly higher than today's levels. In 2100, these scenarios show, on average, between 50% and 250% as much annual

---

[7] The Other Energy Industries sector includes oil refineries, manufacture of solid fuels, coal mining, oil and gas extraction, and other energy-producing industries.

[8] It is estimated that the global radiative forcing of anthropogenic CO_2 is approximately 60% of the total due to all anthropogenic GHGs (IPCC, 2001b).

CO_2 emissions as current rates. Adding together all of the CO_2 emissions projected for the 21st century, the cumulative totals lie in the range of 3,480 to 8,050 GtCO$_2$ (950 to 2,200 GtC) depending on the selected scenario (IPCC, 2001e).

It should be noted that there is potential for confusion about the term 'leakage' since this is widely used in the climate change literature in a spatial sense to refer to the displacement of emissions from one source to another. This report does not discuss leakage of this kind but it does look at the unintended release of CO_2 from storage (which may also be termed leakage). The reader is advised to be aware of the possible ambiguity in the use of the term leakage and to have regard to the context where this word is used in order to clarify the meaning.

1.3 Options for mitigating climate change

As mentioned above, the UN Framework Convention on Climate Change calls for the stabilization of the atmospheric concentration of GHGs but, at present, there is no agreement on what the specific level should be. However, it can be recognized that stabilization of concentrations will only occur once the rate of addition of GHGs to the atmosphere equals the rate at which natural systems can remove them – in other words, when the rate of anthropogenic emissions is balanced by the rate of uptake by natural processes such as atmospheric reactions, net transfer to the oceans, or uptake by the biosphere.

In general, the lower the stabilization target and the higher the level of baseline emissions, the larger the required reduction in emissions below the baseline, and the earlier that it must occur. For example, stabilization at 450 ppmv CO_2 would require emissions to be reduced earlier than stabilization at 650 ppmv, with very rapid emission reductions over the next 20 to 30 years (IPCC, 2000a); this could require the employment of all cost-effective potential mitigation options (IPCC, 2001a). Another conclusion, no less relevant than the previous one, is that the range of baseline scenarios tells us that future economic development policies may impact greenhouse gas emissions as strongly as policies and technologies especially developed to address climate change. Some have argued that climate change is more an issue of economic development, for both developed and developing countries, than it is an environmental issue (Moomaw *et al.*, 1999).

The Third Assessment Report (IPCC, 2001a) shows that, in many of the models that IPCC considered, achieving stabilization at a level of 550 ppmv would require global emissions to be reduced by 7–70% by 2100 (depending upon the stabilization profile) compared to the level of emissions in 2001. If the target were to be lower (450 ppmv), even deeper reductions (55–90%) would be required. For the purposes of this discussion, we will use the term 'deep reductions' to imply net reductions of 80% or more compared with what would otherwise be emitted by an individual power plant or industrial facility.

In any particular scenario, it may be helpful to consider the major factors influencing CO_2 emissions from the supply and use of energy using the following simple but useful identity (after Kaya, 1995):

CO_2 emissions =

$$\text{Population} \times \left(\frac{\text{GDP}}{\text{Population}} \right) \times \left(\frac{\text{Energy}}{\text{GDP}} \right) \times \left(\frac{\text{Emissions}}{\text{Energy}} \right)$$

This shows that the level of CO_2 emissions can be understood to depend directly on the size of the human population, on the level of global wealth, on the energy intensity of the global economy, and on the emissions arising from the production and use of energy. At present, the population continues to rise and average energy use is also rising, whilst the amount of energy required per unit of GDP is falling in many countries, but only slowly (IPCC, 2001d). So achieving deep reductions in emissions will, all other aspects remaining constant, require major changes in the third and fourth factors in this equation, the emissions from energy technology. Meeting the challenge of the UNFCCC's goal will therefore require sharp falls in emissions from energy technology.

A wide variety of technological options have the potential to reduce net CO_2 emissions and/or CO_2 atmospheric concentrations, as will be discussed below, and there may be further options developed in the future. The targets for emission reduction will influence the extent to which each technique is used. The extent of use will also depend on factors such as cost, capacity, environmental impact, the rate at which the technology can be introduced, and social factors such as public acceptance.

1.3.1 Improve energy efficiency

Reductions in fossil fuel consumption can be achieved by improving the efficiency of energy conversion, transport and end-use, including enhancing less energy-intensive economic activities. Energy conversion efficiencies have been increased in the production of electricity, for example by improved turbines; combined heating, cooling and electric-power generation systems reduce CO_2 emissions further still. Technological improvements have achieved gains of factors of 2 to 4 in the energy consumption of vehicles, of lighting and many appliances since 1970; further improvements and wider application are expected (IPCC, 2001a). Further significant gains in both demand-side and supply-side efficiency can be achieved in the near term and will continue to slow the growth in emissions into the future; however, on their own, efficiency gains are unlikely to be sufficient, or economically feasible, to achieve deep reductions in emissions of GHGs (IPCC, 2001a).

1.3.2 Switch to less carbon-intensive fossil fuels

Switching from high-carbon to low-carbon fuels can be cost-effective today where suitable supplies of natural gas are available. A typical emission reduction is 420 kg CO_2 MWh^{-1} for the change from coal to gas in electricity generation; this is about 50% (IPCC, 1996b). If coupled with the introduction of the combined production of heat, cooling and electric power, the reduction in emissions would be even greater. This would make a substantial contribution to emissions reduction from a particular plant but is restricted to plant where supplies of lower carbon fuels are available.

1.3.3   *Increased use of low- and near-zero-carbon energy sources*

Deep reductions in emissions from stationary sources could be achieved by widespread switching to renewable energy or nuclear power (IPCC, 2001a). The extent to which nuclear power could be applied and the speed at which its use might be increased will be determined by that industry's ability to address concerns about cost, safety, long-term storage of nuclear wastes, proliferation and terrorism. Its role is therefore likely to be determined more by the political process and public opinion than by technical factors (IPCC, 2001a).

There is a wide variety of renewable supplies potentially available: commercial ones include wind, solar, biomass, hydro, geothermal and tidal power, depending on geographic location. Many of them could make significant contributions to electricity generation, as well as to vehicle fuelling and space heating or cooling, thereby displacing fossil fuels (IPCC, 2001a). Many of the renewable sources face constraints related to cost, intermittency of supply, land use and other environmental impacts. Between 1992 and 2002, installed wind power generation capacity grew at a rate of about 30% per year, reaching over 31 GW_e by the end of 2002 (Gipe, 2004). Solar electricity generation has increased rapidly (by about 30% per year), achieving 1.1 GW_e capacity in 2001, mainly in small-scale installations (World Energy Assessment, 2004). This has occurred because of falling costs as well as promotional policies in some countries. Liquid fuel derived from biomass has also expanded considerably and is attracting the attention of several countries, for example Brazil, due to its declining costs and co-benefits in creation of jobs for rural populations. Biomass used for electricity generation is growing at about 2.5% per annum; capacity had reached 40 GW_e in 2001. Biomass used for heat was estimated to have capacity of 210 GW_{th} in 2001. Geothermal energy used for electricity is also growing in both developed and developing countries, with capacity of 3 GW_e in 2001 (World Energy Assessment, 2004). There are therefore many options which could make deep reductions by substituting for fossil fuels, although the cost is significant for some and the potential varies from place to place (IPCC, 2001a).

1.3.4   *Sequester CO_2 through the enhancement of natural, biological sinks*

Natural sinks for CO_2 already play a significant role in determining the concentration of CO_2 in the atmosphere. They may be enhanced to take up carbon from the atmosphere. Examples of natural sinks that might be used for this purpose include forests and soils (IPCC, 2000b). Enhancing these sinks through agricultural and forestry practices could significantly improve their storage capacity but this may be limited by land use practice, and social or environmental factors. Carbon stored biologically already includes large quantities of emitted CO_2 but storage may not be permanent.

1.3.5   *CO_2 capture and storage*

As explained above, this approach involves capturing CO_2 generated by fuel combustion or released from industrial processes, and then storing it away from the atmosphere for a very long time. In the Third Assessment Report (IPCC, 2001a) this option was analyzed on the basis of a few, documented projects (e.g., the Sleipner Vest gas project in Norway, enhanced oil recovery practices in Canada and USA, and enhanced recovery of coal bed methane in New Mexico and Canada). That analysis also discussed the large potential of fossil fuel reserves and resources, as well as the large capacity for CO_2 storage in depleted oil and gas fields, deep saline formations, and in the ocean. It also pointed out that CO_2 capture and storage is more appropriate for large sources – such as central power stations, refineries, ammonia, and iron and steel plants – than for small, dispersed emission sources.

The potential contribution of this technology will be influenced by factors such as the cost relative to other options, the time that CO_2 will remain stored, the means of transport to storage sites, environmental concerns, and the acceptability of this approach. The CCS process requires additional fuel and associated CO_2 emissions compared with a similar plant without capture.

Recently it has been recognized that biomass energy used with CO_2 capture and storage (BECS) can yield net removal of CO_2 from the atmosphere because the CO_2 put into storage comes from biomass which has absorbed CO_2 from the atmosphere as it grew (Möllersten *et al.*, 2003; Azar *et al.*, 2003). The overall effect is referred to as 'negative net emissions'. BECS is a new concept that has received little analysis in technical literature and policy discussions to date.

1.3.6   *Potential for reducing CO_2 emissions*

It has been determined (IPCC, 2001a) that the worldwide potential for GHG emission reduction by the use of technological options such as those described above amounts to between 6,950 and 9,500 $MtCO_2$ per year (1,900 to 2,600 MtC per year) by 2010, equivalent to about 25 to 40% of global emissions respectively. The potential rises to 13,200 to 18,500 $MtCO_2$ per year (3,600 to 5,050 MtC per year) by 2020. The evidence on which these estimates are based is extensive but has several limitations: for instance, the data used comes from the 1990s and additional new technologies have since emerged. In addition, no comprehensive worldwide study of technological and economic potential has yet been performed; regional and national studies have generally had different scopes and made different assumptions about key parameters (IPCC, 2001a).

The Third Assessment Report found that the option for reducing emissions with most potential in the short term (up to 2020) was energy efficiency improvement while the near-term potential for CO_2 capture and storage was considered modest, amounting to 73 to 183 $MtCO_2$ per year (20 to 50 MtC per year) from coal and a similar amount from natural gas (see Table TS.1 in IPCC, 2001a). Nevertheless, faced with the longer-term climate challenge described above, and in view of the growing interest in this option, it has become important to analyze the potential of this technology in more depth.

As a result of the 2002 IPCC workshop on CO_2 capture and storage (IPCC, 2002), it is now recognized that the amount of CO_2 emissions which could potentially be captured and stored may be higher than the value given in the Third Assessment Report. Indeed, the emissions reduction may be very significant compared with the values quoted above for the period after 2020. Wider use of this option may tend to restrict the opportunity to use other supply options. Nevertheless, such action might still lead to an increase in emissions abatement because much of the potential estimated previously (IPCC, 2001a) was from the application of measures concerned with end uses of energy. Some applications of CCS cost relatively little (for example, storage of CO_2 from gas processing as in the Sleipner project (Baklid *et al.*, 1996)) and this could allow them to be used at a relatively early date. Certain large industrial sources could present interesting low-cost opportunities for CCS, especially if combined with storage opportunities which generate compensating revenue, such as CO_2 Enhanced Oil Recovery (IEA GHG, 2002). This is discussed in Chapter 2.

1.3.7 *Comparing mitigation options*

A variety of factors will need to be taken into account in any comparison of mitigation options, not least who is making the comparison and for what purpose. The remainder of this chapter discusses various aspects of CCS in a context which may be relevant to decision-makers. In addition, there are broader issues, especially questions of comparison with other mitigation measures. Answering such questions will depend on many factors, including the potential of each option to deliver emission reductions, the national resources available, the accessibility of each technology for the country concerned, national commitments to reduce emissions, the availability of finance, public acceptance, likely infrastructural changes, environmental side-effects, etc. Most aspects of this kind must be considered both in relative terms (e.g., how does this compare with other mitigation options?) and absolute terms (e.g., how much does this cost?), some of which will change over time as the technology advances.

The IPCC (2001a) found that improvements in energy efficiency have the potential to reduce global CO_2 emissions by 30% below year-2000 levels using existing technologies at a cost of less than 30 US$/$tCO_2$ (100 US$/tC). Half of this reduction could be achieved with existing technology at zero or net negative costs[9]. Wider use of renewable energy sources was also found to have substantial potential. Carbon sequestration by

forests was considered a promising near-term mitigation option (IPCC, 2000b), attracting commercial attention at prices of 0.8 to 1.1 US$/$tCO_2$ (3-4 US$/tC). The costs quoted for mitigation in most afforestation projects are presented on a different basis from power generation options, making the afforestation examples look more favourable (Freund and Davison, 2002). Nevertheless, even after allowing for this, the cost of current projects is low.

It is important, when comparing different mitigation options, to consider not just costs but also the potential capacity for emission reduction. A convenient way of doing this is to use Marginal Abatement Cost curves (MACs) to describe the potential capacity for mitigation; these are not yet available for all mitigation options but they are being developed (see, for example, IEA GHG, 2000b). Several other aspects of the comparison of mitigation options are discussed later in this chapter and in Chapter 8.

1.4 Characteristics of CO_2 capture and storage

In order to help the reader understand how CO_2 capture and storage could be used as a mitigation option, some of the key features of the technology are briefly introduced here.

1.4.1 *Overview of the CO_2 capture and storage concept and its development*

Capturing CO_2 typically involves separating it from a gas stream. Suitable techniques were developed 60 years ago in connection with the production of town gas; these involved scrubbing the gas stream with a chemical solvent (Siddique, 1990). Subsequently they were adapted for related purposes, such as capturing CO_2 from the flue gas streams of coal- or gas-burning plant for the carbonation of drinks and brine, and for enhancing oil recovery. These developments required improvements to the process so as to inhibit the oxidation of the solvent in the flue gas stream. Other types of solvent and other methods of separation have been developed more recently. This technique is widely used today for separating CO_2 and other acid gases from natural gas streams[10]. Horn and Steinberg (1982) and Hendriks *et al.* (1989) were among the first to discuss the application of this type of technology to mitigation of climate change, focusing initially on electricity generation. CO_2 removal is already used in the production of hydrogen from fossil fuels; Audus *et al.* (1996) discussed the application of capture and storage in this process as a climate protection measure.

In order to transport CO_2 to possible storage sites, it is compressed to reduce its volume; in its 'dense phase', CO_2 occupies around 0.2% of the volume of the gas at standard temperature and pressure (see Appendix 1 for further information

---

[9] Meaning that the value of energy savings would exceed the technology capital and operating costs within a defined period of time using appropriate discount rates.

[10] The total number of installations is not known but is probably several thousand. Kohl and Nielsen (1997) mention 334 installations using physical solvent scrubbing; this source does not provide a total for the number of chemical solvent plants but they do mention one survey which alone examined 294 amine scrubbing plants. There are also a number of membrane units and other methods of acid gas treatment in use today.

about the properties of CO_2). Several million tonnes per year of CO_2 are transported today by pipeline (Skovholt, 1993), by ship and by road tanker.

In principle, there are many options available for the storage of CO_2. The first proposal of such a concept (Marchetti, 1977) envisaged injection of CO_2 into the ocean so that it was carried into deep water where, it was thought, it would remain for hundreds of years. In order to make a significant difference to the atmospheric loading of greenhouse gases, the amount of CO_2 that would need to be stored in this way would have to be significant compared to the amounts of CO_2 currently emitted to the atmosphere – in other words gigatonnes of CO_2 per year. The only potential storage sites with capacity for such quantities are natural reservoirs, such as geological formations (the capacity of European formations was first assessed by Holloway *et al.*, 1996) or the deep ocean (Cole *et al.*, 1993). Other storage options have also been proposed, as discussed below.

Injection of CO_2 underground would involve similar technology to that employed by the oil and gas industry for the exploration and production of hydrocarbons, and for the underground injection of waste as practised in the USA. Wells would be drilled into geological formations and CO_2 would be injected in the same way as CO_2 has been injected for enhanced oil recovery[11] since the 1970s (Blunt *et al.*, 1993; Stevens and Gale, 2000). In some cases, this could lead to the enhanced production of hydrocarbons, which would help to offset the cost. An extension of this idea involves injection into saline formations (Koide *et al.*, 1992) or into unminable coal seams (Gunter *et al.*, 1997); in the latter case, such injection may sometimes result in the displacement of methane, which could be used as a fuel. The world's first commercial-scale CO_2 storage facility, which began operation in 1996, makes use of a deep saline formation under the North Sea (Korbol and Kaddour, 1995; Baklid *et al.*, 1996).

Monitoring will be required both for purposes of managing the storage site and verifying the extent of CO_2 emissions reduction which has been achieved. Techniques such as seismic surveys, which have developed by the oil and gas industry, have been shown to be adequate for observing CO_2 underground (Gale *et al.*, 2001) and may form the basis for monitoring CO_2 stored in such reservoirs.

Many alternatives to the storage of dense phase CO_2 have been proposed: for example, using the CO_2 to make chemicals or other products (Aresta, 1987), fixing it in mineral carbonates for storage in a solid form (Seifritz, 1990; Dunsmore, 1992), storing it as solid CO_2 ('dry ice') (Seifritz, 1992), as CO_2 hydrate (Uchida *et al.*, 1995), or as solid carbon (Steinberg, 1996). Another proposal is to capture the CO_2 from flue gases using micro-algae to make a product which can be turned into a biofuel (Benemann, 1993).

The potential role of CO_2 capture and storage as a mitigation

option has to be examined using integrated energy system models (early studies by Yamaji (1997) have since been followed by many others). An assessment of the environmental impact of the technology through life cycle analysis was reported by Audus and Freund (1997) and other studies have since examined this further.

The concept of CO_2 capture and storage is therefore based on a combination of known technologies applied to the new purpose of mitigating climate change. The economic potential of this technique to enable deep reductions in emissions was examined by Edmonds *et al.* (2001), and is discussed in more detail in Chapter 8. The scope for further improvement of the technology and for development of new ideas is examined in later chapters, each of which focuses on a specific part of the system.

1.4.2   Systems for CO_2 capture

Figure 1.3 illustrates how CO_2 capture and storage may be configured for use in electricity generation. A conventional fossil fuel-fired power plant is shown schematically in Figure 1.3a. Here, the fuel (e.g., natural gas) and an oxidant (typically air) are brought together in a combustion system; heat from this is used to drive a turbine/generator which produces electricity. The exhaust gases are released to atmosphere.

Figure 1.3b shows a plant of this kind modified to capture CO_2 from the flue gas stream, in other words after combustion. Once it has been captured, the CO_2 is compressed in order to transport it to the storage site. Figure 1.3c shows another variant where CO_2 is removed before combustion (pre-combustion decarbonization). Figure 1.3d represents an alternative where nitrogen is extracted from air before combustion; in other words, pure oxygen is supplied as the oxidant. This type of system is commonly referred to as oxyfuel combustion. A necessary part of this process is the recycling of CO_2 or water to moderate the combustion temperature.

1.4.3   Range of possible uses

The main application examined so far for CO_2 capture and storage has been its use in power generation. However, in other large energy-intensive industries (e.g., cement manufacture, oil refining, ammonia production, and iron and steel manufacture), individual plants can also emit large amounts of CO_2, so these industries could also use this technology. In some cases, for example in the production of ammonia or hydrogen, the nature of the exhaust gases (being concentrated in CO_2) would make separation less expensive.

The main applications foreseen for this technology are therefore in large, central facilities that produce significant quantities of CO_2. However, as indicated in Table 1.1, roughly 38% of emissions arise from dispersed sources such as buildings and, in particular, vehicles. These are generally not considered suitable for the direct application of CO_2 capture because of the economies of scale associated with the capture processes as well as the difficulties and costs of transporting small amounts of

---

[11] For example, there were 40 gas-processing plants in Canada in 2002 separating CO_2 and H_2S from produced natural gas and injecting them into geological reservoirs (see Chapter 5.2.4). There are also 76 Enhanced Oil Recovery projects where CO2 is injected underground (Stevens and Gale, 2000).

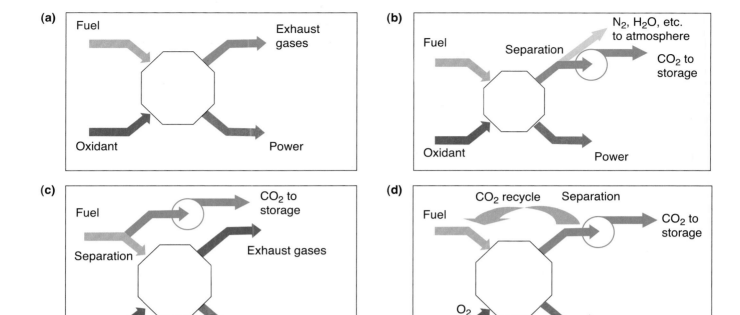

Figure 1.3 a) Schematic diagram of fossil-fuel-based power generation; b) Schematic diagram of post-combustion capture; c) Schematic diagram of pre-combustion capture; d) Schematic diagram of oxyfuel combustion

CO_2. An alternative approach would be to reduce the emissions from dispersed sources by supplying them with an energy carrier with zero net CO_2 emissions from use, such as biofuels, electricity or hydrogen (Johansson *et al.*, 1993). Electricity or hydrogen[12] from fossil fuels could be produced with CO_2 capture and this would avoid most of the CO_2 emissions at the production site (Audus *et al.*, 1996). The cost, applicability and environmental aspects of various applications are discussed later in this report.

1.4.4 Scale of the plant

Some impression of the scale of the plant involved can be gained from considering a coal-fired power plant generating $500MW_e$. This would emit approximately 2.9 $MtCO_2$ per year (0.8 MtC per year) to atmosphere. A comparable plant with CO_2 capture and storage, producing a similar amount of electricity and capturing 85% of the CO_2 (after combustion) and compressing it for transportation, would emit 0.6 $MtCO_2$ per year to the atmosphere (0.16 MtC per year), in other words 80% less than in the case without capture. The latter plant would also send 3.4 $MtCO_2$ per year to storage (0.9 MtC per year). Because of its larger size, the amount of CO_2 generated by the plant with capture and compression is more than the plant without capture (in this example 38% more). This is a result of the energy

requirements of the capture plant and of the CO_2 compressor. The proportion of CO_2 captured (85%) is a level readily achievable with current technology (this is discussed in Chapter 3); it is certainly feasible to capture a higher proportion and designs will vary from case to case. These figures demonstrate the scale of the operation of a CO_2 capture plant and illustrate that capturing CO_2 could achieve deep reductions in emissions from individual power plants and similar installations (IEA GHG, 2000a).

Given a plant of this scale, a pipeline of 300–400 mm diameter could handle the quantities of CO_2 over distances of hundreds of kilometres without further compression; for longer distances, extra compression might be required to maintain pressure. Larger pipelines could carry the CO_2 from several plants over longer distances at lower unit cost. Storage of CO_2, for example by injection into a geological formation, would likely involve several million tonnes of CO_2 per year but the precise amount will vary from site to site, as discussed in Chapters 5 and 6.

1.5 Assessing CCS in terms of environmental impact and cost

The purpose of this section and those that follow is to introduce some of the other issues which are potentially of interest to decision-makers when considering CCS. Answers to some of the questions posed may be found in subsequent chapters, although answers to others will depend on further work and

---

[12] Hydrogen is produced from fossil fuels today in oil refineries and other industrial processes.

local information. When looking at the use of CCS, important considerations will include the environmental and resource implications, as well as the cost. A systematic process of evaluation is needed which can examine all the stages of the CCS system in these respects and can be used for this and other mitigation options. A well-established method of analyzing environmental impacts in a systematic manner is the technique of Life Cycle Analysis (LCA). This is codified in the International Standard ISO 14040 (ISO, 1997). The first step required is the establishment of a system boundary, followed by a comparison of the system with CCS and a base case (reference system) without CCS. The difference will define the environmental impact of CCS. A similar approach will allow a systematic assessment of the resource and/or cost implications of CCS.

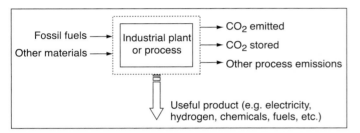

Figure 1.4 System boundary for a plant or process emitting CO_2 (such as a power plant, a hydrogen production plant or other industrial process). The resource and environmental impacts of a CCS system are measured by the changes in total system input and output quantities needed to produce a unit of product.

1.5.1   *Establishing a system boundary*

A generic system boundary is shown in Figure 1.4, along with the flows of materials into and out of the system. The key flow[13] is the product stream, which may be an energy product (such as electricity or heat), or another product with economic value such as hydrogen, cement, chemicals, fuels or other goods. In analyzing the environmental and resource implications of CCS, the convention used throughout this report is to normalize all of the system inputs and outputs to a unit quantity of product (e.g., electricity). As explained later, this concept is essential for establishing the effectiveness of this option: in this particular case, the total amount of CO_2 produced is increased due to the additional equipment and operation of the CCS plant. In contrast, a simple parameter such as the amount of CO_2 captured may be misleading.

Inputs to the process include the fossil fuels used to meet process energy requirements, as well as other materials used by the process (such as water, air, chemicals, or biomass used as a feedstock or energy source). These may involve renewable or non-renewable resources. Outputs to the environment include the CO_2 stored and emitted, plus any other gaseous, liquid or solid emissions released to the atmosphere, water or land. Changes in other emissions – not just CO_2 – may also

---

[13] Referred to as the 'elementary flow' in life cycle analysis.

be important. Other aspects which may be relatively unique to CCS include the ability to keep the CO_2 separate from the atmosphere and the possibility of unpredictable effects (the consequences of climate change, for example) but these are not quantifiable in an LCA.

Use of this procedure would enable a robust comparison of different CCS options. In order to compare a power plant with CCS with other ways of reducing CO_2 emissions from electricity production (the use of renewable energy, for example), a broader system boundary may have to be considered.

1.5.2   *Application to the assessment of environmental and resource impacts*

The three main components of the CO_2 capture, transport and storage system are illustrated in Figure 1.5 as sub-systems within the overall system boundary for a power plant with CCS. As a result of the additional requirements for operating the CCS equipment, the quantity of fuel and other material inputs needed to produce a unit of product (e.g., one MWh of electricity) is higher than in the base case without CCS and there will also be increases in some emissions and reductions in others. Specific details of the CCS sub-systems illustrated in Figure 1.5 are presented in Chapters 3–7, along with the quantification of CCS energy requirements, resource requirements and emissions.

1.5.3   *Application to cost assessment*

The cost of CO_2 capture and storage is typically built up from three separate components: the cost of capture (including compression), transport costs and the cost of storage (including monitoring costs and, if necessary, remediation of any release). Any income from EOR (if applicable) would help to partially offset the costs, as would credits from an emissions trading system or from avoiding a carbon tax if these were to be introduced. The costs of individual components are discussed in Chapters 3 to 7; the costs of whole systems and alternative options are considered in Chapter 8. The confidence levels of cost estimates for technologies at different stages of development and commercialization are also discussed in those chapters.

There are various ways of expressing the cost data (Freund and Davison, 2002). One convention is to express the costs in terms of US\$/tCO_2 avoided, which has the important feature of taking into account the additional energy (and emissions) resulting from capturing the CO_2. This is very important for understanding the full effects on the particular plant of capturing CO_2, especially the increased use of energy. However, as a means of comparing mitigation options, this can be confusing since the answer depends on the base case chosen for the comparison (i.e., what is being avoided). Hence, for comparisons with other ways of supplying energy or services, the cost of systems with and without capture are best presented in terms of a unit of product such as the cost of generation (e.g., US\$ MWh^{-1}) coupled with the CO_2 emissions per unit of electricity generated (e.g., tCO_2 MWh^{-1}). Users can then choose the appropriate base case best suited to their purposes. This is the approach

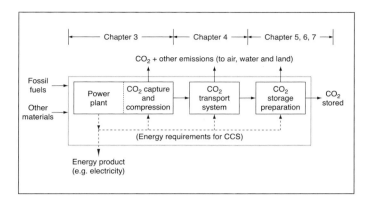

Figure 1.5 System components inside the boundary of Figure 1.4 for the case of a power plant with CO_2 capture and storage. Solid arrows denote mass flows while dashed lines denote energy flows. The magnitude of each flow depends upon the type and design of each sub-system, so only some of the flows will be present or significant in any particular case. To compare a plant with CCS to another system with a similar product, for example a renewables-based power plant, a broader system boundary may have to be used.

used in this report and it is consistent with the treatment of environmental implications described above.

Expressing the cost of mitigation in terms of US\$/tCO_2 avoided is also the approach used when considering mitigation options for a *collection* of plants (such as a national electricity system). This approach is typically found in integrated assessment modelling for policy-related purposes (see Chapter 8). The costs calculated in this way should not be compared with the cost of CO_2-avoided calculated for an *individual* power plant of a particular design as described above because the base case will not be the same. However, because the term 'avoided' is used in both cases, there can be misunderstanding if a clear distinction is not made.

1.5.4 *Other cost and environmental impact issues*

Most of the published studies of specific projects look at particular CO_2 sources and particular storage reservoirs. They are necessarily based on the costs for particular types of plants, so that the quantities of CO_2 involved are typically only a few million tonnes per year. Although these are realistic quantities for the first projects of this kind, they fail to reflect the potential economies of scale which are likely if or when this technology is widely used for mitigation of climate change, which would result in the capture, transport and storage of much greater quantities of CO_2. As a consequence of this greater use, reductions can be expected in costs as a result of both economies of scale and increased experience with the manufacture and operation of most stages of the CCS system. This will take place over a period of several decades. Such effects of 'learning' have been seen in many technologies, including energy technologies, although historically observed rates of improvement and cost reduction are quite variable and have not been accurately predicted for any specific technology (McDonald and Schrattenholzer, 2001).

The construction of any large plant will generate issues relating to environmental impact, which is why impact analyses are required in many countries before the approval of such projects. There will probably be a requirement for gaining a permit for the work. Chapters 3 to 7 discuss in more detail the environmental issues and impacts associated with CO_2 capture, transport and storage. At a power plant, the impact will depend largely on the type of capture system employed and the extra energy required, with the latter increasing the flows of fuel and chemical reagents and some of the emissions associated with generating a megawatt hour of electricity. The construction and operation of CO_2 pipelines will have a similar impact on the environment to that of the more familiar natural gas pipelines. The large-scale transportation and storage of CO_2 could also be a potential hazard, if significant amounts were to escape (see Annex I).

The different storage options may involve different obligations in terms of monitoring and liability. The monitoring of CO_2 flows will take place in all parts of the system for reasons of process control. It will also be necessary to monitor the systems to ensure that storage is safe and secure, to provide data for national inventories and to provide a basis for CO_2 emissions trading.

In developing monitoring strategies, especially for reasons of regulatory compliance and verification, a key question is how long the monitoring must continue; clearly, monitoring will be needed throughout the injection phase but the frequency and extent of monitoring after injection has been completed still needs to be determined, and the organization(s) responsible for monitoring in the long term will have to be identified. In addition, when CO_2 is used, for example, in enhanced oil recovery, it will be necessary to establish the net amount of CO_2 stored. The extent to which the guidelines for reporting emissions already developed by IPCC need to be adapted for this new mitigation option is discussed in Chapter 9.

In order to help understand the nature of the risks, a distinction may usefully be drawn between the slow seepage of CO_2 and potentially hazardous, larger and unintended releases caused by a rapid failure of some part of the system (see Annex I for information about the dangers of CO_2 in certain circumstances). CO_2 disperses readily in turbulent air but seepage from stores under land might have noticeable effects on local ecosystems depending on the amount released and the size of the area affected. In the sea, marine currents would quickly disperse any CO_2 dissolved in seawater. CO_2 seeping from a storage reservoir may intercept shallow aquifers or surface water bodies; if these are sources of drinking water, there could be direct consequences for human activity. There is considerable uncertainty about the potential local ecosystem damage that could arise from seepage of CO_2 from underground reservoirs: small seepages may produce no detectable impact but it is known that relatively large releases from natural CO_2 reservoirs can inflict measurable damage (Sorey *et al.*, 1996). However, if the cumulative amount released from purposeful storage was significant, this could have an impact on the climate. In that case, national inventories would need to take this into account (as discussed in Chapter 9). The likely level of seepage from geological storage reservoirs is the subject of current research described in Chapter 5. Such environmental considerations form the basis for some of the legal barriers to storage of CO_2 which are discussed in Chapters 5 and 6.

The environmental impact of CCS, as with any other energy system, can be expressed as an external cost (IPCC, 2001d) but relatively little has been done to apply this approach to CCS and so it is not discussed further in this report. The results of an application of this approach to CCS can be found in Audus and Freund (1997).

1.6   Assessing CCS in terms of energy supply and CO_2 storage

Some of the first questions to be raised when the subject of CO_2 capture and storage is mentioned are:
- Are there enough fossil fuels to make this worthwhile?
- How long will the CO_2 remain in store?
- Is there sufficient storage capacity and how widely is it available?

These questions are closely related to the minimum time it is necessary to keep CO_2 out of the atmosphere in order to mitigate climate change, and therefore to a fourth, overall, question: 'How long does the CO_2 need to remain in store?' This section suggests an approach that can be used to answer these questions, ending with a discussion of broader issues relating to fossil fuels and other scenarios.

1.6.1   *Fossil fuel availability*

Fossil fuels are globally traded commodities that are available to all countries. Although they may be used for much of the 21st century, the balance of the different fuels may change. CO_2 capture and storage would enable countries, if they wish, to continue to include fossil fuels in their energy mix, even in the presence of severe restrictions on greenhouse gas emissions.

Whether fossil fuels will last long enough to justify the development and large-scale deployment of CO_2 capture and storage depends on a number of factors, including their depletion rate, cost, and the composition of the fossil fuel resources and reserves.

1.6.1.1   *Depletion rate and cost of use*
Proven coal, oil and natural gas reserves are finite, so consumption of these primary fuels can be expected to peak and then decline at some time in the future (IPCC, 2001a). However, predicting the pace at which use of fossil fuels will fall is far from simple because of the many different factors involved. Alternative sources of energy are being developed which will compete with fossil fuels, thereby extending the life of the reserves. Extracting fossil fuels from more difficult locations will increase the cost of supply, as will the use of feedstocks that require greater amounts of processing; the resultant increase in cost will also tend to reduce demand. Restrictions on emissions, whether by capping or tax, would also increase the cost of using

fossil fuels, as would the introduction of CCS. At the same time, improved technology will reduce the cost of using these fuels. All but the last of these factors will have the effect of extending the life of the fossil fuel reserves, although the introduction of CCS would tend to push up demand for them.

1.6.1.2   *Fossil fuel reserves and resources*
In addition to the known reserves, there are significant resources that, through technological advances and the willingness of society to pay more for them, may be converted into commercial fuels in the future. Furthermore, there are thought to be large amounts of non-conventional oil (e.g., heavy oil, tars sands, shales) and gas (e.g., methane hydrates). A quantification of these in the Third Assessment Report (IPCC, 2001a) showed that fully exploiting the known oil and natural gas resources (without any emission control), plus the use of non-conventional resources, would cause atmospheric concentrations of CO_2 to rise above 750 ppmv. In addition, coal resources are even larger than those of oil and gas; consuming all of them would enable the global economy to emit 5 times as much CO_2 as has been released since 1850 (5,200 GtCO_2 or 1,500 GtC) (see Chapter 3 in IPCC, 2001a). A scenario for achieving significant reductions in emissions but without the use of CCS (Berk *et al.*, 2001) demonstrates the extent to which a shift away from fossil fuels would be required to stabilize at 450 ppmv by 2100. Thus, sufficient fossil fuels exist for continued use for decades to come. This means that the availability of fossil fuels does not limit the potential application of CO_2 capture and storage; CCS would provide a way of limiting the environmental impact of the continued use of fossil fuels.

1.6.2   *Is there sufficient storage capacity?*

To achieve stabilization at 550 ppmv, the Third Assessment Report (IPCC, 2001e) showed that, by 2100, the reduction in emissions might have to be about 38 GtCO_2 per year (10 GtC per year)[14] compared to scenarios with no mitigation action. If CO_2 capture and storage is to make a significant contribution towards reducing emissions, several hundreds or thousands of plants would need to be built, each capturing 1 to 5 MtCO_2 per year (0.27–1.4 MtC per year). These figures are consistent with the numbers of plants built and operated by electricity companies and other manufacturing enterprises.

Initial estimates of the capacity of known storage reservoirs (IEA GHG, 2001; IPCC, 2001a) indicate that it is comparable to the amount of CO_2 which would be produced for storage by such plants. More recent estimates are given in Chapters 5 and 6, although differences between the methods for estimating storage capacity demonstrate the uncertainties in these estimates; these issues are discussed in later chapters. Storage outside natural reservoirs, for example in artificial stores or by changing CO_2 into another form (Freund, 2001), does not generally provide

---

[14] This is an indicative value calculated by averaging the figures across the six SRES marker scenarios; this value varies considerably depending on the scenario and the parameter values used in the climate model.

similar capacity for the abatement of emissions at low cost (Audus and Oonk, 1997); Chapter 7 looks at some aspects of this.

The extent to which these reservoirs are within reasonable, cost-competitive distances from the sources of CO_2 will determine the potential for using this mitigation option.

1.6.3   *How long will the CO_2 remain in storage?*

This seemingly simple question is, in fact, a surprisingly complicated one to answer since the mechanisms and rates of release are quite different for different options. In this report, we use the term 'fraction retained' to indicate how much CO_2 remains in store for how long. The term is defined as follows:

- '*Fraction retained*' is the fraction of the cumulative amount of injected CO_2 that is retained in the storage reservoir over a specified period of time, for example a hundred or a million years.

Chapters 5, 6 and 7 provide more information about particular types of storage. Table AI.6 in Annex I provides the relation between leakage of CO_2 and the fraction retained. The above

definition makes no judgement about how the amount of CO_2 retained in storage will evolve over time – if there were to be an escape of CO_2, the rate may not be uniform.

The CO_2 storage process and its relationship to concentrations in the atmosphere can be understood by considering the stocks of stored CO_2 and the flows between reservoirs. Figure 1.6 contains a schematic diagram that shows the major stocks in natural and potential engineered storage reservoirs, and the flows to and from them. In the current pattern of fossil fuel use, CO_2 is released directly to the atmosphere from human sources. The amount of CO_2 released to the atmosphere by combustion and industrial processes can be reduced by a combination of the various mitigation measures described above. These flows are shown as alternative pathways in Figure 1.6.

The flows marked ***CCS*** with a subscript are the *net* tons of carbon dioxide per year that could be placed into each of the three types of storage reservoir considered in this report. Additional emissions associated with the capture and storage process are not explicitly indicated but may be considered as additional sources of CO_2 emission to the atmosphere. The potential release flows from the reservoirs to the atmosphere are indicated by ***R***, with a subscript indicating the appropriate reservoir. In some storage options, the release flows can be very

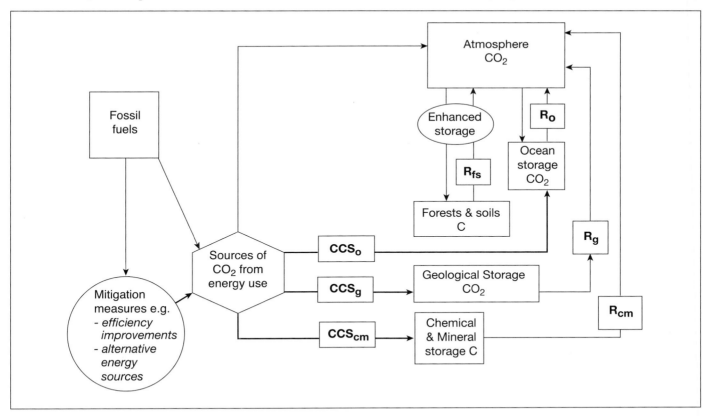

Figure 1.6 Schematic diagram of stocks and flows of CO_2 with net flows of captured CO_2 to each reservoir indicated by the label CCS (these flows exclude residual emissions associated with the process of capture and storage). The release flows from each of the storage reservoirs are indicated by the labels R. The stock in the atmosphere depends upon the difference between the rates at which CO_2 reaches the atmosphere and at which it is removed. Flows to the atmosphere may be slowed by a combination of mitigation options, such as improving energy efficiency or the use of alternatives to fossil fuels, by enhancing biological storage or by storing CCS in geological formations, in the oceans or in chemicals or minerals.

small compared to the flows into those storage reservoirs.

The *amount* in storage at a particular time is determined by the capacity of the reservoir and the past history of additions to, and releases from, the reservoir. The *change* in stocks of CO_2 in a particular storage reservoir over a specified time is determined by the current stock and the relative rates at which the gas is added and released; in the case of ocean storage, the level of CO_2 in the atmosphere will also influence the net rate of release[15]. As long as the *input* storage rate exceeds the *release* rate, CO_2 will accumulate in the reservoir, and a certain amount will be stored away from the atmosphere. Analyses presented in this report conclude that the time frames for different storage options cover a wide range:

- The terrestrial biosphere stores and releases both natural and fossil fuel CO_2 through the global carbon cycle. It is difficult to provide a simple picture of the fraction retained because of the dynamic nature of this process. Typically, however, 99% is stored for decades to centuries, although the average lifetime will be towards the lower end of that range. The terrestrial biosphere at present is a net sink for carbon dioxide but some current biological sinks are becoming net sources as temperatures rise. The annual storage flows and total carbon storage capacity can be enhanced by forestry and soil management practices. Terrestrial sequestration is not explicitly considered in this report but it is covered in IPCC, 2000b.

- Oceans hold the largest amount of mobile CO_2. They absorb and release natural and fossil fuel CO_2 according to the dynamics of the global carbon cycle, and this process results in changes in ocean chemistry. The fraction retained by ocean storage at 3,000 m depth could be around 85% after 500 years. However, this process has not yet been demonstrated at a significant scale for long periods. Injection at shallower depths would result in shorter retention times. Chapter 6 discusses the storage capacity and fractions retained for ocean storage.

- In geological storage, a picture of the likely fraction retained may be gained from the observation of natural systems where CO_2 has been in natural geological reservoirs for millions of years. It may be possible to engineer storage reservoirs that have comparable performance. The fraction retained in appropriately selected and managed geological reservoirs is likely to exceed 99% over 1000 years. However, sudden gas releases from geological reservoirs could be triggered by failure of the storage seal or the injection well, earthquakes or volcanic eruptions, or if the reservoir were accidentally punctured by subsequent drilling activity. Such releases might have significant local effects. Experience with engineered natural-gas-storage facilities and natural CO_2 reservoirs may be relevant to understanding whether such releases might occur. The storage capacity and fraction retained for the various geological storage options are discussed in Chapter 5.

- Mineral carbonation through chemical reactions would provide a fraction retained of nearly 100% for exceptionally long times in carbonate rock. However, this process has not yet been demonstrated on a significant scale for long periods and the energy balance may not be favourable. This is discussed in Chapter 7.

- Converting carbon dioxide into other, possibly useful, chemicals may be limited by the energetics of such reactions, the quantities of chemicals produced and their effective lifetimes. In most cases this would result in very small net storage of CO_2. Ninety-nine per cent of the carbon will be retained in the product for periods in the order of weeks to months, depending on the product. This is discussed in Chapter 7.

1.6.4   How long does the CO_2 need to remain in storage?

In deciding whether a particular storage option meets mitigation goals, it will be important to know both the net storage capacity and the fraction retained over time. Alternative ways to frame the question are to ask 'How long is enough to achieve a stated policy goal?' or 'What is the benefit of isolating a specific amount of CO_2 away from the atmosphere for a hundred or a million years?' Understanding the effectiveness of storage involves the consideration of factors such as the maximum atmospheric concentration of CO_2 that is set as a policy goal, the timing of that maximum, the anticipated duration of the fossil fuel era, and available means of controlling the CO_2 concentration in the event of significant future releases.

The issue for policy is whether CO_2 will be held in a particular class of reservoirs long enough so that it will not increase the difficulty of meeting future targets for CO_2 concentration in the atmosphere. For example, if 99% of the CO_2 is stored for periods that exceed the projected time span for the use of fossil fuels, this should not to lead to concentrations higher than those specified by the policy goal.

One may assess the implications of possible future releases of CO_2 from storage using simulations similar to those developed for generating greenhouse gas stabilization trajectories[16]. A framework of this kind can treat releases from storage as delayed emissions. Some authors examined various ways of assessing unintended releases from storage and found that a delay in emissions in the order of a thousand years may be almost as effective as perfect storage (IPCC, 2001b; Herzog *et al.*, 2003; Ha-Duong and Keith, 2003)[17]. This is true if marginal carbon prices remain constant or if there is a backstop technology that can cap abatement costs in the not too distant

---

[15]   For further discussion of this point, see Chapter 6.

[16]   Such a framework attempts to account for the intergenerational trade-offs between climate impact and the cost of mitigation and aims to select an emissions trajectory (modified by mitigation measures) that maximizes overall welfare (Wigley *et al.*, 1996; IPCC, 2001a).

[17]   For example, Herzog *et al.* (2003) calculated the effectiveness of an ocean storage project relative to permanent storage using economic arguments; given a constant carbon price, the project would be 97% effective at a 3% discount rate; if the price of carbon were to increase at the same rate as the discount rate for 100 years and remain constant thereafter, the project would be 80% effective; for a similar rate of increase but over a 500 year period, effectiveness would be 45%.

future. However, if discount rates decline in the long term, then releases of CO_2 from storage must be lower in order to achieve the same level of effectiveness.

Other authors suggest that the climate impact of CO_2 released from imperfect storage will vary over time, so they expect carbon prices to depend on the method of accounting for the releases. Haugan and Joos (2004) found that there must be an upper limit to the rate of loss from storage in order to avoid temperatures and CO_2 concentrations over the next millennium becoming higher in scenarios with geological CCS than in those without it[18].

Dooley and Wise (2003) examined two hypothetical release scenarios using a relatively short 100-year simulation. They showed that relatively high rates of release from storage make it impossible to achieve stabilization at levels such as 450 ppmv. They imply that higher emissions trajectories are less sensitive to such releases but, as stabilization is not achieved until later under these circumstances, this result is inconclusive.

Pacala (2003) examined unintended releases using a simulation over several hundred years, assuming that storage security varies between the different reservoirs. Although this seemed to suggest that quite high release rates could be acceptable, the conclusion depends on extra CO_2 being captured and stored, and thereby accumulating in the more secure reservoirs. This would imply that it is important for reservoirs with low rates of release to be available.

Such perspectives omit potentially important issues such as the political and economic risk that policies will not be implemented perfectly, as well as the resulting ecological risk due to the possibility of non-zero releases which may preclude the future stabilization of CO_2 concentrations (Baer, 2003). Nevertheless, all methods imply that, if CO_2 capture and storage is to be acceptable as a mitigation measure, there must be an upper limit to the amount of unintended releases.

The discussion above provides a framework for considering the effectiveness of the retention of CO_2 in storage and suggests a potential context for considering the important policy question: 'How long is long enough?' Further discussion of these issues can be found in Chapters 8 and 9.

1.6.5   Time frame for the technology

Discussions of CCS mention various time scales. In this section, we propose some terminology as a basis for the later discussion.

Energy systems, such as power plant and electricity transmission networks, typically have operational lifetimes of 30–40 years; when refurbishment or re-powering is taken into account, the generating station can be supplying electricity for even longer still. Such lifetimes generate expectations which are reflected in the design of the plant and in the rate of return on the investment. The capture equipment could be built and refurbished on a similar cycle, as could the CO_2 transmission system. The operational lifetime of the CO_2 storage reservoir will be determined by its capacity and the time frame over which it can retain CO_2, which cannot be so easily generalized. However, it is likely that the phase of filling the reservoir will be at least as long as the operational lifetime of a power plant[19]. In terms of protecting the climate, we shall refer to this as the **medium term**, in contrast to the **short-term** nature of measures connected with decisions about operating and maintaining such facilities.

In contrast, the mitigation of climate change is determined by longer time scales: for example, the lifetime (or adjustment time) of CO_2 in the atmosphere is often said to be about 100 years (IPCC, 2001c). Expectations about the mitigation of climate change typically assume that action will be needed during many decades or centuries (see, for example, IPCC, 2000a). This will be referred to as the **long term**.

Even so, these descriptors are inadequate to describe the storage of CO_2 as a mitigation measure. As discussed above, it is anticipated that CO_2 levels in the atmosphere would rise, peak and decline over a period of several hundred years in virtually all scenarios; this is shown in Figure 1.7. If there is effective action to mitigate climate change, the peak would occur sooner

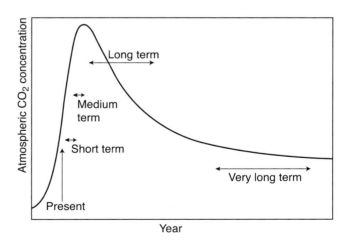

Figure 1.7 The response of atmospheric CO_2 concentrations due to emissions to the atmosphere. Typical values for 'short term', 'medium term', 'long term' and' very long term' are years, decades, centuries, millennia, respectively. In this example, cumulative emissions are limited to a maximum value and concentrations stabilize at 550 ppmv (adapted from Kheshgi, 2003). This figure is indicative and should not be read as prescribing specific values for any of these periods. If the goal were to constrain concentrations in the atmosphere to lower levels, such as 450 ppmv, greater reductions in emission rates would be required.

---

[18] These authors calculated the effectiveness of a storage facility measured in terms of the global warming avoided compared with perfect storage. For a store which annually releases 0.001 of the amount stored, effectiveness is around 60% after 1000 years. This rate of release would be equivalent to a fraction retained of 90% over 100 years or 60% over 500 years. It is likely that, in practice, geological and mineral storage would have lower rates of release than this (see chapters 5 and 7) and hence higher effectiveness – for example, a release rate of 0.01% per year would be equivalent to a fraction retained of 99% over 100 years or 95% over 500 years.

[19] It should be noted that there will not necessarily be a one-to-one correspondence between a CO_2-producing plant and storage reservoir. Given a suitable network for the transport of CO_2, the captured CO_2 from one plant could be stored in different locations during the lifetime of the producing plant.

(and be at a lower level) than if no action is taken. As suggested above, most of the CO_2 must be stored for much longer than the time required to achieve stabilization. We consider this to be the **very long term**, in other words periods of time lasting centuries or millennia. Precisely how long is a subject of much debate at present and this will be explored in later chapters.

1.6.6   Other effects of introducing CCS into scenarios

In view of the economic importance of energy carriers (more than 2 trillion dollars annually, World Energy Assessment, 2004) as well as fossil fuel's contribution to climate forcing (50 to 60% of the total), the decision to invest economic resources in the development of a technology such as CCS may have far-reaching consequences, including implications for equity and sustainable development (these are discussed in the following section). This emphasizes the importance of considering the wider ramifications of such investment.

The implementation of CCS would contribute to the preservation of much of the energy infrastructure established in the last century and may help restrain the cost of meeting the target for emissions reduction. From another perspective, its use may reduce the potential for application of alternative energy sources (Edmonds *et al.*, 2001). As noted in section 1.3, the mitigation of climate change is a complex issue and it seems likely that any eventual solution will involve a portfolio of methods[20]. Even so, there is concern in some quarters that the CO_2 capture and storage option could capture financial resources and the attention of policymakers that would otherwise be spent on alternative measures, although this issue has not been extensively analyzed in the literature.

The possibility of obtaining net negative emissions when coupling biomass energy and CCS may provide an opportunity to reduce CO_2 concentration in the atmosphere if this option is available at a sufficiently large scale. In view of the uncertainty about the safe concentration of CO_2 in the atmosphere, a large-scale option providing net negative emissions could be especially useful in the light of the precautionary principle.

1.6.6.1   Effect of CCS on energy supply and use
All of the SRES scenarios (IPCC, 2000a) show significant consumption of fossil fuels for a long time into the future. One of the consequences of deploying CCS would be a continued use of fossil fuels in the energy mix but the minimization of their effect on the climate system and environment. By enabling countries to access a wider range of energy supplies than would otherwise be the case, energy security will be improved. Such aspects are important when considering climate change policy and sustainable development: as indicated before, decision-makers are likely to balance pure economic effectiveness against other socially relevant issues.

The successful development and implementation of CCS on a large scale might therefore be interpreted by society as a driver for reinforcing socio-economic and behavioural trends that are increasing total energy use, especially in developed countries and within high-income groups in developing countries[21] (IPCC, 2001a).

1.6.6.2   Effect of CCS on technological diversity
The fossil fuel energy system and its infrastructure can be thought of as a technology cluster. Such a phenomenon can be recognized as possibly presenting dangers as well as offering benefits for society. It can lead to specialization as innovations improve on dominant technologies, thereby generating further innovations which help to retain market share. On the other hand, innovations in technologies with small market shares are less valuable and so there is less incentive to improve on those technologies; a minor technology can therefore become trapped by high costs and a small market share. This phenomenon leads to path dependence or technology lock-in (Bulter and Hofkes, 2004; Unruh, 2000). Although CCS has not yet been examined specifically in this respect, it may be that reinforcing the position of the fossil fuel energy system may present barriers to increased technological diversity (a key element in evolutionary change; see Nelson and Winter, 1982).

It could be argued that increasing demand for some alternative energy sources will bring significant additional benefits outside the climate change arena such as rural sector jobs, or a large labour force for maintenance (World Energy Assessment, 2004). It is not possible to forecast the full societal impacts of such technology in its early days, especially as it seems likely that stabilizing atmospheric concentrations of CO_2 will require the full slate of available technologies (including ones not yet developed). The available information is not adequate for predictions of the differences in job creation potential between different mitigation options.

In view of the paucity of literature on these aspects of CCS, this report cannot provide tools for a full quantitative judgment of options; it merely flags some of the other issues that decision-makers will wish to consider. This is further discussed in Chapter 8.

1.6.6.3   Financing of the projects
Compared to a similar plant that releases CO_2 to the atmosphere, a facility with capture and storage will cost more to build and to operate and will be less efficient in its use of primary energy. If regulations are adopted which cause the owners of CO_2-emitting plant to limit emissions, and they choose to use CCS (or any other measure which increases their costs), they will need to find ways to recover the extra costs or accept a lower rate of return on their investment. In circumstances where emissions trading is allowed, companies may, in some cases, reduce the cost of meeting emission targets by buying or selling

---

[20] The optimum portfolio of mitigation measures is likely to be different in different places and at different times. Given the variety of measures available, it seems likely that several will be used in a complementary fashion as part of the portfolio, and that there will not be a single clear 'winner' amongst them.

[21] For example, housing units in many countries are increasing in size, and the intensity of electrical appliance use is increasing. The use of electrical office equipment in commercial buildings is also rising rapidly.

credits. Where the project is located in another Annex I country, it may be possible to fund this through Joint Implementation (JI). The Clean Development Mechanism (CDM) may provide opportunities for developing countries to acquire technology for emission reduction purposes, with some of the costs being borne by external funders who can claim credit for these investments. At the time of writing, it is uncertain whether CCS projects would be covered by the CDM and there are many issues to be considered. The current low value of Certified Emission Reductions is a major barrier to such projects at present (IEA GHG, 2004a). It is possible that some CO_2-EOR projects could be more attractive, especially if the project would also delay the abandonment of a field or prevent job losses. The issue of the longevity of storage has still to be resolved but the longer retention time for geological formations may make it easier for CCS to be accepted than was the case for natural sinks. A number of countries have the potential to host CCS projects involving geological storage under CDM (IEA GHG, 2004a) but the true potential can only be assessed when the underground storage resources have been mapped. The above discussion shows that there are many questions to be answered about the financing of such options, not least if proposed as a project under the flexible mechanisms of the Kyoto Protocol.

1.6.7 *Societal requirements*

Even if CO_2 capture and storage is cost-effective and can be recognized as potentially fulfilling a useful role in energy supply for a climate-constrained world, there will be other aspects that must be addressed before it can be widely used. For example, what are the legal issues that face this technology? What framework needs to be put in place for long-term regulation? Will CO_2 capture and storage gain public acceptance?

1.6.7.1 *Legal issues concerning CCS*
Some legal questions about CCS can be identified and answered relatively easily; for example, the legal issues relating to the process of capturing CO_2 seem likely to be similar to those facing any large chemical plant. Transporting CO_2 through pipelines can probably be managed under current regulatory regimes for domestic and international pipelines. The extent to which the CO_2 is contaminated with other substances, such as compounds of sulphur (see Chapter 4), might alter its classification to that of a hazardous substance, subjecting it to more restrictive regulation. However, the storage of carbon dioxide is likely to pose new legal challenges. What licensing procedure will be required by national authorities for storage in underground reservoirs onshore? It seems likely that factors to be considered will include containment criteria, geological stability, potential hazard, the possibility of interference with other underground or surface activities and agreement on sub-surface property rights, and controls on drilling or mining nearby.

Storage in geological formations below the sea floor will be controlled by different rules from storage under land. The Law of the Sea[22], the London Convention and regional agreements such as the OSPAR Convention[23] will affect storage of CO_2 under the sea but the precise implications have yet to be worked out. This is discussed further in Chapter 5. Ocean storage raises a similar set of questions about the Law of the Sea and the London Convention but the different nature of the activity may generate different responses. These are discussed in Chapter 6.

A further class of legal issues concerns the responsibility for stored carbon dioxide. This is relevant because the CO_2 will have been the subject of a contract for storage, or a contract for emissions reduction, and/or because of the possibility of unintended release. Should society expect private companies to be responsible over centuries for the storage of CO_2? A judgement may have to be made about a reasonable balance between the costs and benefits to current and to future generations. In the case of the very long-term storage of nuclear waste, states have taken on the responsibility for managing storage; the companies that generate the waste, and make a profit from using the nuclear material, pay a fee to the government to take responsibility. In other fields, the deep-well injection of hazardous materials is sometimes the responsibility of governments and sometimes the responsibility of the companies concerned under a licensing system (IEA GHG, 2004b). Rules about insurance and about liability (if there were to be a release of CO_2) will need to be developed so that, even if something happens in the distant future, when the company that stored it is no longer in business, there will be a means of ensuring another organization is capable and willing to accept responsibility.

The information on legal issues presented in this report reflects the best understanding at the time of writing but should not be taken as definitive as the issues have not been tested.

1.6.7.2 *Public acceptance*
Only a few studies have been carried out of public attitudes towards CCS. Such research presents challenges because the public is not familiar with the technology, and may only have a limited understanding of climate change and the possibilities for mitigation. As a result the studies completed to date have had to provide information on CCS (and on climate change) to their subjects. This tends to limit the scale of the study which can be carried out. This issue is examined in more detail in Chapter 5.

What form of public consultation will be needed before approval of a CCS project? Will the public compare CCS with other activities below ground such as the underground storage of natural gas or will CCS be compared to nuclear waste disposal? Will they have different concerns about different forms of storage, such as geological or ocean storage of CO_2? Will the general attitude towards building pipelines affect the development of CO_2 pipelines? These and other issues are the subject of current discussion and investigation.

When a CCS project is proposed, the public and governments will want to be satisfied that storage of carbon dioxide is so

---

[22] The full text of these conventions is accessible on the Internet.
[23] Issues of interest for this report are at the time of writing being discussed in the OSPAR convention that regulates the uses of the North East Atlantic.

secure that emissions will be reduced and also that there will be no significant threat to human health or to ecosystems (Hawkins, 2003). Carbon dioxide transport and storage will have to be monitored to ensure there is little or no release to the atmosphere but monitoring issues are still being debated. For example, can the anticipated low rates of CO_2 release from geological storage be detected by currently available monitoring techniques? Who will do this monitoring (IEA GHG, 2004b)? How long should monitoring continue after injection: for periods of decades or centuries (IEA GHG, 2004c)?

1.7   Implications for technology transfer and sustainable development

1.7.1   *Equity and sustainable development*

The climate change issue involves complex interactions between climatic, environmental, economic, political, institutional, social, scientific, and technological processes. It cannot be addressed in isolation from broader societal goals, such as equity or sustainable development (IPCC, 2001a), or other existing or probable future sources of environmental, economic or social stress. In keeping with this complexity, a multiplicity of approaches has emerged to analyze climate change and related challenges. Many of these incorporate concerns about development, equity, and sustainability, albeit partially and gradually (IPCC, 2001a).

Sustainable development is too complex a subject for a simple summary; the study of this field aims to assess the benefits and trade-offs involved in the pursuit of the multiple goals of environmental conservation, social equity, economic growth, and eradication of poverty (IPCC, 2001a, Chapter 1). Most of the studies only make a first attempt to integrate a number of important sustainable development indicators and only a few have considered the implications for CCS (Turkenburg, 1997). To date, studies have focused on short-term side-effects of climate change mitigation policies (e.g., impact on local air and water quality) but they have also suggested a number of additional indicators to reflect development (e.g., job creation) and social impact (e.g., income distribution). CCS also poses issues relating to long-term liability for possible unintended releases or contamination which may have inter-generational and, in some cases, international consequences[24]. Further studies will be needed to develop suitable answers about CCS. In particular, long-term liability must be shown to be compatible with sustainable development.

There are various viewpoints relating to climate policy: one is based on cost-effectiveness, another on environmental sustainability, and another on equity (Munasinghe and Swart,

2005). Most policies designed to achieve the mitigation of climate change also have other important rationales. They can be related to the objectives of development, sustainability and equity. 'Conventional' climate policy analyses have tended to be driven (directly or indirectly) by the question: what is the cost-effective means of mitigating climate change for the global economy? Typically, these analyses start from a baseline projection of greenhouse gas emissions and reflect a specific set of socio-economic projections. Equity considerations are added to the process, to broaden the discussion from global welfare as a single subject to include the effects of climate change and mitigation policies on existing inequalities, amongst and within nations. The goal here goes beyond providing for basic survival, extending to a standard of living that provides security and dignity for all.

Ancillary effects of mitigation policies may include reductions in local and regional air pollution, as well as indirect effects on transportation, agriculture, land use practices, biodiversity preservation, employment, fuel security, etc. (Krupnick *et al.*, 2000). The concept of 'co-benefits' can be used to capture dimensions of the response to mitigation policies from the equity and sustainability perspectives in a way that could modify the projections produced by those working from the cost-effectiveness perspective. As yet, little analysis has been reported of the option of CCS in these respects.

Will CO_2 capture and storage favour the creation of job opportunities for particular countries? Will it favour technological and financial elitism or will it enhance equity by reducing the cost of energy? In terms of sustainable development, does the maintenance of the current market structures aid those countries that traditionally market fossil fuels, relative to those that import them? Is this something which mitigation policies should be developed to assist? There are no simple answers to these questions but policymakers may want to consider them. However, no analysis of these aspects of CCS is yet available. Furthermore, the mitigation options available will vary from country to country; in each case, policymakers have to balance such ancillary benefits with the direct benefits of the various options in order to select the most appropriate strategy.

1.7.2   *Technology transfer*

Article 4.5 of the UNFCCC requires all Annex I countries to take 'All practicable steps to promote, facilitate and finance, as appropriate, the transfer of, or access to, environmentally sound technologies and know-how to other parties, particularly developing countries, to enable them to implement provisions of the convention.' This applies to CCS as much as it does to any other mitigation option. This was precisely stated in the declaration issued at COP 7 (UNFCCC, 2001). Paragraph 8, item (d) states: 'Cooperating in the development, diffusion and transfer (…) and/or technologies relating to fossil fuels that capture and store GHGs, and encouraging their wider use, and facilitating the participation of the least developed countries and other Parties not included in Annex I in this effort'

In achieving these objectives of the Convention, several key

---

[24] Some legislation is already in place which will influence this: for example both the London Convention (Article X) and its 1996 Protocol (Article 15) contain provisions stating that liability is in accordance with the principles of international law regarding a state's responsibility for damage caused to the environment of other states or to any other area of the environment. Similarly, regional agreements such as the OSPAR Convention incorporate the 'polluter pays' principle (Article 2(b)).

elements will have to be considered (IPCC, 2001a). These are discussed in the IPCC Special Report on Technology Transfer (IPCC, 2000c), which looked into all aspects of the processes affecting the development, application and diffusion of technology. This looks at technology transfer for the purposes of adapting to climate change as well as for mitigation. It looks at processes within countries and between countries, covering hardware, knowledge and practices. Particularly important are the assessment of technology needs, the provision of technology information, capacity building, the creation of an enabling environment, and innovative financing to facilitate technology transfer.

Although no academic examination of CCS in these respects has yet been undertaken, some remarks can be made in general about this mitigation option.

1.7.2.1   Potential barriers

Technology transfer faces several barriers, including intellectual property rights, access to capital, etc. As with any new technology, CCS opens opportunities for proprietary rights. As it will rely on the development and/or integration of technologies, some of which are not yet used for such purposes, there is considerable scope for learning by doing. Several developing countries are already taking an active interest in this option, where they have national resources that would allow them to make use of this technique. For example, Deshun *et al.* (1998) have been looking at the related technique of CO_2-EOR. Some of the key technologies will be developed by particular companies (as is occurring with wind power and solar photovoltaics) but will the intellectual property for CCS be accumulated in the hands of a few? CCS will involve both existing and future technologies, some of which will be proprietary. Will the owners of these rights to be willing to exploit their developments by licensing others to use them? At present it appears to be too early to answer these questions.

Given that the essential parts of CCS systems are based on established technology, it can be expected that it will be accessible to anyone who can afford it and wants to buy it. Several companies currently offer competing methods of capturing CO_2; pipelines for CO_2 and ships are constructed today by companies specializing in this type of equipment; the drilling of injection wells is standard practice in the oil and gas industry, and is carried out by many companies around the world. More specialist skills may be required to survey geological reservoirs; indeed, monitoring of CO_2 underground is a very new application of seismic analysis. However, it is anticipated that, within a short space of time, these will become as widely available as other techniques derived from the international oil and gas industry. Making these technologies available to developing countries will pose similar challenges as those encountered with other modern technological developments. This shows the relevance of the UNFCCC declaration on technology transfer quoted above to ensure that developing countries have access to the option of CO_2 capture and storage.

1.7.2.2   Potential users

CO_2 emissions are rising rapidly in some developing countries; if these countries wish to reduce the rate of increase of emissions, they will want to have access to a range of mitigation options, one of which could be CCS. Initially it seems likely that CCS would be exploited by countries with relevant experience, such as oil and gas production[25], but this may not be the case in other natural resource sectors. Will there be fewer opportunities for the transfer of CCS technology than for other mitigation options where technologies are in the hands of numerous companies? Or will the knowledge and experience already available in the energy sector in certain developing countries provide an opportunity for them to exploit CCS technologies? Will CO_2 capture and storage technologies attract more interest from certain developing countries if applied to biomass sources[26]? If there is a year-round supply of CO_2 from the biomass processing plant and good storage reservoirs within reasonable distance, this could be an important opportunity for technology transfer. As yet there are no answers to these questions.

1.8   Contents of this report

This report provides an assessment of CO_2 capture and storage as an option for the mitigation of climate change. The report does not cover the use of natural sinks to sequester carbon since this issue is covered in the Land Use, Land Use Change and Forestry report (IPCC, 2000b) and in IPCC's Third Assessment Report (IPCC, 2001a).

There are many technical approaches which could be used for capturing CO_2. They are examined in Chapter 3, with the exception of biological processes for fixation of CO_2 from flue gases, which are not covered in this report. The main natural reservoirs which could, in principle, hold CO_2 are geological formations and the deep ocean; they are discussed in Chapters 5 and 6 respectively. Other options for the storage and re-use of CO_2 are examined in Chapter 7.

Chapter 2 considers the geographical correspondence of CO_2 sources and potential storage reservoirs, a factor that will determine the cost-effectiveness of moving CO_2 from the place where it is captured to the storage site. A separate chapter, Chapter 4, is dedicated to transporting CO_2 from capture to storage sites.

The overall cost of this technology and the consequences of including it in energy systems models are described in Chapter 8. Some of the other requirements outlined above, such as legality, applicable standards, regulation and public acceptance, are discussed in detail at the appropriate point in several of the chapters. Governments might also wish to know how this method of emission reduction would be taken into account in national inventories of greenhouse gas emissions. This area is discussed in Chapter 9. Government and industry alike will be interested in the accessibility of the technology, in methods of financing the plant and in whether assistance will be available

---

[25] In 1999, there were 20 developing countries that were each producing more than 1% of global oil production, 14 developing countries that were each producing more than 1% of global gas production, and 7 developing countries producing more than 1% of global coal production (BP, 2003).
[26] For further discussion of using CCS with biomass, see Chapter 2.

from industry, government or supra-national bodies. At present, it is too early in the exploitation of this technology to make confident predictions about these matters. Three annexes provide information about the properties of CO_2 and carbon-based fuels, a glossary of terms and the units used in this report. Gaps and areas for further work are discussed in the chapters and in the Technical Summary to this report.

References

Aresta, M. (ed.), 1987: Carbon dioxide as a source of carbon; biochemical and chemical use. Kluwer, the Hague.

Audus, H. and H. Oonk, 1997: An assessment procedure for chemical utilisation schemes intended to reduce CO_2 emission to atmosphere. *Energy Conversion and Management,* **38**(suppl. Proceedings of the Third International Conference on Carbon Dioxide Removal, 1996), pp S409–414.

Audus, H. and P. Freund, 1997: The costs and benefits of mitigation: a full fuel cycle examination of technologies for reducing greenhouse gas emissions. *Energy Conversion and Management,* **38**, Suppl., pp S595–600.

Audus, H., O. Kaarstad, and M. Kowal, 1996: Decarbonisation of fossil fuels: Hydrogen as an energy carrier. Proceedings of the 11th World Hydrogen Energy Conference, International Association of Hydrogen Energy, published by Schon and Wetzel, Frankfurt, Germany.

Azar, C., K. Lindgren, and B.A. Andersson, 2003: Global energy scenarios meeting stringent CO_2 constraints - cost-effective fuel choices in the transportation sector. *Energy Policy,* **31**, pp. 961–976.

Baer, P., 2003: An issue of scenarios: carbon sequestration as an investment and the distribution of risk. An editorial comment. *Climate Change,* **59**, 283–291.

Baklid, A., R. Korbøl, and G. Owren, 1996: Sleipner Vest CO_2 disposal: CO_2 injection into a shallow underground aquifer. Paper presented at the 1996 SPE Annual Technical Conference, Denver, Colorado, USA. SPE paper 36600, 1–9.

Benemann, J.R., 1993: Utilization of carbon dioxide from fossil fuel burning power plant with biological systems. *Energy Conversion and Management,* **34**(9–11) pp. 999–1004.

Berk, M.M., J.G. van Minnen, B. Metz, and W. Moomaw, 2001: Keeping Our Options Open, Climate Options for the Long Term (COOL) – Global Dialogue synthesis report, RIVM, NOP rapport nr. 410 200 118.

Blunt, M., F.J. Fayers, and F.M. Orr Jr. 1993: Carbon Dioxide in Enhanced Oil Recovery. *Energy Conversion and Management,* **34**(9–11) pp. 1197–1204.

BP, 2003: *BP Statistical Review of World Energy.* London.

Bulter, F.A.G den and M.W. Hofkes, 2004: Technological Transition: a neo-classical economics viewpoint. In *Sciences for Industrial Transformation: views from different disciplines.* X. Olsthoorn and A.Wieczoreck (eds.). Kluwer Academic Publishers, Dordrecht.

Cole, K.H., G.R. Stegen, D. Spencer, 1993: *Energy Conversion and Management,* **34** (9–11), pp, 991–998.

Deshun, L., Y.G. Chen, O. Lihui, 1998: Waste CO_2 capture and utilization for enhanced oil recovery (EOR) and underground storage - a case study in Jilin Oil field, China. In Greenhouse Gas Mitigation - Technologies for Activities Implemented Jointly, Riemer P.W.F., A.Y. Smith, K.V. Thambimuthu, (eds). Pergamon, Oxford.

Dooley, J.J. and M.A. Wise, 2003: Retention of CO_2 in Geologic Sequestration Formations: Desirable Levels, Economic Considerations, and the Implications for Sequestration R&D. Proceedings of the 6th International Conference on Greenhouse Gas Control Technologies. J. Gale and Y. Kaya (eds). *Elsevier Science,* Amsterdam pp. 273–278.

Dunsmore, H.E., 1992: A geological perspective on global warming and the possibility of carbon dioxide removal as calcium carbonate mineral. *Energy Conversion and Management,* **33**(5–8), pp. 565–572.

Edmonds, J.A., P. Freund, J.J. Dooley, 2001: The role of carbon management technologies in addressing atmospheric stabilization of greenhouse gases. Proceedings of the 5th International Conference on Greenhouse Gas Control Technologies, D. Williams, B. Durie, P. McMullan, C. Paulson, A. Smith (eds). CSIRO, Australia, pp. 46–51.

Fletcher Multilaterals, The texts of UNCLOS, the London Convention, OSPAR convention and other treaties can be seen at http://fletcher.tufts.edu/multilaterals.html.

Freund, P. and J.E. Davison, 2002: General overview of costs, IPCC Workshop, Regina.

Freund, P., 2001: Progress in understanding the potential role of CO_2 storage. Proceedings of the 5th International Conference on Greenhouse Gas Control Technologies (GHGT-5), D.J. Williams, R.A. Durie, P. McMullan, C.A.J. Paulson, and A.Y. Smith (eds). CSIRO, 13–16 August 2000, Cairns, Australia, pp. 272–278.

Gale, J., N.P. Christensen, A. Cutler, and T.A. Torp, 2001: Demonstrating the Potential for Geological Storage of CO_2: The Sleipner and GESTCO Projects, *Environmental Geosciences,* **8**(3), pp.160–165.

Gipe, P., 2004: Wind Power: Renewable Energy for Home, **Farm, & Business.** Chelsea Green Publishing Co., USA ISBN 1-931498-14-8.

Gunter, W.D., T. Gentzis, B.A. Rottengusser, R.J.H. Richardson, 1997: Deep coalbed methane in Alberta, Canada: a fuel resource with the potential of zero greenhouse gas emissions. *Energy Conversion and Management,* **38**, Suppl., pp. S217–222.

Ha-Duong, M. and D.W. Keith, 2003: Carbon storage: the economic efficiency of storing CO_2 in leaky reservoirs. *Clean Technologies and Environmental Policy,* **5**, pp.181–189.

Haugan, P.M. and F. Joos, 2004: Metrics to assess the mitigation of global warming by carbon capture and storage in the ocean and in geological reservoirs. *Geophysical Research Letters,* **31**, L18202.

Hawkins, D.G., 2003: Passing gas: policy implications for geologic carbon storage sites. Proceedings of the 6th International Conference on Greenhouse Gas Control Technologies, J. Gale and Y Kaya (eds), *Elsevier Science,* Amsterdam pp. 249–254.

Hendriks, C.A., K. Blok, and W.C. Turkenburg, 1989: The recovery of carbon dioxide from power plants. Proceedings of the Symposium on Climate and Energy, Utrecht, The Netherlands.

Herzog, H.J., K. Caldeira, and J. Reilly, 2003: An issue of permanence: assessing the effectiveness of temporary carbon storage. *Climatic Change,* **59,** pp. 293–310.

Holloway, S., J.P. Heederik, L.G.H. van der Meer, I. Czernichowski-Lauriol, R. Harrison, E. Lindeberg, I.R. Summerfield, C. Rochelle, T. Schwarzkopf, O. Kaarstad, and B. Berger, 1996: The Underground Disposal of Carbon Dioxide, Final Report of JOULE II Project No. CT92-0031, British Geological Survey, Keyworth, Nottingham, UK.

Horn, F.L. and M. Steinberg, 1982: Control of carbon dioxide emissions from a power plant (and use in enhanced oil recovery). *Fuel,* **61,** May 1982.

IEA GHG, 2000a, Leading options for the capture of CO_2 emissions at power stations, Report Ph3/14. IEA Greenhouse Gas R&D Programme, Cheltenham, UK.

IEA GHG, 2000b: The potential of wind energy to reduce CO_2 emissions. Report Ph3/24. IEA Greenhouse Gas R&D Programme, Cheltenham, UK.

IEA GHG, 2001: Putting Carbon back in the Ground. IEA Greenhouse Gas R&D Programme, Cheltenham, UK.

IEA GHG, 2002: Opportunities for early application of CO_2 sequestration technology. Report Ph4/10. IEA Greenhouse Gas R&D Programme, Cheltenham, UK.

IEA GHG, 2004a: Implications of the Clean Development Mechanism for use of CO_2 Capture and Storage, Report Ph4/36. IEA Greenhouse Gas R&D Programme, Cheltenham, UK.

IEA GHG, 2004b: Overview of Long-term Framework for CO_2 Capture and Storage. Report Ph4/35. IEA Greenhouse Gas R&D Programme, Cheltenham, UK.

IEA GHG, 2004c: Overview of Monitoring Requirements for Geological Storage Projects. Report Ph4/29. IEA Greenhouse Gas R&D Programme, Cheltenham, UK.

IEA, 2003: CO_2 emissions from fuel combustion, 1971–2001, OECD/IEA, Paris.

IEA, 2004: Energy Balances of Non-OECD Countries, 2001–2002. OECD/IEA, Paris.

IPCC, 1996a: Climate Change 1995: Impacts, Adaptations and Mitigation of Climate Change: Scientific-Technical Analyses. Contribution of Working Group II to The Second Assessment Report of the Intergovernmental Panel on Climate Change. R.T. Watson, M.C. Zinyowera, and R.H. Moss, (eds.). Cambridge University Press, Cambridge, UK.

IPCC, 1996b: Technologies, Policies, and Measures for Mitigating Climate Change - IPCC Technical Paper I.

IPCC, 2000a: Special Report on Emission Scenarios, Cambridge University Press, Cambridge, UK.

IPCC, 2000b: Land Use, Land-Use Change and Forestry. IPCC Special Report, R.T. Watson, I.R. Noble, B. Bolin, N.H. Ravindranath, D.J. Verardo, and D.J. Dokken (eds.). Cambridge University Press, Cambridge, UK.

IPCC, 2000c: Summary for Policymakers. Methodological and Technological Issues in Technology Transfer. Cambridge University Press, Cambridge, UK.

IPCC, 2001a: Climate Change 2001 - Mitigation. The Third Assessment Report of the Intergovernmental Panel on Climate Change. B. Metz, O. Davidson, R. Swart, and J. Pan (eds.). Cambridge University Press, Cambridge, UK.

IPCC, 2001b: Climate Change 2001. The Third Assessment Report of the Intergovernmental Panel on Climate Change. Cambridge University Press, Cambridge, UK.

IPCC, 2001c: Climate Change 2001: the Scientific Basis. Contribution of Working Group I to the Third Assessment Report of the Intergovernmental Panel on Climate Change. J.T. Houghton, Y. Ding, D.J. Griggs, M. Noguer, P.J. van der Linden, X. Dai, K. Maskell, and C.A. Johnson, (eds.). Cambridge University Press, Cambridge, UK.

IPCC, 2001d: Costing Methodologies. A. Markandya, K. Halsnaes, A. Lanza, Y. Matsuoka, S. Maya, J. Pan, J. Shogren, R Seroa de Motta, and T. Zhang, In: Climate Change 2001: Mitigation. Contribution of Working Group III to the Third Assessment Report of the Intergovernmental Panel on Climate Change. B. Metz, O. Davidson, R. Swart, and J. Pan (eds.). Cambridge University Press, Cambridge, UK.

IPCC, 2001e: Climate Change 2001. Synthesis Report. A contribution of Working Groups I, II and III to The Third Assessment Report of the Intergovernmental Panel on Climate Change. R.T. Watson and the Core Writing Team (eds.). Cambridge University Press, Cambridge, UK

IPCC, 2002: Workshop on Carbon Dioxide Capture and Storage. Proceedings published by ECN, the Netherlands.

ISO, 1997: International Standard ISO 14040: Environmental Management - Life Cycle Assessment - Principles and Framework. International Organisation for Standardisation, Geneva, Switzerland.

Johansson, T.B., H. Kelly, A.K.N. Reddy, R. Williams, 1993: Renewable Fuels and Electricity for a Growing World Economy: Defining and Achieving the Potential, in Renewable Energy - Sources for Fuels and Electricity, T.B. Johansson, H. Kelly, A.K.N. Reddy, R. Williams (eds.). Island Press.

Kaya, Y., 1995: The role of CO_2 removal and disposal. *Energy Conversion and Management,* **36**(6–9) pp. 375–380.

Kheshgi, H.S., 2003: Evasion of CO_2 injected into the ocean in the context of CO_2 stabilisation. Proceedings of the 6th International Conference on Greenhouse Gas Control Technologies, J. Gale, and Y. Kaya (eds), Elsevier Science Ltd, Amsterdam, pp. 811–816.

Kohl, A. and R. Nielsen, 1997: Gas Purification, Gulf Publishing Company, Houston, USA.

Koide, H., Y. Tazaki, Y. Noguchi, S. Nakayama, M. Iijima, K. Ito, Y. Shindo, 1992: Subterranean containment and long-term storage of carbon dioxide in unused aquifers and in depleted natural gas reservoirs. *Energy Conversion and Management.* **33**(5–8), pp. 619–626.

Korbol, R. and A. Kaddour, 1995: Sleipner Vest CO_2 disposal - injection of removed CO_2 into the Utsira formation. *Energy Conversion and Management,* **36**(3–9), pp. 509–512.

Krupnick, A.J., D. Buttraw, A. Markandya, 2000: The Ancillary Benefits and Costs of Climate Change Mitigation: A Conceptual Framework. Paper presented to the Expert Workshop on Assessing the Ancillary Benefits and Costs of Greenhouse Gas Mitigation Strategies, 27–29 March 2000, Washington D.C.

Marchetti, C. and N. Nakicenovic, 1979: The Dynamics of Energy Systems and the Logistic Substitution Model. RR-79-13. Laxenburg, Austria: International Institute for Applied Systems Analysis (IIASA).

Marchetti, C., 1977: On Geo-engineering and the CO_2 problem. *Climate Change*, **1**, pp. 59–68.

Marland, G., A. Brenkert, O. Jos, 1999: CO_2 from fossil fuel burning: a comparison of ORNL and EDGAR estimates of national emissions. *Environmental Science & Policy*, **2**, pp. 265–273.

McDonald, A. and L. Schrattenholzer, 2001: Learning rates for energy technologies. *Energy Policy* **29**, pp. 255–261.

Möllersten, K., J. Yan, and J.R. Moreira, 2003: Promising market niches for biomass energy with CO_2 removal and disposal - Opportunities for energy supply with negative CO_2 emissions, *Biomass and Bioenergy*, **25**, pp. 273–285.

Moomaw, W., K. Ramakrishna, K. Gallagher, and T. Fried, 1999: The Kyoto Protocol: A Blueprint for Sustainability. *Journal of Environment and Development*, **8**, pp. 82–90.

Munasinghe, M. and R. Swart, 2005: Primer on Climate Change and Sustainable Development – Facts, Policy Analysis, and Application, Cambridge University Press, Cambridge, UK.

Nelson, R.R. and S. Winter, 1982: An Evolutionary Theory of Economic Change. Harvard University Press, Cambridge, MA.

Pacala, S.W., 2003: Global Constraints on Reservoir Leakage. Proceedings of the 6th International Conference on Greenhouse Gas Control Technologies. J. Gale and Y. Kaya (eds). *Elsevier Science*, Amsterdam pp. 267–272.

Seifritz, W., 1990: CO_2 disposal by means of silicates. *Nature,* **345**, pp. 486.

Seifritz, W., 1992: The terrestrial storage of CO_2-ice as a means to mitigate the greenhouse effect. Hydrogen Energy Progress IX (C.D.J. Pottier and T.N. Veziroglu (eds), pp. 59–68.

Siddique, Q., 1990: Separation of Gases. Proceedings of 5th Priestley Conference, Roy. Soc. Chem., London, pp. 329.

Skovholt, O., 1993: CO_2 transportation systems. *Energy Conversion and Management*, **34**, 9–11, pp.1095–1103.

Sorey, M.L., C.D. Farrar, W.C. Evans, D.P. Hill, R.A. Bailey, J.W. Hendley, P.H. Stauffer, 1996: Invisible CO_2 Gas Killing Trees at Mammoth Mountain, California. US Geological Survey Fact Sheet, 172–96.

Steinberg, M. 1996: The Carnol process for CO_2 mitigation from power plants and the transportation sector. *Energy Conversion and Management,* **37**(6–8) pp 843–848.

Stevens, S.H. and J. Gale, 2000: Geologic CO_2 sequestration. *Oil and Gas Journal*, May 15th, 40–44.

Turkenburg, W.C., 1997: Sustainable development, climate change and carbon dioxide removal. *Energy Conversion and Management*, **38**, S3–S12.

Uchida, T., T. Hondo, S. Mae, J. Kawabata, 1995: Physical data of CO_2 hydrate. In Direct Ocean Disposal of Carbon Dioxide. N. Handa and T. Ohsumi (eds), Terrapub, Tokyo pp. 45–61.

UNFCCC, 1992: United Nations, New York.

UNFCCC, 2001: Report of the Conference of the Parties on its Seventh Session, held in Marrakech, from 29 October to 10 November 2001, Addendum. FCCC/CP/2001/13/Add.1.

Unruh, G., 2000: Understanding Carbon Lock-in. *Energy Policy*, **28**(12) pp. 817–830.

Wigley, T.M.L., R. Richels, J.A. Edmonds, 1996: Economic and environmental choices in the stabilization of atmospheric CO_2 concentrations. *Nature,* **379**, pp. 240–243.

World Energy Assessment, 2004: Overview: 2004 Update. J. Goldemberg and T.B. Johansson (eds), United Nations Development Programme, New York.

Yamaji, K., 1997: A study of the role of end-of-pipe technologies in reducing CO_2 emissions. *Waste Management,* **17**(5–6) pp. 295–302.

2

Sources of CO$_2$

Coordinating Lead Author
John Gale (United Kingdom)

Lead Authors
John Bradshaw (Australia), Zhenlin Chen (China), Amit Garg (India), Dario Gomez (Argentina), Hans-Holger Rogner (Germany), Dale Simbeck (United States), Robert Williams (United States)

Contributing Authors
Ferenc Toth (Austria), Detlef van Vuuren (Netherlands)

Review Editors
Ismail El Gizouli (Sudan), Jürgen Friedrich Hake (Germany)

Contents

EXECUTIVE SUMMARY

Assessing CO$_2$ capture and storage calls for a comprehensive delineation of CO$_2$ sources. The attractiveness of a particular CO$_2$ source for capture depends on its volume, concentration and partial pressure, integrated system aspects, and its proximity to a suitable reservoir. Emissions of CO$_2$ arise from a number of sources, mainly fossil fuel combustion in the power generation, industrial, residential and transport sectors. In the power generation and industrial sectors, many sources have large emission volumes that make them amenable to the addition of CO$_2$ capture technology. Large numbers of small point sources and, in the case of transport, mobile sources characterize the other sectors, making them less amenable for capture at present. Technological changes in the production and nature of transport fuels, however, may eventually allow the capture of CO$_2$ from energy use in this sector.

Over 7,500 large CO$_2$ emission sources (above 0.1 MtCO$_2$ yr^{-1}) have been identified. These sources are distributed geographically around the world but four clusters of emissions can be observed: in North America (the Midwest and the eastern freeboard of the USA), North West Europe, South East Asia (eastern coast) and Southern Asia (the Indian sub-continent). Projections for the future (up to 2050) indicate that the number of emission sources from the power and industry sectors is likely to increase, predominantly in Southern and South East Asia, while the number of emission sources suitable for capture and storage in regions like Europe may decrease slightly.

Comparing the geographical distribution of the emission sources with geological storage opportunities, it can be seen that there is a good match between sources and opportunities. A substantial proportion of the emission sources are either on top of, or within 300 km from, a site with potential for geological storage. Detailed studies are, however, needed to confirm the suitability of such sites for CO$_2$ storage. In the case of ocean storage, related research suggests that only a small proportion of large emission sources will be close to potential ocean storage sites.

The majority of the emissions sources have concentrations of CO$_2$ that are typically lower than 15%. However, a small proportion (less than 2%) have concentrations that exceed 95%, making them more suitable for CO$_2$ capture. The high-content sources open up the possibility of lower capture costs compared to low-content sources because only dehydration and compression are required. The future proportion of high- and low-content CO$_2$ sources will largely depend on the rate of introduction of hydrogen, biofuels, and the gasification or liquefaction of fossil fuels, as well as future developments in plant sizes.

Technological changes, such as the centralized production of liquid or gaseous energy carriers (e.g., methanol, ethanol or hydrogen) from fossil sources or the centralized production of those energy carriers or electricity from biomass, may allow for CO$_2$ capture and storage. Under these conditions, power generation and industrial emission sources would largely remain unaffected but CO$_2$ emissions from transport and distributed energy-supply systems would be replaced by additional point sources that would be amenable to capture. The CO$_2$ could then be stored either in geological formations or in the oceans. Given the scarcity of data, it is not possible to project the likely numbers of such additional point sources, or their geographical distribution, with confidence (estimates range from 0 to 1,400 GtCO$_2$ (0–380 GtC) for 2050).

According to six illustrative SRES scenarios, global CO$_2$ emissions could range from 29.3 to 44.2 GtCO$_2$ (8–12 GtC) in 2020 and from 22.5 to 83.7 GtCO$_2$ (6–23 GtC) in 2050. The technical potential of CO$_2$ capture associated with these emission ranges has been estimated recently at 2.6–4.9 GtCO$_2$ for 2020 (0.7–1.3 GtC) and 4.9–37.5 GtCO$_2$ for 2050 (1.3–10 GtC). These emission and capture ranges reflect the inherent uncertainties of scenario and modelling analyses. However, there is one trend common to all of the six illustrative SRES scenarios: the general increase of future CO$_2$ emissions in the developing countries relative to the industrialized countries.

2.1 Sources of CO$_2$

This chapter aims to consider the emission sources of CO$_2$ and their suitability for capture and subsequent storage, both now and in the future. In addition, it will look at alternative energy carriers for fossil fuels and at how the future development of this technology might affect the global emission sources of CO$_2$ and the prospects for capturing these emissions.

Chapter 1 showed that the power and industry sectors combined dominate current global CO$_2$ emissions, accounting for about 60% of total CO$_2$ emissions (see Section 1.2.2). Future projections indicate that the share of these sectoral emissions will decline to around 50% of global CO$_2$ emissions by 2050 (IEA, 2002). The CO$_2$ emissions in these sectors are generated by boilers and furnaces burning fossil fuels and are typically emitted from large exhaust stacks. These stacks can be described as large stationary sources, to distinguish them from mobile sources such as those in the transport sector and from smaller stationary sources such as small heating boilers used in the residential sector. The large stationary sources represent potential opportunities for the addition of CO$_2$ capture plants. The volumes produced from these sources are usually large and the plants can be equipped with a capture plant to produce a source of high-purity CO$_2$ for subsequent storage. Of course, not all power generation and industrial sites produce their emissions from a single point source. At large industrial complexes like refineries there will be multiple exhaust stacks, which present an additional technical challenge in terms of integrating an exhaust-gas gathering system in an already congested complex, undoubtedly adding to capture costs (Simmonds *et al.*, 2003).

Coal is currently the dominant fuel in the power sector, accounting for 38% of electricity generated in 2000, with hydro power accounting for 17.5%, natural gas for 17.3%, nuclear for 16.8%, oil for 9%, and non-hydro renewables for 1.6%. Coal is projected to remain the dominant fuel for power generation in 2020 (about 36%), whilst natural-gas generation will become the second largest source, surpassing hydro. The use of biomass as a fuel in the power sector is currently limited. Fuel selection in the industrial sector is largely sector-specific. For example, the use of blast furnaces dominates primary steel production in the iron and steel sector, which primarily uses coal and coke (IEA GHG, 2000b; IPCC, 2001). In the refining and chemical sectors, oil and gas are the primary fuels. For industries like cement manufacture, all fossil fuels are used, with coal dominating in areas like the USA, China and India (IEA GHG, 1999), and oil and gas in countries like Mexico (Sheinbaum and Ozawa, 1998). However, the current trend in European cement manufacture is to use non-fossil fuels: these consist principally of wastes like tyres, sewage sludge and chemical-waste mixtures (IEA GHG, 1999). In global terms, biomass is not usually a significant fuel source in the large manufacturing industries. However, in certain regions of the world, like Scandinavia and Brazil, it is acknowledged that biomass use can be significant (Möllersten *et al.*, 2003).

To reduce the CO_2 emissions from the power and industry sectors through the use of CO_2 capture and storage, it is important to understand where these emissions arise and what their geographical relationship is with respect to potential storage opportunities (Gale, 2002). If there is a good geographical relationship between the large stationary emission sources and potential geological storage sites then it is possible that a significant proportion of the emissions from these sources can be reduced using CO_2 capture and storage. If, however, they are not well matched geographically, then there will be implications for the length and size of the transmission infrastructure that is required, and this could impact significantly on the cost of CO_2 capture and storage, and on the potential to achieve deep reductions in global CO_2 emissions. It may be the case that there are regions of the world that have greater potential for the application of CO_2 capture and storage than others given their source/storage opportunity relationship. Understanding the regional differences will be an important factor in assessing how much of an impact CO_2 capture and storage can have on global emissions reduction and which of the portfolio of mitigation options is most important in a regional context.

Other sectors of the economy, such as the residential and transport sectors, contribute around 30% of global CO_2 emissions and also produce a large number of point source emissions. However, the emission volumes from the individual sources in these sectors tend to be small in comparison to those from the power and industry sectors and are much more widely distributed, or even mobile rather than stationary. It is currently not considered to be technically possible to capture emissions from these other small stationary sources, because there are still substantial technical and economic issues that need to be resolved (IPCC, 2001). However, in the future, the use of low-carbon energy carriers, such as electricity or hydrogen produced from fossil fuels, may allow CO_2 emissions to be captured from the residential and transport sectors as well. Such fuels would most probably be produced in large centralized plants and would be accompanied by capture and storage of the CO_2 co-product. The distributed fuels could then be used for distributed generation in either heaters or fuels cells and in vehicles in the transport sector.

In this scenario, power generation and industrial sources would be unaffected but additional point sources would be generated that would also require storage. In the medium to long term therefore, the development and commercial deployment of such technology, combined with an accelerated shift to low- or zero-carbon fuels in the transport sector, could lead to a significant change in the geographical pattern of CO_2 emissions compared to that currently observed.

2.2   Characterization of CO_2 emission sources

This section presents information on the characteristics of the CO_2 emission sources. It is considered necessary to review the different CO_2 contents and volumes of CO_2 from these sources as these factors can influence the technical suitability of these emissions for storage, and the costs of capture and storage.

2.2.1   Present

2.2.1.1   Source types
The emission sources considered in this chapter include all large stationary sources (>0.1 $MtCO_2$ yr[-1]) involving fossil fuel and biomass use. These sources are present in three main areas: fuel combustion activities, industrial processes and natural-gas processing. The largest CO_2 emissions by far result from the oxidation of carbon when fossil fuels are burned. These emissions are associated with fossil fuel combustion in power plants, oil refineries and large industrial facilities.

For the purposes of this report, large stationary sources are considered to be those emitting over 0.1 $MtCO_2$ yr[-1]. This threshold was selected because the sources emitting less than 0.1 $MtCO_2$ yr[-1] together account for less than 1% of the emissions from all the stationary sources under consideration (see Table 2.1). However, this threshold does not exclude emissions capture at smaller CO_2 sources, even though this is more costly and technically challenging.

Carbon dioxide not related to combustion is emitted from a variety of industrial production processes which transform materials chemically, physically or biologically. Such processes include:

- the use of fuels as feedstocks in petrochemical processes (Chauvel and Lefebvre, 1989; Christensen and Primdahl, 1994);
- the use of carbon as a reducing agent in the commercial production of metals from ores (IEA GHG, 2000; IPCC, 2001);
- the thermal decomposition (calcination) of limestone and dolomite in cement or lime production (IEA GHG, 1999, IPCC 2001);
- the fermentation of biomass (e.g., to convert sugar to alcohol).

In some instances these industrial-process emissions are produced in combination with fuel combustion emissions, a typical example being aluminium production (IEA GHG, 2000).

Table 2.1 Properties of candidate gas streams that can be inputted to a capture process (Sources: Campbell et al., 2000; Gielen and Moriguchi, 2003; Foster Wheeler, 1998; IEA GHG, 1999; IEA GHG, 2002a).

Source	CO_2 concentration % vol (dry)	Pressure of gas stream MPa[a]	CO_2 partial pressure MPa
CO_2 from fuel combustion			
• Power station flue gas:			
Natural gas fired boilers	7 - 10	0.1	0.007 - 0.010
Gas turbines	3 - 4	0.1	0.003 - 0.004
Oil fired boilers	11 - 13	0.1	0.011 - 0.013
Coal fired boilers	12 - 14	0.1	0.012 - 0.014
IGCC[b]: after combustion	12 - 14	0.1	0.012 - 0.014
• Oil refinery and petrochemical plant fired heaters	8	0.1	0.008
CO_2 from chemical transformations + fuel combustion			
• Blast furnace gas:			
Before combustion[c]	20	0.2 - 0.3	0.040 - 0.060
After combustion	27	0.1	0.027
• Cement kiln off-gas	14 - 33	0.1	0.014 - 0.033
CO_2 from chemical transformations before combustion			
• IGCC: synthesis gas after gasification	8 - 20	2 - 7	0.16 - 1.4

[a]   0.1 MPa = 1 bar.
[b]   IGCC: Integrated gasification combined cycle.
[c]   Blast furnace gas also contains significant amounts of carbon monoxide that could be converted to CO_2 using the so-called shift reaction.

A third type of source occurs in natural-gas processing installations. CO_2 is a common impurity in natural gas, and it must be removed to improve the heating value of the gas or to meet pipeline specifications (Maddox and Morgan, 1998).

2.2.1.2 CO_2 content

The properties of those streams that can be inputted to a CO_2 capture process are discussed in this section. In CO_2 capture, the CO_2 partial pressure of the gas stream to be treated is important as well as the concentration of the stream. For practical purposes, this partial pressure can be defined as the product of the total pressure of the gas stream times the CO_2 mole fraction. It is a key variable in the selection of the separation method (this is discussed further in Chapter 3). As a rule of thumb, it can be said that the lower the CO_2 partial pressure of a gas stream, the more stringent the conditions for the separation process.

Typical CO_2 concentrations and their corresponding partial pressures for large stationary combustion sources are shown in Table 2.1, which also includes the newer Integrated Gasification Combined Cycle technology (IGCC). Typically, the majority of emission sources from the power sector and from industrial processes have low CO_2 partial pressures; hence the focus of the discussion in this section. Where emission sources with high partial pressure are generated, for example in ammonia or hydrogen production, these sources require only dehydration and some compression, and therefore they have lower capture costs.

Table 2.1 also provides a summary of the properties of CO_2 streams originating from cement and metal production in which chemical transformations and combustion are combined. Flue gases found in power plants, furnaces in industries, blast furnaces and cement kilns are typically generated at atmospheric pressure and temperatures ranging between 100°C and 200°C, depending on the heat recovery conditions.

Carbon dioxide levels in flue gases vary depending on the type of fuel used and the excess air level used for optimal combustion conditions. Flue gas volumes also depend on these two variables. Natural-gas-fired power generation plants are typically combined cycle gas turbines which generate flue gases with low CO_2 concentrations, typically 3–4% by volume (IEA GHG, 2002a). Coal for power generation is primarily burnt in pulverized-fuel boilers producing an atmospheric pressure flue gas stream with a CO_2 content of up to 14% by volume (IEA GHG, 2002a). The newer and potentially more efficient IGCC technology has been developed for generating electricity from coal, heavy fuel oil and process carbonaceous residues. In this process the feedstock is first gasified to generate a synthesis gas (often referred to as 'syngas'), which is burnt in a gas turbine after exhaustive gas cleaning (Campbell *et al.*, 2000). Current IGCC plants where the synthesis gas is directly combusted in the turbine, like conventional thermal power plants, produce a flue gas with low CO_2 concentrations (up to 14% by volume). At present, there are only fifteen coal- and oil-fired IGCC plants, ranging in size from 40 to 550 MW. They were started up in the 1980s and 1990s in Europe and the USA (Giuffrida *et al.*, 2003). It should be noted that there are conceptual designs in which the CO_2 can be removed before the synthesis gas is combusted, producing a high-concentration, high-pressure CO_2 exhaust gas stream that could be more suitable for storage (see Chapter 3 for more details). However, no such plants have been built or are under construction.

Fossil fuel consumption in boilers, furnaces and in process operations in the manufacturing industry also typically produces flue gases with low CO_2 levels comparable to those in the power

Table 2.2 Typical properties of gas streams that are already input to a capture process (Sources: Chauvel and Lefebvre, 1989; Maddox and Morgan, 1998; IEA GHG, 2002a).

Source	CO_2 concentration % vol	Pressure of gas stream MPa[a]	CO_2 partial pressure MPa
Chemical reaction(s)			
• Ammonia production[b]	18	2.8	0.5
• Ethylene oxide	8	2.5	0.2
• Hydrogen production[b]	15 - 20	2.2 - 2.7	0.3 - 0.5
• Methanol production[b]	10	2.7	0.27
Other processes			
• Natural gas processing	2 - 65	0.9 - 8	0.05 - 4.4

[a]   0.1 MPa = 1 bar
[b]   The concentration corresponds to high operating pressure for the steam methane reformer.

sector. CO_2 concentrations in the flue gas from cement kilns depend on the production process and type of cement produced and are usually higher than in power generation processes (IEA GHG, 1999). Existing cement kilns in developing countries such as China and India are often relatively small. However, the quantity of CO_2 produced by a new large cement kiln can be similar to that of a power station boiler. Integrated steel mills globally account for over 80% of CO_2 emissions from steel production (IEA GHG, 2000b). About 70% of the carbon input to an integrated steel mill is present in the blast furnace gas, which is used as a fuel gas within the steel mill. CO_2 could be captured before or after combustion of this gas. The CO_2 concentration after combustion in air would be about 27% by volume, significantly higher than in the flue gas from power stations. Other process streams within a steel mill may also be suitable candidates for CO_2 capture before or after combustion. For example, the off-gas from an oxygen-steel furnace typically contains 16% CO_2 and 70% carbon monoxide.

The off-gases produced during the fermentation of sugars to ethanol consist of almost pure CO_2 with a few impurities. This gas stream is generated at a rate of 0.76 kg CO_2^{-1} and is typically available at atmospheric pressure (0.1 MPa) (Kheshgi and Prince, 2005).

CO_2 also occurs as an undesirable product that must be removed in some petrochemical processes, particularly those using synthesis gas as an intermediate or as an impurity in natural gas. The properties of the raw gas streams from which CO_2 is customarily removed in some of these industries are shown in Table 2.2. It can be seen from Table 2.1 that the CO_2 partial pressures of flue gases are at least one order of magnitude less than the CO_2 partial pressures of the streams arising from the processes listed in Table 2.2. This implies that CO_2 recovery from fuel combustion streams will be comparatively much more difficult.

2.2.1.3   Scale of emissions

A specific detailed dataset has been developed for CO_2 stationary sources for 2000, giving their geographical distribution by process type and country (IEA GHG, 2002a). The stationary sources of CO_2 in this database comprise power plants, oil refineries, gas-processing plants, cement plants, iron and steel plants and those industrial facilities where fossil fuels are used as feedstock, namely ammonia, ethylene, ethylene oxide and hydrogen. This global inventory contains over 14 thousand emission sources with individual CO_2 emissions ranging from 2.5 tCO_2 yr[-1] to 55.2 MtCO_2 yr[-1]. The information for each single source includes location (city, country and region), annual CO_2 emissions and CO_2 emission concentrations. The coordinates (latitude/longitude) of 74% of the sources are also provided. The total emissions from these 14 thousand sources amount to over 13 GtCO_2 yr[-1]. Almost 7,900 stationary sources with individual emissions greater than or equal to 0.1 MtCO_2 per year have been identified globally. These emissions included over 90% of the total CO_2 emissions from large point sources in 2000. Some 6,000 emission sources with emissions below 0.1 MtCO_2 yr[-1] were also identified, but they represent only a small fraction of the total emissions volume and were therefore excluded from further discussion in this chapter. There are also a number of regional and country-specific CO_2 emission estimates for large sources covering China, Japan, India, North West Europe and Australia (Hibino, 2003; Garg *et al.*, 2002; Christensen *et al.*, 2001, Bradshaw *et al.*, 2002) that can be drawn upon. Table 2.3 summarizes the information concerning large stationary sources according to the type of emission generating process. In the case of the petrochemical and gas-processing industries, the CO_2 concentration listed in this table refers to the stream leaving the capture process. The largest amount of CO_2 emitted from large stationary sources originates from fossil fuel combustion for power generation, with an average annual emission of 3.9 MtCO_2 per source. Substantial amounts of CO_2 arise in the oil and gas processing industries while cement production is the largest emitter from the industrial sector.

In the USA, 12 ethanol plants with a total productive capacity of 5.3 billion litres yr[-1] each produce CO_2 at rates in excess of 0.1 MtCO_2 yr[-1] (Kheshgi and Prince, 2005); in Brazil, where ethanol production totalled over 14 billion litres per year during 2003-2004, the average distillery productive capacity is 180 million litres yr[-1]. The corresponding average fermentation CO_2 production rate is 0.14 MtCO_2 yr[-1], with the largest distillery producing nearly 10 times the average.

Table 2.3 Profile of worldwide large CO$_2$ stationary sources emitting more than 0.1 Mt CO$_2$ per year (Source: IEA GHG, 2002a).

Process	CO$_2$ concentration in gas stream % by vol.	Number of sources	Emissions (MtCO$_2$)	% of total CO$_2$ emissions	Cumulative total CO$_2$ emissions (%)	Average emissions/source (MtCO$_2$ per source)
CO$_2$ from fossil fuels or minerals						
Power						
Coal	12 to 15	2,025	7,984	59.69	59.69	3.94
Natural gas	3	985	759	5.68	65.37	0.77
Natural gas	7 to 10	743	752	5.62	70.99	1.01
Fuel oil	8	515	654	4.89	75.88	1.27
Fuel oil	3	593	326	2.43	78.31	0.55
Other fuels[a]	NA	79	61	0.45	78.77	0.77
Hydrogen	NA	2	3	0.02	78.79	1.27
Natural-gas sweetening						
	NA[b]	NA	50[c]	0.37	79.16	
Cement production						
Combined	20	1175	932	6.97	86.13	0.79
Refineries						
	3 to 13	638	798	5.97	92.09	1.25
Iron and steel industry						
Integrated steel mills	15	180	630[d]	4.71	96.81	3.50
Other processes[d]	NA	89	16	0.12	96.92	0.17
Petrochemical industry						
Ethylene	12	240	258	1.93	98.85	1.08
Ammonia: process	100	194	113	0.84	99.70	0.58
Ammonia: fuel combustion	8	19	5	0.04	99.73	0.26
Ethylene oxide	100	17	3	0.02	99.75	0.15
Other sources						
Non-specified	NA	90	33	0.25	100.00	0.37
		7,584	**13,375**	**100**		**1.76**
CO$_2$ from biomass[e]						
Bioenergy	3 to 8	213	73			0.34
Fermentation	100	90	17.6			**0.2**

[a] Other gas, other oil, digester gas, landfill gas.

[b] A relatively small fraction of these sources has a high concentration of CO$_2$. In Canada, only two plants out of a total of 24 have high CO$_2$ concentrations.

[c] Based on an estimate that about half of the annual worldwide natural-gas production contains CO$_2$ at concentrations of about 4% mol and that this CO$_2$ content is normally reduced from 4% to 2% mol (see Section 3.2.2).

[d] This amount corresponds to the emissions of those sources that have been individually identified in the reference database. The worldwide CO$_2$ emissions, estimated by a top-down approach, are larger than this amount and exceed 1 Gt (Gielen and Moriguchi, 2003).

[e] For North America and Brazil only. All numbers are for 2003, except for power generation from biomass and waste in North America, which is for 2000.

The top 25% of all large stationary CO$_2$ emission sources (those emitting more than 1 MtCO$_2$ per year) listed in Table 2.3 account for over 85% of the cumulative emissions from these types of sources. At the other end of the scale, the lowest 41% (in the 0.1 to 0.5 MtCO$_2$ range) contribute less than 10% (Figure 2.1). There are 330 sources with individual emissions above 10 MtCO$_2$ per year. Of their cumulative emissions, 78% come from power plants, 20% from gas processing and the remainder from iron and steel plants (IEA GHG, 2000b). High-concentration/

high-partial-pressure sources (e.g., from ammonia/hydrogen production and gas processing operations) contribute a relatively low share (<2%) of the emissions from large stationary sources (van Bergen *et al.*, 2004). However, these high-concentration sources could represent early prospects for the implementation of CO$_2$ capture and storage. The costs for capture are lower than for low-concentration/low-partial-pressure sources. If these sources can then be linked to enhanced production schemes in the vicinity (<50km), like CO$_2$-enhanced oil recovery, they could be low-cost options for CO_2 capture and storage (van Bergen *et al.*, 2004). Such sources emit 0.36 $GtCO_2$ yr^{-1} (0.1 GtC yr^{-1}), which equates to 3% of emissions from point sources larger than 0.1 $MtCO_2$ yr^{-1} (IEA GHG, 2002b). The geographical relationship between these high-concentration sources and prospective storage opportunities is discussed in Section 2.4.3. A small number of source streams with high CO_2 concentrations are already used in CO_2-EOR operations in the USA and Canada (Stevens and Gale, 2000).

2.2.2   Future

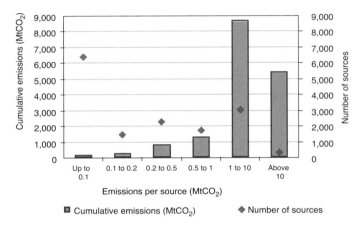

Figure 2.1 Relationship between large stationary source emissions and number of emission sources (Source: IEA GHG, 2002a).

Future anthropogenic CO_2 emissions will be the product of different drivers such as demographic development, socio-economic development, and technological changes (see Chapter 1, Section 1.2.4). Because their future evolution is inherently uncertain and because numerous combinations of different rates of change are quite plausible, analysts resort to scenarios as a way of describing internally consistent, alternative images of how the future might unfold. The IPCC developed a set of greenhouse gas emission scenarios for the period until 2100 (IPCC, 2000). The scenarios show a wide range of possible future worlds and CO_2 emissions (see Figure 2.2), consistent with the full uncertainty range of the underlying literature reported by Morita and Lee (1998). The scenarios are important as they provide a backdrop for determining the baseline for emission reductions that may be achieved with new technologies, including CO_2 capture and storage implemented specially for such purposes.

Technology change is one of the key drivers in long-term scenarios and plays a critical role in the SRES scenarios. Future rates of innovation and diffusion are integral parts of, and vary with, the story lines. Scenario-specific technology change may differ in terms of technology clusters (i.e., the type of technologies used) or rate of diffusion. In the fossil-intensive A1FI scenario, innovation concentrates on the fossil source-to-service chains stretching from exploration and resource

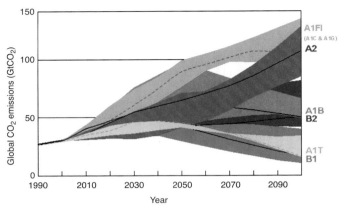

Figure 2.2 Range of annual global CO_2 emission in he SRES scenarios ($GtCO_2$) (Source: IPCC, 2000).

extraction to fuel upgrading/cleaning, transport, conversion and end-use. Alternatively, innovation in the environmentally-oriented B1 scenario focuses on renewable and hydrogen technologies.

The way in which technology change was included in the SRES scenarios depended on the particular model used. Some models applied autonomous performance improvements to fuel utilization, while others included specific technologies with detailed performance parameters. Even models with a strong emphasis on technology reflected new technologies or innovation in a rather generic manner. For example, advanced coal technology could be either an integrated coal gasification combined cycle (IGCC) plant, a pressurized fluidized bed combustion facility or any other, as-yet-unidentified, technology. The main characteristics of advanced coal technology are attractive investment costs, high thermal efficiency, potential multi-production integration and low pollution emissions – features that are prerequisites for any coal technology carrying the "advanced" label.

In general, technological diversity remained a feature in all scenarios, despite the fact that different clusters may dominate more in different scenarios. The trend towards cleaner and more convenient technologies, especially at the level of end-use (including transport), is common to all scenarios. In addition, transport fuels shift broadly towards supply schemes suitable for pre-combustion decarbonization. Centralized non-fossil technologies penetrate the power sector to various extents, while decentralized and home-based renewable and hydrogen-production infrastructures expand in all scenarios, but mostly in the environmentally-conscious and technology-intensive scenarios.

Despite the trend towards cleaner fuels, CO_2 emissions are projected to rise at different rates, at least until 2050. Emission patterns then diverge. Scenario-specific rates of technology change (performance improvements) and technology diffusion lead to different technology mixes, fuel uses and unit sizes. As regards fossil fuel use for power generation and industrial energy supply, the number of large stationary emission sources generally increases in the absence of restrictions on CO_2 emissions and a fundamental change in the characteristics of these emission

Table 2.4 Sectoral and regional distribution of energy-related CO$_2$ emissions in 2000 (MtCO$_2$) (Source: IEA, 2003).

		Public electricity and heat production	Unallocated autoproducers	Other energy industries	Manufacturing industries and construction	Transport	Commercial and public services	Residential	Other sectors	CO$_2$ sectoral approach total
1	Economies in transition	1,118.5	391.4	106.6	521.7	317.1	58.0	312.5	127.7	2,953.6
2	OECD West	1,087.3	132.0	222.8	722.1	1,040.9	175.1	494.6	96.2	3,971.0
3	USA	2,265.1	134.9	272.4	657.9	1,719.9	225.5	371.4	42.7	5,689.7
4	OECD Pacific	509.2	87.0	62.2	301.1	344.4	95.3	75.8	35.7	1,510.5
5	South/East Asia	925.5	104.1	137.9	533.3	451.8	50.9	185.6	39.7	2,428.7
6	Centrally Planned Asia	1,332.2	37.7	138.5	978.4	245.4	72.6	221.4	118.7	3,144.8
7	Middle East	280.6	6.6	118.6	193.0	171.6	16.6	90.8	112.5	990.4
8	Africa	276.3	15.9	40.2	137.7	143.5	5.0	44.5	34.8	697.8
9	Latin America	222.3	37.0	134.5	279.3	396.0	17.9	81.0	41.5	1,209.6
	Sector total	**8,016.9**	**946.5**	**1,233.7**	**4,324.7**	**4,830.6**	**716.8**	**1,877.5**	**649.4**	**22,596.1**

sources is unlikely to occur before 2050. In addition, the ratio of low-concentration to high-concentration emission sources remains relatively stable, with low-concentration sources dominating the emission profile.

In some scenarios, low- or zero-carbon fuels such as ethanol, methanol or hydrogen begin to dominate the transport sector and make inroads into the industrial, residential and commercial sectors after 2050. The centralized production of such fuels could lead to a significant change in the number of high-concentration emission sources and a change in the ratio of low- to high-purity emission sources; this is discussed in more detail in Section 2.5.2.

2.3 Geographical distribution of sources

This section discusses the geographical locations of large point sources discussed in the preceding sections. It is necessary to understand how these sources are geographically distributed across the world in order to assess their potential for subsequent storage.

2.3.1 Present

A picture of the geographical distribution of the sources of CO$_2$ emissions and the potential storage reservoirs helps us to understand the global cost of CO$_2$ mitigation, particularly those components associated with CO$_2$ transport. Geographical information about emission sources can be retrieved from a number of data sets. Table 2.4 shows the sectoral and regional distribution of energy-related CO$_2$ emissions in 2000. As mentioned earlier in this report, over 60% of global CO$_2$ emissions come from the power and industry sectors. Geographically,

these power and industry emissions are dominated by four regions which account for over 90% of the emissions. These regions are: Asia (30%), North America (24%), the transitional economies (13%), and OECD West[1] (12%). All the other regions account individually for less than 6% of the global emissions from the power and industry sectors.

Figure 2.3 shows the known locations of stationary CO$_2$ sources worldwide, as taken from the database referred to in Section 2.2 (IEA GHG, 2002a). North America is the region with the largest number of stationary sources (37%), followed by Asia (24%) and OECD Europe[2] (14%). Figure 2.3 shows three large clusters of stationary sources located in the central and eastern states of the US, in northwestern and central regions of Europe (Austria, Czech Republic, Germany, Hungary, Netherlands and UK) and in Asia (eastern China and Japan with an additional smaller cluster in the Indian subcontinent).

The distribution of stationary CO$_2$ emissions as a proportion of the total stationary emissions for 2000 indicates that the regions that are the largest emitters of CO$_2$ from stationary sources are: Asia at 41% (5.6 GtCO$_2$ yr^{-1}), North America at 20% (2.69 GtCO$_2$ yr^{-1}) and OECD Europe at 13% (1.75 GtCO$_2$ yr^{-1}). All other regions emitted less than 10% of the total CO$_2$ emission from stationary sources in 2000.

A comparison of the estimates of CO$_2$ emissions from the IEA and IEA GHG databases showed that the two sets produced

---

[1] Note: OECD West refers to the following countries: Austria, Belgium, Canada, Denmark, Finland, France, Germany, Greece, Iceland, Ireland, Italy, Luxembourg, Netherlands, Norway, Portugal, Spain, Sweden, Switzerland, Turkey, United Kingdom.

[2] OECD Europe includes the OECD West countries listed above, plus the Czech Republic, Hungary, Iceland, Norway, Poland, Slovak Republic, Switzerland and Turkey.

Figure 2.3 Global distribution of large stationary CO_2 sources (based on a compilation of publicly available information on global emission sources, IEA GHG 2002).

similar estimates for the total of global emissions but that results differed significantly for many countries. Regional differences of this kind have also been noted for other CO_2 emission databases (Marland *et al.*, 1999).

2.3.2 *Future CO_2 emissions and technical capture potentials*

The total CO_2 emissions from fossil fuel combustion in the SRES scenarios provide the upper limit for potential CO_2 capture for this assessment. In fact, the theoretical maximum is even higher because of the possibility of CO_2 capture from biomass. These emissions are also included in the tables of CO_2 emissions and they are therefore potentially available for capture. Obviously, the capture potential that is practical in technical terms is much smaller than the theoretical maximum, and the economic potential[3] is even smaller. Needless to say, it is the economic potential that matters most. This section presents estimates of the technical potential and Chapter 8 will address the economic potential.

Table 2.5 shows the CO_2 emissions by economic sector and major world regions for 2020 and 2050, and for six scenarios[4]. It should be noted that the total CO_2 emissions in Table 2.5 are higher than reported in SRES because emissions from biomass are explicitly included here (as these are potentially available for capture), while they where considered "climate-neutral" in the SRES presentations and therefore not counted as emission releases to the atmosphere. Geographically, the distribution of emission sources is set to change substantially. Between 2000 and 2050, the bulk of emission sources will shift from the OECD countries to the developing regions, especially China, South Asia and Latin America. As to emissions by sector, power generation, transport, and industry will remain the three main sources of CO_2 emissions over the next 50 years. Globally, the projected energy sector emissions will fluctuate around the 40% mark in 2050 (this matches the current figure), emissions from the industry sector will decline and transport sector emissions (i.e., mobile sources) increase. Power generation, which typically represent the bulk of large point sources, will account for about 50% of total emissions by 2050[5].

These emissions form the theoretical maximum potential for CO_2 capture from fossil fuel use. Toth and Rogner (2006) derived a set of capture factors on the basis of the technical or technological feasibility of adding CO_2 capture before, during or after combustion of fossil fuels. Capture factors are defined as the estimated maximum share of emissions for which capture is technically plausible. A detailed assessment of the power plants

---

[3] Economic potential is the amount of reductions in greenhouse gas emissions from a specific option that could be achieved cost-effectively given prevailing circumstances (i.e. a price for CO_2 reductions and the costs of other options).

[4] For the four marker scenarios and the technology-intensive A1T and the fossil-intensive A1FI illustrative scenarios, it is important to note that comparisons between the results of different models are not straightforward. First, the modelling methodologies imply different representations of energy technologies and their future evolutions. Secondly, the sectoral disaggregation and the energy/fuel details vary across the models. Thirdly, there are differences in how countries of the world are grouped together into regions. Tables 2.5 and 2.6 are based on the work by Toth and Rogner (2005) that attempts to create the best possible approximation for the purposes of comparing the regional and sectoral model and scenario results.

[5] As regards the share of emissions across sectors in 2020 (Table 2.5), there is an inherent divergence between scenarios with longer and shorter time horizons. Given the quasi perfect foresight of the underlying models, the SRES scenarios account for resource depletion over a period of a century and, due to the anticipated transition to higher-fuel-cost categories in the longer run, they shift to non-fossil energy sources much earlier than, for example, the IEA scenarios, especially for electricity supply. Consequently, the range for the shares of fossil-sourced power generation is between 43 and 58% for 2020, while the IEA projects a share of 71%. The corresponding sectoral shares in CO_2 emissions mirror the electricity generating mix: the IEA projects 43% for power generation (IEA, 2002) compared to a range of 28 to 32% in the six illustrative SRES scenarios.

Table 2.5 Carbon dioxide emissions from sectors in major world regions in six IPCC SRES scenarios in 2020 and 2050 (IPCC, 2000). Continued on next page.

A1B

Sector	Africa	CPA	EEFSU	LAM	Middle East	USA	P-OECD	S&EA	OECD West	Sector total
Power	2,016	3,193	1,482	1,182	721	1,607	698	2,063	1,244	14,207
Industry	1,046	2,512	1,465	1,689	966	1,122	564	1,834	1,123	12,321
Res/Com	642	1,897	439	566	195	637	238	950	933	6,496
Transport	877	1,008	312	1,502	1,052	2,022	659	1,592	2,175	11,199
Region total	4,580	8,610	3,698	4,938	2,934	5,388	2,159	6,439	5,476	44,222

A1T Sub-Saharan

Sector	Africa	CPA	E Europe	FSU	LAM	ME-N Africa	NAM	P-OECD	PAS	SAS	W. Europe	Sector total
Power	333	2,165	356	705	396	368	2,470	448	1,388	195	1,221	10,045
Industry	358	2,840	208	727	885	465	690	292	954	748	530	8,699
Res/Com	730	2,773	105	352	713	149	771	150	795	690	627	7,855
Refineries	107	211	23	196	282	139	370	75	250	42	219	1,913
Synfuels	59	122	9	22	139	36	127	30	211	38	107	900
Hydrogen	57	145	26	80	57	61	231	74	75	47	177	1,030
Transport	435	1,235	96	578	1,159	837	2,394	450	620	432	1,448	9,684
Region total	2,078	9,491	823	2,661	3,631	2,055	7,053	1,519	4,292	2,192	4,330	40,126

A1FI

Sector	Africa	CPA	EEFSU	LAM	Middle East	USA	Canada	P-OECD	South East Asia	W. Europe	Sector total
Power	427	3,732	2,248	680	370	2,618	181	753	2,546	1,640	15,195
Industry	622	3,498	1,121	695	426	1,418	153	416	1,530	1,384	11,262
Res/Com	135	1,363	582	125	25	755	102	115	488	786	4,477
Transport	456	542	588	977	297	2,210	168	357	1,357	1,345	8,297
Synfuels	10	12	126	2	0	52	3	12	2	21	238
Hydrogen	0	0	0	0	0	0	0	0	0	0	0
Fuel flared	21	11	19	135	74	9	1	1	52	4	327
Region total	1,670	9,159	4,682	2,613	1,192	7,062	608	1,654	5,976	5,181	39,796

Source: Total emissions MtCO$_2$ 2020

CPA = Centrally Planned Asia. EE = Eastern Europe, FSU = Former Soviet Union, LAM = Latin America, P-OECD = Pacific OECD, S&EA = South and Southeast Asia, OECD-West = Western Europe + Canada, Africa, ME = Middle East, PAS = Pacific Asia, SAS = South Asia

Table 2.5 Continued.

A2

Sector	Africa	East Asia	E. Europe	FSU	LAM	Middle East	USA	Canada	P-OECD	South East Asia	South Asia	OECD Europe	Sector total
Power	670	1,616	488	923	1,130	857	3,680	224	689	356	1,282	1,663	13,579
Industry	290	1,786	261	417	625	402	808	111	291	218	708	528	6,444
Res/Com	269	746	118	539	209	434	639	92	155	87	251	644	4,181
Transport	358	606	130	314	1,060	569	2,013	200	406	334	332	1,270	7,592
Others	394	439	112	371	644	538	567	68	247	269	142	532	4,324
Region total	1,981	5,193	1,109	2,563	3,668	2,800	7,706	696	1,788	1,264	2,715	4,638	36,120

B1

Sector	Africa	East Asia	E. Europe	FSU	LAM	Middle East	USA	Canada	P-OECD	South East Asia	South Asia	OECD Europe	Sector total
Power	629	1,148	377	670	1,031	699	2,228	128	477	354	972	1,118	9,829
Industry	259	1,377	210	290	531	362	537	79	205	209	611	355	5,024
Res/Com	283	602	108	471	193	350	511	74	132	79	250	557	3,611
Transport	384	578	136	343	987	509	1,708	172	365	314	370	1,204	7,070
Others	392	413	99	291	591	502	481	55	169	266	164	432	3,856
Region total	1,946	4,118	931	2,064	3,333	2,422	5,466	506	1,348	1,222	2,367	3,665	29,389

B2

Sector	Sub-Saharan Africa	CPA	E. Europe	FSU	LAM	ME-N Africa	NAM	P-OECD	PAS	SAS	W. Europe	Sector total
Power	317	1,451	398	149	338	342	3,317	459	1,017	398	1,234	9,420
Industry	307	2,017	232	956	754	400	993	223	796	634	679	7,990
Res/Com	854	1,936	137	330	462	177	1,213	174	440	929	768	7,420
Refineries	70	241	42	169	223	193	480	98	242	111	271	2,139
Synfuels	30	18	2	32	47	16	126	4	77	12	56	420
Hydrogen	15	274	15	18	24	17	159	31	108	36	119	817
Transport	224	655	105	530	715	506	2,278	384	784	468	1,164	7,812
Region total	1,816	6,591	931	2,184	2,563	1,652	8,566	1,373	3,464	2,589	4,292	36,019

Source: Total emissions MtCO$_2$, 2020

CPA = Centrally Planned Asia. EE = Eastern Europe, FSU = Former Soviet Union, LAM = Latin America, P-OECD = Pacific OECD, S&EA = South and Southeast Asia, OECD-West = Western Europe + Canada, Africa, ME = Middle East, PAS = Pacific Asia, SAS = South Asia

Table 2.5 Continued.

A1B

Sector	Africa	CPA	EEFSU	LAM	Middle East	USA	P-OECD	S&EA	OECD West	Sector total
Power	4,078	2,708	1,276	1,165	840	1,361	588	2,700	1,459	16,174
Industry	2,304	2,555	1,645	2,384	1,635	969	395	3,273	1,038	16,199
Res/Com	2,610	3,297	879	1,074	415	797	236	2,056	1,004	12,369
Transport	4,190	2,082	512	2,841	2,676	2,091	690	4,506	2,278	21,867
Region total	13,182	10,643	4,311	7,465	5,566	5,218	1,909	12,535	5,779	66,609

A1T

Sector	Sub-Sharan Africa	CPA	E. Europe	FSU	LAM	ME-N Africa	NAM	P-OECD	PAS	SAS	W. Europe	Sector total
Power	925	3,831	119	203	788	958	606	107	1,039	745	147	9,469
Industry	1,871	983	77	299	433	614	420	104	521	1,394	278	6,996
Res/Com	774	2,574	70	448	1,576	598	878	116	1,154	1,285	507	9,979
Refineries	71	477	12	395	314	299	263	32	287	137	42	2,330
Synfuels	811	442	137	118	699	22	715	114	515	339	418	4,329
Hydrogen	290	99	37	364	0	647	0	0	151	256	612	2,456
Transport	1,083	4,319	280	1,121	2,106	1,613	2,094	386	1,839	1,545	1,464	17,851
Region total	5,825	12,725	732	2,949	5,917	4,751	4,977	859	5,506	5,702	3,468	53,411

A1FI

Sector	Africa	CPA	EEFSU	LAM	Middle East	USA	Canada	P-OECD	South East Asia	W. Europe	Sector total
Power	4,413	7,598	4,102	2,604	1,409	3,485	240	918	9,530	2,374	36,673
Industry	2,022	4,899	1,066	948	857	1,295	118	337	2,731	1,244	15,517
Res/Com	503	2,093	814	238	70	854	95	112	1,172	854	6,805
Transport	2,680	1,207	1,031	2,173	860	2,753	176	418	4,525	1,516	17,340
Synfuels	259	2,629	2,189	35	0	1,021	50	171	267	418	7,039
Hydrogen	0	0	0	0	0	0	0	0	0	0	0
Fuel flared	50	26	43	102	40	13	3	1	20	6	305
Region total	9,927	18,453	9,246	6,099	3,236	9,421	682	1,958	18,246	6,412	83,679

Source: Total emissions MtCO₂ 2050
CPA = Centrally Planned Asia. EE = Eastern Europe, FSU = Former Soviet Union, LAM = Latin America, P-OECD = Pacific OECD, S&EA = South and Southeast Asia, OECD-West = Western Europe + Canada, Africa, ME = Middle East, PAS = Pacific Asia, SAS = South Asia

Table 2.5 Continued.

A2

Sector	Africa	East Asia	E. Europe	FSU	LAM	Middle East	USA	Canada	P-OECD	South East Asia	South Asia	OECD Europe	Sector total
Power	2,144	3,406	913	1,679	2,621	2,518	4,653	310	1,028	967	3,660	1,766	25,666
Industry	881	2,727	345	725	1,118	899	895	115	276	413	1,627	487	10,506
Res/Com	907	1,451	157	735	325	719	644	95	144	179	599	628	6,582
Transport	1,061	901	193	646	1,547	1,370	1,946	191	378	578	703	1,275	10,788
Others	719	643	106	452	754	904	582	67	142	304	359	429	5,461
Region total	5,713	9,127	1,714	4,237	6,365	6,409	8,719	778	1,967	2,441	6,949	4,585	59,003

B1

Sector	Africa	East Asia	E. Europe	FSU	LAM	Middle East	USA	Canada	P-OECD	South East Asia	South Asia	OECD Europe	Sector total
Power	573	251	104	343	496	662	342	30	82	313	1,243	311	4,749
Industry	556	985	121	235	465	574	319	44	103	250	877	171	4,699
Res/Com	517	465	92	358	242	298	338	52	81	105	455	384	3,389
Transport	959	571	127	466	946	834	976	104	204	390	660	732	6,968
Others	414	280	45	209	378	458	230	29	60	198	253	225	2,779
Region total	3,019	2,551	488	1,612	2,527	2,825	2,205	259	529	1,255	3,488	1,824	22,584

B2 Sub-Saharan

Sector	Africa	CPA	E. Europe	FSU	LAM	ME-N Africa	NAM	P-OECD	PAS	SAS	W. Europe	Sector total
Power	654	1,703	474	576	274	753	2,280	289	762	1,357	936	10,060
Industry	932	1,751	166	685	688	601	708	66	827	1,499	406	8,328
Res/Com	623	1,850	85	386	477	127	1,084	129	661	1,106	610	7,138
Refineries	43	360	14	409	200	85	382	47	244	262	112	2,157
Synfuels	453	139	56	285	326	448	174	50	223	54	97	2,304
Hydrogen	308	1,312	43	278	277	186	319	29	185	444	364	3,743
Transport	572	1,531	145	840	1,230	799	2,577	340	1,014	1,075	1,336	11,459
Region total	3,584	8,645	984	3,458	3,471	2,999	7,524	951	3,917	5,797	3,861	45,189

Notes:

Source: Total emissions MtCO$_2$ 2050.
The division of the world into large economic regions differs between the various models underlying the SRES scenarios. Tables 2.5 and 2.6 consolidate the original model regions at a level that makes model results comparable (although the exact geographical coverage of the regions may vary).
CPA = Centrally Planned Asia. EE = Eastern Europe, FSU = Former Soviet Union, LAM = Latin America, P-OECD = Pacific OECD, S&EA = South and Southeast Asia, OECD-West = Western Europe + Canada, Africa, ME = Middle East, PAS = Pacific Asia, SAS = South Asia

currently in operation around the world and those planned to be built in the near future was conducted, together with a review of industrial boilers in selected regions. Capture factors were established on the basis of installed capacity, fuel type, unit size, and other technical parameters. Outside the energy and industry sectors, there are only very limited prospects for practical CO$_2$ capture because sources in the residential sectors are small, dispersed, and often mobile, and contain only low concentrations. These factors result in lower capture factors.

In the assessment of CO$_2$ capture, perhaps the most important open question is what will happen in the transport sector over the next few decades. If the above average increases in energy use for transport projected by all models in all scenarios involve traditional fossil-fuelled engine technologies, the capture and storage of transport-related CO$_2$ will – though theoretically possible –remain technically meaningless (excess weight, on-board equipment, compression penalty, etc.). However, depending on the penetration rate of hydrogen-based transport technologies, it should be possible to retrofit CO$_2$-emitting hydrogen production facilities with CO$_2$ capture equipment. The transport sector provides a huge potential for indirect CO$_2$ capture but feasibility depends on future hydrogen production technologies.

CO$_2$ capture might also be technically feasible from biomass-fuelled power plants, biomass fermentation for alcohol production or units for the production of biomass-derived hydrogen. It is conceivable that these technologies might play a significant role by 2050 and produce negative emissions across the full technology chain.

The results of applying the capture factors developed by Toth and Rogner (2006) to the CO$_2$ emissions of the SRES scenarios of Table 2.5 are presented in Table 2.6. Depending on the scenario, between 30 and 60% of global power generation emissions could be suitable for capture by 2050 and 30 to 40% of industry emissions could also be captured in that time frame.

The technical potentials for CO$_2$ capture presented here are only the first step in the full carbon dioxide capture and storage chain. The variations across scenarios reflect the uncertainties inherently associated with scenario and modelling analyses. The ranges of the technical capture potential relative to total CO$_2$ emissions are 9–12% (or 2.6–4.9 GtCO$_2$) by 2020 and 21–45% (or 4.7–37.5 GtCO$_2$) by 2050.

2.4　Geographical relationship between sources and storage opportunities

The preceding sections in this chapter have described the geographical distributions of CO$_2$ emission sources. This section gives an overview of the geographic distribution of potential storage sites that are in relative proximity to present-day sites with large point sources.

2.4.1　*Global storage opportunities*

Global assessments of storage opportunities for CO$_2$ emissions involving large volumes of CO$_2$ storage have focused on the options of geological storage or ocean storage, where CO$_2$ is:

- injected and trapped within geological formations at subsurface depths greater than 800 m where the CO$_2$ will be supercritical and in a dense liquid-like form in a geological reservoir, or
- injected into deep ocean waters with the aim of dispersing it quickly or depositing it at great depths on the floor of the ocean with the aim of forming CO$_2$ lakes.

High-level global assessments of both geological and ocean storage scenarios have estimated that there is considerable capacity for CO$_2$ storage (the estimates range from hundreds to tens of thousands of GtCO$_2$). The estimates in the literature of storage capacity in geological formations and in the oceans are discussed in detail in Chapters 5 and 6 respectively and are not discussed further in this chapter.

2.4.2　*Consideration of spatial and temporal relationships*

As discussed in Chapter 5, the aim of geological storage is to replicate the natural occurrence of deep subsurface fluids, where they have been trapped for tens or hundreds of millions of years. Due to the slow migration rates of subsurface fluids observed in nature (often centimetres per year), and even including scenarios where CO$_2$ leakage to the surface might unexpectedly occur, CO$_2$ injected into the geological subsurface will essentially remain geographically close to the location where it is injected. Chapter 6 shows that CO$_2$ injected into the ocean water column does not remain in a static location, but will migrate at relatively rapid speed throughout the ocean as dissolved CO$_2$ within the prevailing circulation of ocean currents. So dissolved CO$_2$ in the water column will not remain where it is injected in the immediate short term (i.e., a few years to some centuries). Deep-ocean lakes of CO$_2$ will, in principle, be more static geographically but will dissolve into the water column over the course of a few years or centuries.

These spatial and temporal characteristics of CO$_2$ migration in geological and ocean storage are important criteria when attempting to make maps of source and storage locations. In both storage scenarios, the possibility of adjoining storage locations in the future and of any possible reciprocal impacts will need to be considered.

2.4.3　*Global geographical mapping of source/storage locations*

To appreciate the relevance of a map showing the geographic distribution of sources and potential storage locations, it is necessary to know the volumes of CO$_2$ emissions and the storage capacity that might be available, and to establish a picture of the types and levels of technical uncertainty associated with the

Table 2.6 CO$_2$ emissions available for capture and storage in 2020 and 2050 from sectors in major world regions under six IPCC SRES scenarios (after Toth and Rogner, 2005). Continued on next page.

Potential CO$_2$ capture in MtCO$_2$ 2020

A1B

Sector	Africa	CPA	EEFSU	LAM	MEA	NAM	P-OECD	S&EA	OECD West	Sector total
Power	117	475	319	165	167	479	185	290	351	2,548
Industry	33	182	168	155	127	156	64	130	159	1,173
Res/Com	6	46	21	16	7	30	12	17	51	207
Transport	0	0	0	0	0	0	0	0	0	0
Region total	156	702	508	337	301	665	261	437	561	3,928

A1T   Sub-Saharan

Sector	Africa	CPA	E. Europe	FSU	LAM	ME-N Africa	NAM	P-OECD	PAS	SAS	W. Europe	Sector total
Power	21	334	78	139	39	110	715	128	164	20	366	2,115
Industry	6	195	18	70	56	57	85	21	35	57	65	664
Res/Com	4	59	4	16	14	4	37	7	12	6	36	200
Refineries	22	54	6	50	71	42	113	23	63	11	67	521
Synfuels	30	74	6	16	85	25	91	23	86	16	81	532
Hydrogen	46	125	24	73	50	56	211	68	65	41	162	919
Transport	0	0	0	0	0	0	0	0	0	0	0	0
Region total	129	840	135	364	315	294	1,251	270	426	150	777	4,950

A1FI

Sector	Africa	CPA	EEFSU	LAM	Middle East	USA	Canada	P-OECD	South East Asia	W. Europe	Sector total
Power	30	607	525	95	90	791	55	226	401	500	3,319
Industry	15	259	144	49	58	189	22	51	104	198	1,091
Res/Com	1	31	26	4	1	36	4	6	7	48	165
Transport	0	0	0	0	0	0	0	0	0	0	0
Synfuels	5	7	89	1	0	37	2	9	1	16	167
Hydrogen	0	0	0	0	0	0	0	0	0	0	0
Fuel flared	0	0	0	0	0	0	0	0	0	0	0
Region total	50	904	785	149	149	1,053	83	292	513	763	4,741

CPA = Centrally Planned Asia. EE = Eastern Europe, FSU = Former Soviet Union, LAM = Latin America, P-OECD = Pacific OECD, S&EA = South and Southeast Asia, OECD-West = Western Europe + Canada, Africa, ME = Middle East, PAS = Pacific Asia, SAS = South Asia

Table 2.6 Continued.

Potential CO₂ capture in MtCO₂ 2020

A2

Sector	Africa	East Asia	E. Europe	FSU	LAM	Middle East	USA	Canada	P-OECD	South East Asia	South Asia	OECD Europe	Sector total
Power	41	241	102	217	150	208	1,111	66	201	60	140	477	3,016
Industry	8	127	26	49	42	48	111	15	35	12	49	68	590
Res/Com	3	25	5	26	6	15	30	4	8	2	5	35	163
Transport	0	0	0	0	0	0	0	0	0	0	0	0	0
Others	0	0	0	0	0	0	0	0	0	0	0	0	0
Region total	51	392	134	292	198	271	1,252	86	244	74	194	579	3,769

B1

Sector	Africa	East Asia	E. Europe	FSU	LAM	Middle East	USA	Canada	P-OECD	South East Asia	South Asia	OECD Europe	Sector total
Power	38	156	81	160	147	174	632	35	126	57	129	304	2,040
Industry	6	79	19	32	35	43	68	10	22	10	45	43	411
Res/Com	3	22	5	22	5	11	22	3	6	2	5	28	134
Transport	0	0	0	0	0	0	0	0	0	0	0	0	0
Others	0	0	0	0	0	0	0	0	0	0	0	0	0
Region total	47	256	105	214	187	228	722	49	155	69	179	375	2,584

B2

Sector	Sub-Saharan Africa	CPA	E. Europe	FSU	LAM	ME-N Africa	NAM	P-OECD	PAS	SAS	W. Europe	Sector total
Power	18	225	82	24	52	100	982	114	153	41	349	2,140
Industry	5	122	19	89	42	50	103	12	19	30	73	565
Res/Com	5	42	5	15	6	4	46	8	6	6	35	178
Refineries	14	60	11	42	56	58	144	29	61	28	81	583
Synfuels	15	11	2	22	28	11	88	3	31	5	42	258
Hydrogen	12	233	14	16	20	16	144	28	92	31	107	712
Transport	0	0	0	0	0	0	0	0	0	0	0	0
Region total	69	693	132	209	204	239	1,507	196	361	140	687	4,437

CPA = Centrally Planned Asia. EE = Eastern Europe, FSU = Former Soviet Union, LAM = Latin America, P-OECD = Pacific OECD, S&EA = South and Southeast Asia, OECD-West = Western Europe + Canada, Africa, ME = Middle East, PAS = Pacific Asia, SAS = South Asia

Table 2.6 Continued.

Potential CO$_2$ Capture in MtCO$_2$, 2050

A1B

Sector	Africa	CPA	EEFSU	LAM	Middle East	NAM	P-OECD	S&EA	OECD West	Sector total
Power	2,167	1,701	831	674	548	1,015	438	1,658	1,092	10,124
Industry	760	931	726	1,015	701	439	165	1,201	481	6,419
Res/Com	222	660	191	128	87	172	68	393	319	2,241
Transport	0	0	0	0	0	0	0	0	0	0
Region total	3,149	3,291	1,747	1,818	1,337	1,627	671	3,253	1,892	18,783

A1T

Sector	Sub-Saharan Africa	CPA	E. Europe	FSU	LAM	ME-N Africa	NAM	P-OECD	PAS	SAS	W. Europe	Sector total
Power	526	2,530	90	127	469	753	477	84	702	423	115	6,296
Industry	329	307	25	110	165	191	139	33	111	288	102	1,799
Res/Com	66	445	16	94	189	126	190	32	238	140	159	1,694
Refineries	37	367	9	304	242	245	216	26	221	98	35	1,799
Synfuels	665	407	126	109	645	20	660	105	449	296	386	3,867
Hydrogen	283	96	36	354	0	630	0	0	147	249	596	2,392
Transport	0	0	0	0	0	0	0	0	0	0	0	0
Region total	1,905	4,154	301	1,098	1,709	1,965	1,681	280	1,867	1,493	1,393	17,846

A1FI

Sector	Africa	CPA	EEFSU	LAM	Middle East	USA	Canada	P-OECD	South East Asia	W. Europe	Sector total
Power	2,369	4,836	2,691	1,486	992	2,677	186	705	5,979	1,862	23,781
Industry	557	1,817	462	332	370	559	53	144	962	569	5,826
Res/Com	37	430	188	27	15	189	23	30	229	279	1,448
Transport	0	0	0	0	0	0	0	0	0	0	0
Synfuels	213	2,425	2,019	32	0	942	46	158	233	385	6,453
Hydrogen	0	0	0	0	0	0	0	0	0	0	0
Fuel flared	0	0	0	0	0	0	0	0	0	0	0
Region total	3,175	9,509	5,360	1,877	1,377	4,367	308	1,038	7,403	3,095	37,508

CPA = Centrally Planned Asia. EE = Eastern Europe, FSU = Former Soviet Union, LAM = Latin America, P-OECD = Pacific OECD, S&EA = South and Southeast Asia, OECD-West = Western Europe + Canada, Africa, ME = Middle East, PAS = Pacific Asia, SAS = South Asia

Table 2.6 Continued.

Potential CO₂ Capture in MtCO₂ 2050

A2

Sector	Africa	East Asia	E. Europe	FSU	LAM	Middle East	USA	Canada	P-OECD	South East Asia	South Asia	OECD Europe	Sector total
Power	1,158	2,080	571	1,110	1,407	1,628	3,569	230	779	631	1,912	1,284	16,359
Industry	257	991	128	286	365	319	384	46	112	139	519	194	3,741
Res/Com	78	293	34	155	41	148	143	21	42	30	113	197	1,295
Transport	0	0	0	0	0	0	0	0	0	0	0	0	0
Others	0	0	0	0	0	0	0	0	0	0	0	0	0
Region total	1,493	3,365	733	1,552	1,812	2,095	4,096	298	933	799	2,544	1,675	21,394

B1

Sector	Africa	East Asia	E. Europe	FSU	LAM	Middle East	USA	Canada	P-OECD	South East Asia	South Asia	OECD Europe	Sector total
Power	266	130	63	218	258	418	221	19	52	185	635	203	2,668
Industry	138	268	40	83	137	196	118	16	36	72	271	64	1,437
Res/Com	44	80	19	69	28	57	69	11	21	16	73	111	598
Transport	0	0	0	0	0	0	0	0	0	0	0	0	0
Others	0	0	0	0	0	0	0	0	0	0	0	0	0
Region total	447	478	121	371	423	671	408	46	110	273	980	377	4,703

B2

Sector	Sub-Saharan Africa	CPA	E. Europe	FSU	LAM	ME-N Africa	NAM	P-OECD	PAS	SAS	W. Europe	Sector total
Power	339	1,067	307	345	164	563	1,710	216	439	673	704	6,526
Industry	166	459	63	248	266	257	225	20	157	238	144	2,243
Res/Com	42	309	18	77	52	16	224	35	102	104	182	1,161
Refineries	22	270	11	306	150	68	305	38	183	183	89	1,625
Synfuels	362	125	51	256	293	403	157	45	189	46	87	2,015
Hydrogen	293	1,246	41	264	263	176	303	27	176	421	345	3,556
Transport	0	0	0	0	0	0	0	0	0	0	0	0
Region total	1,223	3,476	489	1,496	1,187	1,484	2,924	383	1,246	1,665	1,552	17,125

Notes: The division of the world into large economic regions differs in the different models underlying the SRES scenarios. Tables 2.5 and 2.6 consolidate the original model regions at a level that makes model results comparable (although the exact geographical coverage of the regions may vary).

CPA = Centrally Planned Asia. EE = Eastern Europe, FSU = Former Soviet Union, LAM = Latin America, P-OECD = Pacific OECD, S&EA = South and Southeast Asia, OECD-West = Western Europe + Canada, Africa, ME = Middle East, PAS = Pacific Asia, SAS = South Asia

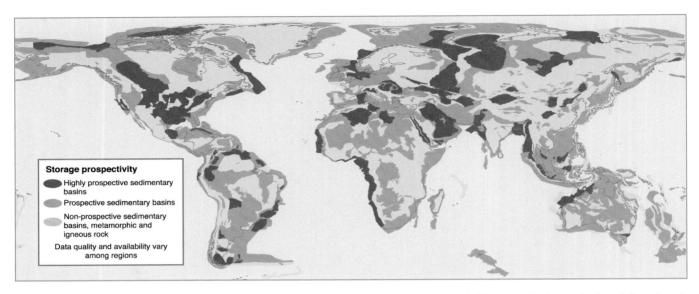

Figure 2.4 Prospective areas in sedimentary basins where suitable saline formations, oil or gas fields, or coal beds may be found. Locations for storage in coal beds are only partly included. Prospectivity is a qualitative assessment of the likelihood that a suitable storage location is present in a given area based on the available information. This figure should be taken as a guide only, because it is based on partial data, the quality of which may vary from region to region, and which may change over time and with new information (Bradshaw and Dance, 2004).

storage sites that will affect their viability as potential solutions. As indicated above in this chapter, there are some 7,500 large stationary sources with emissions in excess of 0.1 MtCO$_2$ yr^{-1} and that number is projected to rise by 2050. The mapping does not take into account the 'capture factors' presented in Section 2.3.2.

2.4.3.1   Geological storage and source location matching
Chapter 5 includes detailed discussions of the geological characteristics of storage sites. Before discussing the global locations for geological storage opportunities, it is necessary to describe some basic fundamentals of geological storage. The world's geological provinces can be allocated to a variety of rock types, but the main ones relevant to geological storage are sedimentary basins that have undergone only minor tectonic deformation and are at least 1000 m thick with adequate reservoir/seal pairs to allow for the injection and trapping of CO$_2$. The petroleum provinces of the world are a subset of the sedimentary basins described above, and are considered to be promising locations for the geological storage of CO$_2$ (Bradshaw *et al.*, 2002). These basins have adequate reservoir/seal pairs, and suitable traps for hydrocarbons, whether liquids or gases. The remaining geological provinces of the world can generally be categorized as igneous (rocks formed from crystallization of molten liquid) and metamorphic (pre-existing rocks formed by chemical and physical alteration under the influence of heat, pressure and chemically active fluids) provinces. These rock types are commonly known as hard-rock provinces, and they will not be favourable for CO$_2$ storage as they are generally not porous and permeable and will therefore not readily transmit fluids. More details on the suitability of sedimentary basins and characterization of specific sites are provided in Chapter 5.

Figure 2.4 shows the 'prospectivity'(see Annex II) of

various parts of the world for the geological storage of CO$_2$. Prospectivity is a term commonly used in explorations for any geological resource, and in this case it applies to CO$_2$ storage space. Prospectivity is a qualitative assessment of the likelihood that a suitable storage location is present in a given area based on the available information. By nature, it will change over time and with new information. Estimates of prospectivity are developed by examining data (if possible), examining existing knowledge, applying established conceptual models and, ideally, generating new conceptual models or applying an analogue from a neighbouring basin or some other geologically similar setting. The concept of prospectivity is often used when it is too complex or technically impossible to assign numerical estimates to the extent of a resource.

Figure 2.4 shows the world's geological provinces broken down into provinces that are thought, at a very simplistic level, to have CO$_2$ storage potential that is either: 1) highly prospective, 2) prospective, or 3) non-prospective (Bradshaw and Dance, 2004). Areas of high prospectivity are considered to include those basins that are world-class petroleum basins, meaning that they are the basins of the world that are producing substantial volumes of hydrocarbons. It also includes areas that are expected to have substantial storage potential. Areas of prospective storage potential are basins that are minor petroleum basins but not world-class, as well as other sedimentary basins that have not been highly deformed. Some of these basins will be highly prospective for CO$_2$ storage and others will have low prospectivity.

Determining the degree of suitability of any of these basins for CO$_2$ storage will depend on detailed work in each area. Areas that are non-prospective are highly deformed sedimentary basins and other geological provinces, mainly containing metamorphic and igneous rocks. Some of these

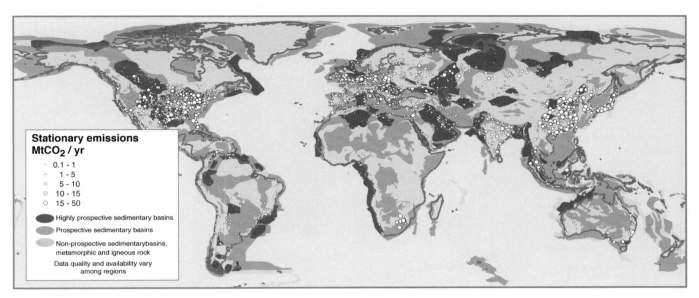

Figure 2.5 Geographical relationship between CO$_2$ emission sources and prospective geological storage sites. The dots indicate CO$_2$ emission sources of 0.1–50 MtCO$_2$ yr^{-1}. Prospectivity is a qualitative assessment of the likelihood that a suitable storage location is present in a given area based on the available information. This figure should be taken as a guide only, because it is based on partial data, the quality of which may vary from region to region, and which may change over time and with new information.

provinces might have some local niche opportunities for CO$_2$ storage, but at this stage they would not be considered suitable for a conventional form of CO$_2$ storage. As Bradshaw and Dance (2004) explain, this map is subject to significant caveats and based on significant assumptions because of the data source from which it was generated. However, it can be used as a general (although not specific) guide at the global scale to the location of areas that are likely to provide opportunities for the geological storage of CO$_2$. Due to the generalized manner in which this map has been created, and the lack of specific or hard data for each of the basins assessed, the 'prospectivity' levels assigned to each category have no meaningful correlative statistical or probabilistic connotation. To achieve a numerical analysis of risk or certainty would require specific information about each and every basin assessed.

Figure 2.5 shows the overlap of the sedimentary basins that are prospective for CO$_2$ storage potential with the current locations of large sources of stationary emissions (IEA GHG, 2002a). The map can be simplistically interpreted to identify areas where large distances might be required to transport emissions from any given source to a geological storage location. It clearly shows areas with local geological storage potential and low numbers of emission sites (for example, South America) as well as areas with high numbers of emission sites and few geological storage options in the vicinity (the Indian sub-continent, for example). This map, however, does not address the relative capacity of any of the given sites to match either large emission sources or small storage capacities. Neither does it address any of the technical uncertainties that could exist at any of the storage sites, or the cost implications for the emission sources of the nature of the emission plant or the purity of the emission sources. Such issues of detailed source-to-store matching are dealt with in Chapter 5.

Figures 2.6, 2.7 and 2.8 show the regional emission clusters for twelve regions of the world and the available storage opportunities within each region. They also compare the relative ranking of the area of available prospective sedimentary basins in a 300 km radius around emission clusters (Bradshaw and Dance, 2004). The 300 km radius was selected because it was considered useful as an indicator of likely transport distances for potentially viable source-to-storage matches (see Chapter 5). Although this data could suggest trends, such as high emissions for China with a small area of prospective sedimentary basins, or a large area of prospective sedimentary basins with low emissions for the Middle East, it is premature to make too many assumptions until detailed assessments are made in each region as to the quality and viability of each sedimentary basin and specific proposed sites. Each basin will have its own technical peculiarities, and because the science of injection and storage of very large volumes of CO$_2$ is still developing, it is premature at this stage to make any substantive comments about the viability of individual sedimentary basins unless there are detailed data sets and assessments (see Chapter 5). These maps do, however, indicate where such detailed geological assessments will be required – China and India, for example – before a comprehensive assessment can be made of the likely worldwide impact of the geological storage of CO$_2$. These maps also show that CO$_2$ storage space is a resource, just like any other resource; some regions will have many favourable opportunities, and others will not be so well-endowed (Bradshaw and Dance, 2004).

Figure 2.9 shows those emission sources with high concentrations (>95%) of CO$_2$, with their proximity to prospective geological storage sites. Clusters of high-concentration sources can be observed in China and North America and to lesser extent in Europe.

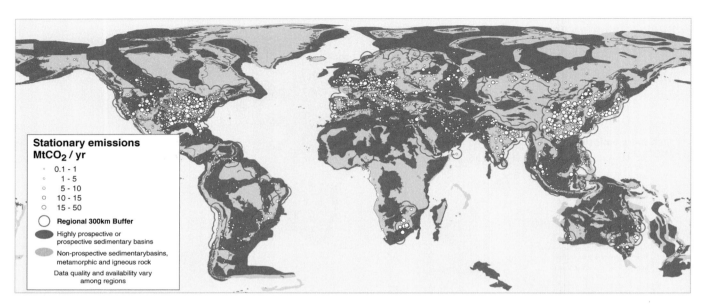

Figure 2.6 Regional emission clusters with a 300 km buffer relative to world geological storage prospectivity (Bradshaw and Dance, 2004).

2.4.3.2 Ocean storage and source-location matching
Due to a lack of publicly available literature, a review of the proximity of large CO_2 point sources and their geographical relationship to ocean storage opportunities on the global scale could not be undertaken. A related study was undertaken that analysed seawater scrubbing of CO_2 from power stations along the coastlines of the world. The study considered the number

of large stationary sources (in this case, power generation plants) on the coastlines of the worldwide that are located within 100 km of the 1500 m ocean floor contour (IEA GHG, 2000a). Eighty-nine potential power generation sources were identified that were close to these deep-water locations. This number represents only a small proportion (< 2%) of the total number of large stationary sources in the power generation

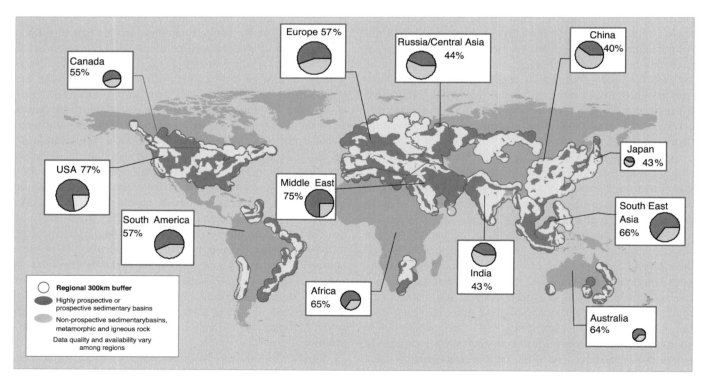

Figure 2.7 Regional storage opportunities determined by using a ratio (percentage) of all prospective areas to non-prospective areas within a 300 km buffer around major stationary emissions. The pie charts show the proportion of the prospective areas (sedimentary basins) in the buffer regions (Bradshaw and Dance, 2004).

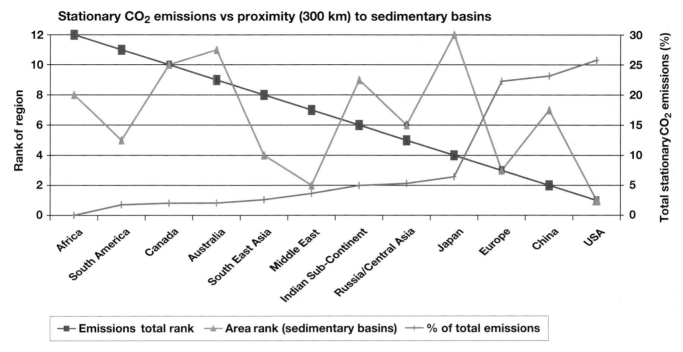

Figure 2.8 Proximity of emissions to sedimentary basins.

sector worldwide (see Section 2.1). A larger proportion of power plants could possibly turn to deep-ocean storage because transport over distances larger than 100 km may prove cost-effective in some cases; nevertheless, this study indicates that a higher fraction of large stationary sources could be more cost-effectively matched to geological storage reservoirs than ocean storage sites. There are many issues that will also need to be addressed when considering deep-ocean storage sites, including jurisdictional boundaries, site suitability, and environmental impact etc., which are discussed in Chapter 6. The spatial and temporal nature of ocean water-column injection may affect the

approach to source and storage matching, as the CO_2 will not remain adjacent to the local region where the CO_2 is injected, and conceivably might migrate across jurisdictional boundaries and into sensitive environmental provinces.

2.5   Alternative energy carriers and CO₂ source implications

As discussed earlier in this chapter, a significant fraction of the world's CO_2 emissions comes from transport, residences, and other small, distributed combustion sources. Whilst it is

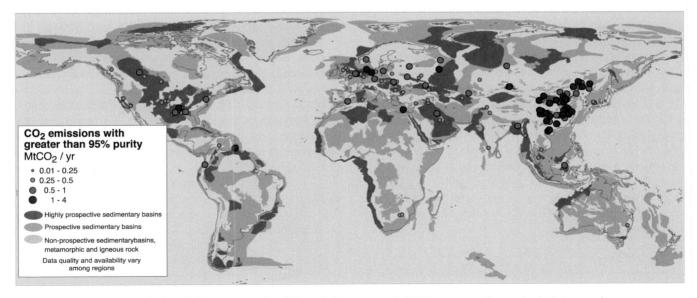

Figure 2.9 Geographical proximity of high-concentration CO_2 emission sources (> 95%) to prospective geological storage sites.

currently not economically feasible to capture and store CO_2 from these small, distributed sources, these emissions could be reduced if the fossil fuels used in these units were replaced with either:

- carbon-free energy carriers (e.g. electricity or hydrogen);
- energy carriers that are less carbon-intensive than conventional hydrocarbon fuels (e.g., methanol, Fischer-Tropsch liquids or dimethyl ether);
- biomass energy that can either be used directly or to produce energy carriers like bioethanol. If the biomass is grown sustainably the energy produced can be considered carbon-neutral.

In the first two cases, the alternative energy carriers can be produced in centralized plants that incorporate CO_2 capture and storage. In the case of biomass, CO_2 capture and storage can also be incorporated into the energy carrier production schemes. The aim of this section is to explore the implications that introducing such alternative energy carriers and energy sources might have for future large point sources of CO_2 emissions.

2.5.1   Carbon-free energy carriers

2.5.1.1   Electricity
The long-term trend has been towards the electrification of the energy economy, and this trend is expected to continue (IPCC, 2000). To the extent that expanded electricity use is a substitute for the direct use of fossil fuels (e.g., in transport, or for cooking or heating applications in households), the result can be less CO_2 emissions if the electricity is from carbon-free primary energy sources (renewable or nuclear) or from distributed generators such as fuel cells powered by hydrogen produced with near-zero fuel-cycle-wide emissions or from large fossil-fuel power plants at which CO_2 is captured and stored.

While, in principle, all energy could be provided by electricity, most energy projections envision that the direct use of fuels will be preferred for many applications (IPCC, 2000). In transport, for example, despite intensive developmental efforts, battery-powered electric vehicles have not evolved beyond niche markets because the challenges of high cost, heavy weight, and long recharging times have not been overcome. Whilst the prospects of current hybrid electric vehicles (which combine fossil fuel and electric batteries) penetrating mass markets seem good, these vehicles do not require charging from centralized electrical grids. The successful development of 'plug-in hybrids' might lead to an expanded role for electricity in transport but such vehicles would still require fuel as well as grid electricity. In summary, it is expected that, although electricity's share of total energy might continue to grow, most growth in large point sources of CO_2 emissions will be the result of increased primary energy demand.

2.5.1.2   Hydrogen
If hydrogen can be successfully established in the market as an energy carrier, a consequence could be the emergence of large new concentrated sources of CO_2 if the hydrogen

is manufactured from fossil fuels in large pre-combustion decarbonization plants with CO_2 capture and storage. Such plants produce a high concentration source of CO_2 (see Chapter 3 for details on system design). Where fossil fuel costs are low and CO_2 capture and storage is feasible, hydrogen manufactured in this way is likely to be less costly than hydrogen produced from renewable or nuclear primary energy sources (Williams, 2003; NRC, 2004). It should be noted that this technology can be utilized only if production sites are within a couple of hundred kilometres of where the hydrogen will be used, since cost-effective, long-distance hydrogen transport represents a significant challenge. Producing hydrogen from fossil fuels could be a step in technological development towards a hydrogen economy based on carbon-free primary energy sources through the establishment of a hydrogen utilization infrastructure (Simbeck, 2003).

Energy market applications for hydrogen include its conversion to electricity electrochemically (in fuel cells) and in combustion applications. Substituting hydrogen for fossil fuel burning eliminates CO_2 emissions at the point of energy use. Much of the interest in hydrogen market development has focused on distributed stationary applications in buildings and on transport. Fuel cells are one option for use in stationary distributed energy systems at scales as small as apartment buildings and even single-family residences (Lloyd, 1999). In building applications, hydrogen could also be combusted for heating and cooking (Ogden and Williams, 1989). In the transport sector, the hydrogen fuel cell car is the focus of intense development activity, with commercialization targeted for the middle of the next decade by several major automobile manufacturers (Burns *et al.*, 2002). The main technological obstacles to the widespread use of fuel cell vehicles are the current high costs of the vehicles themselves and the bulkiness of compressed gaseous hydrogen storage (the only fully proven hydrogen storage technology), which restricts the range between refuelling (NRC, 2004). However, the currently achievable ranges might be acceptable to many consumers, even without storage technology breakthroughs (Ogden *et al.*, 2004).

Hydrogen might also be used in internal combustion engine vehicles before fuel cell vehicles become available (Owen and Gordon, 2002), although efficiencies are likely to be less than with fuel cells. In this case, the range between refuelling would also be less than for hydrogen fuel cell vehicles with the same performance (Ogden et al., 2004). For power generation applications, gas turbines originally designed for natural gas operation can be re-engineered to operate on hydrogen (Chiesa *et al.*, 2003).

Currently, there are a number of obstacles on the path to a hydrogen economy. They are: the absence of cost-competitive fuel cells and other hydrogen equipment and the absence of an infrastructure for getting hydrogen to consumers. These challenges are being addressed in many hydrogen R&D programmes and policy studies being carried out around the world (Sperling and Cannon, 2004). There are also safety concerns because, compared to other fuels, hydrogen has a wide flammability and detonation range, low ignition energy, and high flame speed. However, industrial experience shows that hydrogen can be manufactured and used safely in many applications (NRC, 2004).

There is widespread industrial experience with the production and distribution of hydrogen, mainly for the synthesis of ammonia fertilizer and hydro-treatment in oil refineries. Current global hydrogen production is 45 million t yr^{-1}, the equivalent to 1.4% of global primary energy use in 2000 (Simbeck, 2003). Forty-eight per cent is produced from natural gas, 30% from oil, 18% from coal, and 4% via electrolysis of water. Ammonia production, which consumes about 100,000 MW_t of hydrogen, is growing by 2–4% per year. Oil refinery demand for hydrogen is also increasing, largely because of the ongoing shift to heavier crude oils and regulations limiting the sulphur content of transport fuels. Most hydrogen is currently manufactured via steam methane reforming (SMR), steam reforming of naphtha, and the gasification of petroleum residues and coal. The SMR option is generally favoured due to its lower capital cost wherever natural gas is available at reasonable prices. Nevertheless, there are currently about 75 modern commercial gasification plants making about 20,000 MW_t of hydrogen from coal and oil refinery residues (NETL-DOE, 2002); these are mostly ammonia fertilizer plants and hydrogen plants in oil refineries in China, Europe, and North America. There are currently over 16,000 km of hydrogen pipelines around the world. Most are relatively short and located in industrial areas for large customers who make chemicals, reduce metals, and engage in the hydro-treatment of oil at refineries. The longest pipeline currently in operation is 400 km long and is located in a densely populated area of Europe, running from Antwerp to northern France. The pipeline operates at a pressure of about 60 atmospheres (Simbeck, 2004).

Fossil fuel plants producing hydrogen with CO_2 capture and storage would typically be large, producing volumes of the order of 1000 MW_t (720 t day^{-1})[6] in order to keep the hydrogen costs and CO_2 storage costs low. Per kg of hydrogen, the co-production rate would be about 8 $kgCO_2$ with SMR and 15 $kgCO_2$ with coal gasification, so that the CO_2 storage rates (for plants operated at 80% average capacity factor) would be 1.7 and 3.1 million tonnes per year for SMR and coal gasification plants respectively.

Making hydrogen from fossil fuels with CO_2 capture and storage in a relatively small number of large plants for use in large numbers of mobile and stationary distributed applications could lead to major reductions in fuel-cycle-wide emissions compared to petroleum-based energy systems. This takes into account all fossil fuel energy inputs, including energy for petroleum refining and hydrogen compression at refuelling stations (NRC, 2004; Ogden *et al.*, 2004). No estimates have yet been made of the number of large stationary, concentrated CO_2 sources that could be generated via such hydrogen production systems and their geographical distribution.

2.5.2 *Alternative energy carriers and CO₂ source implications*

Interest in synthetic liquid fuels stems from concerns about both the security of oil supplies (TFEST, 2004) and the expectation that it could possibly be decades before hydrogen can make a major contribution to the energy economy (NRC, 2004).

There is considerable activity worldwide relating to the manufacture of Fischer-Tropsch liquids from stranded natural gas supplies. The first major gas to liquids plant, producing 12,500 barrels per day, was built in Malaysia in 1993. Several projects are underway to make Fischer-Tropsch liquid fuels from natural gas in Qatar at plant capacities ranging from 30,000 to 140,000 barrels per day. Although gas to liquids projects do not typically produce concentrated by-product streams of CO_2, synthetic fuel projects using synthesis gas derived from coal (or other solid feedstocks such as biomass or petroleum residuals) via gasification could produce large streams of concentrated CO_2 that are good candidates for capture and storage. At Sasol in South Africa, coal containing some 20 million tonnes of carbon is consumed annually in the manufacture of synthetic fuels and chemicals. About 32% of the carbon ends up in the products, 40% is vented as CO_2 in dilute streams, and 28% is released as nearly pure CO_2 at a rate of about 20 million tonnes of CO_2 per year. In addition, since 2000, 1.5 million tonnes per year of CO_2 by-product from synthetic methane production at a coal gasification plant in North Dakota (United States) have been captured and transported 300 km by pipeline to the Weyburn oil field in Saskatchewan (Canada), where it is used for enhanced oil recovery (see Chapter 5 for more details). Coal-based synthetic fuel plants being planned or considered in China include six 600,000 t yr^{-1} methanol plants, two 800,000 t yr^{-1} dimethyl ether plants, and two or more large Fischer-Tropsch liquids plants[7]. In the United States, the Department of Energy is supporting a demonstration project in Pennsylvania to make 5,000 barrels/day of Fischer-Tropsch liquids plus 41 MW_e of electricity from low-quality coal.

If synthesis-gas-based energy systems become established in the market, economic considerations are likely to lead, as in the case of hydrogen production, to the construction of large facilities that would generate huge, relatively pure, CO_2 co-product streams. Polygeneration plants, for example plants that could produce synthetic liquid fuels plus electricity, would benefit as a result of economies of scale, economies of scope, and opportunities afforded by greater system operating flexibility (Williams *et al.*, 2000; Bechtel *et al.*, 2003; Larson and Ren, 2003; Celik *et al.*, 2005). In such plants, CO_2 could be captured from shifted synthesis gas streams both upstream and downstream of the synthesis reactor where the synthetic fuel is produced.

With CO_2 capture and storage, the fuel-cycle-wide greenhouse gas emissions per GJ for coal derived synthetic

---

[6] A plant of this kind operating at 80% capacity could support 2 million hydrogen fuel cell cars with a gasoline-equivalent fuel economy of 2.9 L per 100 km driving 14,000 km per year.

[7] Most of the methanol would be used for making chemicals and for subsequent conversion to dimethyl ether, although some methanol will be used for transport fuel. The dimethyl ether would be used mainly as a cooking fuel.

fuels can sometimes be less than for crude oil-derived fuels. For example, a study of dimethyl ether manufacture from coal with CO_2 capture and storage found that fuel-cycle-wide greenhouse gas emissions per GJ ranged from 75 to 97% of the emission rate for diesel derived from crude oil, depending on the extent of CO_2 capture (Celik *et al.*, 2005).

The CO_2 source implications of making synthetic low-carbon liquid energy carriers with CO_2 capture and storage are similar to those for making hydrogen from fossil fuels: large quantities of concentrated CO_2 would be available for capture at point sources. Again, no estimates have yet been made of the number of large stationary sources that could be generated or of their geographical distribution.

2.5.3   *CO_2 source implications of biomass energy production*

There is considerable interest in some regions of the world in the use of biomass to produce energy, either in dedicated plants or in combination with fossil fuels. One set of options with potentially significant but currently uncertain implications for future CO_2 sources is bioenergy with CO_2 capture and storage. Such systems could potentially achieve negative CO_2 emissions. The perceived CO_2 emission benefits and costs of such systems are discussed elsewhere in this report (see Chapters 3 and 8) and are not discussed further here. The aim of this section is to assess the current scale of emissions from biomass energy production, to consider how they might vary in the future, and therefore to consider their impact on the future number, and scale, of CO_2 emission sources.

2.5.3.1   *Bioethanol production*
Bioethanol is the main biofuel being produced today. Currently, the two largest producers of bioethanol are the USA and Brazil. The USA produced 11 billion litres in 2003, nearly double the capacity in 1995. Production is expected to continue to rise because of government incentives. Brazilian production was over 14 billion litres per year in 2003/2004, similar to the level in 1997/1998 (Möllersten *et al.*, 2003). Bioethanol is used directly in internal combustion engines, without modification, as a partial replacement for petroleum-based fuels (the level of replacement in Europe and the USA is 5 to 10%).

Bioethanol plants are a high-concentration source of CO_2 at atmospheric pressure that can be captured and subsequently stored. As can be seen in Table 2.3, the numbers of these plants are significant in the context of high-purity sources, although their global distribution is restricted. These sources are comparable in size to those from ethylene oxide plants but smaller than those from ammonia plants.

Although the trend in manufacture is towards larger production facilities, the scale of future production will be determined by issues such as improvements in biomass production and conversion technologies, competition with other land use, water demand, markets for by-product streams and competition with other transport fuels.

On the basis of the literature currently available, it is not possible to estimate the number of bioethanol plants that will be built in the future or the likely size of their CO_2 emissions.

2.5.3.2   *Biomass as a primary energy source*
A key issue posed by biomass energy production, both with and without CO_2 capture and storage, is that of size. Current biomass energy production plants are much smaller than fossil fuel power plants; typical plant capacities are about 30 MW$_e$, with CO_2 emissions of less than 0.2 MtCO_2 per year. The size of these biomass energy production plants reflects the availability and dispersed nature of current biomass supplies, which are mainly crop and forestry residues.

The prospects for biomass energy production with CO_2 capture and storage might be improved in the future if economies of scale in energy production and/or CO_2 capture and storage can be realized. If, for instance, a CO_2 pipeline network is established in a country or region, then small CO_2 emission sources (including those from biomass energy plants) could be added to any nearby CO_2 pipelines if it is economically viable to do so. A second possibility is that existing large fossil fuel plants with CO_2 capture and storage represent an opportunity for the co-processing of biomass. Co-processing biomass at coal power plants already takes place in a number of countries. However, it must be noted that if biomass is co-processed with a fossil fuel, these plants do not represent new large-scale emissions sources. A third possibility is to build larger biomass energy production plants than the plants typically in place at present. Larger biomass energy production plants have been built or are being planned in a number of countries, typically those with extensive biomass resources. For example, Sweden already has seven combined heat and power plants using biomass at pulp mills, with each plant producing around 130 MW$_e$ equivalent. The size of biomass energy production plants depends on local circumstances, in particular the availability of concentrated biomass sources; pulp mills and sugar processing plants offer concentrated sources of this kind.

Larger plants could also be favoured if there were a shift from the utilization of biomass residues to dedicated energy crops. Several studies have assessed the likely size of future biomass energy production plants, but these studies conflict when it comes to the scale issue. One study, cited in Audus and Freund (2004), surveyed 28 favoured sites using woody biomass crops in Spain and concluded that the average appropriate scale would be in the range 30 to 70 MW$_e$. This figure is based on the fact that transport distances longer than the assumed maximum of 40 km would render larger plants uneconomic. In contrast, another study based on dedicated energy crops in Brazil and the United States estimated that economies of scale outweigh the extra costs of transporting biomass over long distances. This study found that plant capacities of hundreds of MW$_e$ were feasible (Marrison and Larson, 1995). Other studies have come up with similar findings (Dornburg and Faaij, 2001; Hamelinck and Faaij, 2002). A recent study analyzed a variety of options including both electricity and synthetic fuel production and indicated that large plants processing about 1000 MW$_{th}$ of biomass would tend to be preferred for dedicated energy crops in the United States (Greene *et al.*, 2004).

The size of future emission sources from bioenergy options depends to a large degree on local circumstances and the extent to which economic forces and/or public policies will encourage the development of dedicated energy crops. The projections of annual global biomass energy use rise from 12–60 EJ by 2020, to 70–190 EJ per year by 2050, and to 120–380 EJ by 2100 in the SRES Marker Scenarios (IPCC, 2000), showing that many global energy modellers expect that dedicated energy crops may well become more and more important during the course of this century. So if bioenergy systems prove to be viable at scales suitable for CO$_2$ capture and storage, then the negative emissions potential of biomass (see Chapter 8) might, during the course of this century, become globally important. However, it is currently unclear to what extent it will be feasible to exploit this potential, both because of the uncertainties about the scale of bioenergy conversion and the extent to which dedicated biomass energy crops will play a role in the energy economy of the future.

In summary, based on the available literature, it is not possible at this stage to make reliable quantitative statements on number of biomass energy production plants that will be built in the future or the likely size of their CO$_2$ emissions.

2.6   Gaps in knowledge

Whilst it is possible to determine emission source data for the year 2000 (CO$_2$ concentration and point source geographical location) with a reasonable degree of accuracy for most industrial sectors, it is more difficult to predict the future location of emission point sources. Whilst all projections indicate there will be an increase in CO$_2$ emissions, determining the actual locations for new plants currently remains a subjective business.

A detailed description of the storage capacity for the world's sedimentary basins is required. Although capacity estimates have been made, they do not yet constitute a full resource assessment. Such information is essential to establish a better picture of the existing opportunities for storing the CO$_2$ generated at large point sources. At present, only a simplistic assessment is possible based on the limited data about the storage capacity currently available in sedimentary basins.

An analysis of the storage potential in the ocean for emissions from large point sources was not possible because detailed mapping indicating the relationship between storage locations in the oceans and point source emissions has not yet been carefully assessed.

This chapter highlights the fact that fossil fuel-based hydrogen production from large centralized plants will potentially result in the generation of more high-concentration emission sources. However, it is not currently possible to predict with any accuracy the number of these point sources in the future, or when they will be established, because of market development uncertainties surrounding hydrogen as an energy carrier. For example, before high-concentration CO$_2$ sources associated with hydrogen production for energy can

be exploited, cost-effective end-use technologies for hydrogen (e.g., low-temperature fuel cells) must be readily available on the market. In addition, it is expected that it will take decades to build a hydrogen infrastructure that will bring the hydrogen from large centralized sources (where CCS is practical) to consumers.

Synthetic liquid fuels production or the co-production of liquid fuels and electricity via the gasification of coal or other solid feedstocks or petroleum residuals can also lead to the generation of concentrated streams of CO$_2$. It is unclear at the present time to what extent such synthetic fuels will be produced as alternatives to crude-oil-derived hydrocarbon fuels. The co-production options, which seem especially promising, require market reforms that make it possible to co-produce electricity at a competitive market price.

During the course of this century, biomass energy systems might become significant new large CO$_2$ sources, but this depends on the extent to which bioenergy conversion will take place in large plants, and the global significance of this option may well depend critically on the extent to which dedicated energy crops are pursued.

References

Audus, H. and P. Freund, 2004: Climate change mitigation by biomass gasification combined with CO$_2$ capture and storage. Proceedings of 7th International Conference on Greenhouse Gas Control Technologies. E.S. Rubin, D.W. Keith, and C.F. Gilboy (eds.), Vol. 1 pp. 187-200: Peer-Reviewed Papers and Plenary Presentations, Pergamon, 2005

Bechtel Corporation, Global Energy Inc., and Nexant Inc., 2003: Gasification Plant Cost and Performance Optimization, Task 2 Topical Report: Coke/Coal Gasification with Liquids Co-production, prepared for the National Energy Technology Laboratory, US Department of Energy under Contract No. DE-AC26-99FT40342, September.

Bradshaw, J. and T. Dance, 2004: Mapping geological storage prospectivity of CO$_2$ for the world's sedimentary basins and regional source to sink matching. Proceedings of the 7th International Conference on Greenhouse Gas Technologies, Vol. 1; peer reviewed Papers and Plenary Presentations. pp. 583-592. Eds. E.S. Rubin, D.W. Keith and C.F. Gilboy, Pergamon, 2005

Bradshaw, J., B.E. Bradshaw, G. Allinson, A.J. Rigg, V. Nguyen, and L. Spencer, 2002: The Potential for Geological Sequestration of CO$_2$ in Australia: Preliminary findings and implications to new gas field development. *APPEA Journal*, **42**(1), 25-46.

Burns, L., J. McCormick, and C. Borroni-Bird, 2002: Vehicle of change. *Scientific American*, **287**(4), 64-73.

Campbell, P.E., J.T. McMullan, and B.C. Williams, 2000: Concept for a competitive coal fired integrated gasification combined cycle power plant. *Fuel*, **79**(9), 1031-1040.

Celik, F., E.D. Larson, and R.H. Williams, 2005: Transportation Fuel from Coal with Low CO_2 Emissions. Wilson, M., T. Morris, J. Gale and K. Thambimuthu (eds.), Proceedings of 7th International Conference on Greenhouse Gas Control Technologies. Volume II: Papers, Posters and Panel Discussion, pp. 1053-1058, Pergamon, 2005

Chauvel, A. and G. Lefebvre, 1989: Petrochemical Processes, Technical and Economic Characteristics, 1 Synthesis-Gas Derivatives and Major Hydrocarbons, Éditions Technip, Paris, 2001.

Chiesa, P., G. Lozza, and L. Mazzocchi, 2003: Using hydrogen as gas turbine fuel, *Proceedings of ASME Turbo Expo 2003: Power for Land, Sea, and Air*, Atlanta, GA, 16-19 June.

Christensen, N.P., 2001: The GESTCO Project: Assessing European potential for geological storage and CO_2 from fossil fuel combustion. Proceedings of the Fifth International Conference on Greenhouse Gas Control Technologies (GHGT-5), 12-16 August 2000, Cairns, Australia. pp. 260-265.

Christensen, T.S. and I.I. Primdahl, 1994: Improve synthesis gas production using auto thermal reforming. *Hydrocarbon Processing*, 39-46, March, 1994.

Dornburg, V. and A. Faaij, 2001: Efficiency and economy of wood-fired biomass energy systems in relation to scale regarding heat and power generation using combustion and gasification technologies, *Biomass and Biomass energy,* **21** (2): 91-108.

Foster Wheeler, 1998: Solving the heavy fuel oil problem with IGCC technology. *Heat Engineering*, **62**(2), 24-28.

Gale, J., 2002: Overview of CO_2 emissions sources, potential, transport and geographical distribution of storage possibilities. Proceedings of the workshop on CO_2 dioxide capture and storage, Regina, Canada, 18-21 November 2002, pp. 15-29.

Garg, A., M. Kapshe, P.R. Shukla, and D.Ghosh, 2002: Large Point Source (LPS) emissions for India: Regional and sectoral analysis. *Atmospheric Environment,* **36**, pp. 213-224.

Gielen, D.J. and Y. Moriguchi, 2003: Technological potentials for CO_2 emission reduction in the global iron and steel industry. *International Journal of Energy Technology and Policy,* **1**(3), 229-249.

Greene, N., 2004: Growing energy: how biofuels can help end America's growing oil dependence, *NCEP Technical Appendix: Expanding Energy Supply*, in The National Commission on Energy Policy, Ending the Energy Stalemate: A Bipartisan Strategy to Meet America's Energy Challenges, Washington, DC.

Hamelinck, C.N. and A. Faaij, 2002: Future prospects for production of methanol and hydrogen from biomass, *Journal of Power Sources*, **111** (1): 1-22.

Hibino, G., Y. Matsuoka, and M. Kainuma, 2003: AIM/Common Database: A Tool for AIM Family Linkage. In: M. Kainuma, Y. Matsuoka, and T. Morita, (eds.), Climate Policy Assessment: Asia-Pacific Integrated Modelling. Springer-Verlag, Tokyo, Japan. pp. 233-244.

IEA, 2002: World Energy Outlook - 2002. International Energy Agency of the Organisation for Economic Co-operation and Development (OECD/IEA), Paris, France.

IEA GHG, 1999: The Reduction of Greenhouse Gas Emissions from the Cement Industry, PH3/7, May, 112 pp.

IEA GHG, 2000: Greenhouse Gas Emissions from Major Industrial Sources - IV, the Aluminium Industry, PH3/23, April, 80 pp.

IEA GHG, 2000a: Capture of CO_2 using water scrubbing, IEA Report Number PH3/26, July, 150 pp.

IEA GHG, 2000b: Greenhouse Gas Emissions from Major Industrial Sources - III, Iron and Steel Production, PH3/30, September, 130 pp.

IEA GHG, 2002a: Building the Cost Curves for CO_2 Storage, Part 1: Sources of CO_2, PH4/9, July, 48 pp.

IEA GHG, 2002b: Opportunities for Early Application of CO_2 Sequestration Technology, Ph4/10, September, 91 pp.

IPCC, 2000: Emissions Scenarios, a Special Report of IPCC Working Party III, Summary for Policy Makers, 20 pp.

IPCC, 2001: Climate Change 2001: Mitigation, Contribution of Working Group III to the Third Assessment Report of the Intergovernmental Panel on Climate Change, Cambridge University Press, Cambridge, UK. 752 pp, ISBN: 0521015022.

Kheshgi, H.S. and R.C. Prince, 2005: Sequestration of fermentation CO_2 from ethanol production. *Energy*, **30**, 1865-1871.

Larson, E.D., and T. Ren, 2003: Synthetic fuels production by indirect coal liquefaction, *Energy for Sustainable Development*, **VII** (4), 79-102.

Lloyd, A.C. 1999: The Power Plant in your Basement. *Scientific American,* **280**(7), 80-86.

Maddox, R.N. and D.J. Morgan, 1998: Gas Conditioning and Gas Treating, Volume 4: Gas treating and liquid sweetening, Campbell Petroleum Series, OK, USA, 498 pp.

Marland, G., A. Brenkert, and J. Oliver, 1999: CO_2 from fossil fuel burning: a comparison of ORNL and EDGAR estimates of national emissions. *Environmental Science & Policy*, **2**, pp. 265-273.

Marrison, C. and E. Larson, 1995: Cost vs scale for advanced plantation-based biomass energy systems in the USA and Brazil. Proceedings of the Second Biomass Conference of the America, NREL, Golden, Colorado, pp. 1272-1290.

Möllersten, K., J. Yan, and J.R. Moreira, 2003: Potential markets niches for biomass supply with CO_2 capture and storage - Opportunities for energy supply with negative CO_2 emissions, *Biomass and Bioenergy*, **25**, pp 273-285.

Morita, T., and H.-C. Lee, 1998: Appendix to Emissions Scenarios Database and Review of Scenarios. *Mitigation and Adaptation Strategies for Global Change*, 3(2-4), 121-131.

NETL-DOE, 2002: Worldwide gasification database which can be viewed at www.netl.doe.gov/coal/Gasification/index.html.

NRC (Committee on Alternatives and Strategies for Future Hydrogen Production and Use of the National Research Council), 2004: *The Hydrogen Economy - Opportunities, Costs, Barriers, and R&D Needs*, The National Academies Press, Washington, DC, www.nap.edu.

Ogden, J. and R. Williams, 1989: *Solar Hydrogen*, World Resources Institute, Washington, DC.

Ogden, J., R. Williams, and E. Larson, 2004: Societal lifecycle costs of cars with alternative fuels, *Energy Policy*, **32**, 7-27.

Owen and Gordon, N. Owen and R. Gordon, 2002: "CO$_2$ to Hydrogen" Roadmaps for Passenger Cars, a study for the Department for Transport and the Department of Trade and Industry carried out by Ricardo Consulting Engineers Ltd., West Sussex, UK, November.

Simbeck, D.R., 2003: CO$_2$ Capture and Storage, the Essential Bridge to the Hydrogen Economy, Elsevier Science Oxford, UK, July.

Simbeck, D.R., 2004: CO$_2$ Capture and Storage, the Essential Bridge to the Hydrogen Economy, *Energy, 29*: 1633-1641.

Simmonds, S., P. Horst, M.B. Wilkinson, C. Watt and C.A. Roberts, 2003: Proceedings of the 6th International Conference on Greenhouse Gas Control Technologies, J. Gale, Y. Kaya (eds), 1-4 October 2002, Kyoto, Japan, pp. 39-44.

Sperling, D. and J.S. Cannon (eds.), 2004: *The Hydrogen Energy Transition*, Elsevier, St. Louis.

Stevens, S.H. and J. Gale, 2000: Geologic CO$_2$ Sequestration, *Oil and Gas Journal*, May 15th, 40-44.

TFEST (Task Force on Energy Strategies and Technologies), 2003: Transforming coal for sustainability: a strategy for China, *Energy for Sustainable Development*, **VII** (4): 21-30.

Toth, F.L and H-H. Rogner, 2006: Carbon Dioxide Capture: An Assessment of Plausible Ranges, Accepted for publication *International Journal of Global Energy Issues*, **25**, forthcoming.

Van Bergen, F., J. Gale, K.J. Damen, and A.F.B. Wildenborg, 2004: Worldwide selection of early opportunities for CO$_2$-EOR and CO$_2$-ECBM, *Energy, 29* (9-10): 1611-1621.

Williams, R.H. (Convening Lead Author) *et al.*, 2000: Advanced energy supply technologies. In World Energy Assessment: Energy the Challenge of Sustainability, (a study sponsored jointly by the United Nations Development Programme, the United Nations Department of Social and Economic Affairs, and the World Energy Council), published by the Bureau for Development Policy, United Nations Development Programme, New York. Bureau for Development Policy, United Nations Development Program, New York, pp. 273-329.

Williams, R.H., 1998: Fuel decarbonisation for fuel cell applications and sequestration of the separated CO$_2$, in *Eco-Restructuring: Implications for Sustainable Development*, R.W. Ayres (ed.), United Nations University Press, Tokyo, pp. 180-222.

Williams, R.H., 2003: Decarbonised fossil energy carriers and their energy technological competitors, pp. 119-135, in Proceedings of the Workshop on Carbon Capture and Storage of the Intergovernmental Panel on Climate Change, Regina, Saskatchewan, Canada, published by ECN (Energy Research Center of The Netherlands), 18-21 November, 178 pp.

3

Capture of CO$_2$

Coordinating Lead Authors
Kelly (Kailai) Thambimuthu (Australia and Canada), Mohammad Soltanieh (Iran), Juan Carlos Abanades (Spain)

Lead Authors
Rodney Allam (United Kingdom), Olav Bolland (Norway), John Davison (United Kingdom), Paul Feron (The Netherlands), Fred Goede (South Africa), Alyce Herrera (Philippines), Masaki Iijima (Japan), Daniël Jansen (The Netherlands), Iosif Leites (Russian Federation), Philippe Mathieu (Belgium), Edward Rubin (United States), Dale Simbeck (United States), Krzysztof Warmuzinski (Poland), Michael Wilkinson (United Kingdom), Robert Williams (United States)

Contributing Authors
Manfred Jaschik (Poland), Anders Lyngfelt (Sweden), Roland Span (Germany), Marek Tanczyk (Poland)

Review Editors
Ziad Abu-Ghararah (Saudi Arabia), Tatsuaki Yashima (Japan)

Contents

EXECUTIVE SUMMARY

The purpose of CO_2 capture is to produce a concentrated stream that can be readily transported to a CO_2 storage site. CO_2 capture and storage is most applicable to large, centralized sources like power plants and large industries. Capture technologies also open the way for large-scale production of low-carbon or carbon-free electricity and fuels for transportation, as well as for small-scale or distributed applications. The energy required to operate CO_2 capture systems reduces the overall efficiency of power generation or other processes, leading to increased fuel requirements, solid wastes and environmental impacts relative to the same type of base plant without capture. However, as more efficient plants with capture become available and replace many of the older less efficient plants now in service, the net impacts will be compatible with clean air emission goals for fossil fuel use. Minimization of energy requirements for capture, together with improvements in the efficiency of energy conversion processes will continue to be high priorities for future technology development in order to minimize overall environmental impacts and cost.

At present, CO_2 is routinely separated at some large industrial plants such as natural gas processing and ammonia production facilities, although these plants remove CO_2 to meet process demands and not for storage. CO_2 capture also has been applied to several small power plants. However, there have been no applications at large-scale power plants of several hundred megawatts, the major source of current and projected CO_2 emissions. There are three main approaches to CO_2 capture, for industrial and power plant applications. *Post-combustion* systems separate CO_2 from the flue gases produced by combustion of a primary fuel (coal, natural gas, oil or biomass) in air. *Oxy-fuel combustion* uses oxygen instead of air for combustion, producing a flue gas that is mainly H_2O and CO_2 and which is readily captured. This is an option still under development. *Pre-combustion* systems process the primary fuel in a reactor to produce separate streams of CO_2 for storage and H_2 which is used as a fuel. Other industrial processes, including processes for the production of low-carbon or carbon-free fuels, employ one or more of these same basic capture methods. The monitoring, risk and legal aspects associated with CO_2 capture systems appear to present no new challenges, as they are all elements of long-standing health, safety and environmental control practice in industry.

For all of the aforementioned applications, we reviewed recent studies of the performance and cost of commercial or near-commercial technologies, as well as that of newer CO_2 capture concepts that are the subject of intense R&D efforts worldwide. For power plants, current commercial CO_2 capture systems can reduce CO_2 emissions by 80-90% kWh^{-1} (85-95% capture efficiency). Across all plant types the cost of electricity production (COE) increases by 12-36 US$ MWh^{-1} (US$ 0.012-0.036 kWh^{-1}) over a similar type of plant without capture, corresponding to a 40-85% increase for a supercritical pulverized coal (PC) plant, 35-70% for a natural gas combined cycle (NGCC) plant and 20-55% for an integrated gasification combined cycle (IGCC) plant using bituminous coal. Overall the COE for fossil fuel plants with capture, ranges from 43-86 US$ MWh^{-1}, with the cost per tonne of CO_2 ranging from 11-57 US$/$tCO_2$ captured or 13-74 US$/$tCO_2$ avoided (depending on plant type, size, fuel type and a host of other factors). These costs include CO_2 compression but not additional transport and storage costs. NGCC systems typically have a lower COE than new PC and IGCC plants (with or without capture) for gas prices below about 4 US$ GJ^{-1}. Most studies indicate that IGCC plants are slightly more costly without capture and slightly less costly with capture than similarly sized PC plants, but the differences in cost for plants with CO_2 capture can vary with coal type and other local factors. The lowest CO_2 capture costs (averaging about 12 US$/t CO_2 captured or 15 US$/$tCO_2$ avoided) were found for industrial processes such as hydrogen production plants that produce concentrated CO_2 streams as part of the current production process; such industrial processes may represent some of the earliest opportunities for CO_2 Capture and Storage (CCS). In all cases, CO_2 capture costs are highly dependent upon technical, economic and financial factors related to the design and operation of the production process or power system of interest, as well as the design and operation of the CO_2 capture technology employed. Thus, comparisons of alternative technologies, or the use of CCS cost estimates, require a specific context to be meaningful.

New or improved methods of CO_2 capture, combined with advanced power systems and industrial process designs, can significantly reduce CO_2 capture costs and associated energy requirements. While there is considerable uncertainty about the magnitude and timing of future cost reductions, this assessment suggests that improvements to commercial technologies can reduce CO_2 capture costs by at least 20-30% over approximately the next decade, while new technologies under development promise more substantial cost reductions. Realization of future cost reductions, however, will require deployment and adoption of commercial technologies in the marketplace as well as sustained R&D.

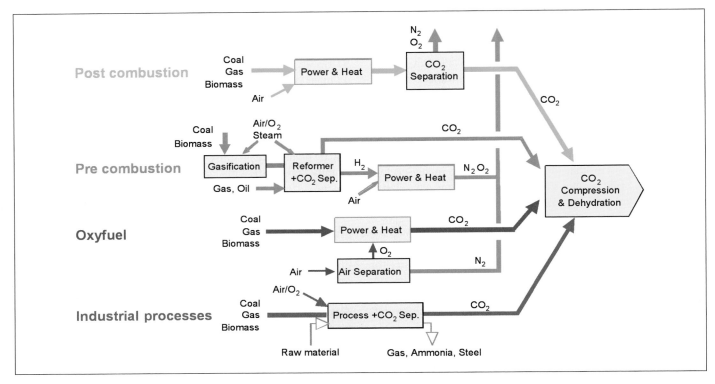

Figure 3.1 CO$_2$ capture systems (adapted from BP).

3.1   Introduction

3.1.1   *The basis for CO$_2$ capture*

The main application of CO$_2$ capture is likely to be at large point sources: fossil fuel power plants, fuel processing plants and other industrial plants, particularly for the manufacture of iron, steel, cement and bulk chemicals, as discussed in Chapter 2.

Capturing CO$_2$ directly from small and mobile sources in the transportation and residential & commercial building sectors is expected to be more difficult and expensive than from large point sources. Small-scale capture is therefore not further discussed in this chapter. An alternative way of avoiding emissions of CO$_2$ from these sources would be by use of energy carriers such as hydrogen or electricity produced in large fossil fuel-based plants with CO$_2$ capture or by using renewable energy sources. Production of hydrogen with CO$_2$ capture is included in this chapter.

The possibility of CO$_2$ capture from ambient air (Lackner, 2003) is not discussed in this chapter because the CO$_2$ concentration in ambient air is around 380 ppm, a factor of 100 or more lower than in flue gas. Capturing CO$_2$ from air by the growth of biomass and its use in industrial plants with CO$_2$ capture is more cost-effective based on foreseeable technologies, and is included in this chapter.

In an analysis of possible future scenarios for anthropogenic greenhouse-gas emissions it is implicit that technological innovations will be one of the key factors which determines our future path (Section 2.5.3). Therefore this chapter deals not only with application of existing technology for CO$_2$ capture, but describes many new processes under development which may result in lower CO$_2$ capture costs in future.

3.1.2   *CO$_2$ capture systems*

There are four basic systems for capturing CO$_2$ from use of fossil fuels and/or biomass:
*   Capture from industrial process streams (described in Section 3.2);
*   Post-combustion capture (described in Section 3.3);
*   Oxy-fuel combustion capture (described in Section 3.4);
*   Pre-combustion capture (described in Section 3.5).

These systems are shown in simplified form in Figure 3.1.

3.1.2.1   *Capture from industrial process streams*
CO$_2$ has been captured from industrial process streams for 80 years (Kohl and Nielsen, 1997), although most of the CO$_2$ that is captured is vented to the atmosphere because there is no incentive or requirement to store it. Current examples of CO$_2$ capture from process streams are purification of natural gas and production of hydrogen-containing synthesis gas for the manufacture of ammonia, alcohols and synthetic liquid fuels. Most of the techniques employed for CO$_2$ capture in the examples mentioned are also similar to those used in pre-combustion capture. Other industrial process streams which are a source of CO$_2$ that is not captured include cement and steel production, and fermentation processes for food and drink production. CO$_2$ could be captured from these streams using

techniques that are common to post-combustion capture, oxy-fuel combustion capture and pre-combustion capture (see below and Section 3.2).

3.1.2.2 Post-combustion capture

Capture of CO$_2$ from flue gases produced by combustion of fossil fuels and biomass in air is referred to as post-combustion capture. Instead of being discharged directly to the atmosphere, flue gas is passed through equipment which separates most of the CO$_2$. The CO$_2$ is fed to a storage reservoir and the remaining flue gas is discharged to the atmosphere. A chemical sorbent process as described in Section 3.1.3.1 would normally be used for CO$_2$ separation. Other techniques are also being considered but these are not at such an advanced stage of development.

Besides industrial applications, the main systems of reference for post-combustion capture are the current installed capacity of 2261 GW$_e$ of oil, coal and natural gas power plants (IEA WEO, 2004) and in particular, 155 GW$_e$ of supercritical pulverized coal fired plants (IEA CCC, 2005) and 339 GW$_e$ of natural gas combined cycle (NGCC) plants, both representing the types of high efficiency power plant technology where CO$_2$ capture can be best applied (see Sections 3.3 and 3.7).

3.1.2.3 Oxy-fuel combustion capture

In oxy-fuel combustion, nearly pure oxygen is used for combustion instead of air, resulting in a flue gas that is mainly CO$_2$ and H$_2$O. If fuel is burnt in pure oxygen, the flame temperature is excessively high, but CO$_2$ and/or H$_2$O-rich flue gas can be recycled to the combustor to moderate this. Oxygen is usually produced by low temperature (cryogenic) air separation and novel techniques to supply oxygen to the fuel, such as membranes and chemical looping cycles are being developed. The power plant systems of reference for oxy-fuel combustion capture systems are the same as those noted above for post-combustion capture systems.

3.1.2.4 Pre-combustion capture

Pre-combustion capture involves reacting a fuel with oxygen or air and/or steam to give mainly a 'synthesis gas (syngas)' or 'fuel gas' composed of carbon monoxide and hydrogen. The carbon monoxide is reacted with steam in a catalytic reactor, called a shift converter, to give CO$_2$ and more hydrogen. CO$_2$ is then separated, usually by a physical or chemical absorption process, resulting in a hydrogen-rich fuel which can be used in many applications, such as boilers, furnaces, gas turbines, engines and fuel cells. These systems are considered to be strategically important (see Section 3.5) but the power plant systems of reference today are 4 GW$_e$ of both oil and coal-based, integrated gasification combined cycles (IGCC) which are around 0.1% of total installed capacity worldwide (3719 GW$_e$; IEA WEO, 2004). Other reference systems for the application of pre-combustion capture include substantially more capacity than that identified above for IGCC in existing natural gas, oil and coal-based syngas/hydrogen production facilities and other types of industrial systems described in more detail in Sections 3.2 and 3.5.

3.1.3 Types of CO$_2$ capture technologies

CO$_2$ capture systems use many of the known technologies for gas separation which are integrated into the basic systems for CO$_2$ capture identified in the last section. A summary of these separation methods is given below while further details are available in standard textbooks.

3.1.3.1 Separation with sorbents/solvents

The separation is achieved by passing the CO$_2$-containing gas in intimate contact with a liquid absorbent or solid sorbent that is capable of capturing the CO$_2$. In the general scheme of Figure 3.2a, the sorbent loaded with the captured CO$_2$ is transported to a different vessel, where it releases the CO$_2$ (regeneration) after being heated, after a pressure decrease or after any other change in the conditions around the sorbent. The sorbent resulting after the regeneration step is sent back to capture more CO$_2$ in a cyclic process. In some variants of this scheme the sorbent is a solid and does not circulate between vessels because the sorption and regeneration are achieved by cyclic changes (in pressure or temperature) in the vessel where the sorbent is contained. A make-up flow of fresh sorbent is always required to compensate for the natural decay of activity and/or sorbent losses. In some situations, the sorbent may be a solid oxide which reacts in a vessel with fossil fuel or biomass producing heat and mainly CO$_2$ (see Section 3.4.6). The spent sorbent is then circulated to a second vessel where it is re-oxidized in air for reuse with some loss and make up of fresh sorbent.

The general scheme of Figure 3.2 governs many important CO$_2$ capture systems, including leading commercial options like chemical absorption and physical absorption and adsorption. Other emerging processes based on new liquid sorbents, or new solid regenerable sorbents are being developed with the aim of overcoming the limitations of the existing systems. One common problem of these CO$_2$ capture systems is that the flow of sorbent between the vessels of Figure 3.2a is large because it has to match the huge flow of CO$_2$ being processed in the power plant. Therefore, equipment sizes and the energy required for sorbent regeneration are large and tend to translate into an important efficiency penalty and added cost. Also, in systems using expensive sorbent materials there is always a danger of escalating cost related to the purchase of the sorbent and the disposal of sorbent residues. Good sorbent performance under high CO$_2$ loading in many repetitive cycles is obviously a necessary condition in these CO$_2$ capture systems.

3.1.3.2 Separation with membranes

Membranes (Figure 3.2b) are specially manufactured materials that allow the selective permeation of a gas through them. The selectivity of the membrane to different gases is intimately related to the nature of the material, but the flow of gas through the membrane is usually driven by the pressure difference across the membrane. Therefore, high-pressure streams are usually preferred for membrane separation. There are many different types of membrane materials (polymeric, metallic, ceramic) that may find application in CO$_2$ capture systems to

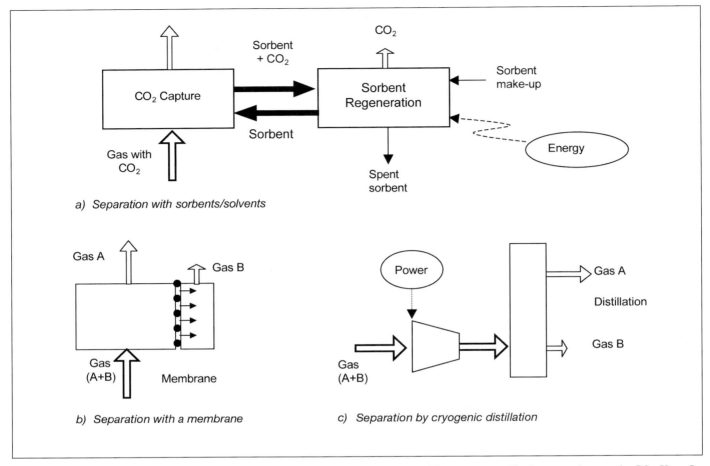

Figure 3.2 General schemes of the main separation processes relevant for CO_2 capture. The gas removed in the separation may be CO_2, H_2 or O_2. In Figures 3.2b and 3.2c one of the separated gas streams (A and B) is a concentrated stream of CO_2, H_2 or O_2 and the other is a gas stream with all the remaining gases in the original gas (A+B).

preferentially separate H_2 from a fuel gas stream, CO_2 from a range of process streams or O_2 from air with the separated O_2 subsequently aiding the production of a highly concentrated CO_2 stream. Although membrane separation finds many current commercial applications in industry (some of a large scale, like CO_2 separation from natural gas) they have not yet been applied for the large scale and demanding conditions in terms of reliability and low-cost required for CO_2 capture systems. A large worldwide R&D effort is in progress aimed at the manufacture of more suitable membrane materials for CO_2 capture in large-scale applications.

3.1.3.3 *Distillation of a liquefied gas stream and refrigerated separation*

A gas can be made liquid by a series of compression, cooling and expansion steps. Once in liquid form, the components of the gas can be separated in a distillation column. In the case of air, this operation is currently carried out commercially on a large scale. Oxygen can be separated from air following the scheme of Figure 3.2c and be used in a range of CO_2 capture systems (oxy-fuel combustion and pre-combustion capture). As in the previous paragraphs, the key issue for these systems is

the large flow of oxygen required. Refrigerated separation can also be used to separate CO_2 from other gases. It can be used to separate impurities from relatively high purity CO_2 streams, for example, from oxy-fuel combustion and for CO_2 removal from natural gas or synthesis gas that has undergone a shift conversion of CO to CO_2.

3.1.4 *Application of CO_2 capture*

The CO_2 capture systems shown in Figure 3.1 can be cross-referenced with the different separation technologies of Figure 3.2, resulting in a capture toolbox. Table 3.1 gives an overview of both current and emerging technologies in this toolbox. In the next sections of this chapter a more detailed description of all these technological options will be given, with more emphasis on the most developed technologies for which the CO_2 capture cost can be estimated most reliably. These leading commercial options are shown in bold in Table 3.1. An overview of the diverse range of emerging options being investigated worldwide for CO_2 capture applications will also be provided. All of these options are aimed at more efficient and lower cost CO_2-capture systems (compared with the leading options). It is important

Table 3.1 Capture toolbox.

Separation task	Process streams[a]		Post-combustion capture		Oxy-fuel combustion capture		Pre-combustion capture	
	CO₂/CH₄		CO₂/N₂		O₂/N₂		CO₂/H₂	
Capture Technologies	Current	Emerging	Current	Emerging	Current	Emerging	Current	Emerging
Solvents (Absorption)	**Physical solvents** Chemical solvents	Improved solvents Novel contacting equipment Improved design of processes	**Chemical solvents**	Improved solvents Novel contacting equipment Improved design of processes	n. a.	Biomimetic solvents, e.g. hemoglobine-derivatives	**Physical solvent** **Chemical solvents**	Improved chemical solvents Novel contacting equipment Improved design of processes
Membranes	Polymeric	Ceramic Facilitated transport Carbon Contactors	Polymeric	Ceramic Facilitated transport Carbon Contactors	Polymeric	Ion transport membranes Facilitated transport	Polymeric	Ceramic Palladium Reactors Contactors
Solid sorbents	Zeolites Activated carbon		Zeolites Activated carbon	Carbonates Carbon based sorbents	Zeolites Activated carbon	Adsorbents for O₂/N₂ separation, Perovskites Oxygen chemical looping	Zeolites Activated carbon Alumina	Carbonates Hydrotalcites Silicates
Cryogenic	Ryan-Holmes process		Liquefaction	Hybrid processes	**Distillation**	Improved distillation	Liquefaction	Hybrid processes

[a] Notes: Processes shown in bold are commercial processes that are currently preferred in most circumstances. Some process streams involve CO₂/H₂ or CO₂/N₂ separations but this is covered under pre-combustion capture and post-combustion capture. The key separation processes are outlined in Section 3.1.3 and described in Sections 3.2-3.5.

to understand that this wide variety of approaches for CO_2 capture will tend to settle with time as the expected benefits (and potential weaknesses) in the technological portfolio of Table 3.1 becomes obvious with new results from current and future research and demonstration projects. Only a few of these options will prove truly cost-effective in the medium to long term.

CO_2 capture may be installed in new energy utilization plants or it may be retrofitted to existing plants. In principle, if CO_2 capture is to be introduced rapidly, it may have to be retrofitted to some existing plants or these plants would have to be retired prematurely and replaced by new plants with capture. Disadvantages of retrofits are:

- There may be site constraints such as availability of land for the capture equipment;
- A long remaining plant life may be needed to justify the large expense of installing capture equipment;
- Old plants tend to have low energy efficiencies. Including CO_2 capture will have a proportionally greater impact on the net output than in high efficiency plants.

To minimize the site constraints, new energy utilization plants could be built 'capture-ready', that is with the process design initially factoring in the changes necessary to add capture and with sufficient space and facilities made available for simple installation of CO_2 capture at a later date. For some types of capture retrofit, for example pre-combustion capture and oxy-fuel combustion, much of the retrofit equipment could be built on a separate site if necessary.

The other barriers could be largely overcome by upgrading or substantially rebuilding the existing plant when capture is retrofitted. For example, old inefficient boilers and steam turbines could be replaced by modern, high-efficiency supercritical boilers and turbines or IGCC plants. As the efficiencies of power generation technologies are increasing, the efficiency of the retrofitted plant with CO_2 capture could be as high as that of the original plant without capture.

3.2 Industrial process capture systems

3.2.1 Introduction

There are several industrial applications involving process streams where the opportunity exists to capture CO_2 in large quantities and at costs lower than from the systems described in the rest of this chapter. Capture from these sources will not be the complete answer to the needs of climate change, since the volumes of combustion-generated CO_2 are much higher, but it may well be the place where the first capture and storage occurs.

3.2.2 Natural gas sweetening

Natural gas contains different concentration levels of CO_2, depending on its source, which must be removed. Often pipeline specifications require that the CO_2 concentration be lowered to around 2% by volume (although this amount varies in different places) to prevent pipeline corrosion, to avoid excess energy for transport and to increase the heating value of the gas. Whilst accurate figures are published for annual worldwide natural gas production (BP, 2004), none seem to be published on how much of that gas may contain CO_2. Nevertheless, a reasonable assumption is that about half of raw natural gas production contains CO_2 at concentrations averaging at least 4% by volume. These figures can be used to illustrate the scale of this CO_2 capture and storage opportunity. If half of the worldwide production of 2618.5 billion m³ of natural gas in 2003 is reduced in CO_2 content from 4 to 2% mol, the resultant amount of CO_2 removed would be at least 50 Mt CO_2 yr⁻¹. It is interesting to note that there are two operating natural gas plants capturing and storing CO_2, BP's In Salah plant in Algeria and a Statoil plant at Sleipner in the North Sea. Both capture about 1 MtCO_2 yr⁻¹ (see Chapter 5). About 6.5 million tCO_2 yr⁻¹ from natural gas sweetening is also currently being used in enhanced oil recovery (EOR) in the United States (Beecy and Kuuskraa, 2005) where in these commercial EOR projects, a large fraction of the injected CO_2 is also retained underground (see Chapter 5).

Depending on the level of CO_2 in natural gas, different processes for natural gas sweetening (i.e., H_2S and CO_2 removal) are available (Kohl and Nielsen, 1997 and Maddox and Morgan, 1998):

- Chemical solvents
- Physical solvents
- Membranes

Natural gas sweetening using various alkanolamines (MEA, DEA, MDEA, etc.; See Table 3.2), or a mixture of them, is the most commonly used method. The process flow diagram for CO_2 recovery from natural gas is similar to what is presented for flue gas treatment (see Figure 3.4, Section 3.3.2.1), except that in natural gas processing, absorption occurs at high pressure, with subsequent expansion before the stripper column, where CO_2 will be flashed and separated. When the CO_2 concentration in natural gas is high, membrane systems may be more economical. Industrial application of membranes for recovery of CO_2 from

natural gas started in the early 1980s for small units, with many design parameters unknown (Noble and Stern, 1995). It is now a well-established and competitive technology with advantages compared to other technologies, including amine treatment in certain cases (Tabe-Mohammadi, 1999). These advantages include lower capital cost, ease of skid-mounted installation, lower energy consumption, ability to be applied in remote areas, especially offshore and flexibility.

3.2.3 Steel production

The iron and steel industry is the largest energy-consuming manufacturing sector in the world, accounting for 10-15% of total industrial energy consumption (IEA GHG, 2000a). Associated CO_2 emissions were estimated at 1442 MtCO_2 in 1995. Two types of iron- and steel-making technologies are in operation today. The integrated steel plant has a typical capacity of 3-5 Mtonnes yr⁻¹ of steel and uses coal as its basic fuel with, in many cases, additional natural gas and oil. The mini-mill uses electric arc furnaces to melt scrap with a typical output of 1 Mtonnes yr⁻¹ of steel and an electrical consumption of 300-350 kWh tonne⁻¹ steel. Increasingly mini-mills blend direct-reduced iron (DRI) with scrap to increase steel quality. The production of direct-reduced iron involves reaction of high oxygen content iron ore with H_2 and CO to form reduced iron plus H_2O and CO_2. As a result, many of the direct reduction iron processes could capture a pure CO_2 stream.

An important and growing trend is the use of new iron-making processes, which can use lower grade coal than the coking coals required for blast furnace operation. A good example is the COREX process (von Bogdandy *et. al*, 1989), which produces a large additional quantity of N_2-free fuel gas which can be used in a secondary operation to convert iron ore to iron. Complete CO_2 capture from this process should be possible with this arrangement since the CO_2 and H_2O present in the COREX top gas must be removed to allow the CO plus H_2 to be heated and used to reduce iron oxide to iron in the secondary shaft kiln. This process will produce a combination of molten iron and iron with high recovery of CO_2 derived from the coal feed to the COREX process.

Table 3.2 Common solvents used for the removal of CO_2 from natural gas or shifted syngas in pre-combustion capture processes.

Solvent name	Type	Chemical name	Vendors
Rectisol	Physical	Methanol	Lurgi and Linde, Germany Lotepro Corporation, USA
Purisol	Physical	N-methyl-2-pyrolidone (NMP)	Lurgi, Germany
Selexol	Physical	Dimethyl ethers of polyethylene glycol (DMPEG)	Union Carbide, USA
Benfield	Chemical	Potassium carbonate	UOP
MEA	Chemical	Monoethanolamine	Various
MDEA	Chemical	Methyldiethylamine	BASF and others
Sulfinol	Chemical	Tetrahydrothiophene 1,1-dioxide (Sulfolane), an alkaloamine and water	Shell

Early opportunities exist for the capture of CO_2 emissions from the iron and steel industry, such as:

- CO_2 recovery from blast furnace gas and recycle of CO-rich top gas to the furnace. A minimum quantity of coke is still required and the blast furnace is fed with a mixture of pure O_2 and recycled top gas. The furnace is, in effect, converted from air firing to oxy-fuel firing with CO_2 capture (see Section 3.4). This would recover 70% of the CO_2 currently emitted from an integrated steel plant (Dongke et al., 1988). It would be feasible to retrofit existing blast furnaces with this process.

- Direct reduction of iron ore, using hydrogen derived from a fossil fuel in a pre-combustion capture step (see Section 3.5) (Duarte and Reich, 1998). Instead of the fuel being burnt in the furnace and releasing its CO_2 to atmosphere, the fuel would be converted to hydrogen and the CO_2 would be captured during that process. The hydrogen would then be used as a reduction agent for the iron ore. Capture rates should be 90-95% according to the design of the pre-combustion capture technique (see Section 3.5).

Other novel process routes for steel making to which CO_2 capture can be applied are currently in the research and development phase (Gielen, 2003; IEA, 2004)

3.2.4 Cement production

Emissions of CO_2 from the cement industry account for 6% of the total emissions of CO_2 from stationary sources (see Chapter 2). Cement production requires large quantities of fuel to drive the high temperature, energy-intensive reactions associated with the calcination of the limestone – that is calcium carbonate being converted to calcium oxide with the evolution of CO_2.

At present, CO_2 is not captured from cement plants, but possibilities do exist. The concentration of CO_2 in the flue gases is between 15-30% by volume, which is higher than in flue gases from power and heat production (3-15% by volume). So, in principle, the post-combustion technologies for CO_2 capture described in Section 3.3 could be applied to cement production plants, but would require the additional generation of steam in a cement plant to regenerate the solvent used to capture CO_2. Oxy-fuel combustion capture systems may also become a promising technique to recover CO_2 (IEA GHG, 1999). Another emerging option would be the use of calcium sorbents for CO_2 capture (see Sections 3.3.3.4 and 3.5.3.5) as calcium carbonate (limestone) is a raw material already used in cement plants. All of these capture techniques could be applied to retrofit, or new plant applications.

3.2.5 Ammonia production

CO_2 is a byproduct of ammonia (NH_3) production (Leites *et al.*, 2003); Two main groups of processes are used:
- Steam reforming of light hydrocarbons (natural gas, liquefied petroleum gas, naphtha)
- Partial oxidation or gasification of heavy hydrocarbons (coal, heavy fuel oil, vacuum residue).

Around 85% of ammonia is made by processes in the steam methane reforming group and so a description of the process is useful. Although the processes vary in detail, they all comprise the following steps:

1. Purification of the feed;
2. Primary steam methane reforming (see Section 3.5.2.1);
3. Secondary reforming, with the addition of air, commonly called auto thermal reforming (see Section 3.5.2.3);
4. Shift conversion of CO and H_2O to CO_2 and H_2;
5. Removal of CO_2;
6. Methanation (a process that reacts and removes trace CO and CO_2);
7. Ammonia synthesis.

The removal of CO_2 as a pure stream is of interest to this report. A typical modern plant will use the amine solvent process to treat 200,000 Nm^3 h^{-1} of gas from the reformer, to produce 72 tonnes h^{-1} of concentrated CO_2 (Apple, 1997). The amount of CO_2 produced in modern plants from natural gas is about 1.27 tCO_2/tNH_3. Hence, with a world ammonia production of about 100 Mtonnes yr^{-1}, about 127 $MtCO_2$ yr^{-1} is produced. However, it should be noted that this is not all available for storage, as ammonia plants are frequently combined with urea plants, which are capable of utilizing 70-90% of the CO_2. About 0.7 $MtCO_2$ yr^{-1} captured from ammonia plants is currently used for enhanced oil recovery in the United States (Beecy and Kuuskraa, 2005) with a large fraction of the injected CO_2 being retained underground (see Chapter 5) in these commercial EOR projects.

3.2.6 Status and outlook

We have reviewed processes – current and potential - that may be used to separate CO_2 in the course of producing another product. One of these processes, natural gas sweetening, is already being used in two industrial plants to capture and store about 2 $MtCO_2$ yr^{-1} for the purpose of climate change mitigation. In the case of ammonia production, pure CO_2 is already being separated. Over 7 $MtCO_2$ yr^{-1} captured from both natural gas sweetening and ammonia plants is currently being used in enhanced oil recovery with some storage (see also Chapter 5) of the injected CO_2 in these commercial EOR projects. Several potential processes for CO_2 capture in steel and cement production exist, but none have yet been applied. Although the total amount of CO_2 that may be captured from these industrial processes is insignificant in terms of the scale of the climate change challenge, significance may arise in that their use could serve as early examples of solutions that can be applied on larger scale elsewhere.

3.3 Post-combustion capture systems

3.3.1 Introduction

Current anthropogenic CO_2 emissions from stationary sources come mostly from combustion systems such as power plants, cement kilns, furnaces in industries and iron and steel production plants (see Chapter 2). In these large-scale processes, the direct firing of fuel with air in a combustion chamber has been (for centuries, as it is today) the most economic technology to extract and use the energy contained in the fuel. Therefore, the strategic importance of post-combustion capture systems becomes evident when confronted with the reality of today's sources of CO_2 emissions. Chapter 2 shows that any attempt to mitigate CO_2 emissions from stationary sources on a relevant scale using CO_2 capture and storage, will have to address CO_2 capture from combustion systems. All the CO_2 capture systems described in this section are aimed at the separation of CO_2 from the flue gases generated in a large-scale combustion process fired with fossil fuels. Similar capture systems can also be applied to biomass fired combustion processes that tend to be used on a much smaller scale compared to those for fossil fuels.

Flue gases or stack gases found in combustion systems are usually at atmospheric pressure. Because of the low pressure, the large presence of nitrogen from air and the large scale of the units, huge flows of gases are generated, the largest example of which may be the stack emissions coming from a natural gas combined cycle power plant having a maximum capacity of around 5 million normal $m^3 h^{-1}$. CO_2 contents of flue gases vary depending on the type of fuel used (between 3% for a natural gas combined cycle to less than 15% by volume for a coal-fired combustion plant See Table 2.1). In principle post-combustion capture systems can be applied to flue gases produced from the combustion of any type of fuel. However, the impurities in the fuel are very important for the design and costing of the complete plant (Rao and Rubin, 2002). Flue gases coming from coal combustion will contain not only CO_2, N_2, O_2 and H_2O, but also air pollutants such as SO_x, NO_x, particulates, HCl, HF, mercury, other metals and other trace organic and inorganic contaminants. Figure 3.3 shows a general schematic of a coal-fired power plant in which additional unit operations are deployed to remove the air pollutants prior to CO_2 capture

in an absorption-based process. Although capture of CO_2 in these flue gases is in principle more problematic and energy intensive than from other gas streams, commercial experience is available at a sufficiently large scale (see Section 3.3.2) to provide the basis for cost estimates for post-combustion CO_2 capture systems (see Section 3.7). Also, a large R&D effort is being undertaken worldwide to develop more efficient and lower cost post-combustion systems (see Section 3.3.3), following all possible approaches for the CO_2 separation step (using sorbents, membranes or cryogenics; see Section 3.1.3).

3.3.2 Existing technologies

There are several commercially available process technologies which can in principle be used for CO_2 capture from flue gases. However, comparative assessment studies (Hendriks, 1994; Riemer and Ormerod, 1995; IEA GHG, 2000b) have shown that absorption processes based on chemical solvents are currently the preferred option for post-combustion CO_2 capture. At this point in time, they offer high capture efficiency and selectivity, and the lowest energy use and costs when compared with other existing post-combustion capture processes. Absorption processes have reached the commercial stage of operation for post-combustion CO_2 capture systems, albeit not on the scale required for power plant flue gases. Therefore, the following paragraphs are devoted to a review of existing knowledge of the technology and the key technical and environmental issues relevant to the application of this currently leading commercial option for CO_2 capture. The fundamentals of the CO_2 separation step using commercial chemical absorption processes are discussed first. The requirements of flue gas pretreatment (removal of pollutants other than CO_2) and the energy requirements for regeneration of the chemical solvent follow.

3.3.2.1 Absorption processes

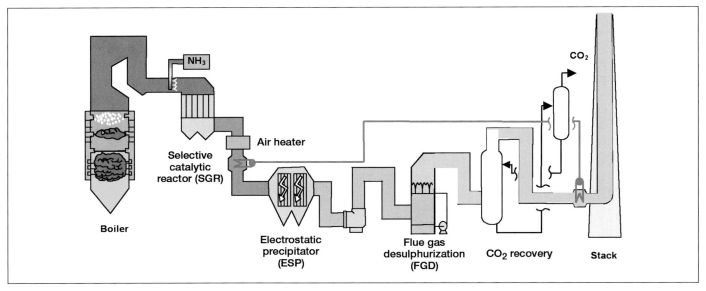

Figure 3.3 Schematic of a pulverized coal-fired power plant with an amine-based CO_2 capture system and other emission controls.

Absorption processes in post-combustion capture make use of the reversible nature of the chemical reaction of an aqueous alkaline solvent, usually an amine, with an acid or sour gas. The process flow diagram of a commercial absorption system is presented in Figure 3.4. After cooling the flue gas, it is brought into contact with the solvent in the absorber. A blower is required to overcome the pressure drop through the absorber. At absorber temperatures typically between 40 and 60°C, CO_2 is bound by the chemical solvent in the absorber. The flue gas then undergoes a water wash section to balance water in the system and to remove any solvent droplets or solvent vapour carried over, and then it leaves the absorber. It is possible to reduce CO_2 concentration in the exit gas down to very low values, as a result of the chemical reaction in the solvent, but lower exit concentrations tend to increase the height of the absorption vessel. The 'rich' solvent, which contains the chemically bound CO_2 is then pumped to the top of a stripper (or regeneration vessel), via a heat exchanger. The regeneration of the chemical solvent is carried out in the stripper at elevated temperatures (100°C–140°C) and pressures not very much higher than atmospheric pressure. Heat is supplied to the reboiler to maintain the regeneration conditions. This leads to a thermal energy penalty as a result of heating up the solvent, providing the required desorption heat for removing the chemically bound CO_2 and for steam production which acts as a stripping gas. Steam is recovered in the condenser and fed back to the stripper, whereas the CO_2 product gas leaves the stripper. The 'lean' solvent, containing far less CO_2 is then pumped back to the absorber via the lean-rich heat exchanger and a cooler to bring it down to the absorber temperature level.

Figure 3.4 also shows some additional equipment needed to maintain the solution quality as a result of the formation of degradation products, corrosion products and the presence of particles. This is generally done using filters, carbon beds and a thermally operated reclaimer. Control of degradation and corrosion has in fact been an important aspect in the development of absorption processes over the past few decades.

The key parameters determining the technical and economic operation of a CO_2 absorption system are:

- *Flue gas flow rate* - The flue gas flow rate will determine the size of the absorber and the absorber represents a sizeable contribution to the overall cost.
- *CO₂ content in flue gas* - Since flue gas is usually at atmospheric pressure, the partial pressure of CO_2 will be as low as 3-15 kPa. Under these low CO_2 partial pressure conditions, aqueous amines (chemical solvents) are the most suitable absorption solvents (Kohl and Nielsen, 1997).
- *CO₂ removal* - In practice, typical CO_2 recoveries are between 80% and 95%. The exact recovery choice is an economic trade-off, a higher recovery will lead to a taller absorption column, higher energy penalties and hence increased costs.
- *Solvent flow rate* - The solvent flow rate will determine the size of most equipment apart from the absorber. For a given solvent, the flow rate will be fixed by the previous parameters and also the chosen CO_2 concentrations within the lean and the rich solutions.
- *Energy requirement* - The energy consumption of the process is the sum of the thermal energy needed to regenerate the solvents and the electrical energy required to operate liquid pumps and the flue gas blower or fan. Energy is also required to compress the CO_2 recovered to the final pressure required for transport and storage.

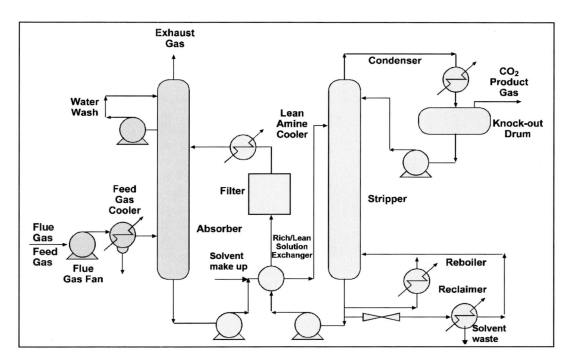

Figure 3.4 Process flow diagram for CO_2 recovery from flue gas by chemical absorption.

• Cooling requirement - Cooling is needed to bring the flue gas and solvent temperatures down to temperature levels required for efficient absorption of CO_2. Also, the product from the stripper will require cooling to recover steam from the stripping process.

The purity and pressure of CO_2 typically recovered from an amine-based chemical absorption process are as follows (Sander and Mariz, 1992):

• CO_2 purity: 99.9% by volume or more (water saturated conditions)
• CO_2 pressure: 50 kPa (gauge)

A further CO_2 purification step makes it possible to bring the CO_2-quality up to food-grade standard. This is required for use in beverages and packaging.

Since combustion flue gases are generally at atmospheric pressure and the CO_2 is diluted, the CO_2 partial pressure is very low. Also, flue gas contains oxygen and other impurities; therefore an important characteristic of an absorption process is in the proper choice of solvent for the given process duty. High CO_2 loading and low heat of desorption energy are essential for atmospheric flue gas CO_2 recovery. The solvents must also have low byproduct formation and low decomposition rates, to maintain solvent performance and to limit the amount of waste materials produced. The important effect of other contaminants on the solvent is discussed in Section 3.3.2.2.

The following three absorption processes are commercially available for CO_2 capture in post-combustion systems:

• The Kerr-McGee/ABB Lummus Crest Process (Barchas and Davis, 1992) - This process recovers CO_2 from coke and coal-fired boilers, delivering CO_2 for soda ash and liquid CO_2 preparations. It uses a 15-20% by weight aqueous MEA (Mono-Ethanolamine) solution. The largest capacity experienced for this process is 800 tCO_2 d^{-1} utilizing two parallel trains (Arnold *et al.*, 1982).

• The Fluor Daniel ® ECONAMINE™ Process (Sander and Mariz, 1992, Chapel *et al.*, 1999) - This process was acquired by Fluor Daniel Inc. from Dow Chemical Company in 1989. It is a MEA-based process (30% by weight aqueous solution) with an inhibitor to resist carbon steel corrosion and is specifically tailored for oxygen-containing gas streams. It has been used in many plants worldwide recovering up to 320 tCO_2 d^{-1} in a single train for use in beverage and urea production.

• The Kansai Electric Power Co., Mitsubishi Heavy Industries, Ltd., KEPCO/MHI Process (Mimura *et al.*, 1999 and 2003) - The process is based upon sterically-hindered amines and already three solvents (KS-1, KS-2 and KS-3) have been developed. KS-1 was commercialized in a urea production application. In this process, low amine losses and low solvent degradation have been noted without the use of inhibitors or additives. As shown in Figure 3.5, the first commercial plant at 200 tCO_2 d^{-1} recovery from a flue gas stream has been operating in Malaysia since 1999 for urea production (equivalent to the emissions from a 10 MWt coal-fired power plant)

The performance of the chemical solvent in the operation is maintained by replacement, filtering and reclaiming, which leads to a consumables requirement. Typical values for the solvent consumption are between 0.2 and 1.6 kg/tCO2. In addition, chemicals are needed to reclaim the amine from the heat stable salt (typically 0.03–0.13 kg NaOH/tCO2) and to remove decomposition products (typically 0.03-0.06 kg activated carbon/tCO2). The ranges are primarily dependent on the absorption process, with KS-1 being at the low end of the range and ECONAMINE ™ at the high end.

3.3.2.2. Flue gas pretreatment

Flue gases from a combustion power plant are usually above 100°C, which means that they need to be cooled down to the temperature levels required for the absorption process. This can be done in a cooler with direct water contact, which also acts as a flue gas wash with additional removal of fine particulates.

In addition to the above, flue gas from coal combustion will contain other acid gas components such as NO_x and SO_x. Flue gases from natural gas combustion will normally only contain NO_x. These acidic gas components will, similar to CO_2, have a chemical interaction with the alkaline solvent. This is not desirable as the irreversible nature of this interaction leads to the formation of heat stable salts and hence a loss in absorption capacity of the solvent and the risk of formation of solids in the solution. It also results in an extra consumption of chemicals to regenerate the solvent and the production of a waste stream such as sodium sulphate or sodium nitrate. Therefore, the pre-removal of NO_x and SO_x to very low values before CO_2

Figure 3.5 CO_2 capture plant in Malaysia using a 200 tonne d^{-1} KEPCO/MHI chemical solvent process (Courtesy of Mitsubishi).

recovery becomes essential. For NO$_x$ it is the NO$_2$ which leads to the formation of heat stable salts. Fortunately, the level of NO$_2$ is mostly less than 10% of the overall NO$_x$ content in a flue gas (Chapel *et al.*, 1999).

The allowable SO$_x$ content in the flue gas is primarily determined by the cost of the solvent - as this is consumed by reaction with SO$_x$. SO$_2$ concentrations in the flue gas are typically around 300-5000 ppm. Commercially available SO$_2$-removal plants will remove up to 98-99%. Amines are relatively cheap chemicals, but even cheap solvents like MEA (with a price around 1.25 US\$ kg^{-1} (Rao and Rubin, 2002) may require SO$_x$ concentrations of around 10 ppm, to keep solvent consumption (around 1.6 kg of MEA/tCO$_2$ separated) and make up costs at reasonable values, which often means that additional flue gas desulphurization is needed. The optimal SO$_2$ content, before the CO$_2$ absorption process is a cost trade-off between CO$_2$-solvent consumption and SO$_2$-removal costs. For the Kerr-Mcgee/ABB Lummus Crest Technology, SO$_2$-removal is typically not justified for SO$_2$ levels below 50 ppm (Barchas and Davis, 1992). For the Fluor Daniel Econamine FG process a maximum of 10 ppm SO$_2$ content is generally set as the feed gas specification (Sander and Mariz, 1992). This can be met by using alkaline salt solutions in a spray scrubber (Chapel *et al.*, 1999). A SO$_2$ scrubber might also double as a direct contact cooler to cool down the flue gas.

Careful attention must also be paid to fly ash and soot present in the flue gas, as they might plug the absorber if contaminants levels are too high. Often the requirements of other flue gas treatment are such that precautions have already been taken. In the case of CO$_2$ recovery from a coal-fired boiler flue gas, the plant typically has to be equipped with a DeNO$_x$ unit, an electrostatic precipitator or a bag house filter and a DeSO$_x$ or flue gas desulphurization unit as part of the environmental protection of the power plant facilities. In some cases, these environmental protection facilities are not enough to carry out deep SO$_x$ removal up to the 1-2 ppm level sometimes needed to minimize solvent consumption and its reclamation from sticking of solvent wastes on reclaimer tube surfaces.

3.3.2.3   *Power generation efficiency penalty in CO$_2$ capture*

A key feature of post-combustion CO$_2$ capture processes based on absorption is the high energy requirement and the resulting efficiency penalty on power cycles. This is primarily due to the heat necessary to regenerate the solvent, steam use for stripping and to a lesser extent the electricity required for liquid pumping, the flue gas fan and finally compression of the CO$_2$ product. Later in this chapter, Sections 3.6 and 3.7 present summaries of CO$_2$ capture energy requirements for a variety of power systems and discuss the environmental and economic implications of these energy demands.

In principle, the thermal energy for the regeneration process can be supplied by an auxiliary boiler in a retrofit situation. Most studies, however, focus on an overall process in which the absorption process is integrated into the power plant. The heat requirement is at such levels that low-pressure steam, for example condensing at 0.3 MPa(g), can be used in the reboiler. The steam required for the regeneration process is then extracted from the steam cycle in the power plant. For a coal-fired power station, low-pressure steam will be extracted prior to the last expansion stage of the steam turbine. For a natural gas fired combined cycle, low-pressure steam will be extracted from the last stage in the heat recovery steam generator. Some of this heat can be recovered by preheating the boiler feed water (Hendriks, 1994). Values for the heat requirement for the leading absorption technologies are between 2.7 and 3.3 GJ/tCO$_2$, depending on the solvent process. Typical values for the electricity requirement are between 0.06 and 0.11 GJ/tCO$_2$ for post-combustion capture in coal- fired power plants and 0.21 and 0.33 GJ/tCO$_2$ for post-combustion capture in natural gas fired combined cycles. Compression of the CO$_2$ to 110 bar will require around 0.4 GJ/tCO$_2$ (IEA GHG, 2004).

Integration of the absorption process with an existing power plant will require modifications of the low-pressure part of the steam cycle, as a sizeable fraction of the steam will be extracted and hence will not be available to produce power (Nsakala *et al.*, 2001, Mimura *et al.*,1995, Mimura *et al.*, 1997). To limit the required modifications, small back-pressure steam turbines using medium pressure steam to drive the flue gas fan and boiler feed water pumps can be used. The steam is then condensed in the reboiler (Mimura *et al.*, 1999). Furthermore, in power plants based on steam cycles more than 50% thermal energy in the steam cycle is disposed off in the steam condenser. If the steam cycle system and CO$_2$ recovery can be integrated, part of the waste heat disposed by the steam condenser can be utilized for regeneration of the chemical solvent.

The reduction of the energy penalty is, nevertheless, closely linked to the chosen solvent system. The IEA Greenhouse Programme (IEA GHG) has carried out performance assessments of power plants with post-combustion capture of CO$_2$, taking into consideration the most recent improvements in post-combustion CO$_2$ capture processes identified by technology licensors (IEA GHG, 2004). In this study, Mitsui Babcock Energy Ltd. and Alstom provided information on the use of a high efficiency, ultra-supercritical steam cycle (29 MPa, 600°C, 620°C reheat) boiler and steam turbine for a coal-fired power plant, while for the NGCC case, a combined cycle using a GE 9FA gas turbine was adopted. Fluor provided information on the Fluor Econamine + process based on MEA, and MHI provided information on KEPCO/MHI process based on the KS-1 solvent for CO$_2$ capture. CO$_2$ leaving these systems were compressed to a pressure of 11 MPa. The overall net power plant efficiencies with and without CO$_2$ capture are shown in Figure 3.6, while Figure 3.7 shows the efficiency penalty for CO$_2$ capture. Overall, results from this study show that the efficiency penalty for post-combustion capture in coal and gas fired plant is lower for KEPCO/MHI's CO$_2$ absorption process. For the purpose of comparison, the performance of power plants with pre-combustion and oxy-fuel capture, based on the same standard set of plant design criteria are also shown in Figures 3.6 and 3.7.

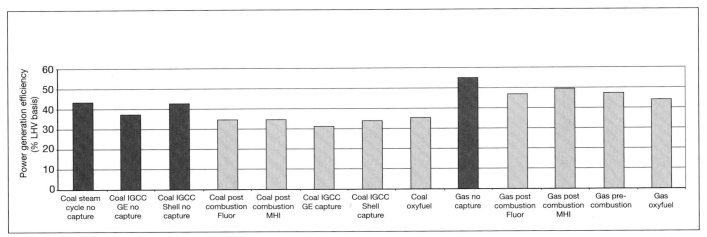

Figure 3.6 Thermal efficiencies of power plants with and without CO$_2$ capture, % LHV-basis (Source data: Davison 2005, IEA GHG 2004, IEA GHG 2003; IEA GHG, 2000b; Dillon *et al.*, 2005).

a. The efficiencies are based on a standard set of plant design criteria (IEA GHG, 2004).
b. The coal steam cycle plants, including the post-combustion capture and oxy-fuel plants, are based on ultra-supercritical steam (29MPa, 600C superheat, 620C reheat). The IGCC and natural gas pre- and post-combustion capture plants are based on GE 9FA gas turbine combined cycles. The natural gas oxy-fuel plant is based on a CO$_2$ recycle gas turbine, as shown in Figure 3.10, with different operating pressures and temperatures but similar mechanical design criteria to that of the 9FA.
c. Data are presented for two types of post-combustion capture solvent: MEA (Fluor plant designs) and KS-1 (MHI plant designs). The solvent desorption heat consumptions are 3.2 and 2.7 MJ/kgCO$_2$ captured respectively for the coal plants and 3.7 and 2.7 MJ kg^{-1} for the natural gas plants.
d. Data are presented for IGCC plants based on two types of gasifier: the Shell dry feed/heat recovery boiler type and the GE (formerly Texaco) slurry feed water quench type.
e. The natural gas pre-combustion capture plant is based on partial oxidation using oxygen.
f. The oxy-fuel plants include cryogenic removal of some of the impurities from the CO$_2$ during compression. Electricity consumption for oxygen production by cryogenic distillation of air is 200 kWh/ tO$_2$ at atmospheric pressure for the coal plant and 320 kWh/ tO$_2$ at 40 bar for the natural gas plant. Oxygen production in the IGCC and natural gas pre-combustion capture plants is partially integrated with the gas turbine compressor, so comparable data cannot be provided for these plants.
g. The percentage CO$_2$ capture is 85–90% for all plants except the natural gas oxy-fuel plant which has an inherently higher percentage capture of 97%.

3.3.2.4   Effluents

As a result of decomposition of amines, effluents will be created, particularly ammonia and heat-stable salts. Rao and Rubin (2002) have estimated these emissions for an MEA-based process based on limited data. In such processes, heat stable salts (solvent decomposition products, corrosion products etc.) are removed from the solution in a reclaimer and a waste stream is created and is disposed of using normal HSE (Health, Safety and Environmental) practices. In some cases, these reclaimer bottoms may be classified as a hazardous waste, requiring special handling (Rao and Rubin, 2002). Also a particle filter and carbon filter is normally installed in the solvent circuit to remove byproducts. Finally, some solvent material will be lost to the environment through evaporation and carry over in the absorber, which is accounted for in the solvent consumption. It is expected that acid gases other than CO$_2$, which are still present in the flue gas (SO$_x$ and NO$_2$) will also be absorbed in the solution. This will lower the concentration of these components further and even the net emissions in some cases depending on the amount of additional energy use for CO$_2$ capture (see Tables 3.4 and 3.5). As SO$_2$-removal prior to CO$_2$-removal is very likely in coal-fired plants, this will lead to the production of a waste or byproduct stream containing gypsum and water from the FGD unit.

3.3.3   Emerging technologies

3.3.3.1   Other absorption process

Various novel solvents are being investigated, with the object of achieving a reduced energy consumption for solvent regeneration (Chakma, 1995; Chakma and Tontiwachwuthikul, 1999; Mimura *et al.*, 1999; Zheng *et al.*, 2003; Cullinane and Rochelle, 2003; Leites, 1998; Erga *et al.*, 1995; Aresta and Dibenedetto, 2003; Bai and Yeh, 1997).

Besides novel solvents, novel process designs are also currently becoming available (Leites *et al.* 2003). Research is also being carried out to improve upon the existing practices and packing types (Aroonwilas *et al.*, 2003). Another area of research is to increase the concentration levels of aqueous MEA solution used in absorption systems as this tends to reduce the size of equipment used in capture plants (Aboudheir *et al.*, 2003). Methods to prevent oxidative degradation of MEA by de-oxygenation of the solvent solutions are also being investigated (Chakravarti *et al.*, 2001). In addition to this, the catalytic removal of oxygen in flue gases from coal firing has been suggested (Nsakala *et al.*, 2001) to enable operation with promising solvents sensitive to oxygen.

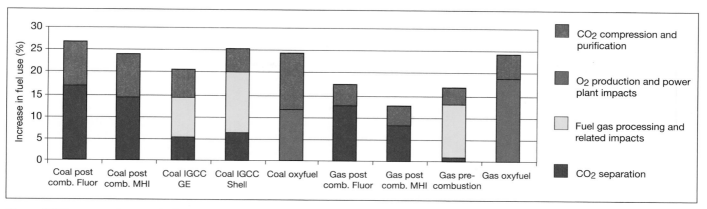

Figure 3.7 Percentage increase in fuel use per kWh of electricity due to CO_2 capture, compared to the same plant without capture (Source data: Davison, 2005; IEA GHG, 2004; IEA GHG, 2003; IEA GHG, 2000b; Dillon *et al.*, 2005).

a. The increase in fuel required to produce a kWh of electricity is calculated by comparing the same type of plant with and without capture. The increase in fuel consumption depends on the type of baseline plant without capture. For example, the increase in energy consumption for a GE IGCC plant with capture compared to a coal steam cycle baseline plant without capture would be 40% as opposed to the lower value shown in the figure that was calculated relative to the same type of baseline plant without capture.

b. The direct energy consumptions for CO_2 separation are lower for pre-combustion capture than for post-combustion capture, because CO_2 is removed from a more concentrated, higher pressure gas, so a physical rather than a chemical solvent can be used.

c. The 'Fuel gas processing and related impacts' category for IGCC includes shift conversion of the fuel gas and the effects on the gas turbine combined cycle of removal of CO_2 from the fuel gas and use of hydrogen as a fuel instead of syngas. For natural gas pre-combustion capture this category also includes partial oxidation/steam reforming of the natural gas.

d. The energy consumption for CO_2 compression is lower in pre-combustion capture than in post-combustion capture because some of the CO_2 leaves the separation unit at elevated pressure.

e. The energy consumption for CO_2 compression in the oxy-fuel processes depends on the composition of the extracted product, namely 75% by volume in the coal-fired plant and 93% by volume in the gas fired plant. Impurities are cryogenically removed from the CO_2 during compression, to give a final CO_2 purity of 96% by volume. The energy consumption of the cryogenic CO_2 separation unit is included in the CO_2 compression power consumption.

f. The 'Oxygen production and power plant impacts' category for oxy-fuel processes includes the power consumption for oxygen production and the impacts of CO_2 capture on the rest of the power plant, that is excluding CO_2 compression and purification. In the coal-fired oxy-fuel plant, the efficiency of the rest of the power plant increases slightly, for example due to the absence of a flue gas desulphurization (FGD) unit. The efficiency of the rest of the gas fired oxy-fuel plant decreases because of the change of working fluid in the power cycle from air to recycled flue gas.

3.3.3.2   Adsorption process

In the adsorption process for flue gas CO_2 recovery, molecular sieves or activated carbons are used in adsorbing CO_2. Desorbing CO_2 is then done by the pressure swing operation (PSA) or temperature swing operation (TSA). Most applications are associated with pressure swing adsorption (Ishibashi *et al.*, 1999 and Yokoyama, 2003). Much less attention has been focused on CO_2 removal via temperature swing adsorption, as this technique is less attractive compared to PSA due to the longer cycle times needed to heat up the bed of solid particles during sorbent regeneration. For bulk separations at large scales, it is also essential to limit the length of the unused bed and therefore opt for faster cycle times.

Adsorption processes have been employed for CO_2 removal from synthesis gas for hydrogen production (see Section 3.5.2.9). It has not yet reached a commercial stage for CO_2 recovery from flue gases. The following main R&D activities have been conducted:

• Study of CO_2 removal from flue gas of a thermal power plant by physical adsorption (Ishibashi *et al.*, 1999);

• Study of CO_2 removal from flue gas of a thermal power plant by a combined system with pressure swing adsorption and a super cold separator (Takamura *et al.*, 1999);

• Pilot tests on the recovery of CO_2 from a coal and oil fired power plant, using pressure temperature swing adsorption (PTSA) and an X-type zeolite as an adsorbent (Yokoyama, 2003).

Pilot test results of coal-fired flue gas CO_2 recovery by adsorption processes show that the energy consumption for capture (blowers and vacuum pumps) has improved from the original 708 kWh/tCO_2 to 560 kWh/tCO_2. An energy consumption of 560 kWh/tCO_2 is equivalent to a loss corresponding to 21% of the energy output of the power plant. Recovered CO_2 purity is about 99.0% by volume using two stages of a PSA and PTSA system (Ishibashi *et al.*, 1999).

It can be concluded that based on mathematical models and data from pilot-scale experimental installations, the design of a full-scale industrial adsorption process might be feasible. A serious drawback of all adsorptive methods is the necessity to treat the gaseous feed before CO_2 separation in an adsorber. Operation at high temperature with other sorbents (see Section 3.3.3.4) can circumvent this requirement (Sircar and Golden, 2001). In many cases gases have to be also cooled and dried, which limits the attractiveness of PSA, TSA or ESA (electric swing adsorption) vis-à-vis capture by chemical absorption described in previous sections. The development of a new generation of materials that would efficiently adsorb CO_2 will undoubtedly enhance the competitiveness of adsorptive separation in a flue gas application.

3.3.3.3 Membranes

Membrane processes are used commercially for CO_2 removal from natural gas at high pressure and at high CO_2 concentration (see Section 3.2.2). In flue gases, the low CO_2 partial pressure difference provides a low driving force for gas separation. The removal of carbon dioxide using commercially available polymeric gas separation membranes results in higher energy penalties on the power generation efficiency compared to a standard chemical absorption process (Herzog *et al.*, 1991, Van der Sluijs *et al.*, 1992 and Feron, 1994). Also, the maximum percentage of CO_2 removed is lower than for a standard chemical absorption processes. Improvements can be made if more selective membranes become available, such as facilitated membranes, described below.

The membrane option currently receiving the most attention is a hybrid membrane – absorbent (or solvent) system. These systems are being developed for flue gas CO_2 recovery. Membrane/solvent systems employ membranes to provide a very high surface area to volume ratio for mass exchange between a gas stream and a solvent resulting in a very compact system. This results in a membrane contactor system in which the membrane forms a gas permeable barrier between a liquid and a gaseous phase. In general, the membrane is not involved in the separation process. In the case of porous membranes, gaseous components diffuse through the pores and are absorbed by the liquid; in cases of non-porous membranes they dissolve in the membrane and diffuse through the membrane. The contact surface area between gas and liquid phase is maintained by the membrane and is independent of the gas and liquid flow rate. The selectivity of the partition is primarily determined by the absorbent (solvent). Absorption in the liquid phase is determined either by physical partition or by a chemical reaction.

The advantages of membrane/solvent systems are avoidance of operational problems occurring in conventional solvent absorption systems (see Section 3.3.2.1) where gas and liquid flows are in direct contact. Operational problems avoided include foaming, flooding entrainment and channelling, and result in the free choice of the gas and liquid flow rates and a fixed interface for mass transfer in the membrane/solvent system. Furthermore, the use of compact membranes result in smaller equipment sizes with capital cost reductions. The choice of a suitable combination of solvent and membrane material is very important. The material characteristics should be such that the transfer of solvent through the membrane is avoided at operating pressure gradients of typically 50–100 kPa,

while the transfer of gas is not hindered. The overall process configuration in terms of unit operations would be very similar to a conventional chemical absorption/desorption process (see Figure 3.4). Membrane/solvent systems can be both used in the absorption as well as in the desorption step. Feron and Jansen (2002) and Falk-Pedersen *et al.* (1999) give examples of suitable membrane/solvent systems.

Research and development efforts have also been reported in the area of facilitated transport membranes. Facilitated transport membranes rely on the formation of complexes or reversible chemical reactions of components present in a gas stream with compounds present in the membrane. These complexes or reaction products are then transported through the membrane. Although solution and diffusion still play a role in the transport mechanism, the essential element is the specific chemical interaction of a gas component with a compound in the membrane, the so-called carrier. Like other pressure driven membrane processes, the driving force for the separation comes from a difference in partial pressure of the component to be transported. An important class of facilitated transport membranes is the so-called supported liquid membrane in which the carrier is dissolved into a liquid contained in a membrane. For CO_2 separations, carbonates, amines and molten salt hydrates have been suggested as carriers (Feron, 1992). Porous membranes and ion-exchange membranes have been employed as the support. Until now, supported liquid membranes have only been studied on a laboratory scale. Practical problems associated with supported liquid membranes are membrane stability and liquid volatility. Furthermore, the selectivity for a gas decreases with increasing partial pressure on the feed side. This is a result of saturation of the carrier in the liquid. Also, as the total feed pressure is increased, the permeation of unwanted components is increased. This also results in a decrease in selectivity. Finally, selectivity is also reduced by a reduction in membrane thickness. Recent development work has focused on the following technological options that are applicable to both CO_2/N_2 and CO_2/H_2 separations:

- Amine-containing membranes (Teramoto *et al.*, 1996);
- Membranes containing potassium carbonate polymer gel membranes (Okabe *et al.*, 2003);
- Membranes containing potassium carbonate-glycerol (Chen *et al.*, 1999);
- Dendrimer-containing membranes (Kovvali and Sirkar, 2001).
- Poly-electrolyte membranes (Quinn and Laciak, 1997);

Facilitated transport membranes and other membranes can also be used in a preconcentration step prior to the liquefaction of CO_2 (Mano *et al.*, 2003).

3.3.3.4 Solid sorbents

There are post-combustion systems being proposed that make use of regenerable solid sorbents to remove CO_2 at relatively high temperatures. The use of high temperatures in the CO_2 separation step has the potential to reduce efficiency penalties with respect to wet-absorption methods. In principle, they all follow the scheme shown in Figure 3.2a, where the combustion flue gas is put in contact with the sorbent in a suitable reactor to allow the gas-solid reaction of CO$_2$ with the sorbent (usually the carbonation of a metal oxide). The solid can be easily separated from the gas stream and sent for regeneration in a different reactor. Instead of moving the solids, the reactor can also be switched between sorption and regeneration modes of operation in a batch wise, cyclic operation. One key component for the development of these systems is obviously the sorbent itself, that has to have good CO$_2$ absorption capacity and chemical and mechanical stability for long periods of operation in repeated cycles. In general, sorbent performance and cost are critical issues in all post-combustion systems, and more elaborate sorbent materials are usually more expensive and will have to demonstrate outstanding performance compared with existing commercial alternatives such as those described in 3.3.2.

Solid sorbents being investigated for large-scale CO$_2$ capture purposes are sodium and potassium oxides and carbonates (to produce bicarbonate), usually supported on a solid substrate (Hoffman *et al.*, 2002; Green *et al.*, 2002). Also, high temperature Li-based and CaO-based sorbents are suitable candidates. The use of lithium-containing compounds (lithium, lithium-zirconia and lithium-silica oxides) in a carbonation-calcination cycle, was first investigated in Japan (Nakagawa and Ohashi, 1998). The reported performance of these sorbents is very good, with very high reactivity in a wide range of temperatures below 700°C, rapid regeneration at higher temperatures and durability in repeated capture-regeneration cycles. This is essential because lithium is an intrinsically expensive material.

The use of CaO as a regenerable CO$_2$ sorbent has been proposed in several processes dating back to the 19th century. The carbonation reaction of CaO to separate CO$_2$ from hot gases (T > 600°C) is very fast and the regeneration of the sorbent by calcining the CaCO$_3$ into CaO and pure CO$_2$ is favoured at T > 900°C (at a partial pressure of CO$_2$ of 0.1 MPa). The basic separation principle using this carbonation-calcination cycle was successfully tested in a pilot plant (40 tonne d^{-1}) for the development of the Acceptor Coal Gasification Process (Curran *et al.*, 1967) using two interconnected fluidized beds. The use of the above cycle for a post-combustion system was first proposed by Shimizu *et al.* (1999) and involved the regeneration of the sorbent in a fluidized bed, firing part of the fuel with O$_2$/CO$_2$ mixtures (see also Section 3.4.2). The effective capture of CO$_2$ by CaO has been demonstrated in a small pilot fluidized bed (Abanades *et al.*, 2004a). Other combustion cycles incorporating capture of CO$_2$ with CaO that might not need O$_2$ are being developed, including one that works at high pressures with simultaneous capture of CO$_2$ and SO$_2$ (Wang *et al.*, 2004). One weak point in all these processes is that natural sorbents (limestones and dolomites) deactivate rapidly, and a large make-up flow of sorbent (of the order of the mass flow of fuel entering the plant) is required to maintain the activity in the capture-regeneration loop (Abanades *et al.*, 2004b). Although the deactivated sorbent may find application in the cement industry and the sorbent cost is low, a range of methods to enhance the activity of Ca-based CO$_2$ sorbents are

being pursued by several groups around the world.

3.3.4 *Status and outlook*

Virtually all the energy we use today from carbon-containing fuels is obtained by directly burning fuels in air. This is despite many decades of exploring promising and more efficient alternative energy conversion cycles that rely on other fuel processing steps prior to fuel combustion or avoiding direct fuel combustion (see pre-combustion capture – Section 3.5). In particular, combustion-based systems are still the competitive choice for operators aiming at large-scale production of electricity and heat from fossil fuels, even under more demanding environmental regulations, because these processes are reliable and well proven in delivering electricity and heat at prices that often set a benchmark for these services. In addition, there is a continued effort to raise the energy conversion efficiencies of these systems through advanced materials and component development. This will allow these systems to operate at higher temperature and higher efficiency.

As was noted in Section 3.1, the main systems of reference for post-combustion capture are the present installed capacity of coal and natural gas power plants, with a total of 970 GW$_e$ subcritical steam and 155 GW$_e$ of supercritical/ultra-supercritical steam-based pulverized coal fired plants, 339 GW$_e$ of natural gas combined cycle, 333 GW$_e$ natural gas steam-electric power plants and 17 GW$_e$ of coal-fired, circulating, fluidized-bed combustion (CFBC) power plants. An additional capacity of 454 GW$_e$ of oil-based power plant, with a significant proportion of these operating in an air-firing mode is also noted (IEA WEO, 2004 and IEA CCC, 2005). Current projections indicate that the generation efficiency of commercial, pulverized coal fired power plants based on ultra-supercritical steam cycles would exceed 50% lower heating value (LHV) over the next decade (IEA, 2004), which will be higher than efficiencies of between 36 and 45% reported for current subcritical and supercritical steam-based plants without capture (see Section 3.7). Similarly, natural gas fired combined cycles are expected to have efficiencies of 65% by 2020 (IEA GHG, 2002b) and up from current efficiencies between 55 and 58% (see Section 3.7). In a future carbon-constrained world, these independent and ongoing developments in power cycle efficiencies will result in lower CO$_2$-emissions per kWh produced and hence a lower loss in overall cycle efficiency when post-combustion capture is applied.

There are proven post-combustion CO$_2$ capture technologies based on absorption processes that are commercially available at present . They produce CO$_2$ from flue gases in coal and gas-fired installations for food/beverage applications and chemicals production in capacity ranges between 6 and 800 tCO$_2$ d^{-1}. They require scale up to 20-50 times that of current unit capacities for deployment in large-scale power plants in the 500 MW$_e$ capacity range (see Section 3.3.2). The inherent limitations of currently available absorption technologies when applied to post-combustion capture systems are well known and their impact on system cost can be estimated relatively accurately for a given application (see Section 3.7). Hence, with the dominant role played by air- blown energy conversion processes in the global energy infrastructure, the availability of post-combustion capture systems is important if CO_2 capture and storage becomes a viable climate change mitigation strategy.

The intense development efforts on novel solvents for improved performance and reduced energy consumption during regeneration, as well as process designs incorporating new contacting devices such as hybrid membrane-absorbent systems, solid adsorbents and high temperature regenerable sorbents, may lead to the use of more energy efficient post-combustion capture systems. However, all these novel concepts still need to prove their lower costs and reliability of operation on a commercial scale. The same considerations also apply to other advanced CO_2 capture concepts with oxy-fuel combustion or pre-combustion capture reviewed in the following sections of this chapter. It is generally not yet clear which of these emerging technologies, if any, will succeed as the dominant commercial technology for energy systems incorporating CO_2 capture.

3.4   Oxy-fuel combustion capture systems

3.4.1   *Introduction*

The oxy-fuel combustion process eliminates nitrogen from the flue gas by combusting a hydrocarbon or carbonaceous fuel in either pure oxygen or a mixture of pure oxygen and a CO_2-rich recycled flue gas (carbonaceous fuels include biomass). Combustion of a fuel with pure oxygen has a combustion temperature of about 3500°C which is far too high for typical power plant materials. The combustion temperature is limited to about 1300-1400°C in a typical gas turbine cycle and to about 1900°C in an oxy-fuel coal-fired boiler using current technology. The combustion temperature is controlled by the proportion of flue gas and gaseous or liquid-water recycled back to the combustion chamber.

The combustion products (or flue gas) consist mainly of carbon dioxide and water vapour together with excess oxygen required to ensure complete combustion of the fuel. It will also contain any other components in the fuel, any diluents in the oxygen stream supplied, any inerts in the fuel and from air leakage into the system from the atmosphere. The net flue gas, after cooling to condense water vapour, contains from about 80-98% CO_2 depending on the fuel used and the particular oxy-fuel combustion process. This concentrated CO_2 stream can be compressed, dried and further purified before delivery into a pipeline for storage (see Chapter 4). The CO_2 capture efficiency is very close to 100% in oxy-fuel combustion capture systems. Impurities in the CO_2 are gas components such as SO_x, NO_x, HCl and Hg derived from the fuel used, and the inert gas components, such as nitrogen, argon and oxygen, derived from the oxygen feed or air leakage into the system. The CO_2 is transported by pipeline as a dense supercritical phase. Inert gases must be reduced to a low concentration to avoid two-phase flow conditions developing in the pipeline systems. The acid gas components may need to be removed to comply

with legislation covering co-disposal of toxic or hazardous waste or to avoid operations or environmental problems with disposal in deep saline reservoirs, hydrocarbon formations or in the ocean. The carbon dioxide must also be dried to prevent water condensation and corrosion in pipelines and allow use of conventional carbon-steel materials.

Although elements of oxy-fuel combustion technologies are in use in the aluminium, iron and steel and glass melting industries today, oxy-fuel technologies for CO_2 capture have yet to be deployed on a commercial scale. Therefore, the first classification between *existing technologies* and *emerging technologies* adopted in post-combustion (Section 3.3) and pre-combustion (Section 3.5) is not followed in this section. However, it is important to emphasize that the key separation step in most oxy-fuel capture systems (O_2 from air) is an 'existing technology' (see Section 3.4.5). Current methods of oxygen production by air separation comprise cryogenic distillation, adsorption using multi-bed pressure swing units and polymeric membranes. For oxy-fuel conversions requiring less than 200 tO_2 d^{-1}, the adsorption system will be economic. For all the larger applications, which include power station boilers, cryogenic air separation is the economic solution (Wilkinson *et al.*, 2003a).

In the following sections we present the main oxy-fuel combustion systems classified according to how the heat of combustion is supplied and whether the flue gas is used as a working fluid (Sections 3.4.2, 3.4.3, 3.4.4). A brief overview of O_2 production methods relevant for these systems is given (Section 3.4.5). In Section 3.4.6, the emerging technology of chemical looping combustion is presented, in which pure oxygen is supplied by a metal oxide rather than an oxygen production process. The section on oxy-fuel systems closes with an overview of the status of the technology (Section 3.4.7).

3.4.2   *Oxy-fuel indirect heating - steam cycle*

In these systems, the oxy-fuel combustion chamber provides heat to a separate fluid by heat transfer through a surface. It can be used for either process heating, or in a boiler with a steam cycle for power generation. The indirect system can be used with any hydrocarbon or carbon-containing fuel.

The application of oxy-fuel indirect heating for CO_2 capture in process heating and power generation has been examined in both pilot-scale trials evaluating the combustion of carbonaceous fuels in oxygen and CO_2-rich recycled flue gas mixtures and engineering assessments of plant conversions as described below.

3.4.2.1   *Oxy-fuel combustion trials*
Work to demonstrate the application of oxy-fuel recycle combustion in process heating and for steam generation for use in steam power cycles have been mostly undertaken in pilot scale tests that have looked at the combustion, heat transfer and pollutant-forming behaviour of natural gas and coal.

One study carried out (Babcock Energy Ltd. *et al.*, 1995) included an oxy-fuel test with flue gas recycle using a 160kW, pulverized coal, low NO$_x$ burner. The system included a heat-transfer test section to simulate fouling conditions. Test conditions included variation in recycle flow and excess O$_2$ levels. Measurements included all gas compositions, ash analysis and tube fouling after a 5-week test run. The work also included a case study on oxy-fuel operation of a 660 MW power boiler with CO$_2$ capture, compression and purification. The main test results were that NO$_x$ levels reduced with increase in recycle rate, while SO$_2$ and carbon in ash levels were insensitive to the recycle rate. Fouling in the convective test section was greater with oxy-fuel firing than with air. High-slagging UK coal had worse slagging when using oxy-fuel firing, the higher excess O$_2$ level lowered carbon in ash and CO concentration.

For the combustion of pulverized coal, other pilot-scale tests by Croiset and Thambimuthu (2000) have reported that the flame temperature and heat capacity of gases to match fuel burning in air occurs when the feed gas used in oxy-fuel combustion has a composition of approximately 35% by volume O$_2$ and 65% by volume of dry recycled CO$_2$ (c.f. 21% by volume O$_2$ and the rest nitrogen in air). In practice, the presence of inerts such as ash and inorganic components in the coal, the specific fuel composition and moisture in the recycled gas stream and the coal feed will result in minor adjustments to this feed mixture composition to keep the flame temperature at a value similar to fuel combustion in air.

At conditions that match O$_2$/CO$_2$ recycle combustion to fuel burning in air, coal burning is reported to be complete (Croiset and Thambimuthu, 2000), with operation of the process at excess O$_2$ levels in the flue gas as low as 1-3% by volume O$_2$, producing a flue gas stream of 95-98% by volume dry CO$_2$ (the rest being excess O$_2$, NO$_x$, SO$_x$ and argon) when a very high purity O$_2$ stream is used in the combustion process with zero leakage of ambient air into the system. No differences were detected in the fly ash formation behaviour in the combustor or SO$_2$ emissions compared to conventional air firing conditions. For NO$_x$ on the other hand, emissions were lower due to zero thermal NO$_x$ formation from the absence of nitrogen in the feed gas - with the partial recycling of NO$_x$ also reducing the formation and net emissions originating from the fuel bound nitrogen. Other studies have demonstrated that the level of NO$_x$ reduction is as high as 75% compared to coal burning in air (Chatel-Pelage *et al.*, 2003). Similar data for natural gas burning in O$_2$/CO$_2$ recycle mixtures report zero thermal NO$_x$ emissions in the absence of air leakage into the boiler, with trace amounts produced as thermal NO$_x$ when residual nitrogen is present in the natural gas feed (Tan *et al.*, 2002).

The above and other findings show that with the application of oxy-fuel combustion in modified utility boilers, the nitrogen-free combustion process would benefit from higher heat transfer rates (McDonald and Palkes, 1999), and if also constructed with higher temperature tolerant materials, are able to operate at higher oxygen concentration and lower flue gas recycle flows – both of which will considerably reduce overall volume flows and size of the boiler.

It should be noted that even when deploying a 2/3 flue gas recycle gas ratio to maintain a 35% by volume O$_2$ feed to a pulverized coal fired boiler, hot recycling of the flue gas prior to CO$_2$ purification and compression also reduces the size of all unit operations in the stream leaving the boiler to 1/5 that of similar equipment deployed in conventional air blown combustion systems (Chatel-Pelage *et al.*, 2003). Use of a low temperature gas purification step prior to CO$_2$ compression (see Section 3.4.2.2) will also eliminate the need to deploy conventional selective catalytic reduction for NO$_x$ removal and flue gas desulphurization to purify the gas, a practice typically adopted in conventional air-blown combustion processes (see Figure 3.3). The overall reduction in flow volumes, equipment scale and simplification of gas purification steps will thus have the benefit of reducing both capital and operating costs of equipment deployed for combustion, heat transfer and final gas purification in process and power plant applications (Marin *et al.*, 2003).

As noted above for pulverized coal, oil, natural gas and biomass combustion, fluidized beds could also be fired with O$_2$ instead of air to supply heat for the steam cycle. The intense solid mixing in a fluidized bed combustion system can provide very good temperature control even in highly exothermic conditions, thereby minimizing the need for flue gas recycling. In principle, a variety of commercial designs for fluidized combustion boilers exist that could be retrofitted for oxygen firing. A circulating fluidized bed combustor with O$_2$ firing was proposed by Shimizu *et al.* (1999) to generate the heat required for the calcination of CaCO$_3$ (see also Section 3.3.3.4). More recently, plans for pilot testing of an oxy-fired circulating fluidized bed boiler have been published by Nsakala *et al.* (2003).

3.4.2.2 *Assessments of plants converted to oxy-fuel combustion*

We now discuss performance data from a recent comprehensive design study for an application of oxy-fuel combustion in a new build pulverized coal fired power boiler using a supercritical steam cycle (see Figure 3.8; Dillon *et al.*, 2005). The overall thermal efficiency on a lower heating value basis is reduced from 44.2% to 35.4%. The net power output is reduced from 677 MW$_e$ to 532 MW$_e$.

Important features of the system include:
- Burner design and gas recycle flow rate have been selected to achieve the same temperatures as in air combustion (compatible temperatures with existing materials in the boiler).
- The CO$_2$-rich flue gas from the boiler is divided into three gas streams: one to be recycled back to the combustor, one to be used as transport and drying gas of the coal feed, and the third as product gas. The first recycle and the product stream are cooled by direct water scrubbing to remove residual particulates, water vapour and soluble acid gases such as SO$_3$ and HCl. Oxygen and entrained coal dust together with the second recycle stream flow to the burners.
- The air leakage into the boiler is sufficient to give a high enough inerts level to require a low temperature inert gas

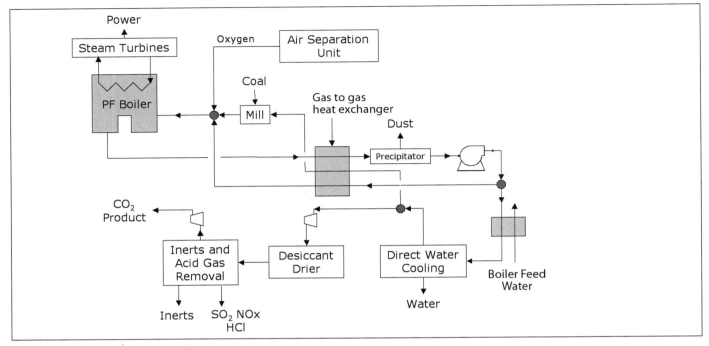

Figure 3.8 Schematic of an oxy-fuel, pulverized coal fired power plant.

removal unit to be installed, even if pure O_2 were used as the oxidant in the boiler. The cryogenic oxygen plant will, in this case, produce 95% O_2 purity to minimize power consumption and capital cost.

- The low temperature (-55°C) CO_2 purification plant (Wilkinson et al., 2003b) integrated with the CO_2 compressor will not only remove excess O_2, N_2, argon but can also remove all NO_x and SO_2 from the CO_2 stream, if high purity CO_2 is required for storage. Significantly, removal of these components before final CO_2 compression eliminates the need to otherwise incorporate upstream NO_x and SO_x removal equipment in the net flue gas stream leaving the boiler. Elimination of N_2 from the flue gas results in higher SO_x concentrations in the boiler and reduced NO_x levels. Suitable corrosion resistant materials of construction must be chosen.
- The overall heat transfer is improved in oxy-fuel firing because of the higher emissivity of the CO_2/H_2O gas mixture in the boiler compared to nitrogen and the improved heat transfer in the convection section. These improvements, together with the recycle of hot flue gas, increase the boiler efficiency and steam generation by about 5%.
- The overall thermal efficiency is improved by running the O_2 plant air compressor and the first and final stages of the CO_2 compressor without cooling, and recovering the compression heat for boiler feed water heating prior to de-aeration.

Engineering studies have also been reported by Simbeck and McDonald (2001b) and by McDonald and Palkes (1999). This work has confirmed that the concept of retrofitting oxy-fuel combustion with CO_2 capture to existing coal-fired power

stations does not have any technical barriers and can make use of existing technology systems.

It has been reported (Wilkinson *et al.*, 2003b) that the application of oxy-fuel technology for the retrofit of power plant boilers and a range of refinery heaters in a refinery complex (Grangemouth refinery in Scotland) is technically feasible at a competitive cost compared to other types of CO_2 capture technologies. In this case, the existing boiler is adapted to allow combustion of refinery gas and fuel oil with highly enriched oxygen and with partial flue gas recycling for temperature control. Oxy-fuel boiler conversions only needed minor burner modifications, a new O_2 injection system and controls, and a new flue gas recycle line with a separate blower. These are cheap and relatively simple modifications and result in an increase in boiler/heater thermal efficiency due to the recycle of hot gas. Modifications to a coal-fired boiler are more complex. In this study, it was found to be more economic to design the air separation units for only 95% O_2 purity instead of 99.5% to comply with practical levels of air leakage into boilers and to separate the associated argon and nitrogen in the CO_2 inert gas removal system to produce a purity of CO_2 suitable for geological storage. After conversion of the boiler, the CO_2 concentration in the flue gas increases from 17 to 60% while the water content increases from 10 to 30%. Impurities (SO_x, NO_x) and gases (excess O_2, N_2, argon) representing about 10% of the stream are separated from CO_2 at low temperature (-55°C). After cooling, compression and drying of the separated or non-recycled flue gas, the product for storage comprises 96% CO_2 contaminated with 2% N_2, 1% argon and less than 1% O_2 and SO_2. Production of ultra-pure CO_2 for storage would also be possible if distillation steps are added to the separation process.

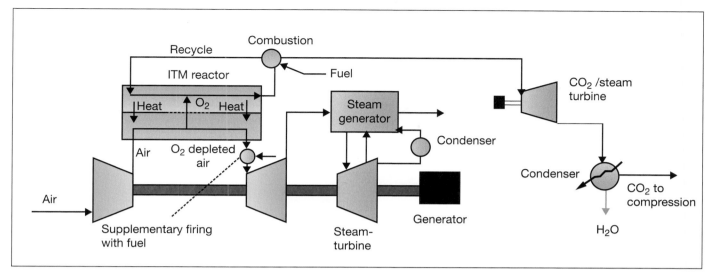

Figure 3.9 Principle flow scheme of the advanced zero emission power plant cycle.

3.4.2.3   *Advanced zero emission power plant*

The advanced zero emission power plant (or AZEP as outlined in Figure 3.9; Griffin *et al.*, 2003) is an indirect heating gas turbine cycle that incorporates a high-temperature oxygen transport membrane, operating at about 800°C -1000°C (see Section 3.4.5.2). This process uses a standard air-based gas turbine in a combined cycle arrangement. Three process steps take place in a reactor system that replaces the combustion chamber of a standard gas turbine: 1) separation of oxygen from hot air using the membrane and transport to the combustion section; 2) combustion and 3) heat exchange from the combustion products to the compressed air.

A net efficiency for advanced zero emission power cycle of around 49–50% LHV is claimed including CO_2 compression for transport. In order to get full advantage of the potential of the most advanced gas turbines, which have inlet temperatures of 1300°C-1400°C, an afterburner fired with natural gas in air may be added behind the reactor system. The efficiency then climbs up to 52% but now 15% of the CO_2 generated by combustion is released at the stack and is not captured.

3.4.3   *Oxy-fuel direct heating - gas turbine cycle*

Oxy-fuel combustion takes place in a pressurized CO_2-rich recirculating stream in a modified gas turbine. The hot gas is expanded in the turbine producing power. The turbine exhaust is cooled to provide heat for a steam cycle and water vapour is condensed by further cooling. The CO_2-rich gas is compressed in the compressor section. The net CO_2-rich combustion product is removed from the system. Only natural gas, light hydrocarbons and syngas (CO + H_2) can be used as fuel.

3.4.3.1   *Cycle description and performance*

Figure 3.10 shows how a gas turbine can be adapted to run with oxy-fuel firing using CO_2 as a working fluid. Exhaust gas leaving the heat recovery steam generator is cooled to condense water. The net CO_2 product is removed and the remaining gas is

recycled to the compressor. Suitable fuels are natural gas, light to medium hydrocarbons or (H_2 + CO) syngas, which could be derived from coal. The use of CO_2 as the working fluid in the turbine will necessitate a complete redesign of the gas turbine (see Section 3.4.3.2). A recent study (Dillon *et al.*, 2005) gives an overall efficiency including CO_2 compression of 45%.

Two typical variants of this configuration are the so-called Matiant and Graz cycles (Mathieu, 2003; Jericha *et al.*, 2003). The Matiant cycle uses CO_2 as the working fluid, and consists of features like intercooled compressor and turbine reheat. The exhaust gas is preheating the recycled CO_2 in a heat exchanger. The CO_2 generated in combustion is extracted from the cycle behind the compressor. The net overall LHV efficiency is expected to be 45-47% and can increase above 50% in a combined cycle configuration similar to that shown in Figure 3.10. The Graz cycle consists of an integrated gas turbine and steam turbine cycle. A net LHV efficiency of above 50% has been calculated for this cycle (Jericha *et al.*, 2003).

A recent comprehensive review of gas turbine cycles with CO_2 capture provides efficiencies of different cycles on a common basis (Kvamsdal *et al.*, 2004).

3.4.3.2   *The CO₂/oxy-fuel gas turbine*

In existing gas turbines the molecular weight of the gases in the compressor and turbine are close to that of air (28.8). In the case of oxy-fuel combustion with CO_2-recycle the compressor fluid molecular weight is about 43 and about 40 in the turbine. The change in working fluid from air to a CO_2-rich gas results in a number of changes in properties that are of importance for the design of the compressor, combustor and the hot gas path including the turbine:

* The speed of sound is 80% of air;
* The gas density is 50% higher than air;
* The specific heat ratio is lower than air resulting in a lower temperature change on adiabatic compression or expansion. An oxy-fuel gas turbine in a combined cycle has a higher optimal pressure ratio, typically 30 to 35 compared to 15

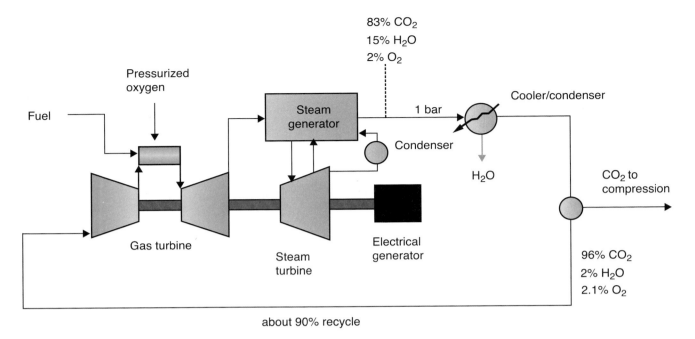

Figure 3.10 Principle of the oxy-fuel gas turbine combined cycle. Exhaust gas is recycled, compressed and used in the combustion chamber to control the temperature entering the turbine.

to 18 used with air in a combined cycle system. With the highest turbine inlet temperature consistent with material limitations, the rather high-pressure ratio results in an exhaust gas temperature of about 600°C, which is optimal for the steam cycle.

These changes in the fundamental properties of the working fluid will have a significant impact on gas turbine components, requiring completely new designs of compressors, combustors (to account for aerodynamic changes and acoustic feedbacks) and hot gas path (O_2 partial pressure must be low in oxy-fuel systems but it is also important to avoid reducing conditions for the materials of the turbine or the change to materials allowing much lower O_2 partial pressures).

3.4.4   Oxy-fuel direct heating - steam turbine cycle

In an oxy-fuel steam turbine cycle, water is pressurized as a liquid and is then evaporated, heated by the direct injection and combustion of a fuel with pure oxygen and expanded in a turbine. Most of the water in the low pressure turbine exhaust gas is cooled and condensed, prior to pumping back to a high pressure while the CO_2 produced from combustion is removed and compressed for pipeline transport. A variant of this cycle in which the heat is provided by burning natural gas fuel in-situ with pure oxygen was proposed by Yantovskii *et al.* (1992).

The direct combustion of fuel and oxygen has been practised for many years in the metallurgical and glass industries where burners operate at near stoichiometric conditions with flame temperatures of up to 3500°C. A water quenched H_2/O_2 burner capable of producing 60 tonne h^{-1}, 6 MPa super heated steam was demonstrated in the mid-1980s (Ramsaier *et al.*, 1985). A

recent development by Clean Energy Systems incorporating these concepts where a mixture of 90 % by volume superheated steam and 10% CO_2 is produced at high temperature and pressure to power conventional or advanced steam turbines is shown in Figure 3.11. The steam is condensed in a low-pressure condenser and recycled, while CO_2 is extracted from the condenser, purified and compressed. (Anderson *et al.*, 2003 and Marin *et al.*, 2003).

Plants of this type require a clean gaseous or liquid fuel and will operate at 20 to 50 MPa pressure. The steam plus CO_2 generator is very compact. Control systems must be very precise as start-up and increase to full flow in a preheated plant can take place in less than 2 seconds. Precise control of this very rapid start was demonstrated (Ramsaier *et al.*, 1985) in a 60 tonne steam h^{-1} unit. The Clean Energy Systems studies claim efficiencies as high as 55% with CO_2 capture depending on the process conditions used.

The Clean Energy Systems technology can be initially applied with current steam turbines (565°C inlet temperature). The main technical issue is clearly the design of the steam turbines which could be used at inlet temperatures up to 1300°C by applying technology similar to that used in the hot path of gas turbines. The combustor itself (the 'gas generator') is adapted from existing rocket engine technology. In 2000, Clean Energy Systems proved the concept with a 110 kW pilot project conducted at the University of California Davis. A 20 MW thermal gas generator was successfully operated in a test run of the order of a few minutes in early 2003. A zero emissions demonstration plant (up to 6 MW electrical) is now on-line. US Department of Energy's National Energy Technology Laboratory designed the reheater (Richards, 2003) and NASA tested it in 2002. Much more technology development and demonstration

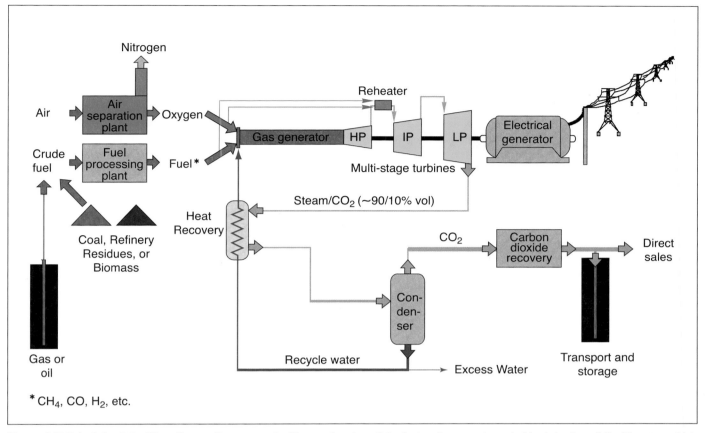

Figure 3.11 Principle of the Clean Energy Systems cycle. The combustion of the fuel and oxygen is cooled by injection of liquid-water, which is recycled in the process.

is needed on this proposed power cycle, but it shows significant potential for low capital cost and high efficiency.

3.4.5 Techniques and improvements in oxygen production

Oxygen is the key requirement for any oxy-fuel combustion system. It is also a key technology for pre-combustion CO_2 capture (see Section 3.5). In the next paragraphs, existing large-scale O_2 production methods are described first, followed by emerging concepts aimed at reducing the energy consumption and cost.

3.4.5.1 Cryogenic oxygen production

The very large quantities of oxygen required for CO_2 capture using the techniques of oxy-fuel combustion and pre-combustion de-carbonization can only be economically produced, at present, by using the established process of oxygen separation from air by distillation at cryogenic temperatures (Latimer, 1967). This is a technology that has been practiced for over 100 years.

In a typical cryogenic air separation plant (Castle, 1991; Figure 3.12), air is compressed to a pressure of 0.5 to 0.6 MPa and purified to remove water, CO_2, N_2O and trace hydrocarbons which could accumulate to dangerous levels in oxygen-rich parts of the plant, such as the reboiler condenser. Two or more switching fixed bed adsorbers are used, which can be

regenerated by either temperature or pressure swing, using in each case, a low pressure waste nitrogen stream. The air is cooled against returning products (oxygen and nitrogen) in a battery of aluminium plate-fin heat exchangers and separated into pure oxygen and nitrogen fractions in a double distillation column, which uses aluminium packing.

Oxygen can be pumped as liquid and delivered as a high-pressure gas at up to 10 MPa. Pumped oxygen plants have largely replaced the oxygen gas compression systems. They have virtually identical power consumptions but in a pumped cycle, a high-pressure air booster compressor provides a means of efficiently vaporizing and heating the liquid oxygen stream to ambient temperature. Current plant sizes range up to 3500 tO_2 d^{-1} and larger single train plants are being designed. Typical power consumption for the delivery of 95% O_2 at low pressure (0.17 MPa, a typical pressure for an oxy-fuel application) is 200 to 240 kWh/tO_2. There are numerous process cycle variations particularly for the production of oxygen at less than 97.5% purity which have been developed to reduce power and capital cost. Note that adsorption and polymeric membrane methods of air separation are only economic for small oxygen production rates.

3.4.5.2 High temperature oxygen ion transport membranes

Ceramic mixed metal oxides have been developed which exhibit simultaneous oxygen ion and electron conduction at

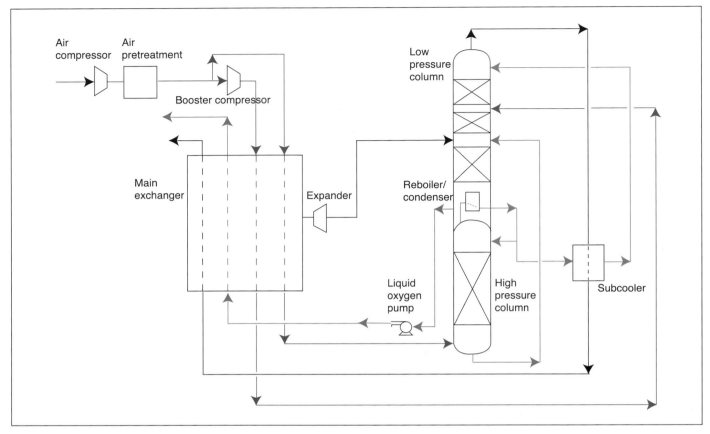

Figure 3.12a Oxygen production by distillation of liquid air.

Figure 3.12b A 3000 t day^{-1} oxygen plant (Courtesy of Air Products).

temperatures above 500°C and preferably above 700°C (Skinner and Kilner 2003; Bouwmeester and Van Laar, 2002; Dyer *et al.*, 2000; Bredesen *et al.*, 2004). Typical crystal structures which exhibit these properties include the perovskites and the brownmillerites. The selectivity of these materials for oxygen is infinite. The oxygen permeability is primarily controlled by the oxygen ion vacancies in the metal oxide lattice. A difference in oxygen partial pressure across the membrane will cause oxygen

molecules to ionize on the ceramic surface and pass into the crystal structure while simultaneously on the permeate side of the membrane, the oxygen ions give up their electrons and leave the ceramic in the region of lower activity. The electron conduction path is through the metal ions in the lattice. Unlike conventional membranes, the flux through the ceramic is a function of the partial pressure ratio. In the technical literature, the engineered structures of these ceramic mixed metal oxides are referred to as *ion transport membranes, ITM* or *oxygen transport membranes, OTM.*

The oxygen transport membrane can be fabricated in the form of plain tubes or as hollow fins on a central collector tube (Armstrong *et al.*, 2002). The finned elements are then mounted in tube sheets within a pressure vessel with high-pressure air flowing over the fins. There are several new concepts that have been proposed for using oxygen transport membranes in power cycles with CO_2 capture. A prime example of an oxy-fuel gas turbine cycle that incorporates an oxygen transport membrane for oxygen production is the advanced zero emission power plant described in Section 3.4.2.3. Another example is found in Sundnes (1998).

Development status
Oxygen transport membrane systems for oxygen production are currently in the early stages of development by at least two consortia receiving research funding from the US Department of Energy and the European Commission. The concept has now

reached the pilot plant stage and projected cost, manufacturing procedures and performance targets for full size systems have been evaluated. Systems capable of large-scale production are projected to be available after industrial demonstration in about 7 years time (Armstrong *et al.*, 2002).

3.4.6   Chemical looping combustion

Originally proposed by Richter and Knoche (1983) and with subsequent significant contributions by Ishida and Jin (1994), the main idea of chemical looping combustion is to split combustion of a hydrocarbon or carbonaceous fuel into separate oxidation and reduction reactions by introducing a suitable metal oxide as an oxygen carrier to circulate between two reactors (Figure 3.13). Separation of oxygen from air is accomplished by fixing the oxygen as a metal oxide. No air separation plant is required. The reaction between fuel and oxygen is accomplished in a second reactor by the release of oxygen from the metal oxide in a reducing atmosphere caused by the presence of a hydrocarbon or carbonaceous fuel. The recycle rate of the solid material between the two reactors and the average solids residence time in each reactor, control the heat balance and the temperature levels in each reactor. The effect of having combustion in two reactors compared to conventional combustion in a single stage is that the CO_2 is not diluted with nitrogen gas, but is almost pure after separation from water, without requiring any extra energy demand and costly external equipment for CO_2 separation.

Possible metal oxides are some oxides of common transition-state metals, such as iron, nickel, copper and manganese (Zafar *et al.*, 2005). The metal/metal oxide may be present in various forms, but most studies so far have assumed the use of particles with diameter 100-500 μm. In order to move particles between the two reactors, the particles are fluidized. This method also ensures efficient heat and mass transfer between the gases and the particles. A critical issue is the long-term mechanical and chemical stability of the particles that have to undergo repeated cycles of oxidation and reduction, to minimize the make-up requirement. When a chemical looping cycle is used in a gas turbine cycle, the mechanical strength for crushing and the filtration system is important to avoid damaging carry-over to the turbine.

The temperature in the reactors, according to available information in the literature, may be in the range 800°C-

1200°C. NO_x formation at these typical operating temperatures will always be low. The fuel conversion in the reduction reactor may not be complete, but it is likely (Cho *et al.*, 2002) that the concentrations of methane and CO when burning natural gas are very small. In order to avoid deposit of carbon in the reduction reactor, it is necessary to use some steam together with the fuel.

The chemical looping principle may be applied either in a gas turbine cycle with pressurized oxidation and reduction reactors, or in a steam turbine cycle with atmospheric pressure in the reactors. In the case of a gas turbine cycle, the oxidation reactor replaces the combustion chamber of a conventional gas turbine. The exothermic oxidation reaction provides heat for increasing the air temperature entering the downstream expansion turbine. In addition, the reduction reactor exit stream may also be expanded in a turbine together with steam production for power generation. The cooled low pressure CO_2 stream will then be compressed to pipeline pressure. Another option is to generate steam using heat transfer surfaces in the oxidation reactor. Current circulating fluidized bed combustion technology operating at atmospheric pressure in both the oxidation and reduction stages necessitates the use of a steam turbine cycle for power generation. Using natural gas as fuel in a chemical looping combustion cycle which supplies a gas turbine combined cycle power plant and delivering CO_2 at atmospheric pressure, the potential for natural gas fuel-to-electricity conversion efficiency is estimated to be in the range 45-50% (Brandvoll and Bolland, 2004). Work on chemical looping combustion is currently in the pilot plant and materials research stage.

3.4.7   Status and outlook

Oxy-fuel combustion applied to furnaces, process heaters, boilers and power generation systems is feasible since no technical barriers for its implementation have been identified. Early use of this capture technology is likely to address applications involving indirect heating in power generation and process heating (Section 3.4.2), since these options involve the minimal modification of technologies and infrastructure that have hitherto been already developed for the combustion of hydrocarbon or carbonaceous fuels in air. However, several novel applications proposed for direct heating in steam turbine cycles or gas turbine cycles for power generation (Sections 3.4.3 and 3.4.4) still require the development of new components such as oxy-fuel combustors, higher temperature tolerant components such as CO_2- and H_2O-based turbines with blade cooling, CO_2 compressors and high temperature ion transport membranes for oxygen separation. As for Chemical Looping Combustion, it is currently still at an early stage of development.

The potential for thermal efficiencies for oxy-fuel cycles with CO_2 capture, assuming the current state of development in power plant technology, is depicted in Figures 3.6 and 3.7. Power generation from pulverized coal fired systems, using supercritical steam conditions presently operate at efficiencies around 45% (LHV), while projections to the 2010-2020 time

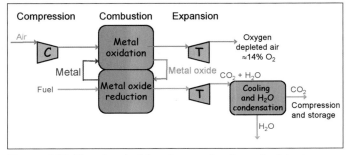

Figure 3.13 The chemical looping combustion principle in a gas turbine cycle.

frame are predicting efficiencies above 50% (IEA, 2004) for plants using ultra-supercritical steam conditions. An increase in efficiency of more than 5% can therefore be expected for future oxy-fuel capture systems based on coal firing that could potentially match the best efficiencies realisable today for pulverized coal-fired plants without CO_2 capture. Similarly, natural gas fired combined cycles will have efficiencies of 65% in 2020 (IEA GHG, 2000b and up from current efficiencies between 55 and 58%), which will enable plant efficiencies for natural gas fired oxy-fuel cycles with CO_2 capture above 50%. The energy penalty for producing oxygen is by far the most important cause for reduced efficiency in an oxy-fuel cycle compared to a conventional power plant.

Current technology development envisages very high efficiency separation of NO_x, SO_x, and Hg, as part of the CO_2 compression and purification system. Improved separation efficiencies of these contaminants are possible based on further process and heat integration in the power cycle.

Current cryogenic oxygen technology is showing continuing cost reduction based on improved compressor efficiencies, more efficient process equipment and larger scale plants. The new high temperature oxygen membrane could significantly improve power generation efficiency and reduce capital cost.

Future oxy-fuel demonstration plants could be based on retrofits to existing equipment such as process heaters and boilers, in order to minimize development costs and achieve early market entry. In this respect, power systems of reference for oxy-fuel combustion capture are mainly the steam-based pulverized coal and natural gas fired plants that currently represent up to 1468 GW_e, or 40% (IEA WEO, 2004) of the existing global infrastructure (see also Section 3.1.2.3). Several demonstration units may be expected within the next few years particularly in Europe, USA, Canada and Australia where active research initiatives are currently underway. As these developments proceed and the technologies achieve market penetration they may become competitive relative to alternate options based on pre- and post-combustion CO_2 capture. A significant incentive to the development of oxy-fuel combustion technology, as well as for pre- and post-combustion capture technologies, is the introduction of environmental requirements and/or fiscal incentives to promote CO_2 capture and storage.

3.5 Pre-combustion capture systems

3.5.1 Introduction

A pre-combustion capture process typically comprises a first stage of reaction producing a mixture of hydrogen and carbon monoxide (syngas) from a primary fuel. The two main routes are to add steam (reaction 1), in which case the process is called 'steam reforming', or oxygen (reaction 2) to the primary fuel. In the latter case, the process is often called 'partial oxidation' when applied to gaseous and liquid fuels and 'gasification' when applied to a solid fuel, but the principles are the same.

Steam reforming
$$C_xH_y + xH_2O \leftrightarrow xCO + (x+y/2)H_2 \quad \Delta H \text{ +ve} \quad (1)$$

Partial oxidation
$$C_xH_y + x/2O_2 \leftrightarrow xCO + (y/2)H_2 \quad \Delta H \text{ -ve} \quad (2)$$

This is followed by the 'shift' reaction to convert CO to CO_2 by the addition of steam (reaction 3):

Water Gas Shift Reaction
$$CO + H_2O \leftrightarrow CO_2 + H_2 \quad \Delta H \text{ -41 kJ mol}^{-1} \quad (3)$$

Finally, the CO_2 is removed from the CO_2/H_2 mixture. The concentration of CO_2 in the input to the CO_2/H_2 separation stage can be in the range 15-60% (dry basis) and the total pressure is typically 2-7 MPa. The separated CO_2 is then available for storage.

It is possible to envisage two applications of pre-combustion capture. The first is in producing a fuel (hydrogen) that is essentially carbon-free. Although the product H_2 does not need to be absolutely pure and may contain low levels of methane, CO or CO_2, the lower the level of carbon-containing compounds, the greater the reduction in CO_2 emissions. The H_2 fuel may also contain inert diluents, such as nitrogen (when air is typically used for partial oxidation), depending on the production process and can be fired in a range of heaters, boilers, gas turbines or fuel cells.

Secondly, pre-combustion capture can be used to reduce the carbon content of fuels, with the excess carbon (usually removed as CO_2) being made available for storage. For example, when using a low H:C ratio fuel such as coal it is possible to gasify the coal and to convert the syngas to liquid Fischer-Tropsch fuels and chemicals which have a higher H:C ratio than coal. In this section, we consider both of these applications.

This section reports on technologies for the production of H_2 with CO_2 capture that already exist and those that are currently emerging. It also describes enabling technologies that need to be developed to enhance the pre-combustion capture systems for power, hydrogen or synfuels and chemicals production or combination of all three.

3.5.2 Existing technologies

3.5.2.1 Steam reforming of gas and light hydrocarbons
Steam reforming is the dominant technology for hydrogen production today and the largest single train plants produce up to 480 tH_2 d^{-1}. The primary energy source is often natural gas, Then the process is referred to as steam methane reforming (SMR), but can also be other light hydrocarbons, such as naphtha. The process begins with the removal of sulphur compounds from the feed, since these are poisons to the current nickel-based catalyst and then steam is added. The reforming reaction (1), which is endothermic, takes place over a catalyst at high temperature (800°C-900°C). Heat is supplied to the reactor tubes by burning part of the fuel (secondary fuel). The reformed gas is cooled in a waste heat boiler which generates the steam needed for the reactions and passed into the CO shift system. Shift reactors in one or two stages are used to convert most of the CO in the syngas to CO_2 (Reaction 3, which is exothermic).

The conventional two-stage CO conversion reduces the CO concentration in syngas (or in hydrogen) down to 0.2-0.3%. High temperature shift reactors operating between 400°C and 550°C and using an iron-chromium catalyst leave between 2% and 3% CO in the exit gas (dry basis). Copper-based catalyst can be used at temperatures from 180°C-350°C and leave from 0.2-1% CO in the exhaust. Lower CO content favours higher CO$_2$ recovery. The gas is then cooled and hydrogen is produced by a CO$_2$/H$_2$ separation step. Until about 30 years ago, the CO$_2$ was removed using a chemical (solvent) absorption process such as an amine or hot potassium carbonate and was rejected to atmosphere as a pure stream from the top of the regenerator. There are many of these plants still in use and the CO$_2$ could be captured readily.

Modern plants, however, use a pressure swing adsorber (PSA), where gases other than H$_2$ are adsorbed in a set of switching beds containing layers of solid adsorbent such as activated carbon, alumina and zeolites (see the fuller description of PSA in Section 3.5.2.9). The H$_2$ exiting the PSA (typically about 2.2 MPa) can have a purity of up to 99.999%, depending on the market need. The CO$_2$ is contained in a stream, from the regeneration cycle, which contains some methane and H$_2$. The stream is used as fuel in the reformer where it is combusted in air and the CO$_2$ ends up being vented to atmosphere in the reformer flue gas. Hence, to capture CO$_2$ from modern SMR plants would require one of the post-combustion processes described above in Section 3.3. Alternatively, the PSA system could be designed not only for high recovery of pure H$_2$ but also to recover pure CO$_2$ and have a fuel gas as the third product stream.

In a design study for a large modern plant (total capacity 720 tH$_2$ d^{-1}), the overall efficiency of making 6.0 MPa H$_2$ from natural gas with CO$_2$ vented that is without CO$_2$ capture, is estimated to be 76%, LHV basis, with emissions of 9.1 kg CO$_2$/kg H$_2$ (IEA GHG, 1996). The process can be modified (at a cost) to provide a nearly pure CO$_2$ co-product. One possibility is to remove most of the CO$_2$ from the shifted, cooled syngas in a 'wet' CO$_2$ removal plant with an appropriate amine solvent. In this case the CO$_2$-deficient syngas exiting the amine scrubber is passed to a PSA unit from which relatively pure H$_2$ is recovered and the PSA purge gases are burned along with additional natural gas to provide the needed reformer heat. The CO$_2$ is recovered from the amine solvent by heating and pressurized for transport. Taking into account the power to compress the CO$_2$ (to 11.2 MPa) reduces the efficiency to about 73% and the emission rate to 1.4 kgCO$_2$/kgH$_2$, while the CO$_2$ removal rate is 8.0 kgCO$_2$/kgH$_2$.

3.5.2.2 Partial oxidation of gas and light hydrocarbons

In the partial oxidation (POX) process (reaction 2), a fuel reacts with pure oxygen at high pressure. The process is exothermic and occurs at high temperatures (typically 1250°C-1400°C). All the heat required for the syngas reaction is supplied by the partial combustion of the fuel and no external heat is required. As with SMR, the syngas will be cooled, shifted and the CO$_2$ removed from the mixture. The comments made on the separation of CO$_2$ from SMR syngas above apply equally to the POX process. POX is a technology in common use today, the efficiency is lower than SMR, but the range of fuels that can be processed is much wider.

For large-scale hydrogen production, the oxygen is supplied from a cryogenic air separation unit (ASU). The high investment and energy consumption of the ASU is compensated by the higher efficiency and lower cost of the gasification process and the absence of N$_2$ (from the air) in the syngas, which reduces the separation costs considerably. However for pre-combustion de-carbonization applications, in which the hydrogen would be used as fuel in a gas turbine, it will be necessary to dilute the H$_2$ with either N$_2$ or steam to reduce flame temperature in the gas turbine combustor and to limit NO$_x$ emission levels. In this case the most efficient system will use air as the oxidant and produce a H$_2$/N$_2$ fuel mixture (Hufton *et al.* 2005)

3.5.2.3 Auto-thermal reforming of gas and light hydrocarbons

The autothermal reforming (ATR) process can be considered as a combination of the two processes described above. The heat required in the SMR reactor is generated by the partial oxidation reaction (2) using air or oxygen, but because steam is supplied to the reactor as well as excess natural gas, the endothermic reforming reaction (1) occurs in a catalytic section of the reactor downstream of the POX burner. The addition of steam enables a high conversion of fuel to hydrogen at a lower temperature. Operating temperatures of the autothermal process are typically 950-1050°C, although this depends on the design of the process. An advantage of the process, compared to SMR, is the lower investment cost for the reactor and the absence of any emissions of CO$_2$ since all heat release is internal, although this is largely offset by investment and operating cost for the oxygen plant. The range of fuels that can be processed is similar to the SMR process, but the feed gas must be sulphur free. CO$_2$ capture is accomplished as described above for the steam methane reforming.

3.5.2.4 Gas heated reformer

Each of the three syngas generation technologies, SMR, ATR and POX produce high temperature gas which must be cooled, producing in each case a steam flow in excess of that required by the reforming and shift reactions. It is possible to reduce this excess production by, for example, using preheated air and a pre-reformer in an SMR plant. Another technique is to use the hot syngas, leaving the primary reactor, as the shell-side heating fluid in a tubular steam/hydrocarbon reforming reactor which can operate in series, or in parallel, with the primary reactor (Abbott *et al.*, 2002). The addition of a secondary gas heated reformer will increase the hydrogen production by up to 33% and eliminate the excess steam production. The overall efficiency is improved and specific capital cost is typically reduced by 15%. Again, CO$_2$ capture is accomplished as described previously for steam methane reforming.

3.5.2.5 Gasification of coal, petroleum residues, or biomass

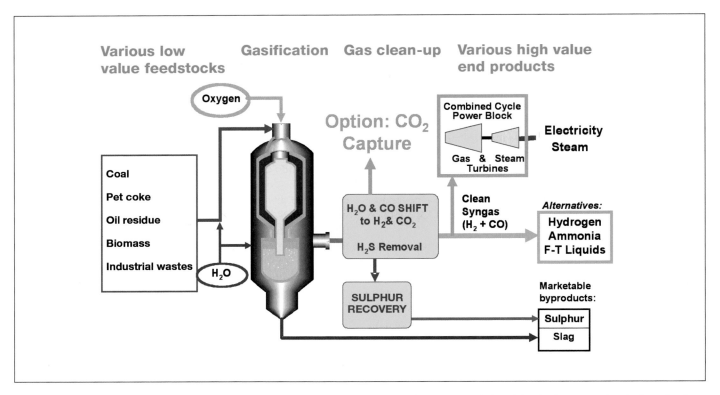

Figure 3.14 Simplified schematic of a gasification process showing options with CO_2 capture and electricity, hydrogen or chemical production.

Gasification (see Figure 3.14) is a chemical process aimed at making high-value products (chemicals, electricity, clean synthetic fuels) out of low-value solid feedstocks such as coal, oil refining residues, or biomass. Gasification is basically partial oxidation (reaction 2), although steam is also supplied to the reactor in most processes. Fixed bed, fluidized bed or entrained flow gasifiers can be used. These can have very different characteristics with respect to oxidant (air or O_2), operating temperature (up to 1350°C), operating pressure (0.1-7 MPa), feed system (dry or water slurry), syngas cooling method (water quench or via radiative and convective heat exchangers) and gas clean-up system deployed. These alternative design options determine the fraction of feedstock converted to syngas, syngas composition and cost. As economics depend strongly on scale, gasification is generally considered to be suitable only for large plants. The gasifier output contains CO, H_2, CO_2 H_2O and impurities (e.g., N_2, COS, H_2S, HCN, NH_3, volatile trace minerals and Hg) that must be managed appropriately.

A worldwide survey of commercial gasification projects identified 128 operating gasification plants with 366 gasifiers producing 42,700 MW_t of syngas (NETL-DOE, 2002 and Simbeck, 2001a). There are also about 24,500 MW_t of syngas projects under development or construction, with 4000-5000 MW_t of syngas added annually. The feedstocks are mainly higher rank coals and oil residues. Most commercial gasification growth for the last 20 years has involved entrained-flow gasifiers, for which there are three competing systems on the market. Recent commercial gasification development has been mainly with industrial ammonia production, industrial polygeneration (in which clean syngas is used to make electricity and steam

along with premium syngas chemicals) and IGCC power plants. Commercial experience with biomass gasification and fluidized bed gasification has been limited.

CO_2 capture technology is well established for gasification systems that make chemicals and synthetic fuels (NETL-DOE, 2002). Gasification-based NH_3 plants (many in China) include making pure H_2 and CO_2 separation at rates up to 3500 tCO_2 d^{-1} per plant. South African plants making Fischer-Tropsch fuels and chemicals and a North Dakota plant making synthetic natural gas (SNG) from coal also produce large streams of nearly pure CO_2. Figure 3.15 shows a picture of the North Dakota gasification plant in which 3.3 $MtCO_2$ yr^{-1} is captured using a refrigerated methanol-based, physical solvent scrubbing process (Rectisol process, see Section 3.5.2.11 and Table 3.2). Most of this captured CO_2 is vented and about 1.5 Mtonnes yr^{-1} of this stream is currently pipelined to the Weyburn, Canada enhanced oil recovery and CO_2 storage project (see Chapter 5).

When CO_2 capture is an objective, O_2-blown and high-pressure systems are preferred because of the higher CO_2 partial pressures. De-carbonization via gasification entails lower energy penalties for CO_2 capture than does post-combustion capture when considering only the separation stage, because the CO_2 can be recovered at partial pressures up to 3 orders of magnitude higher. This greatly reduces CO_2 absorber size, solvent circulation rates and CO_2 stripping energy requirements. However, additional energy penalties are incurred in shifting the CO in the syngas to CO_2 and in other parts of the system (see examples for IGCC plant with CO_2 capture in Figures 3.6 and 3.7). Recent analyses for bituminous coals (see, for example, IEA GHG, 2003) suggest using simple high-pressure

Figure 3.15 North Dakota coal gasification plant with 3.3 MtCO$_2$ yr^{-1} capture using a cold methanol, physical solvent process (cluster of 4 tall columns in the middle of the picture represent the H$_2$S and CO$_2$ capture processes; part of the captured stream is used for EOR with CO$_2$ storage in Weyburn, Saskatchewan, Canada).

entrained-flow gasifiers with water slurry feed and direct water quench followed by 'sour' (sulphur-tolerant) shift reactors and finally co-removal of CO$_2$ and H$_2$S by physical absorption. With sour shifting, hot raw syngas leaving the gasifier requires only one cooling cycle and less processing. Oxygen requirements increase for slurry fed gasifiers and conversion efficiencies decline with higher cycle efficiency losses with quench cooling. Similar trends are also noted with a shift from bituminous to lower rank sub-bituminous coal and lignite (Breton and Amick, 2002). Some analyses (e.g., Stobbs and Clark, 2005) suggest that the advantages of pre-combustion over post-combustion de-carbonization may be small or disappear for low-rank coals converted with entrained-flow gasifiers. High-pressure, fluidized-bed gasifiers may be better suited for use with low-rank coals, biomass and various carbonaceous wastes. Although there are examples of successful demonstration of such gasifiers (e.g., the high temperature Winkler, Renzenbrink *et al.*, 1998), there has been little commercial-scale operating experience.

The H$_2$S in syngas must be removed to levels of tens of ppm for IGCC plants for compliance with SO$_2$ emissions regulations and to levels much less than 1 ppm for plants that make chemicals or synthetic fuels, so as to protect synthesis catalysts. If the CO$_2$ must be provided for storage in relatively pure form, the common practice would be to recover first H$_2$S (which is absorbed more readily than CO$_2$) from syngas (along with a small amount of CO$_2$) in one recovery unit, followed by reduction of H$_2$S to elemental sulphur in a Claus plant and tail gas clean-up, and subsequent recovery of most of the remaining CO$_2$ in a separate downstream unit. An alternative option is to recover sulphur in the form of sulphuric acid (McDaniel and Hormick, 2002). If H$_2$S/CO$_2$ co-storage is allowed, however, it would often be desirable to recover H$_2$S and CO$_2$ in the same physical absorption unit, which would lead to moderate system cost savings (IEA GHG, 2003; Larson and Ren, 2003; Kreutz *et al.*, 2005) especially in light of the typically poor prospects

for selling byproduct sulphur or sulphuric acid. Although co-storage of H$_2$S and CO$_2$ is routinely pursued in Western Canada as an acid gas management strategy for sour natural gas projects (Bachu and Gunter, 2005), it is not yet clear that co-storage would be routinely viable at large scales - a typical gasification-based energy project would involve an annual CO$_2$ storage rate of 1-4 Mtonnes yr^{-1}, whereas the total CO$_2$ storage rate for all 48 Canadian projects is presently only 0.48 Mtonnes yr^{-1} (Bachu and Gunter, 2005).

3.5.2.6 Integrated gasification combined cycle (IGCC) for power generation

In a coal IGCC, syngas exiting the gasifier is cleaned of particles, H$_2$S and other contaminants and then burned to make electricity via a gas turbine/steam turbine combined cycle. The syngas is generated and converted to electricity at the same site, both to avoid the high cost of pipeline transport of syngas (with a heating value only about 1/3 of that for natural gas) and to cost-effectively exploit opportunities for making extra power in the combined cycle's steam turbine using steam from syngas cooling. The main drivers for IGCC development were originally the prospects of exploiting continuing advances in gas turbine technology, the ease of realizing low levels of air-pollutant emissions when contaminants are removed from syngas, and greatly reduced process stream volumes compared to flue gas streams from combustion which are at low pressure and diluted with nitrogen from air.

Since the technology was initially demonstrated in the 1980s, about 4 GW$_e$ of IGCC power plants have been built. Most of this capacity is fuelled with oil or petcoke; less than 1 GW$_e$ of the total is designed for coal (IEA CCC, 2005) and 3 out of 4 plants currently operating on coal and/or petcoke. This experience has demonstrated IGCC load-following capability, although the technology will probably be used mainly in base load applications. All coal-based IGCC projects have been subsidized, whereas only the Italian oil-based IGCC projects have been subsidized. Other polygeneration projects in Canada, the Netherlands and the United States, as well as an oil-based IGCC in Japan, have not been subsidized (Simbeck, 2001a).

IGCC has not yet been deployed more widely because of strong competition from the natural gas combined cycle (NGCC) wherever natural gas is readily available at low prices, because coal-based IGCC plants are not less costly than pulverized coal fired steam-electric plants and because of availability (reliability) concerns. IGCC availability has improved in recent years in commercial-scale demonstration units (Wabash River Energy, 2000; McDaniel and Hornick, 2002). Also, availability has been better for industrial polygeneration and IGCC projects at oil refineries and chemical plants where personnel are experienced with the chemical processes involved. The recent rise in natural gas prices in the USA has also increased interest in IGCC.

Because of the advantages for gasification of CO$_2$ capture at high partial pressures discussed above, IGCC may be attractive for coal power plants in a carbon-constrained world (Karg and Hannemann, 2004). CO$_2$ capture for pre-combustion systems is commercially ready, however, no IGCC plant incorporating CO_2 capture has yet been built. With current technology, average estimates of the energy penalties and the impact of increased fuel use for CO_2 removal are compared with other capture systems in Figures 3.6 and 3.7 and show the prospective potential of IGCC options. The data in Figures 3.6 and 3.7 also show that some IGCC options may be different from others (i.e., slurry fed and quench cooled versus dry feed and syngas cooling) and their relative merits in terms of the capital cost of plant and the delivered cost of power are discussed in Section 3.7.

3.5.2.7 Hydrogen from coal with CO_2 capture

Relative to intensively studied coal IGCC technology with CO_2 capture, there are few studies in the public domain on making H_2 from coal via gasification with CO_2 capture (NRC, 2004; Parsons 2002a, b; Gray and Tomlinson, 2003; Chiesa *et al.*, 2005; Kreutz *et al.*, 2005), even though this H_2 technology is well established commercially, as noted above. With commercial technology, H_2 with CO_2 capture can be produced via coal gasification in a system similar to a coal IGCC plant with CO_2 capture. In line with the design recommendations for coal IGCC plants described above (IEA GHG, 2003), what follows is the description from a design study of a coal H_2 system that produces, using best available technology, 1070 MW_t of H_2 from high-sulphur (3.4%) bituminous coal (Chiesa *et al.*, 2005; Kreutz *et al.*, 2005). In the base case design, syngas is produced in an entrained flow quench gasifier operated at 7 MPa. The syngas is cooled, cleaned of particulate matter, and shifted (to primarily H_2 and CO_2) in sour water gas shift reactors. After further cooling, H_2S is removed from the syngas using a physical solvent (Selexol). CO_2 is then removed from the syngas, again using Selexol. After being stripped from the solvents, the H_2S is converted to elemental S in a Claus unit and a plant provides tail gas clean-up to remove residual sulphur emissions; and the CO_2 is either vented or dried and compressed to 150 atm for pipeline transport and underground storage. High purity H_2 is extracted at 6 MPa from the H_2-rich syngas via a pressure swing adsorption (PSA) unit. The PSA purge gas is compressed and burned in a conventional gas turbine combined cycle, generating 78 MW_e and 39 MW_e of electricity in excess of onsite electricity needs in the without and with CO_2 capture cases, respectively. For this base case analysis, the effective efficiency of H_2 manufacture was estimated to be 64% with CO_2 vented and 61% with CO_2 captured, while the corresponding emission rates are 16.9 $kgCO_2$ and 1.4 $kgCO_2/ kgH_2$, respectively. For the capture case, the CO_2 removal rate was 14.8 $kgCO_2/kgH_2$. Various alternative system configurations were explored. It was found that there are no thermodynamic or cost advantages from increasing the electricity/H_2 output ratio, so this ratio would tend to be determined by relative market demands for electricity and H_2. One potentially significant option for reducing the cost of H_2 with CO_2 capture to about the same level as with CO_2 vented involves H_2S/CO_2 co-capture in a single Selexol unit, as discussed above.

3.5.2.8 Carbon-based fluid fuels and multi-products

As discussed in Chapter 2, clean synthetic high H/C ratio fuels can be made from syngas via gasification of coal or other low H/C ratio feedstocks. Potential products include synthetic natural gas, Fischer-Tropsch diesel/gasoline, dimethyl ether, methanol and gasoline from methanol via the Mobil process. A byproduct is typically a stream of relatively pure CO_2 that can be captured and stored.

Coal derived Fischer-Tropsch synfuels and chemicals have been produced on a commercial scale in South Africa; coal methanol is produced in China and at one US plant; and coal SNG is produced at a North Dakota (US) plant (NETL-DOE, 2002). Since 2000, 1.5 $MtCO_2$ yr^{-1} from the North Dakota synthetic natural gas plant (see Figure 3.15) have been transported by pipeline, 300 km to the Weyburn oil field in Saskatchewan, Canada for enhanced oil recovery with CO_2 storage.

Synfuel manufacture involves O_2-blown gasification to make syngas, gas cooling, gas clean-up, water gas shift and acid gas (H_2S/CO_2) removal. Subsequently cleaned syngas is converted catalytically to fuel in a synthesis reactor and unconverted syngas is separated from the liquid fuel product. At this point either most unconverted gas is recycled to the synthesis reactor to generate additional liquid fuel and the remaining unconverted gas is used to make electricity for onsite needs, or syngas is passed only once through the synthesis reactor, and all unconverted syngas is used for other purposes, for example, to make electricity for sale to the electric grid as well as for onsite use. The latter *once through* option is often more competitive as a technology option (Williams, 2000; Gray and Tomlinson, 2001; Larson and Ren, 2003; Celik *et al.*, 2005).

New slurry-phase synthesis reactors make the once through configuration especially attractive for CO-rich (e.g., coal-derived) syngas by making high once through conversion possible. For once through systems, a water gas shift reactor is often placed upstream of the synthesis reactor to generate the H_2/CO ratio that maximizes synfuel conversion in the synthesis reactor. It is desirable to remove most CO_2 from shifted syngas to maximize synthetic fuel conversion. Also, because synthesis catalysts are extremely sensitive to H_2S and various trace contaminants, these must be removed to very low levels ahead of the synthesis reactor. Most trace metals can be removed at low-cost using an activated carbon filter. CO_2 removal from syngas upstream of the synthesis reactor is a low-cost, partial de-carbonization option, especially when H_2S and CO_2 are co-captured and co-stored as an acid gas management strategy (Larson and Ren, 2003). Further de-carbonization can be realized in once through systems, at higher incremental cost, by adding additional shift reactors downstream of the synthesis reactor, recovering the CO_2, and using the CO_2-depleted, H_2-rich syngas to make electricity or some mix of electricity plus H_2 in a 'polygeneration' configuration (see Figure 3.16). The relative amounts of H_2 and electricity produced would depend mainly on relative demands, as there do not seem to be thermodynamic or cost advantages for particular H_2/electricity production ratios (Chiesa *et al.*, 2005; Kreutz *et al.*, 2005). When syngas is de-carbonized both upstream and downstream of the synthesis reactor (see Figure 3.16) it is feasible to capture and store as CO_2 up to 90% of the carbon in the original feedstock except

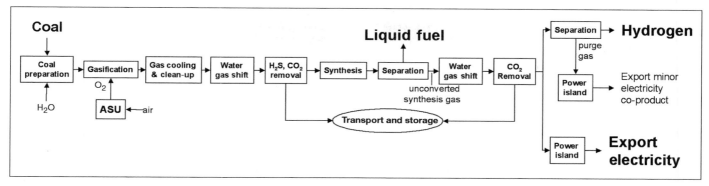

Figure 3.16 Making liquid fuel, electricity and hydrogen from coal via gasification, with CO_2 capture and storage.

that contained in the synthetic fuel produced.

An example of such a system (Celik *et al.*, 2005) is one making 600 MW of dimethyl ether (containing 27% of coal input energy and 20% of coal input carbon) plus 365 MW of electricity (no H_2) from coal. For this system the CO_2 storage rate (equivalent to 74% of C in coal) is 3.8 Mtonnes yr⁻¹ (39% from upstream of the synthesis reactor). The estimated fuel cycle-wide GHG emissions for dimethyl ether are 0.9 times those for crude oil-derived diesel and those for electricity are 0.09 times those for a 43% efficient coal-fired power plant with CO_2 vented.

3.5.2.9   *Pressure swing adsorption*
Pressure Swing Adsorption (PSA) is the system of choice for the purification of syngas, where high purity H_2 is required. However, it does not selectively separate CO_2 from the other waste gases and so for an SMR application the CO_2 concentration in the waste gas would be 40-50% and require further upgrading to produce pure CO_2 for storage. Simultaneous H_2 and CO_2 separation is possible by using an additional PSA section to remove the CO_2 prior to the H_2 separation step, such as the Air Products Gemini Process (Sircar, 1979).

The PSA process is built around adsorptive separations of cyclic character. The cycles consist of two basic steps: adsorption, in which the more adsorbable species are selectively removed from the feed gas and regeneration (desorption), when these species are removed from the adsorbent so that it can be ready for the next cycle. It is possible to obtain useful products during both adsorption and regeneration. The principal characteristic of PSA processes is the use of a decrease in pressure and/or the purge by a less adsorbable gas to clean the adsorbent bed. Apart from adsorption and regeneration, a single commercial PSA cycle consists of a number of additional steps, including co- and counter-current pressurization, pressure equalization and co- and counter-current depressurization. A detailed description of the PSA technique, along with its practical applications can be found elsewhere (Ruthven *et al.*, 1994).

3.5.2.10   *Chemical solvent processes*
Chemical solvents are used to remove CO_2 from syngas at partial pressures below about 1.5 MPa (Astarita *et al.*, 1983) and are similar to those used in post-combustion capture (see Section 3.3.2.1). The solvent removes CO_2 from the shifted syngas by

means of a chemical reaction, which can be reversed by pressure reduction and heating. The tertiary amine methyldiethanolamine (MDEA, see Table 3.2) is widely used in modern industrial processes, due to the high CO_2 loading possible and the low regenerator heating load, relative to other solvents. Hot potassium carbonate (the most common commercial version of which is known as Benfield) was used for CO_2 removal in most hydrogen plants until about 15 years ago.

3.5.2.11   *Physical solvent processes*
Physical solvent (or absorption) processes are mostly applicable to gas streams which have a high CO_2 partial pressure and/or a high total pressure. They are often used to remove the CO_2 from the mixed stream of CO_2 and H_2 that comes from the shift reaction in pre-combustion CO_2 capture processes, such as product from partial oxidation of coal and heavy hydrocarbons.

The leading physical solvent processes are shown in Table 3.2. The regeneration of solvent is carried out by release of pressure at which CO_2 evolves from the solvent, in one or more stages. If a deeper regeneration is required the solvent would be stripped by heating. The process has low energy consumption, as only the energy for pressurizing the solvent (liquid pumping) is required.

The use of high sulphur fossil fuels in a pre-combustion capture process results in syngas with H_2S. Acid gas components must be removed. If transport and storage of mixed CO_2 and H_2S is possible then both components can be removed together. Sulphinol was developed to achieve significantly higher solubilities of acidic components compared to amine solvents, without added problems of excessive corrosion, foaming, or solution degradation. It consists of a mixture of sulpholane (tetrahydrothiophene 1,1-dioxide), an alkanolamine and water in various proportions depending on the duty. If pure CO_2 is required, then a selective process is required using physical solvents - often Rectisol or Selexol. The H_2S must be separated at sufficiently high concentration (generally >50%) to be treated in a sulphur recovery plant.

3.5.2.12   *Effect on other pollutants*
Pre-combustion capture includes reforming, partial oxidation or gasification. In order to maintain the operability of the catalyst of reformers, sulphur (H_2S) has to be removed prior to reforming. In gasification, sulphur can be captured from the

syngas, and in the case when liquid or solid fuels are gasified, particulates, NH_3, COS and HCN are also present in the system that need to be removed. In general, all of these pollutants can be removed from a high-pressure fuel gas prior to combustion, where combustion products are diluted with nitrogen and excess oxygen. In the combustion of hydrogen or a hydrogen-containing fuel gas, NO_x may be formed. Depending upon combustion technology and hydrogen fraction, the rate at which NO_x is formed may vary. If the volumetric fraction of hydrogen is below approximately 50-60%, NO_x formation is at the same level as for natural gas dry low-NO_x systems (Todd and Battista, 2001).

In general, with the exception of H_2S that could be co-removed with CO_2, other pollutants identified above are separated in additional pretreatment operations, particularly in systems that gasify liquid or solid fuels. High temperature pretreatment operations for these multi-pollutants that avoid cooling of the syngas have the advantage of improving the cycle efficiency of the overall gasification process, but these separation processes have not been commercially demonstrated.

Although it is not yet regulated as a 'criteria pollutant', mercury (Hg), is currently the focus of considerable concern as a pollutant from coal power systems. For gasification systems Hg can be recovered from syngas at ambient temperatures at very low-cost, compared to Hg recovery from flue gases (Klett *et al.*, 2002).

3.5.3   Emerging technologies

Emerging options in both natural gas reforming and coal gasification incorporate novel combined reaction/separation systems such as sorption-enhanced reforming and sorption-enhanced water gas shift, membrane reforming and membrane water gas shift. Finally there is a range of technologies that make use of the carbonation of CaO for CO_2 capture.

3.5.3.1   Sorption enhanced reaction
A concept called Sorption Enhanced Reaction (SER) uses a packed bed containing a mixture of a catalyst and a selective adsorbent to remove CO_2 from a high temperature reaction zone, thus driving the reaction to completion. (Hufton *et al.*, 1999). The adsorbent is periodically regenerated by using a pressure swing, or temperature swing adsorption system with steam regeneration (Hufton *et al.*, 2005).

High temperature CO_2 adsorbents such as hydrotalcites (Hufton *et al.*, 1999) or lithium silicate (Nakagawa and Ohashi, 1998) can be mixed with a catalyst to promote either the steam methane reforming reaction (Reaction 1) or water gas shift reaction (Reaction 3) producing pure hydrogen and pure CO_2 in a single process unit. The continuous removal of the CO_2 from the reaction products by adsorption shifts each reaction towards completion.

The SER can be used to produce hydrogen at 400-600°C to fuel a gas turbine combined cycle power generation system. A design study based on a General Electric 9FA gas turbine with hot hydrogen, produced from an air blown ATR with a

sorption enhanced water gas shift reactor, gave a theoretical net efficiency of 48.3% with 90% CO_2 capture at 99% purity and 150 bar pressure (Hufton *et al.*, 2005). The process is currently at the pilot plant stage.

3.5.3.2   Membrane reactors for hydrogen production with CO_2 capture
Inorganic membranes with operating temperatures up to 1000°C offer the possibility of combining reaction and separation of the hydrogen in a single stage at high temperature and pressure to overcome the equilibrium limitations experienced in conventional reactor configurations for the production of hydrogen. The combination of separation and reaction in membrane steam reforming and/or membrane water gas shift offers higher conversion of the reforming and/or shift reactions due to the removal of hydrogen from these equilibrium reactions as shown in Reactions (1) and (3) respectively. The reforming reaction is endothermic and can, with this technique, be forced to completion at lower temperature than normal (typically 500-600°C). The shift reaction being exothermic can be forced to completion at higher temperature (500-600°C).

Another reason to incorporate H_2 separation membranes in the hydrogen production system is that CO_2 is also produced without the need for additional separation equipment. Membrane reactors allow one-step reforming, or a single intermediate water gas shift reaction, with hydrogen separation (the permeate) leaving behind a retentate gas which is predominantly CO_2 and a small amount of non-recovered hydrogen and steam. This CO_2 remains at the relatively high pressure of the reacting system (see Figure 3.17). Condensation of the steam leaves a concentrated CO_2 stream at high pressure, reducing the compression energy for transport and storage. Membrane reforming will benefit from high-pressure operation due to the increased H_2 partial pressure differential across the membrane which is the driving force for hydrogen permeation. Therefore membrane reactors are also seen as a good option for pre-combustion de-carbonization where a low-pressure hydrogen stream for fuel gas and a high-pressure CO_2-rich stream for transport and storage are required. The use of the membrane reformer reactor in a gas turbine combined cycle means that the hydrogen needs to be produced at such pressure that the significant power consumption for the hydrogen compression is avoided. This could be done by increasing the operating pressure of the membrane reactor or by using a sweep gas, for instance steam, at the permeate side of the membrane (Jordal *et al.*, 2003).

For these membrane reactor concepts, a hydrogen selective membrane capable of operating in a high-temperature, high-pressure environment is needed. In the literature a number of membrane types have been reported that have these capabilities and these are listed in Table 3.3. Microporous inorganic membranes based upon surface diffusion separation exhibit rather low separation factors (e.g., H_2/CO_2 separation factor of 15). However, the separation ability of the current commercially available gamma-alumina and silica microporous membranes (which have better separation factors, up to 40) depends upon the stability of the membrane pore size, which is adversely

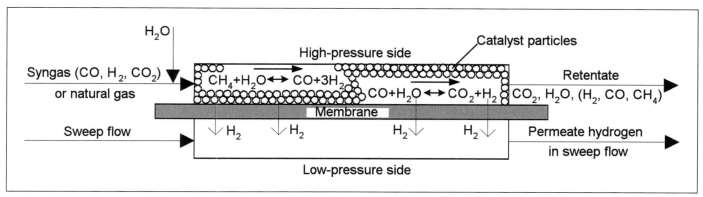

Figure 3.17 Operating principle of a membrane reactor.

Table 3.3 Membrane materials, operating conditions and characteristics for H_2 separation.

	Microporous Ceramic	Microporous Ceramic	Microporous Carbon	Zeolites	Metal
Membrane material	Alumina	Silica	Carbon	Silica (Alumina)	Pd/Ag
Temperature range (°C)	<500	<400	<400	<500 - 700	<600
Pressure range (bar)	>100	>100	10	>100	>100
Pore size distribution (nm)	0.7-2	0.7-2	0.7-2	0.3-0.7	no pores
Separation factors (H_2/CO_2)	15	15	15-25	50	100
Permeability (mol m⁻²s⁻¹Pa⁻¹)	10^{-6}	10^{-6}	10^{-7}	10^{-6}	10^{-7}-10^{-6}
Experim. temp. (°C)	200	200	300-400	300-400	300-400
Pre-clean-up requirements				S	S, HCl, HF (?)
Chemical resistance problem		H_2O	O_2	S	S, HCl, HF
Geometry	Top layer tube	Top layer tube	Top layer tube/fibre	Top layer tube	Top layer tube/plate
Configuration	Cascade/recycle/ once through	Cascade/recycle/ once through	Cascade/recycle/ once through	Once through	Once through
Lifetime	+	-	+	+	0
Costs (US$ m⁻²)	4250	4250	3000?	4000-4250	4000-4250
Scalability	0	0	0	-	0

affected by the presence of steam in the feed streams. The dense ceramic membranes based on inorganic perovskite oxides (also called proton conducting) need high temperatures, higher than 800°C, to achieve practical hydrogen flux rates. Palladium-based dense membranes are also known for their high hydrogen selectivity and permeability over other gases in the temperature range 300°C-600°C that is appropriate for these two reactions. Palladium alloy tubes have been available for several decades, but for CCS applications they are too expensive due to the membrane thickness needed for structural stability and consequently low hydrogen flux rates. In order to be suitable for the target application, a hydrogen separation membrane must have adequate selectivity and flux rate and must be stable in the reducing coal gas or fuel-reforming environment containing steam and hydrogen sulphide.

A number of membrane reactor developments have been reported for hydrogen production with CO_2 capture. Several groups have evaluated methane steam reforming membrane

reactors based on palladium alloy membranes (Middleton *et al.*, 2002, Damle and Dorchak, 2001). These evaluations showed that membrane reactors could achieve 90% CO_2 recovery and that at this moment the projected cost is nearly identical to that for a conventional system. However, a cost-reduction can be achieved by either reducing the material cost of the membrane or by increasing the permeability. Similar evaluations of membrane reactors for the shift conversion and separation of CO_2 from syngas produced from heavy feeds by gasification have been reported (Bracht *et al.*, 1997; Middleton 2002; Lowe *et al.*, 2003). For these gasifier systems the membrane reactors could reduce the costs for capturing CO_2 and the cost reduction would be more significant if they could be made sulphur tolerant.

3.5.3.3 Microchannel reformer
Microreactor technology can be used to produce a SMR, or low temperature air-based POX system using a multichannel plate-

fin heat exchanger, fabricated in stainless steel or high nickel alloy by vacuum brazing or diffusion bonding.

An SMR reactor consists of alternate passages having fins, which are coated with catalyst or porous catalyst insets. Heat is produced by catalytic combustion of fuel gas premixed with air and transferred by conduction to the adjacent passage fed with the steam/hydrocarbon mixture, where the reforming reaction takes place (Babovic *et al.*, 2001). Very compact high efficiency systems can be produced. Although these units are being currently developed by a number of groups for small-scale H_2 production for fuel cell applications, they also show promise in larger H_2 plants.

3.5.3.4 Conversion to hydrogen and carbon

Thermal cracking or pyrolysis of methane is the reaction where methane reacts to carbon and hydrogen through:

Methane pyrolysis:
$$CH_4 \rightarrow C + 2\,H_2 \qquad\qquad (4)$$

The main advantage of the process is that it can potentially yield a clean gas (free of carbon oxides) that could be used directly for power production, but a disadvantage is that the chemical energy from the oxidation of carbon to CO_2 is not released. The cracking reaction is endothermic and so heat has to be supplied to the reaction. If the natural gas is converted fully, the theoretical yield of hydrogen corresponds to 60% of the heating value of the natural gas. The amount of carbon, which can be obtained, corresponds to 49% of the heating value, with the extra 9% of the energy in this calculation being provided as endothermic heat shown by reaction (4) above. Therefore full conversion can be achieved only if heat is supplied from an external source. If full conversion of methane is not achieved, the remaining methane will be combusted to produce heat. There are many different methods under development for reactors based on this principle, including thermal catalytic, thermal non-catalytic and plasma cracking.

In the plasma cracking process natural gas or other hydrocarbons are supplied to a plasma reactor where the hydrocarbons are cracked under pyrolysis conditions (i.e., in absence of oxides, e.g., steam, which can supply oxygen to form CO or CO_2). The plasma arc, for which electricity is used, supplies the heat for the cracking reaction. Advantages of the process are its flexibility with respect to the fuel and the high quality carbon black which can be produced. Two small-scale plasma cracking processes for hydrogen/syngas production have been in development. The Glid Arc process has been developed by the Canadian Synergy Technologies Corporation. The second process is the Kvaerner CB&H process. Kvaerner has reported results for a pilot plant producing 1000 Nm³ hydrogen per hour and 270 kg or 500 kg carbon black using natural gas and aromatic oil respectively (IEA GHG, 2001).

3.5.3.5 Technologies based on calcium oxide

There is a range of pre-combustion systems that make use of the carbonation reaction of CaO at high pressures and temperatures, to further integrate the gasification of the fuel (if solid), the shift reaction, and in-situ CO_2 removal with CaO. The overall reaction aimed in the system is:

Carbonation of calcium oxide:
$$CaO + C + 2\,H_2O \rightarrow CaCO_3 + 2H_2 \qquad\qquad (5)$$

The regeneration of the sorbent produces pure CO_2 when carried out in a separate reactor by calcining $CaCO_3$. A range of systems can be developed under this general reaction scheme depending on the technology adopted for gasification, carbonation-calcination, hydrogen utilization route and storage option for CO_2. The first of these concepts was proposed at the Los Alamos National Laboratory (USA) and is currently under development as the Zero Emission Coal Alliance (ZECA) process. The full system includes (Lackner *et al.*, 2001) a hydro-gasification reactor, solid oxide fuel cell and a technology for mineral carbonation. However, the fuel cell will require more development and mineral carbonation is only at the laboratory investigation stage (see Section 7.2 for a discussion of mineral carbonation).

The HyPrRing process (Lin *et al.*, 2002) is being developed by the Center for Coal Utilization of Japan. It integrates gasification, reforming and *in situ* CO_2 capture in a single reactor at pressures above 12 MPa and temperature above 650°C. Projects in Norway using natural gas and in Germany using brown coal (Bandi *et al.*, 2002) are also underway developing pre-combustion systems using capture of CO_2 with CaO. Finally, General Electric (Rizeq *et al.*, 2002) is developing an innovative system involving the capture of CO_2 in the gasification reactor by a high temperature sorbent and with calcination in a separate reactor by burning part of the fuel with an oxygen carrier.

All these systems are at an early stage of development. Detailed process simulations show that the efficiencies are potentially high because most of the energy employed for sorbent regeneration is effectively transferred to the H_2 generated in reaction (5). The systems are aimed at very large-scale generation of electricity and/or H_2 and cement production (from the deactivated sorbent, CaO). However, many uncertainties remain concerning the performance of the individual units and their practical integration. The main challenge may be the regeneration of the sorbent at very high temperatures (>900°C), to produce a pure stream of CO_2. Another is the operating conditions to achieve sufficient conversion towards hydrogen, without the use of a catalyst for the shift reaction.

3.5.4 Enabling technologies

The performance and cost of a pre-combustion capture system is linked to the availability of the enabling technologies that complete the system. In this section we consider the availability of industrial systems, to produce heat from the de-carbonized fuel and gas turbines and fuel cells to produce power.

3.5.4.1 Use of de-carbonized fuel in industrial systems

The use of hydrogen as a fuel for conventional fired heaters and boilers is considered to be proven and indeed it is practiced at certain industrial sites. There is a very large stock of capital equipment of this type and so the use of hydrogen as a fuel might be considered a valuable technology option in a carbon-constrained world. A study (IEA GHG, 2000c) has looked at the cost of converting an existing refinery to use hydrogen fuel.

3.5.4.2 Use of de-carbonized fuel in gas turbine systems

There is extensive commercial experience with hydrogen-rich fuel gas firing in gas turbines. For example, General Electric reports over 450,000 hours of operating experience with high hydrogen (52-95% by volume) content fuel gas in gas turbines (Shilling and Jones, 2003). Unfortunately, most of that experience is for 'refinery gas' where methane is the other main component of the fuel gas and is utilized in older lower firing temperature gas turbines, not the state-of-the-art over 1300°C gas turbines normally considered for large de-carbonization power plants.

Norsk Hydro and General Electric collaborated to perform full-scale combustion system testing for modern gas turbines firing hydrogen-rich gas with combustion exit temperatures of above 1400°C (Todd and Battista, 2001). The results showed good combustion conditions with low NO_x emission and acceptable hot metal temperatures for mixtures with 54-77% by volume hydrogen with most of the additional gas being nitrogen. Dilution of the hydrogen with nitrogen or steam reduces the NO_x emission.

For pre-combustion capture of CO_2 from natural gas, air-blown gasification or autothermal reforming is usually preferred (IEA GHG, 2000b; Wilkinson and Clarke, 2002). Nitrogen dilution of the hydrogen required for firing in modern gas turbines comes from the gasification air. High-pressure air is usually extracted from the gas turbine to feed the air-blown gasifier, or autothermal reformer to reduce costs and avoid a separate air compressor. The balance between the amount of air withdrawn from the gas turbine and the amount provided from a separate air compressor is determined by the particular characteristics of the gas turbine used. Some gas turbines can accept a higher ratio of expander to compressor flow, allowing greater volumes of dilution gas or smaller air-side draw flow and giving higher power output.

For pre-combustion capture of CO_2 from coal, oxygen-blown gasification is usually preferred (IEA GHG, 2003). Nitrogen dilution of the hydrogen required for firing in modern gas turbines comes from the cryogenic air separation unit (used to make the oxygen; see Section 3.4.5.1). The nitrogen is added to the hydrogen after the gasification, CO shifting and CO_2 capture to reduce the equipment sizes and cost. High-pressure air is usually extracted from the gas turbine to supply a higher than normal pressure cryogenic air separation unit to reduce costs plus air, oxygen and nitrogen compression power. An alternative IGCC scheme that incorporates newly emerging ion transport membranes for oxygen production is also described below in Section 3.5.4.3.

3.5.4.3 Syngas production using oxygen membranes

Oxygen required for a coal-fired IGCC process (Section 3.5.2.6) can be generated in an oxygen transport membrane system by using a heated, high-pressure air stream produced by heating the discharge air from the compressor section of a gas turbine (Allam *et al.*, 2002), typically at 1.6 MPa or 420°C, to the precise inlet temperature of the oxygen transport membrane module which is above 700°C. The oxygen, which permeates to the low-pressure side passes through a heat recovery section and is compressed to the final pressure of use. The O_2 depleted air leaving the oxygen transport membrane module then enters the gas turbine combustor where it is used to burn fuel before entering the gas turbine expander at the required temperature. Note that due to the necessity to have excess air in a gas turbine to limit turbine inlet temperature, removing one mole of oxygen can be compensated by injection of the equivalent thermal capacity of steam to maintain gas turbine power output. Studies have been carried out (Armstrong *et al.*, 2002) to compare oxygen transport membrane and cryogenic oxygen production in an IGCC power plant using coal as fuel. The oxygen plant projected cost was reduced by 35% and the power consumption by 37%. An LHV efficiency of 41.8% without CO_2 capture and compression is reported for this cycle compared to 40.9% when a conventional cryogenic oxygen plant is used.

For autothermal reforming or the partial oxidation of natural gas, if the permeate side of the oxygen transport membrane is exposed to a natural gas plus water vapour stream in the presence of a reforming catalyst, the oxygen will react as it leaves the membrane in an exothermic reaction (Dyer *et al.*, 2001; Carolan *et al.*, 2001), which will provide heat for the endothermic steam/natural gas reforming reaction. The oxygen partial pressure at these highly-reducing, high temperature conditions is extremely low, allowing heated air at close to atmospheric pressure to be used on the feed side of the membrane while producing a H_2 + CO mixture at high pressure from the permeate side. This system can be used to produce H_2 following CO shift reaction and CO_2 removal.

3.5.4.4 Chemical looping gasification/reforming

The chemical looping concept described in 3.4.6 is being considered for reforming of a fuel to produce H_2 and CO (Zafar *et al.*, 2005). When the amount of oxygen brought by the metal oxide into the reduction reactor is below stoichiometric requirements, the chemical reaction with the fuel produces H_2 and CO. The reaction products may subsequently be shifted with steam to yield CO_2 and more H_2.

3.5.4.5 Use of de-carbonized fuel in fuel cells

Fuel cells offer the possibility for highly efficient power production since the conversion process is not controlled by heat to work Carnot cycle restrictions (Blomen and Mugerwa, 1993). In general fuel cells feature the electrochemical oxidation of gaseous fuels directly into electricity, avoiding the mixture of the air and the fuel flows and thus the dilution with nitrogen and excess oxygen of the oxidized products (Campanari, 2002). As a result, the anode outlet stream of a fuel cell already has a very high CO_2 content that simplifies the CO_2 capture subsystem. The fuel is normally natural gas, though some concepts can also be incorporated into coal gasification systems. The systems concepts can be classified into two main groups (Goettlicher, 1999):

- Systems with pre-fuel cell CO_2 capture;
- Systems with post-fuel cell CO_2 capture.

In pre-fuel cell CO_2 capture systems (see Figure 3.18a) the fuel is first converted into hydrogen using steam reforming or coal gasification, followed by the water gas shift conversion. This system approach has been first proposed both for low temperature and for high temperature fuel cells.

The post-fuel cell capture system (see Figure 3.18b) is proposed for high temperature fuel cell systems (Dijkstra and Jansen, 2003). These systems make use of the internal reforming capabilities of the high temperature fuel cells resulting in an anode off-gas that has a high CO_2-content, but also contains H_2O and unconverted CO and H_2. The water can easily be removed by conventional techniques (cooling, knock-out, additional drying). Oxidizing the H_2 and CO from the (SOFC) anode with air will result in a too high dilution of the stream with nitrogen.

Haines (1999) chooses to use an oxygen-transport membrane reactor placed after the SOFC. The anode off-gas is fed to one side of the membrane, the cathode off-gas is fed to the other side of the membrane. The membrane is selective to oxygen, which permeates from the cathode off-gas stream to the anode-off gas. In the membrane unit the H_2 and CO are oxidized. The retenate of the membrane unit consist of CO_2 and water. Finally a concept using a water gas shift membrane reactor has been proposed (Jansen and Dijkstra, 2003).

3.5.5 Status and outlook

This section reviewed a wide variety of processes and fuel conversion routes that share a common objective: to produce a cleaner fuel stream from the conversion of a raw carbonaceous fuel into one that contains little, or none, of the carbon contained in the original fuel. This approach necessarily involves the separation of CO_2 at some point in the conversion process. The resulting H_2-rich fuel can be fed to a hydrogen consuming process, oxidized in a fuel cell, or burned in the combustion chamber of a gas turbine to produce electricity. In systems that operate at high pressure, the energy conversion efficiencies tend to be higher when compared to equivalent systems operating at low pressures following the combustion route, but these efficiency improvements are often obtained at the expense of a higher complexity and capital investment in process plants (see Section 3.7).

In principle, all pre-combustion systems are substantially similar in their conversion routes, allowing for differences that arise from the initial method employed for syngas production from gaseous, liquid or solid fuels and from the subsequent need to remove impurities that originate from the fuel feed to the plant. Once produced, the syngas is first cleaned and then reacted with

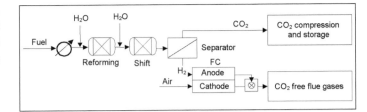

Figure 3.18a Fuel cell system with pre-fuel cell CO_2 capture. The carbon-containing fuel is first completely converted into a mixture of hydrogen and CO_2. Hydrogen and CO_2 are then separated and the H_2-rich fuel is oxidized in the fuel cell to produce electricity. The CO_2 stream is dried and compressed for transport and storage.

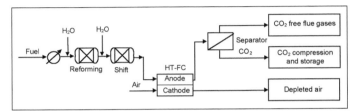

Figure 3.18b Fuel cell system with post-fuel cell CO_2 capture. The carbon-containing fuel is first converted into a syngas. The syngas is oxidized in the fuel cell to produce electricity. At the outlet of the fuel cell CO_2 is separated from the flue gas, dried and compressed for transport and storage.

steam to produce more H_2 and CO_2. The separation of these two gases can be achieved with well-known, commercial absorption-desorption methods, producing a CO_2 stream suitable for storage. Also, intense R&D efforts worldwide are being directed towards the development of new systems that combine CO_2 separation with some of the reaction steps, such as the steam reforming of natural gas or water gas shift reaction stages, but it is not yet clear if these emerging concepts (see Section 3.5.3) will deliver a lower CO_2 capture cost.

In power systems, pre-combustion CO_2 capture in natural gas combined cycles has not been demonstrated. However, studies show that based on current state of the art gas turbine combined cycles, pre-combustion CO_2 capture will reduce the efficiency from 56% LHV to 48% LHV (IEA, 2000b). In natural gas combined cycles, the most significant area for efficiency improvement is the gas turbine and it is expected that by 2020, the efficiency of a natural gas combined cycle could be as high as 65% LHV (IEA GHG, 2000d). For such systems the efficiency with CO_2 capture would equal the current state-of-the-art efficiency for plants without CO_2 capture, that is, 56% LHV.

Integrated Gasification Combined Cycles (IGCC) are large scale, near commercial examples of power systems that can be implemented with heavy oil residues and solid fuels like coal and petroleum coke. For the embryonic coal-fired IGCC technology with the largest unit rated at 331 MW_e, future improvements are expected. A recent study describes improvements potentially realisable for bituminous coals by 2020 that could reduce both energy and cost-of-electricity penalties for CO_2 capture to 13% compared to a same base plant without capture. For such systems the generation efficiency with capture would equal the best efficiency realisable today without CO_2 capture (i.e., 43% LHV; IEA GHG, 2003). Notably, all the innovations considered, with the exception of ion transport membrane technology for air separation (which is motivated by many market drivers other than IGCC needs) involve 'non- breakthrough' technologies, with modest continuing improvements in components that are already established commercially - improvements that might emerge as a natural result of growing commercial experience with IGCC technologies.

All fuel cell types are currently in the development phase. The first demonstration systems are now being tested, with the largest units being at the 1 MW scale. However, it will take at least another 5 to 10 years before these units become commercially available. In the longer term, these highly efficient fuel cell systems are expected to become competitive for power generation. Integrating CO_2 capture in these systems is relatively simple and therefore fuel cell power generation systems offer the prospect of reducing the CO_2 capture penalty in terms of efficiency and capture costs. For instance, for high temperature fuel cell systems without CO_2 capture, efficiencies that exceed 67% are calculated with an anticipated 7% efficiency reduction when CO_2 capture is integrated into the system (Jansen and Dijkstra, 2003). However, fuel cell systems are too small to reach a reasonable level of CO_2 transport cost (IEA GHG, 2002a), but in groups of a total of capacity 100MWe, the cost of CO_2 transport is reduced to a more acceptable level.

Most studies agree that pre-combustion systems may be better suited to implement CO_2 capture at a lower incremental cost compared to the same type of base technology without capture (Section 3.7), but with a key driver affecting implementation being the absolute cost of the carbon emission-free product, or service provided. Pre-combustion systems also have a high strategic importance, because their capability to deliver, in a large scale and at high thermal efficiencies, a suitable mix of electricity, hydrogen and lower carbon-containing fuels or

chemical feedstocks in an increasingly carbon-constrained world.

3.6 Environmental, monitoring, risk and legal aspects of capture systems

The previous sections of this chapter focused on each of the major technologies and systems for CO_2 capture. Here we summarize the major environmental, regulatory and risk issues associated with the use of CO_2 capture technology and the handling of carbon dioxide common to all of these systems. Issues related to the subsequent transport and storage of carbon dioxide are discussed in Chapters 4 to 7.

3.6.1 Emissions and resource use impacts of CO₂ capture systems

3.6.1.1 Overview of emissions from capture systems
Plants with CO_2 capture would produce a stream of concentrated CO_2 for storage, plus in most cases a flue gas or vent gas emitted to the atmosphere and liquid wastes. In some cases solid wastes will also be produced.

The captured CO_2 stream may contain impurities which would have practical impacts on CO_2 transport and storage systems and also potential health, safety and environmental impacts. The types and concentrations of impurities depend on the type of capture process, as shown in Table 3.4, and detailed plant design. The major impurities in CO_2 are well known but there is little published information on the fate of any trace impurities in the feed gas such as heavy metals. If substances are captured along with the CO_2 then their net emissions to the atmosphere will be reduced, but impurities in the CO_2 may result in environmental impacts at the storage site.

CO_2 from most capture processes contains moisture, which has to be removed to avoid corrosion and hydrate formation during transportation. This can be done using conventional

Table 3.4 Concentrations of impurities in dried CO_2, % by volume (Source data: IEA GHG, 2003; IEA GHG, 2004; IEA GHG, 2005).

	SO_2	NO	H_2S	H_2	CO	CH_4	$N_2/Ar/O_2$	Total
COAL FIRED PLANTS								
Post-combustion capture	<0.01	<0.01	0	0	0	0	0.01	0.01
Pre-combustion capture (IGCC)	0	0	0.01-0.6	0.8-2.0	0.03-0.4	0.01	0.03-0.6	2.1-2.7
Oxy-fuel	0.5	0.01	0	0	0	0	3.7	4.2
GAS FIRED PLANTS								
Post-combustion capture	<0.01	<0.01	0	0	0	0	0.01	0.01
Pre-combustion capture	0	0	<0.01	1.0	0.04	2.0	1.3	4.4
Oxy-fuel	<0.01	<0.01	0	0	0	0	4.1	4.1

a. The SO_2 concentration for oxy-fuel and the maximum H_2S concentration for pre-combustion capture are for cases where these impurities are deliberately left in the CO_2, to reduce the costs of capture (see Section 3.6.1.1). The concentrations shown in the table are based on use of coal with a sulphur content of 0.86%. The concentrations would be directly proportional to the fuel sulphur content.

b. The oxy-fuel case includes cryogenic purification of the CO_2 to separate some of the N_2, Ar, O_2 and NO_x. Removal of this unit would increase impurity concentrations but reduce costs.

c. For all technologies, the impurity concentrations shown in the table could be reduced at higher capture costs.

processes and the costs of doing so are included in published costs of CO_2 capture plants.

CO_2 from post-combustion solvent scrubbing processes normally contains low concentrations of impurities. Many of the existing post-combustion capture plants produce high purity CO_2 for use in the food industry (IEA GHG, 2004).

CO_2 from pre-combustion physical solvent scrubbing processes typically contains about 1-2% H_2 and CO and traces of H_2S and other sulphur compounds (IEA GHG, 2003). IGCC plants with pre-combustion capture can be designed to produce a combined stream of CO_2 and sulphur compounds, to reduce costs and avoid the production of solid sulphur (IEA GHG, 2003). Combined streams of CO_2 and sulphur compounds (primarily hydrogen sulphide, H_2S) are already stored, for example in Canada, as discussed in Chapter 5. However, this option would only be considered in circumstances where the combined stream could be transported and stored in a safe and environmentally acceptable manner.

The CO_2-rich gas from oxy-fuel processes contains oxygen, nitrogen, argon, sulphur and nitrogen oxides and various other trace impurities. This gas will normally be compressed and fed to a cryogenic purification process to reduce the impurities concentrations to the levels required to avoid two-phase flow conditions in the transportation pipelines. A 99.99% purity could be produced by including distillation in the cryogenic separation unit. Alternatively, the sulphur and nitrogen oxides could be left in the CO_2 fed to storage in circumstances where that is environmentally acceptable as described above for pre-combustion capture and when the total amount of all impurities left in the CO_2 is low enough to avoid two-phase flow conditions in transportation pipelines.

Power plants with CO_2 capture would emit a CO_2-depleted flue gas to the atmosphere. The concentrations of most harmful substances in the flue gas would be similar to or lower than in the flue gas from plants without CO_2 capture, because CO_2 capture processes inherently remove some impurities and some other impurities have to be removed upstream to enable the CO_2 capture process to operate effectively. For example, post-combustion solvent absorption processes require low concentrations of sulphur compounds in the feed gas to avoid excessive solvent loss, but the reduction in the concentration of an impurity may still result in a higher rate of emissions per kWh of product, depending upon the actual amount removed upstream and the capture system energy requirements. As discussed below (Section 3.6.1.2), the latter measure is more relevant for environmental assessments. In the case of post-combustion solvent capture, the flue gas may also contain traces of solvent and ammonia produced by decomposition of solvent.

Some CO_2 capture systems produce solid and liquid wastes. Solvent scrubbing processes produce degraded solvent wastes, which would be incinerated or disposed of by other means. Post-combustion capture processes produce substantially more degraded solvent than pre-combustion capture processes. However, use of novel post-combustion capture solvents can significantly reduce the quantity of waste compared to MEA

solvent, as discussed in Section 3.3.2.1. The waste from MEA scrubbing would normally be processed to remove metals and then incinerated. The waste can also be disposed of in cement kilns, where the waste metals become agglomerated in the clinker (IEA GHG, 2004). Pre-combustion capture systems periodically produce spent shift and reforming catalysts and these would be sent to specialist reprocessing and disposal facilities.

3.6.1.2   Framework for evaluating capture system impacts

As discussed in Chapter 1, the framework used throughout this report to assess the impacts of CO_2 capture and storage is based on the material and energy flows needed to produce a unit of product from a particular process. As seen earlier in this chapter, CO_2 capture systems require an increase in energy use for their operation. As defined in this report (see Section 1.5 and Figure 1.5), the energy requirement associated with CO_2 capture is expressed as the additional energy required to produce a unit of useful product, such as a kilowatt-hour of electricity (for the case of a power plant). As the energy and resource requirement for CO_2 capture (which includes the energy needed to compress CO_2 for subsequent transport and storage) is typically much larger than for other emission control systems, it has important implications for plant resource requirements and environmental emissions when viewed from the 'systems' perspective of Figure 1.5.

In general, the CCS energy requirement per unit of product can be expressed in terms of the change in net plant efficiency (η) when the reference plant without capture is equipped with a CCS system:[1]

$$\Delta E = (\eta_{ref} / \eta_{ccs}) - 1 \qquad (6)$$

where ΔE is the fractional increase in plant energy input per unit of product and η_{ccs} and η_{ref} are the net efficiencies of the capture plant and reference plant, respectively. The CCS energy requirement directly determines the increases in plant-level resource consumption and environmental burdens associated with producing a unit of useful product (like electricity) while capturing CO_2. In the case of a power plant, the larger the CCS energy requirement, the greater the increases per kilowatt-hour of in-plant fuel consumption and other resource requirements (such as water, chemicals and reagents), as well as environmental releases in the form of solid wastes, liquid wastes and air pollutants not captured by the CCS system. The magnitude of ΔE also determines the magnitude of additional upstream environmental impacts associated with the extraction, storage and transport of additional fuel and other resources consumed at the plant. However, the additional energy for these upstream activities is not normally included in the reported

---

[1] A different measure of the 'energy penalty' commonly reported in the literature is the fractional decrease in plant output (plant derating) for a fixed energy input. This value can be expressed as: $\Delta E^* = 1 - (\eta_{ccs}/\eta_{ref})$. Numerically, ΔE^* is smaller than the value of ΔE given by Equation (6). For example, a plant derating of $\Delta E^* = 25\%$ corresponds to an increase in energy input per kWh of $\Delta E = 33\%$.

energy requirements for CO$_2$ capture systems.[2]

Recent literature on CO$_2$ capture systems applied to electric power plants quantifies the magnitude of CCS energy requirements for a range of proposed new plant designs with and without CO$_2$ capture. As elaborated later in Section 3.7 (Tables 3.7 to 3.15), those data reveal a wide range of ΔE values. For new supercritical pulverized coal (PC) plants using current technology, these ΔE values range from 24-40%, while for natural gas combined cycle (NGCC) systems the range is 11%–22% and for coal-based gasification combined cycle (IGCC) systems it is 14%–25%. These ranges reflect the combined effects of the base plant efficiency and capture system energy requirements for the same plant type with and without capture.

3.6.1.3   Resource and emission impacts for current systems
Only recently have the environmental and resource implications of CCS energy requirements been discussed and quantified for a variety of current CCS systems. Table 3.5 displays the assumptions and results from a recent comparison of three common fossil fuel power plants employing current technology to capture 90% of the CO$_2$ produced (Rubin *et al.*, 2005). Increases in specific fuel consumption relative to the reference plant without CO$_2$ capture correspond directly to the ΔE values defined above. For these three cases, the plant energy requirement per kWh increases by 31% for the PC plant, 16% for the coal-based IGCC plant and 17% for the NGCC plant. For the specific examples used in Table 3.5, the increase in energy consumption for the PC and NGCC plants are in the mid-range of the values for these systems reported later in Tables 3.7 to 3.15 (see also Section 3.6.1.2), whereas the IGCC case is nearer the low end of the reported range for such systems. As a result of the increased energy input per kWh of output, additional resource requirements for the PC plant include proportionally greater amounts of coal, as well as limestone (consumed by the FGD system for SO$_2$ control) and ammonia (consumed by the SCR system for NO$_x$ control). All three plants additionally require more sorbent make-up for the CO$_2$ capture units. Table 3.5 also shows the resulting increases in solid residues for these three cases. In contrast, atmospheric emissions of CO$_2$ decrease sharply as a result of the CCS systems, which also remove residual amounts of other acid gases, especially SO$_2$ in flue gas streams. Thus, the coal combustion system shows a net reduction in SO$_2$ emission rate as a result of CO$_2$ capture. However, because of the reduction in plant efficiency, other air emission rates per kWh increase relative to the reference plants without capture. For the PC and NGCC systems, the increased emissions of ammonia are a result of chemical reactions in the amine-based capture process. Not included in this analysis are the incremental impacts of upstream operations such as mining, processing and transport of fuels and other resources.

Other studies, however, indicate that these impacts, while not insignificant, tend to be small relative to plant-level impacts (Bock *et al.*, 2003).

For the most part, the magnitude of impacts noted above - especially impacts on fuel use and solid waste production - is directly proportional to the increased energy per kWh resulting from the reduction in plant efficiency, as indicated by Equation (6). Because CCS energy requirements are one to two orders of magnitude greater than for other power plant emission control technologies (such as particulate collectors and flue gas desulphurization systems), the illustrative results above emphasize the importance of maximizing overall plant efficiency while controlling environmental emissions.

3.6.1.4   Resource and emission impacts of future systems
The analysis above compared the impacts of CO$_2$ capture for a given plant type based on current technology. The magnitude of actual future impacts, however, will depend on four important factors: (1) the performance of technologies available at the time capture systems are deployed; (2) the type of power plants and capture systems actually put into service; (3) the total capacity of each plant type that is deployed; and, (4) the characteristics and capacity of plants they may be replacing.

Analyses of both current and near-future post-combustion, pre-combustion and oxy-fuel combustion capture technology options reveal that some of the advanced systems currently under development promise to significantly reduce the capture energy requirements - and associated impacts - while still reducing CO$_2$ emissions by 90% or more, as shown in Figure 3.19. Data in this figure was derived from the studies previously reported in Figures 3.6 and 3.7.

The timetable for deploying more efficient plants with CO$_2$ capture will be the key determinant of actual environmental changes. If a new plant with capture replaces an older, less efficient and higher-emitting plant currently in service, the net change in plant-level emission impacts and resource requirements would be much smaller than the values given earlier (which compared identical new plants with and without

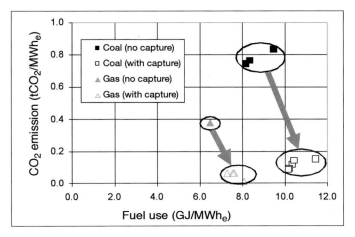

Figure 3.19 Fuel use for a reduction of CO$_2$ emissions from capture plants (data presented from design studies for power plants with and without capture shown in Figures 3.6 and 3.7).

---

[2] Those additional energy requirements, if quantified, could be included by re-defining the system boundary and system efficiency terms in Equation (6) to apply to the full life cycle, rather than only the power plant. Such an analysis would require additional assumptions about the methods of fuel extraction, processing, transport to the power plant, and the associated energy requirements of those activities; as well as the CO$_2$ losses incurred during storage.

Table 3.5 Illustrative impacts of CCS energy requirements on plant-level resource consumption and non-CO_2 emission rates for three current power plant systems. Values shown are mass flow rates in kg per MWh for the capture plant, plus increases over the reference plant rates for the same plant type. See footnotes for additional details. (Source: Rubin et al., 2005)

Capture Plant Parameter [a]	PC [b]		IGCC [c]		NGCC [d]	
	Rate	Increase	Rate	Increase	Rate	Increase
			(All values in kg MWh^{-1})			
Resource consumption						
Fuel	390	93	361	49	156	23
Limestone	27.5	6.8	-	-	-	-
Ammonia	0.80	0.19	-	-	-	-
CCS Reagents	2.76	2.76	0.005	0.005	0.80	0.80
Solid Wastes/byproduct						
Ash/slag	28.1	6.7	34.2	4.7	-	-
FGD residues	49.6	12.2	-	-	-	-
Sulfur	-	-	7.53	1.04	-	-
Spent CCS sorbent	4.05	4.05	0.005	0.005	0.94	0.94
Atmospheric emissions						
CO_2	107	-704	97	-720	43	-342
SO_x	0.001	-0.29	0.33	0.05	-	-
NO_x	0.77	0.18	0.10	0.01	0.11	0.02
NH_3	0.23	0.22	-	-	0.002	0.002

[a] Net power output of all plants is approximately 500 MW. Coal plants use Pittsburgh #8 coal with 2.1%S, 7.2% ash, 5.1% moisture and 303.2 MJ kg^{-1} lower heating value basis (LHV). Natural gas LHV = 59.9 MJ kg^{-1}. All plants capture 90% of potential CO_2 emissions and compress to 13.7 MPa.

[b] PC= Pulverized coal-fired plant; based on a supercritical unit with SCR, ESP and FGD systems, followed by an amine system for CO_2 capture. SCR system assumes 2 ppmv ammonia slip. SO_2 removal efficiency is 98% for reference plant and 99% for capture plant. Net plant efficiency (LHV basis) is 40.9% without CCS and 31.2% with CCS.

[c] IGCC=integrated gasification combined cycle system based on Texaco quench gasifiers (2 + 1 spare), two GE 7FA gas turbines, 3-pressure reheat HRSG. Sulfur removal efficiency is 98% via hydrolyzer plus Selexol system; Sulfur recovery via Claus plant and Beavon-Stretford tailgas unit. Net plant efficiency (LHV basis) is 39.1% without CCS and 33.8% with CCS.

[d] NGCC=natural gas combined cycle plant using two GE 7FA gas turbines and 3-pressure reheat HRSG, with an amine system for CO_2 capture. Net plant efficiency (LHV basis) is 55.8% without CCS and 47.6% with CCS.

capture). For example, the efficiency of a modern coal-based plant with capture is close to many older coal-burning plants currently in service. Replacing the latter with the former would thus reduce CO_2 emissions significantly with little or no net change in plant coal consumption or related solid waste impacts. In some cases, there could in fact be net reductions in other plant emissions, in support of clean air goals. If, however, the deployment of new CCS plants is delayed significantly, older existing plants could well be replaced by modern high-efficiency plants without capture. Such plants also would be built to provide additional capacity in regions with high electricity growth rates, such as in China and other parts of Asia today. A decade or two from now, the fleet of 'existing' plants in those regions would thus look very different from the present. Accordingly, the environmental and resource impacts of additional new plants with CO_2 capture would have to be assessed in the context of the future situation.

Because comparisons of different plant types require a specific context (or scenario) to be meaningful, this chapter has only focused on characterizing the effects of CO_2 capture systems relative to the same type of power plant and not the type of infrastructure it would replace (either currently, or in a future carbon-constrained world). If other systems such as the use of renewable energy, or electricity and synfuels cogenerated from coal, find significant applications, those systems too would require more comprehensive comparative life-cycle assessments of resource use and impacts that are not currently available. Chapter 8, however, assesses overall energy use impacts for illustrative scenarios of CCS deployment in competition with other carbon mitigation options.

3.6.2   *Issues related to the classification of carbon dioxide as a product*

As a current commercial product, carbon dioxide is subject to classification and regulations. The classification of carbon dioxide is dependent on its physical state (gas, liquid or solid), its concentration, impurities present and other criteria established by national legislative classification in different regions of the world. During the capture and concentration process, the quality properties can change the classification of the substance. A detailed assessment of carbon dioxide physical and chemical properties is provided in Annex I.

The environmental, monitoring, risk and legal aspects associated with carbon dioxide handling and storage are well established in the processing industry. However, much larger volumes are targeted for carbon dioxide processing for purposes of CCS than the volumes handled at present. On a local and regional level, additional emergency response and other regulatory measures can be expected in the future, depending on the rate of development of CCS. It is anticipated that human capacity will be developed to assess the monitoring, risk and legal aspects as required by the market.

At present, carbon dioxide typically occurs and is mainly traded as a non-flammable gas (US Department of Transportation classification class 2.2). The classification system of Transport

Dangerous Goods, International Maritime Organization/ International Maritime Dangerous Goods and International Civil Aviation Organization / International Air Transport Association, all classify carbon dioxide in class 2.2, non-flammable, non-corrosive and non-poisonous gases. In US federal regulations, carbon dioxide is not listed as a product in the Clean Water Act (CWA 307 and 311), Clean Air Act (CAA 112) or the Toxics Release Inventory. In other international regulations carbon dioxide is not classified in the European Inventory of Existing Commercial Chemical Substance or other international lists, but in Canada is classified as a compressed gas (class A) on the Canadian Energy Pipeline Association Dangerous Substances List (Hazardous Substances Data Bank, 2002).

3.6.3   *Health and safety risks associated with carbon dioxide processing*

The effects of exposure to carbon dioxide are described in Annex I. However, a risk assessment that includes an understanding of both exposure and effects is required to characterize the risk for various situations associated with carbon dioxide processing (European Chemicals Bureau, 2003); see the following two sections for established risk management practices. The most probable routes of human exposure to carbon dioxide are inhalation or skin contact. The need for a risk-based approach is clear from the following two descriptions. Carbon dioxide and its products of degradation are not legally classified as a toxic substance; is non-hazardous on inhalation, is a non-irritant and does not sensitize or permeate the skin. However, chronic effects on humans follow from long-term exposure to airborne carbon dioxide concentrations of between 0.5 and 1% resulting in metabolic acidosis and increased calcium deposits in soft tissues. The substance is toxic to the cardiovascular system and upper respiratory tract at concentrations above 3%. Sensitive populations to elevated carbon dioxide levels are described in Annex I. The product risk assessment process is therefore necessary as with any other chemical use to determine the risk and establish the necessary risk management processes.

As an asphyxiate carbon dioxide presents the greatest danger. If atmospheric oxygen is displaced such that oxygen concentration is 15-16%, signs of asphyxia will be noted. Skin contact with dry ice has caused serious frostbites and blisters (Hazardous Substances Data Bank, 2002). Protective equipment and clothing required in the processing industries include full face-piece respirators to prevent eye contact and appropriate personal protective clothing to protect the skin from becoming frozen by the liquid.

3.6.4   *Plant design principles and guidelines used by governments, industries and financiers*

New plant facilities like those envisioned for carbon dioxide are subject to design guidelines for the petrochemical industry as determined by relevant authorities. One example is the European Unions' Integrated Pollution Prevention and Control (IPPC) directive requiring the application of the principles of Best Available Technology Not Entailing Excessive Cost (BATNEEC). Carbon dioxide capture and compression processes are listed in several guidelines as gas-processing facilities. Typically the World Bank guidelines and other financial institutions have specific requirements to reduce risk and these require monitoring (World Bank, 1999) which is part of routine plant monitoring to detect accidental releases. Investor guidelines like the World Bank guidelines are particularly important for developing countries where there is less emphasis on monitoring and legislation. National and regional legislation for plant design and specifications from organizations like the US Environmental Protection Agency are available to guide the development of technology.

3.6.5 *Commissioning, good practice during operations and sound management of chemicals*

The routine engineering design, commissioning and start-up activities associated with petrochemical facilities are applicable to the capture and compression of carbon dioxide; for example Hazard Operability studies are conducted on a routine basis for new facilities (Sikdar and Diwekar, 1999).

The management of carbon dioxide and reagents inside factory battery limits will be in accordance with the relevant practices in use for carbon dioxide. For carbon dioxide, US Occupational Health and Safety Act standards and National Institute for Occupational Safety and Health recommendations exist, which are applied widely in industry to guide safe handling of carbon dioxide and the same applies to reagents and catalysts used. Well established and externally audited management systems such as International Standards Organization's ISO 14001 (environment) and ISO 9001 (quality) and Occupational Health and Safety (OHSAS 18000) exist to provide assurance that environment, safety, health and quality management systems are in place (American Institute of Chemical Engineers, 1995). Tools like life-cycle assessment (ISO 14040 series) with the necessary boundary expansion methodology are useful to determine the overall issues associated with a facility and assist with selection of parameters such as energy carriers, operational conditions and materials used in the process. The life-cycle assessment will also indicate if a trouble-free capture system does generate environmental concerns elsewhere in the product life cycle.

3.6.6 *Site closure and remediation*

It is not anticipated that carbon dioxide capture will result in a legacy of polluted sites requiring remediation after plant closure, assuming that standard operating procedures and management practices in the previous section are followed. However, depending on the technology used and the materials procured for operations, waste disposal at the facilities and operation according to a formal management system from construction, operation to the development of site closure plans will largely assist to reduce the risk of a polluted site after closure of operations.

3.7　Cost of CO_2 capture

This section of the report deals with the critical issue of CO_2 capture costs. We begin with an overview of the many factors that affect costs and the ability to compare published estimates on a consistent basis. Different measures of CO_2 capture cost also are presented and discussed. The literature on CO_2 capture costs for currently available technologies is then reviewed, along with the outlook for future costs over the next several decades.

3.7.1　*Factors affecting CO_2 capture cost*

Published estimates for CO_2 capture costs vary widely, mainly as a result of different assumptions regarding technical factors related to plant design and operation (e.g., plant size, net efficiency, fuel properties and load factor), as well as key economic and financial factors such as fuel cost, interest rates and plant lifetime. A number of recent papers have addressed this issue and identified the principal sources of cost differences and variability (Herzog, 1999; Simbeck, 1999; Rubin and Rao, 2003). This section draws heavily on Rubin and Rao (2003) to highlight the major factors affecting the cost of CO_2 capture.

3.7.1.1　*Defining the technology of interest*
Costs will vary with the choice of CO_2 capture technology and the choice of power system or industrial process that generates the CO_2 emissions. In engineering-economic studies of a single plant or CO_2 capture technology, such definitions are usually clear. However, where larger systems are being analyzed, such as in regional, national or global studies of CO_2 mitigation options, the specific technologies assumed for CO_2 production and capture may be unclear or unspecified. In such cases, the context for reported cost results also may be unclear.

3.7.1.2　*Defining the system boundary*
Any economic assessment should clearly define the 'system' whose CO_2 emissions and cost is being characterized. The most common assumption in studies of CO_2 capture is a single facility (most often a power plant) that captures CO_2 and transports it to an off-site storage area such as a geologic formation. The CO_2 emissions considered are those released at the facility before and after capture. Reported costs may or may not include CO_2 transport and storage costs. *The system boundary of interest in this section of the report includes only the power plant or other process of interest and does not include CO_2 transport and storage systems, whose costs are presented in later chapters. CO_2 compression, however, is assumed to occur within the facility boundary and therefore the cost of compression is included in the cost of capture.*[3]

In some studies the system boundary includes emissions of

---

[3] Alternatively, compression costs could be attributed wholly or in part to CO_2 transport and storage. Most studies, however, include compression with capture cost. This also facilitates comparisons of capture technologies that operate at different pressures, and thus incur different costs to achieve a specified final pressure.

CO₂ and other greenhouse gases such as methane (expressed as equivalent CO₂) over the complete fuel cycle encompassing not only the power plant or facility in question, but also the 'upstream' processes of extraction, refining and transport of fuel used at the facility, plus any 'downstream' emissions from the use or storage of captured CO₂. Still larger system boundaries might include all power plants in a utility company's system; all plants in a regional or national grid; or a national economy where power plant and industrial emissions are but one element of the overall energy system being modelled. In each of these cases it is possible to derive a mitigation cost for CO₂, but the results are not directly comparable because they reflect different system boundaries and considerations. Chapter 8 discusses such differences in more detail and presents results for alternative systems of interest.

3.7.1.3   Defining the technology time frame and maturity
Another factor that is often unclear in economic evaluations of CO₂ capture is the assumed time frame and/or level of maturity for the technology under study. Does the cost estimate apply to a facility that would be built today, or at some future time? This is especially problematic in studies of 'advanced' technologies that are still under development and not currently commercial. In most cases, studies of advanced technologies assume that costs apply to an 'nth plant' to be built sometime in the future when the technology is mature. Such estimates reflect the expected benefits of technological learning, but may or may not adequately account for the increased costs that typically occur in the early stages of commercialization. The choice of technology time frame and assumed rate of cost improvements and can therefore make a big difference in CO₂ capture cost estimates.

3.7.1.4   Different cost measures and assumptions
The literature reveals a number of different measures used to characterize CO₂ capture and storage costs, including capital cost, cost of electricity, cost of CO₂ avoided and others. Because some of these measures are reported in the same units (e.g., US dollars per tonne of CO₂) there is great potential for misunderstanding. Furthermore, for any given cost measure, different assumptions about the technical, economic and financial parameters used in cost calculations can also give rise to large differences in reported capture costs. Section 3.7.2 elaborates on some of the common metrics of cost and the parameters they employ.

3.7.2   *Measures of CO₂ capture cost*

We define four common measures of CO₂ capture cost here: capital cost, incremental product cost (such as the cost of electricity), cost of CO₂ avoided and cost of CO₂ captured or removed. Each of these measures provides a different perspective on CO₂ capture cost for a particular technology or system of interest. All of them, however, represent an 'engineering economic' perspective showing the added cost of capturing CO₂ in a particular application. Such measures are required to address larger questions such as which options or strategies to pursue - a topic addressed later in Chapter 8.

3.7.2.1   Capital cost
Capital cost (also known as investment cost or first cost) is a widely used, albeit incomplete, metric of the cost of a technology. It is often reported on a normalized basis (e.g., cost per kW). For CO₂ capture systems, the capital cost is generally assumed to represent the total expenditure required to design, purchase and install the system of interest. It may also include the additional costs of other plant components not needed in the absence of a CO₂ capture device, such as the costs of an upstream gas purification system to protect the capture device. Such costs often arise in complex facilities like a power plant. Thus, the total incremental cost of CO₂ capture for a given plant design is best determined as the difference in total cost between plants with and without CO₂ capture, producing the same amounts of useful (primary) product, such as electricity.

Different organizations employ different systems of accounts to specify the elements of a capital cost estimate. For electric power plants, one widely used procedure is that defined by the Electric Power Research Institute (EPRI, 1993). However, because there is no universally employed nomenclature or system of accounts, capital costs reported by different organizations or authors may not always include the same items. The terms used to report capital costs may further disguise such differences and lead to misunderstandings about what is and is not included. For example, power plant cost studies often report a value of capital cost that does not include the cost of interest during construction or other so-called 'owners costs' that typically add at least 10-20% (sometimes substantially more) to the 'total capital requirement' of a system. Only if a capital cost breakdown is reported can such omissions be discovered. Studies that fail to report the year of a cost estimate introduce further uncertainty that may affect cost comparisons.

3.7.2.2   Incremental product cost
The effect of CO₂ capture on the cost of electricity (or other product) is one of the most important measures of economic impact. Electric power plants, a major source of CO₂ emissions, are of particular interest in this regard. The cost electricity (COE) for a power plant can be calculated as:[4]

$$COE = [(TCR)(FCF) + (FOM)]/[(CF)(8760)(kW)] + VOM + (HR)(FC)$$
(7)

where, COE = levelized cost of electricity (US$ kWh⁻¹), TCR = total capital requirement (US$), FCF = fixed charge factor (fraction yr⁻¹), FOM = fixed operating costs (US$ yr⁻¹), VOM = variable operating costs (US$ kWh⁻¹), HR = net plant heat rate (kJ kWh⁻¹), FC = unit fuel cost (US$ kJ⁻¹), CF = capacity

---
[4] For simplicity, the value of FCF in Equation (7) is applied to the total capital requirement. More detailed calculations of COE based on a year-by-year analysis apply the FCF to the total capital cost excluding owner's costs (such as interest during construction), which are separately accounted for in the years prior to plant start-up.

factor (fraction), 8760 = total hours in a typical year and kW = net plant power (kW). In this chapter, the costs in Equation (7) include only the power plant and capture technologies and not the additional costs of CO_2 transport and storage that are required for a complete system with CCS. The incremental COE is the difference in electricity cost with and without CO_2 capture.[5] Again, the values reported here exclude transport and storage costs. Full CCS costs are reported in Chapter 8.

Equation (7) shows that many factors affect this incremental cost. For example, just as the total capital cost includes many different items, so too do the fixed and variable costs associated with plant operation and maintenance (O&M). Similarly, the fixed charge factor (FCF, also known as the capital recovery factor) reflects assumptions about the plant lifetime and the effective interest rate (or discount rate) used to amortize capital costs.[6] Assumptions about any of the factors in Equation (7) can have a pronounced effect on overall cost results. Nor are these factors all independent of one another. For example, the design heat rate of a new power plant may affect the total capital requirement since high-efficiency plants usually are more costly than lower-efficiency designs.

Finally, because several of the parameter values in Equation (7) may change over the operating life of a facility (such as the capacity factor, unit fuel cost, or variable operating costs), the value of COE also may vary from year to year. To include such effects, an economic evaluation would calculate the net present value (NPV) of discounted costs based on a schedule of year-to-year cost variations, in lieu of the simpler formulation of Equation (7). However, most engineering-economic studies use Equation (7) to calculate a single value of 'levelized' COE over the assumed life of the plant. The levelized COE is the cost of electricity, which, if sustained over the operating life of the plant, would produce the same NPV as an assumed stream of variable year-to-year costs. In most economic studies of CO_2 capture, however, all parameter values in Equation (7) are held constant, reflecting (either implicitly or explicitly) a levelized COE over the life of the plant.[7]

3.7.2.3   Cost of CO_2 avoided

One of the most widely used measures for the cost of CO_2 capture and storage is the 'cost of CO_2 avoided.' This value reflects the average cost of reducing atmospheric CO_2 mass emissions by one unit while providing the same amount of useful product as a 'reference plant' without CCS. For an electric power plant the avoidance cost can be defined as:

Cost of CO_2 avoided (US$/$tCO_2$) =

$$[(COE)_{capture} - (COE)_{ref}] / [(CO_2\ kWh^{-1})_{ref} - (CO_2\ kWh^{-1})_{capture}] \tag{8}$$

where, COE = levelized cost of electricity (US$ kWh^{-1}) as given by Equation (7) and $CO_2\ kWh^{-1}$ = CO_2 mass emission rate (in tonnes) per kWh generated, based on the net plant capacity for each case. The subscripts 'capture' and 'ref' refer to the plant with and without CO_2 capture, respectively. Note that while this equation is commonly used to report a cost of CO_2 avoided for the capture portion of a full CCS system, strictly speaking it should be applied only to a complete CCS system including transport and storage costs (since all elements are required to avoid emissions to the atmosphere).

The choice of the reference plant without CO_2 capture plays a key role in determining the CO_2 avoidance cost. *Here the reference plant is assumed to be a plant of the same type and design as the plant with CO_2 capture.* This provides a consistent basis for reporting the incremental cost of CO_2 capture for a particular type of facility.

Using Equation (8), a cost of CO_2 avoided can be calculated for any two plant types, or any two aggregates of plants. Thus, special care should be taken to ensure that the basis for a reported cost of CO_2 avoided is clearly understood or conveyed. For example, the avoidance cost is sometimes taken as a measure of the cost to society of reducing GHG emissions.[8] In that case, the cost per tonne of CO_2 avoided reflects the average cost of moving from one situation (e.g., the current mix of power generation fuels and technologies) to a different mix of technologies having lower overall emissions. Alternatively, some studies compare individual plants with and without capture (as we do), but assume different types of plants for the two cases. Such studies, for example, might compare a coal-fired plant with capture to an NGCC reference plant without capture. Such cases reflect a different choice of system boundaries and address very different questions, than those addressed here. However, the data presented in this section (comparing the same type of plant with and without capture) can be used to estimate a cost of CO_2 avoided for any two of the systems of interest in a particular situation (see Chapter 8).

3.7.2.4   Cost of CO_2 captured or removed

Another cost measure frequently reported in the literature is based on the mass of CO_2 captured (or removed) rather than emissions avoided. For an electric power plant it can be defined as:

Cost of CO_2 Captured (US$/$tCO_2$) =
$$[(COE)_{capture} - (COE)_{ref}] / (CO_{2,\ captured}\ kWh^{-1}) \tag{9}$$

---

[5] For CO_2 capture systems with large auxiliary energy requirements, the magnitude of incremental cost also depends on whether the plant with capture is assumed to be a larger facility producing the same net output as the reference plant without capture, or whether the reference plant is simply derated to supply the auxiliary energy. While the latter assumption is most common, the former yields a smaller incremental cost due to economy-of-scale effects.

[6] In its simplest form, FCF can be calculated from the project lifetime, n (years), and annual interest rate, i (fraction), by the equation: FCF = i / [$1 - (1 + i)^{-n}$].

[7] Readers not familiar with these economic concepts and calculations may wish to consult a basic economics text, or references such as (EPRI, 1993) or (Rubin, 2001) for more details.

[8] As used here, 'cost' refers only to money spent for technology, fuels and related materials, and not to broader societal measures such as macroeconomic costs or societal damage costs associated with atmospheric emissions. Further discussions and use of the term 'cost of CO_2 avoided' appear in Chapter 8 and in the references cited earlier.

where, $CO_{2, captured}$ kWh^{-1} = total mass of CO_2 captured (in tonnes) per net kWh for the plant with capture. This measure reflects the economic viability of a CO_2 capture system given a market price for CO_2 (as an industrial commodity). If the CO_2 captured at a power plant can be sold at this price (e.g., to the food industry, or for enhanced oil recovery), the COE for the plant with capture would be the same as for the reference plant having higher CO_2 emissions. Numerically, the cost of CO_2 captured is lower than the cost of CO_2 avoided because the energy required to operate the CO_2 capture systems increases the amount of CO_2 emitted per unit of product.

3.7.2.5 Importance of CCS energy requirements

As the energy requirement for CCS is substantially larger than for other emission control systems, it has important implications for plant economics as well as for resource requirements and environmental impacts. The energy 'penalty' (as it is often called) enters cost calculations in one of two ways. Most commonly, all energy needed to operate CCS absorbers, compressors, pumps and other equipment is assumed to be provided within the plant boundary, thus lowering the net plant capacity (kW) and output (kWh, in the case of a power plant). The result, as shown by Equation (7), is a higher unit capital cost (US$ kW^{-1}) and a higher cost of electricity production (US$ kWh^{-1}). Effectively, these higher unit costs reflect the expense of building and operating the incremental capacity needed to operate the CCS system.

Alternatively, some studies - particularly for industrial processes such as hydrogen production - assume that some or all of the energy needed to operate the CCS system is purchased from outside the plant boundary at some assumed price. Still other studies assume that new equipment is installed to generate auxiliary energy on-site. In these cases, the net plant capacity and output may or may not change and may even increase. However, the COE in Equation (7) again will rise due to the increases in VOM costs (for purchased energy) and (if applicable) capital costs for additional equipment. The assumption of purchased power, however, does not guarantee a full accounting of the replacement costs or CO_2 emissions associated with CCS. In all cases, however, the larger the CCS energy requirement, the greater the difference between the costs of CO_2 captured and avoided.

3.7.2.6 Other measures of cost

The cost measures above characterize the expense of adding CO_2 capture to a single plant of a given type and operating profile. A broader modelling framework is needed to address questions involving multiple plants (e.g., a utility system, regional grid, or national network), or decisions about what type of plant to build (and when). Macroeconomic models that include emission control costs as elements of a more complex framework typically yield cost measures such as the change in gross domestic product (GDP) from the imposition of a carbon constraint, along with changes in the average cost of electricity and cost per tonne of CO_2 abated. Such measures are often useful for policy analysis, but reflect many additional assumptions about the structure of an economy as well as the cost of technology. Chapter 8 provides a discussion of macroeconomic modelling as it relates to CO_2 capture costs.

3.7.3 The context for current cost estimates

Recall that CO_2 capture, while practiced today in some industrial applications, is not currently a commercial technology used at large electric power plants, which are the focus of most CCS studies. Thus, cost estimates for CO_2 capture systems rely mainly on studies of hypothetical plants. Published studies also differ significantly in the assumptions used for cost estimation. Equation (7), for example, shows that the plant capacity factor has a major impact on the cost of electric power generation, as do the plant lifetime and discount rate used to compute the fixed charge factor. The COE, in turn, is a key element of CO_2 avoidance cost, Equation (8). Thus, a high plant capacity factor or a low fixed charge rate will lower the cost of CO_2 capture per kWh. The choice of other important parameters, such as the plant size, efficiency, fuel type and CO_2 removal rate will similarly affect the CO_2 capture cost. Less apparent, but often equally important, are assumptions about parameters such as the 'contingency cost factors' embedded in capital cost estimates to account for unspecified costs anticipated for technologies at an early stage of development, or for commercial systems that have not yet been demonstrated for the application, location, or plant scale under study.

Because of the variability of assumptions employed in different studies of CO_2 capture, a systematic comparison of cost results is not straightforward (or even possible in most cases). Moreover, there is no universally 'correct' set of assumptions that apply to all the parameters affecting CO_2 capture cost. For example, the quality and cost of natural gas or coal delivered to power plants in Europe and the United States may differ markedly. Similarly, the cost of capital for a municipal or government-owned utility may be significantly lower than for a privately-owned utility operating in a competitive market. These and other factors lead to real differences in CO_2 capture costs for a given technology or power generation system. Thus, we seek in this report to elucidate the key assumptions employed in different studies of similar systems and technologies and their resulting impact on the cost of CO_2 capture. Analyses comparing the costs of alternative systems on an internally consistent basis (within a particular study) also are highlighted. Nor are all studies equally credible, considering their vintage, data sources, level of detail and extent of peer review. Thus, the approach adopted here is to rely as much as possible on recent peer-reviewed literature, together with other publicly-available studies by governmental and private organizations heavily involved in the field of CO_2 capture. Later, in Chapter 8, the range of capture costs reported here are combined with cost estimates for CO_2 transport and storage to arrive at estimates of the overall cost of CCS for selected power systems and industrial processes.

Table 3.6 Confidence levels for technology and system cost estimates.

Confidence Level	Description
Very High	Mature technology with multiple commercial replications for this application and scale of operation; considerable operating experience and data under a variety of conditions.
High	Commercially deployed in applications similar to the system under study, but at a smaller scale and/or with limited operating experience; no major problems or issues anticipated in this application; commercial guarantees available.
Moderate	No commercial application for the system and/or scale of interest, but technology is commercially deployed in other applications; issues of scale-up, operability and reliability remain to be demonstrated for this application.
Low	Experience and data based on pilot plant or proof-of-concept scale; no commercial applications or full-scale demonstrations; significant technical issues or cost-related questions still to be resolved for this application.
Very Low	A new concept or process not yet tested, or with operational data limited to the laboratory or bench-scale level; issues of large-scale operability, effectiveness, reliability and manufacturability remain to be demonstrated.

3.7.4   Overview of technologies and systems evaluated

Economic studies of CO_2 capture have focused mainly on electric power generation, a major source of CO_2 emissions. To a lesser extent, CO_2 capture from industrial processes also has been subject to economic evaluations, especially processes producing hydrogen, often in combination with other products.

The sections below review and summarize recent estimates of CO_2 capture costs for major systems of interest. Sections 3.7.5 to 3.7.8 focus first on the cost of current CO_2 capture technologies, while Sections 3.7.10 to 3.7.12 go on to discuss improved or 'advanced' technologies promising lower costs in the future. In all cases the system boundary is defined as a single facility at which CO_2 is captured and compressed for delivery to a transport and storage system. To reflect different levels of confidence (or uncertainty) in cost estimates for technologies at different stages of development, the qualitative descriptors shown in Table 3.6 are applied in summarizing published cost estimates.[9] The studies reviewed typically report costs in US dollars for reference years ranging from 2000 to early 2004. Because inflation effects generally have been small during this period no adjustments have been made in summarizing ranges of reported costs.

3.7.5   Post-combustion CO_2 capture cost for electric power plants (current technology)

Most of the world's electricity is currently generated from the combustion of fossil fuels, especially coal and (to an increasing extent) natural gas. Hence, the ability to capture and store the CO_2 emitted by such plants has been a major focus of investigation. This section of the report focuses on the cost of currently available technology for CO_2 capture. Because of the relatively low CO_2 concentration in power plant flue gases, chemical absorption systems have been the dominant technology of interest for post-combustion capture (see Section 3.3.2). However, the cost of CO_2 capture depends not only on

the choice of capture technology, but also - and often more importantly - on the characteristics and design of the overall power plant. For purposes of cost reporting, we distinguish between coal-fired and gas-fired plant designs and between new and existing facilities.

3.7.5.1   New coal-fired power plants
Table 3.7 summarizes the key assumptions and results of recent studies of post-combustion CO_2 capture at new coal-fired power plants. Assumed plant sizes with CO_2 capture range from approximately 300-700 MW net power output. In all cases, CO_2 capture is accomplished using an amine-based absorption system, typically MEA. Capture efficiencies range from 85-95% with the most common value being 90%. The studies employ different assumptions about other key parameters such as the base power plant efficiency, coal properties, coal cost, plant capacity factor, CO_2 product pressure and financial parameters such as the fixed charge factor. All of these factors have a direct influence on total plant cost and the cost of CO_2 capture.

Table 3.7 summarizes several measures of CO_2 capture cost, both in absolute and relative terms. Across the full set of studies, CO_2 capture adds 44-87% to the capital cost of the reference plant (US$ kW^{-1}) and 42-81% to the cost of electricity (US$ MWh^{-1}), while achieving CO_2 reductions of approximately 80-90% per net kWh produced. The cost of CO_2 avoided for these cases varies from 29-51 US$/t$CO_2$. The absolute values of capital cost, COE and incremental cost of electricity in Table 3.7 reflect the different assumptions employed in each study. The result is an incremental COE of 18-38 US$ MWh^{-1} (or US$ 0.018-0.038 kWh^{-1}) for CO_2 capture. The total COE for plants with capture ranges from 62-87 US$ MWh^{-1}. In all cases, a significant portion of the total CO_2 capture cost is due to the energy requirement for CO_2 capture and compression. For the studies in Table 3.7, the plants with CO_2 capture require 24-42% more fuel input per MWh of plant output relative to a similar reference plant without capture. Roughly half the energy is required for solvent regeneration and a third for CO_2 compression.

While many factors contribute to the cost differences observed in Table 3.7, systematic studies of the influence of different factors indicate that the most important sources of variability in reported cost results are assumptions about the

---
[9] These descriptions are used in subsequent tables to characterize systems with CO_2 capture. In most cases the cost estimates for reference plants (without capture) would rank as high (e.g., IGCC power plants) or very high (e.g., PC and NGCC power plants).

Table 3.7 CO_2 capture costs: new pulverized-coal power plants using current technology.

Study Assumptions and Results	SUPERCRITICAL UNITS / BITUMINOUS COALS								SUBCRIT UNITS / LOW RANK COALS		
	Parsons	Parsons	Simbeck	IEA GHG	IEA GHG	Rubin et al.	Range	Range	NETL	Rao & Rubin	Stobbs & Clark
	2002b	2002b	2002	2004	2004	2005	min	max	2002	2002	2005
Reference Plant (without capture)			*		*	*			*	*	*
Boiler type (subcritical, super, ultra)	super	ultra	ultra	ultra	ultra	super			subcritical	subcritical	super
Coal type (bit, sub-bit, lig) and %S	bit, 2.5% S	bit, 2.5% S	bit, 1% S	bit, 1% S	bit, 1% S	bit, 2.1% S			bit, 2.5%S	sub-bit, 0.5%S	lignite
Emission control technologies (SO_2/NO_x)	FGD, SCR	FGD, SCR	FGD, SCR	FGD, SCR	FGD, SCR	FGD, SCR			FGD	FGD, SCR	FGD, SCR, LoTOx
Reference plant net output (MW)	462	506	520	758	754	524	462	758	397	462	424
Plant capacity factor (%)	65	65	80	85	85	75	65	85	85	75	90
Net plant efficiency, LHV (%)	42.2	44.8	44.5	44.0	43.7	40.9	41	45	38.9	36.1	43.4
Coal cost, LHV (US\$ GJ⁻¹)	1.29	0.98	0.98	1.50	1.50	1.25	0.98	1.50	1.03	1.25	0.88
Reference plant emission rate (t CO_2 MWh⁻¹)	0.774	0.736	0.76	0.743	0.747	0.811	0.74	0.81	0.835	0.941	0.883
Capture Plant Design											
CO_2 capture technology	MEA	MEA	MEA	MEA	KS-1	MEA			MEA	MEA	MEA
Net plant output with capture (MW)	329	367	408	666	676	492	329	676	283	326	311.0
Net plant efficiency, LHV (%)	30.1	32.5	34.9	34.8	35.4	31.1	30	35	27.7	25.4	31.8
CO_2 capture system efficiency (%)	90	90	85	87.5	90	90	85	90	95	90	95
CO_2 emission rate after capture (t MWh⁻¹)	0.108	0.101	0.145	0.117	0.092	0.107	0.09	0.15	0.059	0.133	0.060
CO_2 captured (Mt yr⁻¹)	1.830	2.350	2.360	4.061	4.168	3.102	1.83	4.17	2.346	2.580	2.795
CO_2 product pressure (MPa)	8.4	8.4	13.7	11.0	11.0	13.9	8	14	10.3	13.9	13.9
CCS energy requirement (% more input MWh⁻¹)	40	38	28	26	24	31	24	40	40	42	36
CO_2 reduction per kWh (%)	86	86	81	84	88	87	81	88	93	86	93
Cost Results		***	***	**	**	*					***
Cost year basis (constant dollars)	2000	2000	2000	2004	2004	2002			2002	2000	2003
Fixed charge factor (%)	15.5	15.5	12.7	11.0	11.0	14.8	11.0	15.5	14.8	15.0	
Reference plant TCR (US\$ kW⁻¹)	1281	1161	1486	1319	1265	1205	1161	1486	1268	1236	1891
Capture plant TCR (US\$ kW⁻¹)	2219	1943	2578	1894	2007	1936	1894	2578	2373	2163	3252
Incremental TCR for capture (US\$ kW⁻¹)	938	782	1092	575	742	731	575	1092	1105	927	1361
Reference plant COE (US\$ MWh⁻¹)	51.5	51.0	42.9	43.9	42.8	46.1	43	52	42.3	49.2	44.5
Capture plant COE (US\$ MWh⁻¹)	85.6	82.4	70.9	62.4	63.0	74.1	62	86	76.6	87.0	74.3
Incremental COE for capture (US\$ MWh⁻¹)	34.1	31.4	28	18.5	20.2	28	18	34	37.8	37.8	29.8
% increase in capital cost (over ref. plant)	73	67	74	44	59	61	44	74	87	75	72
% increase in COE (over ref. plant)	66	62	65	42	47	61	42	66	81	77	67
Cost of CO_2 captured (US\$/tCO₂)	35	28	34	23	24	29	23	35	31	31	26
Cost of CO_2 avoided (US\$/tCO₂)	51	49	43	29	31	40	29	51	43	47	36
Capture cost confidence level (see Table 3.6)					moderate					moderate	

Notes: All costs in this table are for capture only and do not include the costs of CO_2 transport and storage; see Chapter 8 for total CCS costs. * Reported HHV values converted to LHV assuming LHV/HHV = 0.96 for coal. ** Reported capital costs increased by 8% to include interest during construction. ***Reported capital costs increased by 15% to estimate interest during construction and other owners' costs.

CO_2 capture system energy requirement, power plant efficiency, fuel type, plant capacity factor and fixed charge rate (Rao and Rubin, 2002). In this regard, it is useful to note that the lowest-cost capture systems in Table 3.7 (in terms of COE and cost of CO_2 avoided) come from a recent study (IEA GHG, 2004) that combines an efficient supercritical power plant design using bituminous coal, with high plant utilization, lowest fixed charge rate and more energy-efficient amine system designs, as recently announced by two major vendors (but not yet demonstrated on coal-fired power plants). In contrast, the highest reported COE values are for less efficient subcritical plant designs using low rank coal, combined with lower capacity factors, higher fixed charge rates and employing amine system designs typical of units currently in operation at small power plants.

Recent increases in world coal prices, if sustained, also would affect the levelized COE values reported here. Based on one recent study (IEA GHG, 2004), each 1.00 US\$ GJ^{-1} increase in coal price would increase the COE by 8.2 US\$ MWh^{-1} for a new PC plant without capture and by 10.1 US\$ MWh^{-1} for a plant with capture.

These results indicate that new power plants equipped with CO_2 capture are likely to be high-efficiency supercritical units, which yield lowest overall costs. The worldwide use of supercritical units (without capture) with current usage at 155 GW_e (Section 3.1.2.2), is rapidly increasing in several regions of the world and, as seen in Table 3.7, the preponderance of recent studies of CO_2 capture are based on supercritical units using bituminous coals. For these plants, Table 3.7 shows that capture systems increase the capital cost by 44-74% and the COE by 42-66% (18-34 US\$ MWh^{-1}). The major factors contributing to these ranges were differences in plant size, capacity factor and fixed charge factor. New or improved capture systems and power plant designs that promise to further reduce the costs of CO_2 capture are discussed later in Section 3.7.7. First, however, we examine CO_2 capture costs at existing plants.

3.7.5.2   Existing coal-fired plants

Compared to the study of new plants, CO_2 capture options for existing power plants have received relatively little study to date. Table 3.8 summarizes the assumptions and results of several studies estimating the cost of retrofitting an amine-based CO_2 capture system to an existing coal-fired power plant. Several factors significantly affect the economics of retrofits, especially the age, smaller sizes and lower efficiencies typical of existing plants relative to new builds. The energy requirement for CO_2 capture also is usually higher because of less efficient heat integration for sorbent regeneration. All of these factors lead to higher overall costs. Existing plants not yet equipped with a flue gas desulphurization (FGD) system for SO_2 control also must be retrofitted or upgraded for high-efficiency sulphur capture in addition to the CO_2 capture device. For plants with high NO_x levels, a NO_2 removal system also may be required to minimize solvent loss from reactions with acid gases. Finally, site-specific difficulties, such as land availability, access to plant areas and the need for special ductwork, tend to further increase the capital cost of any retrofit project relative to an equivalent new

plant installation. Nonetheless, in cases where the capital cost of the existing plant has been fully or substantially amortized, Table 3.8 shows that the COE of a retrofitted plant with capture (including all new capital requirements) can be comparable to or lower than that of a new plant, although the incremental COE is typically higher because of the factors noted above.

Table 3.8 further shows that for comparable levels of about 85% CO_2 reduction per kWh, the average cost of CO_2 avoided for retrofits is about 35% higher than for the new plants analyzed in Table 3.7. The incremental capital cost and COE depend strongly on site-specific assumptions, including the degree of amortization and options for providing process energy needs. As with new plants, heat and power for CO_2 capture are usually assumed to be provided by the base (reference) plant, resulting in a sizeable (30 to 40%) plant output reduction. Other studies assume that an auxiliary gas-fired boiler is constructed to provide the CO_2 capture steam requirements and (in some cases) additional power. Low natural gas prices can make this option more attractive than plant output reduction (based on COE), but such systems yield lower CO_2 reductions (around 60%) since the emissions from natural gas combustion are typically not captured. For this reason, the avoided cost values for this option are not directly comparable to those with higher CO_2 reductions.

Also reflected in Table 3.8 is the option of rebuilding an existing boiler and steam turbine as a supercritical unit to gain efficiency improvements in conjunction with CO_2 capture. One recent study (Gibbins *et al.*, 2005) suggests this option could be economically attractive in conjunction with CO_2 capture since the more efficient unit minimizes the cost of capture and yields a greater net power output and a lower COE compared to a simple retrofit. The use of a new and less energy-intensive capture unit yields further cost reductions in this study. Another recent study similarly concluded that the most economical approach to CO_2 capture for an existing coal-fired plant was to combine CO_2 capture with repowering the unit with an ultra-supercritical steam system (Simbeck, 2004). One additional option, repowering an existing unit with a coal gasifier, is discussed later in Section 3.7.6.2.

3.7.5.3   Natural gas-fired power plants

Power plants fuelled by natural gas may include gas-fired boilers, simple-cycle gas turbines, or natural gas combined cycle (NGCC) units. The current operating capacity in use globally is 333 GW_e for gas-fired boilers, 214 GW_e for simple cycle gas turbines and 339 GW_e for NGCC (IEA WEO, 2004). The absence of sulphur and other impurities in natural gas reduces the capital costs associated with auxiliary flue gas clean-up systems required for amine-based CO_2 capture technology. On the other hand, the lower concentration of CO_2 in gas-fired units tends to increase the cost per tonne of CO_2 captured or avoided relative to coal-fired units.

Table 3.9 summarizes the assumptions and cost results of several recent studies of CO_2 capture at gas-fired combined cycle power plants ranging in size from approximately 300-700 MW. Relative to reference plants without capture, to achieve net

Table 3.8 CO$_2$ capture costs: existing pulverized-coal power plants using current technology.

Study Assumptions and Results	Simbeck & McDonald	Alstom et al.	Rao & Rubin	Rao & Rubin	Chen et al.	Chen et al.	Chen et al.	Singh et al.	Gibbins et al.	Range		Gibbins et al.	Gibbins et al.	Chen et al.
	2000	2001	2002	2002	2003	2003	2003	2003	2005	min	max	2006	2006	2003
	AMINE SYSTEM RETROFITS TO EXISTING BOILERS											REPOWERING + CO$_2$ CAPTURE		
Reference Plant (without capture)														
Boiler type (subcritical, super, ultra)	sub	sub *	sub *	sub *	sub *	sub *	sub *	sub-bit	sub			super	super	sub
Coal type (bit, sub-bit, lig) and %S	sub-bit, 0.5%	bit, 2.7%S	sub-bit, 0.5%	sub-bit, 0.5%	sub-bit, 1.1%S	sub-bit, 1.1%S	sub-bit, 1.1%S	sub-bit						
Emission control technologies (SO_2/NO_2)	none	FGD	none	FGD	FGD	FGD	FGD	not reported	not reported			not reported	not reported	FGD
Reference plant size (MW)	292	434	470	470	248	248	248	400	not reported	248	470	not reported	not reported	248
Plant capacity factor (%)	80	67	75	75	80	76 (Capture=80)	76 (Capture=80)	91.3	80	67	91	80	80	80
Net plant efficiency, LHV (%)	36.2	36.2	36.6	36.6	33.1	33.1	33.1		36.0	33	37	43.5	43.5	
Coal cost, LHV (US$ GJ^{-1})	0.98	1.30	1.25	1.25	1.20	1.20	1.20		3.07	0.98	3.07	3.07	3.07	1.20
Reference plant emission rate (t CO_2 MWh^{-1})	0.901	0.908	0.941	0.95	1.004	1.004	1.004	0.925		0.90	1.00			1.004
Capture Plant Design														
CO_2 capture technology	MEA	MEA	MEA	MEA	MEA	MEA	MEA	MEA	MEA			MEA	KS	Selexol
Other equipment included	new FGD	FGD upgrade	New FGD	FGD upgrade	FGD upgrade	FGD upgrade	FGD upgrade	FGD				Advanced supercrit boiler retrofit	Advanced supercrit boiler retrofit	IGCC (Texaco Q) repower +current steam turbine
Net plant size with capture (MW)	294	255	275	275	140	282	282	400		140	400			590
Auxiliary boiler/fuel used? (type, LHV cost)	NG. $4.51 GJ^{-1}	none	none	none	none	NG. $2.59 GJ^{-1}	NG. $5.06 GJ^{-1}	NG. $3.79 GJ^{-1}	none			none	none	none
Net plant efficiency, LHV (%)	25.3	21.3	21.4	21.4	18.7	14.8	14.8	9.4	24.0	19	25	31.5	34.5	32.6
CO_2 capture system efficiency (%)	90	96	90	90	90	90	90	90		90	96	90	90	90
CO_2 emission rate after capture (t MWh^{-1})	0.113	0.059	0.155	0.16	0.177	0.369	0.369	0.324		0.06	0.37			0.099
CO_2 captured (Mt yr^{-1})	2.090	2.228			1.480	1.480	1.480	2.664		1.48	2.66			3.684
CO_2 product pressure (MPa)	13.7	13.9	13.9	13.9	13.9	13.9	13.9		10.0	10	14	10.0	10.0	14.5
CCS energy requirement (% more input MWh^{-1})	43	70	71		77				50	43	77	38	26	
CO_2 reduction per kWh (%)	87 **	94	84	83	82	63	63	65 **		63	94			
Cost Results														
Cost year basis (constant dollars)	1999	n/a	2000	2000	2000	2000	2000	2001						
Fixed charge factor (%)	12.8	13.0	15.0	15.0	14.8	14.8	14.8	9.4	11.8	9.4	15.0	11.8	11.8	15
Reference plant TCR (US$ kW^{-1})	112				0	0	0	0	160	0	160	480	480	0
Capture plant TCR (US$ kW^{-1})	1059	1941			837	647	654	846	1028	647	1941	1282	1170	1493
Incremental TCR for capture (US$ kW^{-1})	947	1602			837	647	654	846	868	647	1602	802	690	1493
Reference plant COE (US$ MWh^{-1})	18.8		18.0	18.0	20.6	20.6	20.6		26.0	18	26	27.0	27.0	21
Capture plant COE (US$ MWh^{-1})	54.3		70.4	66.7	66.8	51.1	62.2		65.0	51	70	58.0	53.0	62.2
Incremental COE for capture (US$ MWh^{-1})	35.5	61.7	52.4	48.7	46.2	30.6	41.7	33.2	39.0	31	62	31.0	26.0	41.2
% increase in capital cost (over ref. plant)														
% increase in COE (over ref. plant)	189		291	271	225	149	203		150	149	291	115	96	196
Cost of CO_2, captured (US$/t$CO_2$)	35	42			31	41	56	40		31	56			
Cost of CO_2, avoided (US$/t$CO_2$)	45	73	67	59	56	48	66	55		45	73			46
Capture cost confidence level (see Table 3.6)	moderate											moderate		

Notes: All costs in this table are for capture only and do not include the costs of CO$_2$, transport and storage; see Chapter 8 for total CCS costs. * Reported HHV values converted to LHV assuming LHV/HHV = 0.96 for coal and 0.90 for natural gas. **Reported capital costs increased by 15% to estimate interest during construction and other owners' costs.

Table 3.9 CO_2 capture costs: natural gas-fired power plants using current technology.

Study Assumptions and Results	Parsons 2002(b) *	NETL 2002	IEA GHG 2004	IEA GHG 2004	CCP 2005	Rubin et al. 2005	Rubin et al. 2005 *	Rubin et al. 2005 *	Range min	Range max
Reference Plant (without capture)										
Plant type (boiler, gas turbine, comb.cycle)	comb.cycle	comb.cycle	comb.cycle	comb.cycle	comb.cycle	comb.cycle	comb.cycle	comb.cycle		
Reference plant size (MW)	509	379	776	776	392	507	507	507	379	776
Plant capacity factor (%)	65	85	85	85	95	75	50	50	50	95
Net plant efficiency, LHV (%)	55.1	57.9	55.6	55.6	57.6	55.8	55.8	55.8	55	58
Fuel cost, LHV (US$ GJ⁻¹)	2.82	3.55	3.00	3.00	2.96	4.44	4.44	4.44	2.82	4.44
Reference plant emission rate (tCO2 MWh⁻¹)	0.364	0.344	0.379	0.379	0.37	0.367	0.367	0.367	0.344	0.379
Capture Plant Design										
CO2 capture technology	MEA	MEA	MEA	KS-1	MEA	MEA	MEA	MEA		
Net plant size with capture (MW)	399	327	662	692	323	432	432	432	323	692
Net plant efficiency, LHV (%)	47.4	49.9	47.4	49.6	47.4	47.6	47.6	47.6	47	50
CO2 capture system efficiency (%)	90	90	85	85	86	90	90	90	85	90
CO2 emission rate after capture (t MWh⁻¹)	0.045	0.040	0.066	0.063	0.063	0.043	0.043	0.043	0.040	0.066
CO2 captured (Mt yr⁻¹)	0.949	0.875	1.844	1.844	1.09	1.099	0.733	0.733	0.733	1.844
CO2 product pressure (MPa)	8.4	10.3	11.0	11.0		13.7	13.7	13.7	8	14
CCS energy requirement (% more input MWh⁻¹)	16	16	15	11	22	17	17	17	11	22
CO2 reduction per kWh (%)	88	88	83	83	83	88	88	88	83	88
Cost Results										
Cost year basis (constant dollars)	2000	2002	2004	2004		2001	2001	2001		
Fixed charge factor (%)			11.0	11.0	11.0	14.8	14.8	14.8	11.0	14.8
Reference plant TCR (US$ kW⁻¹)	549	515	539	539	724	554	554	554	515	724
Capture plant TCR (US$ kW⁻¹)	1099	911	938	958	1261	909	909	909	909	1261
Incremental TCR for capture (US$ kW⁻¹)	550	396	399	419	537	355	355	355	355	550
Reference plant COE (US$ MWh⁻¹)	34.2	34.7	31.3	31.3	34.2	43.1	50	50	31	50
Capture plant COE (US$ MWh⁻¹)	57.9	48.3	44	43.1	51.8	58.9	72	72	43	72
Incremental COE for capture (US$ MWh⁻¹)	23.7	13.6	12.7	11.8	17.6	15.8	22	22	12	24
% increase in capital cost (over ref. plant)	100	77	74	78	74	64	64	64	64	100
% increase in COE (over ref. plant)	69	39	41	38	51	37	44	44	37	69
Cost of CO2, captured (US$/tCO2)	57	38	34	33	46	41	57	57	33	57
Cost of CO2, avoided (US$/tCO2)	74	45	41	37	57	49	68	68	37	74
Capture cost confidence level (see Table 3.6)					moderate					

Notes: All costs in this table are for capture only and do not include the costs of CO_2 transport and storage; see Chapter 8 for total CCS costs. * Reported HHV values converted to LHV assuming LHV/HHV = 0.90 for natural gas.

CO$_2$ reductions (per kWh) of the order of 83-88%, the capital cost per kW increases by 64-100%, while the COE increases by 37-69%, or by 12-24 US\$ MWh^{-1} on an absolute basis. The corresponding cost of CO$_2$ avoided ranges from 37-74 US\$/tCO$_2$, while the CCS energy requirement increases plant fuel consumption per kWh by 11-22%.

As seen earlier in Equations (7) to (9), assumptions about the plant fuel cost have an especially important influence on the COE for gas-fired plants because the contribution of capital costs is relatively low compared to coal plants. The studies in Table 3.9 assume stable gas prices of 2.82-4.44 US\$ GJ^{-1} (LHV basis) over the life of the plant, together with high capacity factors (65-95%) representing base load operation. These assumptions result in relatively low values of COE for both the reference plant and capture plant. Since about 2002, however, natural gas prices have increased significantly in many parts of the world, which has also affected the outlook for future prices. Based on the assumptions of one recent study (IEA GHG, 2004), the COE for an NGCC plant without capture would increase by 6.8 US\$ MWh^{-1} for each 1.00 US\$ GJ^{-1} increase in natural gas price (assuming no change in plant utilization or other factors of production). An NGCC plant with CCS would see a slightly higher increase of 7.3 US\$ MWh^{-1}. The price of natural gas, and its relation to the price of competing fuels like coal, is an important determinant of which type of power plant will provide the lowest cost electricity in the context of a particular situation. However, across a twofold increase in gas price (from 3-6 US\$ GJ^{-1}), the incremental cost of CO$_2$ capture changed by only 2 US\$ MWh^{-1} (US\$ 0.002 kWh^{-1}) with all other factors held constant.

In countries like the US, higher gas prices have also resulted in lower utilization rates (averaging 30-50%) for plants originally designed for base-load operation, but where lower-cost coal plants are available for dispatch. This further raises the average cost of electricity and CO$_2$ capture for those NGCC plants, as reflected in one case in Table 3.9 with a capacity factor of 50%. In other parts of the world, however, lower-cost coal plants may not be available, or gas supply contracts might limit the ability to curtail gas use. Such situations again illustrate that options for power generation with or without CO$_2$ capture should be evaluated in the context of a particular situation or scenario.

Studies of commercial post-combustion CO$_2$ capture applied to simple-cycle gas turbines have been conducted for the special case of retrofitting an auxiliary power generator in a remote location (CCP, 2005). This study reported a relatively high cost of 88 US\$/tCO$_2$ avoided. Studies of post-combustion capture for gas-fired boilers have been limited to industrial applications, as discussed later in Section 3.7.8.

3.7.5.4 Biomass-firing and co-firing systems

Power plants can be designed to be fuelled solely by biomass, or biomass can be co-fired in conventional coal-burning plants. The requirement to reduce net CO$_2$ emissions could lead to an increased use of biomass fuel, because plants that utilize biomass as a primary or supplemental fuel may be able to take credit for the carbon removed from the atmosphere during the

biomass growth cycle. If the biomass carbon released during combustion (as CO$_2$) is then captured and stored, the net quantity of CO$_2$ emitted to the atmosphere could in principle be negative.

The most important factor affecting the economics of biomass use is the cost of the biomass. This can range from a negative value, as in the case of some biomass wastes, to costs substantially higher than coal, as in the case of some purposely-grown biomass fuels, or wastes that have to be collected from diffuse sources. Power plants that use only biomass are typically smaller than coal-fired plants because local availability of biomass is often limited and biomass is more bulky and hence more expensive to transport than coal. The smaller sizes of biomass-fired plants would normally result in lower energy efficiencies and higher costs of CO$_2$ capture. Biomass can be co-fired with coal in larger plants (Robinson *et al.*, 2003). In such circumstances the incremental costs of capturing biomass-derived CO$_2$ should be similar to costs of capturing coal-derived CO$_2$. Another option is to convert biomass into pellets or refined liquid fuels to reduce the cost of transporting it over long distances. However, there are costs and emissions associated with production of these refined fuels. Information on costs of CO$_2$ capture at biomass-fired plants is sparse but some information is given in Section 3.7.8.4. The overall economics of CCS with biomass combustion will depend very much on local circumstances, especially biomass availability and cost and (as with fossil fuels) proximity to potential CO$_2$ storage sites.

3.7.6 Pre-combustion CO$_2$ capture cost for electric power plants (current technology)

Studies of pre-combustion capture for electric power plants have focused mainly on IGCC systems using coal or other solid fuels such as petroleum coke. This section of the report focuses on currently available technology for CO$_2$ capture at such plants. As before, the cost of CO$_2$ capture depends not only on the choice of capture technology, but more importantly on the characteristics and design of the overall power plant, including the fuel type and choice of gasifier. Because IGCC is not widely used for electric power generation at the present time, economic studies of IGCC power plants typically employ design assumptions based on the limited utility experience with IGCC systems and the more extensive experience with gasification in industrial sectors such as petroleum refining and petrochemicals. For oxygen-blown gasifiers, the high operating pressure and relatively high CO$_2$ concentrations achievable in IGCC systems makes physical solvent absorption systems the predominant technology of interest for pre-combustion CO$_2$ capture (see Section 3.5.2.11). For purposes of cost reporting, we again distinguish between new plant designs and the retrofitting of existing facilities.

3.7.6.1 New coal gasification combined cycle power plants

Table 3.10 summarizes the key assumptions and results of several recent studies of CO$_2$ capture costs for new IGCC power plants ranging in size from approximately 400-800 MW

Table 3.10 CO$_2$ capture costs: new IGCC power plants using current technology.

Study Assumptions and Results	NETL	NETL	NETL	Parsons	Simbeck	Nsakala et al.	IEA GHG	IEA GHG	IEA GHG	Rubin et al.	Rubin et al.	Range	
	2002	2002	2002	2002b	2002	2003	2003	2003	2003	2005	2005	min	max
					PLANTS WITH BITUMINOUS COAL FEEDSTOCK								
Reference Plant without capture													
Gasifier name or type	Shell, O$_2$ blown, CGCU	E-gas, O$_2$ blown, CGCU	Texaco quench, O$_2$ blown	E-gas, O$_2$ blown	Texaco quench, O$_2$ blown	Texaco syngas cooler, O$_2$ blown	Texaco quench, O$_2$ blown	Texaco quench, O$_2$ blown	Shell, O$_2$ blown	Texaco quench, O$_2$ blown	Texaco quench, O$_2$ blown		
Fuel type (bit, subbit, lig; other) and %S	Illinois #6	Illinois #6	Illinois #6	bit, 2.5% S	bit, 1% S	bit	bit, 1%S	bit, 1%S	bit, 1%S	bit, 2.1%S	bit, 2.1%S		
Reference plant size (MW)	413	401	571	425	521		827	827	776	527	527	401	827
Plant capacity factor (%)	85	85	65	65	80	80	85	85	85	75	65	65	85
Net plant efficiency, LHV (%)	47.4	46.7	39.1	44.8	44.6		38.0	38.0	43.1	39.1	39.1	38	47
Fuel cost, LHV (US$ GJ^{-1})	1.03	1.03	1.28	1.29	0.98	1.23	1.50	1.50	1.50	1.25	1.25	0.98	1.50
Reference plant emission rate (tCO$_2$ MWh^{-1})	0.682	0.692	0.846	0.718	0.725		0.833	0.833	0.763	0.817	0.817	0.68	0.85
Capture Plant Design													
CO$_2$ capture technology	Selexol	Selexol	Selexol	Selexol	Selexol	Selexol	Selexol	Selexol, NS	Selexol	Selexol	Selexol		
Net plant size, with capture (MW)	351	359	457	404	455		730	742	676	492	492	351	742
Net plant efficiency, LHV (%)	40.1	40.1	31.3	38.5	39.0	31.5	31.5	32.0	34.5	33.8	33.8	31	40
CO$_2$ capture system efficiency (%)	89.2	87.0	89.0	91.0	91.2		85	85	85	90	90	85	91
CO$_2$ emission rate after capture (t MWh^{-1})	0.087	0.105	0.116	0.073	0.065	0.104	0.152	0.151	0.142	0.097	0.097	0.07	0.15
CO$_2$ captured (Mt/yr)	1.803	1.870	2.368	1.379	2.151		4.682	4.728	4.050	2.749	2.383	1.38	4.73
CO$_2$ product pressure (MPa)	14.5	14.5	8.3	8.3			11.0	11.0	11.0	13.7	13.7	8	14
CCS energy requirement (% more input MWh^{-1})	18	16	25	16	14		21	19	25	16	16	14	25
CO$_2$ reduction per kWh (%)	87	85	86	90	91		82	82	81	88	88	81	91
Cost Results							**	**	**				
Cost year basis (constant dollars)	2002	2002	2002	2000	2000		2002	2002	2002	2001	2001		
Fixed charge factor (%)	14.8	14.8	15.0	13.8	13.0		11.0	11.0	11.0	14.8	17.3	11	17
Reference plant TCR (US$ kW^{-1})	1370	1374	1169	1251	1486	1565	1187	1187	1371	1311	1311	1169	1565
Capture plant TCR (US$ kW^{-1})	2270	1897	1549	1844	2067	2179	1495	1414	1860	1748	1748	1414	2270
Incremental TCR for capture (US$ kW^{-1})	900	523	380	593	581	614	308	227	489	437	437	227	900
Reference plant COE (US$ MWh^{-1})	40.6	40.9	43.4	47.7	43.0	53.0	45.0	45.0	48.0	48.3	61	41	61
Capture plant COE (US$ MWh^{-1})	62.9	54.4	59.9	65.8	57.7	71.5	56.0	54.0	63.0	62.6	79	54	79
Incremental COE for capture (US$ MWh^{-1})	22.3	13.5	16.5	18.1	14.7	18.5	11	9	15	14.3	18.2	9	22
% increase in capital cost (over ref. plant)	66	38	33	47	39	39	26	19	36	33	33	19	66
% increase in COE (over ref. plant)	55	33	38	38	34	35	24	20	31	30	30	20	55
Cost of CO$_2$ captured (US$/tCO$_2$)	32	19	18	30	21	21	13	11	19	17	21	11	32
Cost of CO$_2$ avoided (US$/tCO$_2$)	37	23	23	28	22	23	16	13	24	20	25	13	37
Capture cost confidence level (see Table 3.6)						moderate							

Notes: All costs in this table are for capture only and do not include the costs of CO$_2$ transport and storage; see Chapter 8 for total CCS costs. * Reported HHV values converted to LHV assuming LHV/HHV = 0.96 for coal. ** Reported capital costs increased by 8% to include interest during construction. **Reported capital costs increased by 15% to estimate interest during construction and other owners' costs.

Table 3.10. *Continued.*

Study Assumptions and Results	Stobbs & Clark	Stobbs & Clark	Stobbs & Clark	IEA GHG
	2005	2005	2005	2000b
	PLANTS WITH OTHER FEEDSTOCKS			
Reference Plant without capture)				
Gasifier name or type		Texaco quench, O_2 blown	Shell, O_2 blown	O_2 blown, partial oxidation
Fuel type (bit, subbit, lig; other) and %S	bit	Sub-bit	Lignite	Natural gas
Reference plant size (MW)	[No IGCC Reference Plants]			790
Plant capacity factor (%)	90	90	90	90
Net plant efficiency, LHV (%)				56.2
Fuel cost, LHV (US$ GJ⁻¹)	1.90	0.48	0.88	2.00
Reference plant emission rate (tCO₂ MWh⁻¹)				0.370
Capture Plant Design				
CO₂ capture technology	Selexol	Selexol	Selexol	Selexol
Net plant size, with capture (MW)	445	437	361	820
Net plant efficiency, LHV (%)	32.8	27.0	28.3	48.3
CO₂ capture system efficiency (%)	87	92	86	85
CO₂ emission rate after capture (t MWh⁻¹)	0.130	0.102	0.182	0.065
CO₂ captured (Mt/yr)	3.049	4.040	3.183	2.356
CO₂ product pressure (MPa)	13.9	13.9	13.9	11.0
CCS energy requirement (% more input MWh⁻¹)				14
CO₂ reduction per kWh (%)				**82**
Cost Results	***	***	***	**
Cost year basis (constant dollars)	2003	2003	2003	2000
Fixed charge factor (%)				11.0
Reference plant TCR (US$ kW⁻¹)				447
Capture plant TCR (US$ kW⁻¹)	2205	2518	3247	978
Incremental TCR for capture (US$ kW⁻¹)				531
Reference plant COE (US$ MWh⁻¹)				21.6
Capture plant COE (US$ MWh⁻¹)	68.4	62.1	83.9	34.4
Incremental COE for capture (US$ MWh⁻¹)				12.8
% increase in capital cost (over ref. plant)				119
% increase in COE (over ref. plant)				**59**
Cost of CO₂ captured (US$/tCO₂)				35
Cost of CO₂ avoided (US$/tCO₂)	31	33	56	42
Capture cost confidence level (see Table 3.6)	moderate			moderate

Notes: All costs in this table are for capture only and do not include the costs of CO₂ transport and storage; see Chapter 8 for total CCS costs. * Reported HHV values converted to LHV assuming LHV/HHV = 0.96 for coal. ** Reported capital costs increased by 8% to include interest during construction. ***Reported capital costs increased by 15% to estimate interest during construction and other owners' costs.

net power output. While several gasifiers and coal types are represented, most studies focus on the oxygen-blown Texaco quench system,[10] and all but one assume bituminous coals. CO₂ capture efficiencies across these studies range from 85-92% using commercially available physical absorption systems. The energy requirements for capture increase the overall plant heat rate (energy input per kWh) by 16-25%, yielding net CO₂ reductions per kWh of 81-88%. Other study variables that influence total plant cost and the cost of CO₂ capture include the fuel cost, CO₂ product pressure, plant capacity factor and fixed charge factor. Many of the recent studies also include the cost of a spare gasifier to ensure high system reliability.

Table 3.10 indicates that for studies based on the Texaco or E-Gas gasifiers, CO₂ capture adds approximately 20-40% to both the capital cost (US$ kW⁻¹) and the cost of electricity (US$ MWh⁻¹) of the reference IGCC plants, while studies using the Shell gasifier report increases of roughly 30-65%. The total COE reported for IGCC systems ranges from 41-61 US$ MWh⁻¹ without capture and 54-79 US$ MWh⁻¹ with capture. With capture, the lowest COE is found for gasifier systems with quench cooling designs that have lower thermal efficiencies than the more capital-intensive designs with heat recovery systems. Without capture, however, the latter system type has the lowest COE in Table 3.10. Across all studies, the cost of CO₂ avoided ranges from 13-37 US$/tCO₂ relative to an IGCC without capture, excluding transport and storage costs. Part of the reason for this lower incremental cost of CO₂ capture relative to coal combustion plants is the lower average energy requirement for IGCC systems. Another key factor is the smaller gas volume treated in oxygen-blown gasifier systems, which substantially reduces equipment size and cost.

As with PC plants, Table 3.10 again emphasizes the importance of plant financing and utilization assumptions on the calculated cost of electricity, which in turn affects CO₂-capture costs. The lowest COE values in this table are for plants with a low fixed charge rate and high capacity factor, while

---

[10] In 2004, the Texaco gasifier was re-named as the GE gasifier following acquisition by GE Energy (General Electric). However, this report uses the name Texaco, as it is referred to in the original references cited.

substantially higher COE values result from high financing costs and lower plant utilization. Similarly, the type and properties of coal assumed has a major impact on the COE, as seen in a recent Canadian Clean Power Coalition study, which found substantially higher costs for low-rank coals using a Texaco-based IGCC system (Stobbs and Clark, 2005, Table 3.10). EPRI also reports higher IGCC costs for low-rank coals (Holt *et al.*, 2003). On the other hand, where plant-level assumptions and designs are similar across studies, there is relatively little difference in the estimated costs of CO_2 capture based on current commercial technology. Similarly, the several studies in Tables 3.7 and 3.10 that estimate costs for both IGCC and PC plants on an internally consistent basis, all find that IGCC plants with capture have a lower COE than PC plants with capture. There is not yet a high degree of confidence in these cost estimates, however (see Table 3.6).

The costs in Table 3.10 also reflect efforts in some studies to identify least-cost CO_2 capture options. For example, one recent study (IEA GHG, 2003) found that capture and disposal of hydrogen sulphide (H_2S) along with CO_2 can reduce overall capture costs by about 20% (although this may increase transport and storage costs, as discussed in Chapters 4 and 5). The feasibility of this approach depends in a large part on applicable regulatory and permitting requirements. Advanced IGCC designs that may further reduce future CO_2 capture costs are discussed in Section 3.7.7.

3.7.6.2 *Repowering of existing coal-fired plants with IGCC*
For some existing coal-fired power plants, an alternative to the post-combustion capture systems discussed earlier is repowering with an IGCC system. In this case - depending on site-specific circumstances - some existing plant components, such as the steam turbine, might be refurbished and utilized as part of an IGCC plant. Alternatively, the entire combustion plant might be replaced with a new IGCC system while preserving other site facilities and infrastructure.

Although repowering has been widely studied as an option to improve plant performance and increase plant output, there are relatively few studies of repowering motivated by CO_2 capture. Table 3.8 shows results from one recent study (Chen *et al.*, 2003) which reports CO_2 capture costs for IGCC repowering of a 250 MW coal-fired unit that is assumed to be a fully amortized (hence, a low COE of 21 US$ MWh⁻¹). IGCC repowering yielded a net plant capacity of 600 MW with CO_2 capture and a COE of 62-67 US$ MWh⁻¹ depending on whether or not the existing steam turbine can be reused. The cost of CO_2 avoided was 46-51 US$/t$CO_2$. Compared to the option of retrofitting the existing PC unit with an amine-based capture system and retaining the existing boiler (Table 3.8), the COE for IGCC repowering was estimated to be 10-30% lower. These findings are in general agreement with earlier studies by Simbeck (1999). Because the addition of gas turbines roughly triples the gross plant capacity of a steam-electric plant, candidates for IGCC repowering are generally limited to smaller existing units (e.g., 100-300 MW). Taken together with the post-combustion retrofit studies in Table 3.8, the most cost-effective options for existing

plants involve combining CO_2 capture with plant upgrades that increase overall efficiency and net output. Additional studies would be needed to systematically compare the feasibility and cost of IGCC repowering to supercritical boiler upgrades at existing coal-fired plants.

3.7.7 *CO_2 capture cost for hydrogen production and multi-product plants (current technology)*

While electric power systems have been the dominant technologies of interest for CO_2 capture studies, other industrial processes, including hydrogen production and multi-product plants producing a mix of fuels, chemicals and electricity also are of interest. Because CO_2 capture cost depends strongly on the production process in question, several categories of industrial processes are discussed below.

3.7.7.1 *Hydrogen production plants*
Section 3.5 discussed the potential role of hydrogen as an energy carrier and the technological options for its production. Here we examine the cost of capturing CO_2 normally released during the production of hydrogen from fossil fuels. Table 3.11 shows the key assumptions and cost results of recent studies of CO_2 capture costs for plants with hydrogen production rates of 155,000-510,000 Nm³ h⁻¹ (466-1531 MW$_t$), employing either natural gas or coal as a feedstock. The CO_2 capture efficiency for the hydrogen plant ranges from 87-95% using commercially available chemical and physical absorption systems. The CO_2 reduction per unit of product is lower, however, because of the process energy requirements and because of additional CO_2 emitted by an offsite power plant assumed in some of these studies. As hydrogen production requires the separation of H_2 from CO_2, the incremental cost of capture is mainly the cost of CO_2 compression.

At present, hydrogen is produced mainly from natural gas. Two recent studies (see Table 3.11) indicate that CO_2 capture would add approximately 18-33% to the unit cost of hydrogen while reducing net CO_2 emissions per unit of H_2 product by 72-83% (after accounting for the CO_2 emissions from imported electricity). The total cost of hydrogen is sensitive to the cost of feedstock, so different gas prices would alter both the absolute and relative costs of CO_2 capture.

For coal-based hydrogen production, a recent study (NRC, 2004) projects an 8% increase in the unit cost of hydrogen for an 83% reduction in CO_2 emissions per unit of product. Again, this figure includes the CO_2 emissions from imported electricity.

3.7.7.2 *Multi-product plants*
Multi-product plants (also known as polygeneration plants) employ fossil fuel feedstocks to produce a variety of products such as electricity, hydrogen, chemicals and liquid fuels. To calculate the cost of any particular product (for a given rate of return), economic analyses of multi-product plants require that the selling price of all other products be specified over the operating life of the plant. Such assumptions, in addition to

Table 3.11. CO_2 capture costs: Hydrogen and multi-product plants using current or near-commercial technology. (Continued on next page)

Study Assumptions and Results	HYDROGEN AND ELECTRICITY PRODUCTS								
	Simbeck	NRC	NRC	Parsons	Mitretek	Kreutz et al.	Kreutz et al.	Range	
	2005	2004	2004	2002a	2003	2005	2005	min	max
Reference Plant (without capture)	*			*	*				
Plant products (primary/secondary)	H_2	H_2	H_2	H_2+ electricity	H_2+ electricity	H_2+ electricity	H_2+ electricity		
Production process or type	Steam reforming	Steam reforming	Texaco quench, CGCU	Conv E-Gas, CGCU, H_2SO_4 co-product	Texaco quench, CGCU, Claus/Scot sulphur co-product	Texaco quench	Texaco quench		
Feedstock	Natural gas	Natural gas	Coal	Pgh #8 Coal	Coal	Coal	Coal		
Feedstock cost, LHV (US$ GJ^{-1})	5.26	4.73	1,20	0.89	1.03	1.26	1.26	**0.89**	**5.26**
Ref. plant input capacity, LHV (GJ h^{-1})	9848	7235	8861	2627	2954	6706	6706	**2627**	**9848**
Ref plant output capacity, LHV: Fuels (GJ h^{-1})	7504	5513	6004	1419	1579	3853	3853	**1419**	**7504**
Electricity (MW)	-44	-32	-121	38	20	78	78	**-121**	**78**
Net plant efficiency, LHV (%)	74.6	74.6	62.9	59.2	55.9	61.7	61.7	**55.9**	**74.6**
Plant capacity factor (%)	90	90	90	80	85	80	80	**80**	**90**
CO_2 emitted (MtCO₂ yr^{-1})	4.693	3.339	7.399	1.795	2.148	4.215	4.215	**1.80**	**7.40**
Carbon exported in fuels (MtC yr^{-1})	0	0	0	0	0	0	0	**0**	**0**
Total carbon released (kg CO_2 GJ^{-1} products)	81	78	168	164	174	145	145	**78**	**174**
Capture Plant Design									
CO_2 capture/separation technology	Amine scrubber, SMR flue gas	MEA scrubber	Not reported	Selexol	Not reported	Selexol	CO_2 H_2S co-capture, Selexol		
Capture plant input capacity, LHV (GJ h^{-1})	11495	8339	8861	2627	2954	6706	6706	**2627**	**11495**
Capture plant output capacity, LHV: Fuels (GJ h^{-1})	7504	6004	6004	1443	1434	3853	3853	**1434**	**7504**
Electricity (MW)	-129	-91	-187	12	27	39	35	**-187**	**39**
Net plant efficiency, LHV (%)	61.2	68.1	60.2	56.6	51.8	59.5	59.3	**51.8**	**68.1**
CO_2 capture efficiency (%)**	90	90	90	92	87	91	95	**87**	**95**
CO_2 emitted (MtCO₂ yr^{-1})***	1.280	0.604	1.181	0.143	0.279	0.338	0.182	**0.14**	**1.280**
Carbon exported in fuels (MtC yr^{-1})	0	0	0	0	0	0	0	**0.0**	**0**
Total carbon released (kgCO₂ GJ^{-1} products)	23.0	13.5	28.1	13.7	24.5	12.1	6.5	**6.5**	**28.1**
CO_2 captured (MtCO₂ yr^{-1})	4.658	3.378	6.385	1.654	1.869	3.882	4.037	**1.7**	**6.4**
CO_2 product pressure (MPa)	13.7	13.7	13.7	13.4	20	15	15	**13.4**	**20.0**
CCS energy requirement (% more input/GJ plant output)	21.8	9.5	4.5	4.7	7.9	3.6	3.9	**3.6**	**21.8**
CO_2 reduction per unit product (%)	72	83	83	92	86	92	96	**72**	**96**
Cost Results									
Cost year basis (constant dollars)	2003	2000	2000	2000	2000	2002	2002		
Fixed charge rate (%)	20.0	16.0	16.0	14.3	13.0	15.0	15.0	**13.0**	**20.0**
Reference plant TCR (million US$)****	668	469	1192	357	365	887	887	**357**	**1192**
Capture plant TCR (million US$)****	1029	646	1218	415	409	935	872	**409**	**1218**
% increase in capital cost (%)	54.1	37.7	2.2	16.5	11.9	5.4	-1.7	**-1.7**	**54.1**
Ref. plant electricity price (US$ MWh^{-1})	50.0	45.0	45.0	30.8	35.6	46.2	46.2	**30.8**	**50.0**
Capture plant electricity price (US$ MWh^{-1})	50.0	45.0	45.0	30.8	53.6	62.3	60.5	**30.8**	**62.3**
% increase in assumed electricity price	0.0	0.0	0.0	0.0	50.6	34.8	31.0	**0.0**	**50.6**
Ref. plant fuel product cost, LHV (US$ GJ^{-1})	**10.03**	**8.58**	**7.99**	**6.51**	**7.29**	**7.19**	**7.19**	**6.51**	**10.03**
Capture plant fuel product cost, LHV (US$ GJ^{-1})	**13.29**	**10.14**	**8.61**	**7.90**	**8.27**	**7.86**	**7.52**	**7.52**	**13.29**
Increase in fuel product cost (US$ GJ^{-1})	**3.26**	**1.56**	**0.62**	**1.38**	**0.98**	**0.67**	**0.32**	**0.32**	**3.26**
% increase in fuel product cost	**32.5**	**18.2**	**7.7**	**21.1**	**13.4**	**9.3**	**4.5**	**4.5**	**32.5**
Cost of CO_2 captured (US$/$tCO_2$)	38.9	20.7	4.1	8.7	6.0	4.8	2.2	**2.2**	**38.9**
Cost of CO_2 avoided (US$/$tCO_2$)	56.3	24.1	4.4	9.2	6.5	5.0	2.3	**2.3**	**56.3**
Confidence level (see Table 3.6)	high		high			moderate			

Notes: All costs in this table are for capture only and do not include the costs of CO_2 transport and storage; see Chapter 8 for total CCS costs. * Reported HHV values converted to LHV assuming LHV/HHV = 0.96 for coal, 0.846 for hydrogen, and 0.93 for F-T liquids. ** CO_2 capture efficiency = (C in CO_2 captured) /(C in fossil fuel input to plant - C in carbonaceous fuel products of plant) x100; C associated with imported electricity is not included. ***Includes CO_2 emitted in the production of electricity imported by the plant. ****Reported total plant investment values increased by 3.5% to estimate total capital requirement.

those discussed earlier, can significantly affect the outcome of cost calculations when there is not one dominant product at the facility.

Several of the coal-based hydrogen production plants in Table 3.11 also produce electricity, albeit in small amounts (in fact, smaller than the electricity quantities purchased by the stand-alone plants). Most of these studies assume that the value of the electricity product is higher under a carbon capture regime than without CO_2 capture. The result is a 5-33%

increase in hydrogen production cost for CO_2 reductions of 72-96% per unit of product. The case with the lowest incremental product cost and highest CO_2 reduction assumes co-disposal of H_2S with CO_2, thus eliminating the costs of sulphur capture and recovery. As noted earlier (Section 3.7.6.1), the feasibility of this option depends strongly on local regulatory requirements; nor are higher costs for transport and storage reflected in the Table 3.11 cost estimate for this case.

Table 3.11 also presents examples of multi-product plants

Table 3.11. *Continued.*

Study Assumptions and Results	LIQUID FUEL AND ELECTRICITY PRODUCTS									Range	
	Mitretek	Larson/Ren	Larson/Ren	Larson/Ren	Larson/Ren	Celik *et al.*	Celik *et al.*	Celik *et al.*	Celik *et al.*		
	2003	2003	2003	2003	2003	2005	2005	2005	2005	min	max
Reference Plant (without capture)	*										
Plant products (primary/secondary)	F-T liquids + electricity	MeOH +electricity	MeOH +electricity	DME +electricity	DME +electricity	DME + electricity	DME + electricity	DME + electricity	DME + electricity		
Production process or type	Unspecified O₂-blown gasifier, unspecified synthesis reactor	Texaco quench, Liquid phase reactor, Once-through config.	Texaco quench, Liquid phase reactor, Once-through config.	Texaco quench, Liquid phase reactor, Once-through config.	Texaco quench, Liquid phase reactor, Once-through config.	Texaco quench, Liquid phase reactor, Once-through config.	Texaco quench, Liquid phase reactor, Once-through config.	Texaco quench, Liquid phase reactor, Once-through config.	Texaco quench, Liquid phase reactor, Once-through config.		
Feedstock	Coal	Coal	Coal	Coal	Coal	Coal	Coal	Coal	Coal		
Feedstock cost, LHV (US$ GJ^{-1})	1,09	1.00	1.00	1.00	1.00	1.00	1.00	1.00	1.00	1.00	1.09
Ref. plant input capacity, LHV (GJ h^{-1})	16136	9893	9893	8690	8690	7931	7931	7931	7931	7931	16136
Ref plant output capacity, LHV: Fuels (GJ h^{-1})	7161	2254	2254	2160	2160	2161	2161	2161	2161	2160	7161
Electricity (MW)	697	625	625	552	552	490	490	490	490	490	697
Net plant efficiency, LHV (%)	59.9	45.5	45.5	47.7	47.7	49.5	49.5	49.5	49.5	45.5	59.9
Plant capacity factor (%)	90	85	85	85	85	80	80	80	80	80	90
CO₂ emitted (MtCO₂ yr^{-1})	8.067	5.646	5.646	4.895	4.895	4.077	4.077	4.077	4.077	4.08	8.07
Carbon exported in fuels (MtC yr^{-1})	1.190	0.317	0.317	0.334	0.334	0.274	0.274	0.274	0.274	0.27	1.19
Total carbon released (kgCO₂ GJ^{-1} products)	163	203	203	198	198	185	185	185	185	163	203
Capture Plant Design											
CO₂ capture/separation technology	Amine scrubber	Selexol	CO₂ H₂S co-capture. Selexol	Selexol	CO₂ H₂S co-capture. Selexol	CO₂ H₂S co-capture. Rectisol	CO₂ H₂S co-capture. Rectisol	CO₂ H₂S co-capture. Rectisol	CO₂ H₂S co-capture. Rectisol		
Capture plant input capacity, LHV (GJ h^{-1})	16136	9893	9893	8690	Coal	7931	7931	7931	7931	7931	16136
Capture plant output capacity LHV: Fuels (GJ h^{-1})	7242	2254	2254	2160	2160	2161	2160	2160	2160	2160	7242
Electricity (MW)	510	582	577	531	527	469	367	365	353	353	582
Net plant efficiency, LHV (%)	56.3	44.0	43.8	46.9		48.5	43.9	43.8	43.2	43	56
CO₂ capture efficiency (%)**	91	58	63	32	37	36	89	92	97	32	97
CO₂ emitted (MtCO₂ yr^{-1})***	0.733	2.377	2.099	3.320	3.076	2.598	0.390	0.288	0.028	0.03	3.32
Carbon exported in fuels (MtC yr^{-1})	1.2	0.317	0.317	0.294	0.294	0.274	0.274	0.274	0.274	0.274	1.200
Total carbon released (kgCO₂ GJ^{-1} products)	71.7	109.2	101.0	144.9	137.4	134	57	53	43	43	145
CO₂ captured (MtCO₂ yr^{-1})	7.260	3.269	3.547	1.574	1.819	1.479	3.692	3.790	4.021	1.48	7.26
CO₂ product pressure (MPa)	13.8	15	15	15	15	15	15	15	15	14	15
CCS energy requirement. (% more input/GJ plant output)	6.5	3.6	4.0	1.9		2.0	12.8	13.0	14.5	1.9	14.5
CO₂ reduction/unit product (%)	56	46	50	27	31					27	56
Cost Results											
Cost year basis (constant dollars)						2003	2003	2003	2003		
Fixed charge rate (%)	12.7	15.0	15.0	15.0	15.0	15.0	15.0	15.0	15.0	12.7	15.0
Reference plant TCR (million US$)****	2160	1351	1351	1215	1215	1161	1161	1161	1161	1161	2160
Capture plant TCR (million US$)****	2243	1385	1220	1237	1090	1066	1128	1164	1172	1066	2243
% increase in capital cost (%)	3.8	2.6	-9.7	1.8	-10.3	-8.1	-2.8	0.2	0.9	-10.3	3.8
Ref. plant electricity price (US$ MWh^{-1})	35.6	42.9	42.9	42.9	42.9	44.1	44.1	44.1	44.1	35.6	44.1
Capture plant electricity price (US$ MWh^{-1})	53.6	42.9	42.9	42.9	42.9	58.0	58.0	58.0	58.0	42.9	58.0
% increase in assumed elec. price	50.5	0.0	0.0	0.0	0.0	31.5	31.5	31.5	31.5	0.0	50.5
Ref. plant fuel product cost, LHV (US$ GJ^{-1})	**5.58**	**9.12**	**9.12**	**8.68**	**8.68**	**7.41**	**7.41**	**7.41**	**7.41**	**5.6**	**9.1**
Capture plant fuel product cost, LHV (US$ GJ^{-1})	**5.43**	**10.36**	**8.42**	**9.37**	**7.57**	**6.73**	**7.18**	**7.65**	**8.09**	**5.4**	**10.4**
Increase in fuel product cost (US$ GJ^{-1})	**-0.15**	**1.24**	**-0.70**	**0.69**	**-1.11**	**-0.68**	**-0.23**	**0.24**	**0.68**	**-1.1**	**1.2**
% increase in fuel product cost	**-5.7**	**13.6**	**-7.7**	**7.9**	**-12.8**	**-9.2**	**-3.1**	**3.2**	**9.2**	**-12.8**	**13.6**
Cost of CO₂ captured (US$/tCO₂)		12.3	-6.4	13.3	-18.4	-12.4	-1.5	1.5	4.1	-18.4	13.3
Cost of CO₂ avoided (US$/tCO₂)		13.2	-6.9	13.0	-18.3	-13.3	-1.8	1.8	4.8	-18.3	13.2
Confidence level (see Table 3.6)	moderate	moderate	moderate	low to moderate							

Notes: All costs in this table are for capture only and do not include the costs of CO₂ transport and storage; see Chapter 8 for total CCS costs. * Reported HHV values converted to LHV assuming LHV/HHV = 0.96 for coal, 0.846 for hydrogen, and 0.93 for F-T liquids. ** CO₂ capture efficiency = (C in CO₂ captured)/(C in fossil fuel input to plant - C in carbonaceous fuel products of plant) x100; C associated with imported electricity is not included. ***Includes CO₂ emitted in the production of electricity imported by the plant. ****Reported total plant investment values increased by 3.5% to estimate total capital requirement.

producing liquid fuels plus electricity. In these cases the amounts of electricity produced are sizeable compared to the liquid products, so the assumed selling price of electricity has a major influence on the product cost results. So too does the assumption in two of the cases of co-disposal of H$_2$S with CO$_2$ (as described above). For these reasons, the incremental cost of CO$_2$ capture ranges from a 13% decrease to a 13% increase in fuel product cost relative to the no-capture case. Note too that the overall level of CO$_2$ reductions per unit of product is only 27-56%. This is because a significant portion of carbon in the coal feedstock is exported with the liquid fuel products. Nonetheless, an important benefit of these fuel-processing schemes is a reduction (of 30-38%) in the carbon content per unit of fuel energy relative to the feedstock fuel. To the extent these liquid fuels displace other fuels with higher carbon per unit of energy, there is a net benefit in end-use CO$_2$ emissions when the fuels are burned. However, no credit for such reductions is taken in Table 3.11 because the system boundary considered is confined to the fuel production plant.

3.7.8 Capture costs for other industrial processes (current technology)

CO$_2$ can be captured in other industrial processes using the techniques described earlier for power generation. While the costs of capture may vary considerably with the size, type and location of industrial processes, such costs will be lowest for processes or plants having: streams with relatively high CO$_2$ concentrations; process plants that normally operate at high load factors; plants with large CO$_2$ emission rates; and, processes that can utilize waste heat to satisfy the energy requirements of CO$_2$ capture systems. Despite these potential advantages, little detailed work has been carried out to estimate costs of CO$_2$ capture at industrial plants, with most work focused on oil refineries and petrochemical plants. A summary of currently available cost studies appears in Table 3.12.

3.7.8.1 Oil refining and petrochemical plants
Gas-fired process heaters and steam boilers are responsible for the bulk of the CO$_2$ emitted from typical oil refineries and petrochemical plants. Although refineries and petrochemical plants emit large quantities of CO$_2$, they include multiple emission sources often dispersed over a large area. Economies of scale can be achieved by using centralized CO$_2$ absorbers or amine regenerators but some of the benefits are offset by the cost of pipes and ducts. Based on Table 3.14, the cost of capturing and compressing CO$_2$ from refinery and petrochemical plant heaters using current technology is estimated to be 50-60 US$/tCO$_2$ captured. Because of the complexity of these industrial facilities, along with proprietary concerns, the incremental cost of plant products is not normally reported.

High purity CO$_2$ is currently vented to the atmosphere by some gas processing and petrochemical plants, as described in Chapter 2. The cost of CO$_2$ capture in such cases would be simply the cost of drying and compressing the CO$_2$ to the pressure required for transport. The cost would depend on various factors, particularly the scale of operation and the electricity price. Based on 2 MtCO$_2$ yr^{-1} and an electricity price of US$ 0.05 kWh^{-1}, the cost is estimated to be around 10 US$/tCO$_2$ emissions avoided. Electricity accounts for over half of the total cost.

3.7.8.2 Cement plants
As noted in Chapter 2, cement plants are the largest industrial source of CO$_2$ apart from power plants. Cement plants normally burn lower cost high-carbon fuels such as coal, petroleum coke and various wastes. The flue gas typically has a CO$_2$ concentration of 14-33% by volume, significantly higher than at power plants, because CO$_2$ is produced in cement kilns by decomposition of carbonate minerals as well as by fuel combustion. The high CO$_2$ concentration would tend to reduce the specific cost of CO$_2$ capture from flue gas. Pre-combustion capture, if used, would only capture the fuel-related CO$_2$, so would be only a partial solution to CO$_2$ emissions. Oxy-fuel combustion and capture using calcium sorbents are other options, which are described in Sections 3.2.4 and 3.7.11.

3.7.8.3 Integrated steel mills
Integrated steel mills are some of the world's largest emitters of CO$_2$, as described in Chapter 2. About 70% of the carbon introduced into an integrated steel mill is contained in the blast furnace gas in the form of CO$_2$ and CO, each of which comprise about 20% by volume of the gas. The cost of capturing CO$_2$ from blast furnace gas was estimated to be 35 US$/tCO$_2$ avoided (Farla *et al.*, 1995) or 18 US$/tCO$_2$ captured (Gielen, 2003).

Iron ore can be reacted with synthesis gas or hydrogen to produce iron by direct reduction (Cheeley, 2000). Direct reduction processes are already used commercially but further development work would be needed to reduce their costs so as to make them more widely competitive with conventional iron production processes. The cost of capturing CO$_2$ from a direct reduction iron (DRI) production processes was estimated to be 10 US$/tCO$_2$ (Gielen, 2003). CO$_2$ also could be captured from other gases in iron and steel mills but costs would probably be higher as they are more dilute or smaller in scale.

3.7.8.4 Biomass plants
The main large point sources of biomass-derived CO$_2$ are currently wood pulp mills, which emit CO$_2$ from black liquor recovery boilers and bark-fired boilers, and sugar/ethanol mills, which emit CO$_2$ from bagasse-fired boilers. Black liquor is a byproduct of pulping that contains lignin and chemicals used in the pulping process. The cost of post-combustion capture was estimated to be 34 US$/tCO$_2$ avoided in a plant that captures about 1 MtCO$_2$ yr^{-1} (Möllersten *et al.*, 2003). Biomass gasification is under development as an alternative to boilers.

CO$_2$ could be captured from sucrose fermentation and from combustion of sugar cane bagasse at a cost of about 53 US$/tCO$_2$ avoided for a plant capturing 0.6 MtCO$_2$ yr^{-1} avoided (Möllersten *et al.*, 2003). CO$_2$ from sugar cane fermentation has a high purity, so only drying and compression is required. The overall cost is relatively high due to an annual load factor that is lower than that of most power stations and large industrial

Table 3.12. Capture costs: Other industrial processes using current or advanced technology.

Study Assumptions and Cost Results	CURRENT TECHNOLOGY									ADVANCED TECHNOLOGY			
	Farla et al.	IEA GHG	IEA GHG	IEA GHG	Möllersten et al.	Möllersten et al.	Möllersten et al.	CCP	CCP	CCP	CCP	CCP	CCP
	1995	2000c	2000c	2002b	2003	2003	2003	2005	2005	2005	2005	2005	2005
Reference Plant (without capture)													
Industrial process	Iron production	Oil refining petrochemical	Oil refining petrochemical	Pulp mill	Pulp mill	Pulp mill	Ethanol fermentation	Small gas turbines	Refinery heaters & boilers	Small gas turbines	Small gas turbines	Refinery heaters & boilers	Refinery heaters & boilers
Feedstock type	Coke	Refinery gas/ natural gas	Refinery gas/natural gas		Black liquor and bark	Black liquor	Sugar cane	Mixed	NG	Mixed	Mixed	Natural gas	Natural gas
Plant size (specify units)	168 kg s⁻¹ iron	315 kg s⁻¹ crude oil	315 kg s⁻¹ crude oil		17.9 kg s⁻¹ pulp	17.9 kg s⁻¹ pulp	9.1 kg s⁻¹ ethanol	1351 MWt	358 MWt	1351 MWt	1351 MWt	358 MWt	358 MWt
Plant capacity factor (%)	95.3	90	90	90	90.4	90.4	49.3	90.4	98.5	90.4	90.4	98.5	98.5
Feedstock cost (US$ per unit specified)					US$3 GJ⁻¹ LHV	US$3 GJ⁻¹ LHV							
Ref. plant emission rate (kgCO₂ MWh⁻¹)								0.22	0.82	0.22	0.22	0.82	0.82
Capture Plant Design													
CO₂ capture/separation technology	MDEA	MEA	Pre-combustion	Compression only	Amine	Physical solvent							
Location of CO₂ capture	Blast furnace gas	Fired heaters and H₂ plant	Fired heaters and H₂ plant		Boiler	IGCC	Fermentation and bagasse boiler	MEA Baseline (post-comb.)	MEA Baseline (post-comb.)	Membrane Water Gas Shift (pre-comb.)	Flue Gas Recycle & ITM (oxy-fuel)	Very Large-scale ATR (pre-comb.)	Sorption Enhanced Water Gas Shift (pre-comb.)
Capture unit size (specify units)					392 MW fuel	338 MW fuel		1351 MWt	358 MWt	1351 MWt	1351 MWt	358 MWt	358 MWt
CO₂ capture system efficiency (%)	90	95	91	90	90	90	100/90						
Energy source(s) for capture (type +onsite or offsite)													
Are all energy-related CO₂ emissions included?								yes	yes	yes	yes	yes	yes
CO₂ emission rate after capture (kgCO₂ MWh⁻¹)								0.09	0.19	0.09	0.05	0.10	0.14
CO₂ captured (Mt yr⁻¹)	2.795	1.013	1.175	1.970	0.969	0.399	0.560						
CO₂ product pressure (MPa)	11.0	11.0	11.0	8.0	10.0	10.0	10.0	11.0	11.0	11.0	11.0	11.0	11.0
CO₂ reduction per unit of product (%)								60.3	76.5	58.4	75.8	87.4	82.2
Cost Results													
Cost year basis (constant dollars)													
Fixed charge factor (%)					15	15	15	11.0	11.0	11.0	11.0	11.0	11.0
Ref. plant capital cost (US$ per unit capacity)													
Capture plant capital cost (US$ per unit capacity)													
Incremental capital cost (million US$ per kg s⁻¹ CO₂)*	3.8	4.1	4.9	0.3	3.2	1.9	2.6						
Ref. plant cost of product (US$/unit)													
Capture plant cost of product (US$/unit)								10.2	55.1	6.1	6.8	54.2	48.2
Incremental cost of product (US$/unit)								10.2	55.1	6.1	6.8	54.2	48.2
% increase in capital cost (over ref. plant)													
% increase in unit cost of product (over ref. plant)													
Cost of CO₂ captured (US$/tCO₂)	35	50	60	10	34	23	53	55.3	90.9	36.4	38.2	59.0	60.5
Cost of CO₂ avoided (US$/tCO₂)		74	116					78.1	88.2	48.1	41.0	76.0	71.8
Capture cost confidence level (see Table 3.6)				moderate							low		

Notes: All costs in this table are for capture only and do not include the costs of CO₂ transport and storage; see Chapter 8 for total CCS costs. *Capital costs are incremental costs of capture, excluding cost of make-up steam and power generation and also excluding interest during construction and other owner's costs.

plants.

CO$_2$ could be captured at steam-generating plants or power plants that use other biomass byproducts and/or purpose-grown biomass. At present most biomass plants are relatively small. The cost of capturing 0.19 MtCO$_2$ yr^{-1} in a 24 MW biomass-powered IGCC plant, compared to a biomass IGCC plant without capture, is estimated to be about 70 US\$/tCO$_2$ (Audus and Freund, 2005). Larger plants using purpose-grown biomass may be built in the future and biomass can be co-fired with fossil fuels to give economies of scale, as discussed in Chapter 2. Biomass fuels produce similar or slightly greater quantities of CO$_2$ per unit of fuel energy as bituminous coals; thus, the CO$_2$ concentration of flue gases from these fuels will be broadly similar. This implies that the cost of capturing CO$_2$ at large power plants using biomass may be broadly similar to the cost of capturing CO$_2$ in large fossil fuel power plants in cases where plant size, efficiency, load factor and other key parameters are similar. The costs of avoiding CO$_2$ emissions in power plants that use biomass are discussed in more detail in Chapter 8.

3.7.9   *Outlook for future CO$_2$ capture costs*

The following sections focus on 'advanced' technologies that are not yet commercial available, but which promise to lower CO$_2$ capture costs based on preliminary data and design studies. Earlier sections of Chapter 3 discussed some of the efforts underway worldwide to develop lower-cost options for CO$_2$ capture. Some of these developments are based on new process concepts, while others represent improvements to current commercial processes. Indeed, the history of technology innovation indicates that incremental technological change, sustained over many years (often decades), is often the most successful path to substantial long-term improvements in performance and reductions in cost of a technology (Alic *et al.*, 2003). Such trends are commonly represented and quantified in the form of a 'learning curve' or 'experience curve' showing cost reductions as a function of the cumulative adoption of a particular technology (McDonald and Schrattenholzer, 2001). One recent study relevant to CO$_2$ capture systems found that over the past 25 years, capital costs for sulphur dioxide (SO$_2$) and nitrogen oxides (NO$_x$) capture systems at US coal-fired power plants have decreased by an average of 12% for each doubling of installed worldwide capacity (a surrogate for cumulative experience, including investments in R&D) (Rubin *et al.*, 2004a). These capture technologies bear a number of similarities to current systems for CO$_2$ capture. Another recent study (IEA, 2004) suggests a 20% cost reduction for a doubling of the unit capacity of engineered processes due to technological learning. For CCS systems the importance of costs related to energy requirements is emphasized, since reductions in such costs are required to significantly reduce the overall cost of CO$_2$ capture.

At the same time, a large body of literature on technology innovation also teaches us that learning rates are highly uncertain,[11] and that cost estimates for technologies at the early stages of development are often unreliable and overly optimistic (Merrow *et al.*, 1981). Qualitative descriptions of cost trends for advanced technologies and energy systems typically show costs increasing from the research stage through full-scale demonstration; only after one or more full-scale commercial plants are deployed do costs begin to decline for subsequent units (EPRI, 1993; NRC, 2003). Case studies of the SO$_2$ and NO$_x$ capture systems noted above showed similar behaviour, with large (factor of two or more) increases in the cost of early full-scale FGD and SCR installations before costs subsequently declined (Rubin *et al.*, 2004b). Thus, cost estimates for CO$_2$ capture systems should be viewed in the context of their current stage of development. Here we try to provide a perspective on potential future costs that combines qualitative judgments with the quantitative cost estimates offered by technology developers and analysts. The sections below revisit the areas of power generation and other industrial processes to highlight some of the major prospects for CO$_2$ capture cost reductions.

3.7.10   *CO$_2$ capture costs for electric power plants (advanced technology)*

This section first examines oxy-fuel combustion, which avoids the need for CO$_2$ capture by producing a concentrated CO$_2$ stream for delivery to a transport and storage system. Following this we examine potential advances in post-combustion and pre-combustion capture.

3.7.10.1   *Oxy-fuel combustion systems*
It is first important to distinguish between two types of oxy-fuel systems: an oxy-fuel boiler (either a retrofit or new design) and oxy-fuel combustion-based gas turbine cycles. The former are close to demonstration at a commercial scale, while the latter (such as chemical looping combustion systems and novel power cycles using CO$_2$/water as working fluid) are still at the design stage. Table 3.13 summarizes the key assumptions and cost results of several recent studies of CO$_2$ capture costs for oxy-fuel combustion systems applied to new or existing coal-fired units. As discussed earlier in Section 3.4, oxygen combustion produces a flue gas stream consisting primarily of CO$_2$ and water vapour, along with smaller amounts of SO$_2$, nitrogen and other trace impurities. These designs eliminate the capital and operating costs of a post-combustion CO$_2$ capture system, but new costs are incurred for the oxygen plant and other system design modifications. Because oxy-fuel combustion is still under development and has not yet been utilized or demonstrated for large-scale power generation, the design basis and cost estimates for such systems remain highly variable and uncertain. This is reflected in the wide range of oxy-fuel cost estimates in Table 3.13. Note, however, that cost estimates for advanced design

---

[11] In their study of 42 energy-related technologies, McDonald and Schrattenholzer (2001) found learning rates varying from -14% to 34%, with a median value of 16%. These rates represent the average reduction in cost for each doubling of installed capacity. A negative learning rate indicates that costs increased rather than decreased over the period studied.

Table 3.13 Capture costs: Advanced technologies for electric power plants. (continued on next page)

Study Assumptions and Results	OXY-FUEL COMBUSTION							ADVANCED PC	
	Alstom et al. 2001	Singh et al. 2003	Stobbs &Clark 2005	Dillon et al. 2005	Nsakala et al. 2003	Nsakala et al. 2003	Nsakala et al. 2003	Gibbins et al. 2005	Gibbins et al. 2005
Reference Plant (without capture)									
Power plant type	RETROFIT subcrit PC	RETROFIT PC + aux NGCC	RETROFIT PC	New PC	Air-fired CFB	Air-fired CFB	Air-fired CFB	Double reheat supercrit PC	Double reheat supercrit PC
	*				*	*	*		
Fuel type (bit, sub-bit, lig; NG, other) and %S	bit, 2.7%S	sub-bit	lignite	bit	bit, 2.3%S	bit, 2.3%S	bit, 2.3%S		
Reference plant net size (MW)	434	400	300	677	193	193	193		
Plant capacity factor (%)	67	91		85	80	80	80	85	85
Net plant efficiency, LHV (%)				44.2	37.0	37.0	37.0	45.6	45.6
Fuel cost, LHV (US$ GJ^{-1})	1.30			1.50	1.23	1.23	1.23	1.50	1.50
Reference plant emission rate (tCO$_2$ MWh^{-1})	0.908	0.925	0.883	0.722	0.909	0.909	0.909		
Capture Plant Design									
CO$_2$ capture technology	oxy-fuel	oxy-fuel	oxy-fuel	oxy-fuel	oxy-fuel	oxy-fuel with CMB	chemical looping with CMB	MEA	KS-1
Net plant size with capture (MW)	273	400		532	135	197	165		
Net plant efficiency, LHV (%)	23.4		13.7	35.4	25.8	31.3	32.2	34.3	36.5
CO$_2$ capture system efficiency (%)				about 91					
CO$_2$ emission rate after capture (t MWh^{-1})		0.238	0.145	0.085	0.086	0.073	0.005		
CO$_2$ captured (Mt yr^{-1})		2.664							
CO$_2$ product pressure (MPa)	13.9	15	13.7	11				11.0	11.0
CCS energy requirement (% more input MWh^{-1})				25	43	18	15	33	25
CO$_2$ reduction per kWh (%)		**74**	**119**	**88.2**	**90.5**	**92.0**	**99.5**		
Cost Results	**	**							
Cost year basis (constant dollars)		2001	2000		2003	2003	2003		
Fixed charge factor (%)	13.0	9.4		11				11.0	11.0
Reference plant TCR (US$ kW^{-1})	1527	0		1260	1500	1500	1500	1022	1022
Capture plant TCR (US$ kW^{-1})		909	4570	1857	2853	2731	1912	1784	1678
Incremental TCR for capture (US$ kW^{-1})	1198	909		597	1354	1232	413	762	656
Reference plant COE (US$ MWh^{-1})			**44.5**	**44**	**45.3**	**45.3**	**45.3**	**37**	**37**
Capture plant COE (US$ MWh^{-1})			**97.5**	**61.2**	**82.5**	**70.5**	**58.4**	**61**	**57**
Incremental COE for capture (US$ MWh^{-1})	44.5	23.9	53	17.2	37.2	25.2	13.1	24	20
% increase in capital cost (over ref. plant)				47	90	82	28	75	64
% increase in COE (over ref. plant)			**72**	**39**	**82**	**56**	**29**	**65**	**54**
Cost of CO$_2$ captured (US$/tCO$_2$)		29							
Cost of CO$_2$ avoided (US$/tCO$_2$)	54	35	72	27	45	30	14	54	
Capture cost confidence level (see Table 3.6)		low	low		very low	very low	very low	low to moderate	

Notes: All costs in this table are for capture only and do not include the costs of CO$_2$ transport and storage; see Chapter 8 for total CCS costs. * Reported HHV values converted to LHV assuming LHV/HHV = 0.96 for coal. ** Reported value increased by 15% to estimate interest during construction and other owners' costs.

Table 3.13 Continued.

Study Assumptions and Results	ADVANCED NGCC								ADVANCED IGCC					ADVANCED HYBRIDS	
	Simbeck	Parsons	Parsons	CCP	CCP	CCP	CCP	Dillon et al.	Parsons	NETL	NETL	CCP	CCP	NETL	Parsons
	2002	2002b	2002b	2005	2005	2005	2005	2005	2002b	2002	2002	2005	2005	2002	2002b
									*	*	*			*	
Reference Plant (without capture)															
Power plant type	comb. cycle H-class turbine	comb. cycle H-class turbine	comb. cycle H-class turbine					NGCC	E-gas, O_2 water scrubber; H-class turbine	E-gas, O_2, CGCU, Hydraulic air compression	E-gas, O_2, CGCU, Hydraulic air compression with open loop water system	Canada coke gasification	Canada coke gasification	E-gas, O_2, HGCU, "G" GT, SOFC	CHAT SOFC
Fuel type (bit, sub-bit, lig; NG, other) and %S	Nat. gas	Nat. gas	Nat. gas	NG	NG	NG	NG	NG	Illinois #6	Illinois #6	Illinois #6	Coke	Coke	Illinois #6	Nat. gas
Reference plant net size (MW)	480	384	384	392	392	392	392	388	425	326	408	588	588	644	557
Plant capacity factor (%)	80	80	65	95	95	95	95	85	80	85	85	91.3	91.3	85	80
Net plant efficiency, LHV (%)	60.0	59.5	59.5	57.6%	57.6%	57.6%	57.6%	56.0	41.1	43.8	54.9			56.4	66.2
Fuel cost, LHV (US$ GJ⁻¹)	4.86	2.82	2.82	2.96	2.96	2.96	2.96	3.00	1.23	1.03	1.03	2.96	2.96	1.03	2.82
Reference plant emission rate (tCO_2 MWh⁻¹)	0.342	0.338	0.338	0.37	0.37	0.37	0.37	0.371	0.720	0.712	0.568	0.95	0.95	0.572	0.302
Capture Plant Design															
CO_2 capture technology	MEA	MEA	MEA	MEA low-cost/ CCGT-integrated (post-comb.)	Membrane Contactor; KS-1 (post-comb.)	Hydrogen Membrane Reformer (pre-comb.)	Sorption Enhanced Water Gas Shift- Air ATR (pre-comb.)	Oxy-fuel	Selexol	Selexol		IGCC with capture (pre-comb.)	IGCC with advanced capture (pre-comb.)	Selexol	
Net plant size with capture (MW)	413	311	311	345	335	361	424	440	387	312	404	699	734	755	517
Net plant efficiency, LHV (%)	51.7	48.1	48.1	50.6	49.2	53.0	48.2	44.7	33.8	35.2	45.4			49.7	46.1
CO_2 capture system efficiency (%)	85	90	90	86	86	100	90		91.5	92.7	92.7			90	86.8
CO_2 emission rate after capture (t/MWh)	0.06	0.042	0.042	0.06	0.06	0.00	0.04	0.011	0.074	0.065	0.050	0.27	0.28	0.046	0.043
Incremental TCR for capture															
CO_2 captured (Mt yr⁻¹)	0.980	0.669	0.823	1.09	1.09	1.27	1.47		2.074	1.984	1.984	6.80	6.44	3.390	
CO_2 product pressure (MPa)	13.7	8.3	8.3					11	8.3	14.5	14.5			14.5	8.3
CCS energy requirement(% more input MWh-1)	16	24	24					25	22	24	21			13	44
CO_2 reduction per kWh (%)	82	88	88	84.1	83.6	100	87.9	97.0	90	91	91	71.2	71.1	92	86
Cost Results															
Cost year basis (constant dollars)	2001	2000	2000	2005	2005	2005	2005		2000	2002	2002	2005	2005	2002	2000
Fixed charge factor (%)	15.0	15.0		11.0	11.0	11.0	11.0	11	15.0	14.8	14.8	11.0	11.0	14.8	
Reference plant TCR (US$ kW⁻¹)	582	539	496	724	724	724	724	559	1249	1436	881.4	1398	1398	1508	623
Capture plant TCR (US$ kW⁻¹)	1216	1026	943	1002	1225	1058	1089	1034	1698	2189	1450	1919	1823	1822	
Incremental TCR for capture (US$ kW⁻¹)	634	487	447	278	501	334	365	475	449	753	568	521	425	314	
Reference plant COE (US$ MWh⁻¹)	42.9	33.5	30.7	34.2	34.2	34.2	34.2	33.5	41.0	47.0	28.5	32.3	32.3	41.1	
Capture plant COE (US$ MWh⁻¹)	65.9	54.1	48.8	45.1	48.9	43.2	45.4	50.3	53.6	65.5	41.8	42.1	40.5	48.8	
Incremental COE for capture (US$ MWh⁻¹)	23	20.6	18.1	10.9	14.7	9.0	11.2	16.8	12.6	18.5	13.3	9.8	8.2	7.7	
% increase in capital cost (over ref. plant)	109	90	90	38	69	46	50	85	36	52	64	37	30	21	
% increase in COE (over ref. plant)	54	61	59	32	43	26	33	50	31	39	47	30	25	19	
Cost of CO_2 captured (US$/$tCO_2$)	82	48	48	30.2	39.5	22.5	28.2		16	22	20	11	10	13	
Cost of CO_2 avoided (US$/$tCO_2$)		70	61	35.1	47.5	24.4	34.4	47	19	29	26	14	12	15	
Capture cost confidence level (see Table 3.6)	low to moderate	low to moderate	low to moderate	low to very low	low to very low	low to very low	low to very low		low	low	low	low	low	very low	very low

Notes: All costs in this table are for capture only and do not include the costs of CO_2 transport and storage; see Chapter 8 for total CCS costs. * Reported HHV values converted to LHV assuming LHV/HHV = 0.96 for coal and LHV/HHV = 0.90 for natural gas. **Reported value increased by 15% to estimate interest during construction and other owners' costs.

concepts based on oxy-fuel combustion gas turbine cycles are more uncertain at this time than cost estimates for new or retrofitted boilers employing oxy-fuel combustion.

For new plant applications, the data in Table 3.13 indicate that oxy-fuel combustion adds about 30-90% to the capital cost and 30-150% to the COE of a conventional plant, while reducing CO_2 emissions per kWh by 75-100%. Retrofit applications exhibit higher relative costs in cases where the existing plant is wholly or partially amortized. The lowest-cost oxy-fuel system in Table 3.13 is one that employs chemical looping to achieve nearly a 100% reduction in CO_2 emissions. While this concept thus appears promising (see Section 3.4.6), it has yet to be tested and verified at a meaningful scale. Thus cost estimates based on conceptual designs remain highly uncertain at this time.

To judge the potential cost savings of oxy-fuels relative to current CO_2 capture systems, it is useful to compare the costs of alternative technologies evaluated within a particular study based on a particular set of premises. In this regard, the COE for the oxy-fuel retrofit system reported by Alstom *et al.* (2001) in Table 3.13 is 20% lower than the cost of an amine system retrofit (Table 3.13) for the same 255 MW plant, while the cost of CO_2 avoided is 26% lower. In contrast, a recent study by the Canadian Clean Power Coalition (Stobbs and Clark, 2005) reports that the COE for an oxy-fuel system at a large lignite-fired plant (Table 3.13) is 36% higher than for an amine CO_2 capture system, while the cost of CO_2 avoided is more than twice as great. The major source of that cost difference was a specification in the CCPC study that the oxy-fuelled unit also be capable of full air firing. This resulted in a much higher capital cost than for a new unit designed solely for oxy-fuel operation. A more recent study sponsored by IEA GHG (Dillon *et al.*, 2005) found that a large new supercritical coal-fired boiler with oxy-fuel combustion had a COE slightly (2-3%) lower than a state-of-the-art coal plant with post-combustion analyzed in a separate study employing similar assumptions (IEA GHG, 2004). Further cost reductions could be achieved with the successful development of new lower-cost oxygen production technology (see Section 3.4.5). At the current time, the optimum designs of oxy-fuel combustion systems are not yet well established and costs of proposed commercial designs remain uncertain. This is especially true for advanced design concepts that employ components which are not yet available or still in the development stage, such as CO_2 gas turbines or high temperature ceramic membranes for oxygen production.

3.7.10.2 Advanced systems with post-combustion capture

Improvements to current amine-based systems for post-combustion CO_2 capture are being pursued by a number of process developers (Mimura *et al.*, 2003; Muramatsu and Iijima, 2003; Reddy *et al.*, 2003) and may offer the nearest-term potential for cost reductions over the systems currently in use. The newest systems summarized earlier in Table 3.7 reportedly reduce the cost of CO_2 avoided by approximately 20-30% (IEA GHG, 2004). Table 3.13 indicates that additional advances in plant heat integration could further reduce the COE of capture plants by about 5%. These results are consistent with

a recent study by Rao *et al.* (2003), who used expert elicitations and a plant simulation model to quantify the improvements likely achievable by 2015 for four key process parameters: sorbent concentration, regeneration energy requirements, sorbent loss and sorbent cost. The 'most likely' improvement was an 18% reduction in COE, while the 'optimistic' estimates yielded a 36% cost reduction from improvements in just these four parameters. The cost of CO_2 avoided was reduced by similar amounts. Advances in more efficient heat integration (for sorbent regeneration) and higher power plant efficiency could lead to even greater reductions in CO_2 capture cost.

Advances in gas turbine technology produce similar benefits for NGCC systems. Table 3.13 shows several cases based on the H-turbine design. Relative to the cases in Table 3.9, these systems offer higher efficiency and greater CO_2 reductions per kWh. The higher COEs for the advanced NGCC systems reflects the higher natural gas prices assumed in more recent studies.

Table 3.13 indicates that other advanced technologies for post-combustion applications, such as membrane separation systems, may also lower the future cost of CO_2 capture (see Section 3.3.3). Reliable cost estimates for such technologies should await their further development and demonstration.

3.7.10.3 Advanced systems with pre-combustion capture

The cost of gasification-based systems with CO_2 capture also can be expected to fall as a result of continued improvements in gas turbine technology, gasifier designs, oxygen production systems, carbon capture technology, energy management and optimization of the overall facility. One recent study (IEA GHG, 2003) estimates a 20% reduction in the cost of electricity generation from a coal-based IGCC plant with CO_2 capture by 2020. This takes into account improvements in gasification, oxygen production, physical solvent scrubbing and combined cycle processes, but does not take into account any possible radical innovations in CO_2 separation technology. The additional IGCC cases shown in Table 3.13, including recent results of the CO_2 Capture Project (CCP, 2005), foresee similar reductions in the COE of advanced IGCC systems compared to the systems in Table 3.10.

3.7.11 CO_2 capture costs for hydrogen production and multi-product plants (advanced technology)

Table 3.14 shows results of several recent studies that have projected the performance and cost of new or improved ways of producing hydrogen and electricity from fossil fuels.

Compared to the current commercial plants in Table 3.11, the advanced single-product systems with CO_2 capture have hydrogen cost reductions of 16% (for natural gas feedstock) to 26% (for coal feedstock). Additional cases in Table 3.14 show multi-product systems producing hydrogen and electricity. These cases indicate the potential for substantial reductions in the future cost of hydrogen production with CO_2 capture. As before, the results are sensitive to the assumed selling price of co-product electricity. More importantly, these cases assume

Table 3.14 CO$_2$ capture costs: Multi-product plants using advanced technology.

Study Assumptions and Results	Simbeck 2005	NRC 2004	NRC 2004	Parsons 2002a	Mitretek 2003	Mitretek 2003	Mitretek 2003	Range min	Range max
Capture Plant Design	*			*	*	*	*		
Plant products (primary/secondary)	H$_2$	H$_2$	H$_2$	H$_2$+electricity	H$_2$+electricity	H$_2$+electricity	H$_2$+electricity		
Production process or type	Autothermal reforming with O$_2$ provided by ITM	78% efficient ATR/SMR, adv CO$_2$ compressor	Gasifier LHV=75-->80%, Adv ASU, membrane sep, adv CO$_2$ compressor	High-pressure E-gas, HGCU, HTMR, H$_2$SO$_4$ co-product	Advanced E-gas, HGCU, HTMR	Advanced E-gas, HGCU, HTMR, large elec. co-product	Advanced E-gas, HGCU, HTMR, SOFC, large elec. co-product		
Feedstock	Natural gas	Natural gas	Coal	Pgh #8 Coal	Coal	Coal	Coal		
Feedstock cost, LHV (US$ GJ^{-1})	5.26	4.73	1.20	0.89	1.03	1.03	1.03	1	5
Plant capacity factor (%)	90	90	90	80	85	85	85	80	90
CO$_2$ capture/separation technology	Oxy-fuel			Oxy-fuel	Oxy-fuel	Oxy-fuel	Oxy-fuel		
Capture plant input capacity, LHV (GJ h^{-1})	9527	7697	8121	2794	3020	6051	6051	2794	9527
Capture plant output capacity, LHV: Fuels (GJ h^{-1})	7504	6004	6004	1956	1904	1844	1808	1808	7504
Electricity (MW)	–13	–66	–88	7	25	416	519	–88	519
Net plant efficiency, LHV (%)	78.3	74.9	70.0	70.9	66.0	55.2	60.7	55	78
CO$_2$ capture efficiency (%)**	95	90	90	94	100	100	95	90	100
CO$_2$ emitted (MtCO$_2$ yr^{-1})***	0.086	0.505	0.873	0.117	0.000	0.000	0.191	0.000	0.873
Carbon exported in fuels (MtC yr^{-1})	0	0	0	0	0	0	0	0	0
Total carbon released (kgCO$_2$ GJ^{-1} products)	1.46	11.10	19.45	8.45	0.00	0.00	6.96	0.0	19.5
CO$_2$ captured (MtCO$_2$ yr^{-1})	4.074	3.119	5.853	1.855	1.918	3.846	3.652	1.9	5.9
CO$_2$ product pressure (MPa)	13.7	13.7	13.7	13.4	20	20	20	13.4	20.0
Cost Results									
Cost year basis (constant dollars)	2003	2000	2000	2000	2000	2000	2000		
Fixed charge rate (%)	20	16	16	14.3	12.7	12.7	12.7	12.7	20.0
Capture plant TCR (million US$)****	725	441	921	398	441	950	1023	398	1023
Capture plant electricity price (US$ MWh^{-1})	50.0	45.0	45.0	30.8	53.6	53.6	53.6	31	54
Capture plant fuel product cost, LHV (US$ GJ^{-1}) (see Table 3.6)	9.84	8.53	6.39	5.79	6.24	3.27	1.13	1.13	9.84
Capture cost confidence level	low	low	low	low to very low	low to very low	low to very low	very low		

Notes: All costs in this table are for capture only and do not include the costs of CO$_2$ transport and storage; see Chapter 8 for total CCS costs. * Reported HHV values converted to LHV assuming LHV/HHV = 0.96 for coal and 0.846 for hydrogen. ** CO$_2$ capture efficiency = (C in CO$_2$ captured)/(C in fossil fuel input to plant - C in carbonaceous fuel products of plant) x100; C associated with imported electricity is not included. *** Includes CO$_2$ emitted in the production of electricity imported by the plant. ****Reported total plant investment values increased by 3.5% to estimate total capital requirement.

the successful scale-up and commercialization of technologies that have not yet been demonstrated, or which are still under development at relatively small scales, such as solid oxide fuel cells (SOFC). Published cost estimates for these systems thus have a very high degree of uncertainty.

3.7.12   *CO_2 capture costs for other industrial processes (advanced technology)*

As noted earlier, CO_2 capture for industrial processes has not been widely studied. The most extensive analyses have focused on petroleum refineries, especially CO_2 capture options for heaters and other combustion-based processes (see Table 3.12). The use of oxy-fuel combustion offers potential cost savings in several industrial applications. The CO_2 Capture Project reports the cost of capturing CO_2 in refinery heaters and boilers, with an ion transport membrane oxygen plant, to be 31 US$/t$CO_2$ avoided. The cost of pre-combustion capture based on shift and membrane gas separation was predicted to be 41 US$/t$CO_2$ avoided (CCP, 2005).

It also may be possible to apply oxy-fuel combustion to cement plants, but the CO_2 partial pressure in the cement kiln would be higher than normal and the effects of this on the calcination reactions and the quality of the cement product would need to be investigated. The quantity of oxygen required per tonne of CO_2 captured in a cement plant would be only about half as much as in a power plant, because only about half of the CO_2 is produced by fuel combustion. This implies that the cost of CO_2 capture by oxy-fuel combustion at large cement plants would be lower than at power plants, but a detailed engineering cost study is lacking. Emerging technologies that capture CO_2 using calcium-based sorbents, described in Section 3.3.3.4, may be cost competitive in cement plants in the future.

3.7.13   *Summary of CO_2 capture cost estimates*

Table 3.15 summarizes the range of current CO_2 capture costs for the major electric power systems analyzed in this report. These costs apply to case studies of large new plants employing current commercial technologies. For the PC and IGCC systems, the data in Table 3.15 apply only to plants using bituminous coals and the PC plants are for supercritical units only. The cost ranges for each of the three systems reflect differences in the technical, economic and operating assumptions employed in different studies. While some differences in reported costs can be attributed to differences in the CO_2 capture system design, the major sources of variability are differences in the assumed design, operation and financing of the reference plant to which the capture technology is applied (i.e., factors such as plant size, location, efficiency, fuel type, fuel cost, capacity factor and cost of capital). Because no single set of assumptions applies to all situations or all parts of the world, we display the ranges of cost represented by the studies in Tables 3.8, 3.10, 3.11 and 3.12.

For the power plant studies reflected in Table 3.15, current CO_2 capture systems reduce CO_2 emissions per kilowatt-hour by approximately 85-90% relative to a similar plant without capture. The cost of electricity production attributed to CO_2 capture increases by 35-70% for a natural gas combined cycle plant, 40-85% for a new pulverized coal plant and 20-55% for an integrated gasification combined cycle plant. Overall, the COE for fossil fuel plants with capture ranges from 43-86 US$ MWh^{-1}, as compared to 31-61 US$ MWh^{-1} for similar plants without capture. These costs include CO_2 compression but not transport and storage costs. In most studies to date, NGCC systems typically have a lower COE than new PC and IGCC plants (with or without capture) for large base load plants with high capacity factors (75% or more) and gas prices below about 4 US$ GJ^{-1} over the life of the plant. However, for higher gas prices and/ or lower capacity factors, NGCC plants typically have higher COEs than coal-based plants, with or without capture. Recent studies also found that IGCC plants were on average slightly more costly without capture and slightly less costly with capture than similarly sized PC plants. However, the difference in cost between PC and IGCC plants with or without CO_2 capture can vary significantly with coal type and other local factors, such as the cost of capital. Since neither PC nor IGCC systems have yet been demonstrated with CO_2 capture and storage for a large modern power plant (e.g., 500 MW), neither the absolute or relative costs of these systems (nor comparably sized NGCC systems with capture and storage) can be stated with a high degree of confidence at this time, based on the criteria of Table 3.6.

Table 3.15 also shows that the lowest CO_2 capture costs with current technology (as low as 2 US$/t$CO_2$ captured or avoided) were found for industrial processes such as coal-based hydrogen production plants that produce concentrated CO_2 streams as part of the production process. Such industrial processes may represent some of the earliest opportunities for CCS.

Figure 3.20 displays the normalized power plant cost and emissions data from Table 3.15 in graphical form. On this graph, the cost of CO_2 avoided corresponds to the slope of a line connecting any two plants (or points) of interest. While Table 3.15 compares a given capture plant to a similar plant without capture, in some cases comparisons may be sought between a given capture plant and a different type of reference plant. Several cases are illustrated in Figure 3.20 based on either a PC or NGCC reference plant. In each case, the COE and CO_2 emission rate are highly dependent upon technical, economic and financial factors related to the design and operation of the power systems of interest at a particular location. The cost of CO_2 avoided is especially sensitive to these site-specific factors and can vary by an order of magnitude or more when different types of plants are compared. Comparisons of different plant types, therefore, require a specific context and geographical location to be meaningful and should be based on the full COE including CO_2 transport and storage costs. Later, Chapter 8 presents examples of full CCS costs for different plant types and storage options.

In contrast to new plants, CO_2 capture options and costs for existing power plants have not been extensively studied. Current studies indicate that these costs are extremely site-specific and fall into two categories (see Table 3.8). One is the retrofitting of a post-combustion capture system to the existing unit.

Table 3.15 Summary of new plant performance and CO$_2$ capture cost based on current technology.

Performance and Cost Measures	New NGCC Plant Range low	high	Rep. Value	New PC Plant Range low	high	Rep. Value	New IGCC Plant Range low	high	Rep. Value	New Hydrogen Plant Range low	high	Rep. Value	(Units for H$_2$ Plant)
Emission rate without capture (kgCO$_2$ MWh^{-1})	344 -	379	*367*	736 -	811	*762*	682 -	846	*773*	78 -	174	*137*	kgCO$_2$ GJ^{-1} (without capture)
Emission rate with capture (kgCO$_2$ MWh^{-1})	40 -	66	*52*	92 -	145	*112*	65 -	152	*108*	7 -	28	*17*	kgCO$_2$ GJ^{-1} (with capture)
Percent CO$_2$ reduction per kWh (%)	83 -	88	*86*	81 -	88	*85*	81 -	91	*86*	72 -	96	*86*	% reduction/unit of product
Plant efficiency with capture, LHV basis (%)	47 -	50	*48*	30 -	35	*33*	31 -	40	*35*	52 -	68	*60*	Capture plant efficiency (% LHV)
Capture energy requirement (% more input MWh^{-1})	11 -	22	*16*	24 -	40	*31*	14 -	25	*19*	4 -	22	*8*	% more energy input GJ^{-1} product
Total capital requirement without capture (US$ kW^{-1})	515 -	724	*568*	1161 -	1486	*1286*	1169 -	1565	*1326*	*(No unique normalization for multi-product plants)*			Capital requirement without capture
Total capital requirement with capture (US$ kW^{-1})	909 -	1261	*998*	1894 -	2578	*2096*	1414 -	2270	*1825*				Capital requirement with capture
Percent increase in capital cost with capture (%)	64 -	100	*76*	44 -	74	*63*	19 -	66	*37*	-2 -	54	*18*	% increase in capital cost
COE without capture (US$ MWh^{-1})	31 -	50	*37*	43 -	52	*46*	41 -	61	*47*	6.5 -	10.0	*7.8*	H$_2$ cost without capture (US$ GJ^{-1})
COE with capture only (US$ MWh^{-1})	43 -	72	*54*	62 -	86	*73*	54 -	79	*62*	7.5 -	13.3	*9.1*	H$_2$ cost with capture (US$ GJ^{-1})
Increase in COE with capture (US$ MWh^{-1})	12 -	24	*17*	18 -	34	*27*	9 -	22	*16*	0.3 -	3.3	*1.3*	Increase in H$_2$ cost (US$ GJ^{-1})
Percent increase in COE with capture (%)	37 -	69	*46*	42 -	66	*57*	20 -	55	*33*	5 -	33	*15*	% increase in H$_2$ cost
Cost of CO$_2$ captured (US$/tCO$_2$)	33 -	57	*44*	23 -	35	*29*	11 -	32	*20*	2 -	39	*12*	US$/tCO$_2$, captured
Cost of CO$_2$ avoided (US$/tCO$_2$)	37 -	74	*53*	29 -	51	*41*	13 -	37	*23*	2 -	56	*15*	US$/tCO$_2$, avoided
Capture cost confidence level (see Table 3.6)	moderate			moderate			moderate			moderate to high			Confidence Level (see Table 3.6)

Notes: See Section 3.6.1 for calculation of energy requirement for capture plants. Values in italics were adjusted from original reported values as explained below.(a) Ranges and representative values are based on data from Tables 3.8, 3.11, 3.11 and 3.12. All costs in this table are for capture only and do not include the costs of CO$_2$ transport and storage; see Chapter 8 for total CCS costs. (b) All PC and IGCC data are for bituminous coals only at costs of US$1.0-1.5 GJ^{-1} (LHV); all PC plants are supercritical units. (c) NGCC data based on natural gas prices of US$2.8-4.4 GJ^{-1} (LHV basis). (d) Cost are in constant US dollars (approx. year 2002 basis). (e) Power plant sizes range from approximately 400-800 MW without capture and 300-700 MW with capture. (f) Capacity factors vary from 65-85% for coal plants and 50-95% for gas plants (average for each = 80%). (g) Hydrogen plant feedstocks are natural gas (US$ 4.7-5.3 GJ^{-1}) or coal (US$ 0.9-1.3 GJ^{-1}); some plants in data set produce electricity in addition to hydrogen. (h) Fixed charge factors vary from 11-16% for power plants and 13-20% for hydrogen plants. (i) All costs include CO$_2$ compression but not additional CO$_2$ transport and storage costs (see Chapter 8 for full CCS costs).

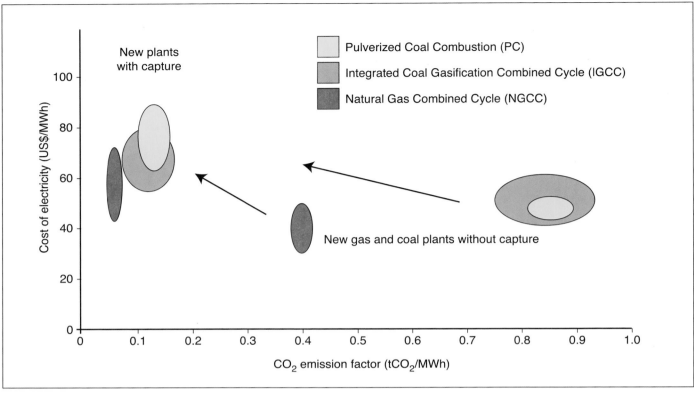

Figure 3.20 Cost of electricity (excluding transport and storage costs) compared to CO_2 emission rate for different reference and capture plants based on current technology. The shaded areas show the Table 3.15 ranges of CO_2 emission rates and levelized cost of electricity (COE) for new PC, IGCC and NGCC plants with and without CO_2 capture. All coal plant data are for bituminous coals only. PC plants are supercritical units only (see Tables 3.7, 3.9, 3.10 and 3.15 for additional assumptions). The cost of CO_2 avoided corresponds to the slope of a line connecting a plant with capture and a reference plant without capture (i.e., the change in electricity cost divided by the change in emission rate). Avoidance costs for the same type of plant with and without capture plant are given in Table 3.15. When comparing different plant types, the reference plant represents the least-cost plant that would 'normally' be built at a particular location in the absence of a carbon constraint. In many regions today, this would be either a PC plant or an NGCC plant. The cost per tonne of CO_2 avoided can be highly variable and depends strongly on the costs and emissions of new plants being considered in a particular situation. See Chapter 8 for the full COE and full cost of CO_2 avoided for different plant types.

The other category combines CO_2 capture with upgrading or repowering the existing plant to significantly improve its efficiency and net power output (see Sections 3.7.4.2 and 3.7.5.2). In general, the latter option appears to be more cost-effective. However, further site-specific studies are required to systematically assess the feasibility and cost of alternative repowering options in conjunction with CO_2 capture for existing power plants.

New or improved methods of CO_2 capture, combined with advanced power systems and industrial process designs, promise to significantly reduce CO_2 capture costs and associated energy requirements. Tables 3.12 to 3.14 summarize the results from recent studies that examine future options. As discussed earlier, there is considerable uncertainty about the magnitude and timing of future cost reductions, as well as the potential for costs to rise above current estimates, especially for technologies still in the early stages of research and development. The current assessment is based on studies of the specific technologies in Tables 3.12 to 3.14 (and the supporting discussions and literature cited in Sections 3.7.9 to 3.7.12), as well as analyses of historical cost trends for related energy and environmental

technologies. This assessment suggests that improvements to current commercial technologies can reduce CO_2 capture costs by at least 20-30% over approximately the next decade, while new technologies under development promise more substantial cost reductions. Achieving future cost reductions, however, will require deployment and adoption of commercial technologies in the marketplace as well as sustained R&D.

3.8   Gaps in knowledge

Gaps in knowledge are related to differences in the stages of development of component technologies for the capture systems reviewed in Sections 3.2 to 3.5. For CO_2 capture from industrial processes, a number of technologies that are commonly used in natural gas sweetening and ammonia production are already used on a commercial scale. For other types of industrial systems capturing CO_2 from steel and cement production, further work is still needed. For CO_2 capture that might be reliant on post-combustion capture or oxy-fuel combustion, options are less well developed, or are available at a smaller scale than those required for applications such as in power generation, where much larger gas flows are handled. For pre-combustion capture many of the required systems have been developed and applied in industry already.

Although many of the component and/or enabling technologies required for CO$_2$ capture in post-combustion, pre-combustion and oxy-fuel combustion are well known, gaps in knowledge are in the practical and/or commercial demonstration of integrated systems. This demonstration is essential to prove the cost of CO$_2$ capture and its use on a large scale, particularly in power generation applications, but also for cement, steel and other large industries. Operating experience is also needed to test system reliability, improved methods of system integration, methods to reduce the energy requirements for CO$_2$ capture, improved process control strategies and the use of optimized functional materials for the implementation of capture processes with advanced, higher efficiency power cycles. As such developments are realized, environmental issues associated with the capture of CO$_2$ and other deleterious pollutants in these systems should also be re-assessed from a perspective involving the whole capture-transport-storage operation.

In an ongoing search to implement existing, new or improved methods of CO$_2$ capture, most capture systems also rely on the application of a range of enabling technologies that influence the attractiveness of a given system. These enabling technologies have their own critical gaps of knowledge. For example, improved processes for the effective removal of sulphur, nitrogen, chlorine, mercury and other pollutants are needed for the effective performance of unit operations for CO$_2$ separation in post- and pre-combustion capture systems, especially when coal is used as the primary fuel. Improved gasification reactors for coals and biomass, the availability of hydrogen-burning gas turbines and fuel cells for stationary power generation also need further development in the pre-combustion route. Combustors and boilers operating at higher temperatures, or a new class of CO$_2$ turbines and compressors, are important requirements for oxy-fuel systems.

With reference to the development of novel CO$_2$ capture and/or other enabling technologies, a wide range of options are currently being investigated worldwide. However, many technical details of the specific processes proposed or under development for these emerging technologies are still not well understood. This makes the assessment of their performance and cost highly uncertain. This is where intense R&D is needed to develop and bring to pilot scale testing the most promising concepts for commercial application. Membranes for H$_2$, CO$_2$ or O$_2$ separation, new sorbents, O$_2$ or CO$_2$ solid carriers and materials for advanced combustors, boilers and turbines all require extensive performance testing. Multi-pollutant emission controls in these novel systems and the impact of fuel impurities and temperature on the functional materials, should also be an area of future work.

References

Abanades, J.C., E.J. Anthony, D. Alvarez, D.Y. Lu, and C. Salvador, 2004a: Capture of CO$_2$ from Combustion Gases in a Fluidised Bed of CaO. *AIChE J*, **50**, No. 7, 1614-1622.

Abanades, J.C., E.S. Rubin and E.J. Anthony, 2004b: Sorbent cost and performance in CO$_2$ capture systems. *Industrial and Engineering Chemistry Research*, **43**, 3462-3466.

Abbot, J., B. Crewdson, and K. Elhius, 2002: Efficient cost effective and environmentally friendly synthesis gas technology for gas to liquids production. IBC Gas to Liquids Conference, London.

Aboudheir, A., P. Tontiwachwuthikul, A. Chakma, and R. Idem, 2003: Kinetics of the reactive absorption of carbon dioxide in high CO$_2$-loaded, concentrated aqueous monoethanolamine solutions. *Chemical Engineering Science* **58**, 5195-5210.

Alic, J.A., D.C. Mowery, and E.S. Rubin, 2003: *U.S. Technology and Innovation Policies: Lessons for Climate Change*. Pew Center on Global Climate Change, Arlington, VA, November.

Allam, R.J., E.P. Foster, V.E. Stein, 2002: Improving Gasification Economics through ITM Oxygen Integration. Proceedings of the Fifth Institution of Chemical Engineers (UK) European Gasification Conference, Noordwijk, The Netherlands.

Alstom Power Inc., ABB Lummus Global Inc., Alstom Power Environmental Systems and American Electric Power, 2001: *Engineering feasibility and economics of CO$_2$ capture on an existing coal-fired power plant*. Report no. PPL-01-CT-09 to Ohio Department of Development, Columbus, OH and US Department of Energy/NETL, Pittsburgh, PA.

American Institute of Chemical Engineers, 1995: Centre for Chemical Process Safety. Guidelines for Technical Planning for On-site Emergencies *Wiley*, New York.

Anderson, R., H. Brandt, S. Doyle, K. Pronske, and F. Viteri, 2003: Power generation with 100% carbon capture and sequestration. Second Annual Conference on Carbon Sequestration, Alexandria, VA.

Apple, M. 1997: Ammonia. Methanol. Hydrogen. Carbon Monoxide. Modern Production Technologies. A Review. Published by *Nitrogen - The Journal of the World Nitrogen and Methanol Industries*. CRU Publishing Ltd.

Aresta, M.A. and A. Dibenedetto, 2003: New Amines for the reversible absorption of carbon dioxide from gas mixtures. *Greenhouse Gas Control Technologies*, Proceedings of the 6th International Conference on Greenhouse Gas Control Technologies (GHGT-6), 1-4 Oct. 2002, Kyoto, Japan, J. Gale and Y. Kaya (eds.), Elsevier Science Ltd, Oxford, UK. 1599-1602.

Armstrong, P.A., D.L. Bennett, E.P. Foster, and V.E. Stein, 2002: Ceramic membrane development for oxygen supply to gasification applications. Proceedings of the Gasification Technologies Conference, San Francisco, CA, USA.

Arnold, D.S., D.A. Barrett and R.H. Isom, 1982: CO$_2$ can be produced from flue gas. *Oil & Gas Journal*, November, 130-136.

Aroonwilas, A., A. Chakma, P. Tontiwachwuthikul, and A. Veawab, 2003: Mathematical Modeling of Mass-Transfer and Hydrodynamics in CO$_2$ Absorbers Packed with Structured Packings, *Chemical Engineering Science,* **58**, 4037-4053.

Astarita, G., D.W. Savage, and A. Bisio, 1983: Gas Treating with Chemical Solvents, Chapter 9 Removal of Carbon Dioxide. *Wiley,* New York.

Audus, H. and P. Freund, 2005: Climate change mitigation by biomass gasification combined with CO_2 capture and storage. Proceedings of 7th International Conference on Greenhouse Gas Control Technologies. E.S. Rubin, D.W. Keith, and C.F. Gilboy (eds.), Vol. 1: Peer-Reviewed Papers and Overviews, E.S. Rubin, D.W. Keith and C.F. Gilboy (eds.), Elsevier Science, Oxford, UK, 187-200.

Babcock Energy Ltd, Air Products Ltd, University of Naples and University of Ulster, 1995: Pulverised coal combustion system for CO_2 capture. Final report 2.1.1, European Commission JOULE II Clean Coal Technology Programme - Powdered Coal Combustion Project.

Babovic, M., A. Gough, P. Leveson, and C. Ramshaw, 2001: Catalytic Plate Reactors for Endo- and Exothermic Reactions. 4th International Conference on Process Intensification for the Chemical Industry, Brugge, Belgium, 10-12 September.

Bachu, S., and W. Gunter, 2005: Overview of Acid Gas Injection in Western Canada. In E.S.Rubin, D.W. Keith, and C.F. Gilboy (eds.), Proceedings of 7th International Conference on Greenhouse Gas Control Technologies. Volume I: Peer Reviewed Papers and Overviews, *Elsevier Science,* Oxford, UK, 443-448.

Bai, H., A.C. Yeh, 1997: Removal of CO_2 Greenhouse Gas by Ammonia Scrubbing. Ind. Eng. Chem. Res, **36** (6), 2490-2493.

Bandi, A., M. Specht, P. Sichler, and N. Nicoloso, 2002: In situ Gas Conditioning in Fuel Reforming for Hydrogen Generation. 5th International Symposium on Gas Cleaning at High Temperature. U.S. DOE National Energy Technology Laboratory, Morgantown, USA.

Barchas, R., R. Davis, 1992: The Kerr-McGee / ABB Lummus Crest Technology for the Recovery of CO_2 from Stack Gases. *Energy Conversion and Management,* **33**(5-8), 333-340.

Beecy, D.J. and Kuuskraa, V.A., 2005: Basic Strategies for Linking CO_2 enhanced oil recovery and storage of CO_2 emissions. In E.S.Rubin, D.W. Keith and C.F. Gilboy (eds.), Proceedings of the 7th International Conference on Greenhouse Gas Control Technologies (GHGT-7), September 5-9, 2004, Vancouver, Canada. Volume I: Peer Reviewed Papers and Overviews, *Elsevier Science,* Oxford, UK, 351-360.

Blomen, L.J.N.J. and M.N. Mugerwa, 1993: Fuel Cell systems, Plenum Press, New York, 1993, ISBN 0-36-44158-6.

Bock, B., R. Rhudy, H. Herzog, M. Klett, J. Davison, D. De la Torre Ugarte, and D.Simbeck, 2003: Economic Evaluation of CO_2 Storage and Sink Options, DOE Research Report DE-FC26-00NT40937, U.S. Department of Energy, Pittsburgh Energy Technology Center, Pittsburgh, PA.

Bouwmeester, H.J.M., L.M.Van Der Haar, 2002: Oxygen permeation through mixed-conducting perovskite oxide membranes. *Ceramic Transactions,* **127**, Materials for Electrochemical Energy Conversion and Storage, 49-57.

BP, 2004: *Statistical Review of World Energy.* Http:\www.bp.com.

Bracht, M, Alderliesten P.T., R. Kloster, R. Pruschek, G. Haupt, E. Xue, J.R.H. Ross, M.K. Koukou, and N. Papayannakos, 1997: Water gas shift membrane reactor for CO_2 control in IGCC systems: techno-economic feasibility study, *Energy Conversion*

and Management, **38** (Suppl.), S159-S164, 1997.

Brandvoll, Ø. and O. Bolland, 2004: Inherent CO_2 capture using chemical looping combustion in a natural gas fired power cycle. ASME Paper No. GT-2002-30129, *ASME Journal of Engineering for Gas Turbines and Power,* **126**, 316-321.

Bredesen, R., K. Jordal and O. Bolland, 2004: High-Temperature Membranes in Power Generation with CO_2 capture. *Journal of Chemical Engineering and Processing,* **43**, 1129-1158.

Breton, D.L. and P. Amick, 2002: Comparative IGCC Cost and Performance for Domestic Coals, Preceedings of the 2002 Gasification Technology Conference, San Francisco, October.

Campanari, S., 2002: Carbon dioxide separation from high temperature fuel cell power plants. *Journal of Power Sources,* **112** (2002), 273-289.

Carolan, M.F., P.N. Dyer, E. Minford, T.F. Barton, D.R. Peterson, A.F. Sammells, D.L. Butt, R.A. Cutler, and D.M. Taylor, 2001: Development of the High Pressure ITM Syngas Process, Proceedings of the 6th Natural Gas Conversion Symposium, Alaska, 17-22 June.

Castle, W.F., 1991: Modern liquid pump oxygen plants: Equipment and performance, Cryogenic Processes and Machinery, AIChE Series No: 294; 89:14-17, 8th Intersociety Cryogenic Symposium, Houston, Texas, USA.

CCP, 2005: Economic and Cost Analysis for CO_2 Capture Costs in the CO_2 Capture Project, Scenarios. In D.C. Thomas (Ed.), Volume 1 - Capture and Separation of Carbon Dioxide from Combustion Sources, Elsevier Science, Oxford, UK.

Celik, F., E.D. Larson, and R.H. Williams, 2005: Transportation Fuel from Coal with Low CO_2 Emissions, Wilson, M., T. Morris, J. Gale and K. Thambimuthu (eds.), Proceedings of 7th International Conference on Greenhouse Gas Control Technologies. Volume II: Papers, Posters and Panel Discussion, *Elsevier Science,* Oxford UK, 1053-1058.

Chakma, A., P. Tontiwachwuthikul, 1999: Designer Solvents for Energy Efficient CO_2 Separation from Flue Gas Streams. Greenhouse Gas Control Technologies. Riemer, P., B. Eliasson, A. Wokaun (eds.), *Elsevier Science, Ltd.,* United Kingdom, 35-42.

Chakma, A., 1995: An Energy Efficient Mixed Solvent for the Separation of CO_2. *Energy Conversion and Management,* **36**(6-9), 427-430.

Chakravarty, S., A. Gupta, B. Hunek, 2001: Advanced technology for the capture of carbon dioxide from flue gases, Presented at First National Conference on Carbon Sequestration, Washington, DC.

Chapel, D.G., C.L. Mariz, and J. Ernest, 1999: Recovery of CO_2 from flue gases: commercial trends, paper No. 340 at the Annual Meeting of the Canadian Society of Chemical Engineering, Saskatoon, Canada, October.

Chatel-Pelage, F., M. Ovidiu, R. Carty, G. Philo, H. Farzan, S. Vecci, 2003: A pilot scale demonstration of oxy-fuel combustion with flue gas recirculation in a pulverised coal-fired boiler, Proceedings 28th International Technical Conference on Coal Utilization & Fuel Systems, Clearwater, Florida, March 10-13.

Cheeley, R., 2000: Combining gasifiers with the MIDREX® direct reduction process, Gasification 4 Conference, Amsterdam, Netherlands, 11-13 April.

Chen, C., A.B. Rao, and E.S. Rubin, 2003: Comparative assessment of

CO$_2$ capture options for existing coal-fired power plants, presented at the Second National Conference on Carbon Sequestration, Alexandria, VA, USA, 5-8 May.

Chen, H., A.S. Kovvali, S. Majumdar, K.K. Sirkar, 1999: Selective CO$_2$ separation from CO$_2$-N$_2$ mixtures by immobilised carbonate-glycerol membranes, *Ind. Eng. Chem.,* **38**, 3489-3498.

Chiesa, P., S. Consonni, T. Kreutz, and R. Williams, 2005: Co-production of hydrogen, electricity and CO$_2$ from coal with commercially ready technology. Part A: Performance and emissions, *International Journal of Hydrogen Energy,* **30** (7): 747-767.

Cho, P., T. Mattisson, and A. Lyngfelt, 2002: Reactivity of iron oxide with methane in a laboratory fluidised bed - application of chemical-looping combustion, 7th International Conference on Circulating Fluidised Beds, Niagara Falls, Ontario, May 5-7, 2002, 599-606.

Croiset, E. and K.V. Thambimuthu, 2000: Coal combustion in O$_2$/CO$_2$ Mixtures Compared to Air. *Canadian Journal of Chemical Engineering,* **78**, 402-407.

Cullinane, J.T. and G. T. Rochelle, 2003: Carbon Dioxide Absorption with Aqueous Potassium Carbonate Promoted by Piperazine, Greenhouse Gas Control Technologies, Vol. II, J. Gale, Y. Kaya, *Elsevier Science, Ltd.,* United Kingdom, 1603-1606.

Curran, G P., C.E. Fink, and E. Gorin, 1967: Carbon dioxide-acceptor gasification process. Studies of acceptor properties. *Adv. Chem. Ser.,* **69**, 141-165.

Damle, A.S. and T.P. Dorchak, 2001: Recovery of Carbon Dioxide in Advanced Fossil Energy Conversion Processes Using a Membrane Reactor, First National Conference on Carbon Sequestration, Washington, DC.

Davison, J.E., 2005: CO$_2$ capture and storage and the IEA Greenhouse Gas R&D Programme. Workshop on CO$_2$ issues, Middelfart, Denmark, 24 May, IEA Greenhouse Gas R&D Programme, Cheltenham, UK.

Dijkstra, J.W. and D. Jansen, 2003: Novel Concepts for CO$_2$ capture with SOFC, Proceedings of the 6th International Conference on Greenhouse Gas Control Technologies (GHGT-6) Volume I, Page 161-166, 1-4 Oct. 2002, Kyoto, Japan, Gale J. and Y. Kaya (eds.), *Elsevier Science Ltd*, Kidlington, Oxford, UK.

Dillon, D.J., R.S. Panesar, R.A.Wall, R.J. Allam, V. White, J. Gibbins, and M.R. Haines, 2005: Oxy-combustion processes for CO$_2$ capture from advanced supercritical PF and NGCC power plant, In: Rubin, E.S., D.W. Keith, and C.F. Gilboy (eds.), Proceedings of 7th International Conference on Greenhouse Gas Control Technologies. Volume I: Peer Reviewed Papers and Overviews, Elsevier Science, Oxford, UK, 211-220.

Dongke, M.A., L. Kong, and W.K. Lu, 1988: Heat and mass balance of oxygen enriched and nitrogen free blast furnace operations with coal injection. I.C.S.T.I. Iron Making Conference Proceedings.

Duarte, P.E. and E. Reich, 1998: A reliable and economic route for coal based D.R.I. production. I.C.S.T.I Ironmaking Conference Proceedings 1998.

Dyer, P.N., C.M. Chen, K.F. Gerdes, C.M. Lowe, S.R. Akhave, D.R. Rowley, K.I. Åsen and E.H. Eriksen, 2001: An Integrated ITM Syngas/Fischer-Tropsch Process for GTL Conversion, 6th Natural Gas Conversion Symposium, Alaska, 17-22 June 2001.

Dyer, P.N., R.E. Richards, S.L. Russek, D.M. Taylor, 2000: Ion transport membrane technology for oxygen separation and syngas production, *Solid State Ionics,* **134** (2000) 21-33.

EPRI, 1993: Technical Assessment Guide, Volume 1: Electricity Supply-1993 (Revision 7), Electric Power Research Institute, Palo Alto, CA, June.

Erga, O., O. Juliussen, H. Lidal, 1995: Carbon dioxide recovery by means of aqueous amines, *Energy Conversion and Management,* **36**(6-9), 387-392.

European Chemicals Bureau, 2003: Technical Guidance Document on Risk Assessment. European Communities. EUR 20418, http://ecb.jrc.it/.

Falk-Pedersen, O., H. Dannström, M. Grønvold, D.-B. Stuksrud, and O. Rønning, 1999: Gas Treatment Using Membrane Gas/Liquid Contractors, Greenhouse Gas Control Technologies. B. Eliasson, P. Riemer and A. Wokaun (eds.), Elsevier Science, Ltd., United Kingdom 115-120.

Farla, J.C., C.A. Hendriks, and K. Blok, 1995: Carbon dioxide recovery from industrial processes, *Climate Change,* **29**, (1995), 439-61.

Feron, P.H.M and A.E. Jansen, 2002: CO$_2$ Separation with polyolefin membrane contactors and dedicated absorption liquids: Performances and prospects, *Separation and Purification Technology,* **27**(3), 231-242.

Feron, P.H.M., 1992: Carbon dioxide capture: The characterisation of gas separation/removal membrane systems applied to the treatment of flue gases arising from power plant generation using fossil fuel. IEA/92/08, IEA Greenhouse Gas R&D programme, Cheltenham, UK.

Feron, P.H.M., 1994: Membranes for carbon dioxide recovery from power plants. In Carbon Dioxide Chemistry: Environmental Issues. J. Paul, C.M. Pradier (eds.), The Royal Society of Chemistry, Cambridge, United Kingdom, 236-249.

Gibbins, J., R.I. Crane, D. Lambropoulos, C. Booth, C.A. Roberts, and M. Lord, 2005: Maximising the effectiveness of post-combustion CO$_2$ capture systems. Proceedings of the 7th International Conference on Greenhouse Gas Control Technologies. Volume I: Peer Reviewed Papers and Overviews, E.S. Rubin, D.W. Keith, and C.F.Gilboy (eds.), Elsevier Science, Oxford, UK, 139-146.

Gielen, D.J., 2003: CO$_2$ removal in the iron and steel industry, *Energy Conversion and Management,* **44** (7), 1027-1037.

Göttlicher, G., 1999: Energetik der Kohlendioxidrückhaltung in Kraftwerken, Fortschritt-Berichte VDI, Reihe 6: Energietechnik Nr. 421, VDI Düsseldorf, Dissertation Universität Essen 1999, ISBN 3-18-342106-2.

Gray, D. and G. Tomlinson, 2001: Coproduction of Ultra Clean Transportation Fuels, Hydrogen, and Electric Power from Coal, Mitretek Technical Report MTR 2001-43, prepared for the National Energy Technology Laboratory, US DOE, July.

Gray, D. and G. Tomlinson, 2003: Hydrogen from Coal. Mitretek Technical Paper MTR-2003-13, prepared for the National Energy Technology Laboratory, US DOE, April.

Green, D.A., B.S. Turk, R.P. Gupta, J.W. Portzer, W.J. McMichael, and D.P. Harrison, 2002: Capture of Carbon Dioxide from flue gas using regenerable sorbents. 19th Annual International Pittsburgh Coal Conference. September 23-27, Pittsburgh, Pennsylvania,

USA.

Griffin, T., S.G. Sundkvist, K. Aasen, and T. Bruun, 2003: Advanced Zero Emissions Gas Turbine Power Plant, ASME Turbo Expo Conference, paper# GT-2003-38120, Atlanta, USA.

Haines, M.R., 1999: Producing Electrical Energy from Natural Gas using a Solid Oxide Fuel Cell. Patent WO 99/10945, 1-14.

Hazardous Substances Data Bank, 2002: US National Library of Medicine, Specialized Information Services: Hazardous Substances Data Bank. Carbon dioxide. 55 pp.

Hendriks, C., 1994: Carbon dioxide removal from coal-fired power plants, Dissertation, Utrecht University, Netherlands, 259 pp.

Herzog, H.J., 1999: The economics of CO_2 capture. Proceedings of the Fourth International Conference on Greenhouse Gas Control Technologies, B. Eliasson, P. Riemer, and A. Wokaun (eds.), 30 August-2 September 1998, Interlaken, Switzerland, Elsevier Science Ltd., Oxford, UK, 101-106.

Herzog, H., D. Golomb, S. Zemba, 1991: Feasibility, modeling and economics of sequestering power plant CO_2 emissions in the deep ocean, *Environmental Progress*, **10**(1), 64-74.

Hoffman, J.S., D.J. Fauth., and H.W. Pennline, 2002: Development of novel dry regenerable sorbents for CO_2 capture. 19th Annual International Pittsburgh Coal Conference. September 23-27, 2002 Pittsburgh, Pennsylvania, USA.

Holt, N., G. Booras, and D. Todd, 2003: Summary of recent IGCC studies of CO_2 for sequestration, Proceedings of Gasification Technologies Conference, October 12-15, San Francisco.

Hufton, J.R., R.J. Allam, R. Chiang, R.P. Middleton, E.L. Weist, and V. White, 2005: Development of a Process for CO_2 Capture from Gas Turbines using a Sorption Enhanced Water Gas Shift Reactor System. Proceedings of 7th International Conference on Greenhouse Gas Control Technologies. Volume I: Peer Reviewed Papers and Overviews, E.S. Rubin, D.W. Keith, and C.F. Gilboy (eds.), Elsevier Science, Oxford, UK, 2005, 253-262.

Hufton, J.R., S. Mayorga, S. Sircar, 1999: Sorption Enhanced Reaction Process for Hydrogen Production *AIChE J*, **45**, 248-254.

IEA WEO, 2004: IEA World Energy Outlook 2004, International Energy Agency, Paris France.

IEA, 2004: Prospects for CO_2 capture and storage, ISBN 92-64-10881-5.

IEA CCC, 2005: IEA CCC (IEA Clean Coal Centre) The World Coal-fired Power Plants Database, Gemini House, Putney, London, United Kingdom.

IEA GHG, 1996: De-carbonisation of fossil fuels, Report PH2/2, March 1996, IEA Greenhouse Gas R&D Programme, Cheltenham, UK.

IEA GHG, 1999: The reduction of greenhouse gas emissions from the cement industry. Report PH3/7, May 1999, IEA Greenhouse Gas R&D Programme, Cheltenham, UK.

IEA GHG, 2000a: Greenhouse gas emissions from major industrial sources III - Iron and Steel Production Report PH3/30, IEA Greenhouse Gas R&D Programme, Cheltenham, UK.

IEA GHG, 2000b: Leading options for the capture of CO_2 emissions at power stations, report PH3/14, Feb. 2000, IEA Greenhouse Gas R&D Programme, Cheltenham, UK.

IEA GHG, 2000c: CO_2 abatement in oil refineries: fired heaters, report PH3/31, Oct. 2000, IEA Greenhouse Gas R&D Programme,

Cheltenham, UK.

IEA GHG, 2000d: Key Components for CO_2 abatement: Gas turbines, report PH3/12 July 2000, IEA Greenhouse Gas R&D Programme, Cheltenham, UK.

IEA GHG, 2001: CO_2 abatement by the use of carbon-rejection processes, report PH3/36, February 2001, IEA Greenhouse Gas R&D Programme, Cheltenham, UK.

IEA GHG, 2002a: Transmission of CO_2 and Energy, report PH4/6, March 2002, IEA Greenhouse Gas R&D Programme, Cheltenham, UK.

IEA GHG, 2002b: Opportunities for early application of CO_2 sequestration technologies, report PH4/10, Sept. 2002, IEA Greenhouse Gas R&D Programme, Cheltenham, UK.

IEA GHG, 2003: Potential for improvements in gasification combined cycle power generation with CO_2 Capture, report PH4/19, IEA Greenhouse Gas R&D Programme, Cheltenham, UK.

IEA GHG, 2004: Improvements in power generation with post-combustion capture of CO_2, report PH4/33, Nov. 2004, IEA Greenhouse Gas R&D Programme, Cheltenham, UK.

IEA GHG, 2005: Retrofit of CO_2 capture to natural gas combined cycle power plants, report 2005/1, Jan. 2005, IEA Greenhouse Gas R&D Programme, Cheltenham, UK.

Ishibashi, M., K. Otake, S. Kanamori, and A. Yasutake, 1999: Study on CO_2 Removal Technology from Flue Gas of Thermal Power Plant by Physical Adsorption Method, Greenhouse Gas Control Technologies. P. Riemer, B. Eliasson, and A. Wokaun (eds.), Elsevier Science, Ltd., United Kingdom, 95-100.

Ishida, M. and H. Jin, 1994: A New Advanced Power-Generation System Using Chemical-Looping Combustion, *Energy*, **19**(4), 415-422.

Jansen, D. and J.W. Dijkstra, 2003: CO_2 capture in SOFC-GT systems, Second Annual Conference on Carbon Sequestration, Alexandria, Virginia USA, May 5-7.

Jericha, H., E. Göttlich, W. Sanz, F. Heitmeir, 2003: Design optimisation of the Graz cycle power plant, ASME Turbo Expo Conference, paper GT-2003-38120, Atlanta, USA.

Jordal, K., R. Bredesen, H.M. Kvamsdal, O. Bolland, 2003: Integration of H_2-separating membrane technology in gas turbine processes for CO_2 sequestration. Proceedings of the 6th International Conference on Greenhouse Gas Control Technologies (GHGT-6), Vol 1, 135-140, 1-4 Oct. 2002, Kyoto, Japan, J. Gale and Y. Kaya (eds.), Elsevier Science Ltd, Oxford, UK.

Karg, J. and F. Hannemann, 2004: IGCC - Fuel-Flexible Technology for the Future, Presented at the Sixth European Gasification Conference, Brighton, UK, May 2004.

Klett, M.G., R.C. Maxwell, and M.D. Rutkowski, 2002: The Cost of Mercury Removal in an IGCC Plant. Final Report for the US Department of Energy National Energy Technology Laboratory, by Parsons Infrastructure and Technology Group Inc., September.

Kohl, A.O. and R.B. Nielsen, 1997: Gas purification, Gulf Publishing Co., Houston, TX, USA.

Kovvali, A.S. and K.K. Sirkar, 2001: Dendrimer liquid membranes: CO_2 separation from gas mixtures, *Ind. Eng. Chem.*, **40**, 2502-2511.

Kreutz, T., R. Williams, P. Chiesa, and S. Consonni, 2005: Co-production of hydrogen, electricity and CO_2 from coal with commercially ready technology. Part B: Economic analysis,

International Journal of Hydrogen Energy, **30** (7): 769-784.

Kvamsdal, H., O. Maurstad, K. Jordal, and O. Bolland, 2004: Benchmarking of gas-turbine cycles with CO_2 capture. Proceedings of 7th International Conference on Greenhouse Gas Control Technologies. Volume I: Peer Reviewed Papers and Overviews, E.S. Rubin, D.W. Keith, and C.F. Gilboy (eds.), Elsevier Science, Oxford, UK, 2005, 233-242.

Lackner, K.S., 2003: Climate change: a guide to CO_2 sequestration, *Science*, **300**, issue 5626, 1677-1678, 13 June.

Lackner, K., H.J. Ziock, D.P. Harrison, 2001: Hydrogen Production from carbonaceous material. United States Patent WO 0142132.

Larson, E.D., and T. Ren, 2003: Synthetic fuels production by indirect coal liquefaction, *Energy for Sustainable Development*, **VII**(4), 79-102.

Latimer, R.E., 1967: Distillation of air. *Chem Eng Progress,* **63**(2), 35-59.

Leites, I.L., D.A. Sama, and N. Lior, 2003: The theory and practice of energy saving in the chemical industry: some methods for reducing thermodynamic irreversibility in chemical technology processes. *Energy,* **28**, N 1, 55-97.

Leites, I.L., 1998: The Thermodynamics of CO_2 solubility in mixtures monoethanolamine with organic solvents and water and commercial experience of energy saving gas purification technology. *Energy Conversion and Management*, **39**, 1665-1674.

Lin, S.Y., Y. Suzuki, H. Hatano, and M. Harada, 2002: Developing an innovative method, HyPr-RING, to produce hydrogen from hydrocarbons, *Energy Conversion and Management,* **43**, 1283-1290.

Lowe, C., V. Francuz, and C. Behrens, 2003: Hydrogen Membrane Selection for a Water Gas Shift Reactor. Second DoE Annual Conference on Carbon Sequestration. May, Arlington, VA.

Maddox, R.N. and D.J. Morgan, 1998: Gas Conditioning and Processing. Volume 4: Gas treating and sulfur recovery. Campbell Petroleum Series, Norman, OK, USA.

Mano, H., S. Kazama, and K. Haraya, 2003: Development of CO_2 separation membranes (1) Polymer membrane, In Greenhouse Gas Control Technologies. J. Gale and Y. Kaya (eds.), Elsevier Science, Ltd., United Kingdom, 1551-1554.

Marin, O., Y. Bourhis, N. Perrin, P. DiZanno, F. Viteri, and R. Anderson, 2003: High efficiency Zero Emission Power Generation based on a high temperature steam cycle, 28th Int. Technical Conference On Coal Utilization and Fuel Systems, Clearwater, FL, March.

Mathieu, P., 2003**:** Mitigation of CO_2 emissions using low and near zero CO_2 emission power plants. Clean Air, *International Journal on Energy for a Clean Environment*, **4**, 1-16.

McDaniel, J.E. and M.J. Hornick, 2002: Tampa Electric Polk Power Station Integrated Gasification Combined Cycle Project, Final Technical Report to the National Energy Technology Laboratory, US Department of Energy, August.

McDonald, A. and L. Schrattenholzer, 2001: Learning rates for energy technologies. *Energy Policy* **29**, pp. 255-261.

McDonald, M. and M. Palkes, 1999: A design study for the application of CO_2/O_2 combustion to an existing 300 MW coal-fired boiler, Proceedings of Combustion Canada 99 Conference-Combustion and Global Climate Change, Calgary, Alberta.

Merrow, E.W., K.E. Phillips and L.W. Myers, 1981: Understanding cost growth and performance shortfalls in pioneer process plants, Rand Publication No. R-2569-DOE, Report to the U.S. Department of Energy by Rand Corporation, Santa Monica, California, September.

Middleton, P., H. Solgaard-Andersen, T. Rostrup-Nielsen T. 2002: Hydrogen Production with CO_2 Capture Using Membrane Reactors. 14th World Hydrogen Energy Conference, June 9-14, Montreal, Canada.

Mimura, T., H. Simayoshi, T. Suda, M. Iijima, S. Mitsuoka, 1997: Development of Energy Saving Technology for Flue Gas Carbon Dioxide Recovery in Power Plant by Chemical Absorption Method and Steam System. *Energy Conversion and Management,* **38**, S57-S62.

Mimura, T., S. Satsumi, M. Iijima, S. Mitsuoka, 1999: Development on Energy Saving Technology for Flue Gas Carbon Dioxide Recovery by the Chemical Absorption Method and Steam System in Power Plant, Greenhouse Gas Control Technologies. P. Riemer, B. Eliasson, A. Wokaun (eds.), Elsevier Science, Ltd., United Kingdom, 71-76.

Mimura, T., S. Shimojo, T. Suda, M. Iijima, S. Mitsuoka, 1995: Research and Development on Energy Saving Technology for Flue Gas Carbon Dioxide Recovery and Steam System in Power Plant, *Energy Conversion and Management*, **36**(6-9), 397-400.

Mimura, T., T. Nojo, M. Iijima, T. Yoshiyama and H. Tanaka, 2003: Recent developments in flue gas CO_2 recovery technology. Greenhouse Gas Control Technologies, Proceedings of the 6th International Conference on Greenhouse Gas Control Technologies (GHGT-6), 1-4 Oct. 2002, Kyoto, Japan, J. Gale and Y. Kaya (eds.), Elsevier Science Ltd, Oxford, UK.

Mitretek, 2003: Hydrogen from Coal, Technical Paper MTR-2003-13, Prepared by D. Gray and G. Tomlinson for the National Energy Technology Laboratory, US DOE, April.

Möllersten, K., J. Yan, and J. Moreira, 2003: Potential market niches for biomass energy with CO_2 capture and storage – opportunities for energy supply with negative CO_2 emissions, *Biomass and Bioenergy,* **25**(2003), 273-285.

Möllersten, K., L. Gao, J. Yan, and M. Obersteiner, 2004: Efficient energy systems with CO_2 capture and storage from renewable biomass in pulp and paper mills, *Renewable Energy,* **29**(2004), 1583-1598.

Muramatsu, E. and M. Iijima, 2003: Life cycle assessment for CO_2 capture technology from exhaust gas of coal power plant. Greenhouse Gas Control Technologies. Proceedings of the 6th International Conference on Greenhouse Gas Control Technologies (GHGT-6), 1-4 Oct. 2002, Kyoto, Japan, J. Gale and Y. Kaya (eds.), Elsevier Science Ltd, Oxford, UK.

Nakagawa, K., T. Ohashi 1998: A novel method of CO_2 capture from high temperature gases, *Journal Electrochem. Soc.*, **145**(4): 1344-1346.

NETL, 2002: Advanced fossil power systems comparison study, Final report prepared for NETL by E.L. Parsons (NETL, Morgantown, WV), W.W. Shelton and J.L. Lyons (EG&G Technical Services, Inc., Morgantown, WV), December.

NETL-DOE, 2002: Worldwide Gasification Database online, Pittsburgh, PA, USA. http://www.netl.doe.gov/coalpower/ gasification/models/dtbs(excel.pdf.

Noble, R. and Stern (eds.), 1995: Membrane Separations Technology, Elsevier Science, Amsterdam, The Netherlands, 718 pp.

NRC, 2003: Review of DOE's Vision 21 Research and Development Program - Phase I, Board on Energy and Environmental Systems of the National Research Council, The National Academies Press, Washington, DC, 97 p.

NRC, 2004: The Hydrogen Economy: Opportunities, Costs, Barriers, and R&D Needs, Prepared by the Committee on Alternatives and Strategies for Future Hydrogen Production and Use, Board on Energy and Environmental Systems of the National Research Council, The National Academies Press, Washington, DC.

Nsakala, N., G. Liljedahl, J. Marion, C. Bozzuto, H. Andrus, and R. Chamberland, 2003: Greenhouse gas emissions control by oxygen firing in circulating fluidised bed boilers. Presented at the Second Annual National Conference on Carbon Sequestration. Alexandria, VA May 5-8, USA.

Nsakala, Y.N., J. Marion, C. Bozzuto, G. Liljedahl, M. Palkes, D. Vogel, J.C. Gupta, M. Guha, H. Johnson, and S. Plasynski, 2001: Engineering feasibility of CO_2 capture on an existing US coal-fired power plant, Paper presented at First National Conference on Carbon Sequestration, Washington DC, May 15-17.

Okabe, K., N. Matsumija, H. Mano, M. Teramoto, 2003: Development of CO_2 separation membranes (1) Facilitated transport membrane, In Greenhouse Gas Control Technologies. J. Gale and Y. Kaya (eds.), Elsevier Science, Ltd., United Kingdom, 1555-1558.

Parsons Infrastructure & Technology Group, Inc., 2002b: *Updated cost and performance estimates for fossil fuel power plants with CO_2 removal.* Report under Contract No. DE-AM26-99FT40465 to U.S.DOE/NETL, Pittsburgh, PA, and EPRI, Palo Alto, CA., December.

Parsons Infrastructure and Technology Group, Inc., 2002a: *Hydrogen Production Facilities: Plant Performance and Cost Comparisons,* Final Report, prepared for the National Energy Technology Laboratory, US DOE, March.

Quinn, R., D.V. Laciak, 1997: Polyelectrolyte membranes for acid gas separations, *Journal of Membrane Science,* **131,** 49-60.

Ramsaier, M., H.J. Sternfeld, K. Wolfmuller, 1985: European Patent 0197 555 A2.

Rao, A.B. and E.S. Rubin, 2002: A technical, economic, and environmental assessment of amine-based CO_2 capture technology for power plant greenhouse gas control. *Environmental Science and Technology, 36,* 4467-4475.

Rao, A.B., E.S. Rubin and M. Morgan, 2003: Evaluation of potential cost reductions from improved CO_2 capture systems. 2nd Annual Conference on Carbon Sequestration, Alexandria, VA, USA, 5-8 May, U.S. Department of Energy, NETL, Pittsburgh, PA.

Reddy, S., J. Scherffius, S. Freguia and C. Roberts, 2003: Fluor's Econamine FG Plus^{SM} technology - an enhanced amine-based CO_2 capture process, 2nd Annual Conference on Carbon Sequestration, Alexandria, VA, USA, 5-8 May, U.S. Department of Energy, National Energy Technology Laboratory, Pittsburgh, PA.

Renzenbrink, W., R. Wischnewski, J. Engelhard, A. Mittelstadt, 1998: High Temperature Winkler (HTW) Coal Gasification: A Fully Developed Process for Methanol and Electricity Production, paper presented at the Gasification Technology Conference, October 1998, San Francisco, CA, USA.

Richards, D., 2003: Dilute oxy-fuel technology for zero emmission power, First International Conference on Industrial Gas Turbine Technologies, Brussels (available on www.came-gt.com).

Richter, H.J., K. Knoche 1983: Reversibility of Combustion processes, Efficiency and Costing - Second Law Analysis of Processes, *ACS Symposium series,* **235,** p. 71-85.

Riemer, P.W.F. and W.G. Ormerod, 1995: International perspectives and the results of carbon dioxide capture disposal and utilisation studies, *Energy Conversion and Management,* **36**(6-9), 813-818.

Rizeq, G., R. Subia, J. West, A. Frydman, and V. Zamansky, 2002: Advanced-Gasification Combustion: Bench-Scale Parametric Study. 19th Annual International Pittsburgh Coal Conference September 23-27, 2002, Pittsburgh, PA, USA.

Robinson, A.L., J.S. Rhodes, and D.W. Keith, 2003: Assessment of potential carbon dioxide reductions due to biomass-coal cofiring in the United States, *Environmental Science and Technology,* **37**(22), 5081-5089.

Rubin, E.S. and A.B. Rao, 2003: Uncertainties in CO_2 capture and sequestration costs, Greenhouse Gas Control Technologies, Proceedings of the 6th International Conference on Greenhouse Gas Control Technologies (GHGT-6), 1-4 Oct. 2002, Kyoto, Japan, J. Gale and Y. Kaya (eds.), Elsevier Science Ltd, Oxford, UK.

Rubin, E.S., 2001: *Introduction to Engineering and the Environment.* McGraw-Hill, Boston, MA, 701 p.

Rubin, E.S., A.B. Rao, and C. Chen, 2005: Comparative Assessments of Fossil Fuel Power Plants with CO_2 Capture and Storage. Proceedings of 7th International Conference on Greenhouse Gas Control Technologies, Volume 1: Peer-Reviewed Papers and Overviews, E.S. Rubin, D.W. Keith and C.F. Gilboy (eds.), Elsevier Science, Oxford, UK, 285-294.

Rubin, E.S., D.A. Hounshell, S. Yeh, M. Taylor, L. Schrattenholzer, K. Riahi, L. Barreto, and S. Rao, 2004b: The Effect of Government Actions on Environmental Technology Innovation: Applications to the Integrated Assessment of Carbon Sequestration Technologies, Final Report of Award No. DE-FG02-00ER63037 from Carnegie Mellon University, Pittsburgh, PA to Office of Biological and Environmental Research, U.S. Department of Energy, Germantown, MD, January, 153 p.

Rubin, E.S., S. Yeh, D.A. Hounshell, and M.R. Taylor, 2004a: Experience Curves for Power Plant Emission Control Technologies, *International Journal of Energy Technology and Policy,* **2,** No.1/2, 52-68, 2004.

Ruthven, D.M., S. Farooq, and K.S. Knaebel, 1994: Pressure Swing Adsorption. VCH, New York, 352 pp.

Sander, M.T., C.L. Mariz, 1992: The Fluor Daniel® Econamine™ FG Process: Past Experience and Present Day Focus, *Energy Conversion Management,* **33**(5-8), 341-348.

Shilling, N. and R. Jones, 2003: The Response of Gas Turbines to a CO_2 Constrained Environment Paper Presented at the Gasification Technology Conference, October 2003, San Francisco, CA, USA, Available at www.gasification.org.

Shimizu, T, T. Hirama, H. Hosoda, K. Kitano, M. Inagaki, and K. Tejima, 1999: A Twin Fluid-Bed Reactor for removal of CO_2 from combustion processes. *IChemE.,* **77**- A, 62-70.

Sikdar, S.K. and U. Diwekar (eds.), 1999: Tools and Methods for

Pollution Prevention. Proceedings of NATO Advanced Research Workshop. NATO Science Series, No. 2: Environmental Security - Vol. 62. Dordrecht: Kluwer. 12-14 October 1998, Prague, Czech Republic.

Simbeck, D.R., 1999: A portfolio selection approach for power plant CO$_2$ capture, separation and R&D options. Proceedings of the 4th International Conference on Greenhouse Gas Control Technologies, 30 Aug. - 2 Sept. 1998, Interlaken, Switzerland, B. Eliasson, P. Riemer and A. Wokaun (eds.), Elsevier Science Ltd., Oxford, UK.

Simbeck, D.R., 2001a: World Gasification Survey: Industrial Trends and Developments. Paper presented at the Gasification Technology Conference, San Francisco, CA, USA, October.www.gasfication.org.

Simbeck, D. R. and M. McDonald, 2001b: Existing coal power plant retrofit CO$_2$ control options analysis, Greenhouse Gas Control Technologies, Proceedings of the 5th International Conference on Greenhouse Gas Control Technologies, 13-16 Aug. 2000, Cairns, Australia, D. Williams et al. (eds.), CSIRO Publishing, Collingwood, Vic., Australia.

Simbeck, D.R., 2002: New power plant CO$_2$ mitigation costs, SFA Pacific, Inc., Mountain View, California, April.

Simbeck, D.R., 2004: Overview and insights on the three basic CO$_2$ capture options, Third Annual Conference on Carbon Capture and Sequestration, Alexandria, Virginia, May.

Simbeck, D.R., 2005: Hydrogen Costs with CO$_2$ Capture. M. Wilson, T. Morris, J. Gale and K. Thambimuthu (eds.): Proceedings of 7th International Conference on Greenhouse Gas Control Technologies. Volume II: Papers, Posters and Panel Discussion, Elsevier Science, Oxford UK, 1059-1066.

Singh, D., E. Croiset, P.L. Douglas and M.A. Douglas, 2003: Techno-economic study of CO$_2$ capture from an existing coal-fired power plant: MEA scrubbing vs. O$_2$/CO$_2$ recycle combustion. *Energy Conversion and Management*, **44**, p. 3073-3091.

Sircar, S., 1979: Separation of multi-component gas mixtures, US Patent No. 4171206, October 16th.

Sircar, S., C.M.A. Golden, 2001: PSA process for removal of bulk carbon dioxide from a wet high-temperature gas. US Patent No. 6322612.

Skinner, S.J.and J.A. Kilner, 2003: Oxygen ion conductors. *Materials Today*, **6**(3), 30-37.

Stobbs, R. and Clark, P., 2005: Canadian Clean Power Coalition: The Evaluation of Options for CO$_2$ Capture From Existing and New Coal-Fired Power Plants, In, Wilson, M., T. Morris, J. Gale and K. Thambimuthu (eds.), Proceedings of 7th International Conference on Greenhouse Gas Control Technologies. Volume II: Papers, Posters and Panel Discussion, Elsevier Science, Oxford, UK, 1187-1192.

Sundnes, A., 1998: Process for generating power and/or heat comprising a mixed conducting membrane reactor. International patent number WO98/55394 Dec. 1998.

Tabe-Mohammadi, A., 1999: A review of the application of membrane separation technology in natural gas treatment, *Sep. Sci. & Tech.*, **34**(10), 2095-2111.

Takamura, Y. Y. Mori, H. Noda, S. Narita, A. Saji, S. Uchida, 1999: Study on CO$_2$ Removal Technology from Flue Gas of Thermal Power Plant by Combined System with Pressure Swing Adsorption and Super Cold Separator. Proceedings of the 5th International Conference on Greenhouse Gas Control Technologies, 13-16 Aug. 2000, Cairns, Australia, D. Williams et al. (eds.), CSIRO Publishing, Collingwood, Vic., Australia.

Tan, Y., M.A., Douglas, E. Croiset, and K.V. Thambimuthu, 2002: CO$_2$ Capture Using Oxygen Enhanced Combustion Strategies for Natural Gas Power Plants, *Fuel*, **81**, 1007-1016.

Teramoto, M., K. Nakai, N. Ohnishi, Q. Huang, T. Watari, H. Matsuyama, 1996: Facilitated transport of carbon dioxide through supported liquid membranes of aqueous amine solutions, *Ind. Eng. Chem.*, **35**, 538-545.

Todd, D.M. and Battista, R.A., 2001: Demonstrated Applicability of Hydrogen Fuel for Gas Turbines, 4th European Gasification Conference 11-13th April, Noordwijk Netherlands.

Van der Sluijs, J.P, C.A. Hendriks, and K. Blok, 1992: Feasibility of polymer membranes for carbon dioxide recovery from flue gases, *Energy Conversion Management,* **33**(5-8), 429-436.

Von Bogdandy, L., W. Nieder, G. Schmidt, U. Schroer, 1989: Smelting reduction of iron ore using the COREX process in power compound systems. *Stahl und Eisen*, **109**(9,10), p 445.

Wabash River Energy Ltd., 2000: Wabash River Coal Gasification Repowering Project, Final Technical Report to the National Energy Technology Laboratory, US Department of Energy, August.

Wang, J., E.J. Anthony, J.C. Abanades, 2004: Clean and efficient use of petroleum coke for combustion and power generation. *Fuel*, **83**, 1341-1348.

Wilkinson, M.B. and Clarke, S.C., 2002: Hydrogen Fuel Production: Advanced Syngas Technology Screening Study.14th World Hydrogen Energy Conference, June 9-14, 2002, Montreal, Canada.

Wilkinson, M.B., J.C. Boden, T. Gilmartin, C. Ward, D.A. Cross, R.J. Allam, and N.W. Ivens, 2003b: CO$_2$ capture from oil refinery process heaters through oxy-fuel combustion, Greenhouse Gas Control Technologies, Proc. of the 6th International Conference on Greenhouse Gas Control Technologies (GHGT-6), 1-4 Oct. 2002, Kyoto, Japan, J. Gale and Y. Kaya (eds.), Elsevier Science Ltd, Oxford, UK. 69-74.

Wilkinson, M.B., M. Simmonds, R.J. Allam, and V. White, 2003a: Oxy-fuel conversion of heaters and boilers for CO$_2$ capture, 2nd Annual Conf on Carbon Sequestration, Virginia (USA), May 2003.

Williams, R.H. (Convening Lead Author), 2000: Advanced energy supply technologies, Chapter 8, 274-329, in *Energy and the Challenge of Sustainability - the World Energy Assessment World Energy Assessment*, 508 pp., UN Development Programme, New York.

World Bank, 1999: Pollution Prevention and Abatement Handbook: Toward Cleaner Production. Washington: The World Bank Group in collaboration with United Nations Industrial Development Organization and United Nations Environment Programme.

Yantovskii, E.I., K.N. Zvagolsky, and V.A. Gavrilenko, 1992: Computer exergonomics of power plants without exhaust gases *Energy Conversion and Management*, **33**, No. 5-8, 405-412.

Yokoyama, T., 2003: Japanese R&D on CO$_2$ Capture. Greenhouse Gas Control Technologies, Proc. of the 6th International Conference on

Greenhouse Gas Control Technologies (GHGT-6), 1-4 Oct. 2002, Kyoto, Japan, J. Gale and Y. Kaya (eds.), Elsevier Science Ltd, Oxford, UK. 13-18.

Zafar, Q., T. Mattisson, and B. Gevert, 2005: Integrated Hydrogen and Power Production with CO_2 Capture Using Chemical-Looping Reforming-Redox Reactivity of Particles of CuO, Mn_2O_3, NiO, and Fe_2O_3 Using SiO_2 as a Support, *Industrial and Engineering Chemistry Research,* **44**(10), 3485-3496.

Zheng, X.Y, Y.-F. Diao, B.-S. He, C.-H. Chen, X.-C. Xu, and W. Feng, 2003: Carbon Dioxide Recovery from Flue Gases by Ammonia Scrubbing. Greenhouse Gas Control Technologies, Proc. of the 6th International Conference on Greenhouse Gas Control Technologies (GHGT-6), 1-4 Oct. 2002, Kyoto, Japan, J. Gale and Y. Kaya (eds.), Elsevier Science Ltd, Oxford, UK. 193-200.

4

Transport of CO$_2$

Coordinating Lead Authors
Richard Doctor (United States), Andrew Palmer (United Kingdom)

Lead Authors
David Coleman (United States), John Davison (United Kingdom), Chris Hendriks (The Netherlands),
Olav Kaarstad (Norway), Masahiko Ozaki (Japan)

Contributing Author
Michael Austell (United Kingdom)

Review Editors
Ramon Pichs-Madruga (Cuba), Svyatoslav Timashev (Russian Federation)

Contents

EXECUTIVE SUMMARY

Transport is that stage of carbon capture and storage that links sources and storage sites. The beginning and end of 'transport' may be defined administratively. 'Transport' is covered by the regulatory framework concerned for public safety that governs pipelines and shipping. In the context of long-distance movement of large quantities of carbon dioxide, pipeline transport is part of current practice. Pipelines routinely carry large volumes of natural gas, oil, condensate and water over distances of thousands of kilometres, both on land and in the sea. Pipelines are laid in deserts, mountain ranges, heavily-populated areas, farmland and the open range, in the Arctic and sub-Arctic, and in seas and oceans up to 2200 m deep.

Carbon dioxide pipelines are not new: they now extend over more than 2500 km in the western USA, where they carry 50 $MtCO_2$ yr^{-1} from natural sources to enhanced oil recovery projects in the west Texas and elsewhere. The carbon dioxide stream ought preferably to be dry and free of hydrogen sulphide, because corrosion is then minimal, and it would be desirable to establish a minimum specification for 'pipeline quality' carbon dioxide. However, it would be possible to design a corrosion-resistant pipeline that would operate safely with a gas that contained water, hydrogen sulphide and other contaminants. Pipeline transport of carbon dioxide through populated areas requires attention be paid to design factors, to overpressure protection, and to leak detection. There is no indication that the problems for carbon dioxide pipelines are any more challenging than those set by hydrocarbon pipelines in similar areas, or that they cannot be resolved.

Liquefied natural gas and petroleum gases such as propane and butane are routinely transported by marine tankers; this trade already takes place on a very large scale. Carbon dioxide is transported in the same way, but on a small scale because of limited demand. The properties of liquefied carbon dioxide are not greatly different from those of liquefied petroleum gases, and the technology can be scaled up to large carbon dioxide carriers. A design study discussed later has estimated costs for marine transport of 1 $MtCO_2$ yr^{-1} by one 22,000 m³ marine tanker over a distance of 1100 km, along with the associated liquefaction, loading and unloading systems.

Liquefied gas can also be carried by rail and road tankers, but it is unlikely that they be considered attractive options for large-scale carbon dioxide capture and storage projects.

4.1 Introduction

CO_2 is transported in three states: gas, liquid and solid. Commercial-scale transport uses tanks, pipelines and ships for gaseous and liquid carbon dioxide.

Gas transported at close to atmospheric pressure occupies such a large volume that very large facilities are needed. Gas occupies less volume if it is compressed, and compressed gas is transported by pipeline. Volume can be further reduced by liquefaction, solidification or hydration. Liquefaction is an established technology for gas transport by ship as LPG (liquefied petroleum gas) and LNG (liquefied natural gas). This existing technology and experience can be transferred to liquid CO_2 transport. Solidification needs much more energy compared with other options, and is inferior from a cost and energy viewpoint. Each of the commercially viable technologies is currently used to transport carbon dioxide.

Research and development on a natural gas hydrate carrying system intended to replace LNG systems is in progress, and the results might be applied to CO_2 ship transport in the future. In pipeline transportation, the volume is reduced by transporting at a high pressure: this is routinely done in gas pipelines, where operating pressures are between 10 and 80 MPa.

A transportation infrastructure that carries carbon dioxide in large enough quantities to make a significant contribution to climate change mitigation will require a large network of pipelines. As growth continues it may become more difficult to secure rights-of-way for the pipelines, particularly in highly populated zones that produce large amounts of carbon dioxide. Existing experience has been in zones with low population densities, and safety issues will become more complex in populated areas.

The most economical carbon dioxide capture systems appear to favour CO_2 capture, first, from pure stream sources such as hydrogen reformers and chemical plants, and then from centralized power and synfuel plants: Chapter 2 discusses this issue in detail. The producers of natural gas speak of 'stranded' reserves from which transport to market is uneconomical. A movement towards a decentralized power supply grid may make CO_2 capture and transport much more costly, and it is easy to envision stranded CO_2 at sites where capture is uneconomic.

A regulatory framework will need to emerge for the low-greenhouse-gas-emissions power industry of the future to guide investment decisions. Future power plant owners may find the carbon dioxide transport component one of the leading issues in their decision-making.

4.2 Pipeline systems

4.2.1 Pipeline transportation systems

CO_2 pipeline operators have established minimum specifications for composition. Box 4.1 gives an example from the Canyon Reef project (Section 4.2.2.1). This specification is for gas for an enhanced oil recovery (EOR) project, and parts of it would not necessarily apply to a CO_2 storage project. A low nitrogen content is important for EOR, but would not be so significant for CCS. A CO_2 pipeline through populated areas might have a lower specified maximum H_2S content.

Dry carbon dioxide does not corrode the carbon-manganese steels generally used for pipelines, as long as the relative humidity is less than 60% (see, for example, Rogers and Mayhew, 1980); this conclusion continues to apply in the presence of N_2, NO_x and SO_x contaminants. Seiersten (2001) wrote:

"The corrosion rate of carbon steel in dry supercritical CO_2 is low. For AISI 1080 values around 0.01 mm yr^{-1} have been measured at 90–120 bar and 160°C–180°C for 200 days. Short-term tests confirm this. In a test conducted at 3°C and 22°C at 140 bar CO_2, and 800 to 1000 ppm H_2S, the corrosion rate for X-60 carbon steel was measured at less than 0.5 μm yr[-1] (0.0005 mm yr[-1]). Field experience also indicates very few problems with transportation of high-pressure dry CO_2 in carbon steel pipelines. During 12 years, the corrosion rate in an operating pipeline amounts to 0.25-2.5 μm yr[-1] (0.00025 to (0.0025 mm yr[-1])".

The water solubility limit in high-pressure CO_2 (500 bar) is 5000 ppm at 75°C and 2000 ppm at 30°C. Methane lowers the solubility limit, and H_2S, O_2 and N_2 may have the same effect.

Corrosion rates are much higher if free water is present; hydrates might also form. Seiersten (2001) measured a corrosion rate of 0.7 mm yr[-1] corrosion rate in 150 to 300 hours exposure at 40°C in water equilibrated with CO_2 at 95 bar, and higher rates at lower pressures. She found little difference between carbon-manganese steel (American Petroleum Institute grade X65) and 0.5 chromium corrosion-resistant alloy. It is unlikely to be practicable to transport wet CO_2 in low-alloy carbon steel pipelines because of this high corrosion rate. If the CO_2 cannot be dried, it may be necessary to build the pipeline of a corrosion-resistant alloy ('stainless steel'). This is an established technology. However the cost of steel has greatly increased recently and this may not be economical.

Once the CO_2 has been dried and meets the transportation criteria, the CO_2 is measured and transported to the final use site. All the pipelines have state-of-the-art metering systems that accurately account for sales and deliveries on to and out of each line, and SCADA (Supervisory Control and Data Acquisition) systems for measuring pressure drops, and redundancies built in to allow for emergencies. In the USA, these pipelines are governed by Department of Transportation regulations. Movement of CO_2 is best accomplished under high pressure: the choice of operating pressure is discussed in an example

below, and the reader is referred to Annex I for a discussion of the physical properties of CO_2.

4.2.2   Existing experience

Table 4.1 lists existing long-distance CO_2 pipelines. Most of the projects listed below are described in greater detail in a report by the UK Department of Trade and Industry (2002). While there are CO_2 pipelines outside the USA, the Permian Basin contains over 90% of the active CO_2 floods in the world (O&GJ, April 15, 2002, EOR Survey). Since then, well over 1600 km of new CO_2 pipelines has been built to service enhanced oil recovery (EOR) in west Texas and nearby states.

4.2.2.1   Canyon Reef
The first large CO_2 pipeline in the USA was the Canyon Reef Carriers, built in 1970 by the SACROC Unit in Scurry County, Texas. Its 352 km moved 12,000 tonnes of anthropogenically produced CO_2 daily (4.4 Mt yr[-1]) from Shell Oil Company gas processing plants in the Texas Val Verde basin.

4.2.2.2   Bravo Dome Pipeline
Oxy Permian constructed this 508 mm (20-inch) line connecting the Bravo Dome CO_2 field with other major pipelines. It is capable of carrying 7.3 $MtCO_2$ yr[-1] and is operated by Kinder Morgan.

4.2.2.3   Cortez Pipeline
Built in 1982 to supply CO_2 from the McElmo Dome in S.E. Colorado, the 762 mm (30-inch), 803 km pipeline carries approximately 20 Mt CO_2 yr[-1] to the CO_2 hub at Denver City, Texas. The line starts near Cortez, Colorado, and crosses the Rocky Mountains, where it interconnects with other CO_2 lines. In the present context, recall that one 1000 MW coal-fired

Box 4.1 Specimen CO_2 quality specifications

The Product delivered by Seller or Seller's representative to Buyer at the Canyon Reef Carriers Delivery Meter shall meet the following specifications, which herein are collectively called 'Quality Specifications':

(a) **Carbon Dioxide**. Product shall contain at least ninety-five mole percent (95%) of Carbon Dioxide as measured at the SACROC delivery meter.

(b) **Water**. Product shall contain no free water, and shall not contain more than 0.48 9 m[-3] in the vapour phase.

(c) **Hydrogen Sulphide**. Product shall not contain more than fifteen hundred (1500) parts per million, by weight, of hydrogen sulphide.

(d) **Total Sulphur**. Product shall not contain more than fourteen hundred and fifty (1450) parts per million, by weight, of total sulphur.

(e) **Temperature**. Product shall not exceed a temperature of 48.9 °C.

(f) **Nitrogen**. Product shall not contain more than four mole percent (4%) of nitrogen.

(g) **Hydrocarbons**. Product shall not contain more than five mole percent (5%) of hydrocarbons and the dew point of Product (with respect to such hydrocarbons) shall not exceed –28.9 °C.

(h) **Oxygen**. Product shall not contain more than ten (10) parts per million, by weight, of oxygen.

(i) **Glycol**. Product shall not contain more than 4 x 10[-5] L m[-3] of glycol and at no time shall such glycol be present in a liquid state at the pressure and temperature conditions of the pipeline.

Table 4.1 Existing long-distance CO$_2$ pipelines (Gale and Davison, 2002) and CO$_2$ pipelines in North America (Courtesy of Oil and Gas Journal).

Pipeline	Location	Operator	Capacity (MtCO$_2$ yr^{-1})	Length (km)	Year finished	Origin of CO$_2$
Cortez	USA	Kinder Morgan	19.3	808	1984	McElmoDome
Sheep Mountain	USA	BP Amoco	9.5	660	-	Sheep Mountain
Bravo	USA	BP Amoco	7.3	350	1984	Bravo Dome
Canyon Reef Carriers	USA	Kinder Morgan	5.2	225	1972	Gasification plants
Val Verde	USA	Petrosource	2.5	130	1998	Val Verde Gas Plants
Bati Raman	Turkey	Turkish Petroleum	1.1	90	1983	Dodan Field
Weyburn	USA & Canada	North Dakota Gasification Co.	5	328	2000	Gasification Plant
Total			49.9	2591		

power station produces about 7 Mt CO$_2$ yr^{-1}, and so one Cortez pipeline could handle the emissions of three of those stations.

The Cortez Pipeline passes through two built-up areas, Placitas, New Mexico (30 km north of Albuquerque, New Mexico) and Edgewood/Moriarty, New Mexico (40 km east of Albuquerque). The line is buried at least 1 m deep and is marked within its right of way. Near houses and built-up areas it is marked more frequently to ensure the residents are aware of the pipeline locations. The entire pipeline is patrolled by air every two weeks, and in built-up areas is frequently patrolled by employees in company vehicles. The public education programme includes the mailing of a brochure describing CO$_2$, signs of a leak and where to report a suspected leak, together with information about the operator and the "one-call" centre.

4.2.2.4 Sheep Mountain Pipeline

BP Oil constructed this 610 mm (24-inch) 772 km line capable of carrying 9.2 MtCO$_2$ yr^{-1} from another naturally occurring source in southeast Colorado. It connects to the Bravo Dome line and into the other major carriers at Denver City and now is operated by Kinder Morgan.

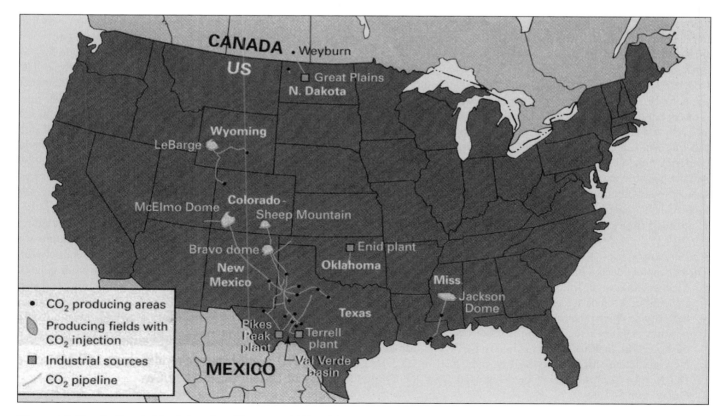

Figure 4.1 CO$_2$ pipelines in North America. (Courtesy of Oil and Gas Journal).

4.2.2.5   Weyburn Pipeline

This 330 km, (305-356 mm diameter) system carries more than 5000 tonne day^{-1} (1.8 Mt yr^{-1}) of CO_2 from the Great Plains Synfuels Plant near Beulah, North Dakota to the Weyburn EOR project in Saskatchewan. The composition of the gas carried by the pipeline is typically CO_2 96%, H_2S 0.9%, CH_4 0.7%, C2+ hydrocarbons 2.3%, CO 0.1%, N_2 less than 300 ppm, O_2 less than 50 ppm and H_2O less than 20 ppm (UK Department of Trade and Industry, 2002). The delivery pressure at Weyburn is 15.2 MPa. There are no intermediate compressor stations. The amount allocated to build the pipeline was 110 US $ million (0.33 x 10^6 US$ km^{-1}) in 1997.

4.2.3   Design

The physical, environmental and social factors that determine the design of a pipeline are summarized in a design basis, which then forms the input for the conceptual design. This includes a system definition for the preliminary route and design aspects for cost-estimating and concept-definition purposes. It is also necessary to consider the process data defining the physical characteristics of product mixture transported, the optimal sizing and pressures for the pipeline, and the mechanical design, such as operating, valves, pumps, compressors, seals, etc. The topography of the pipeline right-of-way must be examined. Topography may include mountains, deserts, river and stream crossings, and for offshore pipelines, the differing challenges of very deep or shallow water, and uneven seabed. It is also important to include geotechnical considerations. For example, is this pipeline to be constructed on thin soil overlaying granite? The local environmental data need to be included, as well as the annual variation in temperature during operation and during construction, potentially unstable slopes, frost heave and seismic activity. Also included are water depth, sea currents, permafrost, ice gouging in Arctic seas, biological growth, aquifers, and other environmental considerations such as protected habitats. The next set of challenges is how the pipeline will accommodate existing and future infrastructure – road, rail, pipeline crossings, military/governmental restrictions and the possible impact of other activities – as well as shipping lanes, rural or urban settings, fishing restrictions, and conflicting uses such as dredging. Finally, this integrated study will serve as the basis for a safety review.

Conceptual design
The conceptual design includes the following components:
- Mechanical design: follows standard procedures, described in detail in (Palmer et al., 2004).
- Stability design: standard methods and software are used to perform stability calculations, offshore (Veritec, 1988) or onshore, though the offshore methods have been questioned. New guidelines for stability will be published in 2005 by Det Norske Veritas and will be designated DNV-RP-F109 On-Bottom Stability
- Protection against corrosion: a well-understood subject of which the application to CO_2 pipelines is described below.

- Trenching and backfilling: onshore lines are usually buried to depth of 1 m. Offshore lines are almost always buried in shallow water. In deeper water pipelines narrower than 400 mm are trenched and sometimes buried to protect them against damage by fishing gear.
- CO_2 pipelines may be more subject to longitudinal running fracture than hydrocarbon gas pipelines. Fracture arresters are installed at intervals of about 500 m.

West (1974) describes the design of the SACROC CO_2 pipeline (Section 4.2.2.1 above). The transportation options examined were:

(i)   a low-pressure CO_2 gas pipeline operating at a maximum pressure of 4.8 MPa;
(ii)   a high-pressure CO_2 gas pipeline operating at a minimum pressure of 9.6 MPa, so that the gas would remain in a dense phase state at all temperatures;
(iii)   a refrigerated liquid CO_2 pipeline;
(iv)   road tank trucks;
(v)   rail tankers, possibly in combination with road tank trucks.

The tank truck and rail options cost more than twice as much as a pipeline. The refrigerated pipeline was rejected because of cost and technical difficulties with liquefaction. The dense phase (Option ii) was 20% cheaper than a low-pressure CO_2 gas pipeline (Option i). The intermediate 4.8 to 9.6 MPa pressure range was avoided so that two-phase flow would not occur. An added advantage of dense-phase transport was that high delivery pressures were required for CO_2 injection.

The final design conforms to the ANSI B31.8 code for gas pipelines and to the DOT regulations applicable at the time. The main 290 km section is 406.4 mm (16 inch) outside diameter and 9.53 mm wall thickness made from grade X65 pipe (specified minimum yield stress of 448 MPa). A shorter 60 km section is 323.85 mm (12.75 inch) outside diameter, 8.74 mm wall thickness, grade X65. Tests showed that dry CO_2 would not corrode the pipeline steel; 304L corrosion-resistant alloy was used for short sections upstream of the glycol dehydrator. The line is buried to a minimum of 0.9 m, and any point on the line is within 16 km of a block valve.

There are six compressor stations, totalling 60 MW, including a station at the SACROC delivery point. The compressor stations are not equally spaced, and the longest distance between two stations is about 160 km. This is consistent with general practice, but some long pipelines have 400 km or more between compressor stations.

Significant nitrogen and oxygen components in CO_2 would shift the boundary of the two-phase region towards higher pressures, and would require a higher operating pressure to avoid two-phase flow.

4.2.4   Construction of land pipelines

Construction planning can begin either before or after rights of way are secured, but a decision to construct will not come before a legal right to construct a pipeline is secured and all governmental regulations met. Onshore and underwater CO_2 pipelines are constructed in the same way as hydrocarbon pipelines, and for both there is an established and well-understood base of engineering experience. Subsection 4.2.5 describes underwater construction.

The construction phases of a land pipeline are outlined below. Some of the operations can take place concurrently.

Environmental and social factors may influence the season of the year in which construction takes place. The land is cleared and the trench excavated. The longest lead items come first: urban areas, river and road crossings. Pipe is received into the pipe yard and welded into double joints (24 m long); transported to staging areas for placement along the pipe route, welded, tested, coated and wrapped, and then lowered into the trench. A hydrostatic test is carried out, and the line is dried. The trench is then backfilled, and the land and the vegetation restored.

4.2.5   *Underwater pipelines*

Most underwater pipelines are constructed by the lay-barge method, in which 12 or 24 m lengths of pipe are brought to a dynamically positioned or anchored barge, and welded one by one to the end of the pipeline. The barge moves slowly forward, and the pipeline leaves the barge over the stern, and passes first over a support structure ('stinger') and then down through the water in a suspended span, until it reaches the seabed. Some lines up to 450 mm diameter are constructed by the reel method, in which the pipeline is welded together onshore, wound onto a reel on a ship, and then unwound from the reel into its final position. Some short lines and lines for shore crossings in shallow water are constructed by various tow and pull methods, in which the line is welded together onshore and then pulled into its final location.

If the design requires that the pipeline be trenched, that is usually done after it has been laid on the seabed, by a jetting sled, a plough or a mechanical cutting device that is pulled along the line. On the other hand, in shore crossings and in very shallow water the trench is often excavated before the pipeline is laid, and that is done by dredgers, backhoes or draglines in soft sediments, or in rock by blasting followed by clamshell excavators. Many shore crossings are drilled horizontally from the shore; this procedure eliminates many uncertainties associated with the surf zone, and reduces the environmental impact of construction.

Underwater connections are made by various kinds of mechanical connection systems, by hyperbaric welding (in air under the local hydrostatic pressure) or by lifting the pipe ends above the surface, welding them together and lowering the connected line to the bottom.

These technologies are established and understood (Palmer and King, 2004). Underwater pipelines up to 1422 mm in diameter have been constructed in many different environments, and pipelines have been laid in depths up to 2200 m. Figure 4.2

plots the diameters and maximum depths of major deepwater pipelines constructed up to 2004. The difficulty of construction is roughly proportional to the depth multiplied by the diameter, and the maximum value of that product has multiplied fourfold since 1980. Still larger and deeper pipelines are technically feasible with today's technology.

4.2.6   *Operations*

Operational aspects of pipelines are divided into three areas: daily operations, maintenance, and health, safety and environment. Operations of a CO_2 pipeline in the USA, for instance, must follow federal operations guidelines (49 CFR 195). Overall operational considerations include training, inspections, safety integration, signs and pipeline markers, public education, damage prevention programmes, communication, facility security and leak detection. Pipelines outside the USA generally have similar regulatory operational requirements.

Personnel form a central part of operations and must be qualified. Personnel are required to be continuously trained and updated on safety procedures, including safety procedures that apply to contractors working on or near the pipeline, as well as to the public.

Operations include daily maintenance, scheduled planning and policies for inspecting, maintaining and repairing all equipment on the line and the pipeline itself, as well as supporting the line and pipeline. This equipment and support includes valves, compressors, pumps, tanks, rights of way, public signs and line markers as well as periodic pipeline flyovers.

Long-distance pipelines are instrumented at intervals so that the flow can be monitored. The monitoring points, compressor stations and block valves are tied back to a central operations centre. Computers control much of the operation, and manual intervention is necessary only in unusual upsets or emergency conditions. The system has inbuilt redundancies to prevent loss of operational capability if a component fails.

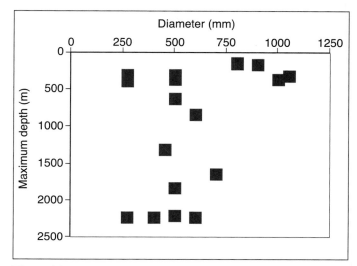

Figure 4.2 Pipelines in deep water.

Pipelines are cleaned and inspected by 'pigs', piston-like devices driven along the line by the gas pressure. Pigs have reached a high level of sophistication, and can measure internal corrosion, mechanical deformation, external corrosion, the precise position of the line, and the development of spans in underwater lines. Further functionality will develop as pig technology evolves, and there is no reason why pigs used for hydrocarbon pipelines should not be used for carbon dioxide.

Pipelines are also monitored externally. Land pipelines are inspected from the air, at intervals agreed between the operator and the regulatory authorities. Inspection from the air detects unauthorized excavation or construction before damage occurs. Currently, underwater pipelines are monitored by remotely operated vehicles, small unmanned submersibles that move along the line and make video records, and in the future, by autonomous underwater vehicles that do not need to be connected to a mother ship by a cable. Some pipelines have independent leak detection systems that find leaks acoustically or by measuring chemical releases, or by picking up pressure changes or small changes in mass balance. This technology is available and routine.

4.3   Ships for CO_2 transportation

4.3.1   Marine transportation system

Carbon dioxide is continuously captured at the plant on land, but the cycle of ship transport is discrete, and so a marine transportation system includes temporary storage on land and a loading facility. The capacity, service speed, number of ships and shipping schedule will be planned, taking into consideration, the capture rate of CO_2, transport distance, and social and technical restrictions. This issue is, of course, not specific to the case of CO_2 transport; CO_2 transportation by ship has a number of similarities to liquefied petroleum gas (LPG) transportation by ship.

What happens at the delivery point depends on the CO_2 storage system. If the delivery point is onshore, the CO_2 is unloaded from the ships into temporary storage tanks. If the delivery point is offshore – as in the ocean storage option – ships might unload to a platform, to a floating storage facility (similar to a floating production and storage facility routinely applied to offshore petroleum production), to a single-buoy mooring or directly to a storage system.

4.3.2   Existing experience

The use of ships for transporting CO_2 across the sea is today in an embryonic stage. Worldwide there are only four small ships used for this purpose. These ships transport liquefied food-grade CO_2 from large point sources of concentrated carbon dioxide such as ammonia plants in northern Europe to coastal distribution terminals in the consuming regions. From these distribution terminals CO_2 is transported to the customers either by tanker trucks or in pressurized cylinders. Design work is ongoing in Norway and Japan for larger CO_2 ships and their

associated liquefaction and intermediate storage facilities.

4.3.3   Design

For the design of hull and tank structure of liquid gas transport ships, such as LPG carriers and LNG carriers, the International Maritime Organization adopted the International Gas Carrier Code in order to prevent the significant secondary damage from accidental damage to ships. CO_2 tankers are designed and constructed under this code.

There are three types of tank structure for liquid gas transport ships: pressure type, low temperature type and semi-refrigerated type. The pressure type is designed to prevent the cargo gas from boiling under ambient air conditions. On the other hand, the low temperature type is designed to operate at a sufficiently low temperature to keep cargo gas as a liquid under the atmospheric pressure. Most small gas carriers are pressure type, and large LPG and LNG carriers are of the low temperature type. The low temperature type is suitable for mass transport because the tank size restriction is not severe. The semi-refrigerated type, including the existing CO_2 carriers, is designed taking into consideration the combined conditions of temperature and pressure necessary for cargo gas to be kept as a liquid. Some tankers such as semi-refrigerated LPG carriers are designed for applicability to the range of cargo conditions between normal temperature/high pressure and low temperature/atmospheric pressure.

Annex I to this report includes the CO_2 phase diagram. At atmospheric pressure, CO_2 is in gas or solid phase, depending on the temperature. Lowering the temperature at atmospheric pressure cannot by itself cause CO_2 to liquefy, but only to make so-called 'dry ice' or solid CO_2. Liquid CO_2 can only exist at a combination of low temperature and pressures well above atmospheric pressure. Hence, a CO_2 cargo tank should be of the pressure-type or semi-refrigerated. The semi-refrigerated type is preferred by ship designers, and the design point of the cargo tank would be around –54 °C per 6 bar to –50°C per 7 bar, which is near the point of CO_2. In a standard design, semi-refrigerated type LPG carriers operate at a design point of –50°C and 7 bar, when transporting a volume of 22,000 m³.

Carbon dioxide could leak into the atmosphere during transportation. The total loss to the atmosphere from ships is between 3 and 4% per 1000 km, counting both boil-off and exhaust from the ship's engines; both components could be reduced by capture and liquefaction, and recapture onshore would reduce the loss to 1 to 2% per 1000 km.

4.3.4   Construction

Carbon dioxide tankers are constructed using the same technology as existing liquefied gas carriers. The latest LNG carriers reach more than 200,000 m³ capacity. (Such a vessel could carry 230 kt of liquid CO_2.) The same type of yards that today build LPG and LNG ships can carry out the construction of a CO_2 tanker. The actual building time will be from one to two years, depending on considerations such as the ship's size.

4.3.5 Operation

4.3.5.1 Loading

Liquid CO_2 is charged from the temporary storage tank to the cargo tank with pumps adapted for high pressure and low temperature CO_2 service. The cargo tanks are first filled and pressurized with gaseous CO_2 to prevent contamination by humid air and the formation of dry ice.

4.3.5.2 Transport to the site

Heat transfer from the environment through the wall of the cargo tank will boil CO_2 and raise the pressure in the tank. It is not dangerous to discharge the CO_2 boil-off gas together with the exhaust gas from the ship's engines, but doing so does, of course, release CO_2 to the air. The objective of zero CO_2 emissions during the process of capture and storage can be achieved by using a refrigeration unit to capture and liquefy boil-off and exhaust CO_2.

4.3.5.3 Unloading

Liquid CO_2 is unloaded at the destination site. The volume occupied by liquid CO_2 in the cargo tanks is replaced with dry gaseous CO_2, so that humid air does not contaminate the tanks. This CO_2 could be recycled and reliquefied when the tank is refilled.

4.3.5.4 Return to port in ballast, and dry-docking

The CO_2 tanker will return to the port for the next voyage. When the CO_2 tanker is in dock for repair or regular inspection, gas CO_2 in cargo tank should be purged with air for safe working. For the first loading after docking, cargo tanks should be fully dried, purged and filled with CO_2 gas.

Ships of similar construction with a combination of cooling and pressure are currently operated for carrying other industrial gases.

4.4 Risk, safety and monitoring

4.4.1 Introduction

There are calculable and perceivable risks for any transportation option. We are not considering perceivable risks because this is beyond the scope of the document. Risks in special cases such as military conflicts and terrorist actions have now been investigated. At least two conferences on pipeline safety and security have taken place, and additional conferences and workshops are planned. However, it is unlikely that these will lead to peer-reviewed journal articles because of the sensitivity of the issue.

Pipelines and marine transportation systems have an established and good safety record. Comparison of CO_2 systems with these existing systems for long-distance pipeline transportation of gas and oil or with marine transportation of oil, yidds that risks should be comparable in terms of failure and accident rates. For the existing transport system these incidents seem to be perceived by the broad community as acceptable in spite of occasional serious pollution incidents such as the *Exxon Valdes* and *Torrey Canyon* disasters (van Bernem and Lubbe, 1997). Because the *consequences* of CO_2 pipeline accidents potentially are of significant concern, stricter regulations for CO_2 pipelines than those for natural gas pipelines currently are in force in the USA.

4.4.2 Land pipelines

Land pipelines are built to defined standards and are subject to regulatory approval. This sometimes includes independent design reviews. Their routes are frequently the subject of public inquiries. The process of securing regulatory approval generally includes approval of a safety plan, of detailed monitoring and inspection procedures and of emergency response plans. In densely populated areas the process of planning, licensing and building new pipelines may be difficult and time-consuming. In some places it may be possible to convert existing hydrocarbon pipelines into CO_2 pipelines.

Pipelines in operation are monitored internally by pigs (internal pipeline inspection devices) and externally by corrosion monitoring and leak detection systems. Monitoring is also done by patrols on foot and by aircraft.

The incidence of failure is relatively small. Guijt (2004) and the European Gas Pipeline Incident Data Group (2002) show that the incidence of failure has markedly decreased. Guijt quotes an incident rate of almost 0.0010 km^{-1} year^{-1} in 1972 falling to below 0.0002 km^{-1} year^{-1} in 2002. Most of the incidents refer to very small pipelines, less than 100 mm in diameter, principally applied to gas distribution systems. The failure incidence for 500 mm and larger pipelines is very much lower, below 0.00005 km^{-1} year^{-1}. These figures include all unintentional releases outside the limits of facilities (such as compressor stations) originating from pipelines whose design pressures are greater than 1.5 MPa. They cover many kinds of incidents, not all of them serious, and there is substantial variation between pipelines, reflecting factors such as system age and inspection frequency.

The corresponding incident figures for western European oil pipelines have been published by CONCAWE (2002). In 1997-2001 the incident frequency was 0.0003 km^{-1} yr^{-1}. The corresponding figure for US onshore gas pipelines was 0.00011 km^{-1} yr^{-1} for the 1986-2002 period, defining an incident as an event that released gas and caused death, inpatient hospitalization or property loss of US$ 50,000: this difference in reporting threshold is thought to account for the difference between European and US statistics (Guijt, 2004).

Lelieveld et al. (2005) examined leakage in 2400 km of the Russian natural gas pipeline system, including compressor stations, valves and machine halls, and concluded that '...overall, the leakage from Russian natural gas transport systems is about 1.4% (with a range of 1.0-2.5%), which is comparable with the amount lost from pipelines in the United States (1.5±0.5%)'. Those numbers refer to total leakage, not to leakage per kilometre.

Gale and Davison (2002) quote incident statistics for CO_2

pipelines in the USA. In the 1990-2002 period there were 10 incidents, with property damage totalling US$ 469,000, and no injuries nor fatalities. The incident rate was 0.00032 $km^{-1} yr^{-1}$. However, unlike oil and gas, CO_2 does not form flammable or explosive mixtures with air. Existing CO_2 pipelines are mainly in areas of low population density, which would also tend to result in lower average impacts. The reasons for the incidents at CO_2 pipelines were relief valve failure (4 failures), weld/gasket/valve packing failure (3), corrosion (2) and outside force (1). In contrast, the principal cause of incidents for natural gas pipelines is outside force, such as damage by excavator buckets. Penetration by excavators can lead to loss of pipeline fluid and sometimes to fractures that propagate great distances. Preventative measures such as increasing the depth of cover and use of concrete barriers above a pipeline and warning tape can greatly reduce the risk. For example, increasing cover from 1 m to 2 m reduces the damage frequency by a factor of 10 in rural areas and by 3.5 in suburban areas (Guijt, 2004).

Carbon dioxide leaking from a pipeline forms a potential physiological hazard for humans and animals. The consequences of CO_2 incidents can be modelled and assessed on a site-specific basis using standard industrial methods, taking into account local topography, meteorological conditions, population density and other local conditions. A study by Vendrig et al. (2003) has modelled the risks of CO_2 pipelines and booster stations. A property of CO_2 that needs to be considered when selecting a pipeline route is the fact that CO_2 is denser than air and can therefore accumulate to potentially dangerous concentrations in low lying areas. Any leak transfers CO_2 to the atmosphere.

If substantial quantities of impurities, particularly H_2S, are included in the CO_2, this could affect the potential impacts of a pipeline leak or rupture. The exposure threshold at which H_2S is immediately dangerous to life or health, according to the National Institute for Occupational Safety and Health, is 100 ppm, compared to 40,000 ppm for CO_2.

If CO_2 is transported for significant distances in densely populated regions, the number of people potentially exposed to risks from CO_2 transportation facilities may be greater than the number exposed to potential risks from CO_2 capture and storage facilities. Public concerns about CO_2 transportation may form a significant barrier to large-scale use of CCS. At present most electricity generation or other fuel conversion plants are built close to energy consumers or sources of fuel supply. New plants with CO_2 capture could be built close to CO_2 storage sites, to minimize CO_2 transportation. However, this may necessitate greater transportation of fuels or electricity, which have their own environmental impacts, potential risks and public concerns. A gathering system would be needed if CO_2 were brought from distributed sources to a trunk pipeline, and for some storage options a distribution system would also be needed: these systems would need to be planned and executed with the same regard for risk outlined here.

4.4.3   Marine pipelines

Marine pipelines are subject to a similar regulatory regime.

The incidence of failure in service is again low. Dragging ships' anchors causes some failures, but that only occurs in shallow water (less than 50 m). Very rarely do ships sink on to pipelines, or do objects fall on to them. Pipelines of 400 mm diameter and larger have been found to be safe from damage caused by fishing gear, but smaller pipelines are trenched to protect them. Damage to underwater pipelines was examined in detail at a conference reported on in Morris and Breaux (1995). Palmer and King (2004) examine case studies of marine pipeline failures, and the technologies of trenching and monitoring. Most failures result from human error. Ecological impacts from a CO_2 pipeline accident have yet to be assessed.

Marine pipelines are monitored internally by inspection devices called 'pigs' (as described earlier in Section 4.2.5), and externally by regular visual inspection from remotely operated vehicles. Some have independent leak detection systems.

4.4.4   Ships

Ship systems can fail in various ways: through collision, foundering, stranding and fire. Perrow's book on accidents (1984) includes many thought-provoking case studies. Many of the ships that he refers to were old, badly maintained and crewed by inadequately trained people. However, it is incorrect to think that marine accidents happen only to poorly regulated 'flag-of-convenience' ships. Gottschalch and Stadler (1990) share Perrow's opinion that many marine accidents can be attributed to system failures and human factors, whereas accidents arising as a consequence of purely technical factors are relatively uncommon.

Ship casualties are well summarized by Lloyds Maritime Information Service. Over 22.5 years between 1978 and 2000, there were 41,086 incidents of varying degrees of severity identified, of which 2,129 were classified as 'serious' (See Table 4.2).

Tankers can be seen to have higher standards than ships in general. Stranding is the source of most of the tanker incidents that have led to public concern. It can be controlled by careful navigation along prescribed routes, and by rigorous standards of operation. LNG tankers are potentially dangerous, but are carefully designed and appear to be operated to very high standards. There have been no accidental losses of cargo from LNG ships. The LNG tanker *El Paso Paul Kaiser* ran aground at 17 knots in 1979, and incurred substantial hull damage, but the LNG tanks were not penetrated and no cargo was lost. There is extensive literature on marine transport of liquefied gas, with a strong emphasis on safety, for example, in Ffooks (1993).

Carbon dioxide tankers and terminals are clearly much less at risk from fire, but there is an asphyxiation risk if collision should rupture a tank. This risk can be minimized by making certain that the high standards of construction and operation currently applied to LPG are also applied to carbon dioxide.

An accident to a liquid CO_2 tanker might release liquefied gas onto the surface of the sea. However, consideration of such an event is a knowledge gap that requires further study. CO_2 releases are anticipated not to have the long-term environmental

Table 4.2 Statistics of serious incidents, depending on the ship type.

Ship type	Number of ships 2000	Serious incidents 1978-2000	Frequency (incidents/ship year)
LPG tankers	982	20	0.00091
LNG tankers	121	1	0.00037
Oil tankers	9678	314	0.00144
Cargo/bulk carriers	21407	1203	0.00250

impacts of crude oil spills. CO_2 would behave differently from LNG, because liquid CO_2 in a tanker is not as cold as LNG but much denser. Its interactions with the sea would be complex: hydrates and ice might form, and temperature differences would induce strong currents. Some of the gas would dissolve in the sea, but some would be released to the atmosphere. If there were little wind and a temperature inversion, clouds of CO_2 gas might lead to asphyxiation and might stop the ship's engines.

The risk can be minimized by careful planning of routes, and by high standards of training and management.

4.5 Legal issues, codes and standards

Transportation of CO_2 by ships and sub-sea pipelines, and across national boundaries, is governed by various international legal conventions. Many jurisdictions/states have environmental impact assessment and strategic environmental assessment legislation that will come into consideration in pipeline building. If a pipeline is constructed across another country's territory (e.g. landlocked states), or if the pipeline is laid in certain zones of the sea, other countries may have the right to participate in the environmental assessment decision-making process or challenge another state's project.

4.5.1 International conventions

Various international conventions could have implications for storage of CO_2, the most significant being the UN Law of the Sea Convention, the London Convention, the Convention on Environmental Impact Assessment in a Transboundary Context (Espoo Convention) and OSPAR (see Chapter 5). The Espoo convention covers environmental assessment, a procedure that seeks to ensure the acquisition of adequate and early information on likely environmental consequences of development projects or activities, and on measures to mitigate harm. Pipelines are subject to environmental assessment. The most significant aspect of the Convention is that it lays down the general obligation of states to notify and consult each other if a project under consideration is likely to have a significant environmental impact across boundaries. In some cases the acceptability of CO_2 storage under these conventions could depend on the method of transportation to the storage site. Conventions that are primarily concerned with discharge and placement rather than transport are discussed in detail in the chapters on ocean and geological storage.

The Basel Convention on the Control of Transboundary

Movements of Hazardous Wastes and their Disposal came into force in 1992 (UNEP, 2000). The Basel Convention was conceived partly on the basis that enhanced control of transboundary movement of wastes will act as an incentive for their environmentally sound management and for the reduction of the volume of movement. However, there is no indication that CO_2 will be defined as a hazardous waste under the convention except in relation to the presence of impurities such as heavy metals and some organic compounds that may be entrained during the capture of CO_2. Adoption of schemes where emissions of SO_2 and NO_x would be included with the CO_2 may require such a review. Accordingly, the Basel Convention does not appear to directly impose any restriction on the transportation of CO_2 (IEA GHG, 2003a).

In addition to the provisions of the Basel Convention, any transport of CO_2 would have to comply with international transport regulations. There are numerous specific agreements, some of which are conventions and others protocols of other conventions that apply depending on the mode of transport. There are also a variety of regional agreements dealing with transport of goods. International transport codes and agreements adhere to the UN Recommendations on the Transport of Dangerous Goods: Model Regulations published by the United Nations (2001). CO_2 in gaseous and refrigerated liquid forms is classified as a non-flammable, non-toxic gas; while solid CO_2 (dry ice) is classified under the heading of miscellaneous dangerous substances and articles. Any transportation of CO_2 adhering to the Recommendations on the Transport of Dangerous Goods: Model Regulations can be expected to meet all relevant agreements and conventions covering transportation by whatever means. Nothing in these recommendations would imply that transportation of CO_2 would be prevented by international transport agreements and conventions (IEA GHG, 2003a).

4.5.2 National codes and standards

The transport of CO_2 by pipeline has been practiced for over 25 years. Internationally adopted standards such as ASME B31.4, Liquid transportation systems for hydrocarbons, liquid petroleum gas, anhydrous ammonia and alcohols' and the widely-applied Norwegian standard (DNV, 2000) specifically mention CO_2. There is considerable experience in the application and use of these standards. Existing standards and codes vary between different countries but gradual unification of these documents is being advanced by such international bodies as ISO and CEN as part of their function. A full review of relevant standards categorized by issues is presented in IEA GHG, 2003b.

Public concern could highlight the issue of leakage of CO_2 from transportation systems, either by rupture or minor leaks, as discussed in Section 4.4. It is possible that standards may be changed in future to address specific public concerns. Odorants are often added to domestic low-pressure gas distribution systems, but not to gas in long-distance pipelines; they could, in principle, be added to CO_2 in pipelines. Mercaptans, naturally present in the Weyburn pipeline system, are the most effective odorants but are not generally suitable for this application because they are degraded by O_2, even at very low concentrations (Katz, 1959). Disulphides, thioethers and ring compounds containing sulphur are alternatives. The value and impact of odorization could be established by a quantitative risk assessment.

4.6   Costs

4.6.1   *Costs of pipeline transport*

The costs of pipelines can be categorized into three items
*   Construction costs
    -   Material/equipment costs (pipe, pipe coating, cathodic protection, telecommunication equipment; possible booster stations)
    -   Installation costs (labour)
*   Operation and maintenance costs
    -   Monitoring costs
    -   Maintenance costs
    -   (Possible) energy costs
*   Other costs (design, project management, regulatory filing fees, insurances costs, right-of-way costs, contingencies allowances)

The pipeline material costs depend on the length of the pipeline, the diameter, the amount of CO_2 to be transported and the quality of the carbon dioxide. Corrosion issues are examined in Section 4.2.2 For costs it is assumed that CO_2 is delivered from the capture system at 10 MPa.

Figure 4.3 shows capital investment costs for pipelines. Investments are higher when compressor station(s) are required to compensate for pressure loss along the pipeline, or for longer pipelines or for hilly terrain. Compressor stations may be avoided by increasing the pipeline diameter and reducing the flow velocity. Reported transport velocity varies from 1 to 5 m s⁻¹. The actual design will be optimized with regard to pipeline diameter, pressure loss (required compressor stations and power) and pipeline wall thickness.

Costs depend on the terrain. Onshore pipeline costs may increase by 50 to 100% or more when the pipeline route is congested and heavily populated. Costs also increase in mountains, in nature reserve areas, in areas with obstacles such as rivers and freeways, and in heavily urbanized areas because of accessibility to construction and additional required safety measures. Offshore pipelines generally operate at higher

pressures and lower temperatures than onshore pipelines, and are often, but not always, 40 to 70% more expensive.

It is cheaper to collect CO_2 from several sources into a single pipeline than to transport smaller amounts separately. Early and smaller projects will face relatively high transport costs, and therefore be sensitive to transport distance, whereas an evolution towards higher capacities (large and wide-spread application) may result in a decrease in transport costs. Implementation of a 'backbone' transport structure may facilitate access to large remote storage reservoirs, but infrastructure of this kind will require large initial upfront investment decisions. Further study is required to determine the possible advantages of such pipeline system.

Figure 4.4 presents onshore and offshore transport costs versus pipeline diameter; where costs are based on investment cost information from various sources. Figure 4.5 gives a cost window for specific transport as function of the flow. Steel is a cost component for both pipelines and ships, and steel prices doubled in the two years up to 2005: this may be temporary.

4.6.2   *Costs of marine transportation systems*

Costs of a marine transport system comprise many cost elements. Besides investments for ships, investments are required for loading and unloading facilities, intermediate storage and liquefaction units. Further costs are for operation (e.g. labour, ship fuel costs, electricity costs, harbour fees), and maintenance. An optimal use of installations and ships in the transport cycle is crucial. Extra facilities (e.g. an expanded storage requirement) have to be created to be able to anticipate on possible disruptions in the transport system.

The cost of marine transport systems is not known in detail at present, since no system has been implemented on a scale required for CCS projects (i.e. in the range of several million tonnes of carbon dioxide handling per year). Designs have been submitted for tender, so a reasonable amount of knowledge is available. Nevertheless, cost estimates vary widely, because CO_2 shipping chains of this size have never been built and economies of scale may be anticipated to have a major impact on the costs.

A ship designed for carrying CO_2 from harbour to harbour may cost about 30-50% more than a similar size semi-refrigerated LPG ship (Statoil, 2004). However, since the density of liquid CO_2 is about 1100 kg m⁻³, CO_2 ships will carry more mass than an equivalent LNG or LPG ship, where the cargo density is about 500 kg m⁻³. The estimated cost of ships of 20 to 30 kt capacity is between 50 and 70 M$ (Statoil, 2004). Another source (IEA GHG, 2004) estimates ship construction costs at US$ 34 million for 10 kt-sized ship, US$ 60 million with a capacity of 30 kt, or US$ 85 million with a capacity of 50 kt. A time charter rate of about 25,000 US$ day⁻¹ covering capital charges, manning and maintenance is not unreasonable for a ship in the 20 kt carrying capacity range.

The cost for a liquefaction facility is estimated by Statoil (2004) at US$ 35 to US$ 50 million for a capacity of 1 Mt per year. The present largest liquefaction unit is 0.35 Mt yr⁻¹.

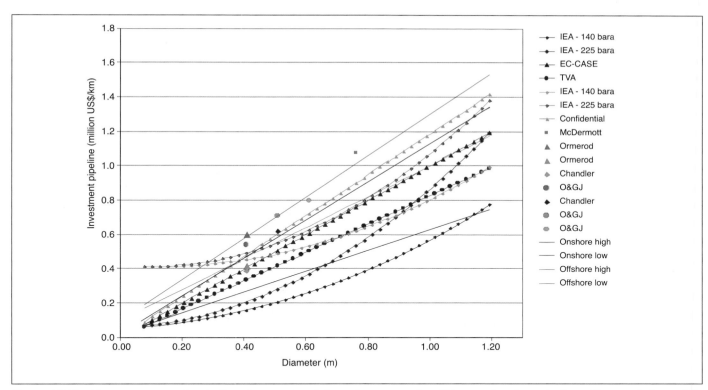

Figure 4.3 Total investment costs for pipelines from various information sources for offshore and onshore pipelines. Costs exclude possible booster stations (IEA GHG, 2002; Hendriks et al., 2005; Bock, 2003; Sarv, 2000; 2001a; 2001b; Ormerod, 1994; Chandler, 2000; O&GJ, 2000).

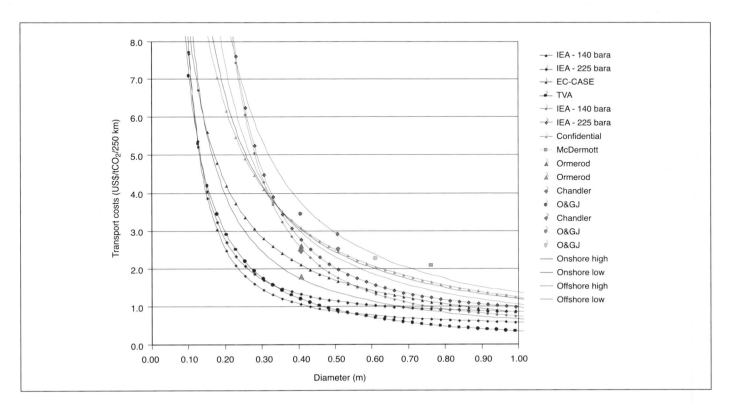

Figure 4.4 Transport costs derived from various information sources for offshore and onshore pipelines. Costs exclude possible booster stations, applying a capital charge rate of 15% and a load factor of 100% (IEA GHG, 2002; Hendriks et al., 2005; Bock, 2003; Sarv, 2000; 2001a; 2001b; Ormerod, 1994; Chandler, 2000; O&GJ, 2000)

IEA GHG (2004) estimates a considerable lower investment for a liquefaction facility, namely US$ 80 million for 6.2 Mt yr^{-1}. Investment costs are reduced to US$ 30 million when carbon dioxide at 100 bar is delivered to the plant. This pressure level is assumed to be delivered from the capture unit. Cost estimates are influenced by local conditions; for example, the absence of sufficient cooling water may call for a more expensive ammonia driven cooling cycle. The difference in numbers also reflects the uncertainty accompanied by scaling up of such facilities

A detailed study (Statoil, 2004) considered a marine transportation system for 5.5 Mt yr^{-1}. The base case had 20 kt tankers with a speed of 35 km h^{-1}, sailing 7600 km on each trip; 17 tankers were required. The annual cost was estimated at US$ 188 million, excluding linquefaction and US$ 300

million, including liquefaction, decreasing to US$ 232 million if compression is allowed (to avoid double counting). The corresponding specific transport costs are 34, 55, and 42 US$ t^{-1}. The study also considered sensitivity to distance: for the case excluding liquefaction, the specific costs were 20 US$ t^{-1} for 500 km, 22 US$ t^{-1} for 1500 km, and 28 US$ t^{-1} for 4500 km.

A study on a comparable ship transportation system carried out for the IEA shows lower costs. For a distance of 7600 km using 30 kt ships, the costs are estimated at 35 US$ t^{-1}. These costs are reduced to 30 US$ tonne^{-1} for 50 kt ships. The IEA study also showed a stronger cost dependency on distance than the Statoil (2004) study.

It should be noted that marine transport induces more associated CO_2 transport emissions than pipelines due to additional energy use for liquefaction and fuel use in ships. IEA GHG (2004) estimated 2.5% extra CO_2 emissions for a transport distance of 200 km and about 18% for 12,000 km. The extra CO_2 emissions for each 1000 km pipelines come to about 1 to 2%.

Ship transport becomes cost-competitive with pipeline transport over larger distances. Figure 4.6 shows an estimate of the costs for transporting 6 Mt yr^{-1} by offshore pipeline and by ship. The break-even distance, i.e. the distance for which the costs per transport mode are the same, is about 1000 km for this example. Transport of larger quantities will shift the break-even distance towards larger distances. However, the cross-over point beyond which ship transportation becomes cheaper than pipeline transportation is not simply a matter of distance alone. It involves many other factors, including loading terminals, pipeline shore crossings, water depth, seabed stability, fuel cost, construction costs, different operating costs in different locations, security, and interaction between land and marine transportation routes.

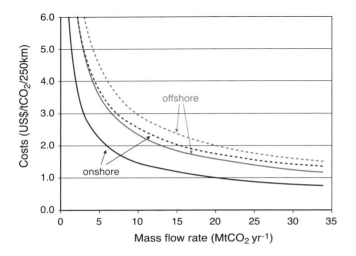

Figure 4.5 Transport costs for onshore and offshore pipelines per 250 km. High (broken lines) and low range (continuous lines) are indicated.

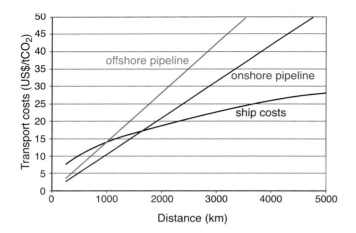

Figure 4.6 Costs, plotted as transportation cost in US$/tCO$_2$ against distance, for onshore and offshore pipelines, and ship transport. The costs include intermediate storage facilities, harbour fees, fuel costs and loading/unloading activities. Costs also include additional costs for liquefaction compared to compression. There is a capital charge factor of 11% for all transport options.

References

Bock, B.R., R. Rhudy, H. Herzog, M. Klett, J. Davison, D.G. de la Torre Ugarte, and D. Simbeck, 2003: Economic Evaluation of CO_2 Storage and Sink Enhancement Options. TVA Public Power Institute, February 2003.

Chandler, H.M. 2000: Heavy Construction Cost Data - 14th Annual Editions. R.S. Means Company, Inc. Kingston, MA, USA.

Concawe, 2002: Western European cross-country oil pipelines 30-year performance statistics, CONCAWE report.

European Gas Pipeline Incident Data Group, 2002: 5th EGIG report 1970-2001 Gas Pipeline Incidents, document EGIG 02.R.0058.

Ffooks, R., 1993: *Natural gas by sea, the development of a new technology.* Royal Institution of Naval Architects, London.

Gale, J. and J. Davison, 2002: Transmission of CO_2 - safety and economic considerations. GHGT-6.

Gottschalch, H. and M. Stadler, 1990: Seefahrtspsychologie (Psychology of navigation). Kasing, Bielefeld.

Guijt, W., 2004: Analyses of incident data show US, European pipelines becoming safer. *Oil and Gas Journal,* January 26, pp.68-73.

Hendriks, C.A., T. Wildenborg, P. Feron, and W. Graus, 2005: Capture and Storage, prepared for EC, DG-ENV, Ecofys Energy and Environment, report nr. M70066.

IEA GHG, 2002: Transmission of CO₂ and Energy, IEA Greenhouse Gas R&D Programme, Report PH4/6, IEA GHG, Cheltenham, UK (March).

IEA GHG, 2003a: Review of International Conventions having Implications for the Storage of Carbon Dioxide in the Ocean and Beneath the Seabed, Report PH4/16, IEA GHG, Cheltenham, UK, 1641 pp.

IEA GHG, 2003b: Barriers to Overcome in Implementation of CO₂ Capture and Storage (2): Rules and Standards for Transmission and Storage of CO₂, Report PH4/23, IEA GHG, Cheltenham, UK, 165 pp.

IEA GHG, 2004: Ship Transport of CO₂, Report PH4/30, IEA GHG, Cheltenham, UK, July 2004-11-16.

Katz, D.L., 1959: Handbook of natural gas engineering. McGraw-Hill, New York, 802 pp.

Lelieveld, J., S. Lechtenböhmer, S.S. Assonov, C.A.M. Brenninkmeijer, C. Dienst, M. Fischedick, and T. Hanke, 2005: Low methane leakage from gas pipelines. *Nature*, **434**, 841-842.

Morris, D. and K. Breaux, 1995: Proceedings of the International Workshop on Damage to Underwater Pipelines, New Orleans, LA, Minerals Management Service.

O&GJ, 2000: Pipeline Economics. *Oil and Gas Journal*, **98**(36), 68-86.

Ormerod, B., 1994: The disposal of carbon dioxide from fossil fuel fired power stations. IEA Greenhouse Gas R&D Programme, Cheltenham, Technical Rep. IEAGHG/SR3, June 1994.

Palmer, A.C. and R.A. King, 2004: Subsea pipeline engineering. Pennwell, Tulsa, OK.

Perrow, C., 1984: Normal accidents. Basic Books, 386 pp.

Rogers, G.F.C. and Y.R. Mayhew, 1980: Engineering thermodynamics and heat transfer. Longman, New York.

Sarv, H. and J. John, 2000: Deep ocean sequestration of captured CO₂. *Technology*, **7S**, 125-135.

Sarv, H., 2001a: Further Technological Evaluation of CO₂ Storage in Deep Oceans. Presented at the 26th International Technical Conference on Coal Utilisation & Fuel Systems, March 5-8, 2001, Clearwater, Florida.

Sarv, H., 2001b: Large-scale CO₂ transportation and deep ocean sequestration - Phase II final report. McDermott Technology Inc., Ohio. Technology Report DE-AC26-98FT40412, 2001.

Seiersten, M., 2001: Material selection for separation, transportation and disposal of CO₂. Proceedings Corrosion 2001, National Association of Corrosion Engineers, paper 01042.

Statoil, 2004: Written communication – O. Kaarstad, Trondheim, Norway, January.

UK Department of Trade and Industry, 2002: Carbon Capture and Storage, report of DTI International Technology Service Mission to the USA and Canada, Advanced Power Generation Technology Forum.

United Nations Environment Programme (UNEP), 2000: Text of the Basel Convention and Decisions of the Conference of the Parties (COP 1 to 5), United Nations Publications, Switzerland.

United Nations, 2001: Recommendations on the Transport of Dangerous Goods: Model Regulations, Twelfth Edition, United Nations Publications ST/SG/AC.10/Rev12, United Nations, New York and Geneva, 732 pp.

Van Bernem, C. and T. Lubbe, 1997: Öl im Meer (Oil in the sea) Wissenschaftliche Buchgesellschaft, Darmstadt.

Vendrig, M., J. Spouge, A. Bird, J. Daycock, and O. Johnsen, 2003: Risk analysis of the geological sequestration of carbon dioxide, Report no. R, Department of Trade and Industry, London, UK.

Veritec, 1988: On-bottom stability design of submarine pipelines. Recommended Practice E305.

West, J.M., 1974: Design and operation of a supercritical CO₂ pipeline-compression system, SACROC unit, Scurry County, Texas. Society of Petroleum Engineers Permian Basin Oil and Gas Recovery Conference, paper SPE 4804.

5

Underground geological storage

Coordinating Lead Authors
Sally Benson (United States), Peter Cook (Australia)

Lead Authors
Jason Anderson (United States), Stefan Bachu (Canada), Hassan Bashir Nimir (Sudan), Biswajit Basu (India), John Bradshaw (Australia), Gota Deguchi (Japan), John Gale (United Kingdom), Gabriela von Goerne (Germany), Wolfgang Heidug (Germany), Sam Holloway (United Kingdom), Rami Kamal (Saudi Arabia), David Keith (Canada), Philip Lloyd (South Africa), Paulo Rocha (Brazil), Bill Senior (United Kingdom), Jolyon Thomson (United Kingdom), Tore Torp (Norway), Ton Wildenborg (Netherlands), Malcolm Wilson (Canada), Francesco Zarlenga (Italy), Di Zhou (China)

Contributing Authors
Michael Celia (United States), Bill Gunter (Canada), Jonathan Ennis King (Australia), Erik Lindeberg (Norway), Salvatore Lombardi (Italy), Curt Oldenburg (United States), Karsten Pruess (United States) andy Rigg (Australia), Scott Stevens (United States), Elizabeth Wilson (United States), Steve Whittaker (Canada)

Review Editors
Günther Borm (Germany), David Hawkins (United States), Arthur Lee (United States)

Contents

EXECUTIVE SUMMARY

Underground accumulation of carbon dioxide (CO_2) is a widespread geological phenomenon, with natural trapping of CO_2 in underground reservoirs. Information and experience gained from the injection and/or storage of CO_2 from a large number of existing enhanced oil recovery (EOR) and acid gas projects, as well as from the Sleipner, Weyburn and In Salah projects, indicate that it is feasible to store CO_2 in geological formations as a CO_2 mitigation option. Industrial analogues, including underground natural gas storage projects around the world and acid gas injection projects, provide additional indications that CO_2 can be safely injected and stored at well-characterized and properly managed sites. While there are differences between natural accumulations and engineered storage, injecting CO_2 into deep geological formations at carefully selected sites can store it underground for long periods of time: it is considered likely that 99% or more of the injected CO_2 will be retained for 1000 years. Depleted oil and gas reservoirs, possibly coal formations and particularly saline formations (deep underground porous reservoir rocks saturated with brackish water or brine), can be used for storage of CO_2. At depths below about 800–1000 m, supercritical CO_2 has a liquid-like density that provides the potential for efficient utilization of underground storage space in the pores of sedimentary rocks. Carbon dioxide can remain trapped underground by virtue of a number of mechanisms, such as: trapping below an impermeable, confining layer (caprock); retention as an immobile phase trapped in the pore spaces of the storage formation; dissolution in the *in situ* formation fluids; and/or adsorption onto organic matter in coal and shale. Additionally, it may be trapped by reacting with the minerals in the storage formation and caprock to produce carbonate minerals. Models are available to predict what happens when CO_2 is injected underground. Also, by avoiding deteriorated wells or open fractures or faults, injected CO_2 will be retained for very long periods of time. Moreover, CO_2 becomes less mobile over time as a result of multiple trapping mechanisms, further lowering the prospect of leakage.

Injection of CO_2 in deep geological formations uses technologies that have been developed for and applied by, the oil and gas industry. Well-drilling technology, injection technology, computer simulation of storage reservoir dynamics and monitoring methods can potentially be adapted from existing applications to meet the needs of geological storage. Beyond conventional oil and gas technology, other successful underground injection practices – including natural gas storage, acid gas disposal and deep injection of liquid wastes – as well as the industry's extensive experience with subsurface disposal of oil-field brines, can provide useful information about designing programmes for long-term storage of CO_2. Geological storage of CO_2 is in practice today beneath the North Sea, where nearly 1 $MtCO_2$ has been successfully injected annually at Sleipner since 1996 and in Algeria at the In-Salah gas field. Carbon dioxide is also injected underground to recover oil. About 30 Mt of non-anthropogenic CO_2 are injected annually, mostly in west Texas, to recover oil from over 50 individual projects, some of which started in the early 1970s. The Weyburn Project

in Canada, where currently 1–2 $MtCO_2$ are injected annually, combines EOR with a comprehensive monitoring and modelling programme to evaluate CO_2 storage. Several more storage projects are under development at this time.

In areas with suitable hydrocarbon accumulations, CO_2-EOR may be implemented because of the added economic benefit of incremental oil production, which may offset some of the costs of CO_2 capture, transport and injection. Storage of CO_2 in coal beds, in conjunction with enhanced coal bed methane (ECBM) production, is potentially attractive because of the prospect of enhanced production of methane, the cleanest of the fossil fuels. This technology, however, is not well developed and a better understanding of injection and storage processes in coals is needed. Carbon dioxide storage in depleted oil and gas reservoirs is very promising in some areas, because these structures are well known and significant infrastructures are already in place. Nevertheless, relatively few hydrocarbon reservoirs are currently depleted or near depletion and CO_2 storage will have to be staged to fit the time of reservoir availability. Deep saline formations are believed to have by far the largest capacity for CO_2 storage and are much more widespread than other options.

While there are uncertainties, the global capacity to store CO_2 deep underground is large. Depleted oil and gas reservoirs are estimated to have a storage capacity of 675–900 $GtCO_2$. Deep saline formations are very likely to have a storage capacity of at least 1000 $GtCO_2$ and some studies suggest it may be an order of magnitude greater than this, but quantification of the upper range is difficult until additional studies are undertaken. Capacity of unminable coal formations is uncertain, with estimates ranging from as little as 3 $GtCO_2$ up to 200 $GtCO_2$. Potential storage sites are likely to be broadly distributed in many of the world's sedimentary basins, located in the same region as many of the world's emission sources and are likely to be adequate to store a significant proportion of those emissions well into the future.

The cost of geological storage of CO_2 is highly site-specific, depending on factors such as the depth of the storage formation, the number of wells needed for injection and whether the project is onshore or offshore – but costs for storage, including monitoring, appear to lie in the range of 0.6–8.3 US$/$tCO_2$ stored. This cost is small compared to present-day costs of CO_2 capture from flue gases, as indicated in Chapter 3. EOR could lead to negative storage costs of 10–16 US$/$tCO_2$ for oil prices of 15–20 US$ per barrel and more for higher oil prices.

Potential risks to humans and ecosystems from geological storage may arise from leaking injection wells, abandoned wells, leakage across faults and ineffective confining layers. Leakage of CO_2 could potentially degrade the quality of groundwater, damage some hydrocarbon or mineral resources, and have lethal effects on plants and sub-soil animals. Release of CO_2 back into the atmosphere could also create local health and safety concerns. Avoiding or mitigating these impacts will require careful site selection, effective regulatory oversight, an appropriate monitoring programme that provides

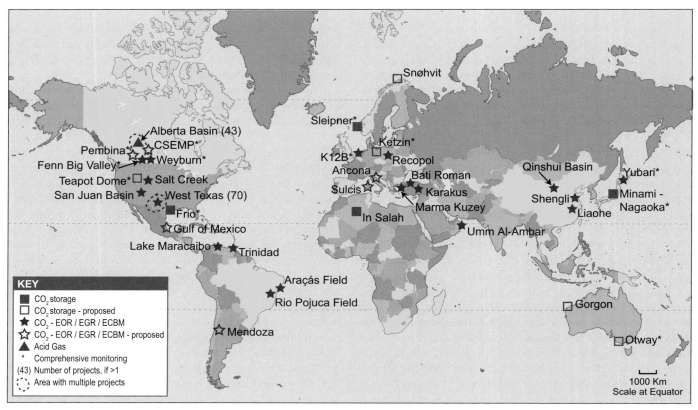

Figure 5.1 Location of sites where activities relevant to CO_2 storage are planned or under way.

early warning that the storage site is not functioning as anticipated and implementation of remediation methods to stop or control CO_2 releases. Methods to accomplish these are being developed and tested.

There are few, if any, national regulations specifically dealing with CO_2 storage, but regulations dealing with oil and gas, groundwater and the underground injection of fluids can in many cases be readily adapted and/or adopted. However, there are no regulations relating specifically to long-term responsibility for storage. A number of international laws that predate any consideration of CO_2 storage are relevant to offshore geological storage; consideration of whether these laws do or do not permit offshore geological storage is under way.

There are gaps in our knowledge, such as regional storage-capacity estimates for many parts of the world. Similarly, better estimation of leakage rates, improved cost data, better intervention and remediation options, more pilot and demonstration projects and clarity on the issue of long-term stewardship all require consideration. Despite the fact that more work is needed to improve technologies and decrease uncertainty, there appear to be no insurmountable technical barriers to an increased uptake of geological storage as an effective mitigation option.

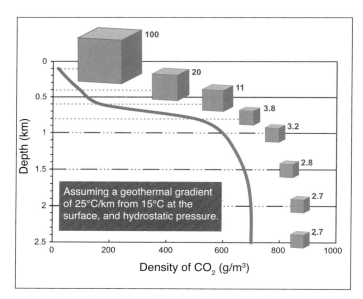

Figure 5.2 Variation of CO_2 density with depth, assuming hydrostatic pressure and a geothermal gradient of 25°C km^{-1} from 15°C at the surface (based on the density data of Angus *et al.*, 1973). Carbon dioxide density increases rapidly at approximately 800 m depth, when the CO_2 reaches a supercritical state. Cubes represent the relative volume occupied by the CO_2 and down to 800 m, this volume can be seen to dramatically decrease with depth. At depths below 1.5 km, the density and specific volume become nearly constant.

5.1 Introduction

5.1.1 What is geological storage?

Capture and geological storage of CO_2 provide a way to avoid emitting CO_2 into the atmosphere, by capturing CO_2 from major stationary sources (Chapter 3), transporting it usually by pipeline (Chapter 4) and injecting it into suitable deep rock formations. This chapter explores the nature of geological storage and considers its potential as a mitigation option.

The subsurface is the Earth's largest carbon reservoir, where the vast majority of the world's carbon is held in coals, oil, gas organic-rich shales and carbonate rocks. Geological storage of CO_2 has been a natural process in the Earth's upper crust for hundreds of millions of years. Carbon dioxide derived from biological activity, igneous activity and chemical reactions between rocks and fluids accumulates in the natural subsurface environment as carbonate minerals, in solution or in a gaseous or supercritical form, either as a gas mixture or as pure CO_2. The engineered injection of CO_2 into subsurface geological formations was first undertaken in Texas, USA, in the early 1970s, as part of enhanced oil recovery (EOR) projects and has been ongoing there and at many other locations ever since.

Geological storage of anthropogenic CO_2 as a greenhouse gas mitigation option was first proposed in the 1970s, but little research was done until the early 1990s, when the idea gained credibility through the work of individuals and research groups (Marchetti, 1977; Baes *et al.*, 1980; Kaarstad, 1992; Koide *et al.*, 1992; van der Meer, 1992; Gunter *et al.*, 1993; Holloway and Savage, 1993; Bachu *et al.*, 1994; Korbol and Kaddour, 1994). The subsurface disposal of acid gas (a by-product of petroleum production with a CO_2 content of up to 98%) in the Alberta Basin of Canada and in the United States provides additional useful experience. In 1996, the world's first large-scale storage project was initiated by Statoil and its partners at the Sleipner Gas Field in the North Sea.

By the late 1990s, a number of publicly and privately funded research programmes were under way in the United States, Canada, Japan, Europe and Australia. Throughout this time, though less publicly, a number of oil companies became increasingly interested in geological storage as a mitigation option, particularly for gas fields with a high natural CO_2 content such as Natuna in Indonesia, In Salah in Algeria and Gorgon in Australia. More recently, coal mining companies and electricity-generation companies have started to investigate geological storage as a mitigation option of relevance to their industry.

In a little over a decade, geological storage of CO_2 has

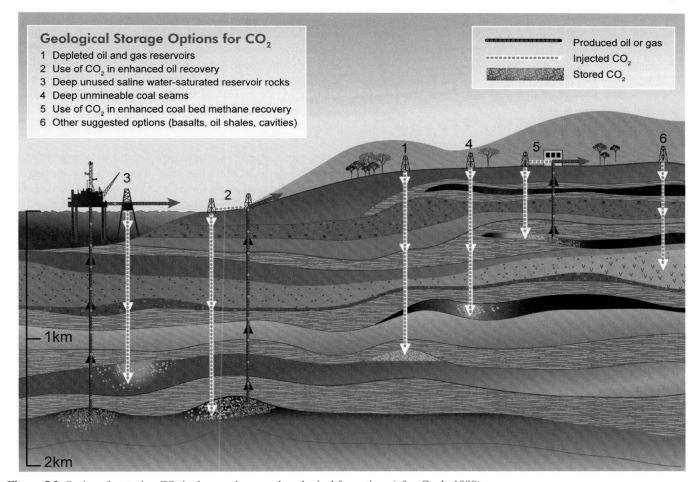

Figure 5.3 Options for storing CO_2 in deep underground geological formations (after Cook, 1999).

grown from a concept of limited interest to one that is quite widely regarded as a potentially important mitigation option (Figure 5.1). There are several reasons for this. First, as research has progressed and as demonstration and commercial projects have been successfully undertaken, the level of confidence in the technology has increased. Second, there is consensus that a broad portfolio of mitigation options is needed. Third, geological storage (in conjunction with CO_2 capture) could help to make deep cuts to atmospheric CO_2 emissions. However, if that potential is to be realized, the technique must be safe, environmentally sustainable, cost-effective and capable of being broadly applied. This chapter explores these issues.

To geologically store CO_2, it must first be compressed, usually to a dense fluid state known as 'supercritical' (see Glossary). Depending on the rate that temperature increases with depth (the geothermal gradient), the density of CO_2 will increase with depth, until at about 800 m or greater, the injected CO_2 will be in a dense supercritical state (Figure 5.2).

Geological storage of CO_2 can be undertaken in a variety of geological settings in sedimentary basins. Within these basins, oil fields, depleted gas fields, deep coal seams and saline formations are all possible storage formations (Figure 5.3).

Subsurface geological storage is possible both onshore and offshore, with offshore sites accessed through pipelines from the shore or from offshore platforms. The continental shelf and some adjacent deep-marine sedimentary basins are potential offshore storage sites, but the majority of sediments of the abyssal deep ocean floor are too thin and impermeable to be suitable for geological storage (Cook and Carleton, 2000). In addition to storage in sedimentary formations, some consideration has been given to storage in caverns, basalt and organic-rich shales (Section 5.3.5).

Fluids have been injected on a massive scale into the deep subsurface for many years to dispose of unwanted chemicals, pollutants or by-products of petroleum production, to enhance the production of oil and gas or to recharge depleted formations (Wilson *et al.*, 2003). The principles involved in such activities are well established and in most countries there are regulations governing these activities. Natural gas has also been injected and stored in the subsurface on a large scale in many parts of the world for many years. Injection of CO_2 to date has been done at a relatively small scale, but if it were to be used to significantly decrease emissions from existing stationary sources, then the injection rates would have to be at a scale similar to other injection operations under way at present.

But what is the world's geological storage capacity and does it occur where we need it? These questions were first raised in Chapter 2, but Section 5.3.8 of this chapter considers geographical matching of CO_2 sources to geological storage sites in detail. Not all sedimentary basins are suitable for CO_2 storage; some are too shallow and others are dominated by rocks with low permeability or poor confining characteristics. Basins suitable for CO_2 storage have characteristics such as thick accumulations of sediments, permeable rock formations saturated with saline water (saline formations), extensive covers of low porosity rocks (acting as seals) and structural simplicity.

While many basins show such features, many others do not.

Is there likely to be sufficient storage capacity to meet the world's needs in the years ahead? To consider this issue, it is useful to draw parallels with the terms 'resources' and 'reserves' used for mineral deposits (McKelvey, 1972). Deposits of minerals or fossil fuels are often cited with very large resource figures, but the 'proven' reserve is only some fraction of the resource. The resource figures are based on the selling price of the commodity, the cost of exploiting the commodity, the availability of appropriate technologies, proof that the commodity exists and whether the environmental or social impact of exploiting the commodity is acceptable to the community. Similarly, to turn technical geological storage capacity into economical storage capacity, the storage project must be economically viable, technically feasible, safe, environmentally and socially sustainable and acceptable to the community. Given these constraints, it is inevitable that the storage capacity that will actually be used will be significantly less than the technical potential. Section 5.3 explores this issue. It is likely that usable storage capacity will exist in many areas where people live and where CO_2 is generated from large stationary sources. This geographical congruence of storage-need and storage-capacity should not come as a surprise, because much of the world's population is concentrated in regions underlain by sedimentary basins (Gunter *et al.*, 2004).

It is also important to know how securely and for how long stored CO_2 will be retained – for decades, centuries, millennia or for geological time? To assure public safety, storage sites must be designed and operated to minimize the possibility of leakage. Consequently, potential leakage pathways must be identified and procedures must be established, to set appropriate design and operational standards as well as monitoring, measurement and verification requirements. Sections 5.4, 5.6 and 5.7 consider these issues.

In this chapter, we primarily consider storage of pure or nearly pure, CO_2. It has been suggested that it may be economically favourable to co-store CO_2 along with H_2S, SO_2 or NO_2. Since only a few scientific studies have evaluated the impacts of these added constituents on storage performance or risks, they are not addressed comprehensively here. Moreover, the limited information gained from practical experience with acid gas injection in Canada is insufficient to assess the impacts of the added components on storage security.

5.1.2   *Existing and planned CO_2 projects*

A number of pilot and commercial CO_2 storage projects are under way or proposed (Figure 5.1). To date, most actual or planned commercial projects are associated with major gas production facilities that have gas streams containing CO_2 in the range of 10–15% by volume, such as Sleipner in the North Sea, Snohvit in the Barents Sea, In Salah in Algeria and Gorgon in Australia (Figure 5.1), as well as the acid gas injection projects in Canada and the United States. At the Sleipner Project, operated by Statoil, more than 7 $MtCO_2$ has been injected into a deep sub-sea saline formation since 1996 (Box 5.1). Existing and planned

Table 5.1 A selection of current and planned geological storage projects.

Project	Country	Scale of Project	Lead organizations	Injection start date	Approximate average daily injection rate	Total storage	Storage type	Geological storage formation	Age of formation	Lithology	Monitoring
Sleipner	Norway	Commercial	Statoil, IEA	1996	3000 t day^{-1}	20 Mt planned	Aquifer	Utsira Formation	Tertiary	Sandstone	4D seismic plus gravity
Weyburn	Canada	Commercial	EnCana, IEA	May 2000	3-5000 t day^{-1}	20 Mt planned	CO_2-EOR	Midale Formation	Mississippian	Carbonate	Comprehensive
Minami-Nagaoka	Japan	Demo	Research Institute of Innovative Technology for the Earth	2002	Max 40 t day^{-1}	10,000 t planned	Aquifer (Sth. Nagaoka Gas Field)	Haizume Formation	Pleistocene	Sandstone	Crosswell seismic + well monitoring
Yubari	Japan	Demo	Japanese Ministry of Economy, Trade and Industry	2004	10 t day^{-1}	200 t Planned	CO_2-ECBM	Yubari Formation (Ishikari Coal Basin)	Tertiary	Coal	Comprehensive
In Salah	Algeria	Commercial	Sonatrach, BP, Statoil	2004	3-4000 t day^{-1}	17 Mt planned	Depleted hydrocarbon reservoirs	Krechba Formation	Carboniferous	Sandstone	Planned comprehensive
Frio	USA	Pilot	Bureau of Economic Geology of the University of Texas	4-13 Oct. 2004	Approx. 177 t day^{-1} for 9 days	1600t	Saline formation	Frio Formation	Tertiary	Brine-bearing sandstone-shale	Comprehensive
K12B	Netherlands	Demo	Gaz de France	2004	100-1000 t day^{-1} (2006+)	Approx 8 Mt	EGR	Rotleigendes	Permian	Sandstone	Comprehensive
Fenn Big Valley	Canada	Pilot	Alberta Research Council	1998	50 t day^{-1}	200 t	CO_2-ECBM	Mannville Group	Cretaceous	Coal	P, T, flow
Recopol	Poland	Pilot	TNO-NITG (Netherlands)	2003	1 t day^{-1}	10 t	CO_2-ECBM	Silesian Basin	Carboniferous	Coal	
Qinshui Basin	China	Pilot	Alberta Research Council	2003	30 t day^{-1}	150 t	CO_2-ECBM	Shanxi Formation	Carboniferous-Permian	Coal	P, T, flow
Salt Creek	USA	Commercial	Anadarko	2004	5-6000 t day^{-1}	27 Mt	CO_2-EOR	Frontier	Cretaceous	Sandstone	Under development
Planned Projects (2005 onwards)											
Snøhvit	Norway	Decided Commercial	Statoil	2006	2000 t day^{-1}		Saline formation	Tubaen Formation	Lower Jurassic	Sandstone	Under development
Gorgon	Australia	Planned Commercial	Chevron	Planned 2009	Approx. 10,000 t day^{-1}		Saline formation	Dupuy Formation	Late Jurassic	Massive sandstone with shale seal	Under development
Ketzin	Germany	Demo	GFZ Potsdam	2006	100 t day^{-1}	60 kt	Saline formation	Stuttgart Formation	Triassic	Sandstone	Comprehensive
Otway	Australia	Pilot	CO2CRC	Planned late 2005	160 t day^{-1} for 2 years	0.1 Mt	Saline fm and depleted gas field	Waarre Formation	Cretaceous	Sandstone	Comprehensive
Teapot Dome	USA	Proposed Demo	RMOTC	Proposed 2006	170 t day^{-1} for 3 months	10 kt	Saline fm and CO_2-EOR	Tensleep and Red Peak Fm	Permian	Sandstone	Comprehensive
CSEMP	Canada	Pilot	Suncor Energy	2005	50 t day^{-1}	10 kt	CO_2-ECBM	Ardley Fm	Tertiary	Coal	Comprehensive
Pembina	Canada	Pilot	Penn West	2005	50 t day^{-1}	50 kt	CO_2-EOR	Cardium Fm	Cretaceous	Sandstone	Comprehensive

storage projects are also listed in Table 5.1.

At the In Salah Gas Field in Algeria, Sonatrack, BP and Statoil inject CO_2 stripped from natural gas into the gas reservoir outside the boundaries of the gas field (Box 5.2). Statoil is planning another project in the Barents Sea, where CO_2 from the Snohvit field will be stripped from the gas and injected into a geological formation below the gas field. Chevron is proposing to produce gas from the Gorgon field off Western Australia, containing approximately 14% CO_2. The CO_2 will be injected into the Dupuy Formation at Barrow Island (Oen, 2003). In The Netherlands, CO_2 is being injected at pilot scale into the almost depleted K12-B offshore gas field (van der Meer *et al.*, 2005).

Forty-four CO_2-rich acid gas injection projects are currently operating in Western Canada, ongoing since the early 1990s (Bachu and Haug, 2005). Although they are mostly small scale, they provide important examples of effectively managing injection of CO_2 and hazardous gases such as H_2S (Section 5.2.4.2).

Box 5.1 The Sleipner Project, North Sea.

The Sleipner Project, operated by Statoil in the North Sea about 250 km off the coast of Norway, is the first commercial-scale project dedicated to geological CO_2 storage in a saline formation. The CO_2 (about 9%) from Sleipner West Gas Field is separated, then injected into a large, deep, saline formation 800 m below the seabed of the North Sea. The Saline Aquifer CO_2 Storage (SACS) project was established to monitor and research the storage of CO_2. From 1995, the IEA Greenhouse Gas R&D Programme has worked with Statoil to arrange the monitoring and research activities. Approximately 1 $MtCO_2$ is removed from the produced natural gas and injected underground annually in the field. The CO_2 injection operation started in October 1996 and, by early 2005, more than 7 $MtCO_2$ had been injected at a rate of approximately 2700 t day^{-1}. Over the lifetime of the project, a total of 20 $MtCO_2$ is expected to be stored. A simplified diagram of the Sleipner scheme is given in Figure 5.4.

The saline formation into which the CO_2 is injected is a brine-saturated unconsolidated sandstone about 800–1000 m below the sea floor. The formation also contains secondary thin shale layers, which influence the internal movement of injected CO_2. The saline formation has a very large storage capacity, on the order of 1–10 $GtCO_2$. The top of the formation is fairly flat on a regional scale, although it contains numerous small, low-amplitude closures. The overlying primary seal is an extensive, thick, shale layer.

This project is being carried out in three phases. Phase-0 involved baseline data gathering and evaluation, which was completed in November 1998. Phase-1 involved establishment of project status after three years of CO_2 injection. Five main project areas involve descriptions of reservoir geology, reservoir simulation, geochemistry, assessment of need and cost for monitoring wells and geophysical modelling. Phase-2, involving data interpretation and model verification, began in April 2000.

The fate and transport of the CO_2 plume in the storage formation has been monitored successfully by seismic time-lapse surveys (Figure 5.16). The surveys also show that the caprock is an effective seal that prevents CO_2 migration out of the storage formation. Today, the footprint of the plume at Sleipner extends over an area of approximately 5 km^2. Reservoir studies and simulations covering hundreds to thousands of years have shown that CO_2 will eventually dissolve in the pore water, which will become heavier and sink, thus minimizing the potential for long-term leakage (Lindeberg and Bergmo, 2003).

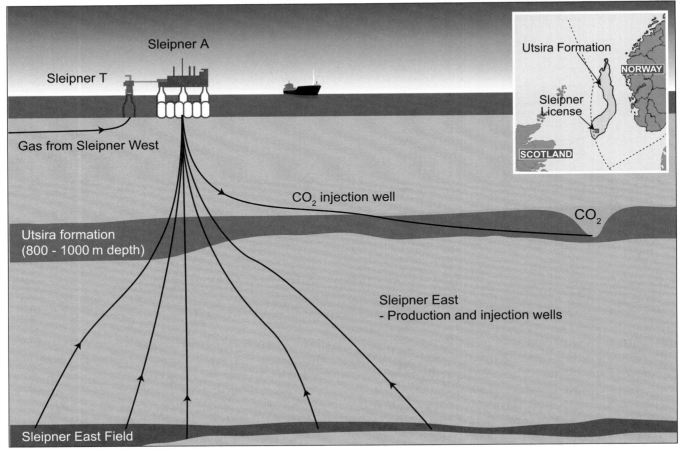

Figure 5.4 Simplified diagram of the Sleipner CO_2 Storage Project. Inset: location and extent of the Utsira formation.

Box 5.2 The In Salah, Algeria, CO_2 Storage Project.

The In Salah Gas Project, a joint venture among Sonatrach, BP and Statoil located in the central Saharan region of Algeria, is the world's first large-scale CO_2 storage project in a gas reservoir (Riddiford *et al.*, 2003). The Krechba Field at In Salah produces natural gas containing up to 10% CO_2 from several geological reservoirs and delivers it to markets in Europe, after processing and stripping the CO_2 to meet commercial specifications. The project involves re-injecting the CO_2 into a sandstone reservoir at a depth of 1800 m and storing up to 1.2 $MtCO_2$ yr^{-1}. Carbon dioxide injection started in April 2004 and, over the life of the project, it is estimated that 17 $MtCO_2$ will be geologically stored. The project consists of four production and three injection wells (Figure 5.5). Long-reach (up to 1.5 km) horizontal wells are used to inject CO_2 into the 5-mD permeability reservoir.

The Krechba Field is a relatively simple anticline. Carbon dioxide injection takes place down-dip from the gas/water contact in the gas-bearing reservoir. The injected CO_2 is expected to eventually migrate into the area of the current gas field after depletion of the gas zone. The field has been mapped with three-dimensional seismic and well data from the field. Deep faults have been mapped, but at shallower levels, the structure is unfaulted. The storage target in the reservoir interval therefore carries minimal structural uncertainty or risk. The top seal is a thick succession of mudstones up to 950 m thick.

A preliminary risk assessment of CO_2 storage integrity has been carried out and baseline data acquired. Processes that could result in CO_2 migration from the injection interval have been quantified and a monitoring programme is planned involving a range of technologies, including noble gas tracers, pressure surveys, tomography, gravity baseline studies, microbiological studies, four-dimensional seismic and geomechanical monitoring.

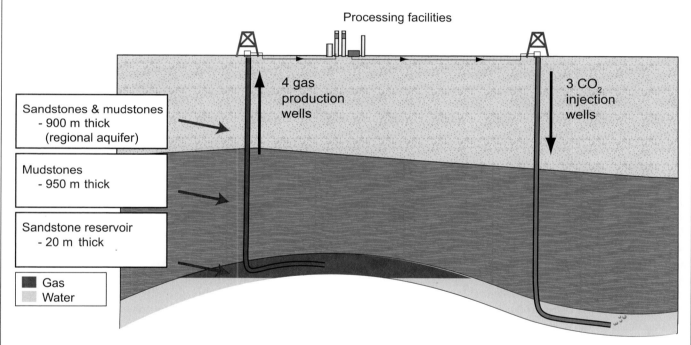

Figure 5.5 Schematic of the In Salah Gas Project, Algeria. One $MtCO_2$ will be stored annually in the gas reservoir. Long-reach horizontal wells with slotted intervals of up to 1.5 km are used to inject CO_2 into the water-filled parts of the gas reservoir.

Opportunities for enhanced oil recovery (EOR) have increased interest in CO_2 storage (Stevens *et al.*, 2001b; Moberg *et al.*, 2003; Moritis, 2003; Riddiford *et al.*, 2003; Torp and Gale, 2003). Although not designed for CO_2 storage, CO_2-EOR projects can demonstrate associated storage of CO_2, although lack of comprehensive monitoring of EOR projects (other than at the International Energy Agency Greenhouse Gas (IEA-GHG) Weyburn Project in Canada) makes it difficult to quantify storage. In the United States, approximately 73 CO_2-EOR operations inject up to 30 $MtCO_2$ yr^{-1}, most of which comes from natural CO_2 accumulations – although approximately 3

$MtCO_2$ is from anthropogenic sources, such as gas processing and fertiliser plants (Stevens *et al.*, 2001b). The SACROC project in Texas was the first large-scale commercial CO_2-EOR project in the world. It used anthropogenic CO_2 during the period 1972 to 1995. The Rangely Weber project (Box 5.6) injects anthropogenic CO_2 from a gas-processing plant in Wyoming.

In Canada, a CO_2-EOR project has been established by EnCana at the Weyburn Oil Field in southern Saskatchewan (Box 5.3). The project is expected to inject 23 $MtCO_2$ and extend the life of the oil field by 25 years (Moberg *et al.*,

Box 5.3 The Weyburn CO_2-EOR Project.

The Weyburn CO_2-enhanced oil recovery (CO_2-EOR) project is located in the Williston Basin, a geological structure extending from south-central Canada into north-central United States. The project aims to permanently store almost all of the injected CO_2 by eliminating the CO_2 that would normally be released during the end of the field life.

The source of the CO_2 for the Weyburn CO_2-EOR Project is the Dakota Gasification Company facility, located approximately 325 km south of Weyburn, in Beulah, North Dakota, USA. At the plant, coal is gasified to make synthetic gas (methane), with a relatively pure stream of CO_2 as a by-product. This CO_2 stream is dehydrated, compressed and piped to Weyburn in southeastern Saskatchewan, Canada, for use in the field. The Weyburn CO_2-EOR Project is designed to take CO_2 from the pipeline for about 15 years, with delivered volumes dropping from 5000 to about 3000 t day^{-1} over the life of the project.

The Weyburn field covers an area of 180 km^2, with original oil in place on the order of 222 million m^3 (1396 million barrels). Over the life of the CO_2-EOR project (20–25 years), it is expected that some 20 MtCO_2 will be stored in the field, under current economic conditions and oil recovery technology. The oil field layout and operation is relatively conventional for oil field operations. The field has been designed with a combination of vertical and horizontal wells to optimize the sweep efficiency of the CO_2. In all cases, production and injection strings are used within the wells to protect the integrity of the casing of the well.

The oil reservoir is a fractured carbonate, 20–27 m thick. The primary upper seal for the reservoir is an anhydrite zone. At the northern limit of the reservoir, the carbonate thins against a regional unconformity. The basal seal is also anhydrite, but is less consistent across the area of the reservoir. A thick, flat-lying shale above the unconformity forms a good regional barrier to leakage from the reservoir. In addition, several high-permeability formations containing saline groundwater would form good conduits for lateral migration of any CO_2 that might reach these zones, with rapid dissolution of the CO_2 in the formation fluids.

Since CO_2 injection began in late 2000, the EOR project has performed largely as predicted. Currently, some 1600 m^3 (10,063 barrels) day^{-1} of incremental oil is being produced from the field. All produced CO_2 is captured and recompressed for reinjection into the production zone. Currently, some 1000 tCO_2 day^{-1} is reinjected; this will increase as the project matures. Monitoring is extensive, with high-resolution seismic surveys and surface monitoring to determine any potential leakage. Surface monitoring includes sampling and analysis of potable groundwater, as well as soil gas sampling and analysis (Moberg *et al.*, 2003). To date, there has been no indication of CO_2 leakage to the surface and near-surface environment (White, 2005; Strutt *et al.*, 2003).

2003; Law, 2005). The fate of the injected CO_2 is being closely monitored through the IEA GHG Weyburn Project (Wilson and Monea, 2005). Carbon dioxide-EOR is under consideration for the North Sea, although there is as yet little, if any, operational experience for offshore CO_2-EOR. Carbon dioxide-EOR projects are also currently under way in a number of countries including Trinidad, Turkey and Brazil (Moritis, 2002). Saudi Aramco, the world's largest producer and exporter of crude oil, is evaluating the technical feasibility of CO_2-EOR in some of its Saudi Arabian reservoirs.

In addition to these commercial storage or EOR projects, a number of pilot storage projects are under way or planned. The Frio Brine Project in Texas, USA, involved injection and storage of 1900 tCO_2 in a highly permeable formation with a regionally extensive shale seal (Hovorka *et al.*, 2005). Pilot projects are proposed for Ketzin, west of Berlin, Germany, for the Otway Basin of southeast Australia and for Teapot Dome, Wyoming, USA (Figure 5.1). The American FutureGen project, proposed for late this decade, will be a geological storage project linked to coal-fired electricity generation. A small-scale CO_2 injection and monitoring project is being carried out by RITE at Nagoaka in northwest Honshu, Japan. Small-scale injection projects to test CO_2 storage in coal have been carried out in Europe (RECOPOL) and Japan (Yamaguchi *et al.*, 2005). A CO_2-enhanced coal bed methane (ECBM) recovery

demonstration project has been undertaken in the northern San Juan Basin of New Mexico, USA (Reeves, 2003a) (Box 5.7). Further CO_2-ECBM projects are under consideration for China, Canada, Italy and Poland (Gale, 2003). In all, some 59 opportunities for CO_2-ECBM have been identified worldwide, the majority in China (van Bergen *et al.*, 2003a).

These projects (Figure 5.1; Table 5.1) demonstrate that subsurface injection of CO_2 is not for the distant future, but is being implemented now for environmental and/or commercial reasons.

5.1.3　Key questions

In the previous section, the point is made that deep injection of CO_2 is under way in a number of places (Figure 5.1). However, if CO_2 storage is to be undertaken on the scale necessary to make deep cuts to atmospheric CO_2 emissions, there must be hundreds, and perhaps even thousands, of large-scale geological storage projects under way worldwide. The extent to which this is or might be, feasible depends on the answers to the key questions outlined below and addressed subsequently in this chapter:

- How is CO_2 stored underground? What happens to the CO_2 when it is injected? What are the physico-chemical and chemical processes involved? What are the geological controls? (Sections 5.2 and 5.3)

- How long can CO_2 remain stored underground? (Section 5.2)
- How much and where can CO_2 be stored in the subsurface, locally, regionally, globally? Is it a modest niche opportunity or is the total storage capacity sufficient to contain a large proportion of the CO_2 currently emitted to the atmosphere? (Section 5.3)
- Are there significant opportunities for CO_2-enhanced oil and gas recovery? (Section 5.3)
- How is a suitable storage site identified and what are its geological characteristics? (see Section 5.4)
- What technologies are currently available for geological storage of CO_2? (Section 5.5)
- Can we monitor CO_2 once it is geologically stored? (Section 5.6)
- Will a storage site leak and what would be the likely consequences? (Sections 5.6 and 5.7)
- Can a CO_2 storage site be remediated if something does go wrong? (Sections 5.6 and 5.7)
- Can a geological storage site be operated safely and if so, how? (Section 5.7)
- Are there legal and regulatory issues for geological storage and is there a legal/regulatory framework that enables it to be undertaken? (Section 5.8)
- What is the likely cost of geological storage of CO_2? (Section 5.9)
- After reviewing our current state of knowledge, are there things that we still need to know? What are these gaps in knowledge? (Section 5.10).

The remainder of this chapter seeks to address these questions.

5.2 Storage mechanisms and storage security

Geological formations in the subsurface are composed of transported and deposited rock grains organic material and minerals that form after the rocks are deposited. The pore space between grains or minerals is occupied by fluid (mostly water, with proportionally minute occurrences of oil and gas). Open fractures and cavities are also filled with fluid. Injection of CO_2 into the pore space and fractures of a permeable formation can displace the *in situ* fluid or the CO_2 may dissolve in or mix with the fluid or react with the mineral grains or there may be some combination of these processes. This section examines these processes and their influence on geological storage of CO_2.

5.2.1 *CO_2 flow and transport processes*

Injection of fluids into deep geological formations is achieved by pumping fluids down into a well (see Section 5.5). The part of the well in the storage zone is either perforated or covered with a permeable screen to enable the CO_2 to enter the formation. The perforated or screened interval is usually on the order of 10–100 m thick, depending on the permeability and thickness of the formation. Injection raises the pressure near the well,

allowing CO_2 to enter the pore spaces initially occupied by the *in situ* formation fluids. The amount and spatial distribution of pressure buildup in the formation will depend on the rate of injection, the permeability and thickness of the injection formation, the presence or absence of permeability barriers within it and the geometry of the regional underground water (hydrogeological) system.

Once injected into the formation, the primary flow and transport mechanisms that control the spread of CO_2 include:

- Fluid flow (migration) in response to pressure gradients created by the injection process;
- Fluid flow in response to natural hydraulic gradients;
- Buoyancy caused by the density differences between CO_2 and the formation fluids;
- Diffusion;
- Dispersion and fingering caused by formation heterogeneities and mobility contrast between CO_2 and formation fluid;
- Dissolution into the formation fluid;
- Mineralization;
- Pore space (relative permeability) trapping;
- Adsorption of CO_2 onto organic material.

The rate of fluid flow depends on the number and properties of the fluid phases present in the formation. When two or more fluids mix in any proportion, they are referred to as miscible fluids. If they do not mix, they are referred to as immiscible. The presence of several different phases may decrease the permeability and slow the rate of migration. If CO_2 is injected into a gas reservoir, a single miscible fluid phase consisting of natural gas and CO_2 is formed locally. When CO_2 is injected into a deep saline formation in a liquid or liquid-like supercritical dense phase, it is immiscible in water. Carbon dioxide injected into an oil reservoir may be miscible or immiscible, depending on the oil composition and the pressure and temperature of the system (Section 5.3.2). When CO_2 is injected into coal beds, in addition to some of the processes listed above, adsorption and desorption of gases (particularly methane) previously adsorbed on the coal take place, as well as swelling or shrinkage of the coal itself (Section 5.3.4).

Because supercritical CO_2 is much less viscous than water and oil (by an order of magnitude or more), migration is controlled by the contrast in mobility of CO_2 and the *in situ* formation fluids (Celia *et al.*, 2005; Nordbotten *et al.*, 2005a). Because of the comparatively high mobility of CO_2, only some of the oil or water will be displaced, leading to an average saturation of CO_2 in the range of 30–60%. Viscous fingering can cause CO_2 to bypass much of the pore space, depending on the heterogeneity and anisotropy of rock permeability (van der Meer, 1995; Ennis-King and Paterson, 2001; Flett *et al.*, 2005). In natural gas reservoirs, CO_2 is more viscous than natural gas, so the 'front' will be stable and viscous fingering limited.

The magnitude of the buoyancy forces that drive vertical flow depends on the type of fluid in the formation. In saline formations, the comparatively large density difference (30–50%) between CO_2 and formation water creates strong buoyancy forces that drive CO_2 upwards. In oil reservoirs, the density difference and buoyancy forces are not as large, particularly if the oil and CO_2 are miscible (Kovscek, 2002). In gas reservoirs, the opposite effect will occur, with CO_2 migrating downwards under buoyancy forces, because CO_2 is denser than natural gas (Oldenburg *et al.*, 2001).

In saline formations and oil reservoirs, the buoyant plume of injected CO_2 migrates upwards, but not evenly. This is because a lower permeability layer acts as a barrier and causes the CO_2 to migrate laterally, filling any stratigraphic or structural trap it encounters. The shape of the CO_2 plume rising through the rock matrix (Figure 5.6) is strongly affected by formation heterogeneity, such as low-permeability shale lenses (Flett *et al.*, 2005). Low-permeability layers within the storage formation therefore have the effect of slowing the upward migration of CO_2, which would otherwise cause CO_2 to bypass deeper parts of the storage formation (Doughty *et al.*, 2001).

As CO_2 migrates through the formation, some of it will dissolve into the formation water. In systems with slowly flowing water, reservoir-scale numerical simulations show that, over tens of years, a significant amount, up to 30% of the injected CO_2, will dissolve in formation water (Doughty *et al.*, 2001). Basin-scale simulations suggest that over centuries, the entire CO_2 plume dissolves in formation water (McPherson and Cole, 2000; Ennis-King *et al.*, 2003). If the injected CO_2 is contained in a closed structure (no flow of formation water), it will take much longer for CO_2 to completely dissolve because of reduced contact with unsaturated formation water. Once CO_2 is dissolved in the formation fluid, it migrates along with the regional groundwater flow. For deep sedimentary basins characterized by low permeability and high salinity, groundwater flow velocities are very low, typically on the order

of millimetres to centimetres per year (Bachu *et al.*, 1994). Thus, migration rates of dissolved CO_2 are substantially lower than for separate-phase CO_2.

Water saturated with CO_2 is slightly denser (approximately 1%) than the original formation water, depending on salinity (Enick and Klara, 1990; Bachu and Adams, 2003). With high vertical permeability, this may lead to free convection, replacing the CO_2-saturated water from the plume vicinity with unsaturated water, producing faster rates of CO_2 dissolution (Lindeberg and Wessel-Berg, 1997; Ennis-King and Paterson, 2003). Figure 5.7 illustrates the formation of convection cells and dissolution of CO_2 over several thousand years. The solubility of CO_2 in brine decreases with increasing pressure, decreasing temperature and increasing salinity (Annex 1). Calculations indicate that, depending on the salinity and depth, 20–60 kgCO_2 can dissolve in 1 m³ of formation fluid (Holt *et al.*, 1995; Koide *et al.*, 1995). With the use of a homogeneous model rather than a heterogeneous one, the time required for complete CO_2 dissolution may be underestimated.

As CO_2 migrates through a formation, some of it is retained in the pore space by capillary forces (Figure 5.6), commonly referred to as 'residual CO_2 trapping', which may immobilize significant amounts of CO_2 (Obdam *et al.*, 2003; Kumar *et al.*, 2005). Figure 5.8 illustrates that when the degree of trapping is high and CO_2 is injected at the bottom of a thick formation, all of the CO_2 may be trapped by this mechanism, even before it reaches the caprock at the top of the formation. While this effect is formation-specific, Holtz (2002) has demonstrated that residual CO_2 saturations may be as high as 15–25% for many typical storage formations. Over time, much of the trapped CO_2 dissolves in the formation water (Ennis-King and

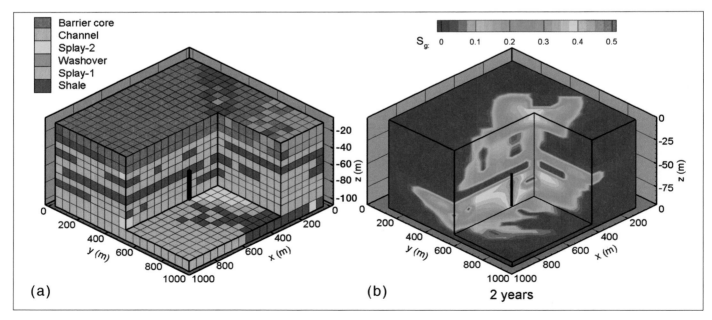

Figure 5.6 Simulated distribution of CO_2 injected into a heterogeneous formation with low-permeability layers that block upward migration of CO_2. (a) Illustration of a heterogeneous formation facies grid model. The location of the injection well is indicated by the vertical line in the lower portion of the grid. (b) The CO_2 distribution after two years of injection. Note that the simulated distribution of CO_2 is strongly influenced by the low-permeability layers that block and delay upward movement of CO_2 (after Doughty and Pruess, 2004).

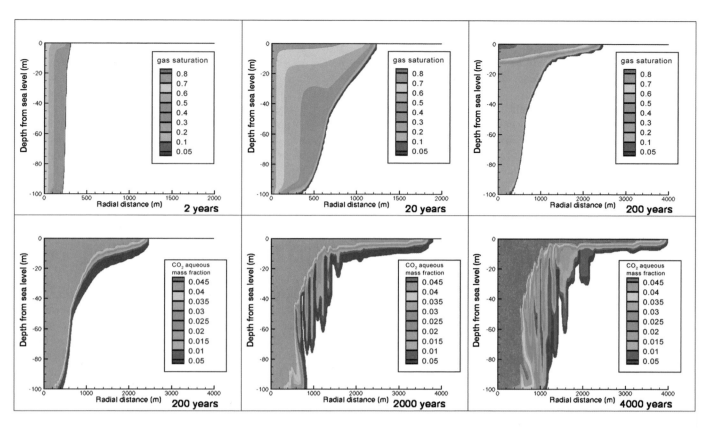

Figure 5.7 Radial simulations of CO_2 injection into a homogeneous formation 100 m thick, at a depth of 1 km, where the pressure is 10 MPa and the temperature is 40°C. The injection rate is 1 $MtCO_2$ yr^{-1} for 20 years, the horizontal permeability is 10^{-13} m^2 (approximately 100 mD) and the vertical permeability is one-tenth of that. The residual CO_2 saturation is 20%. The first three parts of the figure at 2, 20 and 200 years, show the gas saturation in the porous medium; the second three parts of the figure at 200, 2000 and 4000 years, show the mass fraction of dissolved CO_2 in the aqueous phase (after Ennis-King and Paterson, 2003).

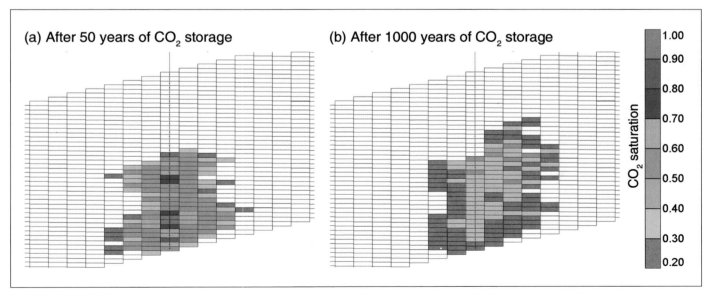

Figure 5.8 Simulation of 50 years of injection of CO_2 into the base of a saline formation. Capillary forces trap CO_2 in the pore spaces of sedimentary rocks. (a) After the 50-year injection period, most CO_2 is still mobile, driven upwards by buoyancy forces. (b) After 1000 years, buoyancy-driven flow has expanded the volume affected by CO_2 and much is trapped as residual CO_2 saturation or dissolved in brine (not shown). Little CO_2 is mobile and all CO_2 is contained within the aquifer (after Kumar et al., 2005).

Paterson, 2003), although appropriate reservoir engineering can accelerate or modify solubility trapping (Keith *et al.*, 2005).

5.2.2   *CO₂ storage mechanisms in geological formations*

The effectiveness of geological storage depends on a combination of physical and geochemical trapping mechanisms (Figure 5.9). The most effective storage sites are those where CO₂ is immobile because it is trapped permanently under a thick, low-permeability seal or is converted to solid minerals or is adsorbed on the surfaces of coal micropores or through a combination of physical and chemical trapping mechanisms.

5.2.2.1   *Physical trapping: stratigraphic and structural*
Initially, physical trapping of CO₂ below low-permeability seals (caprocks), such as very-low-permeability shale or salt beds, is the principal means to store CO₂ in geological formations (Figure 5.3). In some high latitude areas, shallow gas hydrates may conceivably act as a seal. Sedimentary basins have such closed, physically bound traps or structures, which are occupied mainly by saline water, oil and gas. Structural traps include those formed by folded or fractured rocks. Faults can act as permeability barriers in some circumstances and as preferential pathways for fluid flow in other circumstances (Salvi *et al.*, 2000). Stratigraphic traps are formed by changes in rock type caused by variation in the setting where the rocks were deposited. Both of these types of traps are suitable for CO₂ storage, although, as discussed in Section 5.5, care must be taken not to exceed the allowable overpressure to avoid fracturing the caprock or re-activating faults (Streit *et al.*, 2005).

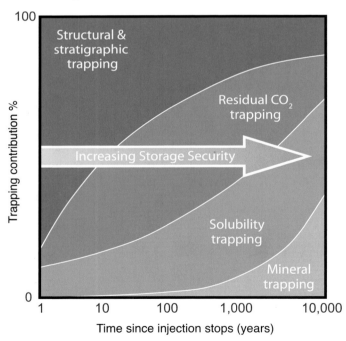

Figure 5.9 Storage security depends on a combination of physical and geochemical trapping. Over time, the physical process of residual CO₂ trapping and geochemical processes of solubility trapping and mineral trapping increase.

5.2.2.2   *Physical trapping: hydrodynamic*
Hydrodynamic trapping can occur in saline formations that do not have a closed trap, but where fluids migrate very slowly over long distances. When CO₂ is injected into a formation, it displaces saline formation water and then migrates buoyantly upwards, because it is less dense than the water. When it reaches the top of the formation, it continues to migrate as a separate phase until it is trapped as residual CO₂ saturation or in local structural or stratigraphic traps within the sealing formation. In the longer term, significant quantities of CO₂ dissolve in the formation water and then migrate with the groundwater. Where the distance from the deep injection site to the end of the overlying impermeable formation is hundreds of kilometres, the time scale for fluid to reach the surface from the deep basin can be millions of years (Bachu *et al.*, 1994).

5.2.2.3   *Geochemical trapping*
Carbon dioxide in the subsurface can undergo a sequence of geochemical interactions with the rock and formation water that will further increase storage capacity and effectiveness. First, when CO₂ dissolves in formation water, a process commonly called solubility trapping occurs. The primary benefit of solubility trapping is that once CO₂ is dissolved, it no longer exists as a separate phase, thereby eliminating the buoyant forces that drive it upwards. Next, it will form ionic species as the rock dissolves, accompanied by a rise in the pH. Finally, some fraction may be converted to stable carbonate minerals (mineral trapping), the most permanent form of geological storage (Gunter *et al.*, 1993). Mineral trapping is believed to be comparatively slow, potentially taking a thousand years or longer. Nevertheless, the permanence of mineral storage, combined with the potentially large storage capacity present in some geological settings, makes this a desirable feature of long-term storage.

Dissolution of CO₂ in formation waters can be represented by the chemical reaction:

$$CO_2\,(g) + H_2O \leftrightarrow H_2CO_3 \leftrightarrow HCO_3^- + H^+ \leftrightarrow CO_3^{2-} + 2H^+$$

The CO₂ solubility in formation water decreases as temperature and salinity increase. Dissolution is rapid when formation water and CO₂ share the same pore space, but once the formation fluid is saturated with CO₂, the rate slows and is controlled by diffusion and convection rates.

CO₂ dissolved in water produces a weak acid, which reacts with the sodium and potassium basic silicate or calcium, magnesium and iron carbonate or silicate minerals in the reservoir or formation to form bicarbonate ions by chemical reactions approximating to:

$$3\ \text{K-feldspar} + 2H_2O + 2CO_2 \leftrightarrow \text{Muscovite} + 6\ \text{Quartz} + 2K^+ + 2HCO_3^-$$

Reaction of the dissolved CO_2 with minerals can be rapid (days) in the case of some carbonate minerals, but slow (hundreds to thousands of years) in the case of silicate minerals.

Formation of carbonate minerals occurs from continued reaction of the bicarbonate ions with calcium, magnesium and iron from silicate minerals such as clays, micas, chlorites and feldspars present in the rock matrix (Gunter *et al.*, 1993, 1997).

Perkins *et al.* (2005) estimate that over 5000 years, all the CO_2 injected into the Weyburn Oil Field will dissolve or be converted to carbonate minerals within the storage formation. Equally importantly, they show that the caprock and overlying rock formations have an even greater capacity for

mineralization. This is significant for leakage risk assessment (Section 5.7) because once CO_2 is dissolved, it is unavailable for leakage as a discrete phase. Modelling by Holtz (2002) suggests more than 60% of CO_2 is trapped by residual CO_2 trapping by the end of the injection phase (100% after 1000 years), although laboratory experiments (Section 5.2.1) suggest somewhat lower percentages. When CO_2 is trapped at residual saturation, it is effectively immobile. However, should there be leakage through the caprock, then saturated brine may degas as it is depressurized, although, as illustrated in Figure 5.7 the tendency of saturated brine is to sink rather than to rise. Reaction of the CO_2 with formation water and rocks may result in reaction products that affect the porosity of the rock and the

Box 5.4 Storage security mechanisms and changes over time.

When the CO_2 is injected, it forms a bubble around the injection well, displacing the mobile water and oil both laterally and vertically within the injection horizon. The interactions between the water and CO_2 phase allow geochemical trapping mechanisms to take effect. Over time, CO_2 that is not immobilized by residual CO_2 trapping can react with *in situ* fluid to form carbonic acid (i.e., H_2CO_3 called solubility trapping – dominates from tens to hundreds of years). Dissolved CO_2 can eventually react with reservoir minerals if an appropriate mineralogy is encountered to form carbon-bearing ionic species (i.e., HCO_3^- and CO_3^{2-} called ionic trapping – dominates from hundreds to thousands of years). Further breakdown of these minerals could precipitate new carbonate minerals that would fix injected CO_2 in its most secure state (i.e., mineral trapping – dominates over thousands to millions of years).

Four injection scenarios are shown in Figure 5.10. Scenarios A, B and C show injection into hydrodynamic traps, essentially systems open to lateral flow of fluids and gas within the injection horizon. Scenario D represents injection into a physically restricted flow regime, similar to those of many producing and depleted oil and gas reservoirs.

In Scenario A, the injected CO_2 is never physically contained laterally. The CO_2 plume migrates within the injection horizon and is ultimately consumed via all types of geochemical trapping mechanisms, including carbonate mineralization. Mineral and ionic trapping dominate. The proportions of CO_2 stored in each geochemical trap will depend strongly on the *in situ* mineralogy, pore space structure and water composition.

In Scenario B, the migration of the CO_2 plume is similar to that of Scenario A, but the mineralogy and water chemistry are such that reaction of CO_2 with minerals is minor and solubility trapping and hydrodynamic trapping dominate.

In Scenario C, the CO_2 is injected into a zone initially similar to Scenario B. However, during lateral migration the CO_2 plume migrates into a zone of physical heterogeneity in the injection horizon. This zone may be characterized by variable porosity and permeability caused by a facies change. The facies change is accompanied by a more reactive mineralogy that causes an abrupt change in path. In the final state, ionic and mineral trapping predominate.

Scenario D illustrates CO_2 injection into a well-constrained flow zone but, similar to Scenario B, it does not have *in-situ* fluid chemistry and mineralogy suitable for ionic or mineral trapping. The bulk of the injected CO_2 is trapped geochemically via solubility trapping and physically via stratigraphic or structural trapping.

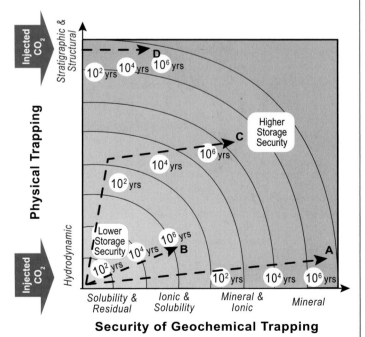

Figure 5.10 Storage expressed as a combination of physical and geochemical trapping. The level of security is proportional to distance from the origin. Dashed lines are examples of million-year pathways, discussed in Box 5.4.

flow of solution through the pores. This possibility has not, however, been observed experimentally and its possible effects cannot be quantified.

Yet another type of fixation occurs when CO_2 is preferentially adsorbed onto coal or organic-rich shales (Section 5.3.4). This has been observed in batch and column experiments in the laboratory, as well as in field experiments at the Fenn Big Valley, Canada and the San Juan Basin, USA (Box 5.7). A rather different form of fixation can occur when CO_2 hydrate is formed in the deep ocean seafloor and onshore in permafrost regions (Koide *et al.*, 1997).

5.2.3 Natural geological accumulations of CO_2

Natural sources of CO_2 occur, as gaseous accumulations of CO_2, CO_2 mixed with natural gas and CO_2 dissolved in formation water (Figure 5.11). These natural accumulations have been studied in the United States, Australia and Europe (Pearce *et al.*, 1996; Allis *et al.*, 2001; Stevens *et al.*, 2003; Watson *et al.*, 2004) as analogues for storage of CO_2, as well as for leakage from engineered storage sites. Production of CO_2 for EOR and other uses provides operational experience relevant to CO_2 capture and storage. There are, of course, differences between natural accumulations of CO_2 and engineered CO_2 storage sites: natural accumulations of CO_2 collect over very long periods of

time and at random sites, some of which might be naturally 'leaky'. At engineered sites, CO_2 injection rates will be rapid and the sites will necessarily be penetrated by injection wells (Celia and Bachu, 2003; Johnson *et al.*, 2005). Therefore, care must be taken to keep injection pressures low enough to avoid damaging the caprock (Section 5.5) and to make sure that the wells are properly sealed (Section 5.5).

Natural accumulations of relatively pure CO_2 are found all over the world in a range of geological settings, particularly in sedimentary basins, intra-plate volcanic regions (Figure 5.11) and in faulted areas or in quiescent volcanic structures. Natural accumulations occur in a number of different types of sedimentary rocks, principally limestones, dolomites and sandstones and with a variety of seals (mudstone, shale, salt and anhydrite) and a range of trap types, reservoir depths and CO_2-bearing phases.

Carbon dioxide fields in the Colorado Plateau and Rocky Mountains, USA, are comparable to conventional natural gas reservoirs (Allis *et al.*, 2001). Studies of three of these fields (McElmo Dome, St. Johns Field and Jackson Dome) have shown that each contains 1600 $MtCO_2$, with measurable leakage (Stevens *et al.*, 2001a). Two hundred Mt trapped in the Pisgah Anticline, northeast of the Jackson Dome, is thought to have been generated more than 65 million years ago (Studlick *et al.*, 1990), with no evidence of leakage, providing additional

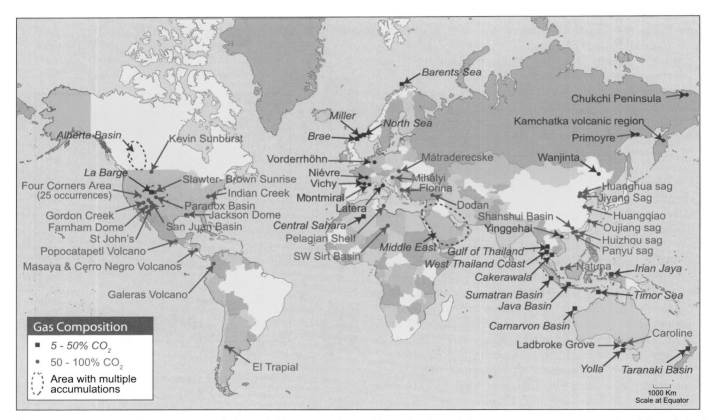

Figure 5.11 Examples of natural accumulations of CO_2 around the world. Regions containing many occurrences are enclosed by a dashed line. Natural accumulations can be useful as analogues for certain aspects of storage and for assessing the environmental impacts of leakage. Data quality is variable and the apparent absence of accumulations in South America, southern Africa and central and northern Asia is probably more a reflection of lack of data than a lack of CO_2 accumulations.

evidence of long-term trapping of CO_2. Extensive studies have been undertaken on small-scale CO_2 accumulations in the Otway Basin in Australia (Watson *et al.*, 2004) and in France, Germany, Hungary and Greece (Pearce *et al.*, 2003).

Conversely, some systems, typically spas and volcanic systems, are leaky and not useful analogues for geological storage. The Kileaua Volcano emits on average 4 $MtCO_2$ yr^{-1}. More than 1200 tCO_2 day^{-1} (438,000 tCO_2 yr^{-1}) leaked into the Mammoth Mountain area, California, between 1990 and 1995, with flux variations linked to seismicity (USGS, 2001b). Average flux densities of 80–160 tCO_2 m^{-2} yr^{-1} are observed near Matraderecske, Hungary, but along faults, the flux density can reach approximately 6600 t m^{-2} yr^{-1} (Pearce *et al.*, 2003). These high seepage rates result from release of CO_2 from faulted volcanic systems, whereas a normal baseline CO_2 flux is of the order of 10–100 gCO_2 m^{-2} day^{-1} under temperate climate conditions (Pizzino *et al.*, 2002). Seepage of CO_2 into Lake Nyos (Cameroon) resulted in CO_2 saturation of water deep in the lake, which in 1987 produced a very large-scale and (for more than 1700 persons) ultimately fatal release of CO_2 when the lake overturned (Kling *et al.*, 1987). The overturn of Lake Nyos (a deep, stratified tropical lake) and release of CO_2 are not representative of the seepage through wells or fractures that may occur from underground geological storage sites. Engineered CO_2 storage sites will be chosen to minimize the prospect of leakage. Natural storage and events such as Lake Nyos are not representative of geological storage for predicting seepage from engineered sites, but can be useful for studying the health, safety and environmental effects of CO_2 leakage (Section 5.7.4).

Carbon dioxide is found in some oil and gas fields as a separate gas phase or dissolved in oil. This type of storage is relatively common in Southeast Asia, China and Australia, less common in other oil and gas provinces such as in Algeria,

Russia, the Paradox Basin (USA) and the Alberta Basin (western Canada). In the North Sea and Barents Sea, a few fields have up to 10% CO_2, including Sleipner and Snohvit (Figure 5.11). The La Barge natural gas field in Wyoming, USA, has 3300 Mt of gas reserves, with an average of 65% CO_2 by volume. In the Appennine region of Italy, many deep wells (1–3 km depth) have trapped gas containing 90% or more CO_2 by volume. Major CO_2 accumulations around the South China Sea include the world's largest known CO_2 accumulation, the Natuna D Alpha field in Indonesia, with more than 9100 $MtCO_2$ (720 Mt natural gas). Concentrations of CO_2 can be highly variable between different fields in a basin and between different reservoir zones within the same field, reflecting complex generation, migration and mixing processes. In Australia's Otway Basin, the timing of CO_2 input and trapping ranges from 5000 years to a million years (Watson *et al.*, 2004).

5.2.4 *Industrial analogues for CO_2 storage*

5.2.4.1 *Natural gas storage*

Underground natural gas storage projects that offer experience relevant to CO_2 storage (Lippmann and Benson, 2003; Perry, 2005) have operated successfully for almost 100 years and in many parts of the world (Figure 5.12). These projects provide for peak loads and balance seasonal fluctuations in gas supply and demand. The Berlin Natural Gas Storage Project is an example of this (Box 5.5). The majority of gas storage projects are in depleted oil and gas reservoirs and saline formations, although caverns in salt have also been used extensively. A number of factors are critical to the success of these projects, including a suitable and adequately characterized site (permeability, thickness and extent of storage reservoir, tightness of caprock, geological structure, lithology, etc.). Injection wells must be properly designed, installed, monitored and maintained and abandoned wells in and near the project must be located and plugged. Finally, taking into account a range of solubility, density and trapping conditions, overpressuring the storage reservoir (injecting gas at a pressure that is well in excess of the in situ formation pressure) must be avoided.

While underground natural gas storage is safe and effective, some projects have leaked, mostly caused by poorly completed or improperly plugged and abandoned wells and by leaky faults (Gurevich *et al.*, 1993; Lippmann and Benson, 2003; Perry, 2005). Abandoned oil and gas fields are easier to assess as natural gas storage sites than are saline formations, because the geological structure and caprock are usually well characterized from existing wells. At most natural gas storage sites, monitoring requirements focus on ensuring that the injection well is not leaking (by the use of pressure measurements and through *in situ* downhole measurements of temperature, pressure, noise/sonic, casing conditions, etc.). Observation wells are sometimes used to verify that gas has not leaked into shallower strata.

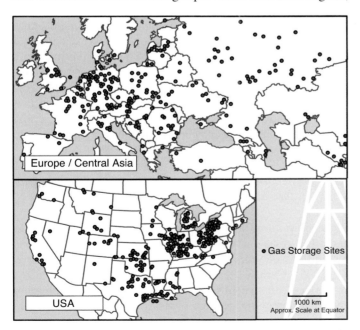

Figure 5.12 Location of some natural gas storage projects.

Box 5.5 The Berlin Natural Gas Storage Facility.

The Berlin Natural Gas Storage Facility is located in central Berlin, Germany, in an area that combines high population density with nature and water conservation reservations. This facility, with a capacity of 1085 million m³, was originally designed to be a reserve natural gas storage unit for limited seasonal quantity equalization. A storage production rate of 450,000 m³ h⁻¹ can be achieved with the existing storage wells and surface facilities. Although the geological and engineering aspects and scale of the facility make it a useful analogue for a small CO_2 storage project, this project is more complex because the input and output for natural gas is highly variable, depending on consumer demand. The risk profiles are also different, considering the highly flammable and explosive nature of natural gas and conversely the reactive nature of CO_2.

The facility lies to the east of the North German Basin, which is part of a complex of basin structures extending from The Netherlands to Poland. The sandstone storage horizons are at approximately 800 m below sea level. The gas storage layers are covered with layers of claystone, anhydrite and halite, approximately 200 m thick. This site has complicated tectonics and heterogeneous reservoir lithologies.

Twelve wells drilled at three sites are available for natural gas storage operation. The varying storage sand types also require different methods of completion of the wells. The wells also have major differences in their production behaviour. The wellheads of the storage wells and of the water disposal wells are housed in 5 m deep cellars covered with concrete plates, with special steel covers over the wellheads to allow for wireline logging. Because of the urban location, a total of 16 deviated storage wells and water disposal wells were concentrated at four sites. Facilities containing substances that could endanger water are set up within fluid-tight concrete enclosures and/or have their own watertight concrete enclosures.

5.2.4.2   Acid gas injection

Acid gas injection operations represent a commercial analogue for some aspects of geological CO_2 storage. Acid gas is a mixture of H_2S and CO_2, with minor amounts of hydrocarbon gases that can result from petroleum production or processing. In Western Canada, operators are increasingly turning to acid gas disposal by injection into deep geological formations. Although the purpose of the acid gas injection operations is to dispose of H_2S, significant quantities of CO_2 are injected at the same time because it is uneconomic to separate the two gases.

Currently, regulatory agencies in Western Canada approve the maximum H_2S fraction, maximum wellhead injection pressure and rate and maximum injection volume. Acid gas is currently injected into 51 different formations at 44 different locations across the Alberta Basin in the provinces of Alberta and British Columbia (Figure 5.13). Carbon dioxide often represents the largest component of the injected acid gas stream, in many cases, 14–98% of the total volume. A total of 2.5 $MtCO_2$ and 2 MtH_2S had been injected in Western Canada by the end of 2003, at rates of 840–500,720 m³ day⁻¹ per site, with an aggregate injection rate in 2003 of 0.45 $MtCO_2$ yr⁻¹ and 0.55 MtH_2S yr⁻¹, with no detectable leakage.

Acid gas injection in Western Canada occurs over a wide range of formation and reservoir types, acid gas compositions and operating conditions. Injection takes place in deep saline formations at 27 sites, into depleted oil and/or gas reservoirs at 19 sites and into the underlying water leg of depleted oil and gas reservoirs at 4 sites. Carbonates form the reservoir at 29 sites and quartz-rich sandstones dominate at the remaining 21 (Figure 5.13). In most cases, shale constitutes the overlying confining unit (caprock), with the remainder of the injection zones being confined by tight limestones, evaporites and anhydrites.

Since the first acid-gas injection operation in 1990, 51 different injection sites have been approved, of which 44 are currently active. One operation was not implemented, three were rescinded after a period of operation (either because injection volumes reached the approved limit or because the gas plant producing the acid gas was decommissioned) and three sites were suspended by the regulatory agency because of reservoir overpressuring.

5.2.4.3   Liquid waste injection

In many parts of the world, large volumes of liquid waste are injected into the deep subsurface every day. For example, for the past 60 years, approximately 9 billion gallons (34.1 million m³) of hazardous waste is injected into saline formations in the United States from about 500 wells each year. In addition, more than 750 billion gallons (2843 million m³) of oil field brines are injected from 150,000 wells each year. This combined annual US injectate volume of about 3000 million m³, when converted to volume equivalent, corresponds to the volume of approximately 2 $GtCO_2$ at a depth of 1 km. Therefore, the experience gained from existing deep-fluid-injection projects is relevant in terms of the style of operation and is of a similar magnitude to that which may be required for geological storage of CO_2.

5.2.5   *Security and duration of CO_2 storage in geological formations*

Evidence from oil and gas fields indicates that hydrocarbons and other gases and fluids including CO_2 can remain trapped for millions of years (Magoon and Dow, 1994; Bradshaw *et al.*, 2005). Carbon dioxide has a tendency to remain in the subsurface (relative to hydrocarbons) via its many physico-chemical immobilization mechanisms. World-class petroleum provinces have storage times for oil and gas of 5–100 million years, others for 350 million years, while some minor petroleum

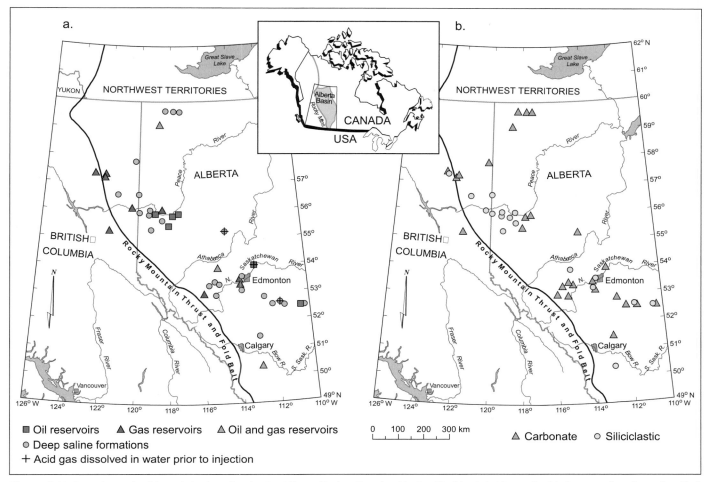

Figure 5.13 Locations of acid gas injection sites in the Alberta Basin, Canada: (a) classified by injection unit; (b) the same locations classified by rock type (from Bachu and Haug, 2005).

accumulations have been stored for up to 1400 million years. However, some natural traps do leak, which reinforces the need for careful site selection (Section 5.3), characterization (Section 5.4) and injection practices (Section 5.5).

5.3 Storage formations, capacity and geographical distribution

In this section, the following issues are addressed: In what types of geological formations can CO_2 be stored? Are such formations widespread? How much CO_2 can be geologically stored?

5.3.1 General site-selection criteria

There are many sedimentary regions in the world (Figures 2.4–2.6 and Figure 5.14) variously suited for CO_2 storage. In general, geological storage sites should have (1) adequate capacity and injectivity, (2) a satisfactory sealing caprock or confining unit and (3) a sufficiently stable geological environment to avoid compromising the integrity of the storage site. Criteria for assessing basin suitability (Bachu, 2000, 2003; Bradshaw *et al.*,

2002) include: basin characteristics (tectonic activity, sediment type, geothermal and hydrodynamic regimes); basin resources (hydrocarbons, coal, salt), industry maturity and infrastructure; and societal issues such as level of development, economy, environmental concerns, public education and attitudes.

The suitability of sedimentary basins for CO_2 storage depends in part on their location on the continental plate. Basins formed in mid-continent locations or near the edge of stable continental plates, are excellent targets for long-term CO_2 storage because of their stability and structure. Such basins are found within most continents and around the Atlantic, Arctic and Indian Oceans. The storage potential of basins found behind mountains formed by plate collision is likely to be good and these include the Rocky Mountain, Appalachian and Andean basins in the Americas, European basins immediately north of the Alps and Carpathians and west of the Urals and those located south of the Zagros and Himalayas in Asia. Basins located in tectonically active areas, such as those around the Pacific Ocean or the northern Mediterranean, may be less suitable for CO_2 storage and sites in these regions must be selected carefully because of the potential for CO_2 leakage (Chiodini *et al.*, 2001; Granieri *et al.*, 2003). Basins located on the edges of plates

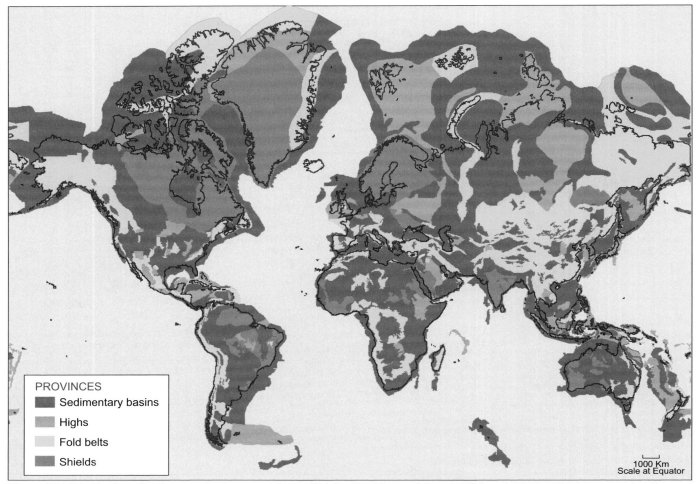

Figure 5.14 Distribution of sedimentary basins around the world (after Bradshaw and Dance, 2005; and USGS, 2001a). In general, sedimentary basins are likely to be the most prospective areas for storage sites. However, storage sites may also be found in some areas of fold belts and in some of the highs. Shield areas constitute regions with low prospectivity for storage. The Mercator projection used here is to provide comparison with Figures 5.1, 5.11 and 5.27. The apparent dimensions of the sedimentary basins, particularly in the northern hemisphere, should not be taken as an indication of their likely storage capacity.

where subduction is occurring or between active mountain ranges, are likely to be strongly folded and faulted and provide less certainty for storage. However, basins must be assessed on an individual basis. For example, the Los Angeles Basin and Sacramento Valley in California, where significant hydrocarbon accumulations have been found, have demonstrated good local storage capacity. Poor CO_2 storage potential is likely to be exhibited by basins that (1) are thin (≤1000 m), (2) have poor reservoir and seal relationships, (3) are highly faulted and fractured, (4) are within fold belts, (5) have strongly discordant sequences, (6) have undergone significant diagenesis or (7) have overpressured reservoirs.

The efficiency of CO_2 storage in geological media, defined as the amount of CO_2 stored per unit volume (Brennan and Burruss, 2003), increases with increasing CO_2 density. Storage safety also increases with increasing density, because buoyancy, which drives upward migration, is stronger for a lighter fluid. Density increases significantly with depth while CO_2 is in gaseous phase, increases only slightly or levels off after passing from the gaseous phase into the dense phase and

may even decrease with a further increase in depth, depending on the temperature gradient (Ennis-King and Paterson, 2001; Bachu, 2003). 'Cold' sedimentary basins, characterized by low temperature gradients, are more favourable for CO_2 storage (Bachu, 2003) because CO_2 attains higher density at shallower depths (700–1000 m) than in 'warm' sedimentary basins, characterized by high temperature gradients where dense-fluid conditions are reached at greater depths (1000–1500 m). The depth of the storage formation (leading to increased drilling and compression costs for deeper formations) may also influence the selection of storage sites.

Adequate porosity and thickness (for storage capacity) and permeability (for injectivity) are critical; porosity usually decreases with depth because of compaction and cementation, which reduces storage capacity and efficiency. The storage formation should be capped by extensive confining units (such as shale, salt or anhydrite beds) to ensure that CO_2 does not escape into overlying, shallower rock units and ultimately to the surface. Extensively faulted and fractured sedimentary basins or parts thereof, particularly in seismically active areas, require careful characterization to be good candidates for CO_2 storage, unless the faults and fractures are sealed and CO_2 injection will not open them (Holloway, 1997; Zarlenga *et al.*, 2004).

The pressure and flow regimes of formation waters in a sedimentary basin are important factors in selecting sites for CO_2 storage (Bachu *et al.*, 1994). Injection of CO_2 into formations overpressured by compaction and/or hydrocarbon generation may raise technological and safety issues that make them unsuitable. Underpressured formations in basins located mid-continent, near the edge of stable continental plates or behind mountains formed by plate collision may be well suited for CO_2 storage. Storage of CO_2 in deep saline formations with fluids having long residence times (millions of years) is conducive to hydrodynamic and mineral trapping (Section 5.2).

The possible presence of fossil fuels and the exploration and production maturity of a basin are additional considerations for selection of storage sites (Bachu, 2000). Basins with little exploration for hydrocarbons may be uncertain targets for CO_2 storage because of limited availability of geological information or potential for contamination of as-yet-undiscovered hydrocarbon resources. Mature sedimentary basins may be prime targets for CO_2 storage because: (1) they have well-known characteristics; (2) hydrocarbon pools and/or coal beds have been discovered and produced; (3) some petroleum reservoirs might be already depleted, nearing depletion or abandoned as uneconomic; (4) the infrastructure needed for CO_2 transport and injection may already be in place. The presence of wells penetrating the subsurface in mature sedimentary basins can create potential CO_2 leakage pathways that may compromise the security of a storage site (Celia and Bachu, 2003). Nevertheless, at Weyburn, despite the presence of many hundreds of existing wells, after four years of CO_2 injection there has been no measurable leakage (Strutt *et al.*, 2003).

5.3.2 *Oil and gas fields*

5.3.2.1 *Abandoned oil and gas fields*
Depleted oil and gas reservoirs are prime candidates for CO_2 storage for several reasons. First, the oil and gas that originally accumulated in traps (structural and stratigraphic) did not escape (in some cases for many millions of years), demonstrating their integrity and safety. Second, the geological structure and physical properties of most oil and gas fields have been extensively studied and characterized. Third, computer models have been developed in the oil and gas industry to predict the movement, displacement behaviour and trapping of hydrocarbons. Finally, some of the infrastructure and wells already in place may be used for handling CO_2 storage operations. Depleted fields will not be adversely affected by CO_2 (having already contained hydrocarbons) and if hydrocarbon fields are still in production, a CO_2 storage scheme can be optimized to enhance oil (or gas) production. However, plugging of abandoned wells in many mature fields began many decades ago when wells were simply filled with a mud-laden fluid. Subsequently, cement plugs were required to be strategically placed within the wellbore, but not with any consideration that they may one day be relied upon to

contain a reactive and potentially buoyant fluid such as CO_2. Therefore, the condition of wells penetrating the caprock must be assessed (Winter and Bergman, 1993). In many cases, even locating the wells may be difficult and caprock integrity may need to be confirmed by pressure and tracer monitoring.

The capacity of a reservoir will be limited by the need to avoid exceeding pressures that damage the caprock (Section 5.5.3). Reservoirs should have limited sensitivity to reductions in permeability caused by plugging of the near-injector region and by reservoir stress fluctuations (Kovscek, 2002; Bossie-Codreanu *et al.*, 2003). Storage in reservoirs at depths less than approximately 800 m may be technically and economically feasible, but the low storage capacity of shallow reservoirs, where CO_2 may be in the gas phase, could be problematic.

5.3.2.2 *Enhanced oil recovery*
Enhanced oil recovery (EOR) through CO_2 flooding (by injection) offers potential economic gain from incremental oil production. Of the original oil in place, 5–40% is usually recovered by conventional primary production (Holt *et al.*, 1995). An additional 10–20% of oil in place is produced by secondary recovery that uses water flooding (Bondor, 1992). Various miscible agents, among them CO_2, have been used for enhanced (tertiary) oil recovery or EOR, with an incremental oil recovery of 7–23% (average 13.2%) of the original oil in place (Martin and Taber, 1992; Moritis, 2003). Descriptions of CO_2-EOR projects are provided in Box 5.3 and Box 5.6, and an illustration is given in Figure 5.15.

Many CO_2 injection schemes have been suggested, including continuous CO_2 injection or alternate water and CO_2 gas injection (Klins and Farouq Ali, 1982; Klins, 1984). Oil displacement by CO_2 injection relies on the phase behaviour of CO_2 and crude oil mixtures that are strongly dependent on reservoir temperature, pressure and crude oil composition. These mechanisms range from oil swelling and viscosity reduction for injection of immiscible fluids (at low pressures) to completely miscible displacement in high-pressure applications. In these applications, more than 50% and up to 67% of the injected CO_2 returns with the produced oil (Bondor, 1992) and is usually separated and re-injected into the reservoir to minimize operating costs. The remainder is trapped in the oil reservoir by various means, such as irreducible saturation and dissolution in reservoir oil that it is not produced and in pore space that is not connected to the flow path for the producing wells.

For enhanced CO_2 storage in EOR operations, oil reservoirs may need to meet additional criteria (Klins, 1984; Taber *et al.*, 1997; Kovscek, 2002; Shaw and Bachu, 2002). Generally, reservoir depth must be more than 600 m. Injection of immiscible fluids must often suffice for heavy- to-medium-gravity oils (oil gravity 12–25 API). The more desirable miscible flooding is applicable to light, low-viscosity oils (oil gravity 25–48 API). For miscible floods, the reservoir pressure must be higher than the minimum miscibility pressure (10–15 MPa) needed for achieving miscibility between reservoir oil and CO_2, depending on oil composition and gravity, reservoir temperature and CO_2 purity (Metcalfe, 1982). To achieve effective removal of the

Box 5.6 The Rangely, Colorado, CO_2-EOR Project.

The Rangely CO_2-EOR Project is located in Colorado, USA and is operated by Chevron. The CO_2 is purchased from the Exxon-Mobil LaBarge natural gas processing facility in Wyoming and transported 283 km via pipeline to the Rangely field. Additional spurs carry CO_2 over 400 km from LaBarge to Lost Soldier and Wertz fields in central Wyoming, currently ending at the Salt Creek field in eastern Wyoming.

The sandstone reservoir of the Rangely field has been CO_2 flooded, by the water alternating gas (WAG) process, since 1986. Primary and secondary recovery, carried out between 1944 and 1986, recovered 1.9 US billion barrels (302 million m^3) of oil (21% of the original oil in place). With use of CO_2 floods, ultimate tertiary recovery of a further 129 million barrels (21 million m^3) of oil (6.8% of original oil in place) is expected. Average daily CO_2 injection in 2003 was equivalent to 2.97 MtCO$_2$ yr^{-1}, with production of 13,913 barrels oil per day. Of the total 2.97 Mt injected, recycled gas comprised around 2.29 Mt and purchased gas about 0.74 Mt. Cumulative CO_2 stored to date is estimated at 22.2 Mt. A simplified flow diagram for the Rangely field is given in Figure 5.15.

The Rangely field, covering an area of 78 km^2, is an asymmetric anticline. A major northeast-to-southwest fault in the eastern half of the field and other faults and fractures significantly influence fluid movement within the reservoir. The sandstone reservoirs have an average gross and effective thickness of 160 m and 40 m, respectively and are comprised of six persistent producing sandstone horizons (depths of 1675–1980 m) with average porosity of 12%. Permeability averages 10 mD (Hefner and Barrow, 1992).

By the end of 2003, there were 248 active injectors, of which 160 are used for CO_2 injection and 348 active producers. Produced gas is processed through two parallel single-column natural-gas-liquids recovery facilities and subsequently compressed to approximately 14.5 MPa. Compressed-produced gas (recycled gas) is combined with purchased CO_2 for reinjection mostly by the WAG process.

Carbon dioxide-EOR operation in the field maintains compliance with government regulations for production, injection, protection of potable water formations, surface use, flaring and venting. A number of protocols have been instituted to ensure containment of CO_2 – for example, pre-injection well-integrity verification, a radioactive tracer survey run on the first injection, injection-profile tracer surveys, mechanical integrity tests, soil gas surveys and round-the-clock field monitoring. Surface release from the storage reservoir is below the detection limit of 170 t yr^{-1} or an annual leakage rate of less than 0.00076% of the total stored CO_2 (Klusman, 2003). Methane leakage is estimated to be 400 t yr^{-1}, possibly due to increased CO_2 injection pressure above original reservoir pressure. The water chemistry portion of the study indicates that the injected CO_2 is dissolving in the water and may be responsible for dissolution of ferroan calcite and dolomite. There is currently no evidence of mineral precipitation that may result in mineral storage of CO_2.

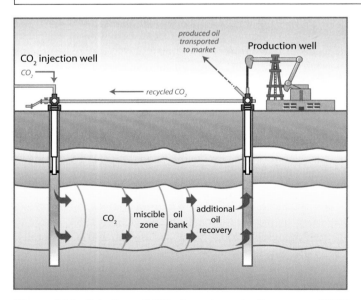

Figure 5.15 Injection of CO_2 for enhanced oil recovery (EOR) with some storage of retained CO_2 (after IEA Greenhouse Gas R&D Programme). The CO_2 that is produced with the oil is separated and re-injected back into the formation. Recycling of produced CO_2 decreases the amount of CO_2 that must be purchased and avoids emissions to the atmosphere.

oil, other preferred criteria for both types of flooding include relatively thin reservoirs (less than 20 m), high reservoir angle, homogenous formation and low vertical permeability. For horizontal reservoirs, the absence of natural water flow, major gas cap and major natural fractures are preferred. Reservoir thickness and permeability are not critical factors.

Reservoir heterogeneity also affects CO_2 storage efficiency. The density difference between the lighter CO_2 and the reservoir oil and water leads to movement of the CO_2 along the top of the reservoir, particularly if the reservoir is relatively homogeneous and has high permeability, negatively affecting the CO_2 storage and oil recovery. Consequently, reservoir heterogeneity may have a positive effect, slowing down the rise of CO_2 to the top of the reservoir and forcing it to spread laterally, giving more complete invasion of the formation and greater storage potential (Bondor, 1992; Kovscek, 2002; Flett *et al.*, 2005).

5.3.2.3 Enhanced gas recovery
Although up to 95% of original gas in place can be produced, CO_2 could potentially be injected into depleted gas reservoirs to enhance gas recovery by repressurizing the reservoir (van der Burgt *et al.*, 1992; Koide and Yamazaki, 2001; Oldenburg *et al.*, 2001). Enhanced gas recovery has so far been implemented only at pilot scale (Gaz de France K12B project, Netherlands,

Table 5.1) and some authors have suggested that CO_2 injection might result in lower gas recovery factors, particularly for very heterogeneous fields (Clemens and Wit, 2002).

5.3.3 Saline formations

Saline formations are deep sedimentary rocks saturated with formation waters or brines containing high concentrations of dissolved salts. These formations are widespread and contain enormous quantities of water, but are unsuitable for agriculture or human consumption. Saline brines are used locally by the chemical industry and formation waters of varying salinity are used in health spas and for producing low-enthalpy geothermal energy. Because the use of geothermal energy is likely to increase, potential geothermal areas may not be suitable for CO_2 storage. It has been suggested that combined geological storage and geothermal energy may be feasible, but regions with good geothermal energy potential are generally less favourable for CO_2 geological storage because of the high degree of faulting and fracturing and the sharp increase of temperature with depth. In very arid regions, deep saline formations may be considered for future water desalinization.

The Sleipner Project in the North Sea is the best available example of a CO_2 storage project in a saline formation (Box 5.1). It was the first commercial-scale project dedicated to geological CO_2 storage. Approximately 1 $MtCO_2$ is removed annually from the produced natural gas and injected underground at Sleipner. The operation started in October 1996 and over the lifetime of the project a total of 20 $MtCO_2$ is expected to be stored. A simplified diagram of the Sleipner scheme is given in Figure 5.4.

The CO_2 is injected into poorly cemented sands about 800–1000 m below the sea floor. The sandstone contains secondary thin shale or clay layers, which influence the internal movement of injected CO_2. The overlying primary seal is an extensive thick shale or clay layer. The saline formation into which CO_2 is injected has a very large storage capacity.

The fate and transport of the Sleipner CO_2 plume has been successfully monitored (Figure 5.16) by seismic time-lapse surveys (Section 5.6). These surveys have helped improve the conceptual model for the fate and transport of stored CO_2. The vertical cross-section of the plume shown in Figure 5.16 indicates both the upward migration of CO_2 (due to buoyancy forces) and the role of lower permeability strata within the formation, diverting some of the CO_2 laterally, thus spreading out the plume over a larger area. The survey also shows that the caprock prevents migration out of the storage formation. The seismic data shown in Figure 5.16 illustrate the gradual growth of the plume. Today, the footprint of the plume at Sleipner extends over approximately 5 km². Reservoir studies and simulations (Section 5.4.2) have shown that the CO_2-saturated brine will eventually become denser and sink, eliminating the potential for long-term leakage (Lindeberg and Bergmo, 2003).

5.3.4 Coal seams

Coal contains fractures (cleats) that impart some permeability to the system. Between cleats, solid coal has a very large number of micropores into which gas molecules from the cleats can diffuse and be tightly adsorbed. Coal can physically adsorb many gases and may contain up to 25 normal m³ (m³ at 1 atm and 0°C) methane per tonne of coal at coal seam pressures. It has a higher affinity to adsorb gaseous CO_2 than methane (Figure 5.17). The volumetric ratio of adsorbable CO_2:CH_4 ranges from as low as one for mature coals such as anthracite, to ten or more for younger, immature coals such as lignite. Gaseous CO_2 injected through wells will flow through the cleat system of the coal, diffuse into the coal matrix and be adsorbed onto the coal micropore surfaces, freeing up gases with lower affinity to coal (i.e., methane).

The process of CO_2 trapping in coals for temperatures and pressures above the critical point is not well understood (Larsen, 2003). It seems that adsorption is gradually replaced by absorption and the CO_2 diffuses or 'dissolves' in coal. Carbon dioxide is a 'plasticizer' for coal, lowering the temperature required to cause the transition from a glassy, brittle structure to a rubbery, plastic structure (coal softening). In one case, the transition temperature was interpreted to drop from about 400°C at 3 MPa to <30°C at 5.5 MPa CO_2 pressure (Larsen, 2003). The transition temperature is dependent on the maturity of the coal, the maceral content, the ash content and the confining stress and is not easily extrapolated to the field. Coal plasticization or softening, may adversely affect the permeability that would allow CO_2 injection. Furthermore, coal swells as CO_2 is adsorbed and/or absorbed, which reduces permeability and injectivity by orders of magnitude or more (Shi and Durucan, 2005) and which may be counteracted by increasing the injection pressures (Clarkson and Bustin, 1997; Palmer and Mansoori, 1998; Krooss et al., 2002; Larsen, 2003). Some studies suggest that the injected CO_2 may react with coal (Zhang et al., 1993), further highlighting the difficulty in injecting CO_2 into low-permeability coal.

If CO_2 is injected into coal seams, it can displace methane, thereby enhancing CBM recovery. Carbon dioxide has been injected successfully at the Allison Project (Box 5.7) and in the Alberta Basin, Canada (Gunter et al., 2005), at depths greater than that corresponding to the CO_2 critical point. Carbon dioxide-ECBM has the potential to increase the amount of produced methane to nearly 90% of the gas, compared to conventional recovery of only 50% by reservoir-pressure depletion alone (Stevens et al., 1996).

Coal permeability is one of several determining factors in selection of a storage site. Coal permeability varies widely and generally decreases with increasing depth as a result of cleat closure with increasing effective stress. Most CBM-producing wells in the world are less than 1000 m deep.

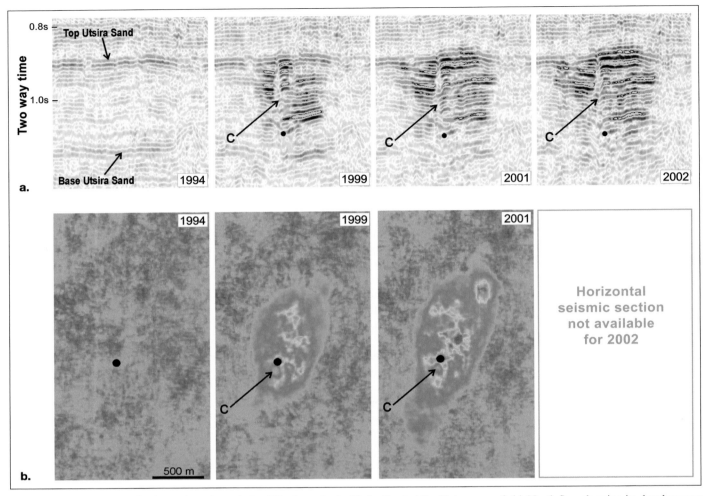

Figure 5.16 (a) Vertical seismic sections through the CO_2 plume in the Utsira Sand at the Sleipner gas field, North Sea, showing its development over time. Note the chimney of high CO_2 saturation (c) above the injection point (black dot) and the bright layers corresponding to high acoustic response due to CO_2 in a gas form being resident in sandstone beneath thin low-permeability horizons within the reservoir. (b) Horizontal seismic sections through the developing CO_2 plume at Sleipner showing its growth over time. The CO_2 plume-specific monitoring was completed in 2001; therefore data for 2002 was not available (courtesy of Andy Chadwick and the CO2STORE project).

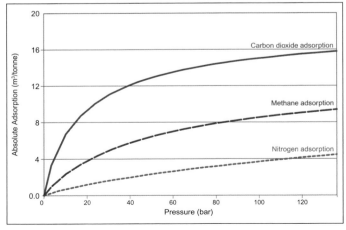

Figure 5.17 Pure gas absolute adsorption in standard cubic feet per tonne (SCF per tonne) on Tiffany Coals at 55°C (after Gasem et al., 2002).

Original screening criteria proposed in selecting favourable areas for CO_2 ECBM (IEA-GHG, 1998) include:

- Adequate permeability (minimum values have not yet been determined);
- Suitable coal geometry (a few, thick seams rather than multiple, thin seams);
- Simple structure (minimal faulting and folding);
- Homogeneous and confined coal seam(s) that are laterally continuous and vertically isolated;
- Adequate depth (down to 1500 m, greater depths have not yet been studied);
- Suitable gas saturation conditions (high gas saturation for ECBM);
- Ability to dewater the formation.

However, more recent studies have indicated that coal rank may play a more significant role than previously thought, owing to the dependence on coal rank of the relative adsorptive capacities

Box 5.7 The Allison Unit CO_2-ECBM Pilot.

The Allison Unit CO_2-ECBM Recovery Pilot Project, located in the northern New Mexico portion of the San Juan Basin, USA, is owned and operated by Burlington Resources. Production from the Allison field began in July 1989 and CO_2 injection operations for ECBM recovery commenced in April 1995. Carbon dioxide injection was suspended in August 2001 to evaluate the results of the pilot. Since this pilot was undertaken purely for the purposes of ECBM production, no CO_2 monitoring programme was implemented.

The CO_2 was sourced from the McElmo Dome in Colorado and delivered to the site through a (then) Shell (now Kinder-Morgan) CO_2 pipeline. The Allison Unit has a CBM resource of 242 million m^3 km^{-2}. A total of 181 million m^3 (6.4 Bcf) of natural CO_2 was injected into the reservoir over six years, of which 45 million m^3 (1.6 Bcf) is forecast to be ultimately produced back, resulting in a net storage volume of 277,000 tCO_2. The pilot consists of 16 methane production wells, 4 CO_2 injection wells and 1 pressure observation well. The injection operations were undertaken at constant surface injection pressures on the order of 10.4 MPa.

The wells were completed in the Fruitland coal, which is capped by shale. The reservoir has a thickness of 13 m, is located at a depth of 950 m and had an original reservoir pressure of 11.5 MPa. In a study conducted under the Coal-Seq Project performed for the US Department of Energy (www.coal-seq.com), a detailed reservoir characterization and modelling of the pilot was developed with the COMET2 reservoir simulator and future field performance was forecast under various operating conditions.

This study provides evidence of significant coal-permeability reduction with CO_2 injection. This permeability reduction resulted in a two-fold reduction in injectivity. This effect compromised incremental methane recovery and project economics. Finding ways to overcome and/or prevent this effect is therefore an important topic for future research. The injection of CO_2 at the Allison Unit has resulted in an increase in methane recovery from an estimated 77% of original gas in place to 95% of the original gas in place within the project area. The recovery of methane was in a proportion of approximately one volume of methane for every three volumes of CO_2 injected (Reeves *et al.*, 2004).

An economic analysis of the pilot indicated a net present value of negative US\$ 627,000, assuming a discount rate of 12% and an initial capital expenditure of US\$ 2.6 million, but not including the beneficial impact of any tax credits for production from non-conventional reservoirs. This was based on a gas price of 2.09 US\$ GJ^{-1} (2.20 US\$/MMbtu) (at the time) and a CO_2 price of 5.19 US\$ t^{-1} (0.30 US\$/Mcf). The results of the financial analysis will change, depending on the cost of oil and gas (the analysis indicated that the pilot would have yielded a positive net present value of US\$2.6 million at today's gas prices) and the cost of CO_2. It was also estimated that if injectivity had been improved by a factor of four (but still using 2.09 US\$ GJ^{-1} (2.20 US\$/MMbtu)), the net present value would have increased to US\$ 3.6 million. Increased injectivity and today's gas prices combined would have yielded a net present value for the pilot of US\$ 15 million or a profit of 34 US\$/tCO_2 retained in the reservoir (Reeves *et al.*, 2003).

of methane and CO_2 (Reeves *et al.*, 2004).

If the coal is never mined or depressurized, it is likely CO_2 will be stored for geological time, but, as with any geological storage option, disturbance of the formation could void any storage. The likely future fate of a coal seam is, therefore, a key determinant of its suitability for storage and in storage site selection and conflicts between mining and CO_2 storage are possible, particularly for shallow coals.

5.3.5   *Other geological media*

Other geological media and/or structures – including basalts, oil or gas shale, salt caverns and abandoned mines – may locally provide niche options for geological storage of CO_2.

5.3.5.1   *Basalts*
Flows and layered intrusions of basalt occur globally, with large volumes present around the world (McGrail *et al.*, 2003). Basalt commonly has low porosity, low permeability and low pore space continuity and any permeability is generally associated

with fractures through which CO_2 will leak unless there is a suitable caprock. Nonetheless, basalt may have some potential for mineral trapping of CO_2, because injected CO_2 may react with silicates in the basalt to form carbonate minerals (McGrail *et al.*, 2003). More research is needed, but in general, basalts appear unlikely to be suitable for CO_2 storage.

5.3.5.2   *Oil or gas rich shale*
Deposits of oil or gas shale or organic-rich shale, occur in many parts of the world. The trapping mechanism for oil shale is similar to that for coal beds, namely CO_2 adsorption onto organic material. Carbon dioxide-enhanced shale-gas production (like ECBM) has the potential to reduce storage costs. The potential for storage of CO_2 in oil or gas shale is currently unknown, but the large volumes of shale suggest that storage capacity may be significant. If site-selection criteria, such as minimum depth, are developed and applied to these shales, then volumes could be limited, but the very low permeability of these shales is likely to preclude injection of large volumes of CO_2.

5.3.5.3   Salt caverns

Storage of CO_2 in salt caverns created by solution mining could use the technology developed for the storage of liquid natural gas and petroleum products in salt beds and domes in Western Canada and the Gulf of Mexico (Dusseault *et al.*, 2004). A single salt cavern can reach more than 500,000 m^3. Storage of CO_2 in salt caverns differs from natural gas and compressed air storage because in the latter case, the caverns are cyclically pressurized and depressurized on a daily-to-annual time scale, whereas CO_2 storage must be effective on a centuries-to-millennia time scale. Owing to the creep properties of salt, a cavern filled with supercritical CO_2 will decrease in volume, until the pressure inside the cavern equalizes the external stress in the salt bed (Bachu and Dusseault, 2005). Although a single cavern 100 m in diameter may hold only about 0.5 Mt of high density CO_2, arrays of caverns could be built for large-scale storage. Cavern sealing is important in preventing leakage and collapse of cavern roofs, which could release large quantities of gas (Katzung *et al.*, 1996). Advantages of CO_2 storage in salt caverns include high capacity per unit volume ($kgCO_2$ m^{-3}), efficiency and injection flow rate. Disadvantages are the potential for CO_2 release in the case of system failure, the relatively small capacity of most individual caverns and the environmental problems of disposing of brine from a solution cavity. Salt caverns can also be used for temporary storage of CO_2 in collector and distributor systems between sources and sinks of CO_2.

5.3.5.4   Abandoned mines

The suitability of mines for CO_2 storage depends on the nature and sealing capacity of the rock in which mining occurs. Heavily fractured rock, typical of igneous and metamorphic terrains, would be difficult to seal. Mines in sedimentary rocks may offer some CO_2-storage opportunities (e.g., potash and salt mines or stratabound lead and zinc deposits). Abandoned coal mines offer the opportunity to store CO_2, with the added benefit of adsorption of CO_2 onto coal remaining in the mined-out area (Piessens and Dusar, 2004). However, the rocks above coal mines are strongly fractured, which increases the risk of gas leakage. In addition, long-term, safe, high-pressure, CO_2-resistant shaft seals have not been developed and any shaft failure could result in release of large quantities of CO_2. Nevertheless, in Colorado, USA, there is a natural gas storage facility in an abandoned coal mine.

5.3.6   Effects of impurities on storage capacity

The presence of impurities in the CO_2 gas stream affects the engineering processes of capture, transport and injection (Chapters 3 and 4), as well as the trapping mechanisms and capacity for CO_2 storage in geological media. Some contaminants in the CO_2 stream (e.g., SO_x, NO_x, H_2S) may require classification as hazardous, imposing different requirements for injection and disposal than if the stream were pure (Bergman *et al.*, 1997). Gas impurities in the CO_2 stream affect the compressibility of the injected CO_2 (and hence the volume needed for storing a given amount) and reduce the capacity for storage in free phase,

because of the storage space taken by these gases. Additionally, depending on the type of geological storage, the presence of impurities may have some other specific effects.

In EOR operations, impurities affect the oil recovery because they change the solubility of CO_2 in oil and the ability of CO_2 to vaporize oil components (Metcalfe, 1982). Methane and nitrogen decrease oil recovery, whereas hydrogen sulphide, propane and heavier hydrocarbons have the opposite effect (Alston *et al.*, 1985; Sebastian *et al.*, 1985). The presence of SO_x may improve oil recovery, whereas the presence of NO_x can retard miscibility and thus reduce oil recovery (Bryant and Lake, 2005) and O_2 can react exothermally with oil in the reservoir.

In the case of CO_2 storage in deep saline formations, the presence of gas impurities affects the rate and amount of CO_2 storage through dissolution and precipitation. Additionally, leaching of heavy metals from the minerals in the rock matrix by SO_2 or O_2 contaminants is possible. Experience to date with acid gas injection (Section 5.2.4.2) suggests that the effect of impurities is not significant, although Knauss *et al.* (2005) suggest that SO_x injection with CO_2 produces substantially different chemical, mobilization and mineral reactions. Clarity is needed about the range of gas compositions that industry might wish to store, other than pure CO_2 (Anheden *et al.*, 2005), because although there might be environmental issues to address, there might be cost savings in co-storage of CO_2 and contaminants.

In the case of CO_2 storage in coal seams, impurities may also have a positive or negative effect, similar to EOR operations. If a stream of gas containing H_2S or SO_2 is injected into coal beds, these will likely be preferentially adsorbed because they have a higher affinity to coal than CO_2, thus reducing the storage capacity for CO_2 (Chikatamarla and Bustin, 2003). If oxygen is present, it will react irreversibly with the coal, reducing the sorption surface and, hence, the adsorption capacity. On the other hand, some impure CO_2 waste streams, such as coal-fired flue gas (i.e., primarily N_2 + CO_2), may be used for ECBM because the CO_2 is stripped out (retained) by the coal reservoir, because it has higher sorption selectivity than N_2 and CH_4.

5.3.7   Geographical distribution and storage capacity estimates

Identifying potential sites for CO_2 geological storage and estimating their capacity on a regional or local scale should conceptually be a simple task. The differences between the various mechanisms and means of trapping (Sections 5.2.2) suggest in principle the following methods:
- For volumetric trapping, capacity is the product of available volume (pore space or cavity) and CO_2 density at *in situ* pressure and temperature;
- For solubility trapping, capacity is the amount of CO_2 that can be dissolved in the formation fluid (oil in oil reservoirs, brackish water or brine in saline formations);
- For adsorption trapping, capacity is the product of coal volume and its capacity for adsorbing CO_2;

- For mineral trapping, capacity is calculated on the basis of available minerals for carbonate precipitation and the amount of CO_2 that will be used in these reactions.

The major impediments to applying these simple methods for estimating the capacity for CO_2 storage in geological media are the lack of data, their uncertainty, the resources needed to process data when available and the fact that frequently more than one trapping mechanism is active. This leads to two situations:

- Global capacity estimates have been calculated by simplifying assumptions and using very simplistic methods and hence are not reliable;
- Country- and region- or basin-specific estimates are more detailed and precise, but are still affected by the limitations imposed by availability of data and the methodology used. Country- or basin-specific capacity estimates are available only for North America, Western Europe, Australia and Japan.

The geographical distribution and capacity estimates are presented below and summarized in Table 5.2.

5.3.7.1 Storage in oil and gas reservoirs

This CO_2 storage option is restricted to hydrocarbon-producing basins, which represent numerically less than half of the sedimentary provinces in the world. It is generally assumed that oil and gas reservoirs can be used for CO_2 storage after their oil or gas reserves are depleted, although storage combined with enhanced oil or gas production can occur sooner. Short of a detailed, reservoir-by-reservoir analysis, the CO_2 storage capacity can and should be calculated from databases of reserves and production (e.g., Winter and Bergman, 1993; Stevens *et al.*, 2001b; Bachu and Shaw, 2003, 2005; Beecy and Kuuskra, 2005).

In hydrocarbon reservoirs with little water encroachment, the injected CO_2 will generally occupy the pore volume previously occupied by oil and/or natural gas. However, not all the previously (hydrocarbon-saturated) pore space will be available for CO_2 because some residual water may be trapped in the pore space due to capillarity, viscous fingering and gravity effects (Stevens *et al.*, 2001c). In open hydrocarbon reservoirs (where pressure is maintained by water influx), in addition to the capacity reduction caused by capillarity and other local effects, a significant fraction of the pore space will be invaded by water, decreasing the pore space available for CO_2 storage,

if repressuring the reservoir is limited to preserve reservoir integrity. In Western Canada, this loss was estimated to be in the order of 30% for gas reservoirs and 50% for oil reservoirs if reservoir repressuring with CO_2 is limited to the initial reservoir pressure (Bachu *et al.*, 2004). The capacity estimates presented here for oil and gas reservoirs have not included any 'discounting' that may be appropriate for water-drive reservoirs because detailed site-specific reservoir analysis is needed to assess the effects of water-drive on capacity on a case-by-case basis.

Many storage-capacity estimates for oil and gas fields do not distinguish capacity relating to oil and gas that has already been produced from capacity relating to remaining reserves yet to be produced and that will become available in future years. In some global assessments, estimates also attribute capacity to undiscovered oil and gas fields that might be discovered in future years. There is uncertainty about when oil and gas fields will be depleted and become available for CO_2 storage. The depletion of oil and gas fields is mostly affected by economic rather than technical considerations, particularly oil and gas prices. It is possible that production from near-depleted fields will be extended if future economic considerations allow more hydrocarbons to be recovered, thus delaying access to such fields for CO_2 storage. Currently few of the world's large oil and gas fields are depleted.

A variety of regional and global estimates of storage capacity in oil and gas fields have been made. Regional and national assessments use a 'bottom-up' approach that is based on field reserves data from each area's existing and discovered oil and gas fields. Although the methodologies used may differ, there is a higher level of confidence in these than the global estimates, for the reasons outlined previously. Currently, this type of assessment is available only for northwestern Europe, United States, Canada and Australia. In Europe, there have been three bottom-up attempts to estimate the CO_2 storage capacity of oil and gas reservoirs covering parts of Europe, but comprising most of Europe's storage capacity since they include the North Sea (Holloway, 1996; Wildenborg *et al.*, 2005b). The methodology used in all three studies was based on the assumption that the total reservoir volume of hydrocarbons could be replaced by CO_2. The operators' estimate of 'ultimately recoverable reserves' (URR) was used for each field where available or was estimated. The underground volume occupied by the URR and the amount of CO_2 that could be stored in that space under reservoir conditions was then calculated. Undiscovered reserves were excluded. For Canada, the assumption was that

Table 5.2 Storage capacity for several geological storage options. The storage capacity includes storage options that are not economical.

Reservoir type	Lower estimate of storage capacity (GtCO$_2$)	Upper estimate of storage capacity (GtCO$_2$)
Oil and gas fields	675[a]	900[a]
Unminable coal seams (ECBM)	3-15	200
Deep saline formations	1000	Uncertain, but possibly 10^4

[a] These numbers would increase by 25% if "undiscovered" oil and gas fields were included in this assessment.

the produced reserves (not the original oil or gas in place) could be replaced by CO_2 (theoretical capacity) for all reservoirs in Western Canada, on the basis of *in situ* pressure, temperature and pore volume. Reduction coefficients were then applied to account for aquifer invasion and all other effects (effective capacity). This value was then reduced for depth (900–3500 m) and size (practical capacity) (Bachu and Shaw, 2005).

The storage potential of northwestern Europe is estimated at more than 40 $GtCO_2$ for gas reservoirs and 7 $GtCO_2$ for oil fields (Wildenborg *et al.*, 2005b). The European estimates are based on all reserves (no significant fields occur above 800 m). Carbon dioxide density was calculated from the depth, pressure and temperature of fields in most cases; where these were not available, a density of 700 kg m^{-3} was used. No assumption was made about the amount of oil recovered from the fields before CO_2 storage was initiated and tertiary recovery by EOR was not included. In Western Canada, the practical CO_2 storage potential in the Alberta and Williston basins in reservoirs with capacity more than 1 $MtCO_2$ each was estimated to be about 1 $GtCO_2$ in oil reservoirs and about 4 $GtCO_2$ in gas reservoirs. The capacity in all discovered oil and gas reservoirs is approximately 10 $GtCO_2$ (Bachu *et al.*, 2004; Bachu and Shaw, 2005). For Canada, the CO_2 density was calculated for each reservoir from the pressure and temperature. The oil and gas recovery was that provided in the reserves databases or was based on actual production. For reservoirs suitable for EOR, an analytical method was developed to estimate how much would be produced and how much CO_2 would be stored (Shaw and Bachu, 2002). In the United States, the total storage capacity in discovered oil and gas fields is estimated to be approximately 98 $GtCO_2$ (Winter and Bergman, 1993; Bergman *et al.*, 1997). Data on production to date and known reserves and resources indicate that Australia has up to 15 $GtCO_2$ storage capacity in gas reservoirs and 0.7 $GtCO_2$ in oil reservoirs. The Australian estimates used field data to recalculate the CO_2 that could occupy the producible volume at field conditions. The total storage capacity in discovered fields for these regions with bottom-up assessments is 170 $GtCO_2$.

Although not yet assessed, it is almost certain that significant storage potential exists in all other oil and gas provinces around the world, such as the Middle East, Russia, Asia, Africa and Latin America.

Global capacity for CO_2-EOR opportunities is estimated to have a geological storage capacity of 61–123 $GtCO_2$, although as practised today, CO_2-EOR is not engineered to maximize CO_2 storage. In fact, it is optimized to maximize revenues from oil production, which in many cases requires minimizing the amount of CO_2 retained in the reservoir. In the future, if storing CO_2 has an economic value, co-optimizing CO_2 storage and EOR may increase capacity estimates. In European capacity studies, it was considered likely that EOR would be attempted at all oil fields where CO_2 storage took place, because it would generate additional revenue. The calculation in Wildenborg *et al.* (2005b) allows for different recovery factors based on API (American Petroleum Institute) gravity of oil. For Canada, all 10,000 oil reservoirs in Western Canada were screened for suitability for EOR on the basis of a set of criteria developed

from EOR literature. Those oil reservoirs that passed were considered further in storage calculations (Shaw and Bachu, 2002).

Global estimates of storage capacity in oil reservoirs vary from 126 to 400 $GtCO_2$ (Freund, 2001). These assessments, made on a top-down basis, include potential in undiscovered reservoirs. Comparable global capacity for CO_2 storage in gas reservoirs is estimated at 800 $GtCO_2$ (Freund, 2001). The combined estimate of total ultimate storage capacity in discovered oil and gas fields is therefore very likely 675–900 $GtCO_2$. If undiscovered oil and gas fields are included, this figure would increase to 900–1200 $GtCO_2$, but the confidence level would decrease.[1]

In comparison, more detailed regional estimates made for northwestern Europe, United States, Australia and Canada indicate a total of about 170 $GtCO_2$ storage capacity in their existing oil and gas fields, with the discovered oil and gas reserves of these countries accounting for 18.9% of the world total (USGS, 2001a). Global storage estimates that are based on proportionality suggest that discovered worldwide oil and gas reservoirs have a capacity of 900 $GtCO_2$, which is comparable to the global estimates by Freund (2001) of 800 $GtCO_2$ for gas (Stevens *et al.*, 2000) and 123 $GtCO_2$ for oil and is assessed as a reliable value, although water invasion was not always taken into account.

5.3.7.2 Storage in deep saline formations

Saline formations occur in sedimentary basins throughout the world, both onshore and on the continental shelves (Chapter 2 and Section 5.3.3) and are not limited to hydrocarbon provinces or coal basins. However, estimating the CO_2 storage capacity of deep saline formations is presently a challenge for the following reasons:

- There are multiple mechanisms for storage, including physical trapping beneath low permeability caprock, dissolution and mineralization;
- These mechanisms operate both simultaneously and on different time scales, such that the time frame of CO_2 storage affects the capacity estimate; volumetric storage is important initially, but later CO_2 dissolves and reacts with minerals;
- Relations and interactions between these various mechanisms are very complex, evolve with time and are highly dependent on local conditions;
- There is no single, consistent, broadly available methodology for estimating CO_2 storage capacity (various studies have used different methods that do not allow comparison);
- Only limited seismic and well data are normally available (unlike data on oil and gas reservoirs).

To understand the difficulties in assessing CO_2 storage capacity in deep saline formations, we need to understand the interplay

---

[1] Estimates of the undiscovered oil and gas are based on the USGS assessment that 30% more oil and gas will be discovered, compared to the resources known today.

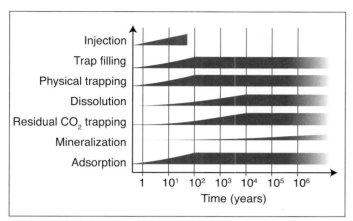

Figure 5.18 Schematic showing the time evolution of various CO_2 storage mechanisms operating in deep saline formations, during and after injection. Assessing storage capacity is complicated by the different time and spatial scales over which these processes occur.

of the various trapping mechanisms during the evolution of a CO_2 plume (Section 5.2 and Figure 5.18). In addition, the storage capacity of deep saline formations can be determined only on a case-by-case basis.

To date, most of the estimates of CO_2 storage capacity in deep saline formations focus on physical trapping and/or dissolution. These estimates make the simplifying assumption that no geochemical reactions take place concurrent with CO_2 injection, flow and dissolution. Some recent work suggests that it can take several thousand years for geochemical reactions to have a significant impact (Xu *et al.*, 2003). The CO_2 storage capacity from mineral trapping can be comparable to the capacity in solution per unit volume of sedimentary rock when formation porosity is taken into account (Bachu and Adams, 2003; Perkins *et al.*, 2005), although the rates and time frames of these two processes are different.

More than 14 global assessments of capacity have been made by using these types of approaches (IEA-GHG, 2004). The range of estimates from these studies is large (200–56,000 $GtCO_2$), reflecting both the different assumptions used to make these estimates and the uncertainty in the parameters. Most of the estimates are in the range of several hundred Gtonnes of CO_2. Volumetric capacity estimates that are based on local, reservoir-scale numerical simulations of CO_2 injection suggest occupancy of the pore space by CO_2 on the order of a few percent as a result of gravity segregation and viscous fingering (van der Meer, 1992, 1995; Krom *et al.*, 1993; Ispen and Jacobsen, 1996). Koide *et al.* (1992) used the areal method of projecting natural resources reserves and assumed that 1% of the total area of the world's sedimentary basins can be used for CO_2 storage. Other studies considered that 2–6% of formation area can be used for CO_2 storage. However, Bradshaw and Dance (2005) have shown there is no correlation between geographic area of a sedimentary basin and its capacity for either hydrocarbons (oil and gas reserves) or CO_2 storage.

The storage capacity of Europe has been estimated as 30–577 $GtCO_2$ (Holloway, 1996; Bøe *et al.*, 2002; Wildenborg *et al.*, 2005b). The main uncertainties for Europe are estimates of

the amount trapped (estimated to be 3%) and storage efficiency, estimated as 2–6% (2% for closed aquifer with permeability barriers; 6% for open aquifer with almost infinite extent), 4% if open/closed status is not known. The volume in traps is assumed to be proportional to the total pore volume, which may not necessarily be correct. Early estimates of the total US storage capacity in deep saline formations suggested a total of up to 500 $GtCO_2$ (Bergman and Winter, 1995). A more recent estimate of the capacity of a single deep formation in the United States, the Mount Simon Sandstone, is 160–800 $GtCO_2$ (Gupta *et al.*, 1999), suggesting that the total US storage capacity may be higher than earlier estimates. Assuming that CO_2 will dissolve to saturation in all deep formations, Bachu and Adams (2003) estimated the storage capacity of the Alberta basin in Western Canada to be approximately 4000 $GtCO_2$, which is a theoretical maximum assuming that all the pore water in the Alberta Basin could become saturated with CO_2, which is not likely. An Australian storage capacity estimate of 740 $GtCO_2$ was determined by a cumulative risked-capacity approach for 65 potentially viable sites from 48 basins (Bradshaw *et al.*, 2003). The total capacity in Japan has been estimated as 1.5–80 $GtCO_2$, mostly in offshore formations (Tanaka *et al.*, 1995).

Within these wide ranges, the lower figure is generally the estimated storage capacity of volumetric traps within the deep saline formations, where free-phase CO_2 would accumulate. The larger figure is based on additional storage mechanisms, mainly dissolution but also mineral trapping. The various methods and data used in these capacity estimates demonstrate a high degree of uncertainty in estimating regional or global storage capacity in deep saline formations. In the examples from Europe and Japan, the maximum estimate is 15 to 50 times larger than the low estimate. Similarly, global estimates of storage capacity show a wide range, 100–200,000 $GtCO_2$, reflecting different methodologies, levels of uncertainties and considerations of effective trapping mechanisms.

The assessment of this report is that it is very likely that global storage capacity in deep saline formations is at least 1000 $GtCO_2$. Confidence in this assessment comes from the fact that oil and gas fields 'discovered' have a global storage capacity of approximately 675–900 $GtCO_2$ and that they occupy only a small fraction of the pore volume in sedimentary basins, the rest being occupied by brackish water and brine. Moreover, oil and gas reservoirs occur only in about half of the world's sedimentary basins. Additionally, regional estimates suggest that significant storage capacity is available. Significantly more storage capacity is likely to be available in deep saline formations. The literature is not adequate to support a robust estimate of the maximum geological storage capacity. Some studies suggest that it might be little more than 1000 $GtCO_2$, while others indicate that the upper figure could be an order of magnitude higher. More detailed regional and local capacity assessments are required to resolve this issue.

5.3.7.3 Storage in coal
No commercial CO_2-ECBM operations exist and a comprehensive realistic assessment of the potential for CO_2

storage in coal formations has not yet been made. Normally, commercial CBM reservoirs are shallower than 1500 m, whereas coal mining in Europe and elsewhere has reached depths of 1000 m. Because CO_2 should not be stored in coals that could be potentially mined, there is a relatively narrow depth window for CO_2 storage.

Assuming that bituminous coals can adsorb twice as much CO_2 as methane, a preliminary analysis of the theoretical CO_2 storage potential for ECBM recovery projects suggests that approximately 60–200 $GtCO_2$ could be stored worldwide in bituminous coal seams (IEA-GHG, 1998). More recent estimates for North America range from 60 to 90 $GtCO_2$ (Reeves, 2003b; Dooley *et al.*, 2005), by including sub-bituminous coals and lignites. Technical and economic considerations suggest a practical storage potential of approximately 7 $GtCO_2$ for bituminous coals (Gale and Freund, 2001; Gale, 2004). Assuming that CO_2 would not be stored in coal seams without recovering the CBM, a storage capacity of 3–15 $GtCO_2$ is calculated, for a US annual production of CBM in 2003 of approximately 0.04 trillion m^3 and projected global production levels of 0.20 trillion m^3 in the future. This calculation assumes that 0.1 $GtCO_2$ can be stored for every Tcf of produced CBM (3.53 $GtCO_2$ for every trillion m^3) and compares well to Gale (2004).

5.3.8   Matching of CO_2 sources and geological storage sites

Matching of CO_2 sources with geological storage sites requires detailed assessment of source quality and quantity, transport and economic and environmental factors. If the storage site is far from CO_2 sources or is associated with a high level of technical uncertainty, then its storage potential may never be realized.

5.3.8.1   Regional studies
Matching sources of CO_2 to potential storage sites, taking into account projections for future socio-economic development, will be particularly important for some of the rapidly developing economies. Assessment of sources and storage sites, together with numerical simulations, emissions mapping and identification of transport routes, has been undertaken for a number of regions in Europe (Holloway, 1996; Larsen *et al.*, 2005). In Japan, studies have modelled and optimized the linkages between 20 onshore emission regions and 20 offshore storage regions, including both ocean storage and geological storage (Akimoto *et al.*, 2003). Preliminary studies have also begun in India (Garg *et al.*, 2005) and Argentina (Amadeo *et al.*, 2005). For the United States, a study that used a Geographic Information System (GIS) and a broad-based economic analysis (Dooley *et al.*, 2005) shows that about two-thirds of power stations are adjacent to potential geological storage locations, but a number would require transportation of hundreds of kilometres.

Studies of Canadian sedimentary basins that include descriptions of the type of data and flow diagrams of the assessment process have been carried out by Bachu (2003).

Results for the Western Canada Sedimentary Basin show that, while the total capacity of oil and gas reservoirs in the basin is several Gtonnes of CO_2, the capacity of underlying deep saline formations is two to three orders of magnitude higher. Most major CO_2 emitters have potential storage sites relatively close by, with the notable exception of the oil sands plants in northeastern Alberta (current CO_2 emissions of about 20 $MtCO_2$ yr^{-1}).

In Australia, a portfolio approach was undertaken for the continent to identify a range of geological storage sites (Rigg *et al.*, 2001; Bradshaw *et al.*, 2002). The initial assessment screened 300 sedimentary basins down to 48 basins and 65 areas. Methodology was developed for ranking storage sites (technical and economic risks) and proximity of large CO_2 emission sites. Region-wide solutions were sought, incorporating an economic model to assess full project economics over 20 to 30 years, including costs of transport, storage, monitoring and Monte Carlo analysis. The study produced three storage estimates:

- Total capacity of 740 $GtCO_2$, equivalent to 1600 years of current emissions, but with no economic barriers considered;
- 'Realistic' capacity of 100–115 $MtCO_2$ yr^{-1} or 50% of annual stationary emissions, determined by matching sources with the closest viable storage sites and assuming economic incentives for storage;
- 'Cost curve' capacity of 20–180 $MtCO_2$ yr^{-1}, with increasing storage capacity depending on future CO_2 values.

5.3.8.2   Methodology and assessment criteria
Although some commonality exists in the various approaches for capacity assessment, each study is influenced by the available data and resources, the aims of the respective study and whether local or whole-region solutions are being sought. The next level of analysis covers regional aspects and detail at the prospect or project level, including screening and selection of potential CO_2 storage sites on the basis of technical, environmental, safety and economic criteria. Finally, integration and analysis of various scenarios can lead to identification of potential storage sites that should then become targets of detailed engineering and economic studies.

The following factors should be considered when selecting CO_2 storage sites and matching them with CO_2 sources (Winter and Bergman, 1993; Bergman *et al.*, 1997; Kovscek, 2002): volume, purity and rate of the CO_2 stream; suitability of the storage sites, including the seal; proximity of the source and storage sites; infrastructure for the capture and delivery of CO_2; existence of a large number of storage sites to allow diversification; known or undiscovered energy, mineral or groundwater resources that might be compromised; existing wells and infrastructure; viability and safety of the storage site; injection strategies and, in the case of EOR and ECBM, production strategies, which together affect the number of wells and their spacing; terrain and right of way; location of population centres; local expertise; and overall costs and economics.

Although technical suitability criteria are initial indicators for identifying potential CO_2 storage sites, once the best candidates have been selected, further considerations will be controlled by economic, safety and environmental aspects. These criteria must be assessed for the anticipated lifetime of the operation, to ascertain whether storage capacity can match supply volume and whether injection rates can match the supply rate. Other issues might include whether CO_2 sources and storage sites are matched on a one-to-one basis or whether a collection and distribution system is implemented, to form an integrated industrial system. Such deliberations affect cost outcomes, as will the supply rates, through economies of scale. Early opportunities for source-storage matching could involve sites where an economic benefit might accrue through the enhanced production of oil or gas (Holtz *et al.*, 2001; van Bergen *et al.*, 2003b).

Assigning technical risks is important for matching of CO_2 sources and storage sites, for five risk factors: storage capacity, injectivity, containment, site and natural resources (Bradshaw *et al.*, 2002, 2003). These screening criteria introduce reality checks to large storage-capacity estimates and indicate which regions to concentrate upon in future detailed studies. The use of 'cost curve' capacity introduces another level of sophistication that helps in identifying how sensitive any storage capacity estimate is to the cost of CO_2. Combining the technical criteria into an economic assessment reveals that costs are quite project-specific.

5.4 Characterization and performance prediction for identified sites

Key goals for geological CO_2 storage site characterization are to assess how much CO_2 can be stored at a potential storage site and to demonstrate that the site is capable of meeting required storage performance criteria (Figure 5.19). Site characterization requires the collection of the wide variety of geological data that are needed to achieve these goals. Much of the data will necessarily be site-specific. Most data will be integrated into geological models that will be used to simulate and predict the performance of the site. These and related issues are considered below.

5.4.1 Characterization of identified sites

Storage site requirements depend greatly upon the trapping mechanism and the geological medium in which storage is proposed (e.g., deep saline formation, depleted oil or gas field or coal seam). Data availability and quality vary greatly between each of these options (Table 5.3). In many cases, oil and gas fields will be better characterized than deep saline formations because a relevant data set was collected during hydrocarbon exploration and production. However, this may not always be the case. There are many examples of deep saline formations whose character and performance for CO_2 storage can be predicted reliably over a large area (Chadwick *et al.*, 2003; Bradshaw *et al.*, 2003).

5.4.1.1 Data types

The storage site and its surroundings need to be characterized in terms of geology, hydrogeology, geochemistry and geomechanics (structural geology and deformation in response to stress changes). The greatest emphasis will be placed on the reservoir and its sealing horizons. However, the strata above the storage formation and caprock also need to be assessed because if CO_2 leaked it would migrate through them (Haidl *et al.*, 2005). Documentation of the characteristics of any particular storage site will rely on data that have been obtained directly from the reservoir, such as core and fluids produced from wells at or near the proposed storage site, pressure transient tests conducted to test seal efficiency and indirect remote sensing measurements such as seismic reflection data and regional hydrodynamic pressure gradients. Integration of all of the different types of data is needed to develop a reliable model that can be used to assess whether a site is suitable for CO_2 storage.

During the site-selection process that may follow an initial screening, detailed reservoir simulation (Section 5.4.2 will be necessary to meaningfully assess a potential storage site. A range of geophysical, geological, hydrogeological and geomechanical information is required to perform the modelling associated with a reservoir simulation. This information must be built into a three-dimensional geological model, populated with known and extrapolated data at an appropriate scale. Examples of the basic types of data and products that may be useful are listed in Table 5.3.

Financial constraints may limit the types of data that can be collected as part of the site characterization and selection process. Today, no standard methodology prescribes how a site must be characterized. Instead, selections about site characterization data will be made on a site-specific basis, choosing those data sets that will be most valuable in the particular geological setting. However, some data sets are likely to be selected for every case. Geological site description from wellbores and outcrops are needed to characterize the storage formation and seal properties. Seismic surveys are needed to define the subsurface geological structure and identify faults or fractures that could create leakage pathways. Formation pressure measurements are needed to map the rate and direction of groundwater flow. Water quality samples are needed to demonstrate the isolation between deep and shallow groundwater.

5.4.1.2 Assessment of stratigraphic factors affecting site integrity

Caprocks or seals are the permeability barriers (mostly vertical but sometimes lateral) that prevent or impede migration of CO_2 from the injection site. The integrity of a seal depends on spatial distribution and physical properties. Ideally, a sealing rock unit should be regional in nature and uniform in lithology, especially at its base. Where there are lateral changes in the basal units of a seal rock, the chance of migration out of the primary reservoir into higher intervals increases. However, if the seal rock is uniform, regionally extensive and thick, then the main issues will be the physical rock strength, any natural or anthropomorphic penetrations (faults, fractures and wells) and

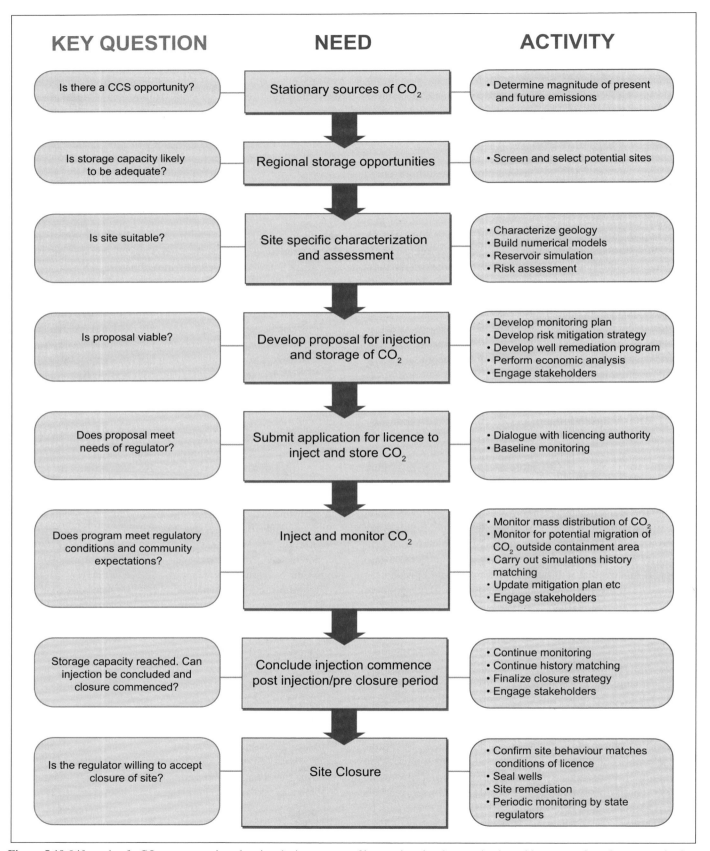

Figure 5.19 Life cycle of a CO_2 storage project showing the importance of integrating site characterization with a range of regulatory, monitoring, economic, risking and engineering issues.

Table 5.3 Types of data that are used to characterize and select geological CO_2 storage sites.

- Seismic profiles across the area of interest, preferably three-dimensional or closely spaced two-dimensional surveys;
- Structure contour maps of reservoirs, seals and aquifers;
- Detailed maps of the structural boundaries of the trap where the CO_2 will accumulate, especially highlighting potential spill points;
- Maps of the predicted pathway along which the CO_2 will migrate from the point of injection;
- Documentation and maps of faults and fault;
- Facies maps showing any lateral facies changes in the reservoirs or seals;
- Core and drill cuttings samples from the reservoir and seal intervals;
- Well logs, preferably a consistent suite, including geological, geophysical and engineering logs;
- Fluid analyses and tests from downhole sampling and production testing;
- Oil and gas production data (if a hydrocarbon field);
- Pressure transient tests for measuring reservoir and seal permeability;
- Petrophysical measurements, including porosity, permeability, mineralogy (petrography), seal capacity, pressure, temperature, salinity and laboratory rock strength testing;
- Pressure, temperature, water salinity;
- In situ stress analysis to determine potential for fault reactivation and fault slip tendency and thus identify the maximum sustainable pore fluid pressure during injection in regard to the reservoir, seal and faults;
- Hydrodynamic analysis to identify the magnitude and direction of water flow, hydraulic interconnectivity of formations and pressure decrease associated with hydrocarbon production;
- Seismological data, geomorphological data and tectonic investigations to indicate neotectonic activity.

potential CO_2-water-rock reactions that could weaken the seal rock or increase its porosity and permeability.

Methods have been described for making field-scale measurements of the permeability of caprocks for formation gas storage projects, based on theoretical developments in the 1950s and 1960s (Hantush and Jacobs, 1955; Hantush, 1960). These use water-pumping tests to measure the rate of leakage across the caprock (Witherspoon *et al.*, 1968). A related type of test, called a pressure 'leak-off' test, can be used to measure caprock permeability and *in situ* stress. The capacity of a seal rock to hold back fluids can also be estimated from core samples by mercury injection capillary pressure (MICP) analysis, a method widely used in the oil and gas industry (Vavra *et al.*, 1992). MICP analysis measures the pressures required to move mercury through the pore network system of a seal rock. The resulting data can be used to derive the height of a column of reservoir rock saturated by a particular fluid (e.g., CO_2) that the sealing strata would be capable of holding back (Gibson-Poole *et al.*, 2002).

5.4.1.3   Geomechanical factors affecting site integrity
When CO_2 is injected into a porous and permeable reservoir rock, it will be forced into pores at a pressure higher than that in the surrounding formation. This pressure could lead to deformation of the reservoir rock or the seal rock, resulting in the opening of fractures or failure along a fault plane. Geomechanical modelling of the subsurface is necessary in any storage site assessment and should focus on the maximum formation pressures that can be sustained in a storage site. As an example, at Weyburn, where the initial reservoir pressure is 14.2 MPa, the maximum injection pressure (90% of fracture pressure) is in the range of 25–27 MPa and fracture pressure is in the range of 29–31 MPa. Coupled geomechanical-geochemical modelling may also be needed to document fracture sealing by precipitation of carbonates in fractures or pores. Modelling these will require knowledge of pore fluid composition, mineralogy,

in situ stresses, pore fluid pressures and pre-existing fault orientations and their frictional properties (Streit and Hillis, 2003; Johnson *et al.*, 2005). These estimates can be made from conventional well and seismic data and leak-off tests, but the results can be enhanced by access to physical measurements of rock strength. Application of this methodology at a regional scale is documented by Gibson-Poole *et al.* (2002).

The efficacy of an oil or gas field seal rock can be characterized by examining its capillary entry pressure and the potential hydrocarbon column height that it can sustain (see above). However, Jimenez and Chalaturnyk (2003) suggest that the geomechanical processes, during depletion and subsequent CO_2 injection, may affect the hydraulic integrity of the seal rock in hydrocarbon fields. Movement along faults can be produced in a hydrocarbon field by induced changes in the pre-production stress regime. This can happen when fluid pressures are substantially depleted during hydrocarbon production (Streit and Hillis, 2003). Determining whether the induced stress changes result in compaction or pore collapse is critical in assessment of a depleted field. If pore collapse occurs, then it might not be possible to return a pressure-depleted field to its original pore pressure without the risk of induced failure. By having a reduced maximum pore fluid pressure, the total volume of CO_2 that can be stored in a depleted field could be substantially less than otherwise estimated.

5.4.1.4   Geochemical factors affecting site integrity
The mixing of CO_2 and water in the pore system of the reservoir rock will create dissolved CO_2, carbonic acid and bicarbonate ions. The acidification of the pore water reduces the amount of CO_2 that can be dissolved. As a consequence, rocks that buffer the pore water pH to higher values (reducing the acidity) facilitate the storage of CO_2 as a dissolved phase (Section 5.2). The CO_2-rich water may react with minerals in the reservoir rock or caprock matrix or with the primary pore fluid. Importantly, it may also react with borehole cements and steels (see discussion

below). Such reactions may cause either mineral dissolution and potential breakdown of the rock (or cement) matrix or mineral precipitation and plugging of the pore system (and thus, reduction in permeability).

A carbonate mineral formation effectively traps stored CO_2 as an immobile solid phase (Section 5.2). If the mineralogical composition of the rock matrix is strongly dominated by quartz, geochemical reactions will be dominated by simple dissolution into the brine and CO_2-water-rock reactions can be neglected. In this case, complex geochemical simulations of rock-water interactions will not be needed. However, for more complex mineralogies, sophisticated simulations, based on laboratory experimental data that use reservoir and caprock samples and native pore fluids, may be necessary to fully assess the potential effects of such reactions in more complex systems (Bachu *et al.*, 1994; Czernichowski-Lauriol *et al.*, 1996; Rochelle *et al.*, 1999, 2004; Gunter *et al.*, 2000). Studies of rock samples recovered from natural systems rich in CO_2 can provide indications of what reactions might occur in the very long term (Pearce *et al.*, 1996). Reactions in boreholes are considered by Crolet (1983), Rochelle *et al.* (2004) and Schremp and Roberson (1975). Natural CO_2 reservoirs also allow sampling of solid and fluid reactants and reaction products, thus allowing formulation of geochemical models that can be verified with numerical simulations, further facilitating quantitative predictions of water-CO_2-rock reactions (May, 1998).

5.4.1.5 Anthropogenic factors affecting storage integrity

As discussed at greater length in Section 5.7.2, anthropogenic factors such as active or abandoned wells, mine shafts and subsurface production can impact storage security. Abandoned wells that penetrate the storage formation can be of particular concern because they may provide short circuits for CO_2 to leak from the storage formation to the surface (Celia and Bachu, 2003; Gasda *et al.*, 2004). Therefore, locating and assessing the condition of abandoned and active wells is an important component of site characterization. It is possible to locate abandoned wells with airborne magnetometer surveys. In most cases, abandoned wells will have metal casings, but this may not be the case for wells drilled long ago or those never completed for oil or gas production. Countries with oil and gas production will have at least some records of the more recently drilled wells, depth of wells and other information stored in a geographic database. The consistency and quality of record keeping of drilled wells (oil and gas, mining exploration and water) varies considerably, from excellent for recent wells to nonexistent, particularly for older wells (Stenhouse *et al.*, 2004).

5.4.2 Performance prediction and optimization modelling

Computer simulation also has a key role in the design and operation of field projects for underground injection of CO_2. Predictions of the storage capacity of the site or the expected incremental recovery in enhanced recovery projects, are vital to

an initial assessment of economic feasibility. In a similar vein, simulation can be used in tandem with economic assessments to optimize the location, number, design and depth of injection wells. For enhanced recovery projects, the timing of CO_2 injection relative to production is vital to the success of the operation and the effect of various strategies can be assessed by simulation. Simulations of the long-term distribution of CO_2 in the subsurface (e.g., migration rate and direction and rate of dissolution in the formation water) are important for the design of cost-effective monitoring programmes, since the results will influence the location of monitoring wells and the frequency of repeat measurements, such as for seismic, soil gas or water chemistry. During injection and monitoring operations, simulation models can be adjusted to match field observations and then used to assess the impact of possible operational changes, such as drilling new wells or altering injection rates, often with the goal of further improving recovery (in the context of hydrocarbon extraction) or of avoiding migration of CO_2 past a likely spill-point.

Section 5.2 described the important physical, chemical and geomechanical processes that must be considered when evaluating a storage project. Numerical simulators currently in use in the oil, gas and geothermal energy industries provide important subsets of the required capabilities. They have served as convenient starting points for recent and ongoing development efforts specifically targeted at modelling the geological storage of CO_2. Many simulation codes have been used and adapted for this purpose (White, 1995; Nitao, 1996; White and Oostrom, 1997; Pruess *et al.*, 1999; Lichtner, 2001; Steefel, 2001; Xu *et al.*, 2003).

Simulation codes are available for multiphase flow processes, chemical reactions and geomechanical changes, but most codes account for only a subset of these processes. Capabilities for a comprehensive treatment of different processes are limited at present. This is especially true for the coupling of multiphase fluid flow, geochemical reactions and (particularly) geomechanics, which are very important for the integrity of potential geological storage sites (Rutqvist and Tsang, 2002). Demonstrating that they can model the important physical and chemical processes accurately and reliably is necessary for establishing credibility as practical engineering tools. Recently, an analytical model developed for predicting the evolution of a plume of CO_2 injected into a deep saline formation, as well as potential CO_2 leakage rates through abandoned wells, has shown good matching with results obtained from the industry numerical simulator ECLIPSE (Celia *et al.*, 2005; Nordbotten *et al.*, 2005b).

A code intercomparison study involving ten research groups from six countries was conducted recently to evaluate the capabilities and accuracy of numerical simulators for geological storage of greenhouse gases (Pruess *et al.*, 2004). The test problems addressed CO_2 storage in saline formations and oil and gas reservoirs. The results of the intercomparison were encouraging in that substantial agreement was found between results obtained with different simulators. However, there were also areas with only fair agreement, as well as some significant discrepancies. Most discrepancies could be traced to differences in fluid property descriptions, such as fluid densities and viscosities and mutual solubility of CO_2 and water. The study concluded that 'although code development work undoubtedly must continue . . . codes are available now that can model the complex phenomena accompanying geological storage of CO_2 in a robust manner and with quantitatively similar results' (Pruess *et al.*, 2004).

Another, similar intercomparison study was conducted for simulation of storage of CO_2 in coal beds, considering both pure CO_2 injection and injection of flue gases (Law *et al.*, 2003). Again, there was good agreement between the simulation results from different codes. Code intercomparisons are useful for checking mathematical methods and numerical approximations and to provide insight into relevant phenomena by using the different descriptions of the physics (or chemistry) implemented. However, establishing the realism and accuracy of physical and chemical process models is a more demanding task, one that requires carefully controlled and monitored field and laboratory experiments. Only after simulation models have been shown to be capable of adequately representing real-world observations can they be relied upon for engineering design and analysis. Methods for calibrating models to complex engineered subsurface systems are available, but validating them requires field testing that is time consuming and expensive.

The principal difficulty is that the complex geological models on which the simulation models are based are subject to considerable uncertainties, resulting both from uncertainties in data interpretation and, in some cases, sparse data sets. Measurements taken at wells provide information on rock and fluid properties at that location, but statistical techniques must be used to estimate properties away from the wells. When simulating a field in which injection or production is already occurring, a standard approach in the oil and gas industry is to adjust some parameters of the geological model to match selected field observations. This does not prove that the model is correct, but it does provide additional constraints on the model parameters. In the case of saline formation storage, history matching is generally not feasible for constraining uncertainties, due to a lack of underground data for comparison. Systematic parameter variation routines and statistical functions should be included in future coupled simulators to allow uncertainty estimates for numerical reservoir simulation results.

Field tests of CO_2 injection are under way or planned in several countries and these tests provide opportunities to validate simulation models. For example, in Statoil's Sleipner project, simulation results have been matched to information on the distribution of CO_2 in the subsurface, based on the interpretation of repeat three-dimensional seismic surveys (Lindeberg *et al.*, 2001; van der Meer *et al.*, 2001; see also Section 5.4.3). At the Weyburn project in Canada, repeat seismic surveys and water chemistry sampling provide information on CO_2 distribution that can likewise be used to adjust the simulation models (Moberg *et al.*, 2003; White *et al.*, 2004).

Predictions of the long-term distribution of injected CO_2, including the effects of geochemical reactions, cannot be directly validated on a field scale because these reactions may take hundreds to thousands of years. However, the simulation of important mechanisms, such as the convective mixing of dissolved CO_2, can be tested by comparison to laboratory analogues (Ennis-King and Paterson, 2003). Another possible route is to match simulations to the geochemical changes that have occurred in appropriate natural underground accumulations of CO_2, such as the precipitation of carbonate minerals, since these provide evidence for the slow processes that affect the long-term distribution of CO_2 (Johnson *et al.*, 2005). It is also important to have reliable and accurate data regarding the thermophysical properties of CO_2 and mixtures of CO_2 with methane, water and potential contaminants such as H_2S and SO_2. Similarly, it is important to have data on relative permeability and capillary pressure under drainage and imbibition conditions. Code comparison studies show that the largest discrepancies between different simulators can be traced to uncertainties in these parameters (Pruess *et al.*, 2004). For sites where few, if any, CO_2-water-rock interactions occur, reactive chemical transport modelling may not be needed and simpler simulations that consider only CO_2-water reactions will suffice.

5.4.3 *Examples of storage site characterization and performance prediction*

Following are examples and lessons learned from two case studies of characterization of a CO_2 storage site: one of an actual operating CO_2 storage site (Sleipner Gas Field in the North Sea) and the other of a potential or theoretical site (Petrel Sub-basin offshore northwest Australia). A common theme throughout these studies is the integration and multidisciplinary approach required to adequately document and monitor any injection site. There are lessons to be learned from these studies, because they have identified issues that in hindsight should be examined prior to any CO_2 injection.

5.4.3.1 *Sleipner*
Studies of the Sleipner CO_2 Injection Project (Box 5.1) highlighted the advantages of detailed knowledge of the reservoir stratigraphy (Chadwick *et al.*, 2003). After the initial CO_2 injection, small layers of low-permeability sediments within the saline formation interval and sandy lenses near the base of the seal were clearly seen to be exercising an important control on the distribution of CO_2 within the reservoir rock (Figure 5.16a,b). Time-lapse three-dimensional seismic imaging of the developing CO_2 plume also identified the need for precision depth mapping of the bottom of the caprock interval. At Sleipner, the top of the reservoir is almost flat at a regional scale. Hence, any subtle variance in the actual versus predicted depth could substantially affect migration patterns and rate. Identification and mapping of a sand lens above what was initially interpreted as the top of the reservoir resulted in a significant change to the predicted migration direction of the CO_2 (Figure 5.16a,b). These results show the benefit of repeated three-dimensional seismic monitoring and integration of monitoring results into

modelling during the injection phase of the project. Refinement of the storage-site characterization continues after injection has started.

5.4.3.2 Petrel Sub-basin

A theoretical case study of the Petrel Sub-basin offshore northwest Australia examined the basin-wide storage potential of a combined hydrodynamic and solution trapping mechanism and identified how sensitive a reservoir simulation will be to the collected data and models built during the characterization of a storage site (Gibson-Poole *et al.*, 2002; Ennis-King *et al.*, 2003). As at Sleipner, the Petrel study identified that vertical permeability and shale beds within the reservoir interval of the geological model strongly influenced the vertical CO_2 migration rate. In the reservoir simulation, use of coarser grids overestimated the dissolution rate of CO_2 during the injection period, but underestimated it during the long-term migration period. Lower values of residual CO_2 saturation led to faster dissolution during the long-term migration period and the rate of complete dissolution depended on the vertical permeability. Migration distance depended on the rate of dissolution and residual CO_2 trapping. The conclusion of the characterization and performance prediction studies is that the Petrel Sub-basin has a regionally extensive reservoir-seal pair suitable for hydrodynamic trapping (Section 5.2). While the characterization was performed on the basis of only a few wells with limited data, analogue studies helped define the characteristics of the formation. Although this is not the ideal situation, performing a reservoir simulation by using geological analogues may often be the only option. However, understanding which elements will be the most sensitive in the simulation will help geoscientists to understand where to prioritize their efforts in data collection and interpretation.

5.5   Injection well technology and field operations

So far in this chapter, we have considered only the nature of the storage site. But once a suitable site is identified, do we have the technology available to inject large quantities of CO_2 (1–10 $MtCO_2$ yr^{-1}) into the subsurface and to operate the site effectively and safely? This section examines the issue of technology availability.

5.5.1   Injection well technologies

As pointed out earlier in this chapter, many of the technologies required for large-scale geological storage of CO_2 already exist. Drilling and completion technology for injection wells in the oil and gas industry has evolved to a highly sophisticated state, such that it is now possible to drill and complete vertical and extended reach wells (including horizontal wells) in deep formations, wells with multiple completions and wells able to handle corrosive fluids. On the basis of extensive oil industry experience, the technologies for drilling, injection, stimulations and completions for CO_2 injection wells exist and are being

practised with some adaptations in current CO_2 storage projects. In a CO_2 injection well, the principal well design considerations include pressure, corrosion-resistant materials and production and injection rates.

The design of a CO_2 injection well is very similar to that of a gas injection well in an oil field or natural gas storage project. Most downhole components need to be upgraded for higher pressure ratings and corrosion resistance. The technology for handling CO_2 has already been developed for EOR operations and for the disposal of acid gas (Section 5.2.4.) Horizontal and extended reach wells can be good options for improving the rate of CO_2 injection from individual wells. The Weyburn field in Canada (Box 5.3) is an example in which the use of horizontal injection wells is improving oil recovery and increasing CO_2 storage. The horizontal injectors reduce the number of injection wells required for field development. A horizontal injection well has the added advantage that it can create injection profiles that reduce the adverse effects of injected-gas preferential flow through high-permeability zones.

The number of wells required for a storage project will depend on a number of factors, including total injection rate, permeability and thickness of the formation, maximum injection pressures and availability of land-surface area for the injection wells. In general, fewer wells will be needed for high-permeability sediments in thick storage formations and for those projects with horizontal wells for injection. For example, the Sleipner Project, which injects CO_2 into a high-permeability, 200-m-thick formation uses only one well to inject 1 $MtCO_2$ yr^{-1} (Korbol and Kaddour, 1994). In contrast, at the In Salah Project in Algeria, CO_2 is injected into a 20-m-thick formation with much lower permeability (Riddiford *et al.*, 2003). Here, three long-reach horizontal wells with slotted intervals over 1 km are used to inject 1 $MtCO_2$ yr^{-1} (Figure 5.5). Cost will depend, to some degree, on the number and completion techniques for these wells. Therefore, careful design and optimization of the number and slotted intervals is important for cost-effective storage projects.

An injection well and a wellhead are depicted in Figure 5.20. Injection wells commonly are equipped with two valves for well control, one for regular use and one reserved for safety shutoff. In acid gas injection wells, a downhole safety valve is incorporated in the tubing, so that if equipment fails at the surface, the well is automatically shut down to prevent back flow. Jarrell *et al.* (2002) recommend an automatic shutoff valve on all CO_2 wells to ensure that no release occurs and to prevent CO_2 from inadvertently flowing back into the injection system. A typical downhole configuration for an injection well includes a double-grip packer, an on-off tool and a downhole shutoff valve. Annular pressure monitors help detect leaks in packers and tubing, which is important for taking rapid corrective action. To prevent dangerous high-pressure buildup on surface equipment and avoid CO_2 releases into the atmosphere, CO_2 injection must be stopped as soon as leaks occur. Rupture disks and safety valves can be used to relieve built-up pressure. Adequate plans need to be in place for dealing with excess CO_2 if the injection well needs to be shut in. Options include having

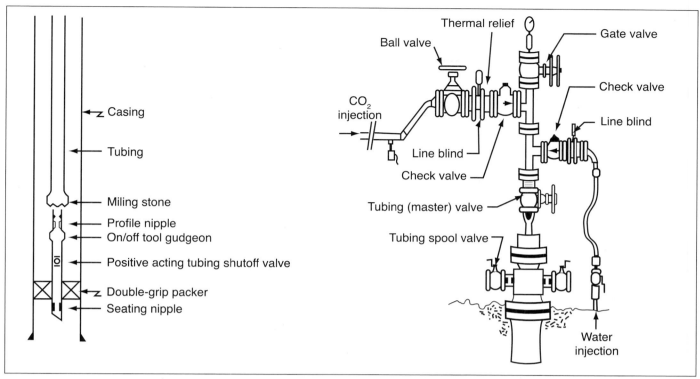

Figure 5.20 Typical CO$_2$ injection well and wellhead configuration.

a backup injection well or methods to safely vent CO$_2$ to the atmosphere.

Proper maintenance of CO$_2$ injection wells is necessary to avoid leakage and well failures. Several practical procedures can be used to reduce probabilities of CO$_2$ blow-out (uncontrolled flow) and mitigate the adverse effects if one should occur. These include periodic wellbore integrity surveys on drilled injection wells, improved blow-out prevention (BOP) maintenance, installation of additional BOP on suspect wells, improved crew awareness, contingency planning and emergency response training (Skinner, 2003).

For CO$_2$ injection through existing and old wells, key factors include the mechanical condition of the well and quality of the cement and well maintenance. A leaking wellbore annulus can be a pathway for CO$_2$ migration. Detailed logging programmes for checking wellbore integrity can be conducted by the operator to protect formations and prevent reservoir cross-flow. A well used for injection (Figure 5.20) must be equipped with a packer to isolate pressure to the injection interval. All materials used in injection wells should be designed to anticipate peak volume, pressure and temperature. In the case of wet gas (containing free water), use of corrosion-resistant material is essential.

5.5.2 Well abandonment procedures

Abandonment procedures for oil, gas and injection wells are designed to protect drinking water aquifers from contamination. If a well remains open after it is no longer in use, brines, hydrocarbons or CO$_2$ could migrate up the well and into shallow drinking water aquifers. To avoid this, many countries

have developed regulations for well 'abandonment' or 'closure' (for example, United States Code of Federal Regulations 40 Part 144 and Alberta Energy and Utilities Board, 2003). These procedures usually require placing cement or mechanical plugs in all or part of the well. Extra care is usually taken to seal the well adjacent to drinking water aquifers. Examples of well abandonment procedures for cased and uncased wells are shown in Figure 5.21. Tests are often required to locate the depth of the plugs and test their mechanical strength under pressure.

It is expected that abandonment procedures for CO$_2$ wells could broadly follow the abandonment methodology used for oil and gas wells and acid-gas disposal wells. However, special care has to be taken to use sealing plugs and cement that are resistant to degradation from CO$_2$. Carbon dioxide-resistant cements have been developed for oil field and geothermal applications. It has been suggested that removing the casing and the liner penetrating the caprock could avoid corrosion of the steel that may later create channels for leakage. The production casing can be removed by pulling or drilling (milling) it out. After removing the casing, a cement plug can be put into the open borehole, as illustrated in Figure 5.21.

The cement plug will act as the main barrier to future CO$_2$ migration. A major issue is related to the sealing quality of the cement plug and the bonding quality with the penetrated caprock. Microchannels created near the wellbore during drilling or milling operations should be sealed with cement. Fluid could also be flushed into the storage reservoir to displace the CO$_2$ and help to improve the cementing quality and bonding to the sealing caprock. Casing protective materials and alternative casing materials, such as composites, should also be evaluated

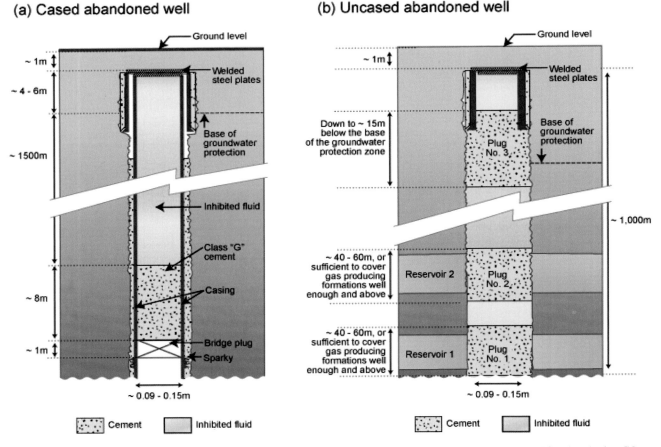

Figure 5.21 Examples of how cased and uncased wells are abandoned today. Special requirements may be developed for abandoning CO_2 storage wells, including use of corrosion-resistant cement plugs and removing all or part of the casing in the injection interval and caprock.

for possible and alternative abandonment procedures. Sealing performance of abandoned wells may need to be monitored for some time after storage operations are completed.

5.5.3   *Injection well pressure and reservoir constraints*

Injectivity characterizes the ease with which fluid can be injected into a geological formation and is defined as the injection rate divided by the pressure difference between the injection point inside the well and the formation. Although CO_2 injectivity should be significantly greater than brine injectivity (because CO_2 has a much lower viscosity than brine), this is not always the case. Grigg (2005) analyzed the performance of CO_2 floods in west Texas and concluded that, in more than half of the projects, injectivity was lower than expected or decreased over time. Christman and Gorell (1990) showed that unexpected CO_2-injectivity behaviour in EOR operations is caused primarily by differences in flow geometry and fluid properties of the oil. Injectivity changes can also be related to insufficiently known relative permeability effects.

To introduce CO_2 into the storage formation, the downhole injection pressure must be higher than the reservoir fluid pressure. On the other hand, increasing formation pressure may induce fractures in the formation. Regulatory agencies

normally limit the maximum downhole pressure to avoid fracturing the injection formation. Measurements of *in-situ* formation stresses and pore fluid pressure are needed for establishing safe injection pressures. Depletion of fluid pressure during production can affect the state of stress in the reservoir. Analysis of some depleted reservoirs indicated that horizontal rock stress decreased by 50–80% of the pore pressure decrease, which increased the possibility of fracturing the reservoir (Streit and Hillis, 2003).

Safe injection pressures can vary widely, depending on the state of stress and tectonic history of a basin. Regulatory agencies have determined safe injection pressures from experience in specific oil and gas provinces. Van der Meer (1996) has derived a relationship for the maximum safe injection pressure. This relationship indicated that for a depth down to 1000 m, the maximum injection pressure is estimated to be 1.35 times the hydrostatic pressure – and this increased to 2.4 for depths of 1–5 km. The maximum pressure gradient allowed for natural gas stored in an aquifer in Germany is 16.8 kPa m^{-1} (Sedlacek, 1999). This value exceeds the natural pressure gradients of formation waters in northeastern Germany, which are on the order of 10.5–13.1 kPa m^{-1}. In Denmark or Great Britain, the maximum pressure gradients for aquifer storage of natural gas do not exceed hydrostatic gradients. In the United States,

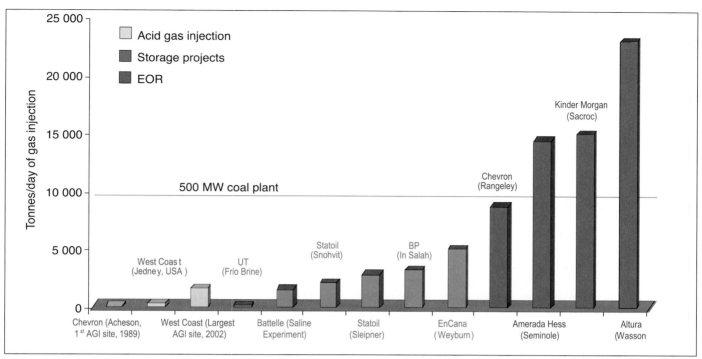

Figure 5.22 Comparison of the magnitude of CO_2 injection activities illustrating that the storage operations from a typical 500-MW coal plant will be the same order of magnitude as existing CO_2 injection operations (after Heinrich *et al.*, 2003).

for industrial waste-water injection wells, injection pressure must not exceed fracture initiation or propagation pressures in the injection formation (USEPA, 1994). For oil and gas field injection wells, injection pressures must not exceed those that would initiate or propagate fractures in the confining units. In the United States, each state has been delegated authority to establish maximum injection pressures. Until the 1990s, many states set state-wide standards for maximum injection pressures; values ranged from 13 to18 kPa m⁻¹. More recently, regulations have changed to require site-specific tests to establish maximum injection pressure gradients. Practical experience in the USEPA's Underground Injection Control Program has shown that fracture pressures range from 11 to 21 kPa m⁻¹.

5.5.4   *Field operations and surface facilities*

Injection rates for selected current CO_2 storage projects in EOR and acid gas injection are compared in Figure 5.22. As indicated, the amount of CO_2 injected from a 500-MW coal-fired power plant would fall within the range of existing experience of CO_2 injection operations for EOR. These examples therefore offer a great deal of insight as to how a geological storage regime might evolve, operate and be managed safely and effectively.

CO_2-EOR operations fall into one of three groups (Jarrell *et al.*, 2002):
- Reservoir management – what to inject, how fast to inject, how much to inject, how to manage water-alternating-gas (WAG), how to maximize sweep efficiency and so on;
- Well management – producing method and remedial work, including selection of workovers, chemical treatment and CO_2 breakthrough;

- Facility management – reinjection plant, separation, metering, corrosion control and facility organization.

Typically, CO_2 is transported from its source to an EOR site through a pipeline and is then injected into the reservoir through an injection well, usually after compression. Before entering the compressor, a suction scrubber will remove any residual liquids present in the CO_2 stream. In EOR operations, CO_2 produced from the production well along with oil and water is separated and then injected back through the injection well.

The field application of CO_2-ECBM technology is broadly similar to that of EOR operations. Carbon dioxide is transported to the CBM field and injected in the coal seam through dedicated injection wells. At the production well, coal-seam gas and formation water is lifted to the surface by electric pumps.

According to Jarrell *et al.* (2002), surface facilities for CO_2-EOR projects include:
- Production systems-fluid separation, gas gathering, production satellite, liquid gathering, central battery, field compression and emergency shutdown systems;
- Injection systems-gas repressurization, water injection and CO_2 distribution systems;
- Gas processing systems-gas processing plant, H_2S removal systems and sulphur recovery and disposal systems.

Jarrell *et al.* (2002) point out that CO_2 facilities are similar to those used in conventional facilities such as for waterfloods. Differences result from the effects of multiphase flow, selection of different materials and the higher pressure that must be handled. The CO_2 field operation setup for the Weyburn Field is shown in Figure 5.23.

It is common to use existing facilities for new CO_2 projects to reduce capital costs, although physical restrictions are always present. Starting a CO_2 flood in an old oil field can affect almost every process and facility (Jarrell *et al.*, 2002); for example, (1) the presence of CO_2 makes the produced water much more corrosive; (2) makeup water from new sources may interact with formation water to create new problems with scale or corrosion; (3) a CO_2 flood may cause paraffins and asphaltenes to precipitate out of the oil, which can cause plugging and emulsion problems; and (4) the potentially dramatic increase in production caused by the flood could cause more formation fines to be entrained in the oil, potentially causing plugging, erosion and processing problems.

5.6   Monitoring and verification technology

What actually happens to CO_2 in the subsurface and how do we know what is happening? In other words, can we monitor CO_2 once it is injected? What techniques are available for monitoring whether CO_2 is leaking out of the storage formation and how sensitive are they? Can we verify that CO_2 is safely and effectively stored underground? How long is monitoring needed? These questions are addressed in this section of the report.

5.6.1   *Purposes for monitoring*

Monitoring is needed for a wide variety of purposes. Specifically, monitoring can be used to:

- Ensure and document effective injection well controls, specifically for monitoring the condition of the injection well and measuring injection rates, wellhead and formation pressures. Petroleum industry experience suggests that leakage from the injection well itself, resulting from improper completion or deterioration of the casing, packers or cement, is one of the most significant potential failure modes for injection projects (Apps, 2005; Perry, 2005);
- Verify the quantity of injected CO_2 that has been stored by various mechanisms;
- Optimize the efficiency of the storage project, including utilization of the storage volume, injection pressures and drilling of new injection wells;
- Demonstrate with appropriate monitoring techniques that CO_2 remains contained in the intended storage formation(s). This is currently the principal method for assuring that the CO_2 remains stored and that performance predictions can be verified;
- Detect leakage and provide an early warning of any seepage or leakage that might require mitigating action.

Figure 5.23 Typical CO_2 field operation setup: Weyburn surface facilities.

In addition to essential elements of a monitoring strategy, other parameters can be used to optimize storage projects, deal with unintended leakage and address regulatory, legal and social issues. Other important purposes for monitoring include assessing the integrity of plugged or abandoned wells, calibrating and confirming performance assessment models (including 'history matching'), establishing baseline parameters for the storage site to ensure that CO_2-induced changes are recognized (Wilson and Monea, 2005), detecting microseismicity associated with a storage project, measuring surface fluxes of CO_2 and designing and monitoring remediation activities (Benson *et al.*, 2004).

Before monitoring of subsurface storage can take place effectively, a baseline survey must be taken. This survey provides the point of comparison for subsequent surveys. This is particularly true of seismic and other remote-sensing technologies, where the identification of saturation of fluids with CO_2 is based on comparative analysis. Baseline monitoring is also a prerequisite for geochemical monitoring, where anomalies are identified relative to background concentrations. Additionally, establishing a baseline of CO_2 fluxes resulting from ecosystem cycling of CO_2, both on diurnal and annual cycles, are useful for distinguishing natural fluxes from potential storage-related releases.

Much of the monitoring technology described below was developed for application in the oil and gas industry. Most of these techniques can be applied to monitoring storage projects in all types of geological formations, although much remains to be learned about monitoring coal formations. Monitoring experience from natural gas storage in saline aquifers can also provide a useful industrial analogue.

5.6.2 *Technologies for monitoring injection rates and pressures*

Measurements of CO_2 injection rates are a common oil field practice and instruments for this purpose are available commercially. Measurements are made by gauges either at the injection wellhead or near distribution manifolds. Typical systems use orifice meters or other devices that relate the pressure drop across the device to the flow rate. The accuracy of the measurements depends on a number of factors that have been described in general by Morrow *et al.* (2003) and specifically for CO_2 by Wright and Majek (1998). For CO_2, accurate estimation of the density is most important for improving measurement accuracy. Small changes in temperature, pressure and composition can have large effects on density. Wright and Majek (1998) developed an oil field CO_2 flow rate system by combining pressure, temperature and differential pressure measurements with gas chromatography. The improved system had an accuracy of 0.6%, compared to 8% for the conventional system. Standards for measurement accuracy vary and are usually established by governments or industrial associations. For example, in the United States, current auditing practices for CO_2-EOR accept flow meter precision of ±4%.

Measurements of injection pressure at the surface and in the formation are also routine. Pressure gauges are installed on most injection wells through orifices in the surface piping near the wellhead. Downhole pressure measurements are routine, but are used for injection well testing or under special circumstances in which surface measurements do not provide reliable information about the downhole pressure. A wide variety of pressure sensors are available and suitable for monitoring pressures at the wellhead or in the formation. Continuous data are available and typically transmitted to a central control room. Surface pressure gauges are often connected to shut-off valves that will stop or curtail injection if the pressure exceeds a predetermined safe threshold or if there is a drop in pressure as a result of a leak. In effect, surface pressures can be used to ensure that downhole pressures do not exceed the threshold of reservoir fracture pressure. A relatively recent innovation, fibre-optic pressure and temperature sensors, is commercially available. Fibre-optic cables are lowered into the wells, connected to sensors and provide real-time formation pressure and temperature measurements. These new systems are expected to provide more reliable measurements and well control.

The current state of the technology is more than adequate to meet the needs for monitoring injection rates, wellhead and formation pressures. Combined with temperature measurements, the collected data will provide information on the state of the CO_2 (supercritical, liquid or gas) and accurate measurement of the amount of CO_2 injected for inventories, reporting and verification, as well as input to modelling. In the case of the Weyburn project, for example, the gas stream is also analyzed to determine the impurities in the CO_2, thus allowing computation of the volume of CO_2 injected.

5.6.3 *Technologies for monitoring subsurface distribution of CO_2*

A number of techniques can be used to monitor the distribution and migration of CO_2 in the subsurface. Table 5.4 summarizes these techniques and how they can be applied to CO_2 storage projects. The applicability and sensitivity of these techniques are somewhat site-specific. Detailed descriptions, including limitations and resolution, are provided in Sections 5.6.3.1 and 5.6.3.2.

5.6.3.1 *Direct techniques for monitoring CO_2 migration*
Direct techniques for monitoring are limited in availability at present. During CO_2 injection for EOR, the injected CO_2 spreads through the reservoir in a heterogeneous manner, because of permeability variations in the reservoir (Moberg *et al.*, 2003). In the case of CO_2-EOR, once the CO_2 reaches a production well, its produced volume can be readily determined. In the case of Weyburn, the carbon in the injected CO_2 has a different isotopic composition from the carbon in the reservoir (Emberley *et al.*, 2002), so the distribution of the CO_2 can be determined on a gross basis by evaluating the arrival of the introduced CO_2 at different production wells. With multiple injection wells in any producing area, the arrival of CO_2 can give only a general indication of distribution in the reservoir.

Table 5.4 Summary of direct and indirect techniques that can be used to monitor CO_2 storage projects.

Measurement technique	Measurement parameters	Example applications
Introduced and natural tracers	Travel time Partitioning of CO_2 into brine or oil Identification sources of CO_2	Tracing movement of CO_2 in the storage formation Quantifying solubility trapping Tracing leakage
Water composition	CO_2, HCO_3^-, CO_3^{2-}· Major ions Trace elements Salinity	Quantifying solubility and mineral trapping Quantifying CO_2-water-rock interactions Detecting leakage into shallow groundwater aquifers
Subsurface pressure	Formation pressure Annulus pressure Groundwater aquifer pressure	Control of formation pressure below fracture gradient Wellbore and injection tubing condition Leakage out of the storage formation
Well logs	Brine salinity Sonic velocity CO_2 saturation	Tracking CO_2 movement in and above storage formation Tracking migration of brine into shallow aquifers Calibrating seismic velocities for 3D seismic surveys
Time-lapse 3D seismic imaging	P and S wave velocity Reflection horizons Seismic amplitude attenuation	Tracking CO_2 movement in and above storage formation
Vertical seismic profiling and crosswell seismic imaging	P and S wave velocity Reflection horizons Seismic amplitude attenuation	Detecting detailed distribution of CO_2 in the storage formation Detection leakage through faults and fractures
Passive seismic monitoring	Location, magnitude and source characteristics of seismic events	Development of microfractures in formation or caprock CO_2 migration pathways
Electrical and electromagnetic techniques	Formation conductivity Electromagnetic induction	Tracking movement of CO_2 in and above the storage formation Detecting migration of brine into shallow aquifers
Time-lapse gravity measurements	Density changes caused by fluid displacement	Detect CO_2 movement in or above storage formation CO_2 mass balance in the subsurface
Land surface deformation	Tilt Vertical and horizontal displacement using interferometry and GPS	Detect geomechanical effects on storage formation and caprock Locate CO_2 migration pathways
Visible and infrared imaging from satellite or planes	Hyperspectral imaging of land surface	Detect vegetative stress
CO_2 land surface flux monitoring using flux chambers or eddycovariance	CO_2 fluxes between the land surface and atmosphere	Detect, locate and quantify CO_2 releases
Soil gas sampling	Soil gas composition Isotopic analysis of CO_2	Detect elevated levels of CO_2 Identify source of elevated soil gas CO_2 Evaluate ecosystem impacts

A more accurate approach is to use tracers (gases or gas isotopes not present in the reservoir system) injected into specific wells. The timing of the arrival of the tracers at production or monitoring wells will indicate the path the CO_2 is taking through the reservoir. Monitoring wells may also be used to passively record the movement of CO_2 past the well, although it should be noted that the use of such invasive techniques potentially creates new pathways for leakage to the surface. The movement of tracers or isotopically distinct carbon (in the CO_2) to production or monitoring wells provides some indication of the lateral distribution of the CO_2 in a storage reservoir. In thick formations, multiple sampling along vertical monitoring or production wells would provide some indication of the vertical distribution of the CO_2 in the formation. With many wells and frequently in horizontal wells, the lack of casing (open hole

completion) precludes direct measurement of the location of CO_2 influx along the length of the well, although it may be possible to run surveys to identify the location of major influx.

Direct measurement of migration beyond the storage site can be achieved in a number of ways, depending on where the migration takes the CO_2. Comparison between baseline surveys of water quality and/or isotopic composition can be used to identify new CO_2 arrival at a specific location from natural CO_2 pre-existing at that site. Geochemical techniques can also be used to understand more about the CO_2 and its movement through the reservoir (Czernichowski-Lauriol *et al.*, 1996; Gunter *et al.*, 2000; Wilson and Monea, 2005). The chemical changes that occur in the reservoir fluids indicate the increase in acidity and the chemical effects of this change, in particular the bicarbonate ion levels in the fluids. At the surface, direct measurement can be undertaken by sampling for CO_2 or tracers in soil gas and near surface water-bearing horizons (from existing water wells or new observation wells). Surface CO_2 fluxes may be directly measurable by techniques such as infrared spectroscopy (Miles *et al.*, 2005; Pickles, 2005; Shuler and Tang, 2005).

5.6.3.2 *Indirect techniques for monitoring CO_2 migration*
Indirect techniques for measuring CO_2 distribution in the subsurface include a variety of seismic and non-seismic geophysical and geochemical techniques (Benson *et al.*, 2004; Arts and Winthaegen, 2005; Hoversten and Gasperikova, 2005). Seismic techniques basically measure the velocity and energy absorption of waves, generated artificially or naturally, through rocks. The transmission is modified by the nature of the rock and its contained fluids. In general, energy waves are generated artificially by explosions or ground vibration. Wave generators and sensors may be on the surface (conventional seismic) or modified with the sensors in wells within the subsurface and the source on the surface (vertical seismic profiling). It is also possible to place both sensors and sources in the subsurface to transmit the wave pulses horizontally through the reservoir (inter-well or cross-well tomography). By taking a series of surveys over time, it is possible to trace the distribution of the CO_2 in the reservoir, assuming the free-phase CO_2 volume at the site is sufficiently high to identify from the processed data. A baseline survey with no CO_2 present provides the basis against which comparisons can be made. It would appear that relatively low volumes of free-phase CO_2 (approximately 5% or more) may be identified by these seismic techniques; at present, attempts are being made to quantify the amount of CO_2 in the pore space of the rocks and the distribution within the reservoir (Hoversten *et al.*, 2003). A number of techniques have been actively tested at Weyburn (Section 5.6.3.3), including time-lapse surface three-dimensional seismic (both 3- and 9-component), at one-year intervals (baseline and baseline plus one and two years), vertical seismic profiling and cross-well (horizontal and vertical) tomography between pairs of wells.

For deep accumulations of CO_2 in the subsurface, where CO_2 density approaches the density of fluids in the storage formation, the sensitivity of surface seismic profiles would suggest that resolution on the order of 2500–10,000 t of free-phase CO_2 can be identified (Myer *et al.*, 2003; White *et al.*, 2004; Arts *et al.*, 2005). At Weyburn, areas with low injection rates (<2% hydrocarbon pore volume) demonstrate little or no visible seismic response. In areas with high injection rates (3–13% hydrocarbon pore volume), significant seismic anomalies are observed. Work at Sleipner shows that the CO_2 plume comprises several distinct layers of CO_2, each up to about 10 m thick. These are mostly beneath the strict limit of seismic resolution, but amplitude studies suggest that layer thicknesses as low as 1 m can be mapped (Arts *et al.*, 2005; Chadwick *et al.*, 2005). Seismic resolution will decrease with depth and certain other rock-related properties, so the above discussion of resolution will not apply uniformly in all storage scenarios. One possible way of increasing the accuracy of surveys over time is to create a permanent array of sensors or even sensors and

energy sources (US Patent 6813566), to eliminate the problems associated with surveying locations for sensors and energy sources.

For CO_2 that has migrated even shallower in the subsurface, its gas-like properties will vastly increase the detection limit; hence, even smaller threshold levels of resolution are expected. To date, no quantitative studies have been performed to establish precise detection levels. However, the high compressibility of CO_2 gas, combined with its low density, indicate that much lower levels of detection should be possible.

The use of passive seismic (microseismic) techniques also has potential value. Passive seismic monitoring detects microseismic events induced in the reservoir by dynamic responses to the modification of pore pressures or the reactivation or creation of small fractures. These discrete microearthquakes, with magnitudes on the order of -4 to 0 on the Richter scale (Wilson and Monea, 2005), are picked up by static arrays of sensors, often cemented into abandoned wells. These microseismic events are extremely small, but monitoring the microseismic events may allow the tracking of pressure changes and, possibly, the movement of gas in the reservoir or saline formation.

Non-seismic geophysical techniques include the use of electrical and electromagnetic and self-potential techniques (Benson *et al.*, 2004; Hoversten and Gasperikova, 2005). In addition, gravity techniques (ground or air-based) can be used to determine the migration of the CO_2 plume in the subsurface. Finally, tiltmeters or remote methods (geospatial surveys from aircraft or satellites) for measuring ground distortion may be used in some environments to assess subsurface movement of the plume. Tiltmeters and other techniques are most applicable in areas where natural variations in the surface, such as frost heave or wetting-drying cycles, do not mask the changes that occur from pressure changes. Gravity measurements will respond to changes in the subsurface brought on by density changes caused by the displacement of one fluid by another of different density (e.g., CO_2 replacing water). Gravity is used with numerical modelling to infer those changes in density that best fit the observed data. The estimations of Benson *et al.* (2004) suggest that gravity will not have the same level of resolution as seismic, with minimum levels of CO_2 needed for detection on the order of several hundred thousand tonnes (an order of magnitude greater than seismic). This may be adequate for plume movement, but not for the early definition of possible leaks. A seabed gravity survey was acquired at Sleipner in 2002 and a repeat survey is planned for 2005. Results from these surveys have not yet been published.

Electrical and electromagnetic techniques measure the conducting of the subsurface. Conductivity changes created by a change in the fluid, particularly the displacement of high conductivity saline waters with low-conductive CO_2, can be detected by electrical or electromagnetic surveys. In addition to traditional electrical or electromagnetic techniques, the self-potential the natural electrical potential of the Earth can be measured to determine plume migration. The injection of CO_2 will enhance fluid flow in the rock. This flow can produce an electrical potential that is measured against a reference electrode. This technique is low cost, but is also of low resolution. It can, however, be a useful tool for measuring the plume movement. According to Hoversten and Gasperikova (2005), this technique will require more work to determine its resolution and overall effectiveness.

5.6.3.3   Monitoring case study: IEA-GHG Weyburn Monitoring and Storage Project

At Weyburn (Box 5.3), a monitoring programme was added to a commercial EOR project to develop and evaluate methods for tracking CO_2. Baseline data was collected prior to CO_2 injection (beginning in late 2000). These data included fluid samples (water and oil) and seismic surveys. Two levels of seismic surveys were undertaken, with an extensive three-dimensional (3D), 3-component survey over the original injection area and a detailed 3D, 9-component survey over a limited portion of the injection area. In addition, vertical seismic profiling and cross-well seismic tomography (between two vertical or horizontal wells) was undertaken. Passive seismic (microseismic) monitoring has recently been installed at the

site. Other monitoring includes surface gas surveys (Strutt *et al.*, 2003) and potable water monitoring (the Weyburn field underlies an area with limited surface water availability, so groundwater provides the major potable water supply). Injected volumes (CO_2 and water) were also monitored. Any leaks from surface facilities are carefully monitored. Additionally, several wells were converted to observation wells to allow access to the reservoir. Subsequently, one well was abandoned, but seismic monitors were cemented into place in the well for passive seismic monitoring to be undertaken.

Since injection began, reservoir fluids have been regularly collected and analyzed. Analysis includes chemical and isotopic analyses of reservoir water samples, as well as maintaining an understanding of miscibility relationships between the oil and the injected CO_2. Several seismic surveys have been conducted (one year and two years after injection of CO_2 was initiated) with the processed data clearly showing the movement of CO_2 in the reservoir. Annual surface analysis of soil gas is also continuing (Strutt *et al.*, 2003), as is analysis of near-surface water. The analyses are being synthesized to gain a comprehensive knowledge of CO_2 migration in the reservoir, to understand

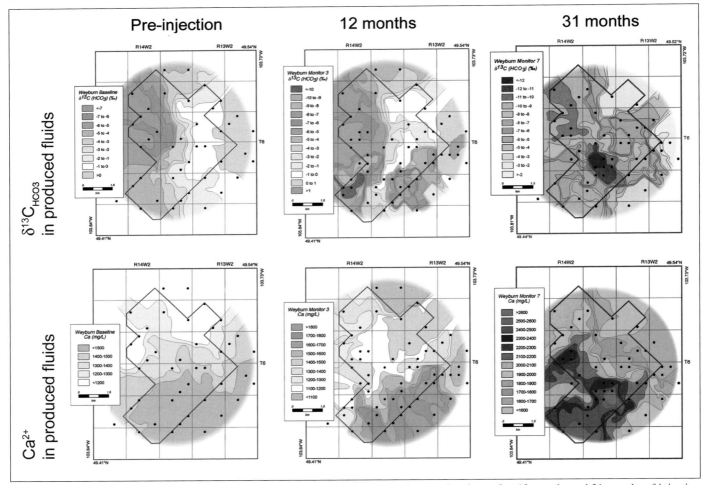

Figure 5.24 The produced water chemistry before CO_2 injection and the produced water chemistry after 12 months and 31 months of injection at Weyburn has been contoured from fluid samples taken at various production wells. The black dots show the location of the sample wells: (a) $\delta^{13}C_{HCO3}$ in the produced water, showing the effect of supercritical CO_2 dissolution and mineral reaction. (b) Calcium concentrations in the produced water, showing the result of mineral dissolution (after Perkins et al., 2005).

geochemical interactions with the reservoir rock and to clearly identify the integrity of the reservoir as a container for long-term storage. Additionally, there is a programme to evaluate the potential role of existing active and abandoned wells in leakage. This includes an analysis of the age of the wells, the use of existing information on cement type and bonding effectiveness and work to better understand the effect of historical and changing fluid chemistry on the cement and steel casing of the well.

The Weyburn summary report (Wilson and Monea, 2005) describes the overall results of the research project, in particular the effectiveness of the seismic monitoring for determining the spread of CO_2 and of the geochemical analysis for determining when CO_2 was about to reach the production wells. Geochemical data also help explain the processes under way in the reservoir itself and the time required to establish a new chemical equilibrium. Figure 5.24 illustrates the change in the chemical composition of the formation water, which forms the basis for assessing the extent to which solubility and mineral trapping will contribute to long-term storage security (Perkins *et al.*, 2005). The initial change in $\delta^{13}C_{HCO3}$ is the result of the supercritical CO_2 dissolving into the water. This change is then muted by the short-term dissolution of reservoir carbonate minerals, as indicated by the increase of calcium concentration, shown in Figure 5.24. In particular, the geochemistry confirms the storage of CO_2 in water in the bicarbonate phase and also CO_2 in the oil phase.

5.6.4 Technologies for monitoring injection well integrity

A number of standard technologies are available for monitoring the integrity of active injection wells. Cement bond logs are used to assess the bond and the continuity of the cement around well casing. Periodic cement bond logs can help detect deterioration in the cemented portion of the well and may also indicate any chemical interaction of the acidized formation fluids with the cement. The initial use of cement bond logs as part of the well-integrity testing can indicate problems with bonding and even the absence of cement.

Prior to converting a well to other uses, such as CO_2 injection, the well usually undergoes testing to ensure its integrity under pressure. These tests are relatively straightforward, with the well being sealed top and bottom (or in the zone to be tested), pressured up and its ability to hold pressure measured. In general, particularly on land, the well will be abandoned if it fails the test and a new well will be drilled, as opposed to attempting any remediation on the defective well.

Injection takes place through a pipe that is lowered into the well and packed off above the perforations or open-hole portion of the well to ensure that the injectant reaches the appropriate level. The pressure in the annulus, the space between the casing and the injection pipe, can be monitored to ensure the integrity of the packer, casing and the injection pipe. Changes in pressure or gas composition in the annulus will alert the operator to problems.

As noted above, the injection pressure is carefully monitored to ensure that there are no problems. A rapid increase in pressure could indicate problems with the well, although industry interpretations suggest that it is more likely to be loss of injectivity in the reservoir.

Temperature logs and 'noise' logs are also often run on a routine basis to detect well failures in natural gas storage projects. Rapid changes in temperature along the length of the wellbore are diagnostic of casing leaks. Similarly, 'noise' associated with leaks in the injection tubing can be used to locate small leaks (Lippmann and Benson, 2003).

5.6.5 Technologies for monitoring local environmental effects

5.6.5.1 Groundwater

If CO_2 leaks from the deep geological storage formation and migrates upwards into overlying shallow groundwater aquifers, methods are available to detect and assess changes in groundwater quality. Of course, it is preferable to identify leakage shortly after it leaks and long before the CO_2 enters the groundwater aquifer, so that measures can be taken to intervene and prevent further migration (see Section 5.7.6). Seismic monitoring methods and potentially others (described in Section 5.6.3.2), can be used to identify leaks before the CO_2 reaches the groundwater zone.

Nevertheless, if CO_2 does migrate into a groundwater aquifer, potential impacts can be assessed by collecting groundwater samples and analyzing them for major ions (e.g., Na, K, Ca, Mg, Mn, Cl, Si, HCO_3^- and SO_4^{2-}), pH, alkalinity, stable isotopes (e.g., ^{13}C, ^{14}C, ^{18}O, ^2H) and gases, including hydrocarbon gases, CO_2 and its associated isotopes (Gunter *et al.*, 1998). Additionally, if shallow groundwater contamination occurs, samples could be analyzed for trace elements such as arsenic and lead, which are mobilized by acidic water (Section 5.5). Methods such as atomic absorption and inductively coupled plasma mass spectroscopy self-potential can be used to accurately measure water quality. Less sensitive field tests or other analytical methods are also available (Clesceri *et al.*, 1998). Standard analytical methods are available to monitor all of these parameters, including the possibility of continuous real-time monitoring for some of the geochemical parameters.

Natural tracers (isotopes of C, O, H and noble gases associated with the injected CO_2) and introduced tracers (noble gases, SF_6 and perfluorocarbons) also may provide insight into the impacts of storage projects on groundwater (Emberley *et al.*, 2002; Nimz and Hudson, 2005). (SF_6 and perfluorocarbons are greenhouse gases with extremely high global warming potentials and therefore caution is warranted in the use of these gases, to avoid their release to the atmosphere.) Natural tracers such as C and O isotopes may be able to link changes in groundwater quality directly to the stored CO_2 by 'fingerprinting' the CO_2, thus distinguishing storage-induced changes from changes in groundwater quality caused by other factors. Introduced tracers such as perfluorocarbons that can be detected at very low concentrations (1 part per trillion) may also be useful for determining whether CO_2 has leaked and is responsible for changes in groundwater quality. Synthetic tracers could be added periodically to determine movement in the reservoir or leakage paths, while natural tracers are present in the reservoir or introduced gases.

5.6.5.2   Air quality and atmospheric fluxes

Continuous sensors for monitoring CO_2 in air are used in a variety of applications, including HVAC (heating, ventilation and air conditioning) systems, greenhouses, combustion emissions measurement and environments in which CO_2 is a significant hazard (such as breweries). Such devices rely on infrared detection principles and are referred to as infrared gas analyzers. These gas analyzers are small and portable and commonly used in occupational settings. Most use non-dispersive infrared or Fourier Transform infrared detectors. Both methods use light attenuation by CO_2 at a specific wavelength, usually 4.26 microns. For extra assurance and validation of real-time monitoring data, US regulatory bodies, such as NIOSH, OSHA and the EPA, use periodic concentration measurement by gas chromatography. Mass spectrometry is the most accurate method for measuring CO_2 concentration, but it is also the least portable. Electrochemical solid state CO_2 detectors exist, but they are not cost effective at this time (e.g., Tamura *et al.*, 2001).

Common field applications in environmental science include the measurement of CO_2 concentrations in soil air, flux from soils and ecosystem-scale carbon dynamics. Diffuse soil flux measurements are made by simple infrared analyzers (Oskarsson *et al.*, 1999). The USGS measures CO_2 flux on Mammoth Mountain, in California (Sorey *et al.*, 1996; USGS, 2001b). Biogeochemists studying ecosystem-scale carbon cycling use data from CO_2 detectors on 2 to 5 m tall towers with wind and temperature data to reconstruct average CO_2 flux over large areas.

Miles *et al.* (2005) concluded that eddy covariance is promising for the monitoring of CO_2 storage projects, both for hazardous leaks and for leaks that would damage the economic viability of geological storage. For a storage project of 100 Mt, Miles *et al.* (2005) estimate that, for leakage rates of 0.01% yr^{-1}, fluxes will range from 1 to 10^4 times the magnitude of typical ecological fluxes (depending on the size of the area over which CO_2 is leaking). Note that a leakage rate of 0.01% yr^{-1} is equivalent to a fraction retained of 90% over 1000 years. This should easily be detectable if background ecological fluxes are measured in advance to determine diurnal and annual cycles. However, with the technology currently available to us, quantifying leakage rates for tracking returns to the atmosphere is likely to be more of a challenge than identifying leaks in the storage reservoir.

Satellite-based remote sensing of CO_2 releases to the atmosphere may also be possible, but this method remains challenging because of the long path length through the atmosphere over which CO_2 is measured and the inherent variability of atmospheric CO_2. Infrared detectors measure average CO_2 concentration over a given path length, so a diffuse or low-level leak viewed through the atmosphere by satellite would be undetectable. As an example, even large CO_2 seeps, such as that at Mammoth Mountain, are difficult to identify today (Martini and Silver, 2002; Pickles, 2005). Aeroplane-based measurement using this same principle may be possible. Carbon dioxide has been measured either directly in the plume by a separate infrared detector or calculated from SO_2 measurements and direct ground sampling of the SO_2: CO_2 ratio for a given volcano or event (Hobbs *et al.*, 1991; USGS, 2001b). Remote-sensing techniques currently under investigation for CO_2 detection are LIDAR (light detection and range-finding), a scanning airborne laser and DIAL (differential absorption LIDAR), which looks at reflections from multiple lasers at different frequencies (Hobbs *et al.*, 1991; Menzies *et al.*, 2001).

In summary, monitoring of CO_2 for occupational safety is well established. On the other hand, while some promising technologies are under development for environmental monitoring and leak detection, measurement and monitoring approaches on the temporal and space scales relevant to geological storage need improvement to be truly effective.

5.6.5.3   Ecosystems

The health of terrestrial and subsurface ecosystems can be determined directly by measuring the productivity and biodiversity of flora and fauna and in some cases (such as at Mammoth Mountain in California) indirectly by using remote-sensing techniques such as hyperspectral imaging (Martini and Silver, 2002; Onstott, 2005; Pickles, 2005). In many areas with natural CO_2 seeps, even those with very low CO_2 fluxes, the seeps are generally quite conspicuous features. They are easily recognized in populated areas, both in agriculture and natural vegetation, by reduced plant growth and the presence of precipitants of minerals leached from rocks by acidic water. Therefore, any conspicuous site could be quickly and easily checked for excess CO_2 concentrations without any large remote-sensing ecosystem studies or surveys. However, in desert environments where vegetation is sparse, direct observation may not be possible. In addition to direct ecosystem observations, analyses of soil gas composition and soil mineralogy can be used to indicate the presence of CO_2 and its impact on soil properties. Detection of elevated concentrations of CO_2 or evidence of excessive soil weathering would indicate the potential for ecosystem impacts.

For aquatic ecosystems, water quality and in particular low pH, would provide a diagnostic for potential impacts. Direct measurements of ecosystem productivity and biodiversity can also be obtained by using standard techniques developed for lakes and marine ecosystems. See Chapter 6 for additional discussion about the impact of elevated CO_2 concentrations on marine environments.

5.6.6 Monitoring network design

There are currently no standard protocols or established network designs for monitoring leakage of CO_2. Monitoring network design will depend on the objectives and requirements of the monitoring programme, which will be determined by regulatory requirements and perceived risks posed by the site (Chalaturnyk and Gunter, 2005). For example, current monitoring for EOR is designed to assess the sweep efficiency of the solvent flood and to deal with health and safety issues. In this regard, the monitoring designed for the Weyburn Project uses seismic surveys to determine the lateral migration of CO_2 over time. This is compared with the simulations undertaken to design the operational practices of the CO_2 flood. For health and safety, the programme is designed to test groundwater for contamination and to monitor for gas buildup in working areas of the field to ensure worker safety. The surface procedure also uses pressure monitoring to ensure that the fracture pressure of the formation is not exceeded (Chalaturnyk and Gunter, 2005).

The Weyburn Project is designed to assess the integrity of an oil reservoir for long-term storage of CO_2 (Wilson and Monea, 2005). In this regard, the demonstrated ability of seismic surveys to measure migration of CO_2 within the formation is important, but in the long term it may be more important to detect CO_2 that has leaked out of the storage reservoir. In this case, the monitoring programme should be designed to achieve the resolution and sensitivity needed to detect CO_2 that has leaked out of the reservoir and is migrating vertically. The use of geochemical monitoring will determine the rate of dissolution of the CO_2 into fluids and the capacity of the minerals within the reservoir to react with the CO_2 and permanently store it. For identification of potential CO_2 leaks, monitoring includes soil gas and groundwater surveys. The soil gas surveys use a grid pattern superimposed on the field to evaluate any change in gas chemistry. Because grid patterns may miss narrow, linear anomalies, the study also looks at the pattern of linear anomalies on the surface that may reflect deeper fault and fracture systems, which could become natural migration pathways.

Current projects, in particular Sleipner and Weyburn, are testing a variety of techniques to determine those that are most effective and least costly. In Western Canada, acid-gas injection wells use pressure monitoring and set maximum wellhead injection pressures to ensure that reservoir fracture pressures are not exceeded. No subsurface monitoring is currently required for these projects. Chalaturnyk and Gunter (2005) suggest that an effectively designed monitoring programme should allow decisions to be made in the future that are based on ongoing interpretation of the data. The data from the programme should also provide the information necessary to decrease uncertainties over time or increase monitoring demand if things develop unexpectedly. The corollary to this is that unexpected changes may result in the requirement of increased monitoring until new uncertainties are resolved.

5.6.7 Long-term stewardship monitoring

The purpose of long-term monitoring is to identify movement of CO_2 that may lead to releases that could impact long-term storage security and safety, as well as trigger the need for remedial action. Long-term monitoring can be accomplished with the same suite of monitoring technologies used during the injection phase. However, at the present time, there are no established protocols for the kind of monitoring that will be required, by whom, for how long and with what purpose. Geological storage of CO_2 may persist over many millions of years. The long duration of storage raises some questions about long-term monitoring – an issue that is also addressed in Section 5.8.

Several studies have attempted to address these issues. Keith and Wilson (2002) have proposed that governments assume responsibility for monitoring after the active phase of the storage project is over, as long as all regulatory requirements have been met during operation. This study did not, however, specify long-term requirements for monitoring. Though perhaps somewhat impractical in terms of implementation, White *et al.* (2003) suggested that monitoring might be required for thousands of years. An alternative point of view is presented by Chow *et al.* (2003) and Benson *et al.* (2004), who suggest that once it has been demonstrated that the plume of CO_2 is no longer moving, further monitoring should not be required. The rationale for this point of view is that long-term monitoring provides little value if the plume is no longer migrating or the cessation of migration can be accurately predicted and verified by a combination of modelling and short- to mid-term monitoring.

If and when long-term monitoring is required, cost-effective, easily deployed methods for monitoring will be preferred. Methods that do not require wells that penetrate the plume will be desirable, because they will not increase the risk of leakage up the monitoring well itself. Technologies are available today, such as 3D seismic imaging, that can provide satisfactory images of CO_2 plume location. While seismic surveys are perceived to be costly, a recent study by Benson *et al.* (2004) suggests that this may be a misconception and indicates that monitoring costs on a discounted basis (10% discount rate) are likely to be no higher than 0.10 US$/t$CO_2$ stored. However, seismic imaging has its limitations, as is evidenced by continued drilling of non-productive hydrocarbon wells, but confidence in its ability to meet most, but not all, of the needs of monitoring CO_2 storage projects is growing. Less expensive and more passive alternatives that could be deployed remotely, such as satellite-based systems, may be desirable, but are not currently able to track underground migration. However, if CO_2 has seeped to the surface, associated vegetative stress can be detected readily in some ecosystems (Martini and Silver, 2002).

Until long-term monitoring requirements are established (Stenhouse *et al.*, 2005), it is not possible to evaluate which technology or combination of technologies for monitoring will be needed or desired. However, today's technology could be deployed to continue monitoring the location of the CO_2 plume over very long time periods with sufficient accuracy to assess the risk of the plume intersecting potential pathways, natural or human, out of the storage site into overlying zones. If CO_2 escapes from the primary storage reservoir with no prospect of remedial action to prevent leakage, technologies are available to monitor the consequent environmental impact on groundwater, soils, ecosystems and the atmosphere.

5.6.8   Verification of CO_2 injection and storage inventory

Verification as a topic is often combined with monitoring such as in the Storage, Monitoring and Verification (SMV) project of the Carbon Capture Project (CCP) or the Monitoring, Mitigation and Verification (MMV) subsection of the DOE-NETL Carbon Sequestration Technology Roadmap and Program Plan (NETL, 2004). In view of this frequently-used combination of terms, there is some overlap in usage between the terms 'verification' and 'monitoring'. For this report, 'verification' is defined as the set of activities used for assessing the amount of CO_2 that is stored underground and for assessing how much, if any, is leaking back into the atmosphere.

No standard protocols have been developed specifically for verification of geological storage. However, experience at the Weyburn and Sleipner projects has demonstrated the utility of various techniques for most if not all aspects of verification (Wilson and Monea, 2005; Sleipner Best Practice Manual, 2004). At the very least, verification will require measurement of the quantity of CO_2 stored. Demonstrating that it remains within the storage site, from both a lateral and vertical migration perspective, is likely to require some combination of models and monitoring. Requirements may be site-specific, depending on the regulatory environment, requirements for economic instruments and the degree of risk of leakage. The oversight for verification may be handled by regulators, either directly or by independent third parties contracted by regulators under national law.

5.7   Risk management, risk assessment and remediation

What are the risks of storing CO_2 in deep geological formations? Can a geological storage site be operated safely? What are the safety concerns and environmental impact if a storage site leaks? Can a CO_2 storage site be fixed if something does go wrong? These questions are addressed in this section of the report.

5.7.1   Framework for assessing environmental risks

The environmental impacts arising from geological storage fall into two broad categories: local environmental effects and global effects arising from the release of stored CO_2 to the atmosphere. Global effects of CO_2 storage may be viewed as the uncertainty in the effectiveness of CO_2 storage. Estimates of the likelihood of release to the atmosphere are discussed below (Section 5.7.3), while the policy implications of potential release from storage are discussed elsewhere (Chapters 1, 8 and 9).

Local health, safety and environmental hazards arise from three distinct causes:
- Direct effects of elevated gas-phase CO_2 concentrations in the shallow subsurface and near-surface environment;
- Effects of dissolved CO_2 on groundwater chemistry;
- Effects that arise from the displacement of fluids by the injected CO_2.

In this section, assessment of possible local and regional environmental hazards is organized by the kind of hazard (e.g., human health and ecosystem hazards are treated separately) and by the underlying physical mechanism (e.g., seismic hazards). For example, the discussion of hazards to groundwater quality includes effects that arise directly from the effect of dissolved CO_2 in groundwater, as well as indirect effects resulting from contamination by displaced brines.

Risks are proportional to the magnitude of the potential hazards and the probability that these hazards will occur. For hazards that arise from locally elevated CO_2 concentrations – in the near-surface atmosphere, soil gas or in aqueous solution – the risks depend on the probability of leakage from the deep storage site to the surface. Thus, most of the hazards described in Section 5.7.4 should be weighted by the probability of release described in Section 5.7.3. Regarding those risks associated with routine operation of the facility and well maintenance, such risks are expected to be comparable to CO_2-EOR operations.

There are two important exceptions to the rule that risk is proportional to the probability of release. First, local impacts will be strongly dependent on the spatial and temporal distribution of fluxes and the resulting CO_2 concentrations. Episodic and localized seepage will likely tend to have more significant impacts per unit of CO_2 released than will seepage that is continuous and or spatially dispersed. Global impacts arising from release of CO_2 to the atmosphere depend only on the average quantity released over time scales of decades to centuries. Second, the hazards arising from displacement, such as the risk of induced seismicity, are roughly independent of the probability of release.

Although we have limited experience with injection of CO_2 for the explicit purpose of avoiding atmospheric emissions, a wealth of closely related industrial experience and scientific knowledge exists that can serve as a basis for appropriate risk management. In addition to the discussion in this section, relevant industrial experience has been described in Sections 5.1 to 5.6.

5.7.2   Processes and pathways for release of CO_2 from geological storage sites

Carbon dioxide that exists as a separate phase (supercritical, liquid or gas) may escape from formations used for geological storage through the following pathways (Figure 5.25):
- Through the pore system in low-permeability caprocks such as shales, if the capillary entry pressure at which CO_2 may enter the caprock is exceeded;
- Through openings in the caprock or fractures and faults;

- Through anthropomorphic pathways, such as poorly completed and/or abandoned pre-existing wells.

For onshore storage sites, CO_2 that has leaked may reach the water table and migrate into the overlying vadose zone. This occurrence would likely include CO_2 contact with drinking-water aquifers. Depending on the mineral composition of the rock matrix within the groundwater aquifer or vadose zone, the reaction of CO_2 with the rock matrix could release contaminants. The US Environmental Protection Agency (USEPA) has witnessed problems with projects designed to replenish groundwater with rainfall wherein mineralized (fixed) contaminants were inadvertently mobilized in concentrations sufficient to cause undesirable contamination.

The vadose zone is only partly saturated with water; the rest of the pore space is filled with soil gas (air). Because it is heavier than air, CO_2 will displace ambient soil gas, leading to concentrations that locally may potentially approach 100% in parts of the vadose zone, even for small leakage fluxes. The dissipating effects of seepage into the surface layer are controlled mostly by pressure-driven flow and diffusion (Oldenburg and Unger, 2003). These occur predominantly in most shallow parts of the vadose zone, leaving the deeper part of the vadose zone potentially subject to accumulation of leaking CO_2. The processes of CO_2 migration in the vadose zone can be modelled, subject to limitations in the characterization of actual complex vadose zone and CO_2 leakage scenarios.

For storage sites that are offshore, CO_2 that has leaked may reach the ocean bottom sediments and then, if lighter than the surrounding water, migrate up through the water column until it reaches the atmosphere. Depending upon the leakage rate, it may either remain as a separate phase or completely dissolve into the water column. When CO_2 dissolves, biological impacts to ocean bottom and marine organisms will be of concern. For those sites where separate-phase CO_2 reaches the ocean surface, hazards to offshore platform workers may be of concern for very large and sudden release rates.

Once through the vadose zone, escaping CO_2 reaches the surface layer of the atmosphere and the surface environment, where humans and other animals can be exposed to it. Carbon dioxide dispersion and mixing result from surface winds and associated turbulence and eddies. As a result, CO_2 concentrations diminish rapidly with elevation, meaning that ground-dwelling animals are more likely to be affected by exposure than are humans (Oldenburg and Unger, 2004). Calm conditions and local topography capable of containing the dense gas will tend to prevent mixing. But such conditions are the exception and in general, the surface layer can be counted on to strongly dilute seeping CO_2. Nevertheless, potential concerns related to buildup of CO_2 concentrations on calm days must be carefully considered in any risk assessment of a CO_2 storage site. Additionally, high subsurface CO_2 concentrations may accumulate in basements, subsurface vaults and other subsurface infrastructures where humans may be exposed to risk.

Carbon dioxide injected into coal seams can escape only if it is in free phase (i.e., not adsorbed onto the coal) via the following pathways (Wo and Liang 2005; Wo *et al.* 2005): flow into surrounding strata during injection when high pressures are used to inject CO_2 into low-permeability coal, either where the cleat system reaches the top of the seam or via hydrofractures induced to improve the contact between the cleat system and CBM production wells; through faults or other natural pathways intersecting the coal seam; via poorly abandoned coal or CBM exploration wells; and through anthropomorphic pathways such

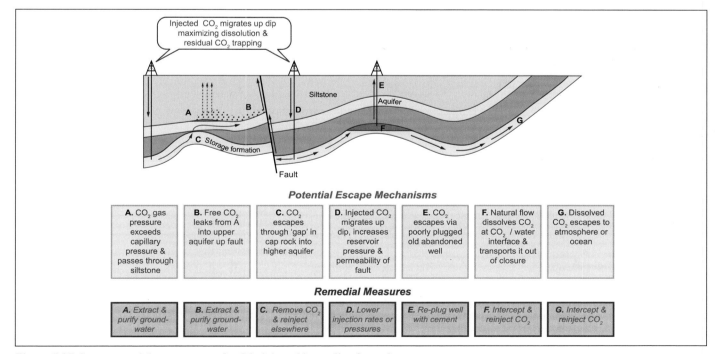

Figure 5.25 Some potential escape routes for CO_2 injected into saline formations.

as coal mines or mining-induced subsidence cracks.

In general, however, CO_2 retained by sorption onto coal will remain confined to the seam even without caprocks, unless the pressure in the coal seam is reduced (e.g., by mining). Changes in pressure and/or temperature lead to changes in the maximum gas content. If the pressure drops markedly, any excess CO_2 may desorb from the coal and flow freely through cleats.

Injection wells and abandoned wells have been identified as one of the most probable leakage pathways for CO_2 storage projects (Gasda *et al.*, 2004; Benson, 2005). When a well is drilled, a continuous, open conduit is created between the land surface and the deep subsurface. If, at the time of drilling, the operator decides that the target formation does not look sufficiently productive, then the well is abandoned as a 'dry hole', in accordance with proper regulatory guidelines. Current guidelines typically require filling sections of the hole with cement (Section 5.5 and Figure 5.21).

Drilling and completion of a well involve not only creation of a hole in the Earth, but also the introduction of engineered materials into the subsurface, such as well cements and well casing. The overall effect of well drilling is replacement of small but potentially significant cylindrical volumes of rock, including low-permeability caprock, with anthropomorphic materials that have properties different from those of the original materials. A number of possible leakage pathways can occur along abandoned wells, as illustrated in Figure 5.26 (Gasda *et al.*, 2004). These include leakage between the cement and the outside of the casing (Figure 5.26a), between the cement and the inside of the metal casing (Figure 5.26b), within the cement plug itself (Figure 5.26c), through deterioration (corrosion) of

the metal casing (Figure 5.26d), deterioration of the cement in the annulus (Figure 5.26e) and leakage in the annular region between the formation and the cement (Figure 5.26f). The potential for long-term degradation of cement and metal casing in the presence of CO_2 is a topic of extensive investigations at this time (e.g., Scherer *et al.*, 2005).

The risk of leakage through abandoned wells is proportional to the number of wells intersected by the CO_2 plume, their depth and the abandonment method used. For mature sedimentary basins, the number of wells in proximity to a possible injection well can be large, on the order of many hundreds. For example, in the Alberta Basin in western Canada, more than 350,000 wells have been drilled. Currently, drilling continues at the rate of approximately 20,000 wells per year. The wells are distributed spatially in clusters, with densities that average around four wells per km^2 (Gasda *et al.*, 2004). Worldwide well densities are provided in Figure 5.27 and illustrate that many areas have much lower well density. Nevertheless, the data provided in Figure 5.27 illustrate an important point made in Section 5.3 – namely that storage security in mature oil and gas provinces may be compromised if a large number of wells penetrate the caprocks. Steps need to be taken to address this potential risk.

5.7.3   *Probability of release from geological storage sites*

Storage sites will presumably be designed to confine all injected CO_2 for geological time scales. Nevertheless, experience with engineered systems suggest a small fraction of operational storage sites may release CO_2 to the atmosphere. No existing studies systematically estimate the probability and magnitude of release across a sample of credible geological storage systems. In the absence of such studies, this section synthesizes the lines of evidence that enable rough quantitative estimates of achievable fractions retained in storage. Five kinds of evidence are relevant to assessing storage effectiveness:

- Data from natural systems, including trapped accumulations of natural gas and CO_2, as well as oil;
- Data from engineered systems, including natural gas storage, gas re-injection for pressure support, CO_2 or miscible hydrocarbon EOR, disposal of acid gases and disposal of other fluids;
- Fundamental physical, chemical and mechanical processes regarding the fate and transport of CO_2 in the subsurface;
- Results from numerical models of CO_2 transport;
- Results from current geological storage projects.

5.7.3.1   *Natural systems*
Natural systems allow inferences about the quality and quantity of geological formations that could be used to store CO_2. The widespread presence of oil, gas and CO_2 trapped in formations for many millions of years implies that within sedimentary basins, impermeable formations (caprocks) of sufficient quality to confine CO_2 for geological time periods are present. For example, the about 200 $MtCO_2$ trapped in the Pisgah Anticline, northeast of the Jackson Dome (Mississippi), is thought to have been generated in Late Cretaceous times, more than 65 million

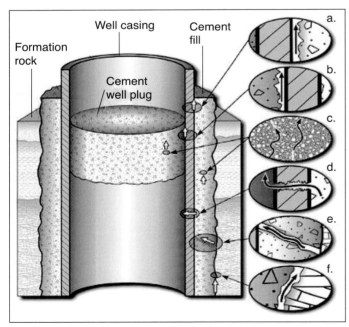

Figure 5.26 Possible leakage pathways in an abandoned well: (a) and (b) between casing and cement wall and plug, respectively; (c) through cement plugs; (d) through casing; (e) through cement wall; and (f) between the cement wall and rock (after Gasda *et al.*, 2004).

years ago (Studlick *et al.*, 1990). Retention times longer than 10 million years are found in many of the world's petroleum basins (Bradshaw *et al.*, 2005). Therefore evidence from natural systems demonstrates that reservoir seals exist that are able to confine CO_2 for millions of years and longer.

5.7.3.2   Engineered systems

Evidence from natural gas storage systems enables performance assessments of engineered barriers (wells and associated management and remediation) and of the performance of natural systems that have been altered by pressure cycling (Lippmann and Benson, 2003; Perry, 2005). Approximately 470 natural gas storage facilities are currently operating in the United States with a total storage capacity exceeding 160 Mt natural gas (Figure 5.12). There have been nine documented incidents of significant leakage: five were related to wellbore integrity, each of which was resolved by reworking the wells; three arose from leaks in caprocks, two of which were remediated and one of which led to project abandonment. The final incident involved early project abandonment owing to poor site selection (Perry, 2005). There are no estimates of the total volumes of gas lost resulting from leakage across all the projects. In one recent serious example of leakage, involving wellbore failure at a facility in Kansas, the total mass released was about 3000 t (Lee, 2001), equal to less than 0.002% of the total gas in storage in the United States and Canada. The capacity-weighted median age of the approximately 470 facilities exceeds 25 years. Given that the Kansas failure was among the worst in the cumulative operating history of gas storage facilities, the average annual release rates, expressed as a fraction of stored gas released per year, are likely below 10^{-5}. While such estimates of the expected (or statistical average) release rates are a useful measure of storage effectiveness, they should not be interpreted as implying that release will be a continuous process.

The performance of natural gas storage systems may be regarded as a lower bound on that of CO_2 storage. One reason for this is that natural gas systems are designed for (and subject to) rapid pressure cycling that increases the probability of caprock leakage. On the other hand, CO_2 will dissolve in pore waters (if present), thereby reducing the risk of leakage. Perhaps the only respect in which gas storage systems present lower risks is that CH_4 is less corrosive than CO_2 to metallic components, such as well casings. Risks are higher in the case of leakage from natural gas storage sites because of the flammable nature of the gas.

5.7.3.3   Fundamental physical, chemical and mechanical processes regarding fate and transport of CO_2 in the subsurface

As described in Section 5.2, scientific understanding of CO_2 storage and in particular performance of storage systems, rests on a large body of knowledge in hydrogeology, petroleum geology, reservoir engineering and related geosciences. Current evaluation has identified a number of processes that alone or in combination can result in very long-term storage. Specifically, the combination of structural and stratigraphic trapping of separate-phase CO_2 below low-permeability caprocks, residual CO_2 trapping, solubility trapping and mineral trapping can create secure storage over geological time scales.

5.7.3.4   Numerical simulations of long-term storage performance

Simulations of CO_2 confinement in large-scale storage projects suggest that, neglecting abandoned wells, the movement of

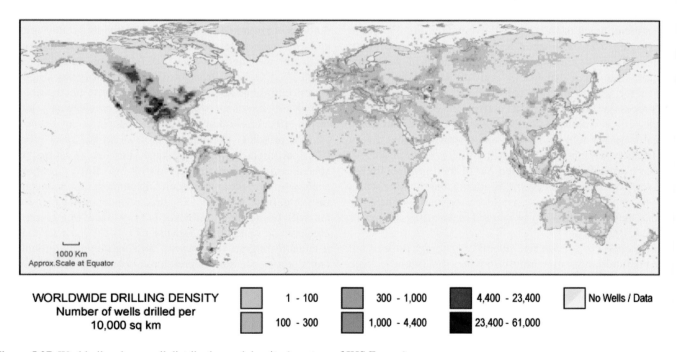

WORLDWIDE DRILLING DENSITY
Number of wells drilled per
10,000 sq km

| 1 - 100 | 300 - 1,000 | 4,400 - 23,400 | No Wells / Data |
| 100 - 300 | 1,000 - 4,400 | 23,400 - 61,000 | |

Figure 5.27 World oil and gas well distribution and density (courtesy of IHS Energy).

CO_2 through the subsurface will be slow. For example, Cawley *et al.* (2005) studied the effect of uncertainties in parameters such as the flow velocity in the aquifer and capillary entry pressure into caprock in their examination of CO_2 storage in the Forties Oilfield in the North Sea. Over the 1000 year time scale examined in their study, Cawley *et al.* (2005) found that less than 0.2% of the stored CO_2 enters into the overlying layers and even in the worse case, the maximum vertical distance moved by any of the CO_2 was less than halfway to the seabed. Similarly, Lindeberg and Bergmo (2003) studied the Sleipner field and found that CO_2 would not begin to migrate into the North Sea for 100,000 years and that even after a million years, the annual rate of release would be about 10^{-6} of the stored CO_2 per year.

Simulations designed to explore the possible release of stored CO_2 to the biosphere by multiple routes, including abandoned wells and other disturbances, have recently become available as a component of more general risk assessment activities (Section 5.7.5). Two studies of the Weyburn site, for example, assessed the probability of release to the biosphere. Walton *et al.* (2005) used a fully probabilistic model, with a simplified representation of CO_2 transport, to compute a probability distribution for the cumulative fraction released to the biosphere. Walton *et al.* found that after 5000 years, the probability was equal that the cumulative amount released would be larger or smaller than 0.1% (the median release fraction) and found a 95% probability that <1% of the total amount stored would be released. Using a deterministic model of CO_2 transport in the subsurface, Zhou *et al.* (2005) found no release to the biosphere in 5000 years. While using a probabilistic model of transport through abandoned wells, they found a statistical mean release of 0.001% and a maximum release of 0.14% (expressed as the cumulative fraction of stored CO_2 released over 5000 years).

In saline formations or oil and gas reservoirs with significant brine content, much of the CO_2 will eventually dissolve in the brine (Figure 5.7), be trapped as a residual immobile phase (Figure 5.8) or be immobilized by geochemical reactions. The time scale for dissolution is typically short compared to the time for CO_2 to migrate out of the storage formation by other processes (Ennis-King and Paterson, 2003; Lindeberg and Bergmo, 2003; Walton *et al.*, 2005). It is expected that many storage projects could be selected and operated so that a very large fraction of the injected CO_2 will dissolve. Once dissolved, CO_2 can eventually be transported out of the injection site by basin-scale circulation or upward migration, but the time scales (millions of years) of such transport are typically sufficiently long that they can (arguably) be ignored in assessing the risk of leakage.

As described in Section 5.1, several CO_2 storage projects are now in operation and being carefully monitored. While no leakage of stored CO_2 out of the storage formations has been observed in any of the current projects, time is too short and overall monitoring too limited, to enable direct empirical conclusions about the long-term performance of geological storage. Rather than providing a direct test of performance, the current projects improve the quality of long-duration performance predictions by testing and sharpening understanding of CO_2 transport and trapping mechanisms.

5.7.3.5 *Assessing the ability of operational geological storage projects to retain CO_2 for long time periods*

Assessment of the fraction retained for geological storage projects is highly site-specific, depending on (1) the storage system design, including the geological characteristics of the selected storage site; (2) the injection system and related reservoir engineering; and (3) the methods of abandonment, including the performance of well-sealing technologies. If the above information is available, it is possible to estimate the fraction retained by using the models described in Section 5.4.2 and risk assessment methods described in Section 5.7.5. Therefore, it is also possible, in principle, to estimate the expected performance of an ensemble of storage projects that adhere to design guidelines such as site selection, seal integrity, injection depth and well closure technologies. Table 5.5 summarizes disparate lines of evidence on the integrity of CO_2 storage systems.

For large-scale operational CO_2 storage projects, assuming that sites are well selected, designed, operated and appropriately monitored, the balance of available evidence suggests the following:
- It is very likely the fraction of stored CO_2 retained is more than 99% over the first 100 years.
- It is likely the fraction of stored CO_2 retained is more than 99% over the first 1000 years.

5.7.4 *Possible local and regional environmental hazards*

5.7.4.1 *Potential hazards to human health and safety*

Risks to human health and safety arise (almost) exclusively from elevated CO_2 concentrations in ambient air, either in confined outdoor environments, in caves or in buildings. Physiological and toxicological responses to elevated CO_2 concentrations are relatively well understood (AI.3.3). At concentrations above about 2%, CO_2 has a strong effect on respiratory physiology and at concentrations above 7–10%, it can cause unconsciousness and death. Exposure studies have not revealed any adverse health effect of chronic exposure to concentrations below 1%.

The principal challenge in estimating the risks posed by CO_2 that might seep from storage sites lies in estimating the spatial and temporal distribution of CO_2 fluxes reaching the shallow subsurface and in predicting ambient CO_2 concentration resulting from a given CO_2 flux. Concentrations in surface air will be strongly influenced by surface topography and atmospheric conditions. Because CO_2 is 50% denser than air, it tends to migrate downwards, flowing along the ground and collecting in shallow depressions, potentially creating much higher concentrations in confined spaces than in open terrain.

Seepage of CO_2 is not uncommon in regions influenced by volcanism. Naturally occurring releases of CO_2 provide a basis for understanding the transport of CO_2 from the vadose zone to the atmosphere, as well as providing empirical data that link CO_2 fluxes into the shallow subsurface with CO_2 concentrations

Table 5.5 Summary of evidence for CO_2 retention and release rates.

Kind of evidence	Average annual fraction released	Representative references
CO_2 in natural formations	The lifetime of CO_2 in natural formations (>10 million yr in some cases) suggests an average release fraction <10^{-7} yr^{-1} for CO_2 trapped in sedimentary basins. In highly fractured volcanic systems, rate of release can be many orders of magnitude faster.	Stevens *et al.*, 2001a; Baines and Worden, 2001
Oil and gas	The presence of buoyant fluids trapped for geological timescales demonstrates the widespread presence of geological systems (seals and caprock) that are capable of confining gasses with release rates <10^{-7} yr^{-1}.	Bradshaw *et al.*, 2005
Natural gas storage	The cumulative experience of natural gas storage systems exceeds 10,000 facility-years and demonstrates that operational engineered storage systems can contain methane with release rates of 10^{-4} to 10^{-6} yr^{-1}.	Lippmann and Benson, 2003; Perry, 2005
Enhanced oil recovery (EOR)	More than 100 $MtCO_2$ has been injected for EOR. Data from the few sites where surface fluxes have been measured suggest that fractional release rates are near zero.	Moritis, 2002; Klusman, 2003
Models of flow through the undisturbed subsurface	Numerical models show that release of CO_2 by subsurface flow through undisturbed geological media (excluding wells) may be near zero at appropriately selected storage sites and is very likely <10^{-6} in the few studies that attempted probabilistic estimates.	Walton *et al.*, 2005; Zhou *et al.*, 2005; Lindeberg and Bergmo, 2003; Cawley *et al.*, 2005
Models of flow through wells	Evidence from a small number of risk assessment studies suggests that average release of CO_2 can be 10^{-5} to 10^{-7} yr^{-1} even in existing oil fields with many abandoned wells, such as Weyburn. Simulations with idealized systems with 'open' wells show that release rates can exceed 10^{-2}, though in practice such wells would presumably be closed as soon as CO_2 was detected.	Walton *et al.*, 2005; Zhou et al., 2005; Nordbotten *et al.*, 2005b
Current CO_2 storage projects	Data from current CO_2 storage projects demonstrate that monitoring techniques are able to detect movement of CO_2 in the storage reservoirs. Although no release to the surface has been detected, little can be concluded given the short history and few sites.	Wilson and Monea, 2005; Arts *et al.*, 2005; Chadwick, *et al.*, 2005

in the ambient air – and the consequent health and safety risks. Such seeps do not, however, provide a useful basis for estimating the spatial and temporal distribution of CO_2 fluxes leaking from a deep storage site, because (in general) the seeps occur in highly fractured volcanic zones, unlike the interiors of stable sedimentary basins, the likely locations for CO_2 storage (Section 5.3).

Natural seeps are widely distributed in tectonically active regions of the world (Morner and Etiope, 2002). In central Italy, for example, CO_2 is emitted from vents, surface degassing and diffuse emission from CO_2-rich groundwater. Fluxes from vents range from less than 100 to more than 430 tCO_2 day^{-1}, which have shown to be lethal to animal and plants. At Poggio dell'Ulivo, for example, a flux of 200 tCO_2 day^{-1} is emitted from diffuse soil degassing. At least ten people have died from CO_2 releases in the region of Lazio over the last 20 years.

Natural and engineered analogues show that it is possible, though improbable, that slow releases from CO_2 storage reservoirs will pose a threat to humans. Sudden, catastrophic releases of natural accumulations of CO_2 have occurred, associated with volcanism or subsurface mining activities. Thus, they are of limited relevance to understanding risks arising from CO_2 stored in sedimentary basins. However, mining or drilling in areas with CO_2 storage sites may pose a long-term risk after site abandonment if institutional knowledge and precautions are not in place to avoid accidentally penetrating a storage formation.

5.7.4.2 *Hazards to groundwater from CO_2 leakage and brine displacement*

Increases in dissolved CO_2 concentration that might occur as CO_2 migrates from a storage reservoir to the surface will alter groundwater chemistry, potentially affecting shallow groundwater used for potable water and industrial and agricultural needs. Dissolved CO_2 forms carbonic acid, altering the pH of the solution and potentially causing indirect effects, including mobilization of (toxic) metals, sulphate or chloride; and possibly giving the water an odd odour, colour or taste. In the worst case, contamination might reach dangerous levels, excluding the use of groundwater for drinking or irrigation.

Wang and Jaffé (2004) used a chemical transport model to investigate the effect of releasing CO_2 from a point source at 100 m depth into a shallow water formation that contained a high concentration of mineralized lead (galena). They found that in weakly buffered formations, the escaping CO_2 could mobilize sufficient dissolved lead to pose a health hazard over a radius of a few hundred metres from the CO_2 source. This analysis represents an extreme upper bound to the risk of metal leaching, since few natural formations have mineral composition so susceptible to the effects of CO_2-mediated leaching and one of the expressed requirements of a storage site is to avoid compromising other potential resources, such as mineral deposits.

The injection of CO_2 or any other fluid deep underground necessarily causes changes in pore-fluid pressures and in the geomechanical stress fields that reach far beyond the volume occupied by the injected fluid. Brines displaced from deep formations by injected CO_2 can potentially migrate or leak through fractures or defective wells to shallow aquifers and contaminate shallower drinking water formations by increasing their salinity. In the worst case, infiltration of saline water into groundwater or into the shallow subsurface could impact wildlife habitat, restrict or eliminate agricultural use of land and pollute surface waters.

As is the case for induced seismicity, the experience with injection of different fluids provides an empirical basis for assessing the likelihood that groundwater contamination will occur by brine displacement. As discussed in Section 5.5 and shown in Figure 5.22, the current site-specific injection rates of fluids into the deep subsurface are roughly comparable to the rates at which CO_2 would be injected if geological storage were adopted for storage of CO_2 from large-scale power plants. Contamination of groundwater by brines displaced from injection wells is rare and it is therefore expected that contamination arising from large-scale CO_2 storage activities would also be rare. Density differences between CO_2 and other fluids with which we have extensive experience do not compromise this conclusion, because brine displacement is driven primarily by the pressure/hydraulic head differential of the injected CO_2, not by buoyancy forces.

5.7.4.3 Hazards to terrestrial and marine ecosystems

Stored CO_2 and any accompanying substances, may affect the flora and fauna with which it comes into contact. Impacts might be expected on microbes in the deep subsurface and on plants and animals in shallower soils and at the surface. The remainder of this discussion focuses only on the hazards where exposures to CO_2 do occur. As discussed in Section 5.7.3, the probability of leakage is low. Nevertheless, it is important to understand the hazards should exposures occur.

In the last three decades, microbes dubbed 'extremophiles', living in environments where life was previously considered impossible, have been identified in many underground habitats. These microorganisms have limited nutrient supply and exhibit very low metabolic rates (D'Hondt *et al.*, 2002). Recent studies have described populations in deep saline formations (Haveman and Pedersen, 2001), oil and gas reservoirs (Orphan *et al.*, 2000) and sediments up to 850 m below the sea floor (Parkes *et al.*, 2000). The mass of subsurface microbes may well exceed the mass of biota on the Earth's surface (Whitman *et al.*, 2001). The working assumption may be that unless there are conditions preventing it, microbes can be found everywhere at the depths being considered for CO_2 storage and consequently CO_2 storage sites may generally contain microbes that could be affected by injected CO_2.

The effect of CO_2 on subsurface microbial populations is not well studied. A low-pH, high-CO_2 environment may favour some species and harm others. In strongly reducing environments, the injection of CO_2 may stimulate microbial communities that would reduce the CO_2 to CH_4; while in other reservoirs, CO_2 injection could cause a short-term stimulation

of Fe(III)-reducing communities (Onstott, 2005). From an operational perspective, creation of biofilms may reduce the effective permeability of the formation.

Should CO_2 leak from the storage formation and find its way to the surface, it will enter a much more biologically active area. While elevated CO_2 concentrations in ambient air can accelerate plant growth, such fertilization will generally be overwhelmed by the detrimental effects of elevated CO_2 in soils, because CO_2 fluxes large enough to significantly increase concentrations in the free air will typically be associated with much higher CO_2 concentrations in soils. The effects of elevated CO_2 concentrations would be mediated by several factors: the type and density of vegetation; the exposure to other environmental stresses; the prevailing environmental conditions like wind speed and rainfall; the presence of low-lying areas; and the density of nearby animal populations.

The main characteristic of long-term elevated CO_2 zones at the surface is the lack of vegetation. New CO_2 releases into vegetated areas cause noticeable die-off. In those areas where significant impacts to vegetation have occurred, CO_2 makes up about 20–95% of the soil gas, whereas normal soil gas usually contains about 0.2–4% CO_2. Carbon dioxide concentrations above 5% may be dangerous for vegetation and as concentration approach 20%, CO_2 becomes phytotoxic. Carbon dioxide can cause death of plants through 'root anoxia', together with low oxygen concentration (Leone *et al.*, 1977; Flower *et al.*, 1981).

One example of plant die-off occurred at Mammoth Mountain, California, USA, where a resurgence of volcanic activity resulted in high CO_2 fluxes. In 1989, a series of small earthquakes occurred near Mammoth Mountain. A year later, 4 ha of pine trees were discovered to be losing their needles and by 1997, the area of dead and dying trees had expanded to 40 ha (Farrar *et al.*, 1999). Soil CO_2 levels above 10–20% inhibit root development and decrease water and nutrient uptake; soil oil-gas testing at Mammoth Mountain in 1994 discovered soil gas readings of up to 95% CO_2 by volume. Total CO_2 flux in the affected areas averaged about 530 t day^{-1} in 1996. Measurements in 2001 showed soil CO_2 levels of 15–90%, with flux rates at the largest affected area (Horseshoe Lake) averaging 90–100 tCO_2 day^{-1} (Gerlach *et al.*, 1999; Rogie *et al.*, 2001). A study of the impact of elevated CO_2 on soils found there was a lower pH and higher moisture content in summer. Wells in the high CO_2 area showed higher levels of silicon, aluminum, magnesium and iron, consistent with enhanced weathering of the soils. Tree-ring data show that CO_2 releases have occurred prior to 1990 (Cook *et al.*, 2001). Data from airborne remote sensing are now being used to map tree health and measure anomalous CO_2 levels, which may help determine how CO_2 affects forest ecosystems (Martini and Silver, 2002).

There is no evidence of any terrestrial impact from current CO_2 storage projects. Likewise, there is no evidence from EOR projects that indicate impacts to vegetation such as those described above. However, no systematic studies have occurred to look for terrestrial impacts from current EOR projects.

Natural CO_2 seepage in volcanic regions, therefore, provides examples of possible impacts from leaky CO_2 storage, although (as mentioned in Section 5.2.3) seeps in volcanic provinces provide a poor analogue to seepage that would occur from CO_2 storage sites in sedimentary basins. As described above, CO_2 seepage can pose substantial hazards. In the Alban Hills, south of Rome (Italy), for example, 29 cows and 8 sheep were asphyxiated in several separate incidents between September 1999 and October 2001 (Carapezza *et al.*, 2003). The measured CO_2 flux was about 60 t day^{-1} of 98% CO_2 and up to 2% H_2S, creating hazardous levels of each gas in localized areas, particularly in low-wind conditions. The high CO_2 and H_2S fluxes resulted from a combination of magmatic activity and faulting.

Human activities have caused detrimental releases of CO_2 from the deep subsurface. In the late 1990s, vegetation died off above an approximately 3-km deep geothermal field being exploited for a 62 MW power plant, in Dixie Valley, Nevada, USA (Bergfeld *et al.*, 2001). A maximum flux of 570 gCO_2 m^{-2} day^{-1} was measured, as compared to a background level of 7 gCO_2 m^{-2} day^{-1}. By 1999, CO_2 flow in the measured area ceased and vegetation began to return.

The relevance of these natural analogues to leakage from CO_2 storage varies. For examples presented here, the fluxes and therefore the risks, are much higher than might be expected from a CO_2 storage facility: the annual flow of CO_2 at the Mammoth Mountain site is roughly equal to a release rate on the order of 0.2% yr^{-1} from a storage site containing 100 $MtCO_2$. This corresponds to a fraction retained of 13.5% over 1000 years and, thus, is not representative of a typical storage site.

Seepage from offshore geological storage sites may pose a hazard to benthic environments and organisms as the CO_2 moves from deep geological structures through benthic sediments to the ocean. While leaking CO_2 might be hazardous to the benthic environment, the seabed and overlying seawater can also provide a barrier, reducing the escape of seeping CO_2 to the atmosphere. These hazards are distinctly different from the environmental effects of the dissolved CO_2 on aquatic life in the water column, which are discussed in Chapter 6. No studies specifically address the environmental effects of seepage from sub-seabed geological storage sites.

5.7.4.4 Induced seismicity

Underground injection of CO_2 or other fluids into porous rock at pressures substantially higher than formation pressures can induce fracturing and movement along faults (see Section 5.5.4 and Healy *et al.*, 1968; Gibbs *et al.*, 1973; Raleigh *et al.*, 1976; Sminchak *et al.*, 2002; Streit *et al.*, 2005; Wo *et al.*, 2005). Induced fracturing and fault activation may pose two kinds of risks. First, brittle failure and associated microseismicity induced by overpressuring can create or enhance fracture permeability, thus providing pathways for unwanted CO_2 migration (Streit and Hillis, 2003). Second, fault activation can, in principle, induce earthquakes large enough to cause damage (e.g., Healy *et al.*, 1968).

Fluid injection into boreholes can induce microseismic activity, as for example at the Rangely Oil Field in Colorado, USA (Gibbs *et al.*, 1973; Raleigh *et al.*, 1976), in test sites such as the drillholes of the German continental deep drilling programme (Shapiro *et al.*, 1997; Zoback and Harjes, 1997) or the Cold Lake Oil Field, Alberta, Canada (Talebi *et al.*, 1998). Deep-well injection of waste fluids may induce earthquakes with moderate local magnitudes (M_L), as suggested for the 1967 Denver earthquakes (M_L of 5.3; Healy *et al.*, 1968; Wyss and Molnar, 1972) and the 1986–1987 Ohio earthquakes (M_L of 4.9; Ahmad and Smith, 1988) in the United States. Seismicity induced by fluid injection is usually assumed to result from increased pore-fluid pressure in the hypocentral region of the seismic event (e.g., Healy *et al.*, 1968; Talebi *et al.*, 1998).

Readily applicable methods exist to assess and control induced fracturing or fault activation (see Section 5.5.3). Several geomechanical methods have been identified for assessing the stability of faults and estimating maximum sustainable pore-fluid pressures for CO_2 storage (Streit and Hillis, 2003). Such methods, which require the determination of *in situ* stresses, fault geometries and relevant rock strengths, are based on brittle failure criteria and have been applied to several study sites for potential CO_2 storage (Rigg *et al.*, 2001; Gibson-Poole *et al.*, 2002).

The monitoring of microseismic events, especially in the vicinity of injection wells, can indicate whether pore fluid pressures have locally exceeded the strength of faults, fractures or intact rock. Acoustic transducers that record microseismic events in monitoring wells of CO_2 storage sites can be used to provide real-time control to keep injection pressures below the levels that induce seismicity. Together with the modelling techniques mentioned above, monitoring can reduce the chance of damage to top seals and fault seals (at CO_2 storage sites) caused by injection-related pore-pressure increases.

Fault activation is primarily dependent on the extent and magnitude of the pore-fluid-pressure perturbations. It is therefore determined more by the quantity and rate than by the kind of fluid injected. Estimates of the risk of inducing significant earthquakes may therefore be based on the diverse and extensive experience with deep-well injection of various aqueous and gaseous streams for disposal and storage. Perhaps the most pertinent experience is the injection of CO_2 for EOR; about 30 $MtCO_2$ yr^{-1} is now injected for EOR worldwide and the cumulative total injected exceeds 0.5 $GtCO_2$, yet there have been no significant seismic effects attributed to CO_2-EOR. In addition to CO_2, injected fluids include brines associated with oil and gas production (>2 Gt yr^{-1}); Floridan aquifer wastewater (>0.5 Gt yr^{-1}); hazardous wastes (>30 Mt yr^{-1}); and natural gas (>100 Mt yr^{-1}) (Wilson *et al.*, 2003).

While few of these cases may precisely mirror the conditions under which CO_2 would be injected for storage (the peak pressures in CO_2-EOR may, for example, be lower than would be used in formation storage), these quantities compare to or exceed, plausible flows of CO_2 into storage. For example, in some cases such as the Rangely Oil Field, USA, current reservoir pressures even exceed the original formation pressure (Raleigh *et al.*, 1976). Thus, they provide a substantial body of empirical data upon which to assess the likelihood of induced seismicity resulting from fluid injection. The fact that only a few individual seismic events associated with deep-well injection have been recorded suggests that the risks are low. Perhaps more importantly, these experiences demonstrate that the regulatory limits imposed on injection pressures are sufficient to avoid significant injection-induced seismicity. Designing CO_2 storage projects to operate within these parameters should be possible. Nevertheless, because formation pressures in CO_2 storage formations may exceed those found in CO_2-EOR projects, more experience with industrial-scale CO_2 storage projects will be needed to fully assess risks of microseismicity.

5.7.4.5 Implications of gas impurity

Under some circumstances, H_2S, SO_2, NO_2 and other trace gases may be stored along with CO_2 (Bryant and Lake, 2005; Knauss *et al.*, 2005) and this may affect the level of risk. For example, H_2S is considerably more toxic than CO_2 and well blow-outs containing H_2S may present higher risks than well blow-outs from storage sites that contain only CO_2. Similarly, dissolution of SO_2 in groundwater creates a far stronger acid than does dissolution of CO_2; hence, the mobilization of metals in groundwater and soils may be higher, leading to greater risk of exposure to hazardous levels of trace metals. While there has not been a systematic and comprehensive assessment of how these additional constituents would affect the risks associated with CO_2 storage, it is worth noting that at Weyburn, one of the most carefully monitored CO_2 injection projects and one for which a considerable effort has been devoted to risk assessment, the injected gas contains approximately 2% H_2S (Wilson and Monea, 2005). To date, most risk assessment studies have assumed that only CO_2 is stored; therefore, insufficient information is available to assess the risks associated with gas impurities at the present time.

5.7.5 Risk assessment methodology

Risk assessment aims to identify and quantify potential risks caused by the subsurface injection of CO_2, where risk denotes a combination (often the product) of the probability of an event happening and the consequences of the event. Risk assessment should be an integral element of risk-management activities, spanning site selection, site characterization, storage system design, monitoring and, if necessary, remediation.

The operation of a CO_2 storage facility will necessarily involve risks arising from the operation of surface facilities such as pipelines, compressors and wellheads. The assessment of such risks is routine practice in the oil and gas industry and available assessment methods like hazard and operability and quantitative risk assessment are directly applicable. Assessment of such risks can be made with considerable confidence, because estimates of failure probabilities and the consequences of failure can be based directly on experience. Techniques used for assessment of operational risks will not, in general, be readily applicable to assessment of risks arising from long-term storage of CO_2 underground. However, they are applicable to the operating phase of a storage project. The remainder of this subsection addresses the long-term risks.

Risk assessment methodologies are diverse; new methodologies arise in response to new classes of problems. Because analysis of the risks posed by geological storage of CO_2 is a new field, no well-established methodology for assessing such risks exists. Methods dealing with the long-term risks posed by the transport of materials through the subsurface have been developed in the area of hazardous and nuclear waste management (Hodgkinson and Sumerling, 1990; North, 1999). These techniques provide a useful basis for assessing the risks of CO_2 storage. Their applicability may be limited, however, because the focus of these techniques has been on assessing the low-volume disposal of hazardous materials, whereas the geological storage of CO_2 is high-volume disposal of a material that involves comparatively mild hazards.

Several substantial efforts are under way to assess the risks posed by particular storage sites (Gale, 2003). These risk assessment activities cover a wide range of reservoirs, use a diversity of methods and consider a very wide class of risks. The description of a representative selection of these risk assessment efforts is summarized in Table 5.6.

The development of a comprehensive catalogue of the risks and of the mechanisms that underlie them, provides a good foundation for systematic risk assessment. Many of the ongoing risk assessment efforts are now cooperating to identify, classify and screen all factors that may influence the safety of storage facilities, by using the features, events and processes (FEP) methodology. In this context, *features* includes a list of parameters, such as storage reservoir permeability, caprock thickness and number of injection wells. *Events* includes processes such as seismic events, well blow-outs and penetration of the storage site by new wells. *Processes* refers to the physical and chemical processes, such as multiphase flow, chemical reactions and geomechanical stress changes that influence storage capacity and security. FEP databases tie information on individual FEPs to relevant literature and allow classification with respect to likelihood, spatial scale, time scale and so on. However, there are alternative approaches.

Most risk assessments involve the use of scenarios that describe possible future states of the storage facility and events that result in leakage of CO_2 or other risks. Each scenario may be considered as an assemblage of selected FEPs. Some risk assessments define a reference scenario that represents the most probable evolution of the system. Variant scenarios are then constructed with alternative FEPs. Various methods are used to structure and rationalize the process of scenario definition in an attempt to reduce the role of subjective judgements in determining the outcomes.

Scenarios are the starting points for selecting and developing mathematical-physical models (Section 5.4.2). Such performance assessment models may include representations of all relevant components including the stored CO_2, the reservoir, the seal, the overburden, the soil and the atmosphere. Many of the fluid-transport models used for risk assessment are derived from (or identical to) well-established models used in the oil and gas or groundwater management industries (Section 5.4.2). The detail or resolution of various components may vary greatly. Some

Table 5.6 Representative selection of risk assessment models and efforts.

Project title	Description and status
Weyburn/ECOMatters	New model, CQUESTRA, developed to enable probabilistic risk assessment. A simple box model is used with explicit representation of transport between boxes caused by failure of wells.
Weyburn/Monitor Scientific	Scenario-based modelling that uses an industry standard reservoir simulation tool (Eclipse3000) based on a realistic model of known reservoir conditions. Initial treatment of wells involves assigning a uniform permeability.
NGCAS/ECL technology	Probabilistic risk assessment using fault tree and FEP (features, events and processes) database. Initial study focused on the Forties oil and gas field located offshore in the North Sea. Concluded that flow through caprock transport by advection in formation waters not important, work on assessing leakage due to well failures ongoing.
SAMARCADS (safety aspects of CO_2 storage)	Methods and tools for HSE risk assessment applied to two storage systems an onshore gas storage facility and an offshore formation.
RITE	Scenario-based analysis of leakage risks in a large offshore formation. Will assess scenarios involving rapid release through faults activated by seismic events.
Battelle	Probabilistic risk assessment of an onshore formation storage site that is intended to represent the Mountaineer site.
GEODISC	Completed a quantitative risk assessment for four sites in Australia: the Petrel Sub-basin; the Dongra depleted oil and gas field; the offshore Gippsland Basin; and, offshore Barrow Island. Also produced a risk assessment report that addressed the socio-political needs of stakeholders.
UK-DTI	Probabilistic risk assessment of failures in surface facilities that uses models and operational data. Assessment of risk of release from geological storage that uses an expert-based Delphi process.

models are designed to allow explicit treatment of uncertainty in input parameters (Saripalli *et al.*, 2003; Stenhouse *et al.*, 2005; Wildenborg *et al.*, 2005a).

Our understanding of abandoned-well behaviour over long time scales is at present relatively poor. Several groups are now collecting data on the performance of well construction materials in high-CO_2 environments and building wellbore simulation models that will couple geomechanics, geochemistry and fluid transport (Scherer *et al.*, 2005; Wilson and Monea, 2005). The combination of better models and new data should enable the integration of physically based predictive models of wellbore performance into larger performance-assessment models, enabling more systematic assessment of leakage from wells.

The parameter values (e.g., permeability of a caprock) and the structure of the performance assessment models (e.g., the processes included or excluded) will both be, in general, uncertain. Risk analysis may or may not treat this uncertainty explicitly. When risks are assessed deterministically, fixed parameter values are chosen to represent the (often unknown) probability distributions. Often the parameter values are selected 'conservatively'; that is, they are selected so that risks are overestimated, although in practice such selections are problematic because the relationship between the parameter value and the risk may itself be uncertain.

Wherever possible, it is preferable to treat uncertainty explicitly. In probabilistic risk assessments, explicit probability distributions are used for some (or all) parameters. Methods such as Monte Carlo analysis are then used to produce probability distributions for various risks. The required probability distributions may be derived directly from data or may involve

formal quantification of expert judgements (Morgan and Henrion, 1999). In some cases, probabilistic risk assessment may require that the models be simplified because of limitations on available computing resources.

Studies of natural and engineered analogues provide a strong basis for understanding and quantifying the health, safety and environmental risks that arise from CO_2 that seeps from the shallow subsurface to the atmosphere. Natural analogues are of less utility in assessing the likelihood of various processes that transport CO_2 from the storage reservoir to the near-surface environment. This is because the geological character of such analogues (e.g., CO_2 transport and seepage in highly fractured zones shaped by volcanism) will typically be very different from sites chosen for geological storage. Engineered analogues such as natural gas storage and CO_2-EOR can provide a basis for deriving quantitative probabilistic models of well performance.

Results from actual risk and assessment for CO_2 storage are provided in 5.7.3.

5.7.6   *Risk management*

Risk management entails the application of a structured process to identify and quantify the risks associated with a given process, to evaluate these, taking into account stakeholder input and context, to modify the process to remove excess risks and to identify and implement appropriate monitoring and intervention strategies to manage the remaining risks.

For geological storage, effective risk mitigation consists of four interrelated activities:

- Careful site selection, including performance and risk

assessment (Section 5.4) and socio-economic and environmental factors;

- Monitoring to provide assurance that the storage project is performing as expected and to provide early warning in the event that it begins to leak (Section 5.6);
- Effective regulatory oversight (Section 5.8);
- Implementation of remediation measures to eliminate or limit the causes and impacts of leakage (Section 5.7.7).

Risk management strategies must use the inputs from the risk assessment process to enable quantitative estimates of the degree of risk mitigation that can be achieved by various measures and to establish an appropriate level of monitoring, with intervention options available if necessary. Experience from natural gas storage projects and disposal of liquid wastes has demonstrated the effectiveness of this approach to risk mitigation (Wilson *et al.*, 2003; Apps, 2005; Perry, 2005).

5.7.7   *Remediation of leaking storage projects*

Geological storage projects will be selected and operated to avoid leakage. However, in rare cases, leakage may occur and remediation measures will be needed, either to stop the leak or to prevent human or ecosystem impact. Moreover, the availability of remediation options may provide an additional level of assurance to the public that geological storage can be safe and effective. While little effort has focused on remediation options thus far, Benson and Hepple (2005) surveyed the practices used to remediate natural gas storage projects, groundwater and soil contamination, as well as disposal of liquid waste in deep geological formations. On the basis of these surveys, remediation options were identified for most of the leakage scenarios that have been identified, namely:

- Leaks within the storage reservoir;
- Leakage out of the storage formation up faults and fractures;
- Shallow groundwater;
- Vadose zone and soil;
- Surface fluxes;
- CO_2 in indoor air, especially basements;
- Surface water.

Identifying options for remediating leakage of CO_2 from active or abandoned wells is particularly important, because they are known vulnerabilities (Gasda *et al.*, 2004; Perry, 2005). Stopping blow-outs or leaks from injection or abandoned wells can be accomplished with standard techniques, such as injecting a heavy mud into the well casing. If the wellhead is not accessible, a nearby well can be drilled to intercept the casing below the ground surface and then pump mud down into the interception well. After control of the well is re-established, the well can be repaired or abandoned. Leaking injection wells can be repaired by replacing the injection tubing and packers. If the annular space behind the casing is leaking, the casing can be perforated to allow injection (squeezing) of cement behind the casing until the leak is stopped. If the well cannot be repaired,

it can be abandoned by following the procedure outlined in Section 5.5.2.

Table 5.7 provides an overview of the remediation options available for the leakage scenarios listed above. Some methods are well established, while others are more speculative. Additional detailed studies are needed to further assess the feasibility of applying these to geological storage projects – studies that are based on realistic scenarios, simulations and field studies.

5.8   Legal issues and public acceptance

What legal and regulatory issues might be involved in CO_2 storage? How do they differ from one country to the next and from onshore to offshore? What international treaties exist that have bearing on geological storage? How does and how will the public view geological storage? These subjects are addressed in this section, which is primarily concerned with geological storage, both onshore and offshore.

5.8.1   *International law*

This section considers the legal position of geological CO_2 storage under international law. Primary sources, namely the relevant treaties, provide the basis for any assessment of the legal position. While States, either individually or jointly, apply their own interpretations to treaty provisions, any determination of the 'correct' interpretation will fall to the International Court of Justice or an arbitral tribunal in accordance with the dispute settlement mechanism under that treaty.

5.8.1.1   *Sources and nature of international obligations*
According to general principles of customary international law, States can exercise their sovereignty in their territories and therefore could engage in activities such as the storage of CO_2 (both geological and ocean) in those areas under their jurisdiction. However, if such storage causes transboundary impacts, States have the responsibility to ensure that activities within their jurisdiction or control do not cause damage to the environment of other States or of areas beyond the limits of national jurisdiction.

More specifically, there exist a number of global and regional environmental treaties, notably those on climate change and the law of the sea and marine environment, which, as presently drafted, could be interpreted as relevant to the permissibility of CO_2 storage, particularly offshore geological storage (Table 5.8).

Before making any assessment of the compatibility of CO_2 storage with the international legal obligations under these treaties, the general nature of such obligations should be recalled – namely that:

- Obligations under a treaty fall only on the Parties to that treaty;
- States take such obligations seriously and so will look to the provisions of such treaties before reaching policy decisions;

Table 5.7. Remediation options for geological CO_2 storage projects (after Benson and Hepple, 2005).

Scenario	Remediation options
Leakage up faults, fractures and spill points	• Lower injection pressure by injecting at a lower rate or through more wells (Buschbach and Bond, 1974); • Lower reservoir pressure by removing water or other fluids from the storage structure; • Intersect the leakage with extraction wells in the vicinity of the leak; • Create a hydraulic barrier by increasing the reservoir pressure upstream of the leak; • Lower the reservoir pressure by creating a pathway to access new compartments in the storage reservoir; • Stop injection to stabilize the project; • Stop injection, produce the CO_2 from the storage reservoir and reinject it back into a more suitable storage structure.
Leakage through active or abandoned wells	• Repair leaking injection wells with standard well recompletion techniques such as replacing the injection tubing and packers; • Repair leaking injection wells by squeezing cement behind the well casing to plug leaks behind the casing; • Plug and abandon injection wells that cannot be repaired by the methods listed above; • Stop blow-outs from injection or abandoned wells with standard techniques to 'kill' a well such as injecting a heavy mud into the well casing. After control of the well is re-established, the recompletion or abandonment practices described above can be used. If the wellhead is not accessible, a nearby well can be drilled to intercept the casing below the ground surface and 'kill' the well by pumping mud down the interception well (DOGGR, 1974).
Accumulation of CO_2 in the vadose zone and soil gas	• Accumulations of gaseous CO_2 in groundwater can be removed or at least made immobile, by drilling wells that intersect the accumulations and extracting the CO_2. The extracted CO_2 could be vented to the atmosphere or reinjected back into a suitable storage site; • Residual CO_2 that is trapped as an immobile gas phase can be removed by dissolving it in water and extracting it as a dissolved phase through groundwater extraction well; • CO_2 that has dissolved in the shallow groundwater could be removed, if needed, by pumping to the surface and aerating it to remove the CO_2. The groundwater could then either be used directly or reinjected back into the groundwate; • If metals or other trace contaminants have been mobilized by acidification of the groundwater, 'pump-and-treat' methods can be used to remove them. Alternatively, hydraulic barriers can be created to immobilize and contain the contaminants by appropriately placed injection and extraction wells. In addition to these active methods of remediation, passive methods that rely on natural biogeochemical processes may also be used.
Leakage into the vadose zone and accumulation in soil gas (Looney and Falta, 2000)	• CO_2 can be extracted from the vadose zone and soil gas by standard vapor extraction techniques from horizontal or vertical wells; • Fluxes from the vadose zone to the ground surface could be decreased or stopped by caps or gas vapour barriers. Pumping below the cap or vapour barrier could be used to deplete the accumulation of CO_2 in the vadose zone; • Since CO_2 is a dense gas, it could be collected in subsurface trenches. Accumulated gas could be pumped from the trenches and released to the atmosphere or reinjected back underground; • Passive remediation techniques that rely only on diffusion and 'barometric pumping' could be used to slowly deplete one-time releases of CO_2 into the vadose zone. This method will not be effective for managing ongoing releases because it is relatively slow; • Acidification of the soils from contact with CO_2 could be remediated by irrigation and drainage. Alternatively, agricultural supplements such as lime could be used to neutralize the soil;
Large releases of CO_2 to the atmosphere	• For releases inside a building or confined space, large fans could be used to rapidly dilute CO_2 to safe levels; • For large releases spread out over a large area, dilution from natural atmospheric mixing (wind) will be the only practical method for diluting the CO_2; • For ongoing leakage in established areas, risks of exposure to high concentrations of CO_2 in confined spaces (e.g. cellar around a wellhead) or during periods of very low wind, fans could be used to keep the rate of air circulation high enough to ensure adequate dilution.
Accumulation of CO_2 in indoor environments with chronic low-level leakage	• Slow releases into structures can be eliminated by using techniques that have been developed for controlling release of radon and volatile organic compounds into buildings. The two primary methods for managing indoor releases are basement/substructure venting or pressurization. Both would have the effect of diluting the CO_2 before it enters the indoor environment (Gadgil *et al.*, 1994; Fischer *et al.*, 1996).
Accumulation in surface water	• Shallow surface water bodies that have significant turnover (shallow lakes) or turbulence (streams) will quickly release dissolved CO_2 back into the atmosphere; • For deep, stably stratified lakes, active systems for venting gas accumulations have been developed and applied at Lake Nyos and Monoun in Cameroon (http://perso.wanadoo.fr/mhalb/nyos/).

Table 5.8 Main international treaties for consideration in the context of geological CO_2 storage (full titles are given in the Glossary).

Treaty	Adoption (Signature)	Entry into Force	Number of Parties/Ratifications
UNFCCC	1992	1994	189
Kyoto Protocol (KP)	1997	2005	132[a]
UNCLOS	1982	1994	145
London Convention (LC)	1972	1975	80
London Protocol (LP)	1996	No	20[a] (26)
OSPAR	1992	1998	15
Basel Convention	1989	1992	162

[a] Several other countries have also announced that their ratification is under way.

- Most environmental treaties contain underlying concepts, such as sustainable development, precautionary approach or principles, that should be taken into account when applying their provisions;
- In terms of supremacy of different treaties, later treaties will supersede earlier ones, but this will depend on *lex specialis,* that is, provisions on a specific subject will supersede general ones (relevant to the relationship between the United Nations Framework Convention on Climate Change (UNFCCC) and its Kyoto Protocol (KP) and the marine treaties);
- Amendment of treaties, if needed to permit CO_2 storage, requires further negotiations, a minimum level of support for their adoption and subsequent entry into force and will amend earlier treaties only for those Parties that have ratified the amendments.

5.8.1.2 Key issues in the application of the marine treaties to CO_2 storage

When interpreting the treaties for the purposes of determining the permissibility of CO_2 storage, particularly offshore geological storage, it is important to bear in mind that the treaties were not drafted to facilitate geological storage but to prohibit marine dumping. Issues to bear in mind include the following:

- Whether storage constitutes 'dumping', that is, it does not if the placement of the CO_2 is 'other than for the purposes of the mere disposal thereof' in accordance with the United Nations Convention on the Law of the Sea (UNCLOS), the London Convention (LC), the London Protocol (LP) and the Convention for the Protection of the Marine Environment of the North-East Atlantic (OSPAR). Alternative scenarios include experiments and storage for the purposes of enhanced oil recovery;
- Whether CO_2 storage can benefit from treaty exemptions concerning wastes arising from the normal operations of offshore installations (LC/LP) or as discharges or emissions from them (OSPAR);
- Is storage in the seabed expressly covered in the treaties or is it limited to the water column (UNCLOS, LC/LP, OSPAR)?
- Is CO_2 (or the substance captured if containing impurities) an 'industrial waste' (LC), 'hazardous waste' (Basel

Convention) or does the process of its storage constitute 'pollution' (UNCLOS) or is it none of these?
- Does the method of the CO_2 reaching the disposal site involve pipelines, vessels or offshore structures (LC/LP, OSPAR)?

5.8.1.3 Literature on geological storage under international law

While it is necessary to look at and interpret the treaty provisions themselves to determine the permissibility of CO_2 storage, secondary sources contain States' or authors' individual interpretations of the treaties.

In their analysis, Purdy and Macrory (2004) conclude that since stored CO_2 does not enter the atmosphere, it will not be classed as an 'emission' for the purposes of the UNFCCC/KP, but as an 'emission reduction'. Emission reductions by CO_2 storage are permitted under the UNFCCC/KP, which allows projects that reduce greenhouse gases at the source. However, the authors consider a potential problem in UNFCCC/KP providing for transparent verification of emission reductions and there could be concerns over permanence, leakage and security.

In terms of marine treaties and in relation to OSPAR, which applies to the North East Atlantic, a report from the OSPAR Group of Jurists and Linguists contains the State Parties' interpretation of OSPAR on the issue of geological (and ocean) offshore storage (OSPAR Commission, 2004). It concludes that, as there is the possibility of pollution or of other adverse environmental effects, the precautionary principle must be applied. More specifically, the report interprets OSPAR as allowing CO_2 placement in the North East Atlantic (including seabed and subsoil) through a *pipeline from land*, provided it does not involve subsequent activities through a vessel or an offshore installation (e.g., an oil or gas platform). The report states, however, that placement from a vessel is prohibited, unless for the purpose of experimentation (which would then require being carried out in accordance with other relevant provisions of OSPAR). In the case of placement in the OSPAR maritime area from an *offshore installation*, this depends upon whether the CO_2 to be stored results from offshore or land-based activities. In the case of offshore-derived CO_2, experimental placement will again be subject to the Convention's provisions, while placement for EOR, climate change mitigation or indeed mere disposal will be strictly subject to authorization or regulation. As regards onshore-derived CO_2, placement only for experimental or EOR purposes will be allowed, subject to the same caveats as for offshore-derived CO_2. The report concludes that, since the applicable OSPAR regime is determined by the method and purpose of placement and not by the effect of placement on the marine environment, the results may well be that placements with different impacts on the environment (for example, placement in the water column and placement in underground strata) may not be distinguished, while different methods of placement having the same impact may be treated differently. A similar analytical exercise concerning the LC/LP has been initiated by Parties to that Convention.

There is uncertainty regarding the extent to which CO_2 storage falls under the jurisdiction of the marine treaties. Some authors argue they will probably not allow such storage or that the LC (globally) and OSPAR (in the North East Atlantic) could significantly restrict geological offshore storage (Lenstra and van Engelenburg, 2002; Bewers, 2003). Specifically regarding the issues raised above, the following propositions have been suggested:

- The long-term storage of CO_2 amounts to 'dumping' under the conventions (Purdy and Macrory, 2004); if CO_2 were to be injected for an industrial purpose, that is, EOR, it would not be considered dumping of waste and would be allowed under the LC (Wall *et al.*, 2005);
- CO_2 captured from an oil or natural gas extraction operation and stored offshore in a geological formation would not be considered 'dumping' under the LC (Wall *et al.*, 2005);
- There remain some ambiguities in the provisions of some conventions, especially in relation to the option of geological storage under the seabed (Ducroux and Bewers, 2005). UNCLOS provides the international legal basis for a range of future uses for the seafloor that could potentially include geological storage of CO_2 (Cook and Carleton, 2000);
- Under the LC, CO_2 might fall under the 'industrial waste' category in the list of wastes prohibited for disposal, while under the LP and OSPAR, it would probably not fall under the categories approved for dumping and should therefore be considered as waste and this is prohibited (Purdy and Macrory, 2004).

If CO_2 is transported *by ship* and then disposed of, either directly from the ship or from an offshore installation, this will be prohibited under the LC/LP (Wall *et al.*, 2005) and OSPAR (Purdy and Macrory, 2004). If CO_2 is transported *by pipeline* to an offshore installation and then disposed of, that would be prohibited under the LC/LP, but not necessarily under OSPAR, where prohibition against dumping applies only to installations carrying out activities concerning hydrocarbons (Purdy and Macrory, 2004). The option of storing CO_2 transported through a pipeline from land appears to remain open under most conventions (Ducroux and Bewers, 2005); the LC/LP apply only to activities that involve ships or platforms and contain no further controls governing pipeline discharges from land-based

sources. Any such discharges would probably be excluded from control by the LC because it would not involve 'disposal at sea' (Wall *et al.*, 2005). Under OSPAR, however, States have general environmental obligations with respect to land-based sources (Purdy and Macrory, 2004) (and discharges from pipelines from land will be regulated, although not prohibited).

5.8.2   *National regulations and standards*

States can regulate subsurface injection and storage of CO_2 within their jurisdiction in accordance with their national rules and regulations. Such rules and regulations could be provided by the mining laws, resource conservation laws, laws on drinking water, waste disposal, oil and gas production, treatment of high-pressurized gases and others. An analysis of existing regulations in North America, Europe, Japan and Australia highlights the lack of regulations that are specifically relevant for CO_2 storage and the lack of clarity relating to post-injection responsibilities (IEA-GHG, 2003; IOGCC, 2005).

Presently, CO_2 is injected into the subsurface for EOR and for disposal of acid gas (Section 5.2.4). Most of these recovery or disposal activities inject relatively small quantities of CO_2 into reasonably well-characterized formations. Generally, the longevity of CO_2 storage underground and the extent of long-term monitoring of the injected fluids are not specified in the regulation of these activities, which are generally regulated under the larger umbrella of upstream oil and gas production and waste disposal regulations that do not specify storage time and need for post-operational monitoring.

In Canada, the practice of deep-well injection of fluids in the subsurface, including disposal of liquid wastes, is legal and regulated. As a result of provincial jurisdiction over energy and mineral resources, there are no generally applicable national laws that specifically regulate deep-well injection of fluids. Onshore CO_2 geological storage would fall under provincial laws and regulations, while storage offshore and in federally administered territories would fall under federal laws and regulations. In the western provinces that are major oil and gas producers, substantive regulations specifically manage the use of injection wells. In Alberta, for example, there are detailed procedural regulations regarding well construction, operation and abandonment, within which specific standards are delineated for five classes of injection wells (Alberta Energy and Utilities Board, 1994). In Saskatchewan, *The Oil and Gas Conservation Regulations 1985* (with Amendments through 2000) prescribe standards for disposal of oil field brine and other wastes. In addition, capture, transport and operational injection of fluids, including acid gas and CO_2, are by and large covered under existing regulations, but no regulations are in place for monitoring the fate of the injected fluids in the subsurface and/or for the post-abandonment stage of an injection operation.

In the United States, the Safe Drinking Water Act regulates most underground injection activities. The USEPA Underground Injection and Control (UIC) Program, created in 1980 to provide minimum standards, helps harmonize regulatory requirements for underground injection activities. The explicit goal of the UIC programme is to protect current and potential sources of public drinking water. The Safe Drinking Water Act expressly prohibits underground injection that 'endangers' an underground source of drinking water. Endangerment is defined with reference to national primary drinking water regulations and adverse human health effects. For certain types or 'classes' of wells, regulations by the USEPA prohibit injection that causes the movement of any contaminant into an underground source of drinking water.

Wells injecting hazardous wastes require the additional development of a no-migration petition to be submitted to the regulators. These petitions place the onus of proof on the project proponent that injected fluid will not migrate from the disposal site for 10,000 years or more. The fluids can exhibit buoyancy effects, as disposed fluids can be less dense than the connate fluids of the receiving formation. Operators are required to use models to demonstrate they can satisfy the 'no-migration' requirement over 10,000 years. Wilson *et al.* (2003) suggests that this process of proving containment could provide a model for long-term storage of CO_2. While detailed requirements exist for siting, constructing and monitoring injection well operation, there are no federal requirements for monitoring or verification of the actual movement of fluids within the injection zone, nor are there general requirements for monitoring in overlying zones to detect leakage. However, there are requirements for ambient monitoring in deep hazardous and industrial waste wells, with the degree of rigour varying from state to state.

Vine (2004) provides an extensive overview of environmental regulations that might affect geological CO_2 storage projects in California. Given that a developer may need to acquire up to 15 permits from federal, state and local authorities, Vine stresses the need for research to quantitatively assess the impacts of regulations on project development.

In Australia, permitting responsibility for onshore oil and gas activities reside with the State Governments, while offshore activities are primarily the responsibility of the Federal Government. A comprehensive assessment of the Australian regulatory regime is under way, but so far only South Australia has adopted legislation regulating the underground injection of gases such as CO_2 for EOR and for storage. Stringent environmental impact assessments are required for all activities that could compromise the quality of surface water or groundwater.

The 25 member states of the European Union (EU) have to ensure that geological storage of CO_2 is in conformity with relevant EU Directives. A number of directives could have an influence on CO_2 geological storage in the EU, notably those on waste (75/442/EEC), landfill (1999/31/EC), water (2000/60/EC), environmental impact assessment (85/337/EEC) and strategic environmental assessment (2001/42/EC). These directives were designed in a situation where CO_2 capture and storage was not taken into account and is not specifically mentioned.

There is one comprehensive Dutch study detailing legal and regulatory aspects of CO_2 underground injection and storage (CRUST Legal Task Force, 2001), including ownership of the stored CO_2, duty of care, liability and claim settlement. It has as its basis the legal situation established by the Dutch Mining

Act of 2003 that covers 'substances' stored underground and unites previously divided regulation of onshore and offshore activities. Storage is defined as 'placing or keeping substances at depth of more than 100 m below the surface of the earth'. Legal interpretation indicates that CO_2 intended for storage would have to be treated as waste, because it was collected with the explicit purpose of disposal.

Regulating CO_2 storage presents a variety of challenges: the scale of the activity, the need to monitor and verify containment and any leakage of a buoyant fluid and the long storage time – all of which require specific regulatory considerations. Additionally, injecting large quantities of CO_2 into saline formations that have not been extensively characterized or may be close to populated areas creates potential risks that will need to be considered. Eventually, linkages between a CO_2 storage programme and a larger national and international CO_2 accounting regime will need to be credibly established.

5.8.3   *Subsurface property rights*

Storage of CO_2 in the subsurface raises several questions: Could rights to pore space be transferred to another party? Who owns CO_2 stored in pore space? How can storage of CO_2 in the pore space be managed so as to assure minimal damage to other property rights (e.g., mineral resources, water rights) sharing the same space? Rights to use subsurface pore space could be granted, separating them from ownership of the surface property. This, for example, appears to apply to most European countries and Canada, whereas in the United States, while there are currently no specific property-rights issues that could govern CO_2 storage, the rights to the subsurface can be severed from the land.

Scale is also an important issue. Simulations have shown that the areal extent of a plume of CO_2 injected from a 1 GW coal-fired power plant over 30 years into a 100-m-thick zone will be approximately 100 km^2 (Rutqvist and Tsang, 2002) and may grow after injection ceases. The approach to dealing with this issue will vary, depending on the legal framework for ownership of subsurface pore space. In Europe, for example, pore space is owned by the State and, therefore, utilization is addressed in the licensing process. In the United States, on the other hand, the determination of subsurface property rights on non-federal lands will vary according to state jurisdiction. In most jurisdictions, the surface owner is entitled to exclusive possession of the space formerly occupied by the subsurface minerals when the minerals are exhausted, that is, the 'pore space'. In other jurisdictions, however, no such precedent exists (Wilson, 2004). Some guidance for answering these questions can be found in the property rights arrangements associated with natural gas storage (McKinnon, 1998).

5.8.4   *Long-term liability*

It is important that liabilities that may apply to a storage project are clear to its proponent, including those liabilities that are applicable after the conclusion of the project. While a White

Paper by the European Commission outlines the general approach to environmental liability (EU, 2000), literature specifically addressing liability regimes for CO_2 storage is sparse. De Figueiredo *et al.* (2005) propose a framework to examine the implications of different types of liability on the viability of geological CO_2 storage and stress that the way in which liability is addressed may have a significant impact on costs and on public perception of CO_2 geological storage.

A number of novel issues arise with CO_2 geological storage. In addition to long-term *in-situ* risk liability, which may become a public liability after project decommissioning, global risks associated with leakage of CO_2 to the atmosphere may need to be considered. Current injection practices do not require any long-term monitoring or verification regime. The cost of monitoring and verification regimes and risk of leakage will be important in managing liability.

There are also considerations about the longevity of institutions and transferability of institutional knowledge. If long-term liability for CO_2 geological storage is transformed into a public liability, can ongoing monitoring and verification be assured and who will pay for these actions? How will information on storage locations be tracked and disseminated to other parties interested in using the subsurface? What are the time frames for storage? Is it realistic (or necessary) to put monitoring or information systems in place for hundreds of years?

Any discussion of long-term CO_2 geological storage also involves intergenerational liability and thus justification of such activities involves an ethical dimension. Some aspects of storage security, such as leakage up abandoned wells, may be realized only over a long time frame, thus posing a risk to future generations. Assumptions on cost, discounting and the rate of technological progress can all lead to dramatically different interpretations of liability and its importance and need to be closely examined.

5.8.5 *Public perception and acceptance*

There is insufficient public knowledge of climate change issues and of the various mitigation options, their potential impact and their practicality. The study of public perceptions and perceived acceptability of CO_2 capture and storage is at an early stage with few studies (Gough *et al.*, 2002; Palmgren *et al.*, 2004; Shackley *et al.*, 2004; Curry *et al.*, 2005; Itaoka *et al.*, 2005). Research on perceptions of CO_2 capture and storage is challenging because of (1) the relatively technical and 'remote' nature of the issue, with few immediate points of connection in the lay public's frame of reference to many key concepts; and (2) the early stage of the technology, with few examples and experiences in the public domain to draw upon as illustrations.

5.8.5.1 *Survey research*
Curry *et al.* (2005) surveyed more than 1200 people representing a general population sample of the United States. They found that less than 4% of the respondents were familiar with the terms *carbon dioxide capture and storage* or *carbon storage*.

Moreover, there was no evidence that those who expressed familiarity were any more likely to correctly identify that the problem being addressed was global warming rather than water pollution or toxic waste. The authors also showed that there was a lack of knowledge of other power generation technologies (e.g., nuclear power, renewables) in terms of their environmental impacts and costs. Eurobarometer (2003) made similar findings across the European Union. The preference of the sample for different methods to address global warming (do nothing, expand nuclear power, continue to use fossil fuels with CO_2 capture and storage, expand renewables, etc.) was quite sensitive to information provided on relative costs and environmental characteristics.

Itaoka *et al.* (2005) conducted a survey of approximately a thousand people in Japan. They found much higher claimed levels of awareness of CO_2 capture and storage (31%) and general support for this mitigation strategy as part of a broader national climate change policy, but generally negative views on specific implementation of CO_2 capture and storage. Ocean storage was viewed most negatively, while offshore geological storage was perceived as the least negative. Part of the sample was provided with more information about CO_2 capture and storage, but this did not appear to make a large difference in the response. Factor analysis was conducted and revealed that four factors were important in influencing public opinion, namely perceptions of the environmental impacts and risks (e.g., leakage), responsibility for reducing CO_2 emissions, the effectiveness of CO_2 capture and storage as a mitigation option and the extent to which it permits the continued use of fossil fuels.

Shackley *et al.* (2004) conducted 212 face-to-face interviews at a UK airport regarding offshore geological storage. They found the sample was in general moderately supportive of the concept of CO_2 capture and storage as a contribution to a 60% reduction in CO_2 emissions in the UK by 2050 (the government's policy target). Provision of basic information on the technology increased the support that was given to it, though just under half of the sample were still undecided or expressed negative views. When compared with other mitigation options, support for CO_2 capture and storage increased slightly, though other options (such as renewable energy and energy efficiency) were strongly preferred. On the other hand, CO_2 capture and storage was much preferred to nuclear power or higher energy bills (no information on price or the environmental impact of other options was provided). When asked, unprompted, if they could think of any negative effects of CO_2 capture and storage, half of the respondents' mentioned leakage, while others mentioned associated potential impacts upon ecosystems and human health. Others viewed CO_2 capture and storage negatively on the grounds it was avoiding the real problem, was short-termist or indicated a reluctance to change.

Huijts (2003) polled 112 individuals living in an area above a gas field in The Netherlands that had experienced two small earthquakes (in 1994 and 2001). She found the sample was mildly positive about CO_2 capture and storage in general terms, but neutral to negative about storage in the immediate neighbourhood. The respondents also thought that the risks and drawbacks were somewhat larger than the benefits to the environment and society. The respondents considered that the personal benefits of CO_2 capture and storage were 'small' or 'reasonably small'. On the basis of her findings, Huijts (2003) observed the storage location could make a large difference to its acceptability; onshore storage below residential areas would probably not be viewed positively, although it has to be borne in mind that the study area had experienced recent earthquakes. Huijts also notes that many respondents (25%) tended to choose a neutral answer to questions about CO_2 capture and storage, suggesting they did not yet have a well-formed opinion.

Palmgren *et al.* (2004) conducted 18 face-to-face interviews in the Pittsburgh, Pennsylvania, USA, area, followed by a closed-form survey administered to a sample of 126 individuals. The study found that provision of more information led the survey respondents to adopt a more negative view towards CO_2 capture and storage. The study also found that, when asked in terms of willingness to pay, the respondents were less favourable towards CO_2 capture and storage as a mitigation option than they were to all the other options provided (which were rated, in descending order, as follows: solar, hydro, wind, natural gas, energy efficiency, nuclear, biomass, geological storage and ocean storage). Ocean storage was viewed more negatively than geological storage, especially after information was provided.

5.8.5.2 Focus-group research

Focus-group research on CO_2 capture and storage was conducted in the UK in 2001 and 2003 (Gough *et al.*, 2002; Shackley *et al.*, 2004). Initial reactions tended to be sceptical; only within the context of the broader discussion of climate change and the need for large cuts in CO_2 emissions, did opinions become more receptive. Typically, participants in these groups were clear that other approaches such as energy efficiency, demand-reduction measures and renewable energy should be pursued as a priority and that CO_2 geological storage should be developed alongside and not as a straight alternative to, these other options. There was general support for use of CO_2 capture and storage as a 'bridging measure' while other zero or low carbon energy technologies are developed or as an emergency stop-gap option if such technologies are not developed in time. There was a moderate level of scepticism among participants towards both government and industry and what may motivate their promotion of CO_2 storage, but there was also some distrust of messages promoted by environmental groups. Levels of trust in key institutions and the role of the media were perceived to have a major influence on how CO_2 capture and storage would be received by the public, a point also made by Huijts (2003).

5.8.5.3 Implications of the research

The existing research described above has applied different methodologies, research designs and terminology, making direct comparisons impossible. Inconsistencies in results have arisen concerning the effect of providing more detailed information to respondents and the evaluation of CO_2 capture and storage in general terms and in comparison with other low-

carbon mitigation options. Explanations for these differences might include the extent of concern expressed regarding future climate change. Representative samples in the USA and EU (Curry *et al.*, 2005) and most of the smaller samples (Shackley *et al.*, 2004; Itaoka *et al.*, 2005) find moderate to high levels of concern over climate change, whereas respondents in the Palmgren *et al.* (2004) study rated climate change as the least of their environmental concerns. A further explanation of the difference in perceptions might be the extent to which perceptions of onshore and offshore geological storage have been distinguished in the research.

From this limited research, it appears that at least three conditions may have to be met before CO_2 capture and storage is considered by the public as a credible technology, alongside other better known options: (1) anthropogenic global climate change has to be regarded as a relatively serious problem; (2) there must be acceptance of the need for large reductions in CO_2 emissions to reduce the threat of global climate change; (3) the public has to accept this technology as a non-harmful and effective option that will contribute to the resolution of (1) and (2). As noted above, many existing surveys have indicated fairly widespread concern over the problem of global climate change and a prevailing feeling that the negative impact outweighs any positive effects (e.g., Kempton *et al.*, 1995; Poortinga and Pidgeon, 2003). On the other hand, some survey and focus-group research suggests that widespread acceptance of the above factors amongst the public – in particular the need for large reduction in CO_2 emissions – is sporadic and variable within and between national populations. Lack of knowledge and uncertainty regarding the economic and environmental characteristics of other principal mitigation options have also been identified as an impediment to evaluating the CO_2 capture and storage option (Curry *et al.*, 2005).

Acceptance of the three conditions does not imply support for CO_2 capture and storage. The technology may still be rejected by some as too 'end of pipe', treating the symptoms not the cause, delaying the point at which the decision to move away from the use of fossil fuels is taken, diverting attention from the development of renewable energy options and holding potential long-term risks that are too difficult to assess with certainty. Conversely, there may be little realization of the practical difficulties in meeting existing and future energy needs from renewables. Acceptance of CO_2 capture and storage, where it occurs, is frequently 'reluctant' rather than 'enthusiastic' and in some cases reflects the perception that CO_2 capture and storage might be required because of failure to reduce CO_2 emissions in other ways. Furthermore, several of the studies above indicate that an 'in principle' acceptance of the technology can be very different from acceptance of storage at a specific site.

5.8.5.4 Underground storage of other fluids

Given minimal experience with storage of CO_2, efforts have been made to find analogues that have similar regulatory (and hence public acceptance) characteristics (Reiner and Herzog, 2004). Proposals for underground natural gas storage schemes have generated public opposition in some localities, despite similar facilities operating close by without apparent concern (Gough *et al.*, 2002). Concern regarding the effects of underground natural gas storage upon local property prices and difficult-to-assess risks appear in one case to have been taken up and possibly amplified by the local media. Public opposition to onshore underground storage is likely to be heightened by accidents such as the two deaths from explosions in 2001 in Hutchinson, Kansas (USA), when compressed natural gas escaped from salt cavern storage facilities (Lee, 2001). However, throughout the world today, many hundreds of natural gas storage sites are evidently acceptable to local communities. There has also been a study of the Underground Injection Control programme in the United States, because of the perceived similarity of the governing regulatory regime (Wilson *et al.*, 2003).

5.9   Costs of geological storage

How much will geological storage cost? What are the major factors driving storage costs? Can costs be offset by enhanced oil and gas production? These questions are covered in this section. It starts with a review of the cost elements and factors that affect storage costs and then presents estimated costs for different storage options. The system boundary for the storage costs used here is the delivery point between the transport system and the storage site facilities. It is generally expected that CO_2 will be delivered as a dense fluid (liquid or supercritical) under pressure at this boundary. The costs of capture, compression and transport to the site are excluded from the storage costs presented here. The figures presented are levelized costs, which incorporate economic assumptions such as the project lifetime, discount rates and inflation (see Section 3.7.2). They incorporate both capital and operating costs.

5.9.1   *Cost elements for geological storage*

The major capital costs for CO_2 geological storage are drilling wells, infrastructure and project management. For some storage sites, there may be in-field pipelines to distribute and deliver CO_2 from centralized facilities to wells within the site. Where required, these are included in storage cost estimates. For enhanced oil, gas and coal bed methane options, additional facilities may be required to handle produced oil and gas. Reuse of infrastructure and wells may reduce costs at some sites. At some sites, there may be additional costs for remediation work for well abandonment that are not included in existing estimates. Operating costs include manpower, maintenance and fuel. The costs for licensing, geological, geophysical and engineering feasibility studies required for site selection, reservoir characterization and evaluation before storage starts are included in the cost estimates. Bock *et al.* (2003) estimate these as US$ 1.685 million for saline formation and depleted oil and gas field storage case studies in the United States. Characterization costs will vary widely from site to site, depending on the extent of pre-existing data, geological complexity of the storage formations and caprock and risks of leakage. In addition, to some degree, economies of scale may

lower the cost per tonne of larger projects; this possibility has not been considered in these estimates.

Monitoring of storage will add further costs and is usually reported separately from the storage cost estimates in the literature. These costs will be sensitive to the regulatory requirements and duration of monitoring. Over the long term, there may be additional costs for remediation and for liabilities.

The cost of CO_2 geological storage is site-specific, which leads to a high degree of variability. Cost depends on the type of storage option (e.g., oil or gas reservoir, saline formation), location, depth and characteristics of the storage reservoir formation and the benefits and prices of any saleable products. Onshore storage costs depend on the location, terrain and other geographic factors. The unit costs are usually higher offshore, reflecting the need for platforms or sub-sea facilities and higher operating costs, as shown in separate studies for Europe (Hendriks *et al.*, 2002) and Australia (Allinson *et al.*, 2003). The equipment and technologies required for storage are already widely used in the energy industries, so that costs can be estimated with confidence.

5.9.2   *Cost estimates*

There are comprehensive assessments of storage costs for the United States, Australia and Europe (Hendriks *et al.*, 2002; Allinson *et al.*, 2003; Bock *et al.*, 2003). These are based on representative geological characteristics for the regions. In some cases, the original cost estimates include compression and pipeline costs and corrections have been made to derive storage costs (Table 5.9). These estimates include capital, operating and site characterization costs, but exclude monitoring costs, remediation and any additional costs required to address long-term liabilities.

The storage option type, depth and geological characteristics affect the number, spacing and cost of wells, as well as the facilities cost. Well and compression costs both increase with depth. Well costs depend on the specific technology, the location, the scale of the operation and local regulations. The cost of wells is a major component; however, the cost of individual wells ranges from about US$ 200,000 for some onshore sites (Bock *et al.*, 2003) to US$ 25 million for offshore horizontal wells (Table 5.10; Kaarstad, 2002). Increasing storage costs with depth have been demonstrated (Hendriks *et al.*, 2002). The geological characteristics of the injection formation are another major cost driver, that is, the reservoir thickness, permeability and effective radius that affect the amount and rate of CO_2 injection and therefore the number of wells needed. It is more costly to inject and store other gases (NO_x, SO_x, H_2S) with CO_2 because of their corrosive and hazardous nature, although the capture cost may be reduced (Allinson *et al.*, 2003).

Table 5.9 Compilation of CO_2 storage cost estimates for different options.

Option type	On or offshore	Location	US$/tCO$_2$ stored			Comments	Nature of Midpoint value
			Low	Mid	High		
Saline formation	Onshore	Australia	0.2	0.5	5.1	Statistics for 20 sites[a]	Median
Saline formation	Onshore	Europe	1.9	2.8	6.2	Representative range[b]	Most likely value
Saline formation	Onshore	USA	0.4	0.5	4.5	Low/base/high cases for USA[c]	Base case for average parameters
Saline formation	Offshore	Australia	0.5	3.4	30.2	Statistics for 34 sites[a]	Median
Saline formation	Offshore	N. Sea	4.7	7.7	12.0	Representative range[b]	Most likely value
Depleted oil field	Onshore	USA	0.5	1.3	4.0	Low/base/high cases for USA[c]	Base case for average parameters
Depleted gas field	Onshore	USA	0.5	2.4	12.2	Low/base/high cases for USA[c]	Base case for average parameters
Disused oil or gas field	Onshore	Europe	1.2	1.7	3.8	Representative range[b]	Most likely value
Disused oil or gas field	Offshore	N. Sea	3.8	6.0	8.1	Low/base/high cases for USA[c]	Most likely value

Note: The ranges and low, most likely (mid), high values reported in different studies were calculated in different ways. The estimates exclude monitoring costs.

a. Figures from Allinson *et al.*, (2003) are statistics for multiple cases from different sites in Australia. Low is the minimum value, most likely is median, high is maximum value of all the cases. The main determinants of storage costs are rate of injection and reservoir characteristics such as permeability, thickness, reservoir depth rather than reservoir type (such as saline aquifer, depleted field, etc.). The reservoir type could be high or low cost depending on these characteristics. The figures are adjusted to exclude compression and transport costs.

b. Figures from Hendriks *et al.*, (2002) are described as a representative range of values for storage options 1000-3000 m depth. The full range of costs is acknowledged to be larger than shown. The figures are converted from Euros to US$.

c. Bock *et al.*, (2003) define a base case, low- and high-cost cases from analysis of typical reservoirs for US sites. Each case has different depth, reservoir, cost and oil/gas price parameters. The figures are adjusted to exclude compression and transport costs.

Table 5.10 Investment costs for industry CO_2 storage projects.

Project	Sleipner	Snøhvit
Country	Norway	Norway
Start	1996	2006
Storage type	Aquifer	Aquifer
Annual CO_2 injection rate (MtCO$_2$ yr^{-1})	1	0.7
Onshore/Offshore	Offshore	Offshore
Number of wells	1	1
Pipeline length (km)	0	160
Capital Investment Costs (US$ million)		
Capture and Transport	79	143
Compression and dehydration	79	70
Pipeline	none	73
Storage	15	48
Drilling and well completion	15	25
Facilities	a	12
Other	a	11
Total capital investment costs (US$ million)	94	191
Operating Costs (US$ million)		
Fuel and CO_2 tax	7	
References	Torp and Brown, 2005	Kaarstad, 2002

a No further breakdown figures are available. Subset of a larger system of capital and operating costs for several processes, mostly natural gas and condensate processing.

5.9.3 Cost estimates for CO_2 geological storage

This section reviews storage costs for options without benefits from enhanced oil or gas production. It describes the detailed cost estimates for different storage options.

5.9.3.1 Saline formations

The comprehensive review by Allinson *et al.,* (2003), covering storage costs for more than 50 sites around Australia, illustrates the variability that might occur across a range of sites at the national or regional scale. Onshore costs for 20 sites have a median cost of 0.5 US\$/tCO$_2$ stored, with a range of 0.2–5.1 US\$/tCO$_2$ stored. The 37 offshore sites have a median value of 3.4 US\$/tCO$_2$ stored and a range of 0.5–30.2 US\$/tCO$_2$ stored. This work includes sensitivity studies that use Monte Carlo analyses of estimated costs to changes in input parameters. The main determinants of storage costs are reservoir and injection characteristics such as permeability, thickness and reservoir depth, that affect injection rate and well costs rather than option type (such as saline formation or depleted field).

Bock *et al.* (2003) have made detailed cost estimates on a series of cases for storage in onshore saline formations in the United States. Their assumptions on geological characteristics are based on a statistical review of more than 20 different formations. These formations represent wide ranges in depth (700–1800 m), thickness, permeability, injection rate and well numbers. The base-case estimate for average characteristics has a storage cost of 0.5 US\$/tCO$_2$ stored. High- and low-cost cases representing a range of formations and input parameters are 0.4–4.5 US\$/tCO$_2$ stored. This illustrates the variability resulting from input parameters.

Onshore storage costs for saline formations in Europe for depths of 1000–3000 m are 1.9–6.2 US\$/tCO$_2$, with a most likely value of 2.8 US\$/tCO$_2$ stored (Hendriks *et al.*, 2002). This study also presents estimated costs for offshore storage over the same depth range. These estimates cover reuse of existing oil and gas platforms (Hendriks *et al.*, 2002). The range is 4.7–12.0 US\$/tCO$_2$ stored, showing that offshore costs are higher than onshore costs.

5.9.3.2 Disused oil and gas reservoirs

It has been shown that storage costs in disused oil and gas fields in North America and Europe are comparable to those for saline formations (Hendriks *et al.*, 2002; Bock *et al.*, 2003). Bock *et al.* (2003) present costs for representative oil and gas reservoirs in the Permian Basin (west Texas, USA). For disused gas fields, the base-case estimate has a storage cost of 2.4 US\$/tCO$_2$ stored, with low and high cost cases of 0.5 and 12.2 US\$/tCO$_2$ stored. For depleted oil fields, the base-case cost estimate is 1.3 US\$/tCO$_2$ stored, with low- and highcost cases of 0.5 and 4.0 US\$/tCO$_2$ stored. Some reduction in these costs may be possible by reusing existing wells in these fields, but remediation of abandoned wells would increase the costs if required.

In Europe, storage costs for onshore disused oil and gas fields at depths of 1000–3000 m are 1.2–3.8 US\$/tCO$_2$ stored. The most likely value is 1.7 US\$/tCO$_2$ stored. Offshore oil

and gas fields at the same depths have storage costs of 3.8–8.1 US\$/tCO$_2$ stored (most likely value is 6.0 US\$/tCO$_2$ stored). The costs depend on the depth of the reservoir and reuse of platforms. Disused fields may benefit from reduced exploration and monitoring costs.

5.9.3.3 Representative storage costs

The different studies for saline formations and disused oil and gas fields show a very wide range of costs, 0.2–30.0 US\$/tCO$_2$ stored, because of the site-specific nature of the costs. This reflects the wide range of geological parameters that occur in any region or country. In effect, there will be multiple sites in any geographic area with a cost curve, providing increasing storage capacity with increasing cost.

The extensive Australian data set indicates that storage costs are less than 5.1 US\$/tCO$_2$ stored for all the onshore sites and more than half the offshore sites. Studies for USA and Europe also show that storage costs are generally less than 8 US\$/tCO$_2$, except for high-cost cases for offshore sites in Europe and depleted gas fields in the United States. A recent study suggests that 90% of European storage capacity could be used for costs less that 2 US\$/tCO$_2$ (Wildenborg *et al.*, 2005b).

Assessment of these cost estimates indicates that there is significant potential for storage at costs in the range of 0.5–8 US\$/tCO$_2$ stored, estimates that are based on the median, base case or most likely values presented for the different studies (Table 5.9). These exclude monitoring costs, well remediation and longer term costs.

5.9.3.4 Investment costs for storage projects

Some information is available on the capital and operating costs of industry capture and storage projects (Table 5.10). At Sleipner, the incremental capital cost for the storage component comprising a horizontal well to inject 1 MtCO$_2$ yr^{-1} was US\$ 15 million (Torp and Brown, 2005). Note that at Sleipner, CO$_2$ had to be removed from the natural gas to ready it for sale on the open market. The decision to store the captured CO$_2$ was at least in part driven by a 40 US\$/tCO$_2$ tax on offshore CO$_2$ emissions. Details of the energy penalty and levelized costs are not available. At the planned Snohvit project, the estimated capital costs for storage are US\$ 48 million for injection of 0.7 MtCO$_2$ yr^{-1} (Kaarstad, 2002). This data set is limited and additional data on the actual costs of industry projects is needed.

5.9.4 Cost estimates for storage with enhanced oil and gas recovery

The costs of CO$_2$ geological storage may be offset by additional revenues for production of oil or gas, where CO$_2$ injection and storage is combined with enhanced oil or gas recovery or ECBM. At present, in commercial EOR and ECBM projects that use CO$_2$ injection, the CO$_2$ is purchased for the project and is a significant proportion of operating costs. The economic benefits from enhanced production make EOR and ECBM potential early options for CO$_2$ geological storage.

5.9.4.1   Enhanced oil recovery

The costs of onshore CO_2-flooding EOR projects in North America are well documented (Klins, 1984; Jarrell *et al.*, 2002). Carbon dioxide EOR projects are business ventures to increase oil recovery. Although CO_2 is injected and stored, this is not the primary driver and EOR projects are not optimized for CO_2 storage.

The commercial basis of conventional CO_2-EOR operations is that the revenues from incremental oil compensate for the additional costs incurred (including purchase of CO_2) and provide a return on the investment. The costs differ from project to project. The capital investment components are compressors, separation equipment and H_2S removal, well drilling and well conversions and completions. New wells are not required for some projects. Operating costs are the CO_2 purchase price, fuel costs and field operating costs.

In Texas, the cost of CO_2 purchase was 55–75% of the total cost for a number of EOR fields (averaging 68% of total costs) and is a major investment uncertainty for EOR. Tax and fiscal incentives, government regulations and oil and gas prices are the other main investment uncertainties (e.g., Jarrell *et al.*, 2002).

The CO_2 price is usually indexed to oil prices, with an indicative price of 11.7 US$/$tCO_2$ (0.62 US$/Mscf) at a West Texas Intermediate oil price of 18 US$ per barrel, 16.3 US$/$tCO_2$ at 25 US$ per barrel of oil and 32.7 US$/$tCO_2$ at 50 US$ per barrel of oil (Jarrell *et al.*, 2002). The CO_2 purchase price indicates the scale of benefit for EOR to offset CO_2 storage costs.

5.9.4.2   Cost of CO_2 storage with enhanced oil recovery

Recent studies have estimated the cost of CO_2 storage in EOR sites (Bock *et al.*, 2003; Hendriks *et al.*, 2002). Estimates of CO_2 storage costs for onshore EOR options in North America have been made by Bock *et al.* (2003). Estimates for a 2-$MtCO_2$ yr^{-1} storage scenario are based on assumptions and parameters from existing EOR operations and industry cost data. These include estimates of the effectiveness of CO_2-EOR, in terms of CO_2 injected for each additional barrel of oil. The methodology for these estimates of storage costs is to calculate the break-even CO_2 price (0.3 tCO_2).

Experience from field operations across North America provides information about how much of the injected CO_2 remains in the oil reservoir during EOR. An average of 170 standard m^3 CO_2 of new CO_2 is required for each barrel of enhanced oil production, with a range of 85 (0.15 tCO_2) to 227 (0.4 tCO_2) standard m^3 (Bock *et al.*, 2003). Typically, produced CO_2 is separated from the oil and reinjected back underground, which reduces the cost of CO_2 purchases.

The base case for a representative reservoir at a depth of 1219 m, based on average EOR parameters in the United States with an oil price of 15 US$ per barrel, has a net storage cost of –14.8 US$/$tCO_2$ stored. Negative costs indicate the amount of cost reduction that a particular storage option offers to the overall capture and storage system. Low- and high-cost cases representing a range of CO_2 effectiveness, depth, transport distance and oil price are –92.0 and +66.7 US$/$tCO_2$ stored. The low-cost case assumes favourable assumptions for all parameters (effectiveness, reservoir depth, productivity) and a 20 US$ per barrel oil price. Higher oil prices, such as the 50 US$ per barrel prices of 2005, will considerably change the economics of CO_2-EOR projects. No published studies are available for these higher oil prices.

Other estimates for onshore EOR storage costs all show potential at negative net costs. These include a range of –10.5 to +10.5 US$/$tCO_2$ stored for European sites (Hendriks *et al.*, 2002). These studies show that use of CO_2 enhanced oil recovery for CO_2 storage can be a lower cost option than saline formations and disused oil and gas fields.

At present, there are no commercial offshore EOR operations and limited information is available on CO_2 storage costs for EOR options in offshore settings. Indicative storage cost estimates for offshore EOR are presented by Hendriks *et al.* (2002). Their range is –10.5 to +21.0 US$/$tCO_2$ stored. For the North Sea Forties Field, it has been shown that CO_2-flooding EOR is technically attractive and could increase oil recovery, although at present it is not economically attractive as a stand-alone EOR project (Espie *et al.*, 2003). Impediments are the large capital requirement for adapting facilities, wells and flowlines, as well as tax costs and CO_2 supply. It is noted that the economics will change with additional value for storage of CO_2.

The potential benefit of EOR can be deduced from the CO_2 purchase price and the net storage costs for CO_2-EOR storage case studies. The indicative value of the potential benefit from enhanced oil production to CO_2 storage is usually in the range of 0–16 US$/$tCO_2$. In some cases, there is no benefit from EOR. The maximum estimate of the benefit ranges up to $92 per tonne of CO_2 for a single case study involving favourable parameters. In general, higher benefits will occur at high-oil-price scenarios similar to those that have occurred since 2003 and for highly favourable sites, as shown above. At 50 US$ per barrel of oil, the range may increase up to 30 US$/$tCO_2$.

5.9.4.3   Cost of CO_2 storage with enhanced gas recovery

CO_2-enhanced gas recovery is a less mature technology than EOR and it is not in commercial use. Issues are the cost of CO_2 and infrastructure, concerns about excessive mixing and the high primary recovery rates of many gas reservoirs. Cost estimates show that CO_2-EGR (enhanced gas recovery) can provide a benefit of 4–16 US$/$tCO_2$, depending on the price of gas and the effectiveness of recovery (Oldenburg *et al.*, 2002).

5.9.4.4   Cost of CO_2 storage with enhanced coal bed methane

The injection of CO_2 for ECBM production is an immature technology not yet in commercial use. In CO_2-ECBM, the revenues from the produced gas could offset the investment costs and provide a source of income for investors. Cost data are based on other types of CBM operations that are in use.

There is significant uncertainty in the effectiveness of CO_2 storage in coal beds in conjunction with ECBM, because there is no commercial experience. The suggested metric for CO_2 retention is 1.5–10 m^3 of CO_2 per m^3 of produced methane. The revenue benefit of the enhanced production will depend on gas prices.

Well costs are a major factor in ECBM because many wells are required. In one recent study for an ECBM project (Schreurs, 2002), the cost per production well was given as approximately US$750,000 per well, plus 1500 US$ m^{-1} of in-seam drilling. The cost of each injection well was approximately US$430,000.

The IEA-GHG (1998) developed a global cost curve for CO_2-ECBM, with storage costs ranging from –20 to +150 US$/tCO$_2$. It concluded that only the most favourable sites, representing less than 10% of global capacity, could have negative costs. Estimates of onshore CO_2-ECBM storage costs in the United States have been made by using the approach described for EOR (Bock *et al.*, 2003). They estimate the effectiveness of ECBM in terms of CO_2 injected for incremental gas produced, ranging from 1.5 to 10 units (base case value of 2) of CO_2 per unit of enhanced methane. Other key inputs are the gas well production rate, the ratio of producers to injectors, well depth and the number of wells. The base case, storing 2.1 MtCO$_2$ per year for a representative reservoir at 610 m depth in a newly built facility, requires 270 wells. The assumed gas price is US$1.90 per GJ (US$2.00 per Mbtu). It has a net storage cost of –8.1 US$/tCO$_2$ stored. Low- and high-cost cases representing a range of parameters are –26.4 and +11.1 US$/tCO$_2$ stored. The range of these estimates is comparable to other estimates – for example, those for Canada (Wong *et al.*, 2001) and Europe (Hendriks *et al.*, 2002), 0 to +31.5 US$/tCO$_2$. Enhanced CBM has not been considered in detail for offshore situations and cost estimates are not available.

Only one industrial-scale CO_2-ECBM demonstration project has taken place to date, the Allison project in the United States and it is no longer injecting CO_2 (Box 5.7). One analysis of the Allison project, which has extremely favourable geological characteristics, suggests the economics of ECBM in the United States are dubious under current fiscal conditions and gas prices (IEA-GHG, 2004). The economic analyses suggest this would be commercial, with high gas prices about 4 US$ per GJ) and a credit of 12–18 US$/tCO$_2$. Alternatively, Reeves (2005) used detailed modelling and economic analysis to show a break-even gas price of US$2.44 per GJ (US$2.57 per Mbtu), including costs of 5.19 US$/tCO$_2$ for CO_2 purchased at the field.

5.9.5   *Cost of monitoring*

While there has been extensive discussion of possible monitoring strategies in the literature and technologies that may be applicable, there is limited information on monitoring costs. These will depend on the monitoring strategy and technologies used and how these are adapted for the duration of storage projects. Some of the technologies likely to be used are already in widespread use in the oil and gas and CBM industries. The costs of individual technologies in current use are well constrained.

Repeated use of seismic surveys was found to be an effective monitoring technology at Sleipner. Its applicability will vary between options and sites. Seismic survey costs are highly variable, according to the technology used, location and terrain and complexity. Seismic monitoring costs have been reviewed for an onshore storage project for a 1000 MW power plant with a 30-year life (Myer *et al.*, 2003). Assuming repeat surveys at five-year intervals during the injection period, monitoring costs are estimated as 0.03 US$/tCO$_2$, suggesting that seismic monitoring may represent only a small fraction of overall storage costs. No discounting was used to develop this estimate.

Benson *et al.* (2005) have estimated life-cycle monitoring costs for two scenarios: (1) storage in an oil field with EOR and (2) storage in a saline formation. For these scenarios, no explicit leakage was considered. If leakage were to occur, the 'enhanced' monitoring programme should be sufficient to detect and locate the leakage and may be sufficient to quantify leakage rates as well. For each scenario, cost estimates were developed for the 'basic' and 'enhanced' monitoring package. The basic monitoring package included periodic seismic surveys, microseismicity, wellhead pressure and injection-rate monitoring. The enhanced package included all of the elements of the 'basic' package and added periodic well logging, surface CO_2 flux monitoring and other advanced technologies. For the basic monitoring package, costs for both scenarios are 0.05 US$/tCO$_2$, based on a discount rate of 10% (0.16–0.19 US$/tCO$_2$ undiscounted). The cost for the enhanced monitoring package is 0.069–0.085 US$/tCO$_2$ (0.27–0.30 US$/tCO$_2$ undiscounted). The assumed duration of monitoring includes the 30-year period of injection, as well as further monitoring after site closure of 20 years for EOR sites and 50 years for saline formations. Increasing the duration of monitoring to 1000 years increased the discounted cost by 10%. These calculations are made assuming a discount rate of 10% for the first 30 years and a discount rate of 1% thereafter.

5.9.6   *Cost of remediation of leaky storage projects*

No estimates have been made regarding the costs of remediation for leaking storage projects. Remediation methods listed in Table 5.7 have been used in other applications and, therefore, could be extrapolated to CO_2 storage sites. However, this has not been done yet.

5.9.7   *Cost reduction*

There is little literature on cost-reduction potential for CO_2 geological storage. Economies of scale are likely to be important (Allinson *et al.*, 2003). It is also anticipated that further cost reduction will be achieved with application of learning from early storage projects, optimization of new projects and application of advanced technologies, such as horizontal and multilateral wells, which are now widely used in the oil and gas industry.

5.10   Knowledge gaps

Knowledge regarding CO_2 geological storage is founded on basic knowledge in the earth sciences, on the experience of the oil and gas industry (extending over the last hundred years or more) and on a large number of commercial activities involving the injection and geological storage of CO_2 conducted over the past 10–30 years. Nevertheless, CO_2 storage is a new technology and many questions remain. Here, we summarize what we know now and what gaps remain.

1. Current storage capacity estimates are imperfect:
   - There is need for more development and agreement on assessment methodologies.
   - There are many gaps in capacity estimates at the global, regional and local levels.
   - The knowledge base for geological storage is for the most part based on Australian, Japanese, North American and west European data.
   - There is a need to obtain much more information on storage capacity in other areas, particularly in areas likely to experience the greatest growth in energy use, such as China, Southeast Asia, India, Russia/Former Soviet Union, Eastern Europe, the Middle East and parts of South America and southern Africa.

2. Overall, storage science is understood, but there is need for greater knowledge of particular mechanisms, including:
   - The kinetics of geochemical trapping and the long-term impact of CO_2 on reservoir fluids and rocks.
   - The fundamental processes of CO_2 adsorption and CH_4 desorption on coal during storage operations.

3. Available information indicates that geological storage operations can be conducted without presenting any greater risks for health and the local environment than similar operations in the oil and gas industry, when carried out at high-quality and well-characterized sites. However, confidence would be further enhanced by increased knowledge and assessment ability, particularly regarding:
   - Risks of leakage from abandoned wells caused by material and cement degradation.
   - The temporal variability and spatial distribution of leaks that might arise from inadequate storage sites.
   - Microbial impacts in the deep subsurface.
   - Environmental impact of CO_2 on the marine seafloor.
   - Methods to conduct end-to-end quantitative assessment of risks to human health and the local environment.

4. There is strong evidence that storage of CO_2 in geological storage sites will be long term; however, it would be beneficial to have:
   - Quantification of potential leakage rates from more storage sites.
   - Reliable coupled hydrogeological-geochemical-geo–mechanical simulation models to predict long-term storage performance accurately.

- Reliable probabilistic methods for predicting leakage rates from storage sites.
- Further knowledge of the history of natural accumulations of CO_2.
- Effective and demonstrated protocols for achieving desirable storage duration and local safety.

5. Monitoring technology is available for determining the behaviour of CO_2 at the surface or in the subsurface; however, there is scope for improvement in the following areas:
   - Quantification and resolution of location and forms of CO_2 in the subsurface, by geophysical techniques.
   - Detection and monitoring of subaquatic CO_2 seepage.
   - Remote-sensing and cost-effective surface methods for temporally variable leak detection and quantification, especially for dispersed leaks.
   - Fracture detection and characterization of leakage potential.
   - Development of appropriate long-term monitoring approaches and strategies.

6. Mitigation and remediation options and technologies are available, but there is no track record of remediation for leaked CO_2. While this could be seen as positive, some stakeholders suggest it might be valuable to have an engineered (and controlled) leakage event that could be used as a learning experience.

7. The potential cost of geological storage is known reasonably well, but:
   - There are only a few experience-based cost data from non-EOR CO_2 storage projects.
   - There is little knowledge of regulatory compliance costs.
   - There is inadequate information on monitoring strategies and requirements, which affect costs.

8. The regulatory and responsibility or liability framework for CO_2 storage is yet to be established or unclear. The following issues need to be considered:
   - The role of pilot and demonstration projects in developing regulations.
   - Approaches for verification of CO_2 storage for accounting purposes.
   - Approaches to regulatory oversight for selecting, operating and monitoring CO_2 storage sites, both in the short and long term.
   - Clarity on the need for and approaches to long-term stewardship.
   - Requirements for decommissioning a storage project.

Additional information on all of these topics would improve technologies and decrease uncertainties, but there appear to be no insurmountable technical barriers to an increased uptake of geological storage as a mitigation option.

References

Ahmad, M.U. and J.A. Smith, 1988: Earthquakes, injection wells and the Perry Nuclear Power Plant, Cleveland, Ohio. *Geology,* **16,** 739–742.

Akimoto, K., H. Kotsubo, T. Asami, X. Li, M. Uno, T. Tomoda and T. Ohsumi, 2003: Evaluation of carbon sequestrations in Japan with a mathematical model. Proceedings of the 6th International Conference on Greenhouse Gas Control Technologies (GHGT-6), J. Gale and Y. Kaya (eds.), 1-4 October 2002, Kyoto, Japan, v.I, 913–918.

Alberta Energy and Utilities Board, 1994: Injection and disposal wells, Guide #51, Calgary, AB, http://eub.gov.ab.ca/bbs/products/guides/g51-1994.pdf.

Alberta Energy and Utilities Board, 2003: Well abandonment guide, August 2003 incorporating errata to August 2004, http://www.eub.gov.ab.ca/bbs/products/guides/g20.pdf.

Allinson, W.G, D.N. Nguyen and J. Bradshaw, 2003: The economics of geological storage of CO_2 in Australia, *APPEA Journal,* **623**.

Allis, R., T. Chidsey, W. Gwynn, C. Morgan, S. White, M. Adams and J. Moore, 2001: Natural CO_2 reservoirs on the Colorado Plateau and southern Rocky Mountains: Candidates for CO_2 sequestration. Proceedings of the First National Conference on Carbon Sequestration, 14–17 May 2001, DOE NETL, Washington, DC.

Alston, R.B., G.P. Kokolis and C.F. James, 1985: CO_2 minimum miscibility pressure: A correlation for impure CO_2 streams and live oil systems. *Society of Petroleum Engineers Journal,* **25**(2), 268–274.

Amadeo, N., H. Bajano, J. Comas, J.P. Daverio, M.A. Laborde, J.A. Poggi and D.R. Gómez, 2005: Assessment of CO_2 capture and storage from thermal power plants in Argentina. Proceedings of the 7th International Conference on Greenhouse Gas Technologies (GHGT-7), September 5–9, 2004, Vancouver, Canada, v.I, 243-252.

Angus, S., B. Armstrong and K.M. de Reuck, 1973: International Thermodynamic Tables of the Fluid State Volume 3. Carbon Dioxide. IUPAC Division of Physical Chemistry, Pergamon Press, London, pp. 266–359.

Anheden, M., A. Andersson, C. Bernstone, S. Eriksson, J. Yan, S. Liljemark and C. Wall, 2005: CO_2 quality requirement for a system with CO_2 capture, transport and storage. Proceedings of the 7th International Conference on Greenhouse Gas Technologies (GHGT-7), September 5–9, 2004, Vancouver, Canada, v.II, 2559-2566.

Apps, J., 2005: The Regulatory Climate Governing the Disposal of Liquid Wastes in Deep Geologic Formations: a Paradigm for Regulations for the Subsurface Disposal of CO_2, Carbon Dioxide Capture for Storage in Deep Geologic Formations - Results from the CO_2 Capture Project, v.2: Geologic Storage of Carbon Dioxide with Monitoring and Verification, S.M. Benson (ed.), *Elsevier Science,* London, pp. 1163–1188.

Arts, R. and P. Winthaegen, 2005: Monitor options for CO_2 storage, Carbon Dioxide Capture for Storage in Deep Geologic Formations - Results from the CO_2 Capture Project, v.2: Geologic Storage of Carbon Dioxide with Monitoring and Verification, S.M. Benson (ed.), Elsevier Science, London. pp. 1001–1013.

Arts, R., A. Chadwick and O. Eiken, 2005: Recent time-lapse seismic data show no indication of leakage at the Sleipner CO_2-injection site. Proceedings of the 7th International Conference on Greenhouse Gas Technologies (GHGT-7), September 5–9, 2004, Vancouver, Canada, v.I, 653-662.

Bachu, S., 2000: Sequestration of carbon dioxide in geological media: Criteria and approach for site selection. *Energy Conservation and Management,* **41**(9), 953–970.

Bachu, S., 2003: Screening and ranking of sedimentary basins for sequestration of CO_2 in geological media. *Environmental Geology,* **44**(3), 277–289.

Bachu, S. and J.J. Adams, 2003: Sequestration of CO_2 in geological media in response to climate change: Capacity of deep saline aquifers to sequester CO_2 in solution. *Energy Conversion and Management,* **44**(20), 3151–3175.

Bachu, S. and M. Dusseault, 2005: Underground injection of carbon dioxide in salt beds. Proceedings of the Second International Symposium on Deep Well Injection, C-F. Tsang and J. Apps (eds.), 22–24 October 2003, Berkeley, CA, In press.

Bachu, S. and K. Haug, 2005: In-situ characteristics of acid -gas injection operations in the Alberta basin, western Canada: Demonstration of CO_2 geological storage, Carbon Dioxide Capture for Storage in Deep Geologic Formations - Results from the CO_2 Capture Project, v. 2: Geologic Storage of Carbon Dioxide with Monitoring and Verification, S.M. Benson (ed.), Elsevier, London, pp. 867–876.

Bachu, S. and J.C. Shaw, 2003: Evaluation of the CO_2 sequestration capacity in Alberta's oil and gas reservoirs at depletion and the effect of underlying aquifers. *Journal of Canadian Petroleum Technology,* **42**(9), 51–61.

Bachu, S. and J.C. Shaw, 2005: CO_2 storage in oil and gas reservoirs in western Canada: Effect of aquifers, potential for CO_2-flood enhanced oil recovery and practical capacity. Proceedings of the 7th International Conference on Greenhouse Gas Control Technologies (GHGT-7), September 5–9, 2004, Vancouver, Canada, v.I, 361-370.

Bachu, S., W.D. Gunter and E.H. Perkins, 1994: Aquifer disposal of CO_2: hydrodynamic and mineral trapping, *Energy Conversion and Management,* **35**(4), 269–279.

Bachu, S., J.C. Shaw and R.M. Pearson, 2004: Estimation of oil recovery and CO_2 storage capacity in CO_2 EOR incorporating the effect of underlying aquifers. SPE Paper 89340, presented at the Fourteenth SPE/DOE Improved Oil Recovery Symposium, Tulsa, OK, April 17–21, 2004, 13 pp.

Baes, C.F., S.E. Beall, D.W. Lee and G. Marland, 1980: The collection, disposal and storage of carbon dioxide. In: Interaction of Energy and Climate, W. Bach, J. Pankrath and J. William (eds.), 495–519, D. Reidel Publishing Co.

Baines, S.J. and R.H. Worden, 2001: Geological CO_2 disposal: Understanding the long-term fate of CO_2 in naturally occurring accumulations. Proceedings of the 5th International Conference on Greenhouse Gas Control Technologies (GHGT-5), D.J. Williams, R.A. Durie, P. McMullan, C.A.J. Paulson and A. Smith (eds.), 13–16 August 2000, Cairns, Australia, CSIRO Publishing, Collingwood, Victoria, Australia, pp. 311–316.

Beecy, D. and V.A. Kuuskra, 2005: Basin strategies for linking CO_2 enhanced oil recovery and storage of CO_2 emissions. Proceedings of the 7th International Conference on Greenhouse Gas Control Technologies (GHGT-7), September 5–9, 2004, Vancouver, Canada, v.I, 351-360.

Benson, S.M., 2005: Lessons learned from industrial and natural analogs for health, safety and environmental risk assessment for geologic storage of carbon dioxide. Carbon Dioxide Capture for Storage in Deep Geologic Formations - Results from the CO_2 Capture Project, v. 2: Geologic Storage of Carbon Dioxide with Monitoring and Verification, S.M. Benson (ed.), Elsevier, London, pp. 1133–1141.

Benson, S.M. and R.P. Hepple, 2005: Prospects for early detection and options for remediation of leakage from CO_2, storage projects, Carbon Dioxide Capture for Storage in Deep Geologic Formations - Results from the CO_2 Capture Project, v. 2: Geologic Storage of Carbon Dioxide with Monitoring and Verification, S.M. Benson (ed.), Elsevier, London, pp. 1189–1204.

Benson, S.M., E. Gasperikova and G.M. Hoversten, 2004: Overview of monitoring techniques and protocols for geologic storage projects, IEA Greenhouse Gas R&D Programme Report.

Benson, S.M., E. Gasperikova and G.M. Hoversten, 2005: Monitoring protocols and life-cycle costs for geologic storage of carbon dioxide. Proceedings of the 7th International Conference on Greenhouse Gas Control Technologies (GHGT-7), September 5–9, 2004, Vancouver, Canada, v.II, 1259-1266.

Bøe, R., C. Magnus, P.T. Osmundsen and B.I. Rindstad, 2002: CO_2 point sources and subsurface storage capacities for CO_2 in aquifers in Norway. Norsk Geologische Undersogelske, Trondheim, Norway, NGU Report 2002.010, 132 pp.

Bergfeld, D., F. Goff and C.J. Janik, 2001: Elevated carbon dioxide flux at the Dixie Valley geothermal field, Nevada; relations between surface phenomena and the geothermal reservoir. *Chemical Geology*, **177**(1–2), 43–66.

Bergman, P.D. and E.M. Winter, 1995: Disposal of carbon dioxide in aquifers in the US. *Energy Conversion and Management*, **36**(6), 523–526.

Bergman, P.D., E.M. Winter and Z-Y. Chen, 1997: Disposal of power plant CO_2 in depleted oil and gas reservoirs in Texas. *Energy Conversion and Management*, **38**(Suppl.), S211–S216.

Bewers, M., 2003: Review of international conventions having implications for ocean storage of carbon dioxide. International Energy Agency, Greenhouse Gas Research and Development Programme, Cheltenham, UK, March 2003.

Bock, B., R. Rhudy, H. Herzog, M. Klett, J. Davison, D. De la Torre Ugarte and D. Simbeck, 2003: Economic Evaluation of CO_2 Storage and Sink Options. DOE Research Report DE-FC26-00NT40937.

Bondor, P.L., 1992: Applications of carbon dioxide in enhanced oil recovery. *Energy Conversion and Management*, **33**(5), 579–586.

Bossie-Codreanu, D., Y. Le-Gallo, J.P. Duquerroix, N. Doerler and P. Le Thiez, 2003: CO_2 sequestration in depleted oil reservoirs. Proceedings of the 6th International Conference on Greenhouse Gas Control Technologies (GHGT-6), J. Gale and Y. Kaya (eds.), 1–4 October 2002, Kyoto, Japan, Pergamon, v.I, 403–408.

Bradshaw, J.B. and T. Dance, 2005: Mapping geological storage prospectivity of CO_2 for the world sedimentary basins and regional source to sink matching. Proceedings of the 7th International Conference on Greenhouse Gas Control Technologies (GHGT-7), September 5–9, 2004, Vancouver, Canada, v.I, 583-592.

Bradshaw, J.B., E. Bradshaw, G. Allinson, A.J. Rigg, V. Nguyen and A. Spencer, 2002: The potential for geological sequestration of CO_2 in Australia: preliminary findings and implications to new gas field development. *Australian Petroleum Production and Exploration Association Journal*, **42**(1), 24–46.

Bradshaw, J., G. Allinson, B.E. Bradshaw, V. Nguyen, A.J. Rigg, L. Spencer and P. Wilson, 2003: Australia's CO_2 geological storage potential and matching of emissions sources to potential sinks. Proceedings of the 6th International Conference on Greenhouse Gas Control Technologies (GHGT-6), J. Gale and Y. Kaya (eds.), 1–4 October 2002, Kyoto, Japan, Pergamon, v.I, 633–638.

Bradshaw, J., C. Boreham and F. la Pedalina, 2005: Storage retention time of CO_2 in sedimentary basins: Examples from petroleum systems. Proceedings of the 7th International Conference on Greenhouse Gas Control Technologies (GHGT-7), September 5–9, 2004, Vancouver, Canada, v.I, 541-550.

Brennan, S.T. and R.C. Burruss, 2003: Specific Sequestration Volumes: A Useful Tool for CO_2 Storage Capacity Assessment. USGS OFR 03-0452 available at http://pubs.usgs.gov/of/2003/of03-452/.

Bryant, S. and L. Lake, 2005: Effect of impurities on subsurface CO_2 storage processes, Carbon Dioxide Capture for Storage in Deep Geologic Formations - Results from the CO_2 Capture Project, v. 2: Geologic Storage of Carbon Dioxide with Monitoring and Verification, S.M. Benson (ed.), Elsevier, London. pp. 983–998.

Buschbach, T.C. and D.C. Bond, 1974: Underground storage of natural gas in Illinois - 1973, *Illinois Petroleum*, 101, Illinois State Geological Survey.

Carapezza, M. L., B. Badalamenti, L. Cavarra and A. Scalzo, 2003: Gas hazard assessment in a densely inhabited area of Colli Albani Volcano (Cava dei Selci, Roma). *Journal of Volcanology and Geothermal Research*, **123**(1–2), 81–94.

Cawley, S., M. Saunders, Y. Le Gallo, B. Carpentier, S. Holloway, G.A. Kirby, T. Bennison, L. Wickens, R. Wikramaratna, T. Bidstrup, S.L.B. Arkley, M.A.E. Browne and J.M. Ketzer, 2005, The NGCAS Project - Assessing the potential for EOR and CO_2 storage at the Forties Oil field, Offshore UK - Results from the CO_2 Capture Project, v.2: Geologic Storage of Carbon Dioxide with Monitoring and Verification, S.M. Benson (ed.), Elsevier Science, London, pp. 1163–1188.

Celia, M.A. and S. Bachu, 2003: Geological sequestration of CO_2: Is leakage avoidable and acceptable? Proceedings of the 6th International Conference on Greenhouse Gas Control Technologies (GHGT-6), J. Gale and Y. Kaya (eds.), 1–4 October, Kyoto Japan, Pergamon, v. 1, pp. 477–482.

Celia, M.A., S. Bachu, J.M. Nordbotten, S.E. Gasda and H.K. Dahle, 2005: Quantitative estimation of CO_2 leakage from geological storage: Analytical models, numerical models and data needs. Proceedings of 7th International Conference on Greenhouse Gas Control Technologies. (GHGT-7), September 5–9, 2004, Vancouver, Canada, v.I, 663-672.

Chadwick, R.A., P. Zweigel, U. Gregersen, G.A. Kirby, S. Holloway and P.N. Johannesen, 2003: Geological characterization of CO_2 storage sites: Lessons from Sleipner, northern North Sea. Proceedings of the 6th International Conference on Greenhouse Gas Control Technologies (GHGT-6), J. Gale and Y. Kaya (eds.), 1–4 October 2002, Kyoto, Japan, Pergamon, v.I, 321–326.

Chadwick, R.A., R. Arts and O. Eiken, 2005: 4D seismic quantification of a growing CO_2 plume at Sleipner, North Sea. In: A.G. Dore and B. Vining (eds.), Petroleum Geology: North West Europe and Global Perspectives - Proceedings of the 6th Petroleum Geology Conference. Petroleum Geology Conferences Ltd. Published by the Geological Society, London, 15pp (in press).

Chalaturnyk, R. and W.D. Gunter, 2005: Geological storage of CO_2: Time frames, monitoring and verification. Proceedings of the 7th International Conference on Greenhouse Gas Control Technologies (GHGT-7), September 5–9, 2004, Vancouver, Canada, v.I, 623-632.

Chikatamarla, L. and M.R. Bustin, 2003: Sequestration potential of acid gases in Western Canadian Coals. Proceedings of the 2003 International Coalbed Methane Symposium, University of Alabama, Tuscaloosa, AL, May 5–8, 2003, 16 pp.

Chiodini, G., F. Frondini, C. Cardellini, D. Granieri, L. Marini and G. Ventura, 2001: CO_2 degassing and energy release at Solfatara volcano, Campi Flegrei, Italy. *Journal of Geophysical Research*, **106**(B8), 16213–16221.

Christman, P.G. and S.B. Gorell, 1990: Comparison of laboratory and field-observed CO_2 tertiary injectivity. *Journal of Petroleum Technology*, February 1990.

Chow, J.C., J.G. Watson, A. Herzog, S.M. Benson, G.M. Hidy, W.D. Gunter, S.J. Penkala and C.M. White, 2003: Separation and capture of CO_2 from large stationary sources and sequestration in geological formations. Air and Waste Management Association (AWMA) *Critical Review Papers,* **53**(10), October 2003.http://www.awma.org/journal/past-issue.asp?month=10&year=2003.

Clarkson, C.R. and R.M. Bustin, 1997: The effect of methane gas concentration, coal composition and pore structure upon gas transport in Canadian coals: Implications for reservoir characterization. Proceedings of International Coalbed Methane Symposium, 12–17 May 1997, University of Alabama, Tuscaloosa, AL, pp. 1–11.

Clemens, T. and K. Wit, 2002: CO_2 enhanced gas recovery studied for an example gas reservoir, SPE 77348, presented at the SPE Annual Technical Meeting and Conference, San Antonio, Texas, 29 September - 2 October 2002.

Clesceri, L.S., A.E. Greenberg and A.D. Eaton (eds.), 1998: Standard Methods for the Examination of Water and Wastewater, 20th Edition. American Public Health Association, Washington, DC, January 1998.

Cook, P.J., 1999: Sustainability and nonrenewable resources. *Environmental Geosciences,* **6**(4), 185–190.

Cook, P.J. and C.M. Carleton (eds.), 2000: Continental Shelf Limits: The Scientific and Legal Interface. Oxford University Press, New York, 360 pp.

Cook, A.C., L. J. Hainsworth, M.L. Sorey, W.C. Evans and J.R. Southon, 2001: Radiocarbon studies of plant leaves and tree rings from Mammoth Mountain, California: a long-term record of magmatic CO_2 release. *Chemical Geology*, **177**(1–2),117–131.

Crolet, J.-L., 1983: Acid corrosion in wells (CO_2, H_2S): Metallurgical aspects. *Journal of Petroleum Technology*, August 1983, 1553–1558.

CRUST Legal Task Force, 2001: Legal aspects of underground CO_2 storage. Ministry of Economic Affairs, the Netherlands. Retrieved from www.CO2-reductie.nl. on August 19, 2003.

Curry, T., D. Reiner, S. Ansolabehere and H. Herzog, 2005: How aware is the public of carbon capture and storage? In E.S. Rubin, D.W. Keith and C.F. Gilboy (Eds.), Proceedings of 7th International Conference on Greenhouse Gas Control Technologies (GHGT-7), September 5–9, 2004, Vancouver, Canada, v.I, 1001-1010.

Czernichowski-Lauriol, I., B. Sanjuan, C. Rochelle, K. Bateman, J. Pearce and P. Blackwell, 1996: Analysis of the geochemical aspects of the underground disposal of CO_2. In: Deep Injection Disposal of Hazardous and Industrial Wastes, Scientific and Engineering Aspects, J.A. Apps and C.-F. Tsang (eds.), Academic Press, ISBN 0-12-060060-9, pp. 565–583.

D'Hondt, S., S. Rutherford and A.J. Spivack, 2002: Metabolic activity of subsurface life in deep-sea sediments. *Science*, **295**, 2067–2070.

DOGGR (California Department of Oil, Gas and Geothermal Resources), 1974: Sixtieth Annual Report of the State Oil and Gas Supervisor. Report No. PR06, pp. 51–55.

Dooley, J.J., R.T. Dahowski, C.L. Davidson, S. Bachu, N. Gupta and J. Gale, 2005: A CO_2 storage supply curve for North America and its implications for the deployment of carbon dioxide capture and storage systems. Proceedings of the 7th International Conference on Greenhouse Gas Control Technologies (GHGT-7), September 5–9, 2004, Vancouver, Canada, v.I, 593-602.

Doughty, C. and K. Pruess, 2004: Modeling Supercritical Carbon Dioxide Injection in Heterogeneous Porous Media, *Vadose Zone Journal,* **3**(3), 837–847.

Doughty, C., K. Pruess, S.M. Benson, S.D. Hovorka, P.R. Knox and C.T. Green, 2001: Capacity investigation of brine-bearing sands of the Frio Formation for geologic sequestration of CO_2. Proceedings of First National Conference on Carbon Sequestration, 14–17 May 2001, Washington, D.C., United States Department of Energy, National Energy Technology Laboratory, CD-ROM USDOE/NETL-2001/1144, Paper P.32, 16 pp.

Ducroux, R. and J.M. Bewers, 2005: Acceptance of CCS under international conventions and agreements, IEA GHG Weyburn CO_2 Monitoring and Storage Project Summary Report 2000-2004, M. Wilson and M. Monea (eds.), Proceedings of the 7th International Conference on Greenhouse Gas Control Technologies (GHGT-7), September 5–9, 2004, Vancouver, Canada, v.II, 1467-1474.

Dusseault, M.B., S. Bachu and L. Rothenburg, 2004: Sequestration of CO_2 in salt caverns. *Journal of Canadian Petroleum Technology*, **43**(11), 49–55.

Emberley, S., I. Hutcheon, M. Shevalier, K. Durocher, W.D. Gunter and E.H. Perkins, 2002: Geochemical monitoring of rock-fluid interaction and CO_2 storage at theWeyburn CO_2 - injection enhanced oil recovery site, Saskatchewan, Canada. Proceedings of the 6th International Conference on Greenhouse Gas Control Technologies (GHGT-6), J. Gale and Y. Kaya (eds.), 1–4 October 2002, Kyoto, Japan, Pergamon, v.I, pp. 365–370.

Enick, R.M. and S.M. Klara, 1990: CO_2 solubility in water and brine under reservoir conditions. *Chemical Engineering Communications*, **90**, 23–33.

Ennis-King, J. and L. Paterson, 2001: Reservoir engineering issues in the geological disposal of carbon dioxide. Proceedings of the 5th International Conference on Greenhouse Gas Control Technologies (GHGT-5), D. Williams, D. Durie, P. McMullan, C. Paulson and A. Smith (eds.), 13–16 August 2000, Cairns, Australia, CSIRO Publishing, Collingwood, Victoria, Australia, pp. 290–295.

Ennis-King, J.P. and L. Paterson, 2003: Role of convective mixing in the long-term storage of carbon dioxide in deep saline formations. Presented at Society of Petroleum Engineers Annual Technical Conference and Exhibition, Denver, Colorado, 5–8 October 2003, SPE paper no. 84344.

Ennis-King, J, C.M. Gibson-Poole, S.C. Lang and L. Paterson, 2003: Long term numerical simulation of geological storage of CO_2 in the Petrel sub-basin, North West Australia. Proceedings of the 6th International Conference on Greenhouse Gas Control Technologies (GHGT-6), J. Gale and Y. Kaya (eds.), 1–4 October 2002, Kyoto, Japan, Pergamon, v.I, 507–511.

Espie, A.A, P.J Brand, R.C. Skinner, R.A. Hubbard and H.I. Turan, 2003: Obstacles to the storage of CO_2 through EOR in the North Sea. Proceedings of the 6th International Conference on Greenhouse Gas Control Technologies (GHGT-6), J. Gale and Y. Kaya (eds.), 1–4 October 2002, Kyoto, Japan, Pergamon, v.I, 213–218.

EU, 2000: White Paper on Environmental Liability. COM(2000) 66 final, 9 February 2000. European Union Commission, Brussels. http://http://aei.pitt.edu/archive/00001197/01/environment_liability_wp_COM_2000_66.pdf

Eurobarometer, 2003: Energy Issues, Options and Technologies: A Survey of Public Opinion in Europe. Energy DG, European Commission, Brussels, Belgium.

Farrar, C.D., J.M. Neil and J.F. Howle, 1999: Magmatic carbon dioxide emissions at Mammoth Mountain, California. U.S. Geological Survey Water-Resources Investigations Report 98-4217, Sacramento, CA.

Figueiredo, M.A. de, H.J. Herzog and D.M. Reiner, 2005: Framing the long-term liability issue for geologic storage carbon storage in the United States. *Mitigation and Adaptation Strategies for Global Change*. In press.

Fischer, M.L., A.J. Bentley, K.A. Dunkin, A.T. Hodgson, W.W. Nazaroff, R.G. Sextro and J.M. Daisy, 1996: Factors affecting indoor air concentrations of volatile organic compounds at a site of subsurface gasoline contamination, *Environmental Science and Technology*, **30**(10), 2948–2957.

Flett, M.A., R.M. Gurton and I.J. Taggart, 2005: Heterogeneous saline formations: Long-term benefits for geo-sequestration of greenhouse gases. Proceedings of the 7th International Conference on Greenhouse Gas Control Technologies (GHGT-7), September 5–9, 2004, Vancouver, Canada, v.I, 501–510.

Flower, F.B., E.F. Gilman and I.A.Leon, 1981: Landfill Gas, What It Does To Trees And How Its Injurious Effects May Be Prevented. *Journal of Arboriculture*, **7**(2), 43–52.

Freund, P., 2001: Progress in understanding the potential role of CO_2 storage. Proceedings of the 5th International Conference on Greenhouse Gas Control Technologies (GHGT-5), D.J. Williams, R.A. Durie, P. McMullan, C.A.J. Paulson and A.Y. Smith (eds.), 13–16 August 2000, Cairns, Australia, pp. 272–278.

Gadgil,A.J.,Y.C. Bonnefous and W.J. Fisk, 1994: Relative effectiveness of sub-slab pressurization and depressurization systems for indoor radon mitigation: Studies with an experimentally verified numerical model, *Indoor Air,* **4**, 265–275.

Gale, J., 2003: Geological storage of CO_2: what's known, where are the gaps and what more needs to be done. Proceedings of the 6th International Conference on Greenhouse Gas Control Technologies (GHGT-6), J. Gale and Y. Kaya (eds.), 1–4 October 2002, Kyoto, Japan, Pergamon, v.I, 207–212.

Gale, J.J., 2004: Using coal seams for CO_2 sequestration. Geologica Belgica, 7(1–2), In press.

Gale, J. and P. Freund, 2001: Coal-bed methane enhancement with CO_2 sequestration worldwide potential. *Environmental Geosciences*, **8**(3), 210–217.

Garg, A., D. Menon-Choudhary, M. Kapshe and P.R. Shukla, 2005: Carbon dioxide capture and storage potential in India. Proceedings of the 7th International Conference on Greenhouse Gas Control Technologies (GHGT-7), September 5–9, 2004, Vancouver, Canada.

Gasda, S.E., S. Bachu and M.A. Celia, 2004: The potential for CO_2 leakage from storage sites in geological media: analysis of well distribution in mature sedimentary basins. *Environmental Geology*, **46**(6–7), 707–720.

Gasem, K.A.M., R.L. Robinson and S.R. Reeves, 2002: Adsorption of pure methane, nitrogen and carbon dioxide and their mixtures on San Juan Basin coal. U.S. Department of Energy Topical Report, Contract No. DE-FC26-OONT40924, 83 pp.

Gerlach, T.M., M.P. Doukas, K.A. McGee and R. Kessler, 1999: Soil efflux and total emission rates of magmatic CO_2 at the Horseshoe Lake tree kill, Mammoth Mountain, California, 1995–1999. *Chemical Geology*, **177**, 101–116.

Gibbs, J.F., J.H. Healy, C.B. Raleigh and J. Coakley, 1973: Seismicity in the Rangely, Colorado area: 1962–1970, *Bulletin of the Seismological Society of America*, **63**, 1557–1570.

Gibson-Poole, C.M., S.C. Lang, J.E. Streit, G.M. Kraishan and R.R Hillis, 2002: Assessing a basin's potential for geological sequestration of carbon dioxide: an example from the Mesozoic of the Petrel Sub-basin, NW Australia. In: M. Keep and S.J. Moss (eds.) The Sedimentary Basins of Western Australia 3, Proceedings of the Petroleum Exploration Society of Australia Symposium, Perth, Western Australia, 2002, pp. 439–463.

Gough, C., I. Taylor and S. Shackley, 2002: Burying carbon under the sea: an initial exploration of public opinion. *Energy & Environment,* **13**(6), 883–900.

Granieri, D., G. Chiodini, W. Marzocchi and R. Avino, 2003: Continuous monitoring of CO_2 soil diffuse degassing at Phlegraean Fields (Italy): influence of environmental and volcanic parameters. *Earth and Planetary Science Letters,* **212**(1–2), 167–179.

Grigg, R.B., 2005: Long-term CO_2 storage: Using petroleum industry experience, Carbon Dioxide Capture for Storage in Deep Geologic Formations - Results from the CO_2 Capture Project, v. 2: Geologic Storage of Carbon Dioxide with Monitoring and Verification, S.M. Benson (ed.), Elsevier, London, pp. 853–866.

Gunter, W.D., E.H. Perkins and T.J. McCann, 1993: Aquifer disposal of CO_2-rich gases: reaction design for added capacity. *Energy Conversion and Management,* **34**, 941–948.

Gunter, W.D., B. Wiwchar and E.H. Perkins, 1997: Aquifer disposal of CO_2-rich greenhouse gases: Extension of the time scale of experiment for CO_2-sequestering reactions by geochemical modelling. *Mineralogy and Petrology,* **59**, 121–140.

Gunter, W.D., S. Wong, D.B. Cheel and G. Sjostrom, 1998: Large CO_2 sinks: their role in the mitigation of greenhouse gases from an international, national (Canadian) and provincial (Alberta) perspective. *Applied Energy,* **61**, 209–227.

Gunter, W.D., E.H. Perkins and I. Hutcheon, 2000: Aquifer disposal of acid gases: Modeling of water-rock reactions for trapping acid wastes. *Applied Geochemistry,* **15**, 1085–1095.

Gunter, W.D., S. Bachu and S. Benson, 2004: The role of hydrogeological and geochemical trapping in sedimentary basins for secure geological storage for carbon dioxide. In: Geological Storage of Carbon Dioxide: Technology. S. Baines and R.H. Worden (eds.), Special Publication of Geological Society, London, UK. Special Publication 233, pp. 129–145.

Gunter, W.D., M.J. Mavor and J.R. Robinson, 2005: CO_2 storage and enhanced methane production: field testing at Fenn-Big Valley, Alberta, Canada, with application. Proceedings of the 7th International Conference on Greenhouse Gas Control Technologies (GHGT-7), September 5–9, 2004, Vancouver, Canada, v.I, 413-422.

Gupta, N., B. Sass, J. Sminchak and T. Naymik, 1999: Hydrodynamics of CO_2 disposal in a deep saline formation in the midwestern United States. Proceedings of the 4th International Conference on Greenhouse Gas Control Technologies (GHGT-4), B. Eliasson, P.W.F. Riemer and A. Wokaun (eds.), 30 August to 2 September 1998, Interlaken, Switzerland, Pergamon, 157–162.

Gurevich, A.E., B.L. Endres, J.O. Robertson Jr. and G.V. Chilingar, 1993: Gas migration from oil and gas fields and associated hazards. *Journal of Petroleum Science and Engineering,* **9**, 223–238.

Haidl, F.M., S.G. Whittaker, M. Yurkowski, L.K. Kreis, C.F. Gilboy and R.B. Burke, 2005: The importance of regional geological mapping in assessing sites of CO_2 storage within intracratonic basins: Examples from the IEA Weyburn CO_2 monitoring and storage project, Proceedings of the 7th International Conference on Greenhouse Gas Control Technologies (GHGT-7), September 5–9, 2004, Vancouver, Canada, v.I, 751-760.

Hantush, M.S., 1960: Modifications to the theory of leaky aquifers, *Journal of Geophysical Research,* **65**(11), 3713–3725.

Hantush, M.S. and C.E. Jacobs, 1955: Non-steady radial flow to an infinite leaky aquifer. *Transactions of the American Geophysical Union,* **2**, 519–524.

Haveman, S.A. and K. Pedersen, 2001: Distribution of culturable microorganisms in Fennoscandian Shield groundwater. *FEMS Microbiology Ecology,* **39**(2), 129–137.

Healy, J.H., W.W. Ruby, D.T. Griggs and C.B. Raleigh, 1968: The Denver earthquakes, *Science,* **161**, 1301–1310.

Hefner, T. A. and K.T. Barrow, 1992: AAPG Treatise on Petroleum Geology. Structural Traps VII, pp. 29–56.

Heinrich, J.J., H.J. Herzog and D.M. Reiner, 2003: Environmental assessment of geologic storage of CO_2. Second National Conference on Carbon Sequestration, 5–8 May 2003, Washington, DC.

Hendriks, C., W. Graus and F. van Bergen, 2002: Global carbon dioxide storage potential and costs. Report Ecofys & The Netherland Institute of Applied Geoscience TNO, Ecofys Report EEP02002, 63 pp.

Hobbs, P.V., L.F. Radke, J.H. Lyons, R.J. Ferek and D.J. Coffman, 1991: Airborne measurements of particle and gas emissions from the 1990 volcanic eruptions of Mount Redoubt. *Journal of Geophysical Research,* **96**(D10), 18735–18752.

Hodgkinson, D.P. and T.J. Sumerling, 1990: A review of approaches to scenario analysis for repository safety assessment. Proceedings of the Paris Symposium on Safety Assessment of Radioactive Waste Repositories, 9–13 October 1989, OECD Nuclear Energy Agency: 333–350.

Holloway, S. (ed.), 1996: The underground disposal of carbon dioxide. Final report of Joule 2 Project No. CT92-0031. British Geological Survey, Keyworth, Nottingham, UK, 355 pp.

Holloway, S., 1997: Safety of the underground disposal of carbon dioxide. *Energy Conversion and Management,* **38**(Suppl.), S241–S245.

Holloway, S. and D. Savage, 1993: The potential for aquifer disposal of carbon dioxide in the UK. *Energy Conversion and Management,* **34**(9–11), 925–932.

Holt, T., J. L. Jensen and E. Lindeberg, 1995: Underground storage of CO_2 in aquifers and oil reservoirs. *Energy Conversion and Management,* **36**(6–9), 535–538.

Holtz, M.H., 2002: Residual gas saturation to aquifer influx: A calculation method for 3-D computer reservoir model construction. SPE Paper 75502, presented at the SPE Gas Technologies Symposium, Calgary, Alberta, Canada. April 2002.

Holtz, M.H., P.K. Nance and R.J. Finley, 2001: Reduction of greenhouse gas emissions through CO_2 EOR in Texas. *Environmental Geosciences,* **8**(3) 187–199.

Hoversten, G.M. and E. Gasperikova, 2005: Non Seismic Geophysical Approaches to Monitoring, Carbon Dioxide Capture for Storage in Deep Geologic Formations - Results from the CO_2 Capture Project, v. 2: Geologic Storage of Carbon Dioxide with Monitoring and Verification, S.M. Benson (ed.), Elsevier Science, London. pp. 1071–1112.

Hoversten, G. M., R. Gritto, J. Washbourne and T.M. Daley, 2003: Pressure and Fluid Saturation Prediction in a Multicomponent Reservoir, using Combined Seismic and Electromagnetic Imaging. *Geophysics*, (in press Sept–Oct 2003).

Hovorka, S.D., C. Doughty and M.H. Holtz, 2005: Testing Efficiency of Storage in the Subsurface: Frio Brine Pilot Experiment, Proceedings of the 7th International Conference on Greenhouse Gas Control Technologies (GHGT-7), Vancouver, Canada. September 5–9, 2004, v.II, 1361-1366.

Huijts, N. 2003: Public Perception of Carbon Dioxide Storage, Masters Thesis, Eindhoven University of Technology, The Netherlands.

IEA-GHG, 1998: Enhanced Coal Bed Methane Recovery with CO_2 Sequestration, IEA Greenhouse Gas R&D Programme, Report No. PH3/3, August, 139 pp.

IEA-GHG, 2003: Barriers to Overcome in Implementation of CO_2 Capture and Storage (2):Rules and Standards for the Transmission and Storage of CO_2, IEA Greenhouse Gas R&D Programme, Report No. PH4/23. Cheltenham, U.K.

IEA-GHG, 2004: A Review of Global Capacity Estimates for the Geological Storage of Carbon Dioxide, IEA Greenhouse Gas R&D Programme Technical Review (TR4), March 23, 2004, 27 pp.

IOGCC (Interstate Oil and Gas Compact Commission), 2005: Carbon Capture and Storage: A Regulatory Framework for States. Report to USDOE, 80 pp.

Ispen, K.H. and F.L. Jacobsen, 1996: The Linde structure, Denmark: an example of a CO_2 depository with a secondary chalk cap rock. *Energy and Conversion and Management*, 37(6–8), 1161–1166.

Itaoka, K., A. Saito and M. Akai, 2005: Public acceptance of CO_2 capture and storage technology: A survey of public opinion to explore influential factors. Proceedings of the 7th International Conference on Greenhouse Gas Control Technologies (GHGT-7), September 5–9, 2004, Vancouver, Canada, v.I, p.1011.

Jarrell, P.M., C.E. Fox, M.H. Stein and S.L. Webb, 2002: Practical Aspects of CO_2 Flooding. SPE Monograph Series No. 22, Richardson, TX, 220 pp.

Jimenez, J.A and R.J. Chalaturnyk, 2003: Are disused hydrocarbon reservoirs safe for geological storage of CO_2? Proceedings of the 6th International Conference on Greenhouse Gas Control Technologies (GHGT-6), J. Gale and Y. Kaya (eds.), 1–4 October 2002, Kyoto, Japan, Pergamon, v.I, 471–476.

Johnson, J.W., J.J. Nitao and J.P. Morris, 2005: Reactive transport modeling of cap rock integrity during natural and engineered CO_2 storage, Carbon Dioxide Capture for Storage in Deep Geologic Formations - Results from the CO_2 Capture Project, v. 2: Geologic Storage of Carbon Dioxide with Monitoring and Verification, S.M. Benson, (ed.), Elsevier, London, pp. 787–814.

Kaarstad, O., 1992: Emission-free fossil energy from Norway. *Energy Conversion and Management*, 33(5–8), 619–626.

Kaarstad, O., 2002: Geological storage including costs and risks, in saline aquifers, Proceedings of workshop on Carbon Dioxide Capture and Storage, Regina Canada, 2002.

Katzung, G., P. Krull and F. Kühn, 1996: Die Havarie der UGS-Sonde Lauchstädt 5 im Jahre 1988 - Auswirkungen und geologische Bedingungen. *Zeitschrift für Angewandte Geologie*, 42, 19–26.

Keith, D.W. and M. Wilson, 2002: Developing recommendations for the management of geologic storage of CO_2 in Canada. University of Regina, PARC, Regina, Saskatchewan.

Keith, D., H. Hassanzadeh and M. Pooladi-Darvish, 2005: Reservoir Engineering To Accelerate Dissolution of Stored CO_2 In Brines. Proceedings of the 7th International Conference on Greenhouse Gas Control Technologies (GHGT-7), September 5–9, 2004, Vancouver, Canada, v.II, 2163-2168.

Kempton, W., J. Boster and J. Hartley, 1995: Environmental Values in American Culture. MIT Press, Boston, MA, 320 pp.

Kling, G.W., M.A. Clark, H.R. Compton, J.D. Devine, W.C. Evans, A.M. Humphrey, E.J. Doenigsberg, J.P. Lockword, M.L. Tuttle and G.W. Wagner, 1987: The lake gas disaster in Cameroon, West Africa, *Science*, **236**, 4798, 169–175.

Klins, M.A., 1984: Carbon Dioxide Flooding, D. Reidel Publishing Co., Boston, MA, 267 pp.

Klins, M.A. and S.M. Farouq Ali, 1982: Heavy oil production by carbon dioxide injection. *Journal of Canadian Petroleum Technology*, 21(5), 64–72.

Klusman, R.W., 2003: A geochemical perspective and assessment of leakage potential for a mature carbon dioxide-enhanced oil recovery project and as a prototype for carbon dioxide sequestration; Rangely field, Colorado. *American Association of Petroleum Geologists Bulletin*, **87**(9), 1485–1507.

Knauss, K.G., J.W. Johnson and C.I Steefel, 2005: Evaluation of the impact of CO_2, co-contaminant gas, aqueous fluid and reservoir rock interactions on the geologic sequestration of CO_2. *Chemical Geology,* Elsevier, **217**, 339–350.

Koide, H. and K. Yamazaki, 2001: Subsurface CO_2 disposal with enhanced gas recovery and biogeochemical carbon recycling. *Environmental Geosciences*, **8**(3), 218–224.

Koide, H.G., Y. Tazaki, Y. Noguchi, S. Nakayama, M. Iijima, K. Ito and Y. Shindo, 1992: Subterranean containment and long-term storage of carbon dioxide in unused aquifers and in depleted natural gas reservoirs. *Energy Conversion and Management*, **33**(5–8), 619–626.

Koide, H.G., M. Takahashi and H. Tsukamoto, 1995: Self-trapping mechanisms of carbon dioxide. *Energy Conversion and Management*, **36**(6–9), 505–508.

Koide, H., M. Takahashi, Y. Shindo, Y. Tazaki, M. Iijima, K. Ito, N. Kimura and K. Omata, 1997: Hydrate formation in sediments in the sub-seabed disposal of CO_2. *Energy-The International Journal*, **22**(2/3), 279–283.

Korbol, R. and A. Kaddour, 1994: Sleipner West CO_2 disposal: injection of removed CO_2 into the Utsira formation. *Energy Conversion and Management*, **36**(6–9), 509–512.

Kovscek, A.R., 2002: Screening criteria for CO_2 storage in oil reservoirs. *Petroleum Science and Technology*, **20**(7–8), 841–866.

Krom, T.D., F.L. Jacobsen and K.H. Ipsen, 1993: Aquifer based carbon dioxide disposal in Denmark: capacities, feasibility, implications and state of readiness. *Energy Conversion and Management*, **34**(9–11), 933–940.

Krooss, B.M., F. van Bergen, Y. Gensterblum, N. Siemons, H.J.M. Pagnier and P. David, 2002: High-pressure methane and carbon dioxide adsorption on dry and moisture-equilibrated Pennsylvanian coals. *International Journal of Coal Geology*, **51**(2), 69–92.

Kumar, A., M.H. Noh, K. Sepehrnoori, G.A. Pope, S.L. Bryant and L.W. Lake, 2005: Simulating CO_2 storage in deep saline aquifers, Carbon Dioxide Capture for Storage in Deep Geologic Formations - Results from the CO_2 Capture Project, v.2: Geologic Storage of Carbon Dioxide with Monitoring and Verification, S.M. Benson, (ed.), Elsevier, London. pp. 977–898.

Larsen, J.W., 2003: The effects of dissolved CO_2 on coal structure and properties. *International Journal of Coal Geology,* **57,** 63–70.

Larsen, M., N.P. Christensen, B. Reidulv, D. Bonijoly, M. Dusar, G. Hatziyannis, C. Hendriks, S. Holloway, F. May and A. Wildenborg, 2005: Assessing European potential for geological storage of CO_2 - the GESTCO project. Proceedings of the 7th International Conference on Greenhouse Gas Control Technologies (GHGT-7), September 5–9, 2004, Vancouver, Canada.

Law, D. (ed.), 2005: Theme 3: CO_2 Storage Capacity and Distribution Predictions and the Application of Economic Limits. In: IEA GHG Weyburn CO_2 Monitoring and Storage Project Summary Report 2000–2004, M. Wilson and M. Monea (eds.), Proceedings of the 7th International Conference on Greenhouse Gas Control Technologies (GHGT7), Volume III, p 151–209.

Law, D.H.-S., L.G.H. van der Meer and W.D. Gunter, 2003: Comparison of numerical simulators for greenhouse gas storage in coal beds, Part II: Flue gas injection. Proceedings of the 6th International Conference on Greenhouse Gas Control Technologies (GHGT-6), J. Gale and Y. Kaya (eds.), 1–4 October 2002, Kyoto, Japan, Pergamon, v.I, 563–568.

Lee, A.M., 2001: The Hutchinson Gas Explosions: Unravelling a Geologic Mystery, Kansas Bar Association, 26th Annual KBA/KIOGA Oil and Gas Law Conference, v1, p3-1 to 3-29.

Lenstra, W.J. and B.C.W. van Engelenburg, 2002: Legal and policy aspects: impact on the development of CO_2 storage. Proceedings of IPCC Working Group III: Mitigation of Climate Change Workshop on Carbon Dioxide Capture and Storage, Regina, Canada, 18–21, November, 2002.

Leone, I.A., F.B. Flower, J.J. Arthur and E.F. Gilman, 1977: Damage To Woody Species By Anaerobic Landfill Gases. *Journal of Arboriculture,* **3**(12), 221–225.

Lichtner, P.C., 2001: FLOTRAN User's Manual. Los Alamos National Laboratory Report LA-UR-01-2349, Los Alamos, NM, 2001.

Lindeberg, E. and P. Bergmo, 2003: The long-term fate of CO_2 injected into an aquifer. Proceedings of the 6th International Conference on Greenhouse Gas Control Technologies (GHGT-6), J. Gale and Y. Kaya (eds.), 1–4 October 2002, Kyoto, Japan, Pergamon, v.I, 489–494.

Lindeberg, E. and D. Wessel-Berg, 1997: Vertical convection in an aquifer column under a gas cap of CO_2. *Energy Conversion and Management,* **38**(Suppl.), S229–S234.

Lindeberg, E., A. Ghaderi, P. Zweigel and A. Lothe, 2001: Prediction of CO_2 dispersal pattern improved by geology and reservoir simulation and verified by time lapse seismic, Proceedings of 5th International Conference on Greenhouse Gas Control Technologies, D.J. Williams, R.A. Durie, P. McMullan, C.A.J. Paulson and A.Y. Smith (eds.), CSIRO, Melbourne, Australia. pp. 372–377.

Lippmann, M.J. and S.M. Benson, 2003: Relevance of underground natural gas storage to geologic sequestration of carbon dioxide. Department of Energy's Information Bridge, http://www.osti.gov/dublincore/ecd/servlets/purl/813565-MVm7Ve/native/813565.pdf, U.S. Government Printing Office (GPO).

Looney, B. and R. Falta, 2000: Vadose Zone Science and Technology Solutions: Volume II, Batelle Press, Columbus, OH.

Magoon, L.B. and W.G. Dow, 1994: The petroleum system. American Association of Petroleum Geologists, *Memoir* **60,** 3–24.

Marchetti, C., 1977: On Geoengineering and the CO_2 Problem. Climatic Change, 1, 59–68.

Martin, F.D. and J. J. Taber, 1992: Carbon dioxide flooding. *Journal of Petroleum Technology,* **44**(4), 396–400.

Martini, B. and E. Silver, 2002: The evolution and present state of tree-kills on Mammoth Mountain, California: tracking volcanogenic CO_2 and its lethal effects. Proceedings of the 2002 AVIRIS Airborne Geoscience Workshop, Jet Propulsion Laboratory, California Institute of Technology, Pasadena, CA.

May, F., 1998: Thermodynamic modeling of hydrothermal alteration and geoindicators for CO_2-rich waters. *Zeitschrift der Deutschen Geologischen Gesellschaft,* **149,** 3, 449–464.

McGrail, B.P., S.P. Reidel and H.T. Schaef, 2003: Use and features of basalt formations for geologic sequestration. Proceedings of the 6th International Conference on Greenhouse Gas Control Technologies (GHGT-6), J. Gale and Y. Kaya (eds.), 1–4 October 2002, Kyoto, Japan, Pergamon, v.II, 1637–1641.

McKelvey, V.E., 1972: Mineral resource estimates and public policy. *American Scientist,* **60**(1), 32–40.

McKinnon, R.J., 1998: The interplay between production and underground storage rights in Alberta, *The Alberta Law Review,* **36**(400).

McPherson, B.J.O.L. and B.S. Cole, 2000: Multiphase CO_2 flow, transport and sequestration in the Powder River basin, Wyoming, USA. *Journal of Geochemical Exploration,* **69–70**(6), 65–70.

Menzies, R.T., D.M., Tratt, M.P. Chiao and C.R. Webster, 2001: Laser absorption spectrometer concept for globalscale observations of atmospheric carbon dioxide. 11th Coherent Laser Radar Conference, Malvern, United Kingdom.

Metcalfe, R.S., 1982: Effects of impurities on minimum miscibility pressures and minimum enrichment levels for CO_2 and rich gas displacements. *SPE Journal,* **22**(2), 219–225.

Miles, N., K. Davis and J. Wyngaard, 2005: Detecting Leaks from CO_2 Reservoirs using Micrometeorological Methods, Carbon Dioxide Capture for Storage in Deep Geologic Formations - Results from the CO_2 Capture Project, v. 2: Geologic Storage of Carbon Dioxide with Monitoring and Verification, S.M. Benson (ed.), Elsevier Science, London. pp.1031–1044.

Moberg, R., D.B. Stewart and D. Stachniak, 2003: The IEA Weyburn CO_2 Monitoring and Storage Project. Proceedings of the 6th International Conference on Greenhouse Gas Control Technologies (GHGT-6), J. Gale and Y. Kaya (eds.), 1–4 October 2002, Kyoto, Japan, 219–224.

Morgan, M.G. and M. Henrion, 1999: Uncertainty: A guide to dealing with uncertainty in quantitative risk and policy analysis. Cambridge University Press, New York, NY.

Moritis, G., 2002: Enhanced Oil Recovery, *Oil and Gas Journal,* **100**(15), 43–47.

Moritis, G., 2003: CO_2 sequestration adds new dimension to oil, gas production. *Oil and Gas Journal,* **101**(9), 71–83.

Morner, N.A. and G. Etiope, 2002: Carbon degassing from the lithosphere. *Global and Planetary Change,* **33,** 185–203.

Morrow, T.B., D.L. George and M.G. Nored, 2003: Operational factors that affect orifice meter accuracy: Key findings from a multi-year study. Flow Control Network.

Myer, L.R., G.M. Hoversten and E. Gasperikova, 2003: Sensitivity and cost of monitoring geologic sequestration using geophysics. Proceedings of the 6th International Conference on Greenhouse Gas Control Technologies (GHGT-6), J. Gale and Y. Kaya (eds.), 1–4 October 2002, Kyoto, Japan. *Pergamon,* **1,** 377–382.

NETL, 2004: Carbon Sequestration Technology Roadmap and Program Plan – 2004. US Department of Energy – National Energy Technology Laboratory Report, April 2004, http://www.fe.doe. gov/programs/sequestration/publications/programplans/2004/ SequestrationRoadmap4-29-04.pdf

Nimz, G.J. and G.B. Hudson, 2005: The use of noble gas isotopes for monitoring leakage of geologically stored CO_2, Carbon Dioxide Capture for Storage in Deep Geologic Formations—Results from the CO_2 Capture Project, v. 2: Geologic Storage of Carbon Dioxide with Monitoring and Verification S.M. Benson (ed.), Elsevier Science, London,. pp. 1113–1130.

Nitao, J.J., 1996: The NUFT code for modeling nonisothermal, multiphase, multicomponent flow and transport in porous media. EOS, *Transactions of the American Geophysical Union,* **74**(3), 3.

Nordbotten, J.M., M.A. Celia and S. Bachu, 2005a: Injection and storage of CO_2 in deep saline aquifers: Analytical solution for CO_2 plume evolution during injection. *Transport in Porous Media,* **58,** 339–360, DOI 10.1007/s11242-004-0670-9.

Nordbotten, J.M., M.A. Celia and S. Bachu, 2005b: Semi-analytical solution for CO_2 leakage through an abandoned well. *Environmental Science and Technology,* **39**(2), 602–611.

North, D.W., 1999: A perspective on nuclear waste. *Risk Analysis,* **19,** 751–758.

Obdam, A., L.G.H. van der Meer, F. May, C. Kervevan, N. Bech and A. Wildenborg, 2003: Effective CO_2 storage capacity in aquifers, gas fields, oil fields and coal fields. Proceedings of the 6th International Conference on Greenhouse Gas Control Technologies (GHGT-6), J. Gale and Y. Kaya (eds.), 1–4 October 2002, Kyoto, Japan, Pergamon, v.I, 339–344.

Oen, P. M., 2003: The development of the Greater Gorgon Gas Fields. *The APPEA Journal 2003,* **43**(2), 167–177.

Oil and Gas Conservation Regulations, 1985 (with amendments through 2000): Saskatchewan Industry and Resources, 70 pp.

Oldenburg, C.M. and A.J. Unger, 2003: On leakage and seepage from geologic carbon sequestration sites: unsaturated zone attenuation. *Vadose Zone Journal,* **2,** 287–296.

Oldenburg, C.M. and A.J.A. Unger, 2004: Coupled subsurface-surface layer gas transport for geologic carbon sequestration seepage simulation. *Vadose Zone Journal,* **3,** 848–857.

Oldenburg, C.M., K. Pruess and S. M. Benson, 2001: Process modeling of CO_2 injection into natural gas reservoirs for carbon sequestration and enhanced gas recovery. *Energy and Fuels,* **15,** 293–298.

Oldenburg, C.M., S.H. Stevens and S.M. Benson, 2002: Economic Feasibility of Carbon Sequestration with Enhanced Gas Recovery (CSEGR). Proceedings of the 6th International Conference on Greenhouse Gas Control Technologies (GHGT-6), J. Gale and Y. Kaya (eds.), 1–4 October 2002, Kyoto, Japan, Pergamon, v.I, 691–696.

Onstott, T., 2005: Impact of CO_2 injections on deep subsurface microbial ecosystems and potential ramifications for the surface biosphere, Carbon Dioxide Capture for Storage in Deep Geologic Formations - Results from the CO_2 Capture Project, v. 2: Geologic Storage of Carbon Dioxide with Monitoring and Verification, S.M. Benson (ed.), Elsevier Science, London, pp. 1217–1250.

Orphan, V.J., L.T. Taylor, D. Hafenbradl and E.F. Delong, 2000: Culture-dependent and culture-independent characterization of microbial assemblages associated with high-temperature petroleum reservoirs. *Applied and Environmental Microbiology,* **66**(2), 700–711.

Oskarsson, N., K. Palsson, H. Olafsson and T. Ferreira, 1999: Experimental monitoring of carbon dioxide by low power IR-sensors; Soil degassing in the Furnas volcanic centre, Azores. *Journal of Volcanology and Geothermal Research,* **92**(1–2), 181–193.

OSPAR Commission, 2004: Report from the Group of Jurists and Linguists on the placement of carbon dioxide in the OSPAR maritime area. Annex 12 to 2004 Summary Record.

Palmer, I. and J. Mansoori, 1998: How permeability depends on stress and pore pressure in coalbeds: a new model. *SPE Reservoir Evaluation & Engineering,* **1**(6), 539–544.

Palmgren, C., M. Granger Morgan, W. Bruine de Bruin and D. Keith, 2004: Initial public perceptions of deep geological and oceanic disposal of CO_2. *Environmental Science and Technology.* In press.

Parkes, R.J., B.A. Cragg and P. Wellsbury, 2000: Recent studies on bacterial populations and processes in subseafloor sediments: a review. *Hydrogeology Journal,* **8**(1), 11–28.

Pearce, J.M., S. Holloway, H. Wacker, M.K. Nelis, C. Rochelle and K. Bateman, 1996: Natural occurrences as analogues for the geological disposal of carbon dioxide. *Energy Conversion and Management,* **37**(6–8), 1123–1128.

Pearce, J.M., J. Baker, S. Beaubien, S. Brune, I. Czernichowski-Lauriol, E. Faber, G. Hatziyannis, A. Hildebrand, B.M. Krooss, S. Lombardi, A. Nador, H. Pauwels and B.M. Schroot, 2003: Natural CO_2 accumulations in Europe: Understanding the long-term geological processes in CO_2 sequestration. Proceedings of the 6th International Conference on Greenhouse Gas Control Technologies (GHGT-6), J. Gale and Y. Kaya (eds.), 1–4 October 2002, Kyoto, Japan, Pergamon, v.I, 417–422

Perkins, E., I. Czernichowski-Lauriol, M. Azaroual and P. Durst, 2005: Long term predictions of CO_2 storage by mineral and solubility trapping in the Weyburn Midale Reservoir. Proceedings of the 7th International Conference on Greenhouse Gas Control Technologies (GHGT-7), September 5–9, 2004, Vancouver, Canada, v.II, 2093-2096.

Perry, K.F., 2005: Natural gas storage industry experience and technology: Potential application to CO_2 geological storage, Carbon Dioxide Capture for Storage in Deep Geologic Formations— Results from the CO_2 Capture Project, v. 2: Geologic Storage of Carbon Dioxide with Monitoring and Verification, S.M. Benson (ed.), Elsevier Science, London, pp. 815–826.

Pickles, W.L., 2005: Hyperspectral geobotanical remote sensing for CO_2, Carbon Dioxide Capture for Storage in Deep Geologic Formations - Results from the CO_2 Capture Project, v.2: Geologic Storage of Carbon Dioxide with Monitoring and Verification, S.M. Benson (ed.), Elsevier Science, London, pp. 1045–1070.

Piessens, K. and M. Dusar, 2004: Feasibility of CO_2 sequestration in abandoned coal mines in Belgium. *Geologica Belgica*, 7-3/4. In press.

Pizzino, L., G. Galli, C. Mancini, F. Quattrocchi and P. Scarlato, 2002: Natural gas hazard (CO_2, ^{222}Rn) within a quiescent volcanic region and its relations with tectonics; the case of the Ciampino-Marino area, Alban Hills Volcano, Italy. *Natural Hazards*, **27**(3), 257–287.

Poortinga, W. and N. Pidgeon, 2003: Public Perceptions of Risk, Science and Governance. Centre for Environmental Risk, University of East Anglia, Norwich, UK, 60 pp.

Pruess, K., C. Oldenburg and G. Moridis, 1999: TOUGH2 User's Guide, Version 2.0, Lawrence Berkeley National Laboratory Report LBNL-43134, Berkeley, CA, November, 1999.

Pruess, K., J. García, T. Kovscek, C. Oldenburg, J. Rutqvist, C. Steefel and T. Xu, 2004: Code Intercomparison Builds Confidence in Numerical Simulation Models for Geologic Disposal of CO_2. *Energy*, 2003.

Purdy, R. and R. Macrory, 2004: Geological carbon sequestration: critical legal issues. Tyndall Centre Working Paper 45.

Raleigh, C.B., J.D. Healy and J.D. Bredehoeft, 1976: An experiment in earthquake control of Rangely, Colorado. *Science*, **191**, 1230–1237.

Reeves, S., 2003a: Coal-Seq project update: field studies of ECBM recovery/CO_2 sequestration in coal seams. Proceedings of the 6th International Conference on Greenhouse Gas Control Technologies (GHGT-6), J. Gale and Y. Kaya (eds.), 1–4 October 2002, Kyoto, Japan, Pergamon, v.I, 557–562.

Reeves, S.R., 2003b: Assessment of CO_2 Sequestration and ECBM Potential of US Coalbeds, Topical Report for US Department of Energy by Advanced Resources International, Report No. DE-FC26-00NT40924, February 2003.

Reeves, S.R., 2005: The Coal-Seq project: Key results from field, laboratory and modeling studies. Proceedings of the 7th International Conference on Greenhouse Gas Control Technologies (GHGT-7), September 5–9, 2004, Vancouver, Canada, v.II, 1399-1406.

Reeves, S., A. Taillefert, L. Pekot and C. Clarkson, 2003: The Allison Unit CO_2-ECBM Pilot: A Reservoir Modeling Study. DOE Topical Report, February, 2003.

Reeves, S., D. Davis and A. Oudinot, 2004: A Technical and Economic Sensitivity Study of Enhanced Coalbed Methane Recovery and Carbon Sequestration in Coal. DOE Topical Report, March, 2004.

Reiner, D.M. and H.J. Herzog, 2004: Developing a set of regulatory analogs for carbon sequestration. *Energy*, **29**(9/10): 1561–1570.

Riddiford, F.A., A. Tourqui, C.D. Bishop, B. Taylor and M. Smith, 2003: A cleaner development: The In Salah Gas Project, Algeria. Proceedings of the 6th International Conference on Greenhouse Gas Control Technologies (GHGT-6), J. Gale and Y. Kaya, (eds.), 1–4 October 2002, Kyoto, Japan, v.I, 601–606.

Rigg, A., G. Allinson, J. Bradshaw, J. Ennis-King, C.M. Gibson-Poole, R.R. Hillis, S.C. Lang and J.E. Streit, 2001: The search for sites for geological sequestration of CO_2 in Australia: A progress report on GEODISC. *APPEA Journal*, **41**, 711–725.

Rochelle, C.A., J.M. Pearce and S. Holloway, 1999: The underground sequestration of carbon dioxide: containment by chemical reactions. In: Chemical Containment of Waste in the Geosphere, Geological Society of London Special Publication No. 157, 117–129.

Rochelle, C.A., I. Czernichowski-Lauriol and A.E. Milodowski, 2004: The impact of chemical reactions on CO_2 storage in geological formations, a brief review. In: Geological Storage of Carbon Dioxide for Emissions Reduction: Technology, S.J. Baines and R.H. Worden (eds.). Geological Society Special Publication, Bath, UK.

Rogie, J.D., D.M. Kerrick, M.L. Sorey, G. Chiodini and D.L. Galloway, 2001: Dynamics of carbon dioxide emission at Mammoth Mountain, California. *Earth and Planetary Science Letters*, **188**, 535–541.

Rutqvist, J. and C-F. Tsang, 2002: A study of caprock hydromechanical changes associated with CO_2 injection into a brine formation. *Environmental Geology*, **42**, 296–305.

Salvi, S., F. Quattrocchi, M. Angelone, C.A. Brunori, A. Billi, F. Buongiorno, F. Doumaz, R. Funiciello, M. Guerra, S. Lombardi, G. Mele, L. Pizzino and F. Salvini, 2000: A multidisciplinary approach to earthquake research: implementation of a Geochemical Geographic Information System for the Gargano site, Southern Italy. Natural Hazard, 20(1), 255–278.

Saripalli, K.P., N.M. Mahasenan and E.M. Cook, 2003: Risk and hazard assessment for projects involving the geological sequestration of CO_2. Proceedings of the 6th International Conference on Greenhouse Gas Control Technologies (GHGT-6), J. Gale and Y. Kaya (eds.), 1–4 October 2002, Kyoto, Japan, Pergamon, v.I, 511–516.

Scherer, G.W., M.A. Celia, J-H. Prevost, S. Bachu, R. Bruant, A. Duguid, R. Fuller, S.E. Gasda, M. Radonjic and W. Vichit-Vadakan, 2005: Leakage of CO_2 through Abandoned Wells: Role of Corrosion of Cement, Carbon Dioxide Capture for Storage in Deep Geologic Formations—Results from the CO_2 Capture Project, v. 2: Geologic Storage of Carbon Dioxide with Monitoring and Verification, Benson, S.M. (Ed.), Elsevier Science, London, pp. 827–850.

Schremp, F.W. and G.R. Roberson, 1975: Effect of supercritical carbon dioxide (CO_2) on construction materials. *Society of Petroleum Engineers Journal*, June 1975, 227–233.

Schreurs, H.C.E., 2002: Potential for geological storage of CO_2 in the Netherlands. Proceedings of the 6th International Conference on Greenhouse Gas Control Technologies (GHGT-6), J. Gale and Y. Kaya (eds.), 1–4 October 2002, Kyoto, Japan, Pergamon, v.I, 303–308.

Sebastian, H.M., R.S. Wenger and T.A. Renner, 1985: Correlation of minimum miscibility pressure for impure CO_2 streams. *Journal of Petroleum Technology*, **37**(12), 2076–2082.

Sedlacek, R., 1999: Untertage Erdgasspeicherung in Europa. Erdol, Erdgas, Kohle 115, 573–540.

Shackley, S., C. McLachlan and C. Gough, 2004: The public perception of carbon dioxide capture and storage in the UK: Results from focus groups and a survey, *Climate Policy*. In press.

Shapiro, S.A., E. Huenges and G. Borm, 1997: Estimating the crust permeability from fluid-injection-induced seismic emission at the KTB site. *Geophysical Journal International*, 131, F15–F18.

Shaw, J. C. and S. Bachu, 2002: Screening, evaluation and ranking of oil reserves suitable for CO_2 flood EOR and carbon dioxide sequestration. *Journal of Canadian Petroleum Technology*, **41**(9), 51–61.

Shi, J-Q. and S. Durucan, 2005: A numerical simulation study of the Allison Unit CO_2-ECBM pilot: the effect of matrix shrinkage and swelling on ECBM production and CO_2 injectivity. Proceedings of the 7th International Conference on Greenhouse Gas Control Technologies (GHGT-7), September 5–9, 2004, Vancouver, Canada, v.I, 431-442.

Shuler, P. and Y. Tang, 2005: Atmospheric CO_2 monitoring systems, Carbon Dioxide Capture for Storage in Deep Geologic Formations—Results from the CO_2 Capture Project, v. 2: Geologic Storage of Carbon Dioxide with Monitoring and Verification, S.M. Benson (ed.), Elsevier Science, London, pp. 1015–1030.

Skinner, L., 2003: CO_2 blowouts: An emerging problem. World Oil, 224(1).

Sleipner Best Practice Manual, 2004: S. Holloway, A. Chadwick, E. Lindeberg, I. Czernichowski-Lauriol and R. Arts (eds.), Saline Aquifer CO_2 Storage Project (SACS). 53 pp.

Sminchak, J., N. Gupta, C. Byrer and P. Bergman, 2002: Issues related to seismic activity induced by the injection of CO_2 in deep saline aquifers. *Journal of Energy & Environmental Research*, **2**, 32–46.

Sorey, M. L., W.C. Evans, B.M. Kennedy, C.D. Farrar, L.J. Hainsworth and B. Hausback, 1996: Carbon dioxide and helium emissions from a reservoir of magmatic gas beneath Mammoth Mountain, California. *Journal of Geophysical Research*, **103**(B7), 15303–15323.

Steefel C. I., 2001: CRUNCH. Lawrence Livermore National Laboratory, Livermore, CA. 76 pp.

Stenhouse, M., M. Wilson, H. Herzog, M. Kozak and W. Zhou, 2004: Regulatory Issues Associated with Long-term Storage and Sequestration of CO_2. IEA Greenhouse Gas Report, 34–35.

Stenhouse, M., W. Zhou, D. Savage and S. Benbow, 2005: Framework methodology for long-term assessment of the fate of CO_2 in the Weyburn Field, Carbon Dioxide Capture for Storage in Deep Geologic Formations—Results from the CO_2 Capture Project, v. 2: Geologic Storage of Carbon Dioxide with Monitoring and Verification, Benson, S.M. (Ed.), Elsevier Science, London, pp. 1251–1262.

Stevens, S. H., J.A. Kuuskraa and R.A. Schraufnagel, 1996: Technology spurs growth of U.S. coalbed methane. *Oil and Gas Journal*, **94**(1), 56–63.

Stevens, S.H., V.K. Kuuskraa and J. Gale, 2000: Sequestration of CO_2 in depleted oil and gas fields: Global capacity and barriers to overcome. Proceedings of the 5th International Conference on Greenhouse Gas Control Technologies (GHGT5), Cairns, Australia, 13–16 August, 2000.

Stevens, S.H., C.E. Fox and L.S. Melzer, 2001a: McElmo dome and St. Johns natural CO_2 deposits: Analogs for geologic sequestration. Proceedings of the 5th International Conference on Greenhouse Gas Control Technologies (GHGT-5), D.J. Williams, R.A. Durie, P. McMullan, C.A.J. Paulson and A.Y. Smith (eds.), 13–16 August 2000, Cairns, Australia, CSIRO Publishing, Collingwood, Victoria, Australia, 317–321.

Stevens, S. H., V.A. Kuuskra and J.J. Gale, 2001b: Sequestration of CO_2 in depleted oil and gas fields: global capacity, costs and barriers. Proceedings of the 5th International Conference on Greenhouse Gas Control Technologies (GHGT-5), D.J. Williams, R.A. Durie, P. McMullan, C.A.J. Paulson and A.Y. Smith (eds.), 13–16 August 2000, Cairns, Australia, CSIRO Publishing, Collingwood, Victoria, Australia, pp. 278–283.

Stevens, S.H., V.A. Kuuskra, J. Gale and D. Beecy, 2001c: CO_2 injection and sequestration in depleted oil and gas fields and deep coal seams: worldwide potential and costs. *Environmental Geosciences, **8**(3), 200–209.

Stevens, S.H., C. Fox, T. White, S. Melzer and C. Byrer, 2003: Production operations at natural CO_2 Fields: Technologies for geologic sequestration. Proceedings of the 6th International Conference on Greenhouse Gas Control Technologies (GHGT-6), J. Gale and Y. Kaya (eds.), 1–4 October 2002, Kyoto, Japan, Pergamon,.v.I, 429–433.

Streit, J.E. and R.R. Hillis, 2003: Building geomechanical models for the safe underground storage of carbon dioxide in porous rock. Proceedings of the 6th International Conference on Greenhouse Gas Control Technologies (GHGT-6), J. Gale and Y. Kaya (eds.), 1–4 October 2002, Kyoto, Japan, Pergamon, Amsterdam, v.I., 495–500.

Streit, J., A. Siggins and B. Evans, 2005: Predicting and monitoring geomechanical effects of CO_2 injection, Carbon Dioxide Capture for Storage in Deep Geologic Formations—Results from the CO_2 Capture Project, v. 2: Geologic Storage of Carbon Dioxide with Monitoring and Verification, S.M. Benson (ed.), Elsevier Science, London, pp. 751–766.

Strutt, M.H, S.E. Beaubien, J.C. Beabron, M. Brach, C. Cardellini, R. Granieri, D.G. Jones, S. Lombardi, L. Penner, F. Quattrocchi and N. Voltatorni, 2003: Soil gas as a monitoring tool of deep geological sequestration of carbon dioxide: preliminary results from the EnCana EOR project in Weyburn, Saskatchewan (Canada). Proceedings of the 6th International Conference on Greenhouse Gas Control Technologies (GHGT-6), J. Gale and Y. Kaya (eds.), 1–4 October 2002, Kyoto, Japan, Pergamon, Amsterdam, v.I., 391–396.

Studlick, J.R.J., R.D. Shew, G.L. Basye and J.R. Ray, 1990: A giant carbon dioxide accumulation in the Norphlet Formation, Pisgah Anticline, Mississippi. In: Sandstone Petroleum Reservoirs, J.H. Barwis, J.G. McPherson and J.R.J. Studlick (eds.), Springer Verlag, New York, 181–203.

Taber, J.J., F.D. Martin and R.S. Seright, 1997: EOR screening criteria revisited - part 1: introduction to screening criteria and enhanced recovery fields projects. *SPE Reservoir Engineering,* **12**(3), 189–198.

Talebi, S., T.J. Boone and J.E. Eastwood, 1998: Injection induced microseismicity in Colorado shales. *Pure and Applied Geophysics,* **153**, 95–111.

Tamura, S., N. Imanaka, M. Kamikawa and G. Adachi, 2001: A CO_2 sensor based on a Sc^{3+} conducting $Sc_{1/3}Zr_2(PO_4)_3$ solid electrolyte. *Sensors and Actuators B,* **73**, 205–210.

Tanaka, S., H. Koide and A. Sasagawa, 1995: Possibility of underground CO_2 sequestration in Japan. *Energy Conversion and Management,* **36**(6–9), 527–530.

Torp, T. and K.R. Brown, 2005: CO_2 underground storage costs as experienced at Sleipner and Weyburn. Proceedings of the 7th International Conference on Greenhouse Gas Control Technologies (GHGT-7), September 5–9, 2004, Vancouver, Canada, v.I, 531-540.

Torp, T.A. and J. Gale, 2003: Demonstrating storage of CO_2 in geological reservoirs: the Sleipner and SACS projects. Proceedings of the 6th International Conference on Greenhouse Gas Control Technologies (GHGT-6), J. Gale and Y. Kaya (eds.), 1–4 October 2002, Kyoto, Japan, Pergamon, Amsterdam, v.I, 311–316.

USEPA, 1994: Determination of Maximum Injection Pressure for Class I Wells. Region 5 -- Underground Injection Control Section Regional Guidance #7.

U.S. Geological Survey, 2001a: U.S. Geological Survey World Petroleum Assessment 2000 - Description and Results. U.S. Geological Survey Digital Data Series - DDS-60. http://greenwood. cr.usgs.gov/energy/WorldEnergy/DDS-60/.

U.S. Geological Survey, 2001b: U.S. Geological Survey, On-line factsheet 172-96 Version 2. Invisible Gas Killing Trees at Mammoth Mountain California. http://wrgis.wr.usgs. gov/fact-sheet/fs172-96/.

Van Bergen, F., H.J.M. Pagnier, L.G.H. van der Meer, F.J.G. van den Belt, P.L.A. Winthaegen and R.S. Westerhoff, 2003a: Development of a field experiment of CO_2 storage in coal seams in the Upper Silesian Basin of Poland (RECOPOL). Proceedings of the 6th International Conference on Greenhouse Gas Control Technologies (GHGT-6), J. Gale and Y. Kaya (eds.), 1–4 October 2002, Kyoto, Japan, Pergamon, v.I, 569–574.

Van Bergen, F., A.F.B. Wildenborg, J. Gale and K.J. Damen, 2003b: Worldwide selection of early opportunities for CO_2-EOR and CO_2-ECBM. Proceedings of the 6th International Conference on Greenhouse Gas Control Technologies (GHGT-6), J. Gale and Y. Kaya (eds.), 1–4 October 2002, Kyoto, Japan, Pergamon, v.I, 639–644.

Van der Burgt, M.J., J. Cantle and V.K. Boutkan, 1992: Carbon dioxide disposal from coal-based IGCC's in depleted gas fields. *Energy Conversion and Management,* **33**(5–8), 603–610.

Van der Meer, L.G.H., 1992: Investigation regarding the storage of carbon dioxide in aquifers in the Netherlands. *Energy Conversion and Management,* **33**(5–8), 611–618.

Van der Meer, L.G.H., 1995: The CO_2 storage efficiency of aquifers. *Energy Conversion and Management,* **36**(6–9), 513–518.

Van der Meer L.G.H., 1996: Computer modeling of underground CO_2 storage. *Energy Conversion and Management,* **37**(6–8), 1155–1160.

Van der Meer, L.G.H., R.J. Arts and L. Paterson, 2001: Prediction of migration of CO_2 after injection into a saline aquifer: reservoir history matching of a 4D seismic image with a compositional gas/water model. Proceedings of the 5th International Conference on Greenhouse Gas Control Technologies (GHGT-5), D.J. Williams, R.A. Durie, P. McMullan, C.A.J. Paulson and A.Y. Smith (eds.), 2001, CSIRO, Melbourne, Australia, 378–384.

Van der Meer, L.G.H., J. Hartman, C. Geel and E. Kreft, 2005: Re-injecting CO_2 into an offshore gas reservoir at a depth of nearly 4000 metres sub-sea. Proceedings of the 7th International Conference on Greenhouse Gas Control Technologies (GHGT-7), September 5–9, 2004, Vancouver, Canada, v.I, 521-530.

Vavra, C.L., J.G. Kaldi and R.M. Sneider, 1992: Geological applications of capillary pressure: a review. *American Association of Petroleum Geologists Bulletin,* **76**(6), 840–850.

Vine, E., 2004: Regulatory constraints to carbon sequestration in terrestrial ecosystems and geological formations: a California perspective. *Mitigation and Adaptation Strategies for Global Change,* **9**, 77–95.

Wall, C., C. Bernstone. and M. Olvstam, 2005: International and European legal aspects on underground geological storage of CO_2, Proceedings of the 7th International Conference on Greenhouse Gas Control Technologies (GHGT-7), v.I, 971-978.

Walton, F.C., J.C Tait, D. LeNeveu and M.I. Sheppard, 2005: Geological storage of CO_2: A statistical approach to assessing performance and risk. Proceedings of the 7th International Conference on Greenhouse Gas Control Technologies (GHGT-7), September 5–9, 2004, Vancouver, Canada, v.I, 693-700.

Wang, S. and P.R. Jaffé, 2004: Dissolution of Trace Metals in Potable Aquifers due to CO_2 Releases from Deep Formations. Energy Conversion and Management. In press.

Watson, M.N., C.J. Boreham and P.R. Tingate, 2004: Carbon dioxide and carbonate elements in the Otway Basin: implications for geological storage of carbon dioxide. *The APPEA Journal,* **44**(1), 703–720.

White, C.M., B.R. Strazisar, E.J. Granite, J.S. Hoffman and H.W. Pennline, 2003: Separation and capture of CO_2 from large stationary sources and sequestration in geological formations--coalbeds and deep saline aquifers, Air and Waste Management Association (AWMA) Critical Review Papers, http://www.awma.org/journal/ShowAbstract.asp?Year=2003&PaperID=1066, June 2003.

White, D. (ed.), 2005: Theme 2: Prediction, Monitoring and Verification of CO_2 Movements. In: IEA GHG Weyburn CO_2 Monitoring and Storage Project Summary Report 2000-2004, M. Wilson and M. Monea (eds.), Proceedings of the 7th International Conference on Greenhouse Gas Control Technologies (GHGT-7), Volume III, p 73–148.

White, D.J., G. Burrowes, T. Davis, Z. Hajnal, K. Hirsche, I. Hutcheon, E. Majer, B. Rostron and S. Whittaker, 2004: Greenhouse gas sequestration in abandoned oil reservoirs: The International Energy Agency Weyburn pilot project. *GSA Today*, **14**, 4–10.

White, M.D. and M. Oostrom, 1997: STOMP, Subsurface Transport Over Multiple Phases. Pacific Northwest National Laboratory Report PNNL-11218, Richland, WA, October 1997.

White, S.P., 1995: Multiphase Non-Isothermal Transport of Systems of Reacting Chemicals. *Water Resources Research,* **32**(7), 1761–1772.

Whitman, W.B., D.C. Coleman and W.J. Wiebe, 2001: Prokaryotes: The unseen majority. *Proceedings of the National Academy of Sciences U.S.A.,* **95**(12), 6578–6583.

Wildenborg, A.F.B., A.L. Leijnse, E. Kreft, M.N. Nepveu, A.N.M. Obdam, B. Orlic, E.L. Wipfler, B. van der Grift, W. van Kesteren, I. Gaus, I. Czernichowski-Lauriol, P. Torfs and R. Wojcik, 2005a: Risk assessment methodology for CO_2 sequestration scenario approach, Carbon Dioxide Capture for Storage in Deep Geologic Formations—Results from the CO_2 Capture Project, v. 2: Geologic Storage of Carbon Dioxide with Monitoring and Verification, S.M. Benson (ed.), Elsevier Science, London, pp. 1293–1316.

Wildenborg, T., J. Gale, C. Hendriks, S. Holloway, R. Brandsma, E. Kreft and A. Lokhorst, 2005b: Cost curves for CO_2 storage: European sector. Proceedings of the 7th International Conference on Greenhouse Gas Control Technologies (GHGT-7), September 5–9, 2004, Vancouver, Canada, v.I, 603-610.

Wilson, E., 2004: Managing the Risks of Geologic Carbon Sequestration: A Regulatory and Legal Analysis. Doctoral Dissertation, Engineering and Public Policy, Carnegie Mellon, Pittsburgh, PA, U.S.A.

Wilson, E., T. Johnson and D. Keith, 2003: Regulating the ultimate sink: managing the risks of geologic CO_2 Storage. Environmental Science and Technology, 37, 3476–3483.

Wilson, M. and M. Monea, 2005: IEA GHG Weyburn Monitoring and Storage Project, Summary Report, 2000-2004. Petroleum Technology Research Center, Regina SK, Canada. In: Proceedings of the 7th International Conference on Greenhouse Gas Control Technologies (GHGT-7), Vol. III, September 5–9, Vancouver, Canada

Winter, E.M. and P.D. Bergman, 1993: Availability of depleted oil and gas reservoirs for disposal of carbon dioxide in the United States. Energy Conversion and Management, **34**(9–11), 1177–1187.

Witherspoon, P.A., I. Javendal, S.P. Neuman and R.A. Freeze, 1968: Interpretation of aquifer gas storage conditions from water pumping tests. American Gas Association.

Wo, S. and J-T. Liang, 2005: CO_2 storage in coalbeds: CO_2/N_2 injection and outcrop seepage modeling, Carbon Dioxide Capture for Storage in Deep Geologic Formations—Results from the CO_2 Capture Project, v. 2: Geologic Storage of Carbon Dioxide with Monitoring and Verification, S.M. Benson (ed.), Elsevier Science, London, pp. 897–924.

Wo, S., J-T. Liang and L.R. Myer, 2005: CO_2 storage in coalbeds: Risk assessment of CO_2 and methane leakage, Carbon Dioxide Capture for Storage in Deep Geologic Formations—Results from the CO_2 Capture Project, v. 2: Geologic Storage of Carbon Dioxide with Monitoring and Verification, S.M. Benson (ed.), Elsevier Science, London. pp. 1263–1292.

Wong, S., W.D. Gunter and J. Gale, 2001: Site ranking for CO_2-enhanced coalbed methane demonstration pilots. Proceedings of the 5th International Conference on Greenhouse Gas Control Technologies (GHGT-5), D.J. Williams, R.A. Durie, P. McMullan, C.A.J. Paulson and A. Smith (eds.), 13–16 August 2000, Cairns, Australia, CSIRO Publishing, Collingwood, Victoria, Australia, pp. 543–548.

Wright, G. and Majek, 1998: Chromatograph, RTU Monitoring of CO_2 Injection. *Oil and Gas Journal,* July 20, 1998.

Wyss, M. and P. Molnar, 1972: Efficiency, stress drop, apparent stress, effective stress and frictional stress of Denver, Colorado, earthquakes. *Journal of Geophysical Research,* **77**, 1433–1438.

Xu, T., J.A. Apps and K. Pruess, 2003: Reactive geochemical transport simulation to study mineral trapping for CO_2 disposal in deep arenaceous formations. *Journal of Geophysical Research,* **108**(B2), 2071–2084.

Yamaguchi, S., K. Ohga, M. Fujioka and S. Muto, 2005: Prospect of CO_2 sequestration in Ishikari coal mine, Japan. Proceedings of the 7th International Conference on Greenhouse Gas Control Technologies (GHGT-7), 5–9 September 2004, Vancouver, Canada, v.I, 423-430.

Zarlenga F., R. Vellone, G.P. Beretta, C. Calore, M.A. Chiaramonte, D. De Rita, R. Funiciello, G. Gambolati, G. Gianelli, S. Grauso, S. Lombardi, I. Marson, S. Persoglia, G. Seriani and S. Vercelli, 2004: Il confinamento geologico della CO_2: Possibilità e problematiche aperte in Italia. Energia e Innovazione, In press (In Italian).

Zhang, C.J., M. Smith, M. and B.J. McCoy, 1993: Kinetics of supercritical fluid extraction of coal: Physical and chemical processes. In: Supercritical Fluid Engineering Science: Fundamentals and Applications, E. Kiran and J.F. Brennecke (eds.), American Chemical Society, Washington, DC, pp. 363–379.

Zhou, W., M.J. Stenhouse, R. Arthur, S. Whittaker, D.H.-S. Law, R. Chalaturnyk and W. Jazwari, 2005: The IEA Weyburn CO_2 monitoring and storage project—Modeling of the long-term migration of CO_2 from Weyburn. Proceedings of the 7th International Conference on Greenhouse Gas Control Technologies (GHGT-7), September 5–9, 2004, Vancouver, Canada, v.I, 721-730. Volume 1: Peer-Reviewed Papers and Plenary Presentations, Elsevier, UK.

Zoback, M.D. and H.P. Harjes, 1997: Injection-induced earthquakes and crustal stress at 9 km depth at the KTB deep drilling site, Germany. *Journal of Geophysical Research,* **102**, 18477–18491.

6

Ocean storage

Coordinating Lead Authors
Ken Caldeira (United States), Makoto Akai (Japan)

Lead Authors
Peter Brewer (United States), Baixin Chen (China), Peter Haugan (Norway), Toru Iwama (Japan), Paul Johnston (United Kingdom), Haroon Kheshgi (United States), Qingquan Li (China), Takashi Ohsumi (Japan), Hans Pörtner (Germany), Chris Sabine (United States), Yoshihisa Shirayama (Japan), Jolyon Thomson (United Kingdom)

Contributing Authors
Jim Barry (United States), Lara Hansen (United States)

Review Editors
Brad De Young (Canada), Fortunat Joos (Switzerland)

Contents

EXECUTIVE SUMMARY

Captured CO_2 could be deliberately injected into the ocean at great depth, where most of it would remain isolated from the atmosphere for centuries. CO_2 can be transported via pipeline or ship for release in the ocean or on the sea floor. There have been small-scale field experiments and 25 years of theoretical, laboratory, and modelling studies of intentional ocean storage of CO_2, but ocean storage has not yet been deployed or thoroughly tested.

The increase in atmospheric CO_2 concentrations due to anthropogenic emissions has resulted in the oceans taking up CO_2 at a rate of about 7 $GtCO_2yr^{-1}$ (2 $GtCyr^{-1}$). Over the past 200 years the oceans have taken up 500 $GtCO_2$ from the atmosphere out of 1300 $GtCO_2$ total anthropogenic emissions. Anthropogenic CO_2 resides primarily in the upper ocean and has thus far resulted in a decrease of pH of about 0.1 at the ocean surface with virtually no change in pH deep in the oceans. Models predict that the oceans will take up most CO_2 released to the atmosphere over several centuries as CO_2 is dissolved at the ocean surface and mixed with deep ocean waters.

The Earth's oceans cover over 70% of the Earth's surface with an average depth of about 3,800 metres; hence, there is no practical physical limit to the amount of anthropogenic CO_2 that could be placed in the ocean. However, the amount that is stored in the ocean on the millennial time scale depends on oceanic equilibration with the atmosphere. Over millennia, CO_2 injected into the oceans at great depth will approach approximately the same equilibrium as if it were released to the atmosphere. Sustained atmospheric CO_2 concentrations in the range of 350 to 1000 ppmv imply that $2,300 \pm 260$ to $10,700 \pm 1,000$ Gt of anthropogenic CO_2 will eventually reside in the ocean.

Analyses of ocean observations and models agree that injected CO_2 will be isolated from the atmosphere for several hundreds of years and that the fraction retained tends to be larger with deeper injection. Additional concepts to prolong CO_2 retention include forming solid CO_2 hydrates and liquid CO_2 lakes on the sea floor, and increasing CO_2 solubility by, for example, dissolving mineral carbonates. Over centuries, ocean mixing results in loss of isolation of injected CO_2 and exchange with the atmosphere. This would be gradual from large regions of the ocean. There are no known mechanisms for sudden or catastrophic release of injected CO_2.

Injection up to a few $GtCO_2$ would produce a measurable change in ocean chemistry in the region of injection, whereas injection of hundreds of $GtCO_2$ would eventually produce measurable change over the entire ocean volume.

Experiments show that added CO_2 can harm marine organisms. Effects of elevated CO_2 levels have mostly been studied on time scales up to several months in individual organisms that live near the ocean surface. Observed phenomena include reduced rates of calcification, reproduction, growth, circulatory oxygen supply and mobility as well as increased mortality over time. In some organisms these effects are seen in response to small additions of CO_2. Immediate mortality is expected close to injection points or CO_2 lakes. Chronic effects may set in with small degrees of long-term CO_2 accumulation, such as might result far from an injection site, however, long-term chronic effects have not been studied in deep-sea organisms.

CO_2 effects on marine organisms will have ecosystem consequences; however, no controlled ecosystem experiments have been performed in the deep ocean. Thus, only a preliminary assessment of potential ecosystem effects can be given. It is expected that ecosystem consequences will increase with increasing CO_2 concentration, but no environmental thresholds have been identified. It is also presently unclear, how species and ecosystems would adapt to sustained, elevated CO_2 levels.

Chemical and biological monitoring of an injection project, including observations of the spatial and temporal evolution of the resulting CO_2 plume, would help evaluate the amount of materials released, the retention of CO_2, and some of the potential environmental effects.

For water column and sea floor release, capture and compression/liquefaction are thought to be the dominant cost factors. Transport (i.e., piping, and shipping) costs are expected to be the next largest cost component and scale with proximity to the deep ocean. The costs of monitoring, injection nozzles etc. are expected to be small in comparison.

Dissolving mineral carbonates, if found practical, could cause stored carbon to be retained in the ocean for 10,000 years, minimize changes in ocean pH and CO_2 partial pressure, and may avoid the need for prior separation of CO_2. Large amounts of limestone and materials handling would be required for this approach.

Several different global and regional treaties on the law of the sea and marine environment could be relevant to intentional release of CO_2 into the ocean but the legal status of intentional carbon storage in the ocean has not yet been adjudicated.

It is not known whether the public will accept the deliberate storage of CO_2 in the ocean as part of a climate change mitigation strategy. Deep ocean storage could help reduce the impact of CO_2 emissions on surface ocean biology but at the expense of effects on deep-ocean biology.

6.1 Introduction and background

6.1.1 Intentional storage of CO_2 in the ocean

This report assesses what is known about intentional storage of carbon dioxide in the ocean by inorganic strategies that could be applied at industrial scale. Various technologies have been envisioned to enable and increase ocean CO_2 storage (Figure 6.1). One class of options involves storing a relatively pure stream of carbon dioxide that has been captured and compressed. This CO_2 can be placed on a ship, injected directly into the ocean, or deposited on the sea floor. CO_2 loaded on ships could either be dispersed from a towed pipe or transported to fixed platforms feeding a CO_2 lake on the sea floor. Such CO_2 lakes must be

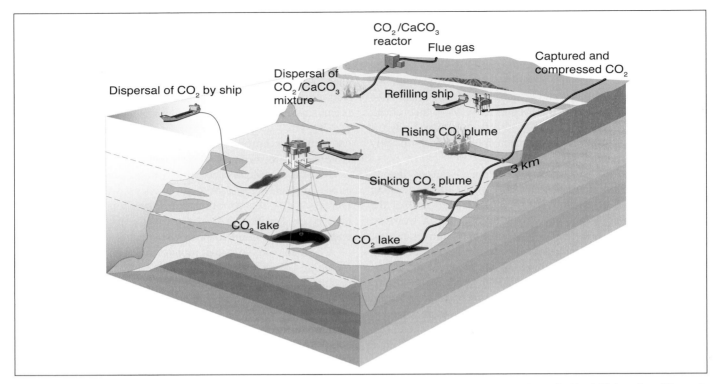

Figure 6.1 Illustration of some of the ocean storage strategies described in this chapter (Artwork courtesy Sean Goddard, University of Exeter.)

deeper than 3 km where CO_2 is denser than sea water. Any of these approaches could in principle be used in conjunction with neutralization with carbonate minerals.

Research, development and analysis of ocean CO_2 storage concepts has progressed to consider key questions and issues that could affect the prospects of ocean storage as a response option to climate change (Section 6.2). Accumulated understanding of the ocean carbon cycle is being used to estimate how long CO_2 released into the oceans will remain isolated from the atmosphere. Such estimates are used to assess the effectiveness of ocean storage concepts (Section 6.3).

Numerical models of the ocean indicate that placing CO_2 in the deep ocean would isolate most of the CO_2 from the atmosphere for several centuries, but over longer times the ocean and atmosphere would equilibrate. Relative to atmospheric release, direct injection of CO_2 into the ocean could reduce maximum amounts and rates of atmospheric CO_2 increase over the next several centuries. Direct injection of CO_2 in the ocean would not reduce atmospheric CO_2 content on the millennial time scale (Table 6.1; Figures 6.2 and 6.3; Hoffert *et al.*, 1979; Kheshgi *et al.*, 1994).

Table 6.1 Amount of additional CO_2 residing in the ocean after atmosphere-ocean equilibration for different atmospheric stabilization concentrations. The uncertainty range represents the influence of climate sensitivity to a CO_2 doubling in the range of 1.5 °C to 4.5 °C (Kheshgi et al., 2005; Kheshgi 2004a). This table considers the possibility of increased carbon storage in the terrestrial biosphere. Such an increase, if permanent, would allow a corresponding increase in total cumulative emissions. This table does not consider natural or engineered dissolution of carbonate minerals, which would increase ocean storage of anthropogenic carbon. The amount already in the oceans exceeds 500 $GtCO_2$ (= 440 $GtCO_2$ for 1994 (Sabine et al., 2004) plus CO_2 absorption since that time). The long-term amount of CO_2 stored in the deep ocean is independent of whether the CO_2 is initially released to the atmosphere or the deep ocean.

Atmospheric CO_2 stabilization concentration (ppmv)	Total cumulative ocean + atmosphere CO_2 release ($GtCO_2$)	Amount of anthropogenic CO_2 stored in the ocean in equilibrium ($GtCO_2$)
350	2880 ± 260	2290 ± 260
450	5890 ± 480	4530 ± 480
550	8350 ± 640	6210 ± 640
650	10,460 ± 750	7540 ± 750
750	12,330 ± 840	8630 ± 840
1000	16,380 ± 1000	10,730 ± 1000

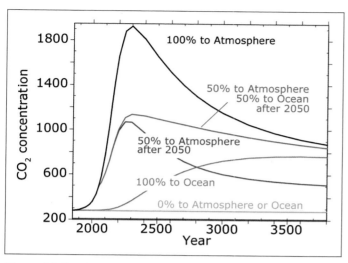

Figure 6.2 Simulated atmospheric CO_2 resulting from CO_2 release to the atmosphere or injection into the ocean at 3,000 m depth (Kheshgi and Archer, 2004). Emissions follow a logistic trajectory with cumulative emissions of 18,000 GtCO$_2$. Illustrative cases include 100% of emissions released to the atmosphere leading to a peak in concentration, 100% of emissions injected into the ocean, and no emissions (i.e., other mitigation approaches are used). Additional cases include atmospheric emission to year 2050, followed by either 50% to atmosphere and 50% to ocean after 2050 or 50% to atmosphere and 50% by other mitigation approaches after 2050. Ocean injection results in lower peak concentrations than atmospheric release but higher than if other mitigation approaches are used (e.g., renewables or permanent storage).

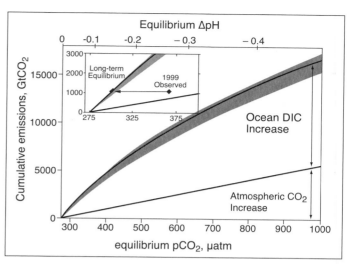

Figure 6.3 Equilibrium partitioning of CO_2 between the ocean and atmosphere. On the time scale of millennia, complete mixing of the oceans leads to a partitioning of cumulative CO_2 emissions between the oceans and atmosphere with the bulk of emissions eventually residing in the oceans as dissolved inorganic carbon. The ocean partition depends nonlinearly on CO_2 concentration according to carbonate chemical equilibrium (Box 6.1) and has limited sensitivity to changes in surface water temperature (shown by the grey area for a range of climate sensitivity of 1.5 to 4.5°C for CO_2 doubling) (adapted from Kheshgi et al., 2005; Kheshgi, 2004a). ΔpH evaluated from pCO_2 of 275 ppm. This calculation is relevant on the time scale of several centuries, and does not consider changes in ocean alkalinity that increase ocean CO_2 uptake over several millennia (Archer et al., 1997).

There has been limited experience with handling CO_2 in the deep sea that could form a basis for the development of ocean CO_2 storage technologies. Before they could be deployed, such technologies would require further development and field testing. Associated with the limited level of development, estimates of the costs of ocean CO_2 storage technologies are at a primitive state, however, the costs of the actual dispersal technologies are expected to be low in comparison to the costs of CO_2 capture and transport to the deep sea (but still non-negligible; Section 6.9). Proximity to the deep sea is a factor, as the deep oceans are remote to many sources of CO_2 (Section 6.4). Ocean storage would require CO_2 transport by ship or deep-sea pipelines. Pipelines and drilling platforms, especially in oil and gas applications, are reaching ever-greater depths, yet not on the scale or to the depth relevant for ocean CO_2 storage (Chapter 4). No insurmountable technical barrier to storage of CO_2 in the oceans is apparent.

Putting CO_2 directly into the deep ocean means that the chemical environment of the deep ocean would be altered immediately, and in concepts where release is from a point, change in ocean chemistry would be greater proximate to the release location. Given only rudimentary understanding of deep-sea ecosystems, only a limited and preliminary assessment of potential ecosystem effects can be given (Section 6.7).

Technologies exist to monitor deep-sea activities (Section 6.6). Practices for monitoring and verification of ocean storage

would depend on which, as of yet undeveloped, ocean storage technology would potentially be deployed, and on environmental impacts to be avoided.

More carbon dioxide could be stored in the ocean with less of an effect on atmospheric CO_2 and fewer adverse effects on the marine environment if the alkalinity of the ocean could be increased, perhaps by dissolving carbonate minerals in sea water. Proposals based on this concept are discussed primarily in Section 6.2.

For ocean storage of CO_2, issues remain regarding environmental consequences, public acceptance, implications of existing laws, safeguards and practices that would need to be developed, and gaps in our understanding of ocean CO_2 storage (Sections 6.7, 6.8, and 6.10).

6.1.2 Relevant background in physical and chemical oceanography

The oceans, atmosphere, and plants and soils are the primary components of the global carbon cycle and actively exchange carbon (Prentice *et al.*, 2001). The oceans cover 71% of the Earth's surface with an average depth of 3,800 m and contain roughly 50 times the quantity of carbon currently contained in the atmosphere and roughly 20 times the quantity of carbon currently contained in plants and soils. The ocean contains so much CO_2 because of its large volume and because CO_2 dissolves in sea water to form various ionic species (Box 6.1).

The increase in atmospheric CO_2 over the past few centuries has been driving CO_2 from the atmosphere into the oceans. The oceans serve as an important sink of CO_2 emitted to the atmosphere taking up on average about 7 $GtCO_2$ yr^{-1} (2 GtC yr^{-1}) over the 20 years from 1980 to 2000 with ocean uptake over the past 200 years estimated to be > 500 $GtCO_2$ (135 GtC) (Prentice *et al.*, 2001; Sabine *et al.*, 2004). On average, the anthropogenic CO_2 signal is detectable to about 1000 m depth; its near absence in the deep ocean is due to the slow exchange between ocean surface and deep –sea waters.

Ocean uptake of anthropogenic CO_2 has led to a perturbation of the chemical environment primarily in ocean surface waters. Increasing ocean CO_2 concentration leads to decreasing carbonate ion concentration and increasing hydrogen ion activity (Box 6.1). The increase in atmospheric CO_2 from about 280 ppm in 1800 to 380 ppm in 2004 has caused an average decrease across the surface of the oceans of about 0.1 pH units ($\Delta pH \approx -0.1$) from an initial average surface ocean pH of about 8.2. Further increase in atmospheric CO_2 will result in a further change in the chemistry of ocean surface waters that will eventually reach the deep ocean (Figure 6.4). The anthropogenic perturbation of ocean chemistry is greatest in the upper ocean where biological activity is high.

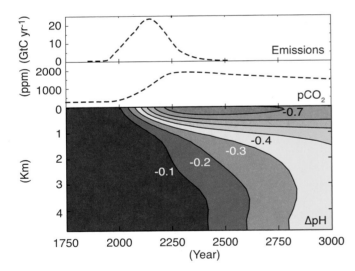

Figure 6.4 Simulated ocean pH changes from CO_2 release to the atmosphere. Modelled atmospheric CO_2 change and horizontally averaged ΔpH driven by a CO_2 emissions scenario: historic atmospheric CO_2 up to 2000, IS92a from 2000 to 2100, and logistic curve extending beyond 2100 with 18,000 $GtCO_2$ (Moomaw et al., 2001) cumulative emissions from 2000 onward (comparable to estimates of fossil-fuel resources – predominantly coal; Caldeira and Wickett, 2003). Since year 1800, the pH of the surface of the oceans has decreased about 0.1 pH units (from an initial average surface ocean pH of about 8.2) and CO_3^{2-} has decreased about 40 μmol kg^{-1}. There are a number of pH scales used by ocean chemists and biologists to characterize the hydrogen ion content of sea water, but ΔpH computed on different scales varies little from scale to scale (Brewer et al., 1995).

Most carbon dioxide released to either the atmosphere or the ocean will eventually reside in the ocean, as ocean chemistry equilibrates with the atmosphere. Thus, stabilization of atmospheric CO_2 concentration at levels above the natural level of 280 ppm implies long-term addition of carbon dioxide to the ocean. In equilibrium, the fraction of an increment of CO_2 released that will reside in the ocean depends on the atmospheric CO_2 concentration (Table 6.1; Figure 6.3; Kheshgi *et al.*, 2005; Kheshgi, 2004a).

The capacity of the oceans to absorb CO_2 in equilibrium with the atmosphere is a function of the chemistry of sea water. The rate at which this capacity can be brought into play is a function of the rate of ocean mixing. Over time scales of decades to centuries, exchange of dissolved inorganic carbon between ocean surface waters and the deep ocean is the primary barrier limiting the rate of ocean uptake of increased atmospheric CO_2. Over many centuries (Kheshgi, 2004a), changes in dissolved inorganic carbon will mix throughout the ocean volume with the oceans containing most of the cumulative CO_2 emissions to the atmosphere/ocean system (Table 6.1; Figure 6.3). Over longer times (millennia), dissolution of $CaCO_3$ causes an even greater fraction of released CO_2 (85–92%) to reside in the ocean (Archer *et al.*, 1997).

Both biological and physical processes lead to the observed distribution of pH and its variability in the world ocean (Figure 6.6). As they transit from the Atlantic to Pacific Basins, deep ocean waters accumulate about 10% more dissolved inorganic carbon dioxide, primarily from the oxidation of sinking organic matter (Figure 6.7).

6.2 Approaches to release of CO_2 into the ocean

6.2.1 Approaches to releasing CO_2 that has been captured, compressed, and transported into the ocean

6.2.1.1 Basic approach
The basic concept of intentional CO_2 storage in the ocean is to take a stream of CO_2 that has been captured and compressed (Chapter 3), and transport it (Chapter 4) to the deep ocean for release at or above the sea floor. (Other ocean storage approaches are discussed in Sections 6.2.2 and 6.2.3.) Once released, the CO_2 would dissolve into the surrounding sea water, disperse and become part of the ocean carbon cycle.

Marchetti (1977) first proposed injecting liquefied CO_2 into the waters flowing over the Mediterranean sill into the mid-depth North Atlantic, where the CO_2 would be isolated from the atmosphere for centuries. This concept relies on the slow exchange of deep ocean waters with the surface to isolate CO_2 from the atmosphere. The effectiveness of ocean storage will depend on how long CO_2 remains isolated from the atmosphere. Over the centuries and millennia, CO_2 released to the deep ocean will mix throughout the oceans and affect atmospheric CO_2 concentration. The object is to transfer the CO_2 to deep waters because the degree of isolation from the atmosphere generally increases with depth in the ocean. Proposed methods

Box 6.1. Chemical properties of CO_2

The oceans absorb large quantities of CO_2 from the atmosphere principally because CO_2 is a weakly acidic gas, and the minerals dissolved in sea water have created a mildly alkaline ocean. The exchange of atmospheric CO_2 with ocean surface waters is determined by the chemical equilibrium between CO_2 and carbonic acid H_2CO_3 in sea water, the partial pressure of CO_2 (pCO_2) in the atmosphere and the rate of air/sea exchange. Carbonic acid dissociates into bicarbonate ion HCO_3^-, carbonate ion CO_3^{2-}, and hydronium ion H^+ by the reactions (see Annex AI.3):

$$CO_2 \text{ (g)} + H_2O \leftrightarrow H_2CO_3 \text{ (aq)} \leftrightarrow HCO_3^- + H^+ \leftrightarrow CO_3^{2-} + 2H^+ \qquad (1)$$

Total dissolved inorganic carbon (DIC) is the sum of carbon contained in H_2CO_3, HCO_3^-, and CO_3^{2-}. The atmospheric concentration of CO_2 in equilibrium with surface water can be calculated from well-known chemical equilibria that depend on ocean total dissolved inorganic carbon, alkalinity, temperature and salinity (Zeebe and Wolf-Gladrow, 2001). The partial pressure of CO_2 in the ocean mixed layer equilibrates with the atmosphere on a time scale of about one year.

The ocean is a highly buffered system, that is the concentration of the chemical species whose equilibrium controls pH is significantly higher than the concentrations of H^+ or OH^-. The pH of sea water is the base–10 log of activity of H^+. Total Alkalinity (TAlk) is the excess of alkaline components, and is defined as the amount of strong acid required to bring sea water to the 'equivalence point' at which the HCO_3^- and H_2CO_3 contributions are equal (Dickson, 1981).

The principal effect of adding CO_2 to sea water is to form bicarbonate ion, for example,

$$CO_2 + H_2O + CO_3^{2-} \rightarrow 2HCO_3^-. \qquad (2)$$

In addition, some CO_2 undergoes simple reaction with water, for example,

$$CO_2 + H_2O \leftrightarrow H^+ + HCO_3^-. \qquad (3)$$

In either case, Total Alkalinity does not change. The combined reactions lower both ocean pH, and carbonate ion concentration. For current ocean composition, CO_2 that is added to sea water is partitioned primarily into HCO_3^- with the net reaction resulting in the generation of H^+ and thus decreasing pH and making sea water more acidic; adding CO_2 thereby decreases the concentration of CO_3^{2-}.

Total Alkalinity is increased when, for example, alkaline minerals such as $CaCO_3$ are dissolved in sea water through the reaction,

$$CaCO_3 \text{ (s)} \leftrightarrow Ca^{2+} + CO_3^{2-} \qquad (4)$$

which releases 2 mole-equivalents of Total Alkalinity and 1 mol of Dissolved Inorganic Carbon for each mole of $CaCO_3$ dissolved. Increasing TAlk more than DIC leads to a decrease in the partial pressure of CO_2 as seen in Figure 6.5. Because most Dissolved Inorganic Carbon is in the form of HCO_3^-, the main effect of dissolving $CaCO_3$ in surface waters is (see Kheshgi, 1995)

$$CaCO_3 \text{ (s)} + CO_2 \text{ (g)} + H_2O \leftrightarrow Ca^{2+} + 2HCO_3^- \qquad (5)$$

thereby shifting CO_2 from the atmosphere to the oceans in equilibrium, neutralizing the effect of CO_2 on pH.

Ocean surface waters are super-saturated with respect to CaCO3, allowing the growth of corals and other organisms that produce shells or skeletons of carbonate minerals. In contrast, the deepest ocean waters have lower pH and lower CO_3^{2-} concentrations, and are thus undersaturated with respect to $CaCO_3$. Marine organisms produce calcium carbonate particles in the surface ocean that settle and dissolve in undersaturated regions of the deep oceans.

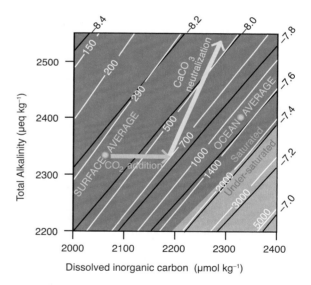

Figure 6.5 Composition diagram for ocean surface waters at 15°C (adapted from Baes, 1982). The white lines denote compositions with the same value of pCO_2 (in ppm); the black lines denote compositions with the same pH. The tan shaded region is undersaturated and the green shaded region is supersaturated with respect to calcite at atmospheric pressure (calcite solubility increases with depth). Surface water and average ocean compositions are also indicated. Adding CO_2 increases Dissolved Inorganic Carbon (DIC) without changing Total Alkalinity (TAlk); dissolving $CaCO_3$ increases both DIC and TAlk, with 2 moles of TAlk added for each mole of DIC added.

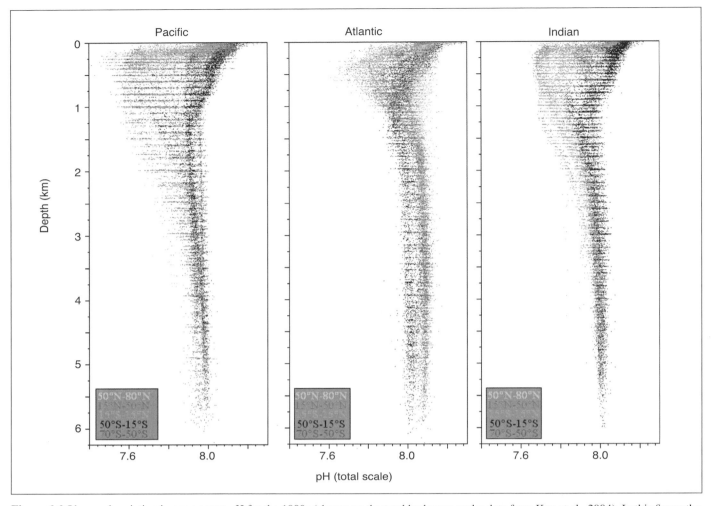

Figure 6.6 Observed variation in open ocean *p*H for the 1990s (shown on the total hydrogen scale; data from Key et al., 2004). In this figure the oceans are separated into separate panels. The three panels are on the same scale and coloured by latitude band to illustrate the large north-south changes in the *p*H of intermediate waters. Pre-industrial surface values would have been about 0.1 *p*H units greater than in the 1990s.

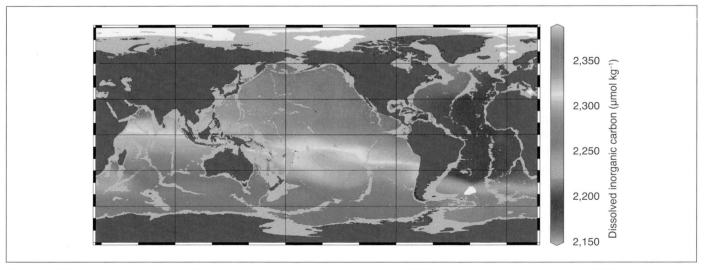

Figure 6.7 Natural variation in total dissolved inorganic carbon concentration at 3000 m depth (data from Key et al., 2004). Ocean carbon concentrations increase roughly 10% as deep ocean waters transit from the North Atlantic to the North Pacific due to the oxidation of organic carbon in the deep ocean.

would inject the CO_2 below the thermocline[1] for more effective storage.

Depending on the details of the release and local sea floor topography, the CO_2 stream could be engineered to dissolve in the ocean or sink to form a lake on the sea floor. CO_2, dissolved in sea water at high concentrations can form a dense plume or sinking current along an inclined sea floor. If release is at a great enough depth, CO_2 liquid will sink and could accumulate on the sea floor as a pool containing a mixture of liquid and hydrate. In the short-term, fixed or towed pipes appear to be the most viable methods for oceanic CO_2 release, relying on technology that is already largely commercially available.

6.2.1.2 *Status of development*

To date, injection of CO_2 into sea water has only been investigated in the laboratory, in small-scale *in-situ* experiments, and in models. Larger-scale *in-situ* experiments have not yet been carried out.

An international consortium involving engineers, oceanographers and ecologists from 15 institutions in the United States, Norway, Japan and Canada proposed an *in-situ* experiment to help evaluate the feasibility of ocean carbon storage as a means of mitigating atmospheric increases. This was to be a collaborative study of the physical, chemical, and biological changes associated with direct injection of CO_2 into the ocean (Adams *et al.*, 2002). The proposed CO_2 Ocean Sequestration Field Experiment was to inject less than 60 tonnes of pure liquid carbon dioxide (CO_2) into the deep ocean near Keahole Point on the Kona coast of the Island of Hawaii. This would have been the largest intentional CO_2 release into the ocean water column. The test was to have taken place in water about 800 m deep, over a period of about two weeks during the summer of 2001. Total project cost was to have been roughly US$ 5 million. A small steel pipeline, about 4 cm in diameter, was to have been deployed from a ship down to the injection depth, with a short section of pipeline resting on the sea floor to facilitate data collection. The liquid CO_2 was to have been dispersed through a nozzle, with CO_2 droplets briefly ascending from the injection point while dissolving into the sea water. However, the project met with opposition from environmental organizations and was never able to acquire all of the necessary permits within the prescribed budget and schedule (de Figueiredo, 2002).

Following this experience, the group developed a plan to release 5.4 tonnes of liquefied CO_2 at a depth of 800 metres off the coast of Norway, and monitor its dispersion in the Norwegian Sea. The Norwegian Pollution Control Authority granted a permit for the experiment. The Conservative Party environment minister in Norway's coalition government, Børge Brende, decided to review the Norwegian Pollution Control Authorities' initial decision. After the public hearing procedure and subsequent decision by the Authority to confirm their initial permit, Brende said, 'The possible future use of the sea as storage for CO_2 is controversial. … Such a deposit could be in defiance of international marine laws and the ministry therefore had to reject the application.' The Norwegian Environment ministry subsequently announced that the project would not go ahead (Giles, 2002).

Several smaller scale scientific experiments (less than 100 litres of CO_2) have however been executed (Brewer *et al.*, 1999, Brewer *et al.*, 2005) and the necessary permits have also been issued for experiments within a marine sanctuary.

6.2.1.3 *Basic behaviour of CO_2 released in different forms*

The near-field behaviour of CO_2 released into the ocean depends on the physical properties of CO_2 (Box 6.2) and the method for CO_2 release. Dissolved CO_2 increases the density of sea water (e.g., Bradshaw, 1973; Song, *et al.*, 2005) and this affects transport and mixing. The near field may be defined as that region in which it is important to take effects of CO_2-induced density changes on the fluid dynamics of the ocean into consideration. The size of this region depends on the scale and design of CO_2 release (Section 6.2.1.4).

CO_2 plume dynamics depend on the way in which CO_2 is released into the ocean water column. CO_2 can be initially in the form of a gas, liquid, solid or solid hydrate. All of these forms of CO_2 would dissolve in sea water, given enough time (Box 6.1). The dissolution rate of CO_2 in sea water is quite variable and depends on the form (gas, liquid, solid, or hydrate), the depth and temperature of disposal, and the local water velocities. Higher flow rates increase the dissolution rate.

Gas. CO_2 could potentially be released as a gas above roughly 500 m depth (Figure 6.8). Below this depth, pressures are too great for CO_2 to exist as a gas. The gas bubbles would be less dense than the surrounding sea water so tend to rise towards the surface, dissolving at a radial speed of about 0.1 cm hr^{-1} (0.26 to 1.1 μmol cm^{-2} s^{-1}; Teng *et al.*, 1996). In waters colder than about 9°C, a CO_2 hydrate film could form on the bubble wall. CO_2 diffusers could produce gaseous CO_2 bubbles that are small enough to dissolve completely before reaching the surface.

Liquid. Below roughly 500 m depth, CO_2 can exist in the ocean as a liquid. Above roughly 2500 m depth CO_2 is less dense than sea water, so liquid CO_2 released shallower than 2500 m would tend to rise towards the surface. Because most ocean water in this depth range is colder than 9°C, CO_2 hydrate would tend to form on the droplet wall. Under these conditions, the radius of the droplet would diminish at a speed of about 0.5 cm hr^{-1} (= 3 μmol cm^{-2} s^{-1}; Brewer *et al.*, 2002). Under these conditions a 0.9 cm diameter droplet would rise about 400 m in an hour before dissolving completely; 90% of its mass would be lost in the first 200 m (Brewer *et al.*, 2002). Thus, CO_2 diffusers could be designed to produce droplets that will dissolve within roughly 100 m of the depth of release. If the droplet reached approximately 500 m depth, it would become a gas bubble.

CO_2 is more compressible than sea water; below roughly

---

[1] The thermocline is the layer of the ocean between about 100 and 1000 m depth that is stably stratified by large temperature and density gradients, thus inhibiting vertical mixing. Vertical mixing rates in the thermocline can be about 1000 times less than those in the deep sea. This zone of slow mixing would act as a barrier to slow degassing of CO_2 released in the deep ocean to the atmosphere.

Box 6.2 Physical properties of CO_2.

The properties of CO_2 in sea water affect its fate upon release to the deep-sea environment. The conditions under which CO_2 can exist in a gas, liquid, solid hydrate, or aqueous phase in sea water are given in Figure 6.8 (see Annex I).

At typical pressures and temperatures that exist in the ocean, pure CO_2 would be a gas above approximately 500 m and a liquid below that depth. Between about 500 and 2700 m depth, liquid CO_2 is lighter than sea water. Deeper than 3000 m, CO_2 is denser than sea water. The buoyancy of CO_2 released into the ocean determines whether released CO_2 rises or falls in the ocean column (Figure 6.9). In the gas phase, CO_2 is lighter than sea water and rises. In the liquid phase CO_2 is a highly compressible fluid compared to sea water. A fully formed crystalline CO_2 hydrate is denser than sea water and will form a sinking mass (Aya *et al.*, 2003); hydrate formation can thus aid ocean CO_2 storage by more rapid transport to depth, and by slowing dissolution. It may also create a nuisance by impeding flow in pipelines or at injectors.

The formation of a solid CO_2 hydrate (Sloan, 1998) is a dynamic process (Figure 6.10; Brewer *et al.*, 1998, 1999, 2000) and the nature of hydrate nucleation in such systems is imperfectly understood. Exposed to an excess of sea water, CO_2 will eventually dissolve forming an aqueous phase with density higher than surrounding sea water. Release of dense or buoyant CO_2 – in a gas, liquid, hydrate or aqueous phase – would entrain surrounding sea water and form plumes that sink, or rise, until dispersed.

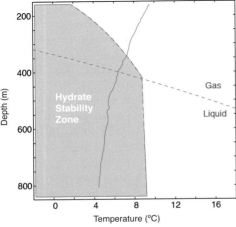

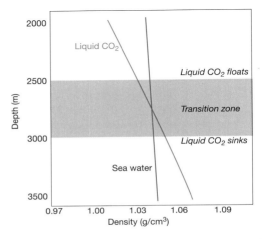

Figure 6.8 CO_2 sea water phase diagram. CO_2 is stable in the liquid phase when temperature and pressure (increasing with ocean depth) fall in the region below the blue curve; a gas phase is stable under conditions above the blue dashed line. In contact with sea water and at temperature and pressure in the shaded region, CO_2 reacts with sea water to from a solid ice-like hydrate $CO_2 \cdot 6H_2O$. CO_2 will dissolve in sea water that is not saturated with CO_2. The red line shows how temperature varies with depth at a site off the coast of California; liquid and hydrated CO_2 can exist below about 400 m (Brewer et al., 2004).

Figure 6.9 Shallower than 2500 m, liquid CO_2 is less dense than sea water, and thus tends to float upward. Deeper than 3000 m, liquid CO_2 is denser than sea water, and thus tends to sink downwards. Between these two depths, the behaviour can vary with location (depending mostly on temperature) and CO_2 can be neutrally buoyant (neither rises nor falls). Conditions shown for the northwest Atlantic Ocean.

Figure 6.10 Liquid CO_2 released at 3600 metres initially forms a liquid CO_2 pool on the sea floor in a small deep ocean experiment (upper picture). In time, released liquid CO_2 reacts with sea water to form a solid CO_2 hydrate in a similar pool (lower picture).

3000 m, liquid CO_2 is denser than the surrounding sea water and sinks. CO_2 nozzles could be engineered to produce large droplets that would sink to the sea floor or small droplets that would dissolve in the sea water before contacting the sea floor. Natural ocean mixing and droplet motion are expected to prevent concentrations of dissolved CO_2 from approaching saturation, except near liquid CO_2 that has been intentionally placed in topographic depressions on the sea floor.

Solid. Solid CO_2 is denser than sea water and thus would tend to sink. Solid CO_2 surfaces would dissolve in sea water at a speed of about 0.2 cm hr^{-1} (inferred from Aya *et al.*, 1997). Thus small quantities of solid CO_2 would dissolve completely before reaching the sea floor; large masses could potentially reach the sea floor before complete dissolution.

Hydrate. CO_2 hydrate is a form of CO_2 in which a cage of water molecules surrounds each molecule of CO_2. It can form in average ocean waters below about 400 m depth. A fully formed crystalline CO_2 hydrate is denser than sea water and will sink (Aya *et al.*, 2003). The surface of this mass would dissolve at a speed similar to that of solid CO_2, about 0.2 cm hr^{-1} (0.47 to 0.60 μm s^{-1}; Rehder *et al.*, 2004; Teng *et al.*, 1999), and thus droplets could be produced that either dissolve completely in the sea water or sink to the sea floor. Pure CO_2 hydrate is a hard crystalline solid and will not flow through a pipe; however a paste-like composite of hydrate and sea water may be extruded (Tsouris *et al.*, 2004), and this will have a dissolution rate intermediate between those of CO_2 droplets and a pure CO_2 hydrate.

6.2.1.4 Behaviour of injected CO_2 in the near field: CO_2-rich plumes

As it leaves the near field, CO_2 enriched water will reside at a depth determined by its density. The oceans are generally stably stratified with density increasing with depth. Parcels of water tend to move upward or downward until they reach water of the same density, then there are no buoyancy forces to induce further motion.

The dynamics of CO_2-rich plumes determine both the depth at which the CO_2 leaves the near-field environment and the amount of initial dilution (and consequently the amount of pH change). When CO_2 is released in any form into seawater, the CO_2 can move upward or downward depending on whether the CO_2 is less or more dense than the surrounding seawater. Drag forces transfer momentum from the CO_2 droplets to the surrounding water column producing motion in the adjacent water, initially in the direction of droplet motion. Simultaneously, the CO_2 dissolves into the surrounding water, making the surrounding water denser and more likely to sink. As the CO_2-enriched water moves, it mixes with surrounding water that is less enriched in CO_2, leading to additional dilution and diminishing the density contrast between the CO_2-enriched water and the surrounding water.

CO_2 releases could be engineered to produce CO_2 plumes with different characteristics (Chen *et al.*, 2003; Sato and Sato, 2002; Alendal and Drange, 2001; Crounse *et al.*, 2001; Drange *et al.*, 2001; Figure 6.11). Modelling studies indicate that

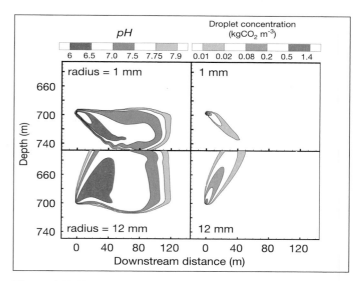

Figure 6.11 Simulated CO_2 enriched sea water plumes (left panels; indicated by pH) and CO_2 droplet plumes (right panels; indicated by kgCO$_2$ m^{-3}) created by injecting 1 cm and 12 cm liquid CO_2 droplets (top and bottom panels, respectively) into the ocean from fixed nozzles (elapsed time is 30 min; injection rate is 1.0 kgCO$_2$ s^{-1}; ocean current speed is 5 cm s^{-1}; Alendal and Drange, 2001). By varying droplet size, the plume can be made to sink (top panels) or rise (bottom panels).

releases of small droplets at slow rates produce smaller plumes than release of large droplets at rapid rates. Where CO_2 is denser than seawater, larger droplet sizes would allow the CO_2 to sink more deeply. CO_2 injected at intermediate depths could increase the density of CO_2-enriched sea water sufficiently to generate a sinking plume that would carry the CO_2 into the deep ocean (Liro *et al.*, 1992; Haugan and Drange, 1992). Apparent coriolis forces would operate on such a plume, turning it towards the right in the Northern Hemisphere and towards the left in the Southern Hemisphere (Alendal *et al.*, 1994). The channelling effects of submarine canyons or other topographic features could help steer dense plumes to greater depth with minimal dilution (Adams *et al.*, 1995).

6.2.1.5 Behaviour of injected CO_2 in the far field

The far field is defined as the region in which the concentration of added CO_2 is low enough such that the resulting density increase does not significantly affect transport, and thus CO_2 may be considered a passive tracer in the ocean. Typically, this would apply within a few kilometres of an injection point in midwater, but if CO_2 is released at the sea floor and guided along topography, concentration may remain high and influence transport for several tens of kilometres. CO_2 is transported by ocean currents and undergoes further mixing and dilution with other water masses (Alendal and Drange, 2001). Most of this mixing and transport occurs along surfaces of nearly constant density, because buoyancy forces inhibit vertical mixing in a stratified fluid. Over time, a release of CO_2 becomes increasingly diluted but affects ever greater volumes of water.

The concept of ocean injection from a moving ship towing a trailing pipe was developed in order to minimize the local

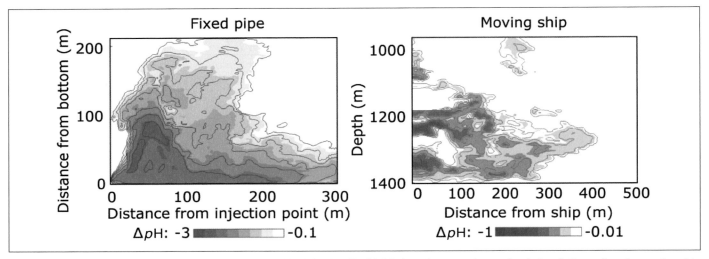

Figure 6.12 Simulated plumes (Chen *et al.*, 2005) created by injecting liquid CO_2 into the ocean from a fixed pipe (left panel) and a moving ship (right panel) at a rate of 100 kg s^{-1} (roughly equal to the CO_2 from a 500 MWe coal-fired power plant). Left panel: injection at 875 m depth (12 m from the sea floor) with an ocean current speed of 2.3 cm s^{-1}. Right panel: injection at 1340 m depth from a ship moving at a speed of 3 m s^{-1}. Note difference in *p*H scales; maximum *p*H perturbations are smaller in the moving ship simulation.

environmental impacts by accelerating the dissolution and dispersion of injected liquid CO_2 (Ozaki, 1997; Minamiura *et al.*, 2004). A moving ship could be used to produce a sea water plume with relatively dilute initial CO_2 concentrations (Figures 6.12 and 6.13). In the upper ocean where CO_2 is less dense than seawater, nozzles engineered to produce mm-scale droplets would generate CO_2 plumes that would rise less than 100 m.

Ocean general circulation models have been used to predict changes in ocean chemistry resulting from the dispersion of

injected CO_2 for hypothetical examples of ocean storage (e.g., Orr, 2004). Wickett *et al.* (2003) estimated that injection into the deep ocean at a rate of 0.37 GtCO_2 yr^{-1} (= 0.1 GtC yr^{-1}) for 100 years would produce a ΔpH < -0.3 over a volume of sea water equivalent to 0.01% or less of total ocean volume (Figure 6.14). In this example, for each GtCO_2 released to the deep ocean, less than about 0.0001%, 0.001% and 0.01% of

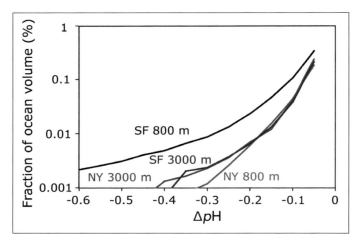

Figure 6.14 Estimated volume of *p*H perturbations at basin scale (Wickett *et al.*, 2003). Simulated fraction of global ocean volume with a ΔpH less than the amount shown on the horizontal axis, after 100 years of simulated injection at a rate of 0.37 GtCO_2 yr^{-1} (= 0.1 GtC yr^{-1}) at each of four different points (two different depths near New York City and San Francisco). Model results indicate, for example, that injecting CO_2 at this rate at a single location for 100 years could be expected to produce a volume of sea water with a ΔpH < -0.3 units in 0.01% or less of total ocean volume (0.01% of the ocean is roughly 10^5 km^3). As with other simulations of direct CO_2 injection in the ocean, results for the upper ocean (e.g., 800 m) tend to be more site-specific than are results for the deep ocean (e.g., 3000 m).

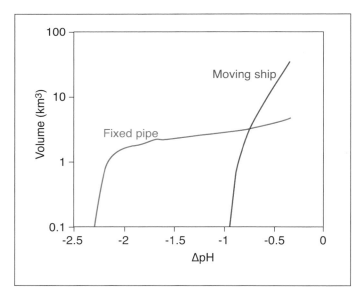

Figure 6.13 Volume of water with a ΔpH less than the value shown on the horizontal axis for the simulations shown in Figure 6.12 corresponding to CO_2 releases from a 500 MW$_e$ power plant. The fixed pipe simulation produces a region with ΔpH < -1, however, the moving ship disperses the CO_2 more widely, largely avoiding *p*H changes of this magnitude.

the ocean volume has ΔpH of less than –0.3, –0.2, and –0.1 pH units respectively. Caldeira and Wickett (2005) predicted volumes of water undergoing a range of pH changes for several atmospheric emission and carbon stabilization pathways, including pathways in which direct injection of CO_2 into the deep ocean was assumed to provide either 10% or 100% of the total atmospheric CO_2 mitigation effort needed to stabilize atmospheric CO_2 according to the WRE550 pathway. This assumed a CO_2 production scenario in which all known fossil-fuel resources were ultimately combusted. Simulations in which ocean injection provided 10% of the total mitigation effort, resulted in significant changes in ocean pH in year 2100 over roughly 1% of the ocean volume (Figure 6.15). By year 2300, injection rates have slowed but previously injected carbon has spread through much of the ocean resulting in an additional 0.1 pH unit reduction in ocean pH over most of the ocean volume compared to WRE550.

6.2.1.6 *Behaviour of CO_2 lakes on the sea floor*
Long-term storage of carbon dioxide might be more effective if CO_2 were stored on the sea floor in liquid or hydrate form below 3000 metres, where CO_2 is denser than sea water (Box 6.2; Ohsumi, 1995; Shindo *et al.*, 1995). Liquid carbon dioxide could be introduced at depth to form a lake of CO_2 on the sea floor (Ohsumi, 1993). Alternatively, CO_2 hydrate could be created in an apparatus designed to produce a hydrate pile or pool on the sea floor (Saji *et al.*, 1992). To date, the concept of CO_2 lakes on the sea floor has been investigated only in the laboratory, in small-scale (tens of litres) *in-situ* experiments and

in numerical models. Larger-scale *in-situ* experiments have not yet been carried out.

Liquid or hydrate deposition of CO_2 on the sea floor could increase isolation, however in the absence of a physical barrier the CO_2 would dissolve into the overlying water (Mori and Mochizuki, 1998; Haugan and Alendal, 2005). In this aspect, most sea floor deposition proposals can be viewed as a means of 'time-delayed release' of CO_2 into the ocean. Thus, many issues relevant to sea floor options, especially the far-field behaviour, are discussed in sections relating to CO_2 release into the water column (e.g., Section 6.2.1.5).

CO_2 released onto the sea floor deeper than 3 km is denser than surrounding sea water and is expected to fill topographic depressions, accumulating as a lake of CO_2 over which a thin hydrate layer would form. This hydrate layer would retard dissolution, but it would not insulate the lake from the overlying water. The hydrate would dissolve into the overlying water (or sink to the bottom of the CO_2 lake), but the hydrate layer would be continuously renewed through the formation of new crystals (Mori, 1998). Laboratory experiments (Aya *et al.*, 1995) and small deep ocean experiments (Brewer *et al.*, 1999) show that deep-sea storage of CO_2 would lead to CO_2 hydrate formation (and subsequent dissolution).

Predictions of the fate of large-scale CO_2 lakes rely on numerical simulations because no large-scale field experiments have yet been performed. For a CO_2 lake with an initial depth of 50 m, the time of complete dissolution varies from 30 to 400 years depending on the local ocean and sea floor environment. The time to dissolve a CO_2 lake depends on its depth, complex

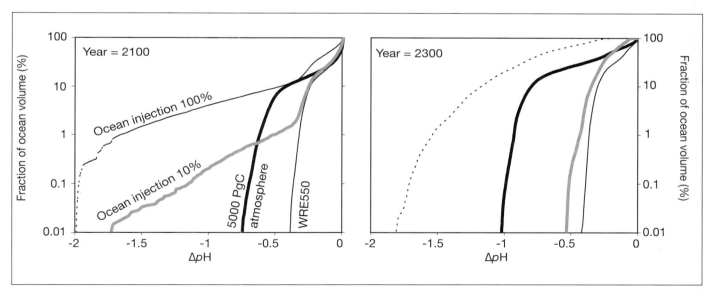

Figure 6.15 Estimated volume of pH perturbations at global scale for hypothetical examples in which injection of CO_2 into the ocean interior provides 100% or 10% of the mitigation effort needed to move from a logistic emissions curve cumulatively releasing 18,000 $GtCO_2$ (=5000 GtC) to emissions consistent with atmospheric CO_2 stabilization at 550 ppm according to the WRE550 pathway (Wigley *et al.*, 1996). The curves show the simulated fraction of ocean volume with a pH reduction greater than the amount shown on the horizontal axis. For the 10% case, in year 2100, injection rates are high and about 1% of the ocean volume has significant pH reductions; in year 2300, injection rates are low, but previously injected CO_2 has decreased ocean pH by about 0.1 unit below the value produced by a WRE550 atmospheric CO_2 pathway in the absence of CO_2 release directly to the ocean (Caldeira and Wickett, 2005).

dynamics of the ocean bottom boundary layer and its turbulence characteristics, mechanism of CO_2 hydrate dissolution, and properties of CO_2 in solution (Haugan and Alendal, 2005). The lifetime of a CO_2 lake would be longest in relatively confined environments, such as might be found in some trenches or depressions with restricted flow (Ohgaki and Akano, 1992). Strong flows have been observed in trenches (Nakashiki, 1997). Nevertheless, simulation of CO_2 storage in a deep trench (Kobayashi, 2003) indicates that the bottom topography can weaken vertical momentum and mass transfer, slowing the CO_2 dissolution rate. In a quiescent environment, transport would be dominated by diffusion. Double-diffusion in the presence of strong stratification may produce long lake lifetimes. In contrast, the flow of sea water across the lake surface would increase mass transfer and dissolution. For example, CO_2 lake lifetimes of >10,000 yr for a 50 m thick lake can be estimated from the dissolution rate of 0.44 cm yr^{-1} for a quiescent, purely diffusive system (Ohsumi, 1997). Fer and Haugan (2003) found that a mean horizontal velocity of 0.05 m s^{-1} would cause the CO_2 lake to dissolve >25 times more rapidly (12 cm yr^{-1}). Furthermore, they found that an ocean bottom storm with a horizontal velocity of 0.20 m s^{-1} could increase the dissolution rate to 170 cm yr^{-1}.

6.2.2 *CO_2 storage by dissolution of carbonate minerals*

Over thousands of years, increased sea water acidity resulting from CO_2 addition will be largely neutralized by the slow natural dissolution of carbonate minerals in sea-floor sediments and on land. This neutralization allows the ocean to absorb more CO_2 from the atmosphere with less of a change in ocean pH, carbonate ion concentration, and pCO_2 (Archer *et al.*, 1997, 1998). Various approaches have been proposed to accelerate carbonate neutralization, and thereby store CO_2 in the oceans by promoting the dissolution of carbonate minerals[2]. These approaches (e.g., Kheshgi, 1995; Rau and Caldeira, 1999) do not entail initial separate CO_2 capture and transport steps. However, no tests of these approaches have yet been performed at sea, so inferences about enhanced ocean CO_2 storage, and effects on ocean pH are based on laboratory experiments (Morse and Mackenzie, 1990; Morse and Arvidson, 2002), calculations (Kheshgi, 1995), and models (Caldeira and Rau, 2000).

Carbonate neutralization approaches attempt to promote reaction (5) (in Box 6.1) in which limestone reacts with carbon dioxide and water to form calcium and bicarbonate ions in solution. Accounting for speciation of dissolved inorganic carbon in sea water (Kheshgi, 1995), for each mole of $CaCO_3$ dissolved there would be 0.8 mole of additional CO_2 stored in sea water in equilibrium with fixed CO_2 partial pressure (i.e., about 2.8 tonnes of limestone per tonne CO_2). Adding alkalinity

to the ocean would increase ocean carbon storage, both in the near term and on millennial time scales (Kheshgi, 1995). The duration of increased ocean carbon storage would be limited by eventual $CaCO_3$ sedimentation, or reduced $CaCO_3$ sediment dissolution, which is modelled to occur through natural processes on the time scale of about 6,000 years (Archer *et al.*, 1997, 1998).

Carbonate minerals have been proposed as the primary source of alkalinity for neutralization of CO_2 acidity (Kheshgi 1995; Rau and Caldeira, 1999). There have been many experiments and observations related to the kinetics of carbonate mineral dissolution and precipitation, both in fresh water and in sea water (Morse and Mackenzie, 1990; Morse and Arvidson, 2002). Carbonate minerals and other alkaline compounds that dissolve readily in surface sea water (such as Na_2CO_3), however, have not been found in sufficient quantities to store carbon in the ocean on scales comparable to fossil CO_2 emissions (Kheshgi, 1995). Carbonate minerals that are abundant do not dissolve in surface ocean waters. Surface ocean waters are typically oversaturated with respect to carbonate minerals (Broecker and Peng, 1982; Emerson and Archer, 1990; Archer, 1996), but carbonate minerals typically do not precipitate in sea water due to kinetic inhibitions (Morse and Mackenzie, 1990).

To circumvent the problem of oversaturated surface waters, Kheshgi (1995) considered promoting reaction (5) by calcining limestone to form CaO, which is readily soluble. If the energy for the calcining step was provided by a CO_2-emission-free source, and the CO_2 released from $CaCO_3$ were captured and stored (e.g., in a geologic formation), then this process would store 1.8 mole CO_2 per mole CaO introduced into the ocean. If the CO_2 from the calcining step were not stored, then a net 0.8 mole CO_2 would be stored per mole CaO. However, if coal without CO_2 capture were used to provide the energy for calcination, and the CO_2 produced in calcining was not captured, only 0.4 mole CO_2 would be stored net per mole lime (CaO) to the ocean, assuming existing high-efficiency kilns (Kheshgi, 1995). This approach would increase the ocean sink of CO_2, and does not need to be connected to a concentrated CO_2 source or require transport to the deep sea. Such a process would, however, need to avoid rapid re-precipitation of $CaCO_3$, a critical issue yet to be addressed.

Rau and Caldeira (1999) proposed extraction of CO_2 from flue gas via reaction with crushed limestone and seawater. Exhaust gases from coal-fired power plants typically have 15,000 ppmv of CO_2 – over 400 times that of ambient air. A carbonic acid solution formed by contacting sea water with flue gases would accelerate the dissolution of calcite, aragonite, dolomite, limestone, and other carbonate-containing minerals, especially if minerals were crushed to increase reactive surface area. The solution of, for example, Ca^{2+} and dissolved inorganic carbon (primarily in the form of HCO_3^-) in sea water could then be released back into the ocean, where it would be diluted by additional seawater. Caldeira and Rau (2000) estimate that dilution of one part effluent from a carbonate neutralization reactor with 100 parts ambient sea water would result, after equilibration with the atmosphere, in a 10% increase in the

---

[2] This approach is fundamentally different than the carbonate mineralization approach assessed in Chapter 7. In that approach CO_2 is stored by reacting it with non-carbonate minerals to form carbonate minerals. In this approach, carbonate minerals are dissolved in the ocean, thereby increasing ocean alkalinity and increasing ocean storage of CO_2. This approach could also make use of non-carbonate minerals, if their dissolution would increase ocean alkalinity.

calcite saturation state, which they contend would not induce precipitation. This approach does not rely on deep-sea release, avoiding the need for energy to separate, transport and inject CO_2 into the deep ocean. The wastewater generated by this carbonate-neutralization approach has been conjectured to be relatively benign (Rau and Caldeira, 1999). For example, the addition of calcium bicarbonate, the primary constituent of the effluent, has been observed to promote coral growth (Marubini and Thake, 1999). This approach will not remove all the CO_2 from a gas stream, because excess CO_2 is required to produce a solution that is corrosive to carbonate minerals. If greater CO_2 removal were required, this approach could be combined with other techniques of CO_2 capture and storage.

Process wastewater could be engineered to contain different ratios of added carbon and calcium, and different ratios of flue gas CO_2 to dissolved limestone (Caldeira and Wickett, 2005). Processes involving greater amounts of limestone dissolution per mole CO_2 added lead to a greater CO_2 fraction being retained. The effluent from a carbonate-dissolution reactor could have the same pH, pCO_2, or $[CO_3^{2-}]$ as ambient seawater, although processing costs may be reduced by allowing effluent composition to vary from these values (Caldeira and Rau, 2000). Elevation in Ca^{2+} and bicarbonate content from this approach is anticipated to be small relative to the already existing concentrations in sea water (Caldeira and Rau, 2000), but effects of the new physicochemical equilibria on physiological performance are unknown. Neutralization of carbon acidity by dissolution of carbonate minerals could reduce impacts on marine ecosystems associated with pH and CO_3^{2-} decline (Section 6.7).

Carbonate neutralization approaches require large amounts of carbonate minerals. Sedimentary carbonates are abundant with estimates of 5×10^{17} tonnes (Berner *et al.*, 1983), roughly 10,000 times greater than the mass of fossil-fuel carbon. Nevertheless, up to about 1.5 mole of carbonate mineral must be dissolved for each mole of anthropogenic CO_2 permanently stored in the ocean (Caldeira and Rau, 2000); therefore, the mass of $CaCO_3$ used would be up to 3.5 times the mass of CO_2 stored. Worldwide, 3 Gt $CaCO_3$ is mined annually (Kheshgi, 1995). Thus, large-scale deployment of carbonate neutralization approaches would require greatly expanded mining and transport of limestone and attendant environmental impacts. In addition, impurities in dissolved carbonate minerals may cause deleterious effects and have yet to be studied.

6.2.3 *Other ocean storage approaches*

Solid hydrate. Water reacts with concentrated CO_2 to form a solid hydrate ($CO_2 \cdot 6H_2O$) under typical ocean conditions at quite modest depths (Løken and Austvik, 1993; Holdren and Baldwin, 2001). Rehder *et al.* (2004) showed that the hydrate dissolves rapidly into the relatively dilute ocean waters. The density of pure CO_2 hydrate is greater than seawater, and this has led to efforts to create a sinking plume of released CO_2 in the ocean water column. Pure CO_2 hydrate is a hard crystalline solid and thus will not flow through a pipe, and so some form of

CO_2 slurry is required for flow assurance (Tsouris *et al.*, 2004).

Water-$CaCO_3$-CO_2 emulsion. Mineral carbonate could be used to physically emulsify and entrain CO_2 injected in sea water (Swett *et al.* 2005); a 1:1 CO_2:$CaCO_3$ emulsion of CO_2 in water could be stabilized by pulverized limestone ($CaCO_3$). The emulsion plume would have a bulk density of 40% greater than that of seawater. Because the emulsion plume is heavier than seawater, the $CaCO_3$ coated CO_2 slurries may sink all the way to the ocean floor. It has been suggested that the emulsion plume would have a pH that is at least 2 units higher than would a plume of liquid CO_2. Carbonate minerals could be mined on land, and then crushed, or fine-grained lime mud could be extracted from the sea floor. These fine-grain carbonate particles could be suspended in sea water upstream from the CO_2-rich plume emanating from the direct CO_2 injection site. The suspended carbonate minerals could then be transported with the ambient sea water into the plume, where the minerals could dissolve, increasing ocean CO_2 storage effectiveness and diminishing the pH impacts of direct injection.

Emplacement in carbonate sediments. Murray *et al.* (1997) have suggested emplacement of CO_2 into carbonate sediments on the sea floor. Insofar as this CO_2 remained isolated from the ocean, this could be categorized as a form of geological storage (Chapter 5).

Dry ice torpedoes. CO_2 could be released from a ship as dry ice at the ocean surface (Steinberg,1985). One costly method is to produce solid CO_2 blocks (Murray *et al.*, 1996). With a density of 1.5 t m^{-3}, these blocks would sink rapidly to the sea floor and could potentially penetrate into the sea floor sediment.

Direct flue-gas injection. Another proposal is to take a power plant flue gas, and pump it directly into the deep ocean without any separation of CO_2 from the flue gas, however costs of compression are likely to render this approach infeasible.

6.3 Capacity and fractions retained

6.3.1 *Capacity*

The physical capacity for storage of CO_2 in the ocean is large relative to fossil-fuel resources. The degree to which this capacity will be utilized may be based on factors such as cost, equilibrium pCO_2, and environmental consequences.

Storage capacity for CO_2 in the ocean can be defined relative to an atmospheric CO_2 stabilization concentration. For example, roughly 2,300 to 10,700 GtCO_2 (above the natural pre-industrial background) would be added to the ocean in equilibrium with atmospheric CO_2 stabilization concentrations, ranging from 350 ppmv to 1000 ppmv, regardless of whether the CO_2 is initially released to the ocean or the atmosphere (Table 6.1, Figure 6.3; Kheshgi *et al.*, 2005; Sorai and Ohsumi, 2005). The capacity of the ocean for CO_2 storage could be increased with the addition of alkalinity to the ocean (e.g., dissolved limestone).

6.3.2 *Measures of fraction retained*

Effectiveness of ocean CO_2 storage has been reported in a variety of ways. These different ways of reporting result in very different numerical values (Box 6.3).

Over several centuries, CO_2 released to the deep ocean would be transported to the ocean surface and interact with the atmosphere. The CO_2-enriched water would then exchange CO_2 with the atmosphere as it approaches chemical equilibrium. In this chemical equilibrium, most of the injected CO_2 remains in the ocean even though it is no longer isolated from the atmosphere (Table 6.1; Figure 6.3). CO_2 that has interacted with the atmosphere is considered to be part of the natural carbon cycle, much in the way that CO_2 released directly to the atmosphere is considered to be part of the natural carbon cycle. Such CO_2 cannot be considered to be isolated from the atmosphere in a way that can be attributable to an ocean storage project.

Loss of isolation of injected CO_2 does not mean loss of all of the injected CO_2 to the atmosphere. In chemical equilibrium with an atmosphere containing 280 ppm CO_2, about 85% of any carbon injected would remain the ocean. If atmospheric CO_2 partial pressures were to approach 1000 ppm, about 66% of the injected CO_2 would remain in the ocean after equilibration with the atmosphere (Table 6.1). Thus, roughly 1/5 to 1/3 of the CO_2 injected into the ocean will eventually reside in the atmosphere, with this airborne fraction depending on the long-term atmosphere-ocean CO_2 equilibrium (Kheshgi, 1995, 2004b). The airborne fraction is the appropriate measure to quantify the effect of ocean storage on atmospheric composition.

6.3.3   Estimation of fraction retained from ocean observations

Observations of radiocarbon, CFCs, and other tracers indicate the degree of isolation of the deep sea from the atmosphere (Prentice *et al.*, 2001). Radiocarbon is absorbed by the oceans from the atmosphere and is transported to the deep-sea, undergoing radioactive decay as it ages. Radiocarbon age (Figure 6.16) is not a perfect indicator of time since a water

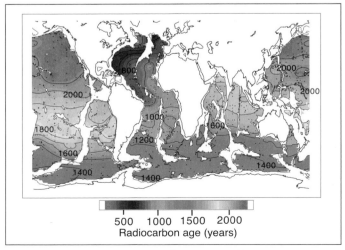

Figure 6.16 Map of radiocarbon (^{14}C) age at 3500 m (Matsumoto and Key, 2004).

parcel last contacted the atmosphere because of incomplete equilibration with the atmosphere (Orr, 2004). Taking this partial equilibration into account, the age of North Pacific deep water is estimated to be in the range of 700 to 1000 years. Other basins, such as the North Atlantic, have characteristic overturning times of 300 years or more. This data suggests that, generally, carbon injected in the deep ocean would equilibrate with the atmosphere over a time scale of 300 to 1000 years.

6.3.4   Estimation of fraction retained from model results

Ocean models have been used to predict the isolation of injected CO_2 from the atmosphere. Many models are calibrated using ocean radiocarbon data, so model-based estimates of retention of injected CO_2 are not completely independent of the estimates based more directly on observations (Section 6.3.3).

A wide number of studies have used three-dimensional ocean general circulation models to study retention of CO_2 injected into the ocean water column (Bacastow and Stegen, 1991; Bacastow et al., 1997; Nakashiki and Ohsumi, 1997; Dewey *et al.*, 1997, 1999; Archer *et al.*, 1998; Xu *et al.*, 1999; Orr, 2004; Hill *et al.*, 2004). These modelling studies generally confirm inferences based on simpler models and considerations of ocean chemistry and radiocarbon decay rates. In ocean general circulation simulations performed by seven modelling groups (Orr, 2004), CO_2 was injected for 100 years at each of seven different locations and at three different depths. Model results indicate that deeper injections will be isolated from the atmosphere for longer durations. Figure 6.17 shows the effect of injection depth on retained fraction for the mean of seven ocean sites (Orr, 2004). Ranges of model results indicate some uncertainty in forecasts of isolation of CO_2 released to the deep ocean, although for all models the time extent of CO_2 isolation is longer for deeper CO_2 release, and isolation is nearly complete for 100 years following CO_2 release at 3000 m depth (Figure 6.18 and 6.19). However, present-day models disagree as to the degassing time scale for particular locations (Figure 6.19). There seems to be no simple and robust correlation of CO_2 retention other than depth of injection (Caldeira *et al.*, 2002), however, there is some indication that the mean fraction retained for stored carbon is greater in the Pacific Ocean than the Atlantic Ocean, but not all models agree on this. Model results indicate that for injection at 1500 m depth, the time scale of the partial CO_2 degassing is sensitive to the location of the injection, but at 3000 m, results are relatively insensitive to injection location. Model results have been found to be sensitive to differences in numerical schemes and model parameterizations (Mignone *et al.*, 2004).

6.4   Site selection

6.4.1   Background

There are no published papers specifically on site selection for intentional ocean storage of CO_2; hence, we can discuss only general factors that might be considered when selecting sites for

Box 6.3 Measures of the fraction of CO_2 retained in storage

Different measures have been used to describe how effective intentional storage of carbon dioxide in the ocean is to mitigate climate change (Mueller *et al.*, 2004). Here, we illustrate several of these measures using schematic model results reported by Herzog *et al.* (2003) for injection of CO_2 at three different depths (Figure 6.17).

Fraction retained (see Chapter 1) is the fraction of the cumulative amount of injected CO_2 that is retained in the storage reservoir over a specified period of time, and thereby does not have the opportunity to affect atmospheric CO_2 concentration (Mignone *et al.*, 2004). The retained fraction approaches zero (Figure 6.17) over long times, indicating that nearly all injected CO_2 will interact with the atmosphere (although a small amount would interact first with carbonate sediments).

Airborne Fraction is the fraction of released CO_2 that adds to atmospheric CO_2 content (Kheshgi and Archer, 2004). For atmospheric release, airborne fraction is initially one and decays to roughly 0.2 (depending on atmospheric CO_2 concentration) as the added CO_2 is mixed throughout the ocean, and decays further to about 0.08 as CO_2 reacts with sediments (Archer *et al.*, 1997). For deep-sea release, airborne fraction is initially zero and then approaches that of atmospheric release. Note that the asymptotic airborne fraction depends on the concentration of CO_2 of surface waters (Figure 6.3).

Fraction retained is used throughout this report to indicate how long the CO_2 is stored. In addition the following measures can be used to compare the effectiveness of ocean carbon storage with other options, for example:

• The **Net Present Value** (NPV) approach (Herzog et al., 2003) considers temporary storage to be equivalent to delayed emission of CO_2 to the atmosphere. The value of delaying CO_2 emissions depends on the future costs of CO_2 emission and economic discount rates. There is economic value to temporary storage (i.e., delayed emission) if the cost of CO_2 emissions increases at a rate that is less than the discount rate (Herzog et al., 2003).

• The **Global-Warming Potential** (GWP) is a measure defined by the IPCC to compare the climatic effect of different greenhouse-gas emissions. It is computed by accumulating the radiative climate forcing of a greenhouse-gas emission over a specified time horizon. This measure has been applied to compare the radiative forcing from oceanic and atmospheric releases of carbon dioxide (Kheshgi et al., 1994, Ramaswamy et al., 2001). Haugan and Joos (2004) propose a modification to the GWP approach that compares the climate effects of the airborne fraction of a CO_2 release to the ocean with those from a release to the atmosphere. Table 6.2 compares these measures for results from a schematic model at three depths.

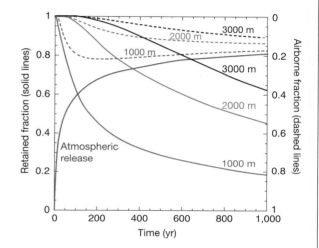

Figure 6.17 Fraction of carbon in the ocean from injection at three different depths and the atmosphere illustrated with results from a schematic model (Herzog et al., 2003). Calculations assume a background 280 ppm of CO_2 in the atmosphere.

Table 6.2 Evaluation of measures described in the text illustrated using schematic model results shown in Figure 6.17. For the Net Present Value measure, the percentage represents the discount rate minus the rate of increase in the cost of CO_2 emission. (If these are equal, the Net Present Value of temporary carbon storage is zero) Two significant digits shown for illustration exceed the accuracy of model results.

Measure		Atmospheric release	Injection depth		
			1000 m	**2000 m**	**3000 m**
Effective	at 20 years	0	0.96	1.00	1.00
Retained	at 100 years	0	0.63	0.97	1.00
Fraction	at 500 years	0	0.28	0.65	0.85
Airborne	at 20 years	0.61	0.03	6×10^{-6}	7×10^{-10}
Fraction	at 100 years	0.40	0.19	0.02	9×10^{-4}
	at 500 years	0.24	0.20	0.12	0.06
Net Present	5% per year	0	0.95	1.00	1.00
Value (constant	1% per year	0	0.72	0.95	0.99
emissions cost)	0.2% per year	0	0.41	0.72	0.85
Global	20 year horizon	1	0.01	1×10^{-6}	6×10^{-10}
Warming	100 year horizon	1	0.21	0.01	4×10^{-4}
Potential	500 year horizon	1	0.56	0.20	0.06

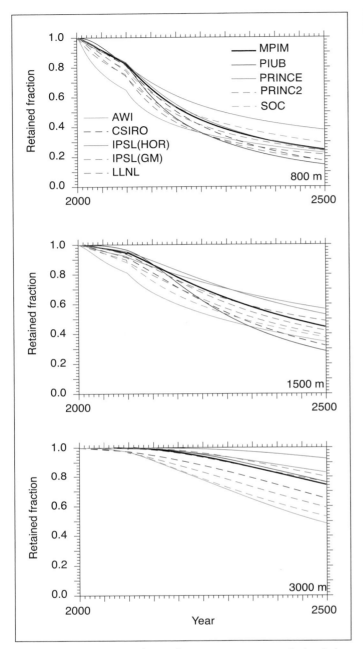

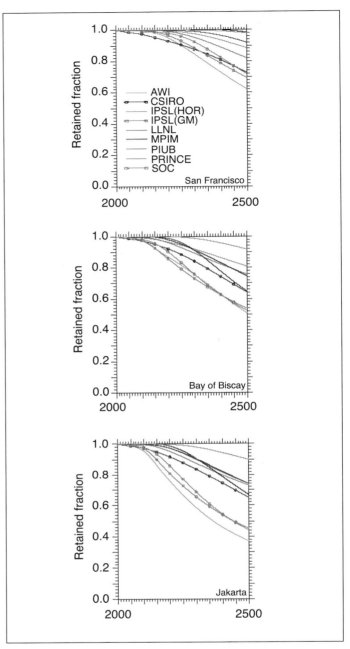

Figure 6.18 Results are shown for seven ocean general circulation models at three different depths averaged over seven injection locations (Orr, 2004). The percentage efficiency shown is the retained fraction for an injection at a constant rate from 2000 to 2100. Models agree that deeper injection isolates CO_2 from the atmosphere longer than shallower injection. For release at 3000 m, most of the added carbon was still isolated from the atmosphere at the end of the 500 year simulations.

Figure 6.19 Comparison of storage results for three injection locations (at 3000 m depth) in ten ocean model simulations (Orr, 2004). Models differ on predictions of CO_2 fraction retained for release in different oceans.

6.4.2   Water column release

ocean storage. Among these considerations are environmental consequences, costs, safety, and international issues (including cross border transport). Because environmental consequences, costs, and social and political issues are addressed in other parts of this report, here we briefly consider site selection factors that enhance the fraction retained or reduce the costs.

Large point sources of CO_2 located near deep water would generally be the most cost effective settings in which to carry out direct CO_2 injection (Figure 6.21; Section 6.9). While models indicate that site-specific differences exist, they do not yet agree on the ranking of potential sites for effectiveness of direct injection CO_2 operations (Orr, 2004).

6.4.3 CO_2 lakes on the sea floor

CO_2 lakes must be on the sea floor at a depth below 3000 m (Figures 6.20 and 6.21), because the liquid CO_2 must be denser than surrounding sea water (Box 6.2).

These ocean general circulation model calculations did not consider interactions with $CaCO_3$ sediments or marine biota. Increased CO_2 concentrations in the ocean promote dissolution of $CaCO_3$ sediments, which would tend to increase predicted CO_2 retention. This has been modelled for the deep sea with results of greater retention for release in the Atlantic because of high $CaCO_3$ inventory in Atlantic sediments (Archer et al., 1998).

Preliminary numerical simulations of ocean CO_2 injection predict increased oceanic retention of injected CO_2 with concurrent global warming due to weaker overturning and a more stratified ocean (Jain and Cao, 2005). Some evidence indicates recent increases in stratification in all major ocean basins (e.g., Joos, 2003; McPhaden and Zhang, 2002; Palmer et al., 2004; Stramma et al., 2004).

6.4.4   *Limestone neutralization*

The amounts of sea water and limestone required to neutralize the acidity of added CO_2 indicate that limestone neutralization would be most suitable for CO_2 point sources located near both the ocean and large deposits of limestone (Rau and Caldeira, 1999).

6.5   Injection technology and operations

6.5.1   *Background*

The development of ocean storage technology is generally at a conceptual stage; thus, we will only discuss general principles. There has been limited engineering analysis and experimental studies of these conceptual technologies for ocean storage (Nihous, 1997), and no field-testing. No operational experience exists. Various technology concepts have been proposed to improve isolation from the atmosphere or diminish environmental consequences of CO_2 injected into the ocean. Further research and development would be needed to make technologies available, but no major technical barriers are apparent.

6.5.2   *Water column release*

Dispersal of liquid CO_2 at a depth of 1000 m or deeper is technologically feasible. Since liquid CO_2 may be relatively easily transported to appropriate depths, the preferred release mode is thought at this time to be as a liquid or dense gas phase (achieved by compression beyond its critical point, 72.8 bar at 31°C). The pipes that would carry this CO_2 to the deep ocean would be similar to the pipes that have been used commercially on land to transport CO_2 for use in CO_2 enhanced oil recovery projects (Ozaki *et al.*, 1997). Models (Liro *et al.*, 1992, Drange and Haugan, 1992) predict that, with a properly designed diffuser, nearly all the CO_2 would dissolve in the ocean within a 100 m of the injection depth. Then, this CO_2-rich water would be diluted as it disperses, primarily horizontally along surfaces of constant density.

Water column injection schemes typically envision minimizing local changes to ocean chemistry by producing a

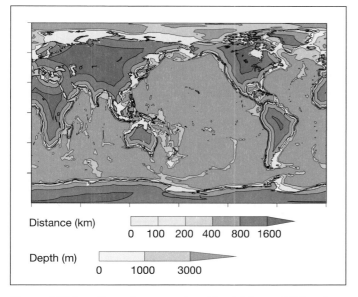

Distance (km)

0   100   200   400   800   1600

Depth (m)

0   1000   3000

Figure 6.20 Locations of ocean water at least 1 km and 3 km deep. Distance over land to water that is at least 3 km deep (Caldeira and Wickett, 2005). In general, land areas with the lightest colours would be the most-cost effective land-based settings for a CO_2-injection operation. However, each potential site would need to be evaluated prior to deployment.

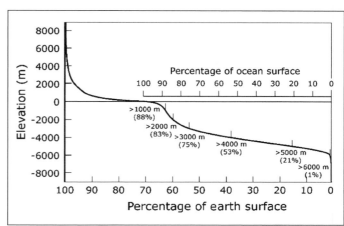

Figure 6.21 Relationship between depth and sea floor area. Flow in ocean bottom boundary layers would need to be taken into account when selecting a site for a CO_2 lake. Bottom friction and turbulence can enhance the dissolution rate and vertical transport of dissolved CO_2 and lead to a short lifetime for the lake (Section 6.2.1.6). It has been suggested that CO_2 lakes would be preferentially sited in relatively restricted depressions or in trenches on sea floor (Ohsumi, 1995).

relatively dilute initial injection through a series of diffusers or by other means. Dilution would reduce exposure of organisms to very low *p*H (very high CO_2) environments (Section 6.7).

One set of options for releasing CO_2 to the ocean involves transporting liquid CO_2 from shore to the deep ocean in a pipeline. This would not present any major new problems in design, 'according to petroleum engineers and naval architects speaking at one of the IEA Greenhouse Gas R&D Programme ocean storage workshops' (Ormerod *et al.*, 2002). The oil industry has been making great advances in undersea offshore technology, with projects routinely working at depths greater than 1000 m. The oil and the gas industry already places pipes on the bottom of the sea in depths down to 1600 m, and design studies have shown 3000 m to be technically feasible (Ormerod *et al.*, 2002). The 1 m diameter pipe would have the capacity to transport 70,000 tCO_2 day^{-1}, enough for CO_2 captured from 3 GW_e of a coal-fired electric power plant (Ormerod *et al.*, 2002). Liro *et al.* (1992) proposed injecting liquid CO_2 at a depth of about 1000 m from a manifold lying near the ocean bottom to form a rising droplet plume. Nihous *et al.* (2002) proposed injecting liquid CO_2 at a depth of below 3000 m from a manifold lying near the ocean bottom and forming a sinking droplet plume. Engineering work would need to be done to assure that, below 500 m depth, hydrates do not form inside the discharged pipe and nozzles, as this could block pipe or nozzle flow.

CO_2 could be transported by tanker for release from a stationary platform (Ozaki *et al.*, 1995) or through a towed pipe (Ozaki *et al.*, 2001). In either case, the design of CO_2 tankers would be nearly identical to those that are now used to transport liquid petroleum gas (LPG). Cooling would be used, in order to reduce pressure requirements, with design conditions of –55 degrees C and 6 bar pressure (Ormerod *et al.*, 2002). Producing a dispersed initial concentration would diminish the magnitude of the maximum *p*H excursion. This would probably involve designing for the size of the initial liquid CO_2 droplet and the turbulent mixing behind the towed pipe (Tsushima *et al.*, 2002). Diffusers could be designed so that CO_2 droplets would dissolve completely before they reach the liquid-gas phase boundary.

CO_2 hydrate is about 15% denser than sea water, so it tends to sink, dissolving into sea water over a broad depth horizon (Wannamaker and Adams, 2002). Kajishima *et al.* (1997) and Saito *et al.* (2001) investigated a proposal to create a dense CO_2-seawater mixture at a depth of between 500 and 1000 m to form a current sinking along the sloping ocean bottom. Another proposal (Tsouris *et al.*, 2004; West *et al.*, 2003) envisions releasing a sinking CO_2-hydrate/seawater slurry at between 1000 and 1500 m depth. This sinking plume would dissolve as it sinks, potentially distributing the CO_2 over kilometres of vertical distance, and achieving some fraction of the CO_2 retained in deep storage despite the initial release into intermediate waters. The production of a hydrate/seawater slurry has been experimentally demonstrated at sea (Tsouris *et al.*, 2004). Tsouris *et al.* (2004) have carried out a field experiment at 1000 m ocean depth in which rapid mixing of sea water with CO_2 in a capillary nozzle to a neutrally buoyant composite paste takes place. This would enhance ocean retention time compared

to that from creation of a buoyant plume. Aya *et al.* (2004) have shown that a rapidly sinking plume of CO_2 can be formed by release of a slurry combining cold liquid and solid CO_2 with a hydrate skin. This would effectively transfer ship released CO_2 at shallow ocean depth to the deep ocean without the cost of a long pipe. In all of these schemes the fate of the CO_2 is to be dissolved into the ocean, with increased depth of dissolution, and thus increased retention.

6.5.3   *Production of a CO_2 lake*

Nakashiki (1997) investigated several different kinds of discharge pipes that could be used from a liquid CO_2 tanker to create a CO_2 lake on the sea floor. They concluded that a 'floating discharge pipe' might be the best option because it is simpler than the alternatives and less likely to be damaged by wind and waves in storm conditions.

Aya *et al.* (2003) proposed creating a slurry of liquid CO_2 mixed with dry ice and releasing into the ocean at around 200 to 500 m depth. The dry ice is denser that the surrounding sea water and would cause the slurry to sink. An *in situ* experiment carried out off the coast of California found that a CO_2 slurry and dry ice mass with initial diameter about 8.0 cm sank approximately 50 metres within two minutes before the dry ice melted (Aya *et al.*, 2003). The initial size of CO_2 slurry and dry ice is a critical factor making it possible to sink more than 3000 m to the sea floor. To meet performance criteria, the dry ice content would be controlled with a system consisting of a main power engine, a compressor, a condenser, and some pipe systems.

6.6   Monitoring and verification

6.6.1   *Background*

Monitoring (Figure 6.22) would be done for at least two different purposes: (1) to gain specific information relating to a particular CO_2 storage operation and (2) to gain general scientific understanding. A monitoring program should attempt to quantify the mass and distribution of CO_2 from each point source and could record related biological and geochemical parameters. These same issues may relate to monitoring of potential leakages from subsea geologic storage, or for verification that such leakage does not occur. Monitoring protocols for submarine sewage disposal for example are already well established, and experience may be drawn from that.

6.6.2   *Monitoring amounts and distributions of materials released*

6.6.2.1   *Monitoring the near field*
It appears that there is no serious impediment to verifying plant compliance with likely performance standards for flow through a pipe. Once CO_2 is discharged from the pipe then the specific monitoring protocols will depend upon whether the plume is buoyant or sinking. Fixed location injections present fewer

verification difficulties than moving ship options.

For ocean injection from large point sources on land, verifying compliance involves above ground inspection of facilities for verification of flow and the CO_2 purity being consistent with environmental regulations (e.g., trace metal concentrations, etc.). For a power plant, flue gases could be monitored for flow rate and CO_2 partial pressure, thus allowing a full power plant carbon audit.

There are a variety of strategies for monitoring release of CO_2 into the ocean from fixed locations. Brewer *et al.* (2005) observed a plume of CO_2-rich sea water emanating from a small-scale experimental release at 4 km depth with an array of pH and

conductivity sensors. Measurements of ocean pH and current profiles at sufficiently high temporal resolution could be used to evaluate the rate of CO_2 release, local CO_2 accumulation and net transport away from the site (Sundfjord *et al.*, 2001). Undersea video cameras can monitor the point of release to observe CO_2 flow. The very large sound velocity contrast between liquid CO_2 (about 300 m s^{-1}) and sea water (about 1,500 m s^{-1}) offers the potential for very efficient monitoring of the liquid CO_2 phase using acoustic techniques (e.g., sonar).

The placement of CO_2 directly in a lake on the sea floor can be verified, and the quantity and loss rate determined by a combination of acoustic, pH, and velocity measurements, and by direct inspection with underwater vehicles. Undersea vehicles, tethered or autonomous, could play a prominent role in monitoring and verification. Autonomous vehicles have been developed that can be programmed to efficiently follow a variety of complex trajectories over large areas (Simonetti, 1998), but accurate pH sensing in a rapidly changing pressure and temperature field has yet to be demonstrated. Deep-sea pH monitoring from tethered vehicles has been shown to be very precise (Brewer *et al.*, 2004), and these vehicles can routinely collect precisely located samples for later analysis.

6.6.2.2 Monitoring the far field
It will be possible to monitor the far field distributions of injected CO_2 using a combination of shipboard measurements and modelling approaches. The ability to identify pH plumes in the ocean has been well demonstrated (Figure 6.23). Current analytical techniques for measuring total CO_2 in the ocean are accurate to about ±0.05% (Johnson *et al.*, 1998). Thus, measurable changes could be seen with the addition of approximately 90 tonnes of CO_2 per km^3. In other words,

Figure 6.22 Schematic of possible approaches for monitoring the injection of CO_2 into the deep ocean via a pipeline. The grey region represents a plume of high CO_2/low pH water extending from the end of the pipeline. Two sets of chemical, biological and current sensors and two underwater cameras are shown at the end of the pipeline. An array of moored sensors to monitor the direction and magnitude of the resulting plume can be seen around the pipe and are also located along the pipeline to monitor for possible leaks. A shore-based facility provides power to the sensors and for obtaining real-time data and an autonomous underwater vehicle maps the near-field distribution of the plume. A towed undulating pumping system monitors at distances of more than a few kilometres from the injection site. The towed system could provide much greater measurement accuracy and precision, but would also be able to provide measurements over large areas in a relatively short period of time. Moored systems are used to monitor the plume between mapping cruises. These moorings have surface buoys and make daily transmissions back to the monitoring facility via satellite. The very far-field distributions are monitored with hydrographic section cruises conducted every 2–5 years using standard discrete sampling approaches. These approaches provide the accuracy and precision required to detect the small CO_2 signals that add to background variations.

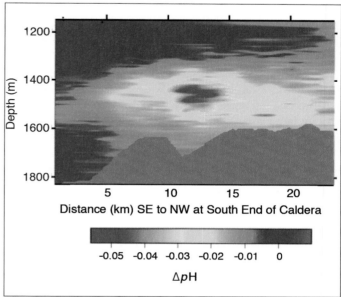

Figure 6.23 Measurements showing the ability to measure chemical effects of a natural CO_2 plume. Profiles for pH were taken in June 1999 near the Axial Volcano at 46°N 130°W, in the ocean near Portland, Oregon, United States.

1 $GtCO_2$ could be detected even if it were dispersed over 10^7 km^3 (i.e., 5000 km x 2000 km x 1 km), if the dissolved inorganic carbon concentrations in the region were mapped out with high-density surveys before the injection began.

Variability in the upper ocean mixed layer would make it difficult to directly monitor small changes in CO_2 in waters shallower than the annual maximum mixed-layer depth. Seasonal mixing from the surface can extend as deep as 800 m in some places, but is less than 200 m in most regions of the ocean. Below the seasonal mixed layer, however, periodic ship-based surveys (every 2 to 5 years) could quantify the expansion of the injection plume.

We do not have a direct means of measuring the evasion of carbon stored in the ocean to the atmosphere. In most cases of practical interest the flux of stored CO_2 from the ocean to atmosphere will be small relative to natural variability and the accuracy of our measurements. Operationally, it would be impossible to differentiate between carbon that has and has not interacted with the atmosphere. The use of prognostic models in evaluating the long-term fate of the injected CO_2 is critical for properly attributing the net storage from a particular site.

Given the natural background variability in ocean carbon concentrations, it would be extremely difficult, if not impossible, to measure CO_2 injected very far from the injection source. The attribution of a signal to a particular point source would become increasingly difficult if injection plumes from different locations began to overlap and mix. In some parts of the ocean it would be difficult to assign the rise in CO_2 to intentional ocean storage as opposed to CO_2 from atmospheric absorption.

6.6.3 Approaches and technologies for monitoring environmental effects

Techniques now being used for field experiments could be used to monitor some near field consequences of direct CO_2 injection (Section 6.7). For example, researchers (Barry *et al.*, 2004, 2005; Carman *et al.*, 2004; Thistle *et al.*, 2005) have been developing experimental means for observing the consequences of elevated CO_2 on organisms in the deep ocean. However, such experiments and studies typically look for evidence of acute toxicity in a narrow range of species (Sato, 2004; Caulfield *et al.*, 1997; Adams *et al.*, 1997; Tamburri *et al.*, 2000). Sub-lethal effects have been studied by Kurihara *et al.* (2004). Process studies, surveys of biogeochemical tracers, and ocean bottom studies could be used to evaluate changes in ecosystem structure and dynamics both before and after an injection.

It is less clear how best to monitor the health of broad reaches of the ocean interior (Sections 6.7.3 and 6.7.4). Ongoing long-term surveys of biogeochemical tracers and deep-sea biota could help to detect long-term changes in deep-sea ecology.

6.7 Environmental impacts, risks, and risk management

6.7.1 Introduction to biological impacts and risk

Overall, there is limited knowledge of deep-sea population and community structure and of deep-sea ecological interactions (Box 6.4). Thus the sensitivities of deep ocean ecosystems to intentional carbon storage and the effects on possibly unidentified goods and services that they may provide remain largely unknown.

Most ocean storage proposals seek to minimize the volume of water with high CO_2 concentrations either by diluting the CO_2 in a large volume of water or by isolating the CO_2 in a small volume (e.g., in CO_2 lakes). Nevertheless, if deployed widely, CO_2 injection strategies ultimately will produce large volumes of water with somewhat elevated CO_2 concentrations (Figure 6.15). Because large amounts of relatively pure CO_2 have never been introduced to the deep ocean in a controlled experiment, conclusions about environmental risk must be based primarily on laboratory and small-scale *in-situ* experiments and extrapolation from these experiments using conceptual and mathematical models. Natural analogues (Box 6.5) can be relevant, but differ significantly from proposed ocean engineering projects.

Compared to the surface, most of the deep sea is stable and varies little in its physiochemical factors over time (Box 6.4). The process of evolutionary selection has probably eliminated individuals apt to endure environmental perturbation. As a result, deep-sea organisms may be more sensitive to environmental disturbance than their shallow water cousins (Shirayama, 1997).

Ocean storage would occur deep in the ocean where there is virtually no light and photosynthesizing organisms are lacking, thus the following discussion primarily addresses CO_2 effects on heterotrophic organisms, mostly animals. The diverse fauna that lives in the waters and sediments of the deep ocean can be affected by ocean CO_2 storage, leading to change in ecosystem composition and functioning. Thus, the effects of CO_2 need to be identified at the level of both the individual (physiological) and the ecosystem.

As described in Section 6.2, introduction of CO_2 into the ocean either directly into sea water or as a lake on the sea floor would result in changes in dissolved CO_2 near to and down current from a discharge point. Dissolving CO_2 in sea water (Box 6.1; Table 6.3) increases the partial pressure of CO_2 (pCO_2, expressed as a ppm fraction of atmospheric pressure, equivalent to μatm), causes decreased pH (more acidic) and decreased CO_3^{2-} concentrations (less saturated). This can lead to dissolution of $CaCO_3$ in sediments or in shells of organisms. Bicarbonate (HCO_3^-) is then produced from carbonate (CO_3^{2-}).

The spatial extent of the waters with increased CO_2 content and decreased pH will depend on the amount of CO_2 released and the technology and approach used to introduce that CO_2 into the ocean. Table 6.3 shows the amount of sea water needed to dilute each tonne of CO_2 to a specified ΔpH reduction. Further dilution would reduce the fraction of ocean at one ΔpH

Box 6.4 Relevant background in biological oceanography.

Photosynthesis produces organic matter in the ocean almost exclusively in the upper 200 m where there is both light and nutrients (e.g., PO_4, NO_3, NH_4^+, Fe). Photosynthesis forms the base of a marine food chain that recycles much of the carbon and nutrients in the upper ocean. Some of this organic matter ultimately sinks to the deep ocean as particles and some of it is mixed into the deep ocean as dissolved organic matter. The flux of organic matter from the surface ocean provides most of the energy and nutrients to support the heterotrophic ecosystems of the deep ocean (Gage and Tyler, 1991). With the exception of the oxygen minimum zone and near volcanic CO_2 vents, most organisms living in the deep ocean live in low and more or less constant CO_2 levels.

At low latitudes, oxygen consumption and CO_2 release can produce a zone at around 1000 m depth characterized by low O_2 and high CO_2 concentrations, known as the 'oxygen minimum zone'. Bacteria are the primary consumers of organic matter in the deep ocean. They obtain energy predominately by consuming dissolved oxygen in reactions that oxidize organic carbon into CO_2. In the oxygen minimum layer, sea water pH may be less than 7.7, roughly 0.5 pH units lower than average pH of natural surface waters (Figure 6.6).

At some locations near the sea floor, especially near submarine volcanic CO_2 sources, CO_2 concentrations can fluctuate greatly. Near deep-sea hydrothermal vents CO_2 partial pressures ($p$$CO_2$, expressed as a ppm fraction of atmospheric pressure, equivalent to μatm) of up to 80,000 ppm have been observed. These are more than 100 times the typical value for deep-sea water. Typically, these vents are associated with fauna that have adapted to these conditions over evolutionary time. For example, tube worms can make use of high CO_2 levels for chemosynthetic CO_2 fixation in association with symbiotic bacteria (Childress *et al.*, 1993). High CO_2 levels (up to a $p$$CO_2$ of 16,000 ppm; Knoll *et al.*, 1996) have been observed in ocean bottom waters and marine sediments where there are high rates organic matter oxidation and low rates of mixing with the overlying seawater. Under these conditions, high CO_2 concentrations are often accompanied by low O_2 concentrations. Near the surface at night, respiratory fluxes in some relatively confined rock pools of the intertidal zone can produce high CO_2 levels. These patterns suggest that in some environments, organisms have evolved to tolerate relatively wide pH oscillations and/or low pH values.

Deep-sea ecosystems generally depend on sinking particles of organic carbon made by photosynthesis near the ocean surface settling down through the water. Most species living in the deep sea display very low metabolic rates (Childress, 1995), especially in oxygen minimum layers (Seibel *et al.*, 1997). Organisms living in the deep seawaters have adapted to the energy-limited environment by conserving energy stores and minimizing energy turnover. As a result of energy limitations and cold temperatures found in the deep sea, biological activities tend to be extremely low. For example, respiration rates of rat-tail fish are roughly 0.1% that of their shallow-water relatives. Community respiration declines exponentially with depth along the California margin, however, rapid turnover of large quantities of organic matter has been observed on the ocean floor (Mahaut *et al.*, 1995; Smith and Demopoulos, 2003). Thus, biological activity of some animals living on the deep sea floor can be as great as is found in relatives living on the sea floor in shallow waters.

Deep-sea ecosystems may take a long time to recover from disturbances that reduce population size. Organisms have adapted to the energy-limited environment of the deep sea by limiting investment in reproduction, thus most deep-sea species produce few offspring. Deep-sea species tend to invest heavily in each of their eggs, making them large and rich in yolk to provide the offspring with the resources they will need for survival. Due to their low metabolic rates, deep-sea species tend to grow slowly and have much longer lifespans than their upper-ocean cousins. For example, on the deep-sea floor, a bivalve less than 1 cm across can be more than 100 years old (Gage, 1991). This means that populations of deep-sea species will be more greatly affected by the loss of individual larvae than would upper ocean species. Upon disturbance, recolonization and community recovery in the deep ocean follows similar patterns to those in shallow waters, but on much longer time scales (several years compared to weeks or months in shallow waters, Smith and Demopoulos, 2003).

The numbers of organisms living on the sea floor per unit area decreases exponentially with depth, probably associated with the diminishing flux of food with depth. On the sea floor of the deepest ocean and of the upper ocean, the fauna can be dominated by a few species. Between 2000 and 3000 m depth ecosystems tend to have high species diversity with a low number of individuals, meaning that each species has a low population size (Snelgrove and Smith, 2002). The fauna living in the water column appear to be less diverse than that on the sea floor, probably due to the relative uniformity of vast volumes of water in the deep ocean.

Box 6.5 Natural analogues and Earth history.

There are several examples of natural systems with strong CO_2 sources in the ocean, and fluid pools toxic to marine life that may be examined to better understand possible physical and biological effects of active CO_2 injection.

Most natural environments that are heavily enriched in CO_2 (or toxic substances) host life forms that have adapted to these special conditions on evolutionary time scales. During Earth history much of the oceans may have hosted life forms specialized on elevated pCO_2, which are now extinct. This limits the use of natural analogues or Earth history to predict and generalize effects of CO_2 injection on most extant marine life.

- *Venting of carbon dioxide-rich fluids:* Hydrothermal vents, often associated with mid-ocean-ridge systems, often release CO_2 rich fluids into the ocean and can be used to study CO_2 behaviour and effects. For example, Sakai *et al.* (1990) observed buoyant hydrate forming fluids containing 86–91% CO_2 (with H_2S, and methane etc. making up the residual) released from the sea floor at 1335–1550 m depth from a hydrothermal vent field. These fluids would be similar to a heavily contaminated industrial CO_2 source. These fluids arise from the reaction of sea water with acid and intermediate volcanic rocks at high temperature; they are released into sea water of 3.8°C. A buoyant hydrate-coated mass forms at the sea floor, which then floats upwards dissolving into the ocean water. Sea floor venting of aqueous fluids, rich in CO_2 and low in pH (3.5–4.4), is also to be found in some hydrothermal systems (Massoth *et al.*, 1989; Karl, 1995).

  Near volcanic vents, deep-sea ecosystems can be sustained by a geochemical input of chemical energy and CO_2. While there has been extensive investigation of these sites, and the plumes emanating from them, this has not yet been in the context of analogues for industrial CO_2 storage effects. Such an investigation would show how a fauna has evolved to adapt to a high-CO_2 environment; it would not show how biota adapted to normal ocean water would respond to increased CO_2 concentrations.*

- *Deep saline brine pools:* The ocean floor is known to have a large number of highly saline brine pools that are anoxic and toxic to marine life. The salty brines freely dissolve, but mixing into the overlying ocean waters is impeded by the stable stratification imparted by the high density of the dissolving brines. The Red Sea contains many such brine pools (Degens and Ross, 1969; Anschutz *et al.*, 1999), some up to 60 km² in area, filled with high-temperature hyper-saline, anoxic, brine. Animals cannot survive in these conditions, and the heat and salt that are transported across the brine-seawater interface form a plume into the surrounding bottom water. Hydrothermal sources resupply brine at the bottom of the brine pool (Anschutz and Blanc, 1996). The Gulf of Mexico contains numerous brine pools. The largest known is the Orca Basin, where a 90 km² brine pool in 2250 m water depth is fed by drainage from exposed salt deposits. The salt is toxic to life, but biogeochemical cycles operate at the interface with the overlying ocean (van Cappellen *et al.*, 1998). The Mediterranean also contains numerous large hypersaline basins (MEDRIFF Consortium, 1995).

  Taken together these naturally occurring brine pools provide examples of vast volumes of soluble, dense, fluids, hostile to marine life, on the sea floor. The number, volume, and extent of these pools exceed those for scenarios for CO_2 lake formation yet considered. There has been little study of the impact of the plumes emanating from these sources. These could be examined to yield information that may be relevant to environmental impacts of a lake of CO_2 on the ocean floor.

- *Changes over geological time:* In certain times in Earth's geological past the oceans may have contained more dissolved inorganic carbon and/or have had a lower pH.

  There is evidence of large-scale changes in calcifying organism distributions in the oceans in the geological record that may be related in changes in carbonate mineral saturation states in the surface ocean. For example, Barker and Elderfield (2002) demonstrated that glacial-interglacial changes in the shell weights of several species of planktonic foraminifera are negatively correlated with atmospheric CO_2 concentrations, suggesting a causal relationship.

  Cambrian CO_2 levels (i.e., about 500 million years ago) were as high as 5000 ppm and mean values decreased progressively thereafter (see. Dudley, 1998; Berner, 2002). Two to three times higher than extant ocean calcium levels ensured that calcification of, for example, coral reefs was enabled in paleo-oceans despite high CO_2 levels (Arp *et al.*, 2001). High performance animal life appeared in the sea only after atmospheric CO_2 began to diminish. The success of these creatures may have depended on the reduction of atmospheric CO_2 levels (reviewed by Pörtner *et al.*, 2004, 2005).

  CO_2 is also thought to have been a potential key factor in the late Permian/Triassic mass extinction, which affected corals, articulate brachiopods, bryozoans, and echinoderms to a larger extent than molluscs, arthropods and chordates (Knoll *et al.*, 1996; Berner, 2002; Bambach *et al.*, 2002). Pörtner *et al.* (2004) hypothesized that this may be due to the corrosive effect of CO_2 on heavily calcified skeletons. CO_2 excursions would have occurred in the context of large climate oscillations. Effects of temperature oscillations, hypoxia events and CO_2 excursions probably contributed to extinctions (Pörtner *et al.*, 2005, see section 6.7.3).

Table 6.3 Relationships between ΔpH, changes in pCO$_2$, and dissolved inorganic carbon concentration calculated for mean deep-sea conditions. Also shown are volumes of water needed to dilute 1 tCO$_2$ to the specified ΔpH, and the amount of CO$_2$ that, if uniformly distributed throughout the ocean, would produce this ΔpH.

pH change ΔpH	Increase in CO$_2$ partial pressure ΔpCO$_2$ (ppm)	Increase in dissolved inorganic carbon ΔDIC (μmol kg^{-1})	Seawater volume to dilute 1 tCO$_2$ to ΔpH (m^3)	GtCO$_2$ to produce ΔpH in entire ocean volume
0	0	0	-	-
-0.1	150	30	656,000	2000
-0.2	340	70	340,000	3800
-0.3	580	100	232,000	5600
-0.5	1260	160	141,000	9200
-1	5250	400	54,800	24,000
-2	57,800	3,260	6800	190,000
-3	586,000	31,900	700	1,850,000

while increasing the volume of water experiencing a lesser ΔpH. Further examples indicating the spatial extent of ocean chemistry change from added CO$_2$ are represented in Figures 6.11, 6.12, 6.13, 6.14, and 6.15.

On evolutionary time scales most extant animal life has adapted to, on average, low ambient CO$_2$ levels. Accordingly, extant animal life may rely on these low pCO$_2$ values and it is unclear to what extent species would be able to adapt to permanently elevated CO$_2$ levels. Exposure to high CO$_2$ levels and extremely acidic water can cause acute mortality, but more limited shifts in CO$_2$, pH, and carbonate levels can be tolerated at least temporarily. Studies of shallow water organisms have identified a variety of physiological mechanisms by which changes in the chemical environment can affect fauna. These mechanisms should also apply to organisms living in the deep ocean. However, knowing physiological mechanisms alone does not enable full assessment of impacts at ecosystem levels. Long-term effects, for intervals greater than the duration of the reproduction cycle or the lifespan of an individual, may be overlooked, yet may still drastically change an ecosystem.

Species living in the open ocean are exposed to low and relatively constant CO$_2$ levels, and thus may be sensitive to CO$_2$ exposure. In contrast, species dwelling in marine sediments, especially in the intertidal zone, are regularly exposed to CO$_2$ fluctuations and thus may be better adapted to high and variable CO$_2$ concentrations. Physiological mechanisms associated with CO$_2$ adaptation have been studied mostly in these organisms. They respond to elevated CO$_2$ concentrations by transiently diminishing energy turnover. However, such responses are likely become detrimental during long-term exposure, as reduced metabolism involves a reduction in physical activity, growth, and reproduction. Overall, marine invertebrates appear more sensitive than fish (Pörtner *et al.*, 2005).

CO$_2$ effects have been studied primarily in fish and invertebrates from shallow waters, although some of these cover wide depth ranges down to below 2000 m or are adapted to cold temperatures (e.g., Langenbuch and Pörtner, 2003, 2004). Some *in situ* biological experiments used CO$_2$ in the deep ocean (See Box 6.6).

6.7.2 Physiological effects of CO$_2$

6.7.2.1 Effects of CO$_2$ on cold-blooded water breathing animals

Hypercapnia is the condition attained when an organism (or part thereof) is surrounded by high concentrations of CO$_2$. Under these conditions, CO$_2$ enters the organisms by diffusion across body and especially respiratory surfaces and equilibrates with all body compartments. This internal accumulation of CO$_2$ will be responsible for most of the effects observed in animals (reviewed by Pörtner and Reipschläger, 1996, Seibel and Walsh, 2001, Ishimatsu *et al.*, 2004, 2005; Pörtner *et al.*, 2004, 2005). Respiratory distress, narcosis, and mortality are the most obvious short-term effects at high CO$_2$ concentrations, but lower concentrations may have important effects on longer time scales. The CO$_2$ level to which an organism has acclimated may affect its acute critical CO$_2$ thresholds, however, the capacity to acclimate has not been investigated to date.

6.7.2.2 Effects of CO$_2$ versus pH

Typically, tolerance limits to CO$_2$ have been characterized by changes in ocean pH or pCO$_2$ (see Shirayama, 1995; Auerbach *et al.*, 1997). However, changes in molecular CO$_2$, carbonate, and bicarbonate concentrations in ambient water and body fluids may each have specific effects on marine organisms (Pörtner and Reipschläger, 1996). In water breathers like fish or invertebrates CO$_2$ entry causes immediate disturbances in acid-base status, which need to be compensated for by ion exchange mechanisms. The acute effect of CO$_2$ accumulation is more severe than that of the reduction in pH or carbonate-ion concentrations. For example, fish larvae are more sensitive to low pH and high CO$_2$ than low pH and low CO$_2$ (achieved by addition of HCl with pCO$_2$ levels kept low by aeration; Ishimatsu *et al.*, 2004).

CO$_2$ added to sea water will change the hydrogen ion concentration (pH). This change in hydrogen ion concentration may affect marine life through mechanisms that do not directly involve CO$_2$. Studies of effects of lowered pH (without concomitant CO$_2$ accumulation) on aquatic organisms have a

Box 6.6 *In-situ* observations of the response of deep-sea biota to added CO_2.

In-situ experiments concerning the sensitivity of deep and shallow-living marine biota to elevated carbon dioxide levels have been limited in scope. Significant CO_2 effects have been observed in experiments, consistent with the mechanisms of CO_2 action reported in Section 6.7.2. Some animals avoid CO_2 plumes, others do not.

Studies evaluating the behaviour and survival of deep-sea animals exposed to liquid CO_2 or to CO_2-rich sea water have been performed on the continental slope and rise off California. Experiments in which about 20–70 kg of liquid CO_2 were released in small corrals on the sea floor at 3600 m depth were used to measure the response of animals that came in contact with liquid CO_2, and to the dissolution plume emanating from CO_2 pools (Barry *et al.*, 2004). Larger bottom-living animals collected from the sea floor were held in cages and placed at distances of 1–50 m from CO_2 pools. In addition, organisms living in the sediment were collected at a range of distances from CO_2 pools, both before CO_2 release and 1–3 months later.

The response of animals to direct contact with liquid CO_2 varied among species. Sea cucumbers (holothurians like *Scotoplanes* sp.) and brittle stars (ophiuroids, unidentified species) died immediately after contact with liquid CO_2 (Barry *et al.*, 2005). A few individuals (<5 individuals) of deep-sea fish (grenadiers, *Coryphaenoides armatus*) that approached CO_2 pools and made contact with the fluid turned immediately and swam out of view. Other deep-sea experiments (Tamburri *et al.* 2000) evaluating the behavioural response of animals to a saturated CO_2 / sea water solution have shown that some scavenger species (deep-sea hagfish) will not avoid acidic, CO_2-rich seawater if chemical cues from decaying bait are also present. In fact, hagfish would maintain contact with the CO_2-rich / bait-scented plume long enough to be apparently 'narcotized' by the CO_2.

Survival rates of abyssal animals exposed to CO_2 dissolution plumes in these experiments varied with the range of pH perturbation and the distance from the CO_2 source. Abyssal animals held in cages or inhabiting sediments that were near (<1 m) CO_2 pools, and which were exposed episodically to large pH reduction (1–1.5 pH units) experienced high rates of mortality (>80%). Animals affected included small (meio-)fauna (flagellates, amoebae, nematodes; Barry *et al.*, 2004) and larger (macro and mega-)fauna (Ampeliscid amphipod species, invertebrates like holothurians, echinoids, and fish like macrourids). Other fish like eelpout (zoarcids), however, all survived month-long exposure to episodic pH shifts of about –1.0 pH units. Animals held further (3–10 m) from CO_2 pools were exposed to mild episodic pH reductions (about 0.1 – 0.2 pH units) exhibited mortality rates were (about 20–50%) higher than at control sites (Barry *et al.*, 2005).

It is unknown whether mortality was caused primarily by short-term exposure to large pH / CO_2 shifts or by chronic, milder pH perturbations. Tidal variation in current direction resulted in a highly variable exposure to pH perturbations

with the most intense exposure to dissolution plumes when the current was flowing directly towards the study animals. During other tidal periods there was often no pH reduction, increasing the difficulty of interpreting these experiments.

Three controlled *in-situ* experiments were carried out at 2000 m in the Kumano Trough using a specially designed chamber (Figure 6.24; Ishida *et al.* 2005) to address the impact of 5,000 and 20,000 ppm rises in $p$$CO_2$ (with resulting pHs of 6.8 and 6.3) on the abundance and diversity of bacteria and of small animals (nano- and meiobenthos). Significant impacts of elevated $p$$CO_2$ on meiobenthic organisms could not be found except for one case where the abundance of foraminifera decreased significantly within 3 days at 20,000 ppm. The abundance of nanobenthos decreased significantly in most cases, whereas the abundance of bacteria increased at 20,000 ppm (Figure 6.25).

In-situ studies of short-term effects of elevated CO_2 concentrations on deep-sea megafauna have been conducted using CO_2 released naturally from the Loihi Seamount (Hawaii) at depths of 1200 to 1300 m (Vetter and Smith, 2005). A submersible was used to manipulate baited traps and bait parcels in Loihi's CO_2 plume to explore the effects of elevated CO_2 on typical deep-sea scavengers. Vent-specialist shrimp were attracted to the bait and proved to be pre-adapted to the high CO_2 levels found close to volcanic vents. Free swimming, amphipods, synaphobranchid eels, and hexanchid sharks avoided open bait parcels placed in the CO_2 plumes

Figure 6.24 Experimental chamber going to the sea floor (Ishida *et al.* 2004). The bottom part houses a chamber that penetrates into the sediment. The top part houses electronics, pumps, valves, and water bags, that are used to control the CO_2 concentration inside the chamber, and to sample sea water in the chamber at designated times. At the time of recovery, the bottom of the chamber is closed, weights are released, and the system returns to the surface of the ocean using buoyancy provided by the glass bulbs (yellow structures around the top).

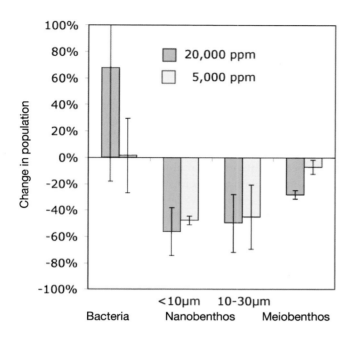

Figure 6.25 Preliminary investigations into the change of bacteria, nanobenthos and meiobenthos abundance after exposure to 20,000 and 5,000 ppm CO_2 for 77 to 375 hr during three experiments carried out at 2,000 m depth in Nankai Trough, north-western Pacific. Error bars represent one standard deviation (Ishida et al. 2005).

long history, with an emphasis on freshwater organisms (Wolff *et al.*, 1988). Observed consequences of lowered water *p*H (at constant pCO_2) include changes in production/productivity patterns in algal and heterotrophic bacterial species, changes in biological calcification/ decalcification processes, and acute and sub-acute metabolic impacts on zooplankton species, ocean bottom species, and fish. Furthermore, changes in the *p*H of marine environments affect: (1) the carbonate system, (2) nitrification (Huesemann *et al.*, 2002) and speciation of nutrients such as phosphate, silicate and ammonia (Zeebe and Wolf-Gladrow, 2001), and (3) speciation and uptake of essential and toxic trace elements. Observations and chemical calculations show that low *p*H conditions generally decrease the association of metals with particles and increase the proportion of biologically available free metals (Sadiq, 1992; Salomons and Forstner, 1984). Aquatic invertebrates take up both essential and non-essential metals, but final body concentrations of metals vary widely across invertebrates. In the case of many trace metals, enhanced bioavailability is likely to have toxicological implications, since free forms of metals are of the greatest toxicological significance (Rainbow, 2002).

6.7.2.3 Acute CO_2 sensitivity: oxygen transport in squid and fish

CO_2 accumulation and uptake can cause anaesthesia in many animal groups. This has been observed in deep-sea animals close to hydrothermal vents or experimental CO_2 pools. A narcotic effect of high, non-determined CO_2 levels was observed in deep-sea hagfish after CO_2 exposure *in situ* (Tamburri *et al.*, 2000). Prior to anaesthesia high CO_2 levels can exert rapid effects on oxygen transport processes and thereby contribute to acute CO_2 effects including early mortality.

Among invertebrates, this type of CO_2 sensitivity may be highest in highly complex, high performance organisms like squid (reviewed by Pörtner *et al.*, 2004). Blue-blooded squid do not possess red blood cells (erythrocytes) to protect their extracellular blood pigment (haemocyanin) from excessive *p*H fluctuations. Acute CO_2 exposure causes acidification of the blood, will hamper oxygen uptake and binding at the gills and reduce the amount of oxygen carried in the blood, limiting performance, and at high concentrations could cause death. Less oxygen is bound to haemocyanin in squid than is bound to haemoglobin in bony fish (teleosts). Jet-propulsion swimming of squid demands a lot of oxygen. Oxygen supply is supported by enhanced oxygen binding with rising blood *p*H (and reduced binding of oxygen with falling *p*H – a large Bohr effect[3]). Maximizing of oxygen transport in squid thus occurs by means of extracellular *p*H oscillations between arterial and venous blood. Therefore, finely controlled extracellular pH changes are important for oxygen transport. At high CO_2 concentrations, animals can asphyxiate because the blood cannot transport enough oxygen to support metabolic functions. In the most active open ocean squid (*Illex illecebrosus*), model calculations predict acute lethal effects with a rise in pCO_2 by 6500 ppm and a 0.25 unit drop in blood *p*H. However, acute CO_2 sensitivity varies between squid species. The less active coastal squid (*Loligo pealei*) is less sensitive to added CO_2.

In comparison to squid and other invertebrates, fish (teleosts) appear to be less sensitive to added CO_2, probably due to their lower metabolic rate, presence of red blood cells (erythrocytes containing haemoglobin) to carry oxygen, existence of a venous oxygen reserve, tighter epithelia, and more efficient acid-base regulation. Thus, adult teleosts (bony fish) exhibit a larger degree of independence from ambient CO_2. A number of tested shallow-water fish have shown relatively high tolerance to added CO_2, with short-term lethal limits of adult fish at a pCO_2 of about 50,000 to 70,000 ppm. European eels (*Anguilla anguilla*) displayed exceptional tolerance of acute hypercapnia up to 104,000 ppm (for review see Ishimatsu *et al.*, 2004, Pörtner *et al.*, 2004). The cause of death in fish involves a depression of cardiac functions followed by a collapse of oxygen delivery to tissues (Ishimatsu *et al.*, 2004). With mean lethal CO_2 levels of 13,000 to 28,000 ppm, juveniles are more sensitive to acute CO_2 stress than adults. In all of these cases, the immediate cause of death appears to be entry of CO_2 into the organism (and not primarily some other *p*H-mediated effect).

---

[3] The Bohr Effect is an adaptation in animals to release oxygen in the oxygen starved tissues in capillaries where respiratory carbon dioxide lowers blood pH. When blood *p*H decreases, the ability of the blood pigment to bind to oxygen decreases. This process helps the release of oxygen in the oxygen-poor environment of the tissues. Modified after ISCID Encyclopedia of Science and Philosophy. 2004. International Society for Complexity, Information, and Design. 12 October 2004 <http://www.iscid.org/encyclopedia/Bohr_Effect>.

Fish may be able to avoid contact to high CO_2 exposure because they possess highly sensitive CO_2 receptors that could be involved in behavioural responses to elevated CO_2 levels (Yamashita *et al.*, 1989). However, not all animals avoid low pH and high concentrations of CO_2; they may actively swim into CO_2-rich regions that carry the odour of potential food (e.g., bait; Tamburri *et al.*, 2000, Box 6.6).

Direct effects of dissolved CO_2 on diving marine air breathers (mammals, turtles) can likely be excluded since they possess higher pCO_2 values in their body fluids than water breathers and gas exchange is minimized during diving. They may nonetheless be indirectly affected through potential CO_2 effects on the food chain (see 6.7.5).

6.7.2.4   Deep compared with shallow acute CO_2 sensitivity

Deep-sea organisms may be less sensitive to high CO_2 levels than their cousins in surface waters, but this is controversial. Fish (and cephalopods) lead a sluggish mode of life with reduced oxygen demand at depths below 300 to 400 m. Metabolic activity of pelagic animals, including fish and cephalopods, generally decreases with depth (Childress, 1995; Seibel *et al.*, 1997). However, Seibel and Walsh (2001) postulated that deep-sea animals would experience serious problems in oxygen supply under conditions of increased CO_2 concentrations. They refer to midwater organisms that may not be representative of deep-sea fauna; as residents of so-called 'oxygen minimum layers' they have special adaptations for efficient extraction of oxygen from low-oxygen waters (Sanders and Childress, 1990; Childress and Seibel, 1998).

6.7.2.5   Long-term CO_2 sensitivity

Long-term impacts of elevated CO_2 concentrations are more pronounced on early developmental than on adult stages of marine invertebrates and fish. Long-term depression of physiological rates may, over time scales of several months, contribute to enhanced mortality rates in a population (Shirayama and Thornton, 2002, Langenbuch and Pörtner, 2004). Prediction of future changes in ecosystem dynamics, structure and functioning therefore requires data on sub-lethal effects over the entire life history of organisms.

The mechanisms limiting performance and long-term survival under moderately elevated CO_2 levels are even less clear than those causing acute mortality. However, they appear more important since they may generate impacts in larger ocean volumes during widespread distribution of CO_2 at moderate levels on long time scales. In animals relying on calcareous exoskeletons, physical damage may occur under permanent CO_2 exposure through reduced calcification and even dissolution of the skeleton, however, effects of CO_2 on calcification processes in the deep ocean have not been studied to date. Numerous studies have demonstrated the sensitivity of calcifying organisms living in surface waters to elevated CO_2 levels on longer time scales (Gattuso *et al.* 1999, Reynaud *et al.*, 2003, Feeley *et al.*, 2004 and refs. therein). At least a dozen laboratory and field studies of corals and coralline algae have suggested reductions in calcification rates by 15–85% with

a doubling of CO_2 (to 560 ppmv) from pre-industrial levels. Shirayama and Thornton (2002) demonstrated that increases in dissolved CO_2 levels to 560 ppm cause a reduction in growth rate and survival of shelled animals like echinoderms and gastropods. These findings indicate that previous atmospheric CO_2 accumulation may already be affecting the growth of calcifying organisms, with the potential for large-scale changes in surface-ocean ecosystem structure. Due to atmospheric CO_2 accumulation, global calcification rates could decrease by 50% over the next century (Zondervan *et al.*, 2001), and there could be significant shifts in global biogeochemical cycles. Despite the potential importance of biogeochemical feedback induced by global change, our understanding of these processes is still in its infancy even in surface waters (Riebesell, 2004). Much less can be said about potential ecosystem shifts in the deep sea (Omori *et al.*, 1998).

Long-term effects of CO_2 elevations identified in individual animal species affects processes in addition to calcification (reviewed by Ishimatsu *et al.*, 2004, Pörtner and Reipschläger, 1996, Pörtner *et al.*, 2004, 2005). In these cases, CO_2 entry into the organism as well as decreased water pH values are likely to have been the cause. Major effects occur through a disturbance in acid-base regulation of several body compartments. Falling pH values result and these affect many metabolic functions, since enzymes and ion transporters are only active over a narrow pH range. pH decreases from CO_2 accumulation are counteracted over time by an accumulation of bicarbonate anions in the affected body compartments (Heisler, 1986; Wheatly and Henry, 1992, Pörtner *et al.*, 1998; Ishimatsu *et al.* 2004), but compensation is not always complete. Acid-base relevant ion transfer may disturb osmoregulation due to the required uptake of appropriate counter ions, which leads to an additional NaCl load of up to 10% in marine fish in high CO_2 environments (Evans, 1984; Ishimatsu *et al.*, 2004). Long-term disturbances in ion equilibria could be involved in mortality of fish over long time scales despite more or less complete compensation of acidification.

Elevated CO_2 levels may cause a depression of aerobic energy metabolism, due to incomplete compensation of the acidosis, as observed in several invertebrate examples (reviewed by Pörtner *et al.* 2004, 2005). In one model organism, the peanut worm *Sipunculus nudus*, high CO_2 levels caused metabolic depression of up to 35% at 20,000 ppm pCO_2. A central nervous mechanism also contributed, indicated by the accumulation of adenosine in the nervous tissue under 10,000 ppm pCO_2. Adenosine caused metabolic depression linked to reduced ventilatory activity even more so when high CO_2 was combined with oxygen deficiency (anoxia; Lutz and Nilsson, 1997). Studies addressing the specific role of adenosine or other neurotransmitters at lower CO_2 levels or in marine fish during hypercapnia are not yet available.

The depression of metabolism observed under high CO_2 concentrations in marine invertebrates also includes inhibition of protein synthesis – a process that is fundamental to growth and reproduction. A CO_2 induced reduction of water pH to 7.3 caused a 55% reduction in growth of Mediterranean mussels (Michaelidis *et al.* 2005; for review see Pörtner *et al.* 2004,

2005). Fish may also grow slowly in high CO_2 waters. Reduced growth was observed in juvenile white sturgeon (Crocker and Cech, 1996). In this case, the stimulation of ventilation and the associated increase in oxygen consumption indicated a shift in energy budget towards maintenance metabolism, which occurred at the expense of growth. This effect was associated with reductions in foraging activity. A harmful influence of CO_2 on reproductive performance was found in two species of marine copepods (*Acartia steuri, Acartia erythrea*) and sea urchins (*Hemicentrotus purcherrimus, Echinometra mathaei*). While survival rates of adult copepods were not affected during 8 days at pCO_2 up to 10,000 ppm, egg production and hatching rates of eggs were significantly reduced concomitant to an increased mortality of young-stage larvae seen at water pH 7.0 (Kurihara *et al.*, 2004). In both sea urchin species tested, fertilization rates decreased with pCO_2 rising above 1000 ppm (below water pH 7.6; Kurihara *et al.*, 2004). Hatching and survival of fish larvae also declined with water pCO_2 and exposure time in all examined species (Ishimatsu *et al.*, 2004).

6.7.3   *From physiological mechanisms to ecosystems*

CO_2 effects propagate from molecules via cells and tissues to whole animals and ecosystems (Figure 6.26; Table 6.4). Organisms are affected by chemistry changes that modulate crucial physiological functions. The success of a species can depend on effects on the most sensitive stages of its life cycle (e.g., egg, larvae, adult). Effects on molecules, cells, and tissues thus integrate into whole animal effects (Pörtner *et al.*, 2004), affecting growth, behaviour, reproduction, and development of eggs and larvae. These processes then determine the ecological success (fitness) of a species, which can also depend on complex interaction among species. Differential effects of chemistry changes on the various species thus affect the entire ecosystem. Studies of CO_2 susceptibility and affected mechanisms in individual species (Figure 6.26) support development of a cause and effect understanding for an entire ecosystem's response to changes in ocean chemistry, but need to be complemented by field studies of ecosystem consequences.

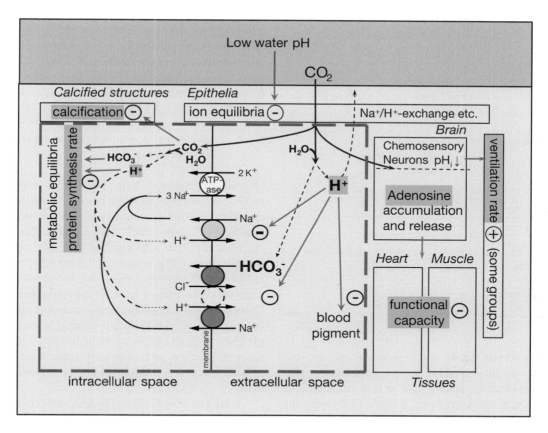

Figure 6.26 Effects of added CO_2 at the scale of molecule to organism and associated changes in proton (H^+), bicarbonate (HCO_3^-) and carbonate (CO_3^{2-}) levels in a generalized and simplified marine invertebrate or fish. The blue region on top refers to open water; the tan region represents the organism. Generalized cellular processes are depicted on the left and occur in various tissues like brain, heart or muscle; depression of these processes has consequences (depicted on the right and top). Under CO_2 stress, whole animal functions, like growth, behaviours or reproduction are depressed (adopted from Pörtner et al., 2005, – or + denotes a depression or stimulation of the respective function). Black arrows reflect diffusive movement of CO_2 between compartments. Red arrows reflect effective factors, CO_2, H^+, HCO_3^- that modulate functions. Shaded areas indicate processes relevant for growth and energy budget.

Table 6.4 Physiological and ecological processes affected by CO_2 (note that listed effects on phytoplankton are not relevant in the deep sea, but may become operative during large-scale mixing of CO_2). Based on reviews by Heisler, 1986, Wheatly and Henry, 1992, Claiborne et al., 2002, Langdon et al., 2003 Shirayama, 2002, Kurihara et al., 2004, Ishimatsu et al., 2004, 2005, Pörtner et al. 2004, 2005, Riebesell, 2004, Feeley et al., 2004 and references therein.

Affected processes	Organisms tested
Calcification	• Corals • Calcareous benthos and plankton
Acid-base regulation	• Fish • Sipunculids • Crustaceans
Mortality	• Scallops • Fish • Copepods • Echinoderms/gastropods • Sipunculids
N-metabolism	• Sipunculids
Protein biosynthesis	• Fish • Sipunculids • Crustaceans
Ion homeostasis	• Fish, crustaceans • Sipunculids
Growth	• Crustaceans • Scallops • Mussels • Fish • Echinoderms/gastropods
Reproductive performance	• Echinoderms • Fish • Copepods
Cardio-respiratory functions	• Fish
Photosynthesis	• Phytoplankton
Growth and calcification	
Ecosystem structure	
Feedback on biogeo-chemical cycles (elemental stoichiometry C:N:P, DOC exudation)	

Tolerance thresholds likely vary between species and phyla, but still await quantification for most organisms. Due to differential sensitivities among and within organisms, a continuum of impacts on ecosystems is more likely than the existence of a well-defined threshold beyond which CO_2 cannot be tolerated. Many species may be able to tolerate transient CO_2 fluctuations, but may not be able to settle and thrive in areas where CO_2 levels remain permanently elevated. At concentrations that do not cause acute mortality, limited tolerance may include reduced capacities of higher functions, that is added CO_2 could reduce the capacity of growth and reproduction, or hamper resistance to infection (Burnett, 1997).

It could also reduce the capacity to attack or escape predation, which would have consequences for the organism's food supply and thus overall fitness with consequences for the rest of the ecosystem.

Complex organisms like animals proved to be more sensitive to changing environmental conditions like temperature extremes than are simpler, especially unicellular, organisms (Pörtner, 2002). It is not known whether animals are also more sensitive to extremes in CO_2. CO_2 affects many physiological mechanisms that are also affected by temperature and hypoxia (Figure 6.26). Challenges presented by added CO_2 could lower long-term resistance to temperature extremes and thus narrow zoogeographical distribution ranges of affected species (Reynaud *et al.*, 2003, Pörtner *et al.*, 2005).

At the ecosystem level, few studies carried out in surface oceans report that species may benefit under elevated CO_2 levels. Riebesell (2004) summarized observations in surface ocean mesocosms under glacial (190 ppm) and increased CO_2 concentrations (790 ppm). High CO_2 concentrations caused higher net community production of phytoplankton. Diatoms dominated under glacial and elevated CO_2 conditions, whereas *Emiliania huxleyi* dominated under present CO_2 conditions. This example illustrates how species that are less sensitive to added CO_2 could become dominant in a high CO_2 environment, in this case due to stimulation of photosynthesis in resource limited phytoplankton species (Riebesell 2004). These conclusions have limited applicability to the deep sea, where animals and bacteria dominate. In animals, most processes are expected to be depressed by high CO_2 and low pH levels (Table 6.4).

6.7.4   *Biological consequences for water column release scenarios*

Overall, extrapolation from knowledge mostly available for surface oceans indicates that acute CO_2 effects (e.g., narcosis, mortality) will only occur in areas where pCO_2 plumes reach significantly above 5000 ppm of atmospheric pressure (in the most sensitive squid) or above 13,000 or 40,000 ppm for juvenile or adult fish, respectively. Such effects are thus expected at CO_2 increases with $\Delta pH < -1.0$ for squid. According to the example presented in Figure 6.12, a towed pipe could avoid pH changes of this magnitude, however a fixed pipe without design optimization would produce a volume of several km^3 with this pH change for an injection rate of 100 kg s^{-1}. Depending on the scale of injection such immediate effects may thus be chosen to be confined to a small region of the ocean (Figures 6.13 and 6.14).

Available knowledge of CO_2 effects and underlying mechanisms indicate that effects on marine fauna and their ecosystems will likely set in during long-term exposure to pCO_2 of more than 400 to 500 ppm or associated moderate pH changes (by about 0.1–0.3 units), primarily in marine invertebrates (Pörtner *et al.* 2005) and, possibly, unicellular organisms. For injection at a rate of 0.37 GtCO$_2$ yr^{-1} for 100 years (Figure 6.14), such critical pH shifts would occur in less than 1% of the total ocean volume by the end of this period. However, baseline *p*H shifts by 0.2 to 0.4 *p*H-units expected from the WRE550 stabilization scenario already reach that magnitude of change. Additional long-term repeated large-scale global injection of 10% of the CO_2 originating from 18,000 $GtCO_2$ fossil fuel would cause an extension of these *p*H shifts from the surface ocean to significantly larger (deeper) fractions of the ocean by 2100 to 2300 (Figure 6.15). Finally, large-scale ocean disposal of all of the CO_2 would lead to *p*H decreases of more than 0.3 and associated long-term effects in most of the ocean. Expected effects will include a reduction in the productivity of calcifying organisms leading to higher ratios of non-calcifiers over calcifiers (Pörtner *et al.*, 2005).

Reduced capacities for growth, productivity, behaviours, and reduced lifespan imply a reduction in population densities and productivities of some species, if not reduced biodiversity. Food chain length and composition may be reduced associated with reduced food availability for high trophic levels. This may diminish resources for local or global fisheries. The suggested scenarios of functional depression also include a CO_2 induced reduction in tolerance to thermal extremes, which may go hand in hand with reduced distribution ranges as well as enhanced geographical distribution shifts. All of these expectations result from extrapolations of current physiological and ecological knowledge and require verification in experimental field studies. The capacity of ecosystems to compensate or adjust to such CO_2 induced shifts is also unknown. Continued research efforts could identify critical mechanisms and address the potential for adaptation on evolutionary time scales.

6.7.5 Biological consequences associated with CO_2 lakes

Strategies that release liquid CO_2 close to the sea floor will be affecting two ecosystems: the ecosystem living on the sea floor, and deep-sea ecosystem living in the overlying water. Storage as a topographically confined 'CO_2 lake' would limit immediate large-scale effects of CO_2 addition, but result in the mortality of most organisms under the lake that are not able to flee and of organisms that wander into the lake. CO_2 will dissolve from the lake into the bottom water, and this will disperse around the lake, with effects similar to direct release of CO_2 into the overlying water. According to the scenarios depicted in Figures 6.11 and 6.12 for CO_2 releases near the sea floor, *p*H reductions expected in the near field are well within the scope of those expected to exert significant effect on marine biota, depending on the length of exposure.

6.7.6 Contaminants in CO_2 streams

The injection of large quantities of CO_2 into the deep ocean will itself be the topic of environmental concern, so the matter of possible small quantities of contaminants in the injected material is of additional but secondary concern. In general there are already stringent limits on contaminants in CO_2 streams due to human population concerns, and technical pipeline considerations. The setting of any additional limits for ocean disposal cannot be addressed with any certainty at this time.

There are prohibitions in general against ocean disposal; historical concerns have generally focused on heavy metals, petroleum products, and toxic industrial chemicals and their breakdown products.

A common contaminant in CO_2 streams is H_2S. There are very large sources of H_2S naturally occurring in the ocean: many marine sediments are anoxic and contain large quantities of sulphides; some large ocean basins (the Black Sea, the Cariaco Trench etc.) are anoxic and sulphidic. As a result ocean ecosystems that have adapted to deal with sulphide and sulphur-oxidizing bacteria are common throughout the worlds oceans. Nonetheless the presence of H_2S in the disposal stream would result in a lowering of local dissolved oxygen levels, and have an impact on respiration and performance of higher marine organisms.

6.7.7 Risk management

There is no peer-reviewed literature directly addressing risk management for intentional ocean carbon storage; however, there have been risk management studies related to other uses of the ocean. Oceanic CO_2 release carries no expectation of risk of catastrophic atmospheric degassing such as occurred at Lake Nyos (Box 6.7). Risks associated with transporting CO_2 to depth are discussed in Chapter 4 (Transport).

It may be possible to recover liquid CO_2 from a lake on the ocean floor. The potential reversibility of the production of CO_2 lakes might be considered a factor that diminishes risk associated with this option.

6.7.8 Social aspects; public and stakeholder perception

The study of public perceptions and perceived acceptability of intentional CO_2 storage in the ocean is at an early stage and comprises only a handful of studies (Curry *et al.*, 2005; Gough *et al.*, 2002; Itaoka *et al.*, 2004; Palmgren *et al.*, 2004). Issues crosscutting public perception of both geological and ocean storage are discussed in Section 5.8.5.

All studies addressing ocean storage published to date have shown that the public is largely uninformed about ocean carbon storage and thus holds no well-developed opinion. There is very little awareness among the public regarding intentional or unintentional ocean carbon storage. For example, Curry *et al.* (2005) found that the public was largely unaware of the role of the oceans in absorbing anthropogenic carbon dioxide released to the atmosphere. In the few relevant studies conducted thus far, the public has expressed more reservations regarding ocean carbon CO_2 storage than for geological CO_2 storage.

Education can affect the acceptance of ocean storage options. In a study conducted in Tokyo and Sapporo, Japan (Iatoka *et al*, 2004), when members of the public, after receiving some basic information, were asked to rate ocean and geologic storage options on a 1 to 5 scale (1 = no, 5 = yes) the mean rating for dilution-type ocean storage was 2.24, lake-type ocean storage was rated at 2.47, onshore geological storage was rated at 2.57, and offshore geological storage was rated at

Box 6.7 Lake Nyos and deep-sea carbon dioxide storage.

About 2 million tonnes of CO_2 gas produced by volcanic activity were released in one night in 1986 by Lake Nyos, Cameroon, causing the death of at least 1700 people (Kling *et al.*, 1994). Could CO_2 released in the deep sea produce similar catastrophic release at the ocean surface?

Such a catastrophic degassing involves the conversion of dissolved CO_2 into the gas phase. In the gas phase, CO_2 is buoyant and rises rapidly, entraining the surrounding water into the rising plume. As the water rises, CO_2 bubbles form more readily. These processes can result in the rapid release of CO_2 that has accumulated in the lake over a prolonged period of magmatic activity.

Bubbles of CO_2 gas can only form in sea water shallower than about 500 m when the partial pressure of CO_2 in sea water exceeds the ambient total pressure. Most release schemes envision CO_2 release deeper than this. CO_2 released below 3000 m would tend to sink and then dissolve into the surrounding seawater. CO_2 droplets released more shallowly generally dissolve within a few 100 vertical metres of release.

The resulting waters are too dilute in CO_2 to produce partial CO_2 pressures exceeding total ambient pressure, thus CO_2 bubbles would not form. Nevertheless, if somehow large volumes of liquid CO_2 were suddenly transported above the liquid-gas phase boundary, there is a possibility of a self-accelerating regime of fluid motion that could lead to rapid degassing at the surface. The disaster at Lake Nyos was exacerbated because the volcanic crater confined the CO_2 released by the lake; the open ocean surface does not provide such topographic confinement. Thus, there is no known mechanism that could produce an unstable volume of water containing 2 $MtCO_2$ at depths shallower than 500 m, and thus no mechanism known by which ocean carbon storage could produce a disaster like that at Lake Nyos.

2.75. After receiving additional information from researchers, the mean rating for dilution-type and lake-type ocean storage increased to 2.42 and 2.72, respectively, while the mean ratings for onshore and offshore geologic storage increased to 2.65 and 2.82, respectively. In a similar conducted study in Pittsburgh, USA, Palmgren *et al.* (2004) found that when asked to rate ocean and geologic storage on a 1 to 7 scale (1 = completely oppose, 7 = completely favour) respondents' mean rating was about 3.2 for ocean storage and about 3.5 for geological storage. After receiving information selected by the researchers, the respondents changed their ratings to about 2.4 for ocean storage and 3.0 for geological storage. Thus, in the Itaoka *et al.* (2004) study the information provided by the researchers increased the acceptance of all options considered whereas in the Study of Palmgren *et al.* (2004) the information provided by the researchers decreased the acceptance of all options considered. The differences could be due to many causes, nevertheless, they suggest that the way information is provided by researchers could affect whether the added information increases or decreases the acceptability of ocean storage options.

Gough *et al.* (2002) reported results from discussions of carbon storage from two unrepresentative focus groups comprising a total of 19 people. These focus groups also preferred geological storage to ocean storage; this preference appeared to be based, 'not primarily upon concerns for the deep-sea ecological environment', but on 'the lack of a visible barrier to prevent CO_2 escaping' from the oceans. Gough *et al.* (2002) notes that 'significant opposition' developed around a proposed ocean CO_2 release experiment in the Pacific Ocean (see Section 6.2.1.2).

6.8 Legal issues

6.8.1 *International law*

Please refer to Sections 5.8.1.1 (*Sources and nature of international obligations*) and 5.8.1.2 (*Key issues in the application of the treaties to CO_2 storage*) for the general position of both geological and ocean storage of CO_2 under international law. It is necessary to look at and interpret the primary sources, the treaty provisions themselves, to determine the permissibility or otherwise of ocean storage. Some secondary sources, principally the 2004 OSPAR Jurists Linguists' paper containing the States Parties' interpretation of the Convention (considered in detail in Section 5.8.1.3) and conference papers prepared for the IEA workshop in 1996, contain their authors' individual interpretations of the treaties.

McCullagh (1996) considered the international legal control of ocean storage, and found that, whilst the UN Framework Convention on Climate Change (UNFCCC) encourages the use of the oceans as a reservoir for CO_2, the UN Convention on the Law of the Sea (UNCLOS) is ambiguous in its application to ocean storage. Whilst ocean storage will reduce CO_2 emissions and combat climate change, to constitute an active use of sinks and reservoirs as required by the UNFCCC, ocean storage would need to be the most cost-effective mitigation option. As for UNCLOS, it is unclear whether ocean storage will be allowable in all areas of the ocean, but provisions on protecting and preserving the marine environment will be applicable if CO_2 is deemed to be 'pollution' under the Convention (which will be so, as the large quantity of CO_2 introduced is likely to cause harm to living marine resources). In fulfilling their obligation to prevent, reduce and control pollution of the marine environment, states must act so as not to transfer damage or hazards from one area to another or transform one type of pollution into another, a requirement that could be relied upon by proponents and opponents alike.

Churchill (1996) also focuses on UNCLOS in his assessment of the international legal issues, and finds that the consent of the coastal state would be required if ocean storage occurred in that state's territorial sea (up to 12 miles from the coast). In that state's Exclusive Economic Zone (up to 200 miles), the storage of CO_2 via a vessel or platform (assuming it constituted 'dumping' under the Convention) would again require the consent of that state. Its discretion is limited by its obligation to have due regard to the rights and duties of other states in the Exclusive Economic Zone under the Convention, by other treaty obligations (London and OSPAR) and the Convention's general duty on parties not to cause damage by pollution to other states' territories or areas beyond their national jurisdiction. He finds that it is uncertain whether the definition of 'dumping' would apply to use of a pipeline system from land for ocean storage, but, in any event, concludes that the discharge of CO_2 from a pipeline will, in many circumstances, constitute pollution and thus require the coastal state to prevent, reduce and control such pollution from land-based sources. But ocean storage by a pipeline from land into the Exclusive Economic Zone will not fall within the rights of a coastal or any other state and any conflict between them will be resolved on the basis of equity and in the light of all the relevant circumstances, taking into account the respective importance of the interests involved to the parties as well as to the international community as a whole. He finds that coastal states do have the power to regulate and control research in their Exclusive Economic Zones, although such consent is not normally withheld except in some cases.

As for the permissibility of discharge of CO_2 into the high seas (the area beyond the Exclusive Economic Zone open to all states), Churchill (1996) concludes that this will depend upon whether the activity is a freedom of the high sea and is thus not prohibited under international law, and finds that the other marine treaties will be relevant in this regard.

Finally, the London Convention is considered by Campbell (1996), who focuses particularly on the 'industrial waste' definition contained in Annex I list of prohibited substances, but does not provide an opinion upon whether CO_2 is covered by that definition 'waste materials generated by manufacturing or processing operations', or indeed the so-called reverse list exceptions to this prohibition, or to the general prohibition under the 1996 Protocol.

6.8.2   National laws

6.8.2.1   Introduction

CO_2 ocean storage, excluding injection from vessels, platforms or other human-made structures into the subseabed to which the Assessment made in Section 5.8 applies, is categorized into the following two types according to the source of injection of the CO_2 (land or sea) and its destination (sea): (1) injection from land (via pipe) into the seawater; (2) injection from vessels, platforms or other human-made structures into sea water (water column, ocean floor).

States are obliged to comply with the provisions of international law mentioned above in Section 6.8.1, in particular treaty law to which they are parties. States have to implement their international obligations regarding CO_2 ocean storage either by enacting relevant national laws or revising existing ones. There have been a few commentaries and papers on the assessment of the legal position of ocean storage at national level. However, the number of countries covered has been quite limited. Summaries of the assessment of the national legal issues having regard to each type of storage mentioned above to be considered when implementing either experimental or fully-fledged ocean storage of CO_2 are provided below.

With regard to the United States, insofar as CO_2 from a fossil-fuel power plant is considered industrial waste, it would be proscribed under the Ocean Dumping Ban Act of 1988. The Marine Protection, Research, and Sanctuaries Act of 1972 (codified as 33 U.S.C. 1401–1445, 16 U.S.C. 1431–1447f, 33 U.S.C. 2801–2805), including the amendments known as the Ocean Dumping Ban Act of 1988, has the aim of regulating intentional ocean disposal of materials, while authorizing related research. The Ocean Dumping Ban Act of 1988 placed a ban on ocean disposal of sewage sludge and industrial wastes after 31 December 1991.

The US Environmental Protection Agency (US EPA) specified protective criteria for marine waters, which held pH to a value between 6.5 and 8.5, with a limit on overall excursion of no more than 0.2 pH units outside the naturally occurring range (see: Train, 1979). Much of the early work on marine organisms reflected concerns about the dumping of industrial acid wastes (e.g., acid iron wastes from TiO_2 manufacture) into marine waters. For the most part, however, these studies failed to differentiate between true pH effects and the effects due to CO_2 liberated by the introduction of acid into the test systems.

6.8.2.2   Injection from land (via pipe) into seawater

States can regulate the activity of injection within their jurisdiction in accordance with their own national rules and regulations. Such rules and regulations would be provided by, if any, the laws relating to the treatment of high-pressure gases, labour health and safety, control of water pollution, dumping at sea, waste disposal, biological diversity, environmental impact assessment etc. It is, therefore, necessary to check whether planned activities of injection fall under the control of relevant existing rules and regulations.

6.8.2.3   Injection from vessels, platforms or other humanmade structures into sea water (water column, ocean floor)

It is necessary to check whether the ocean storage of CO_2 is interpreted as 'dumping' of 'industrial waste' by relevant national laws, such as those on dumping at sea or waste disposal, because this could determine the applicability of the London Convention and London Protocol (see Section 6.8.1). Even if ocean storage is not prohibited, it is also necessary to check whether planned activities will comply with the existing relevant classes of rules and regulations, if any, mentioned above.

Table 6.5 Ocean storage cost estimate for CO_2 transport and injection at 3000 m depth from a floating platform. Scenario assumes three pulverized coal fired power plants with a net generation capacity of 600 MWe each transported either 100 or 500 km by a CO_2 tanker ship of 80,000 m³ capacity to a single floating discharge platform.

Ship transport distance	100 km	500 km
Onshore CO_2 Storage (US$/t$CO_2$ shipped)	3.3	3.3
Ship transport to injection platform (US$/t$CO_2$ shipped)	2.9	4.2
Injection platform, pipe and nozzle (US$/t$CO_2$ shipped)	5.3	5.3
Ocean storage cost (US$/t$CO_2$ shipped)	11.5	12.8
Ocean storage cost (US$/t$CO_2$ net stored)	11.9	13.2

6.9.   Costs

6.9.1   Introduction

Studies on the engineering cost of ocean CO_2 storage have been published for cases where CO_2 is transported from a power plant located at the shore by either ship to an offshore injection platform or injection ship (Section 6.9.2), or pipeline running on the sea floor to an injection nozzle (Section 6.9.3). Costs considered in this section include those specific to ocean storage described below and include the costs of handling of CO_2 and transport of CO_2 offshore, but not costs of onshore transport (Chapter 4).

6.9.2   Dispersion from ocean platform or moving ship

Costs have been estimated for ship transport of CO_2 to an injection platform, with CO_2 injection from a vertical pipe into mid- to deep ocean water, or a ship trailing an injection pipe (Akai *et al.*, 2004; IEA-GHG, 1999; Ozaki, 1997; Akai *et al.*, 1995; Ozaki *et al.*, 1995). In these cases, the tanker ship transports liquid CO_2 at low temperature (–55 to –50°C) and high pressure (0.6 to 0.7 MPa).

Table 6.5 shows storage costs for cases (Akai *et al.*, 2004) of ocean storage using an injection platform. In these cases, CO_2 captured from three power plants is transported by a CO_2 tanker ship to a single floating discharge platform for injection at a depth of 3000 m. The cost of ocean storage is the sum of three major components: tank storage of CO_2 onshore awaiting shipping; shipping of CO_2; and the injection platform pipe and nozzle. The sum of these three components is 11.5 to 12.8 US$/t$CO_2$ shipped 100 to 500 km. Assuming an emission equal to 3% of shipped CO_2 from boil-off and fuel consumption, the estimated cost is 11.9 to 13.2 US$/t$CO_2$ net stored.

Liquid CO_2 could be delivered by a CO_2 transport ship to the injection area and then transferred to a CO_2 injection ship, which would tow a pipe injecting the CO_2 into the ocean at a depth of 2,000 to 2,500 m. Estimated cost of ocean storage (Table 6.6) is again the sum of three major components: tank storage of CO_2 onshore awaiting shipping; shipping of CO_2; and the injection ship, pipe and nozzle (Table 6.6; Akai *et al.*, 2004). The sum of these three components is 13.8 to 15.2 US$/t$CO_2$ shipped 100 to 500 km. Assuming an emission equal to 3% of shipped CO_2 from boil-off and fuel consumption, the estimated cost is 14.2 to 15.7 US$/t$CO_2$ net stored.

6.9.3   Dispersion by pipeline extending from shore into shallow to deep water

Compared with the ship transport option (6.9.2), pipeline transport of CO_2 is estimated to cost less for transport over shorter distances (e.g., 100 km) and more for longer distances (e.g., 500 km), since the cost of ocean storage via pipeline scales with pipeline length.

The cost for transporting CO_2 from a power plant located at the shore through a pipeline running on the sea floor to an injection nozzle has been estimated by IEA-GHG (1994) and Akai *et al.* (2004). In the recent estimate of Akai *et al.* (2004), CO_2 captured from a pulverized coal fired power plant with a net generation capacity of 600 MW$_e$ is transported either 100 or 500 km by a CO_2 pipeline for injection at a depth of 3000 m at a cost of 6.2 US$/t$CO_2$ net stored (100 km case) to 31.1 US$/t$CO_2$ net stored (500 km case).

There are no published cost estimates specific to the production of a CO_2 lake on the sea floor; however, it might be reasonable to assume that there is no significant difference between the cost of CO_2 lake production and the cost of water column injection given this dominance of pipeline costs.

Table 6.6 Ocean storage cost estimate for CO_2 transport and injection at 2000-2500 m depth from a moving ship.

Ship transport distance	100 km	500 km
Onshore CO_2 Storage (US$/t$CO_2$ shipped)	2.2	2.2
Ship transport to injection ship(US$/t$CO_2$ shipped)	3.9	5.3
Injection ship, pipe and nozzle (US$/t$CO_2$ shipped)	7.7	7.7
Ocean storage cost (US$/t$CO_2$ shipped)	13.8	15.2
Ocean storage cost (US$/t$CO_2$ net stored)	14.2	15.7

6.9.4   Cost of carbonate neutralization approach

Large-scale deployment of carbonate neutralization would require a substantial infrastructure to mine, transport, crush, and dissolve these minerals, as well as substantial pumping of seawater, presenting advantages for coastal power plants near carbonate mineral sources.

There are many trade-offs to be analyzed in the design of an economically optimal carbonate-neutralization reactor along the lines of that described by Rau and Caldeira (1999). Factors to be considered in reactor design include water flow rate, gas flow rate, particle size, pressure, temperature, hydrodynamic conditions, purity of reactants, gas-water contact area, etc. Consideration of these factors has led to preliminary cost estimates for this concept, including capture, transport, and energy penalties, of 10 to 110 US\$/tCO_2 net stored (Rau and Caldeira, 1999).

6.9.5   Cost of monitoring and verification

The cost of a monitoring and verification program could involve deploying and maintaining a large array of sensors in the ocean. Technology exists to conduct such monitoring, but a significant fraction of the instrument development and production is limited to research level activities. No estimate of costs for near-field monitoring for ocean storage have been published, but the costs of limited near-field monitoring would be small compared to the costs of ocean storage in cases of the scale considered in 6.9.2 and 6.9.3. Far field monitoring can benefit from international research programs that are developing global monitoring networks.

6.10   Gaps in knowledge

The science and technology of ocean carbon storage could move forward by addressing the following major gaps:

- *Biology and ecology:* Studies of the response of biological systems in the deep sea to added CO_2, including studies that are longer in duration and larger in scale than yet performed.

- Research facilities: Research facilities where ocean storage concepts (e.g., release of CO_2 from a fixed pipe or ship, or carbonate-neutralization approaches) can be applied and their effectiveness and impacts assessed in situ at small-scale on a continuing basis for the purposes of both scientific research and technology development.

- *Engineering:* Investigation and development of technology for working in the deep sea, and the development of pipes, nozzles, diffusers, etc., which can be deployed in the deep sea with assured flow and be operated and maintained cost-effectively.

- *Monitoring:* Development of techniques and sensors to detect CO_2 plumes and their biological and geochemical consequences.

References

Adams, E., D. Golomb, X. Zhang, and H.J. Herzog, 1995: Confined release of CO_2 into shallow seawater. *Direct Ocean Disposal of Carbon Dioxide*. N. Handa, (ed.), Terra Scientific Publishing Company, Tokyo, pp. 153-161.

Adams, E., J. Caulfield, H.J. Herzog, and D.I. Auerbach, 1997: Impacts of reduced *p*H from ocean CO_2 disposal: Sensitivity of zooplankton mortality to model parameters. *Waste Management,* **17**(5-6), 375-380.

Adams, E., M. Akai, G. Alendal, L. Golmen, P. Haugan, H.J. Herzog, S. Matsutani, S. Murai, G. Nihous, T. Ohsumi, Y. Shirayama, C. Smith, E. Vetter, and C.S. Wong, 2002: International Field Experiment on Ocean Carbon Sequestration (Letter). *Environmental Science and Technology,* **36**(21), 399A.

Akai, M., N. Nishio, M. Iijima, M. Ozaki, J. Minamiura, and T. Tanaka, 2004: Performance and Economic Evaluation of CO_2 Capture and Sequestration Technologies. Proceedings of the Seventh International Conference on Greenhouse Gas Control Technologies.

Akai, M., T. Kagajo, and M. Inoue, 1995: Performance Evaluation of Fossil Power Plant with CO_2 Recovery and Sequestering System. *Energy Conversion and Management,* **36**(6-9), 801-804.

Alendal, G. and H. Drange, 2001: Two-phase, near field modelling of purposefully released CO_2 in the ocean. *Journal of Geophysical Research-Oceans,* **106**(C1), 1085-1096.

Alendal, G., H. Drange, and P.M. Haugan, 1994: Modelling of deep-sea gravity currents using an integrated plume model. The Polar Oceans and Their Role in Shaping the Global Environment: The Nansen Centennial Volume, O.M. Johannessen, R.D. Muench, and J.E. Overland (eds.) *AGU Geophysical Monograph*, 85, American Geophysical Union, pp. 237-246.

Anschutz, P. and G. Blanc, 1996: Heat and salt fluxes in the Atlantis II deep (Red Sea). *Earth and Planetary Science Letters*, 142, 147-159.

Anschutz, P., G. Blanc, F. Chatin, M. Geiller, and M.-C. Pierret, 1999: Hydrographic changes during 20 years in the brine-filled basins of the Red Sea. Deep-Sea Research Part I **46**(10) 1779-1792.

Archer, D.E., 1996: An atlas of the distribution of calcium carbonate in sediments of the deep-sea. *Global Biogeochemical Cycles,* **10**(1), 159-174.

Archer, D.E., H. Kheshgi, and E. Maier-Reimer, 1997: Multiple timescales for neutralization of fossil fuel CO_2. *Geophysical Research Letters*, **24**(4), 405-408.

Archer, D.E., H. Kheshgi, and E. Maier-Reimer, 1998: Dynamics of fossil fuel neutralization by Marine $CaCO_3$. *Global Biogeochemical Cycles,* **12**(2), 259-276.

Arp, G., A. Reimer, and J. Reitner, 2001: Photosynthesis-induced biofilm calcification and calcium concentrations in Phanerozoic oceans. *Science*, 292, 1701-1704.

Auerbach, D.I., J.A. Caulfield, E.E. Adams, and H.J. Herzog, 1997: Impacts of Ocean CO_2 Disposal on Marine Life: I. A toxicological assessment integrating constant-concentration laboratory assay data with variable-concentration field exposure. *Environmental Modelling and Assessment,* **2**(4), 333-343.

Aya, I., K. Yamane, and N. Yamada, 1995: Simulation experiment of CO_2 storage in the basin of deep-ocean. *Energy Conversion and Management*, **36**(6-9), 485-488.

Aya, I, R. Kojima, K. Yamane, P. G. Brewer, and E. T. Peltzer, 2004: *In situ* experiments of cold CO_2 release in mid-depth. *Energy*, **29**(9-10), 1499-1509.

Aya, I., K. Yamane, and H. Nariai, 1997: Solubility of CO_2 and density of CO_2 hydrate at 30MPa. *Energy*, **22**(2-3), 263-271.

Aya, I., R. Kojima, K. Yamane, P. G. Brewer, and E. T. Pelter, III, 2003: *In situ* experiments of cold CO_2 release in mid-depth. Proceedings of the International Conference on Greenhouse Gas Control Technologies, 30th September-4th October, Kyoto, Japan.

Bacastow, R.B. and G.R. Stegen, 1991: Estimating the potential for CO_2 sequestration in the ocean using a carbon cycle model. Proceedings of OCEANS '91. Ocean Technologies and Opportunities in the Pacific for the 90's, 1-3 Oct. 1991, Honolulu, USA, 1654-1657.

Bacastow, R.B., R.K. Dewey, and G.R. Stegen, 1997: Effectiveness of CO_2 sequestration in the pre- and post-industrial oceans. *Waste Management*, **17**(5-6), 315-322.

Baes, C. F., 1982: Effects of ocean chemistry and biology on atmospheric carbon dioxide. Carbon Dioxide Review. W.C. Clark (ed.), Oxford University Press, New York, pp. 187-211.

Bambach, R.K., A.H. Knoll, and J.J. Sepkowski, jr., 2002: Anatomical and ecological constraints on Phanerozoic animal diversity in the marine realm. Proceedings of the National Academy of Sciences, **99**(10), 6845-6859.

Barker, S. and H. Elderfield, 2002: Foraminiferal calcification response to glacial-interglacial changes in atmospheric CO_2. *Science*, **297**, 833-836.

Barry, J.P., K.R. Buck, C.F. Lovera, L.Kuhnz, P.J. Whaling, E.T. Peltzer, P. Walz, and P.G. Brewer, 2004: Effects of direct ocean CO_2 injection on deep-sea meiofauna. *Journal of Oceanography*, **60**(4), 759-766.

Barry, J.P. K.R. Buck, C.F. Lovera, L.Kuhnz, and P.J. Whaling, 2005: Utility of deep-sea CO_2 release experiments in understanding the biology of a high CO_2 ocean: effects of hypercapnia on deep-sea meiofauna. *Journal of Geophysical Research-Oceans*, in press.

Berner, R. A., A. C. Lasaga, and R. M. Garrels, 1983: The carbonate-silicate geochemical cycle and its effect on atmospheric carbon dioxide over the past 100 million years. *American Journal of Science* **283**, 641-683.

Berner, R.A., 2002: Examination of hypotheses for the Permo-Triassic boundary extinction by carbon cycle modeling. Proceedings of the National Academy of Sciences, **99**(7), 4172-4177.

Bradshaw, A., 1973: The effect of carbon dioxide on the specific volume of seawater. *Limnology and Oceanography*, **18**(1), 95-105.

Brewer, P.G., D.M. Glover, C. Goyet, and D.K. Shafer, 1995: The pH of the North-Atlantic Ocean - improvements to the global-model for sound-absorption in seawater. *Journal of Geophysical Research-Oceans*, **100**(C5), 8761-8776.

Brewer, P.G., E. Peltzer, I. Aya, P. Haugan, R. Bellerby, K. Yamane, R. Kojima, P. Walz, and Y. Nakajima, 2004: Small scale field study of an ocean CO_2 plume. *Journal of Oceanography*, **60**(4), 751-758.

Brewer, P.G., E.T. Peltzer, G. Friederich, and G. Rehder, 2002: Experimental determination of the fate of a CO_2 plume in seawater. *Environmental Science and Technology*, **36**(24), 5441-5446.

Brewer, P.G., E.T. Peltzer, G. Friederich, I. Aya, and K. Yamane, 2000: Experiments on the ocean sequestration of fossil fuel CO_2: pH measurements and hydrate formation. *Marine Chemistry*, **72**(2-4), 83-93.

Brewer, P.G., F.M. Orr, Jr., G. Friederich, K.A. Kvenvolden, and D.L. Orange, 1998: Gas hydrate formation in the deep-sea: *In situ* experiments with controlled release of methane, natural gas and carbon dioxide. *Energy and Fuels*, **12**(1), 183-188.

Brewer, P.G., G. Friederich, E.T. Peltzer, and F.M. Orr, Jr., 1999: Direct experiments on the ocean disposal of fossil fuel CO_2. *Science*, **284**, 943-945.

Brewer, P.G., E.T. Peltzer, P. Walz, I. Aya, K. Yamane, R. Kojima, Y. Nakajima, N. Nakayama, P. Haugan, and T. Johannessen, 2005: Deep ocean experiments with fossil fuel carbon dioxide: creation and sensing of a controlled plume at 4 km depth. *Journal of Marine Research*, **63**(1), 9-33.

Broecker, W.S. and T.-H. Peng, 1982: Tracers in the Sea. Eldigio Press, Columbia University, Palisades, New York, 690 pp.

Burnett, L.E., 1997: The challenges of living in hypoxic and hypercapnic aquatic environments. *American Zoologist*, **37**(6), 633-640.

Caldeira, K. and G.H. Rau, 2000: Accelerating carbonate dissolution to sequester carbon dioxide in the ocean: Geochemical implications. *Geophysical Research Letters*, **27**(2), 225-228.

Caldeira, K. and M.E. Wickett, 2003: Anthropogenic carbon and ocean pH. *Nature*, **425**, 365-365.

Caldeira, K. and M.E. Wickett, 2005: Ocean chemical effects of atmospheric and oceanic release of carbon dioxide. *Journal of Geophysical Research-Oceans*, **110**.

Caldeira, K., M.E. Wickett, and P.B. Duffy, 2002: Depth, radiocarbon and the effectiveness of direct CO_2 injection as an ocean carbon sequestration strategy. *Geophysical Research Letters*, **29**(16), 1766.

Campbell, J.A., 1996: Legal, jurisdictional and policy issues - 1972 London Convention. Ocean Storage of CO_2, Workshop 3, International links and Concerns, IEA Greenhouse Gas R&D Programme, Cheltenham, UK, pp.127-131.

Carman, K.R., D. Thistle, J. Fleeger, and J. P. Barry, 2004: The influence of introduced CO_2 on deep-sea metazoan meiofauna. *Journal of Oceanography*, **60**(4), 767-772.

Caulfield, J.A., E.E. Adams, D.I. Auerbach, and H.J. Herzog, 1997: Impacts of Ocean CO_2 Disposal on Marine Life: II. Probabilistic plume exposure model used with a time-varying dose-response model, Environmental Modelling and Assessment, **2**(4), 345-353.

Chen, B., Y. Song, M. Nishio, and M. Akai, 2003: Large-eddy simulation on double-plume formation induced by CO_2 Dissolution in the ocean. *Tellus* (B), **55**(2), 723-730.

Chen, B., Y. Song, M. Nishio, and M. Akai, 2005: Modelling of CO_2 dispersion from direct injection of CO_2 in the water column. *Journal of Geophysical Research - Oceans*, **110**.

Childress, J.J. and B.A. Seibel, 1998: Life at stable low oxygen levels: adaptations of animals to oceanic oxygen minimum layers. *Journal of Experimental Biology*, **201**(8), 1223-1232.

Childress, J.J., 1995: Are there physiological and biochemical adaptations of metabolism in deep-sea animals? *Trends in Ecology and Evolution*, **10**(1), 30-36.

Childress, J.J., R. Lee, N.K. Sanders, H. Felbeck, D. Oros, A. Toulmond, M.C.K. Desbruyeres II, and J. Brooks, 1993: Inorganic carbon uptake in hydrothermal vent tubeworms facilitated by high environmental pCO_2. *Nature*, **362**, 147-149.

Churchill, R., 1996: International legal issues relating to ocean Storage of CO_2: A focus on the UN Convention on the Law of the Sea. Ocean Storage of CO_2, Workshop 3, International links and Concerns, IEA Greenhouse Gas R&D Programme, Cheltenham, UK, pp. 117-126.

Claiborne, J.B., S.L. Edwards, and A.I. Morrison-Shetlar, 2002: Acid-base regulation in fishes: Cellular and molecular mechanisms. *Journal of Experimental Zoology*, **293**(3), 302-319.

Crocker, C.E., and J.J. Cech, 1996: The effects of hypercapnia on the growth of juvenile white sturgeon, Acipenser transmontanus. *Aquaculture*, **147**(3-4), 293-299.

Crounse, B., E. Adams, S. Socolofsky, and T. Harrison, 2001: Application of a double plume model to compute near field mixing for the international field experiment of CO_2 ocean sequestration. Proceedings of the 5th International Conference on Greenhouse Gas Control Technologies, August 13th -16th 2000, Cairns Australia, CSIRO pp. 411-416.

Curry, T., D. Reiner, S. Ansolabehere, and H. Herzog, 2005: How aware is the public of carbon capture and storage? E.S. Rubin, D.W. Keith and C.F. Gilboy (eds.), Proceedings of 7th International Conference on Greenhouse Gas Control Technologies (GHGT-7), September 5-9, 2004, Vancouver, Canada.

De Figueiredo, M.A., D.M. Reiner, and H.J. Herzog, 2002: Ocean carbon sequestration: A case study in public and institutional perceptions. Proceedings of the Sixth International Conference on Greenhouse Gas Control Technologies, September 30th-October 4th Kyoto, Japan.

Degens, E.T. and D.A. Ross, 1969: Hot Brines and Recent Heavy Metal Deposits in the Red Sea. Springer-Verlag, New York, 600 pp.

Dewey, R.K., G.R. Stegen and R. Bacastow, 1997: Far-field impacts associated with ocean disposal of CO_2. *Energy and Management*, **38** (Supplement1), S349-S354.

Dewey, R., and G. Stegen, 1999: The dispersion of CO_2 in the ocean: consequences of basin-scale variations in turbulence levels. Greenhouse Gas Control Technologies. Eliasson, B., P. Riemer, A. Wokaun, (eds.), *Elsevier Science Ltd.*, Oxford, pp. 299-304.

Dickson, A.G., 1981: An exact definition of total alkalinity and a procedure for the estimation of alkalinity and total CO_2 from titration data. Deep-Sea Research Part A **28**(6), 609-623.

Drange, H., and P.M. Haugan, 1992: Disposal of CO_2 in sea-water. *Nature*, **357**, 547.

Drange, H., G. Alendal, and O.M. Johannessen, 2001: Ocean release of fossil fuel CO_2: A case study. *Geophysical Research Letters*, **28**(13), 2637-2640.

Dudley, R., 1998: Atmospheric oxygen, giant Palaeozoic insects and the evolution of aerial locomotor performance. *Journal of Experimental Biology*, **201**(8), 1043-1050.

Emerson, S. and D. Archer, 1990: Calcium carbonate preservation in the ocean. Philosophical Transactions of the Royal Society of London (Series A), **331**, 29-41.

Evans, D.H., 1984: The roles of gill permeability and transport mechanisms in euryhalinity. Fish Physiology. W.S. Haar and D.J. Randall (eds.), *Academic Press*, New York, pp. 239-283.

Feely, R.A., C.L. Sabine, K. Lee, W. Berelson, J. Kleypas, V.J. Fabry, and F.J. Millero, 2004: Impact of anthropogenic CO_2 on the $CaCO_3$ system in the oceans. *Science*, **305**, 362-366.

Fer, I. and P. M. Haugan, 2003: Dissolution from a liquid CO_2 lake disposed in the deep ocean. *Limnology and Oceanography*, **48**(2), 872-883.

Gage, J.D. and P.A. Tyler, 1991: Deep-Sea Biology: A Natural History of Organisms at the Deep-sea Floor. Cambridge University Press, Cambridge, 504 pp.

Gattuso J.-P., D. Allemand and M. Frankignoulle, 1999: Interactions between the carbon and carbonate cycles at organism and community levels in coral reefs: a review on processes and control by the carbonate chemistry. *Am. Zool.*, **39**(1): 160-183.

Giles, J., 2002. Norway sinks ocean carbon study. *Nature* **419**, page 6.

Gough, C., I. Taylor, and S. Shackley, 2002: Burying carbon under the sea: an initial exploration of public opinion. *Energy & Environment*, **13**(6), 883-900.

Haugan, P.M. and F. Joos, 2004: Metrics to assess the mitigation of global warming by carbon capture and storage in the ocean and in geological reservoirs. *Geophysical Research Letters*, 31, L18202, doi:10.1029/2004GL020295.

Haugan, P.M. and G. Alendal, 2005: Turbulent diffusion and transport from a CO_2 lake in the deep ocean. *Journal of Geophysical Research-Oceans*, **110**.

Haugan, P.M. and H. Drange, 1992: Sequestration of CO_2 in the deep ocean by shallow injection. *Nature*, **357**, 318-320.

Heisler, N. (ed.), 1986: Acid-base Regulation in Animals. Elsevier Biomedical Press, Amsterdam, 491 pp.

Herzog, H., K. Caldeira, and J. Reilly, 2003: An issue of permanence: assessing the effectiveness of ocean carbon sequestration. *Climatic Change*, **59**(3), 293-310.

Hill, C., V. Bognion, M. Follows, and J. Marshall, 2004: Evaluating carbon sequestration efficiency in an ocean model using adjoint sensitivity analysis. *Journal of Geophysical Research-Oceans*, **109**, C11005, doi:10.1029/2002JC001598.

Hoffert, M.I., Y.-C. Wey, A.J. Callegari, and W.S. Broecker, 1979: Atmospheric response to deep-sea injections of fossil-fuel carbon dioxide. Climatic Change, **2**(1), 53-68.

Holdren, J.P., and S.F. Baldwin, 2001: The PCAST energy studies: toward a national consensus on energy research, development, demonstration, and deployment policy. Annual Review of Energy and the Environment, **26**, 391-434.

Huesemann, M.H., A.D. Skillman, and E.A. Crecelius, 2002: The inhibition of marine nitrification by ocean disposal of carbon dioxide. *Marine Pollution Bulletin*, **44**(2), 142-148.

Ishida, H., Y. Watanabe, T. Fukuhara, S. Kaneko, K. Firisawa, and Y. Shirayama, 2005: *In situ* enclosure experiment using a benthic chamber system to assess the effect of high concentration of CO_2 on deep-sea benthic communities. *Journal of Oceanography*, in press.

Ishimatsu, A., M. Hayashi, K.-S. Lee, T. Kikkawa, and J. Kita, 2005: Physiological effects on fishes in a high-CO_2 world. *Journal of Geophysical Research - Oceans*, **110**.

Ishimatsu, A., T. Kikkawa, M. Hayashi, K.-S. Lee, and J. Kita, 2004: Effects of CO_2 on marine fish: larvae and adults. *Journal of Oceanography*, **60**(4), 731-742.

Itaoka, K., A. Saito, and M. Akai, 2004: Public Acceptance of CO_2 capture and storage technology: A survey of public opinion to explore influential factors. Proceedings of the 7th International Conference on Greenhouse Gas Control Technologies (GHGT-7), September 5-9, 2004, Vancouver, Canada.

Jain, A.K. and L. Cao, 2005: Assessing the effectiveness of direct injection for ocean carbon sequestration under the influence of climate change, *Geophysical Research Letters*, **32**.

Johnson, K.M., A.G. Dickson, G. Eischeid, C. Goyet, P. Guenther, F.J. Millero, D. Purkerson, C.L. Sabine, R.G. Schottle, D.W.R. Wallace, R.J. Wilke, and C.D. Winn, 1998: Coulometric total carbon dioxide analysis for marine studies: Assessment of the quality of total inorganic carbon measurements made during the US Indian Ocean CO_2 Survey 1994-1996. *Marine Chemistry*, **63**(1-2), 21-37.

Joos, F., G.K. Plattner, T.F. Stocker, A. Körtzinger, and D.W.R. Wallace, 2003: Trends in marine dissolved oxygen: implications for ocean circulation changes and the carbon budget. EOS Transactions, *American Geophysical Union*, **84** (21), 197, 201.

Kajishima, T., T. Saito, R. Nagaosa, and S. Kosugi, 1997: GLAD: A gas-lift method for CO_2 disposal into the ocean. *Energy*, **22**(2-3), 257-262.

Karl, D.M., 1995: Ecology of free-living, hydrothermal vent microbial communities. In: The Microbiology of Deep-Sea Hydrothermal Vents, D.M. Karl, (ed.), CRC Press, Boca Raton, pp. 35-125.

Key, R.M., A. Kozyr, C.L. Sabine, K. Lee, R. Wanninkhof, J. Bullister, R.A. Feely, F. Millero, C. Mordy, and T.-H. Peng. 2004: A global ocean carbon climatology: Results from GLODAP. Global Biogeochemical Cycles, **18**, GB4031.

Kheshgi, H.S. and D. Archer, 2004: A nonlinear convolution model for the evasion of CO_2 injected into the deep ocean. *Journal of Geophysical Research-Oceans*, **109**.

Kheshgi, H.S., 1995: Sequestering atmospheric carbon dioxide by increasing ocean alkalinity. *Energy*, **20**(9), 915-922.

Kheshgi, H.S., 2004a: Ocean carbon sink duration under stabilization of atmospheric CO_2: a 1,000-year time-scale. *Geophysical Research Letters*, **31**, L20204.

Kheshgi, H.S., 2004b: Evasion of CO_2 injected into the ocean in the context of CO_2 stabilization. *Energy*, 29 (9-10), 1479-1486.

Kheshgi, H.S., B.P. Flannery, M.I. Hoffert, and A.G. Lapenis, 1994: The effectiveness of marine CO_2 disposal. *Energy*, **19**(9), 967-975.

Kheshgi, H.S., S.J. Smith, and J.A. Edmonds, 2005: Emissions and Atmospheric CO_2 Stabilization: Long-term Limits and Paths, Mitigation and Adaptation. *Strategies for Global Change*, **10**(2), pp. 213-220.

Kling, G.W., W.C. Evans, M.L. Tuttle, and G. Tanyileke, 1994: Degassing of Lake Nyos. *Nature*, **368**, 405-406.

Knoll, A.K., R.K. Bambach, D.E. Canfield, and J.P. Grotzinger, 1996: Comparative Earth history and late Permian mass extinction. *Science*, **273**, 452-457.

Kobayashi, Y., 2003: BFC analysis of flow dynamics and diffusion from the CO_2 storage in the actual sea bottom topography. *Transactions of the West-Japan Society of Naval Architects*, **106**, 19-31.

Kurihara, H., S. Shimode, and Y. Shirayama, 2004: Sub-lethal effects of elevated concentration of CO_2 on planktonic copepods and sea urchins. *Journal of Oceanography*, **60**(4), 743-750.

Langdon, C., W.S. Broecker, D.E. Hammond, E. Glenn, K. Fitzsimmons, S.G. Nelson, T.H. Peng, I. Hajdas, and G. Bonani, 2003: Effect of elevated CO_2 on the community metabolism of an experimental coral reef. *Global Biogeochemical Cycles*, **17**.

Langenbuch, M. and H.O. Pörtner, 2003: Energy budget of Antarctic fish hepatocytes (Pachycara brachycephalum and Lepidonotothen kempi) as a function of ambient CO_2: pH dependent limitations of cellular protein biosynthesis? *Journal of Experimental Biology*, **206** (22), 3895-3903.

Langenbuch, M. and H.O. Pörtner, 2004: High sensitivity to chronically elevated CO_2 levels in a eurybathic marine sipunculid. *Aquatic Toxicology*, **70** (1), 55-61.

Liro, C., E. Adams, and H. Herzog, 1992: Modelling the releases of CO_2 in the deep ocean. *Energy Conversion and Management*, **33**(5-8), 667-674.

Løken, K.P., and T. Austvik, 1993: Deposition of CO_2 on the seabed in the form of hydrates, Part-II. *Energy Conversion and Management*, **34**(9-11), 1081-1087.

Lutz, P.L. and G.E. Nilsson, 1997: Contrasting strategies for anoxic brain survival - glycolysis up or down. *Journal of Experimental Biology*, **200**(2), 411-419.

Mahaut, M.-L., M. Sibuet, and Y. Shirayama, 1995: Weight-dependent respiration rates in deep-sea organisms. *Deep-Sea Research (Part I)*, **42** (9), 1575-1582.

Marchetti, C., 1977: On geoengineering and the CO_2 problem. *Climate Change*, **1**(1), 59-68.

Marubini, F. and B. Thake, 1999: Bicarbonate addition promotes coral growth. *Limnol. Oceanog.* **44**(3a): 716-720.

Massoth, G.J., D.A. Butterfield, J.E. Lupton, R E. McDuff, M.D. Lilley, and I.R. Jonasson, 1989: Submarine venting of phase-separated hydrothermal fluids at axial volcano, Juan de Fuca Ridge. *Nature*, **340**, 702-705.

Matsumoto, K. and R.M. Key, 2004: Natural radiocarbon distribution in the deep ocean. Global environmental change in the ocean and on land, edited by M. Shiyomi, H. Kawahata and others, Terra Publishing Company, Tokyo, Japan, pp. 45-58,

McCullagh, J., 1996: International legal control over accelerating ocean storage of carbon dioxide. Ocean Storage of CO_2, Workshop 3, International links and Concerns. IEA Greenhouse Gas R&D Programme, Cheltenham, UK, pp. 85-115.

McPhaden, M.J., and D. Zhang, 2002: Slowdown of the meridional overturning circulation in the upper Pacific Ocean. *Nature*, **415**, 603-608.

MEDRIFF Consortium, 1995: Three brine lakes discovered in the seafloor of the eastern Mediterranean. EOS Transactions, American Geophysical Union **76**, 313-318.

Michaelidis, B., C. Ouzounis, A. Paleras, and H.O. Pörtner, 2005: Effects of long-term moderate hypercapnia on acid-base balance and growth rate in marine mussels (Mytilus galloprovincialis). Marine Ecology Progress Series **293**, 109-118.

Mignone, B.K., J.L. Sarmiento, R.D. Slater, and A. Gnanadesikan, 2004: Sensitivity of sequestration efficiency to mixing processes in the global ocean, *Energy*, **29**(9-10), 1467-1478.

Minamiura, J., H. Suzuki, B. Chen, M. Nishio, and M. Ozaki, 2004: CO_2 Release in Deep Ocean by Moving Ship. Proceedings of the 7th International Conference on Greenhouse Gas Control Technologies, 5th-9th September 2004, Vancouver, Canada.

Moomaw, W., J.R. Moreira, K. Blok, D.L. Greene, K. Gregory, T. Jaszay, T. Kashiwagi, M. Levine, M. McFarland, N. Siva Prasad, L. Price, H.-H. Rogner, R. Sims, F. Zhou, and P. Zhou, 2001: Technological and Economic Potential of Greenhouse Gas Emission Reduction. B. Metz *et al.* (eds.), Climate Change 2001: Mitigation, Contribution of Working Group III to the Third Assessment Report of the Intergovernmental Panel on Climate Change, Cambridge University Press, Cambridge, UK, 2001, pp 167-277.

Mori, Y.H. and T. Mochizuki, 1998: Dissolution of liquid CO_2 into water at high pressures: a search for the mechanism of dissolution being retarded through hydrate-film formation. *Energy Conversion and Management*, **39**(7), 567-578.

Mori, Y.H., 1998: Formation of CO_2 hydrate on the surface of liquid CO_2 droplets in water - some comments on a previous paper. *Energy Conversion and Management*, **39**(5-6) 369-373.

Morse, J.W. and F.T. Mackenzie, 1990: Geochemistry of Sedimentary Carbonates. *Elsevier*, Amsterdam, 707 pp.

Morse, J.W. and R.S. Arvidson, 2002: Dissolution kinetics of major sedimentary carbonate minerals. *Earth Science Reviews*, **58** (1-2), 51-84.

Mueller, K., L. Cao, K. Caldeira, and A. Jain, 2004: Differing methods of accounting ocean carbon sequestration efficiency. *Journal of Geophysical Research-Oceans*, **109**, C12018, doi:10.1029/2003JC002252.

Murray, C.N., and T.R.S. Wilson, 1997: Marine carbonate formations: their role in mediating long-term ocean-atmosphere carbon dioxide fluxes - A review. *Energy Conversion and Management*, **38** (Supplement 1), S287-S294.

Murray, C.N., L. Visintini, G. Bidoglio, and B. Henry, 1996: Permanent storage of carbon dioxide in the marine environment: The solid CO_2 penetrator. *Energy Conversion and Management*, 37(6-8), 1067-1072.

Nakashiki, N., 1997: Lake-type storage concepts for CO_2 disposal option. *Waste Management*, **17**(5-6), 361-367.

Nakashiki, N., and T. Ohsumi, 1997: Dispersion of CO_2 injected into the ocean at the intermediate depth. *Energy Conversion and Management*, **38** (Supplement 1) S355-S360.

Nihous, G.C., 1997: Technological challenges associated with the sequestration of CO_2 in the ocean. *Waste Management*, **17**(5-6), 337-341.

Nihous, G.C., L. Tang, and S.M. Masutani, 2002: A sinking plume model for deep CO_2 discharge, In Proceedings of the 6th International Conference on Greenhouse Gas Control Technologies, 30th September-4th October, Kyoto, Japan.

Ohgaki, K. and T. Akano, 1992: CO_2 Storage in the Japan deep trench and utilization of gas hydrate. *Energy and Resources*, **13**(4), 69-77.

Ohsumi, T., 1993: Prediction of solute carbon dioxide behaviour around a liquid carbon dioxide pool on deep ocean basin. *Energy Conversion and Management*, **33**(5-8), 685-690.

Ohsumi, T., 1995: CO_2 storage options in the deep sea. *Marine Technology Society Journal*, **29**(3), 58-66.

Ohsumi, T., 1997: CO_2 Storage Options in the Deep-sea, *Marine Tech. Soc. J.*, **29**(3), 58-66.

Omori, M., C.P., Norman, and T. Ikeda, 1998: Oceanic disposal of CO_2: potential effects on deep-sea plankton and micronekton- A review. *Plankton Biology and Ecology*, **45**(2), 87-99.

Ormerod, W.G., P. Freund, A. Smith, and J. Davison, 2002: Ocean Storage of CO_2, International Energy Agency, Greenhouse Gas R&D Programme, ISBN 1 898373 30 2.

Orr, J.C., 2004: Modelling of ocean storage of CO_2---The GOSAC study, Report PH4/37, International Energy Agency, Greenhouse Gas R&D Programme, Cheltenham, UK, 96 pp.

Ozaki, M., 1997: CO_2 injection and dispersion in mid-ocean by moving ship. *Waste Management*, **17**(5-6), 369-373.

Ozaki, M., J. Minamiura, Y. Kitajima, S. Mizokami, K. Takeuchi, and K. Hatakenka, 2001: CO_2 ocean sequestration by moving ships. *Journal of Marine Science and Technology*, **6**, 51-58.

Ozaki, M., K. Sonoda, Y.Fujioka, O. Tsukamoto, and M. Komatsu, 1995: Sending CO_2 into deep ocean with a hanging pipe from floating platform. *Energy Conversion and Management*, **36**(6-9), 475-478.

Ozaki, M., K. Takeuchi, K. Sonoda, and O. Tsukamoto, 1997: Length of vertical pipes for deep-ocean sequestration of CO_2 in rough seas. *Energy*, **22**(2-3), 229-237.

Palmer, M.D., H.L. Bryden, J.L. Hirschi, and J. Marotzke, 2004: Observed changes in the South Indian Ocean gyre circulation, 1987-2002. Geophysical Research Letters, **31**(15) L15303, doi:10.1029/2004GL020506.

Palmgren, C., M. Granger Morgan, W. Bruine de Bruin and D. Keith, 2004: Initial public perceptions of deep geological and oceanic disposal of CO_2. *Environmental Science and Technology*, **38**(24), 6441-6450.

Pörtner, H.O. and A. Reipschläger, 1996: Ocean disposal of anthropogenic CO_2: physiological effects on tolerant and intolerant animals. Ocean Storage of CO_2- Environmental Impact. B. Ormerod, M. Angel (eds.), Massachusetts Institute of Technology and International Energy Agency, Greenhouse Gas R&D Programme, Boston/Cheltenham, pp. 57-81.

Pörtner, H.O., 2002: Climate change and temperature dependent biogeography: systemic to molecular hierarchies of thermal tolerance in animals. *Comparative Biochemistry and Physiology(A)*, **132**(4), 739-761.

Pörtner, H.O., A. Reipschläger, and N. Heisler, 1998: Metabolism and acid-base regulation in Sipunculus nudus as a function of ambient carbon dioxide. *Journal of Experimental Biology*, **201**(1), 43-55.

Pörtner, H.O., M. Langenbuch, and A. Reipschläger, 2004: Biological impact of elevated ocean CO_2 concentrations: lessons from animal physiology and Earth history? *Journal of Oceanography*, **60**(4): 705-718.

Pörtner, H.O., M. Langenbuch, and B. Michaelidis, 2005: Effects of CO_2 on marine animals: Interactions with temperature and hypoxia regimes. *Journal of Geophysical Research - Oceans*, **110**, doi:10.1029/2004JC002561.

Prentice, C., G. Farquhar, M. Fasham, M. Goulden, M. Heimann, V. Jaramillo, H. Kheshgi, C.L. Quéré, R. Scholes, and D. Wallace, 2001: The carbon cycle and atmospheric CO_2. Climate Change 2001: The Scientific Basis: Contribution of WGI to the Third Assessment Report of the IPCC. J.T. Houghton *et al.*, (eds.), Cambridge University Press, New York, pp. 183-237.

Rainbow, P.S., 2002: Trace metal concentrations in aquatic invertebrates: why and so what? *Environmental Pollution*, **120**(3), 497-507.

Ramaswamy, V., O. Boucher, J. Haigh, D. Hauglustaine, J. Haywood, G. Myhre, T. Nakajima, G. Y. Shi, and S. Solomon, 2001: Radiative forcing of climate change. In Climate Change 2001: The Scientific Basis: Contribution of WGI to the Third Assessment Report of the IPCC. J.T. Houghton *et al.*, (eds.), Cambridge University Press, New York, pp. 349-416.

Rau, G. H. and K. Caldeira, 1999: Enhanced carbonate dissolution: A means of sequestering waste CO_2 as ocean bicarbonate. *Energy Conversion and Management*, **40**(17), 1803-1813.

Rehder, G., S.H. Kirby, W.B. Durham, L.A. Stern, E.T. Peltzer, J. Pinkston, and P.G. Brewer, 2004: Dissolution rates of pure methane hydrate and carbon dioxide hydrate in under-saturated sea water at 1000 m depth. *Geochimica et Cosmochimica Acta*, **68**(2), 285-292.

Reynaud, S., N. Leclercq, S. Romaine-Lioud, C. Ferrier-Pagès, J. Jaubert, and J.P. Gattuso, 2003: Interacting effects of CO_2 partial pressure and temperature on photosynthesis and calcification in a scleratinian coral. *Global Change Biology*, **9**(1) 1-9.

Riebesell, U., 2004: Effects of CO_2 enrichment on marine plankton. *Journal of Oceanography*, **60**(4), 719-729.

Sabine, C.L., R.A. Feely, N. Gruber, R.M. Key, K. Lee, J.L. Bullister, R. Wanninkhof, C.S. Wong, D.W.R. Wallace, B. Tilbrook, F.J. Millero, T.H. Peng, A. Kozyr, T. Ono, and A.F. Rios, 2004: The oceanic sink for anthropogenic CO_2. *Science*, **305**, 367-371.

Sadiq, M., 1992: Toxic Metal Chemistry in Marine Environments. Marcel Dekker Inc., New York, 390 pp.

Saito, T., S. Kosugi, T. Kajishima, and K. Tsuchiya, 2001: Characteristics and performance of a deep-ocean disposal system for low-purity CO_2 gas via gas lift effect. *Energy and Fuels*, **15**(2), 285-292.

Saji, A., H. Yoshida, M. Sakai, T. Tanii, T. Kamata, and H. Kitamura, 1992: Fixation of carbon dioxide by hydrate-hydrate. *Energy Conversion and Management*, **33**(5-8), 634-649.

Sakai, H., T. Gamo, E-S. Kim, M. Tsutsumi, T. Tanaka, J. Ishibashi, H. Wakita, M. Yamano, and T. Omori, 1990: Venting of carbon dioxide-rich fluid and hydrate formation in mid-Okinawa trough backarc basin. *Science*, **248**, 1093-1096.

Salomons, W. and U. Forstner, 1984: Metals in the Hydrocycle. Springer-Verlag, Heidelberg, 349 pp.

Sanders, N.K. and J.J. Childress, 1990: A comparison of the respiratory function of the hemocyanins of vertically migrating and non-migrating oplophorid shrimps. *Journal of Experimental Biology*, **152**(1), 167-187.

Sato, T., 2004: Numerical Simulation of Biological Impact Caused by Direct Injection of Carbon Dioxide in the ocean. *Journal of Oceanography*, **60**, 807-816.

Sato, T., and K. Sato, 2002: Numerical Prediction of the Dilution Process and its Biological Impacts in CO_2 Ocean Sequestration. *Journal of Marine Science and Technology*, **6**(4), 169-180.

Seibel, B.A. and P.J. Walsh, 2001: Potential impacts of CO_2 injections on deep-sea biota. *Science*, **294**, 319-320.

Seibel, B.A., E.V. Thuesen, J.J. Childress, and L.A. Gorodezky, 1997: Decline in pelagic cephalopod metabolism with habitat depth reflects differences in locomotory efficiency. *Biological Bulletin*, **192**, (2) 262-278.

Shindo, Y., Y. Fujioka, and H. Komiyama, 1995: Dissolution and dispersion of CO_2 from a liquid CO_2 pool in the deep ocean. *International Journal of Chemical Kinetics*, **27**(11), 1089-1095.

Shirayama, Y. and H. Thornton, 2005: Effect of increased atmospheric CO_2 on shallow-water marine benthos. *Journal of Geophysical Research-Oceans*, **110**.

Shirayama, Y., 1995: Current status of deep-sea biology in relation to the CO_2 disposal. Direct Ocean Disposal of Carbon Dioxide. N. Handa, T. Ohsumi, (eds.), Terra Scientific Publishing Company, Tokyo, pp. 253-264.

Shirayama, Y., 1997: Biodiversity and biological impact of ocean disposal of carbon dioxide. *Waste Management,* **17**(5-6), 381-384.

Simonetti, P., 1998: Low-cost, endurance ocean profiler. *Sea Technology*, **39**(2), 17-21.

Sloan, E.D., 1998. Clathrate Hydrates of Natural Gases. 2nd ed. Marcel Dekker Inc., New York, 705 pp.

Smith, C.R., and A.W. Demopoulos, 2003: Ecology of the deep Pacific Ocean floor. In Ecosystems of the World, Volume 28: Ecosystems of the Deep Ocean. P.A. Tyler, (ed.), Elsevier, Amsterdam, pp. 179-218.

Snelgrove, P.V.R. and C.R. Smith, 2002: A riot of species in an environmental calm: The paradox of the species-rich deep-sea floor. Oceanography and Marine Biology: An Annual Review, **40**, 311-342.

Song, Y., B. Chen, M. Nishio, and M. Akai, 2005: The study on density change of carbon dioxide seawater solution at high pressure and low temperature. *Energy*, **30**(11-12) 2298-2307.

Sorai, M. and T. Ohsumi, 2005: Ocean uptake potential for carbon dioxide sequestration. *Geochemical Journal*, **39**(1) 29-45.

Steinberg, M., 1985: Recovery, disposal, and reuse of CO_2 for atmospheric control. *Environmental Progress*, **4**, 69-77.

Stramma, L., D. Kieke, M. Rhein, F. Schott, I. Yashayaev, and K. P. Koltermann, 2004: Deep water changes at the western boundary of the subpolar North Atlantic during 1996 to 2001. Deep-Sea Research Part I, **51**(8), 1033-1056.

Sundfjord, A., A. Guttorm, P.M. Haugan, and L. Golmen, 2001: Oceanographic criteria for selecting future sites for CO_2 sequestration. Proceedings of the 5th International Conference on Greenhouse Gas Control Technologies, August 13th-16th 2000, Cairns Australia, CSIRO pp. 505-510.

Swett, P., D. Golumn, E. Barry, D. Ryan and C. Lawton, 2005: Liquid carbon dioxide/pulverized limestone globulsion delivery system for deep ocean storage. Proceedings, Seventh International Conference on Greenhouse Gas Control Technologies.

Tamburri, M.N., E.T. Peltzer, G.E. Friederich, I. Aya, K. Yamane, and P.G. Brewer, 2000: A field study of the effects of CO_2 ocean disposal on mobile deep-sea animals. *Marine Chemistry*, **72**(2-4), 95-101.

Teng, H., A. Yamasaki, and Y. Shindo, 1996: The fate of liquid CO_2 disposed in the ocean. *International Energy*, **21**(9), 765-774.

Teng, H., A. Yamasaki, and Y. Shindo, 1999: The fate of CO_2 hydrate released in the ocean. *International Journal of Energy Research*, **23**(4), 295-302.

Thistle, D., K.R. Carman, L. Sedlacek, P.G. Brewer, J.W. Fleeger, and J.P. Barry, 2005: Deep-ocean, sediment-dwelling animals are sensitive to sequestered carbon dioxide. *Marine Ecology Progress Series*, **289**, 1-4.

Train, R.E., 1979: Quality criteria for water, Publ Castlehouse Publications Ltd. UK. 256pp

Tsouris, C., P.G. Brewer, E. Peltzer, P. Walz, D. Riestenberg, L. Liang, and O.R. West, 2004: Hydrate composite particles for ocean carbon sequestration: field verification. *Environmental Science and Technology*, **38**(8), 2470-2475.

Tsushima, S., S. Hirai, H. Sanda, and S. Terada, 2002: Experimental studies on liquid CO_2 injection with hydrate film and highly turbulent flows behind the releasing pipe, In Proceedings of the Sixth International Conference on Greenhouse Gas Control Technologies, Kyoto, pp. 137.

Van Cappellen, P., E. Viollier, A. Roychoudhury, L. Clark, E. Ingall, K. Lowe, and T. Dichristina, 1998: Biogeochemical cycles of manganese and iron at the oxic-anoxic transition of a stratified marine basin (Orca Basin, Gulf of Mexico). *Environmental Science and Technology*, **32**(19), 2931-2939.

Vetter, E.W. and C.R. Smith, 2005: Ecological effects of deep-ocean CO_2 enrichment: Insights from natural high-CO_2 habitats. *Journal of Geophysical Research*, **110**.

Wannamaker, E.J. and E.E. Adams, 2002: Modelling descending carbon dioxide injections in the ocean. Proceedings of the 6th International Conference on Greenhouse Gas Control Technologies, 30th September-4th October, Kyoto, Japan.

West, O.R., C. Tsouris, S. Lee, S.D. Mcallum, and L. Liang, 2003: Negatively buoyant CO_2-hydrate composite for ocean carbon sequestration. *AIChE Journal*, **49**(1), 283-285.

Wheatly, M.G. and R.P. Henry, 1992: Extracellular and intracellular acid–base regulation in crustaceans. *Journal of Experimental Zoology*, **263**(2): 127-142.

Wickett, M.E., K. Caldeira, and P.B. Duffy, 2003: Effect of horizontal grid resolution on simulations of oceanic CFC-11 uptake and direct injection of anthropogenic CO_2. *Journal of Geophysical Research*, **108**.

Wigley, T.M.L., R. Richels, and J.A. Edmonds, 1996: Economic and environmental choices in the stabilization of atmospheric CO_2 concentrations. *Nature*, **379**, 240-243.

Wolff, E.W., J. Seager, V.A. Cooper, and J. Orr, 1988: *Proposed environmental quality standards for list II substances in water:* pH. Report ESSL TR259 Water Research Centre, Medmenham, UK. 66 pp.

Xu, Y., J. Ishizaka, and S. Aoki, 1999: Simulations of the distributions of sequestered CO_2 in the North Pacific using a regional general circulation model. *Energy Conversion and Management*, **40**(7), 683-691.

Yamashita, S., R.E. Evans, and T.J. Hara, 1989: Specificity of the gustatory chemoreceptors for CO_2 and H^+ in rainbow trout (Oncorhynchus mykiss). *Canadian Special Publication of Fisheries and Aquatic Sciences*, **46**(10), 1730-1734.

Zeebe, R.E. and D. Wolf-Gladrow, 2001: CO_2 in Seawater Equilibrium, Kinetics, Isotopes. *Elsevier Oceanography Series*, 65, Amsterdam, 346 pp.

Zondervan, I., R.E. Zeebe, B. Rost, and U. Riebesell, 2001: Decreasing marine biogenic calcification: A negative feedback on rising atmospheric pCO_2. *Global Biogeochemical Cycles*, **15**.

7

Mineral carbonation and industrial uses of carbon dioxide

Coordinating Lead Author
Marco Mazzotti (Italy and Switzerland)

Lead Authors
Juan Carlos Abanades (Spain), Rodney Allam (United Kingdom), Klaus S. Lackner (United States),
Francis Meunier (France), Edward Rubin (United States), Juan Carlos Sanchez (Venezuela), Katsunori
Yogo (Japan), Ron Zevenhoven (Netherlands and Finland)

Review Editors
Baldur Eliasson (Switzerland), R.T.M. Sutamihardja (Indonesia)

Contents

EXECUTIVE SUMMARY

This Chapter describes two rather different options for carbon dioxide (CO_2) storage: (i) the fixation of CO_2 in the form of inorganic carbonates, also known as 'mineral carbonation' or 'mineral sequestration', and (ii) the industrial utilization of CO_2 as a technical fluid or as feedstock for carbon containing chemicals.

In the case of mineral carbonation (see Section 7.2), captured CO_2 is reacted with metal-oxide bearing materials, thus forming the corresponding carbonates and a solid byproduct, silica for example. Natural silicate minerals can be used in artificial processes that mimic natural weathering phenomena, but also alkaline industrial wastes can be considered. The products of mineral carbonation are naturally occurring stable solids that would provide storage capacity on a geological time scale. Moreover, magnesium and calcium silicate deposits are sufficient to fix the CO_2 that could be produced from the combustion of all fossil fuels resources. To fix a tonne of CO_2 requires about 1.6 to 3.7 tonnes of rock. From a thermodynamic viewpoint, inorganic carbonates represent a lower energy state than CO_2; hence the carbonation reaction is exothermic and can theoretically yield energy. However, the kinetics of natural mineral carbonation is slow; hence all currently implemented processes require energy intensive preparation of the solid reactants to achieve affordable conversion rates and/or additives that must be regenerated and recycled using external energy sources. The resulting carbonated solids must be stored at an environmentally suitable location. The technology is still in the development stage and is not yet ready for implementation. The best case studied so far is the wet carbonation of the natural silicate olivine, which costs between 50 and 100 US\$/tCO_2 stored and translates into a 30-50% energy penalty on the original power plant. When accounting for the 10-40% energy penalty in the capture plant as well, a full CCS system with mineral carbonation would need 60-180% more energy than a power plant with equivalent output without CCS.

The industrial use of CO_2 (see Section 7.3) as a gas or a liquid or as feedstock for the production of chemicals could contribute to keeping captured CO_2 out of the atmosphere by storing it in anthropogenic carbon products. Industrial uses provide a carbon sink, as long as the pool size keeps growing and the lifetime of the compounds produced is long. Neither prerequisite is fulfilled in practice, since the scale of CO_2 utilization is small compared to anthropogenic CO_2 emissions, and the lifetime of the chemicals produced is too short with respect to the scale of interest in CO_2 storage. Therefore, the contribution of industrial uses of captured CO_2 to the mitigation of climate change is expected to be small.

7.1   Introduction

This chapter deals with: (i) the fixation of CO_2 in the form of inorganic carbonates, also known as 'mineral carbonation' or 'mineral sequestration' that is discussed in Section 7.2, and (ii) the industrial uses of CO_2 as a technical fluid or as feedstock for carbon containing chemicals, which is the subject of Section 7.3.

7.2   Mineral carbonation

7.2.1   *Definitions, system boundaries and motivation*

Mineral carbonation is based on the reaction of CO_2 with metal oxide bearing materials to form insoluble carbonates, with calcium and magnesium being the most attractive metals. In nature such a reaction is called silicate weathering and takes place on a geological time scale. It involves naturally occurring silicates as the source of alkaline and alkaline-earth metals and consumes atmospheric CO_2. This chapter deals, however, with so-called mineral carbonation, where high concentration CO_2 from a capture step (see Chapter 3) is brought into contact with metal oxide bearing materials with the purpose of fixing the CO_2 as carbonates (Seifritz, 1990; Dunsmore, 1992; Lackner *et al.*, 1995). Suitable materials may be abundant silicate rocks, serpentine and olivine minerals for example, or on a smaller-scale alkaline industrial residues, such as slag from steel production or fly ash. In the case of silicate rocks, carbonation can be carried out either *ex-situ* in a chemical processing plant after mining and pretreating the silicates, or *in-situ*, by injecting CO_2 in silicate-rich geological formations or in alkaline aquifers. Industrial residues on the other hand can be carbonated in the same plant where they are produced. It is worth noting that products of *in-situ* mineral carbonation and geological storage may be similar for the fraction of the CO_2 injected for geological

storage that reacts with the alkaline or alkaline-earth metals in the cap rock leading to 'mineral trapping' (see Chapter 5.2.2).

In terms of material and energy balances, mineral carbonation can be schematized as illustrated in Figure 7.1, which applies to a power plant with CO_2 capture and subsequent storage through mineral carbonation. With respect to the same scheme for a power plant with capture and either geological or ocean storage (see Figure 1.4) two differences can be observed. First, there is an additional material flux corresponding to the metal oxide bearing materials; this is present as input and also as output, in the form of carbonates, silica, non-reacted minerals and for some input minerals product water. Secondly, for the same usable energy output, the relative amounts of fossil fuels as input and of energy rejected as lower grade heat are different. *In-situ* carbonation is an operation similar to geological storage, while *ex-situ* carbonation involves processing steps requiring additional energy input that are difficult to compensate for with the energy released by the carbonation reaction. Given the similarities of *in-situ* carbonation with geological storage, this chapter will focus on *ex-situ* mineral carbonation. With present technology there is always a net demand for high grade energy to drive the mineral carbonation process that is needed for: (i) the preparation of the solid reactants, including mining, transport, grinding and activation when necessary; (ii) the processing, including the equivalent energy associated with the use, recycling and possible losses of additives and catalysts; (iii) the disposal of carbonates and byproducts. The relative importance of the three items differs depending on the source of the metal oxides, for example whether they are natural silicates or industrial wastes.

Despite this potential energy penalty, interest in mineral carbonation stems from two features that make it unique among the different storage approaches, namely the abundance of metal oxide bearing materials, particularly of natural silicates, and the permanence of storage of CO_2 in a stable solid form. However,

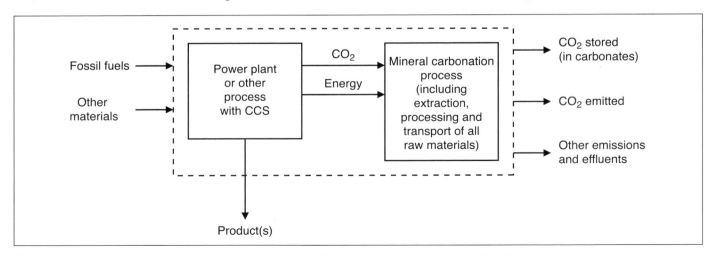

Figure 7.1 Material and energy balances through the system boundaries for a power plant with CO_2 capture and storage through mineral carbonation. The fossil fuel input provides energy both to the power plant that produces CO_2 and to the mineralization process (either directly or indirectly via the power plant). The 'other materials' input serves all processes within the system boundaries and includes the metal oxide bearing materials for mineralization. The 'other emissions' output is made up of the byproducts of the mineralization reaction - silica and possibly water - as well as of non-reacted input materials.

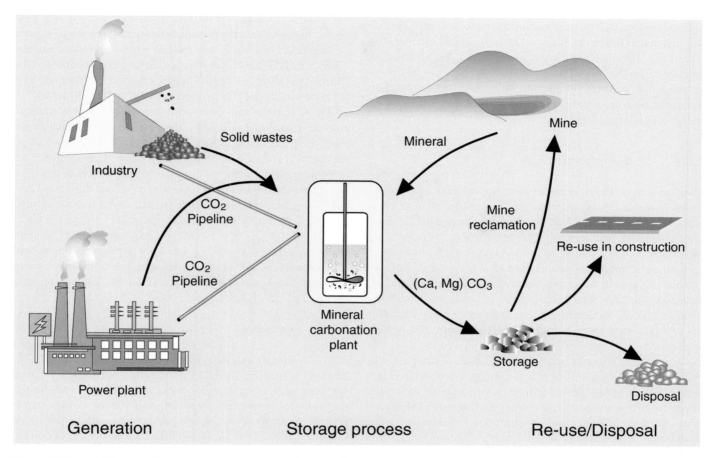

Figure 7.2 Material fluxes and process steps associated with the ex-situ mineral carbonation of silicate rocks or industrial residues (Courtesy Energy Research Centre of the Netherlands (ECN)).

mineral carbonation is today still an immature technology. Studies reported in the literature have not yet reached a level where a thorough assessment of the technology, potential, costs and impacts is possible.

7.2.2   Chemistry of mineral carbonation

When CO_2 reacts with metal oxides (indicated here as MO, where M is a divalent metal, e.g., calcium, magnesium, or iron) the corresponding carbonate is formed and heat is released according to the following chemical reaction:

$$MO + CO_2 \rightarrow MCO_3 + heat \qquad (1)$$

The amount of heat depends on the specific metal and on the material containing the metal oxide. In general this is a large fraction (up to 46% in the case of calcium oxide) of the heat released by the upstream combustion process forming CO_2 (393.8 kJ $mol^{-1}CO_2$ for combustion of elemental carbon). In the case of a few natural silicates the following exothermic chemical reactions take place (in all cases heat values are given per unit mol of CO_2 and standard conditions 25°C and 0.1 MPa, Robie *et al.* 1978):

Olivine:

$$Mg_2SiO_4 + 2CO_2 \rightarrow 2MgCO_3 + SiO_2$$
$$+ 89 \text{ kJ mol}^{-1}CO_2 \qquad (2a)$$

Serpentine:

$$Mg_3Si_2O_5(OH)_4 + 3\,CO_2 \rightarrow 3MgCO_3 + 2SiO_2 + 2H_2O$$
$$+ 64 \text{ kJ mol}^{-1}CO_2 \qquad (2b)$$

Wollastonite:

$$CaSiO_3 + CO_2 \rightarrow CaCO_3 + SiO_2 + 90 \text{ kJ mol}^{-1}CO_2 \qquad (2c)$$

Since the reaction releases heat, the formation of carbonates is thermodynamically favoured at low temperature, whereas at high temperature (above 900°C for calcium carbonate and above 300°C for magnesium carbonate, at a CO_2 partial pressure of one bar) the reverse reaction, that is calcination, is favoured. The representative member of the olivine family considered in the first reaction above is forsterite, which is iron-free. In nature most olivines contain some iron that can form iron oxides or siderite ($FeCO_3$).

Even at the low partial pressure of atmospheric CO_2 and at ambient temperature, carbonation of metal oxide bearing minerals occurs spontaneously, though on geological time scales (Robie *et al.*, 1978; Lasaga and Berner, 1998). Limitations arise from the formation of silica or carbonate layers on the mineral surface during carbonation that tend to hinder further reaction and to limit conversion (Butt *et al.*, 1996) and from the rate of CO_2 uptake from the gas phase in the case of aqueous reactions. The challenge for mineral carbonation is to find ways to accelerate carbonation and to exploit the heat of reaction within the environmental constraints, for example with minimal energy and material losses.

7.2.3   *Sources of metal oxides*

Most processes under consideration for mineral carbonation focus on metal oxide bearing material that contains alkaline-earth metals (such as calcium and magnesium) as opposed to alkali metals (such as sodium and potassium) whose corresponding carbonates are very soluble in water. Oxides and hydroxides of calcium and magnesium would be the ideal source materials, but because of their reactivity they are also extremely rare in nature. Therefore, suitable metal oxide bearing minerals may be silicate rocks or alkaline industrial residues, the former being abundant but generally difficult to access and the latter scarcer but easily available.

Among silicate rocks, mafic and ultramafic rocks are rocks that contain high amounts of magnesium, calcium and iron and have a low content of sodium and potassium. Some of their main mineral constituents are olivines, serpentine, enstatite ($MgSiO_3$), talc ($Mg_3Si_4O_{10}(OH)_2$) and wollastonite. Although molar abundances of magnesium and calcium silicates in the Earth's crust are similar, rocks containing magnesium silicate exhibit a higher MgO concentration (up to 50% by weight, corresponding to a theoretical CO_2 storage capacity of 0.55 kg CO_2/kg rock), than rocks containing calcium silicates, for example basalts, that have CaO content of about 10% by weight only (with a theoretical CO_2 storage capacity of 0.08 kg CO_2/kg rock) (Goff and Lackner, 1998). Deposits of wollastonite, the most calcium-rich silicate, are much rarer than those of magnesium-rich silicates.

Serpentine and olivine are mainly found in ophiolite belts – geological zones where colliding continental plates lead to an uplifting of the earth's crust (Coleman 1977). For example, considering ultramafic deposits containing serpentine and olivine in the Eastern United States and in Puerto Rico, it was found that they have R_{CO2} values between 1.97 and 2.51, depending on purity and type (the R_{CO2} is the ratio of the mass of mineral needed to the mass of CO_2 fixed when assuming complete conversion of the mineral upon carbonation, that is the reciprocal of the theoretical CO_2 storage capacity introduced above). Peridotites and serpentinites exceed the total Mg requirement to neutralize the CO_2 from all worldwide coal resources estimated at 10,000 Gt (Lackner *et al.*, 1995). Specific ore deposits identified in two studies in the USA and Puerto Rico add to approximately $300GtCO_2$ (Goff and Lackner, 1998; Goff *et al.*, 2000). This should be compared to CO_2 emissions of about 5.5 $GtCO_2$ in the United States and about 24 $GtCO_2$/yr[-1] worldwide. No comprehensive mapping of the worldwide storage potential in ophiolite belts has been reported. However, their total surface exposure is estimated to be of the order of 1000 km by 100 km (Goff *et al.*, 2000). It is well known however that magnesium silicate reserves are present in all continents, but since they tend to follow present or ancient continental boundaries, they are not present in all countries. The feasibility of their use for *ex-situ* or *in-situ* mineral carbonation is yet to be established (Brownlow, 1979; Newall *et al.*, 2000).

On a smaller-scale, industrial wastes and mining tailings provide sources of alkalinity that are readily available and reactive. Even though their total amounts are too small to substantially reduce CO_2 emissions, they could help introduce the technology. Waste streams of calcium silicate materials that have been considered for mineral carbonation include pulverized fuel ash from coal fired power plants (with a calcium oxide content up to 65% by weight), bottom ash (about 20% by weight CaO) and fly ash (about 35% by weight CaO) from municipal solid waste incinerators, de-inking ash from paper recycling (about 35% by weight CaO), stainless steel slag (about 65% by weight CaO and MgO) and waste cement (Johnson, 2000; Fernández Bertos *et al.*, 2004; Iizuka *et al.*, 2004).

7.2.4   *Processing*

7.2.4.1   *Mining and mine reclamation*
Mining serpentine would not differ substantially from conventional mining of other minerals with similar properties, for example copper ores. Serpentine and olivine are both mined already, although rarely on the scale envisioned here (Goff and Lackner, 1998; Goff *et al.*, 2000). Like in other mining operations, disposal of tailings and mine reclamation are important issues to consider. Tailing disposal depends on the material characteristics – particle size and cohesion, moisture content and chemical stability against natural leaching processes – and these depend in turn on the specific process. It is likely that carbonation plants will be located near the metal oxide bearing material, either the factory producing the residues to be treated or the silicate mine, to avoid transport of solid materials (see Figure 7.2).

Economies of scale applying to today's mining technology suggest a minimum mining operation of 50,000 to 100,000 tonnes day[-1] (Hartman, 1992), which translates into a minimum mineable volume of about 0.3 km^3 for a mine with a 30 year life. This is a rather small size for ophiolite ore bodies, which are often kilometres wide and hundreds of meters thick (Goff and Lackner, 1998; Goff *et al.*, 2000; Newall *et al.*, 2000). Since coal, in contrast to ophiolite bodies, occurs in thin seams and is buried under substantial overburden, it has been argued that a typical above ground coal mine must move more material (Lackner *et al.*, 1995) and disturb a far larger area (Ziock and Lackner, 2000) for the same amount of carbon atoms treated than the equivalent ophiolite mine, assuming maximum conversion of the mineral to carbonate (one carbon atom yields one CO_2 molecule upon combustion, which has to be fixed in one molecule of carbonate).

Serpentine can take many different forms, from decorative stones to chrysotile asbestos (O'Hanley, 1996). The possibility of encountering asbestos requires adequate precautions. With current best practice it would reportedly not be an obstacle

(Newall et al., 2000). Moreover, since the asbestos form of serpentine is the most reactive, reaction products are expected to be asbestos free (O'Connor et al., 2000). Mineral carbonation could therefore remediate large natural asbestos hazards that occur in certain areas, in California for example (Nichols, 2000).

7.2.4.2   Mineral pretreatment

Mineral pretreatment, excluding the chemical processing steps, involves crushing, grinding and milling, as well as some mechanical separation, for example magnetic extraction of magnetite (Fe_3O_4).

7.2.4.3   CO₂ pre-processing

Mineral carbonation requires little CO_2 pre-processing. If CO_2 is pipelined to the disposal site, the constraints on pipeline operations are likely to exceed pre-processing needs for mineral carbonation. The current state of research suggests that CO_2 should be used at a pressure similar to the pipeline pressure, thus requiring minimal or no compression (Lackner, 2002; O'Connor et al., 2002). Purity demands in carbonation are minimal; acidic components of the flue gas could pass through the same process as they would also be neutralized by the base and could probably be disposed of in a similar manner. Most carbonation processes would preheat CO_2, typically to between 100°C and 150°C for aqueous processes, whereas in gas-solid reactions temperatures could reach 300°C to 500°C (Butt et al., 1996).

7.2.4.4   Carbonation reaction engineering

The simplest approach to mineral carbonation would be the reaction of gaseous CO_2 with particulate metal oxide bearing material at suitable temperature and pressure levels. Unfortunately, such direct gas-solid reactions are too slow to be practical in the case of the materials mentioned in Section 7.2.3 (Newall et al., 2000) and are only feasible at reasonable pressures for refined, rare materials like the oxides or hydroxides of calcium and magnesium (Butt and Lackner, 1997; Bearat et al., 2002; Zevenhoven and Kavaliauskaite, 2004). As a result, mineral carbonation without refined materials cannot directly capture CO_2 from flue gases, but could possibly in the case of pressurized CO_2 rich gases from IGCC plants.

Since the direct fixation of carbon dioxide on solid unrefined material particles seems at present not feasible, the alternative requires the extraction of the metal from the solid. This can be accomplished by suspending the solid material in an aqueous solution and by letting it dissolve and release metal ions, for example calcium or magnesium ions. These ions come in contact with carbonic acid (H_2CO_3) that is formed in the same solution upon carbon dioxide dissolution. Conditions can be achieved where the carbonate and the byproducts – silica in the case of silicate carbonation for example – precipitate. This involves proper choice of the operating parameters of this single-step or multi-step process – particularly temperature, concentration of possible additives and CO_2 pressure (that controls the carbonic acid concentration in solution). At the end of the operation a suspension of fine particles of carbonate, byproducts and non-reacted solid materials remains. These have to be separated by filtration and drying from the solution from which residual metal ions and additives are to be quantitatively recovered.

This wet process scheme is currently in the research phase and has to overcome three major hurdles to become cost-effective and to be considered as a viable option for carbon storage: (i) acceleration of the overall rate of the process, which may be limited by the dissolution rate of the metal oxide bearing material; (ii) elimination of the interference between the concomitant metal oxide dissolution and carbonate precipitation; (iii) complete recovery of all the chemical species involved, if additives are used.

Mineral carbonation starting from natural silicates is a slow process that can be kinetically enhanced by raising the temperature, although thermodynamics are a limiting factor. In aqueous systems, this is typically kept below 200°C, since high temperature favours gaseous CO_2 over precipitated carbonates. It is believed that the metal oxide dissolution constitutes the rate-limiting step and most research efforts have been devoted to finding ways to speed up the metal extraction from the solid input materials. This could be achieved either by activating the mineral to make it more labile and reactive, or by enhancing the metal oxide extraction through the presence of additives or catalysts in solution. Activation can take different forms, namely heat-treatment at 650°C for serpentine (Barnes et al., 1950; Drägulescu et al., 1972; O'Connor et al., 2000) and ultra-fine (attrition) grinding for olivine and wollastonite (O'Connor et al., 2002; Kim and Chung, 2002). The energy cost of activation has been estimated to be of 300 kWh t^{-1} of mineral and 70–150 kWh t^{-1} of mineral for thermal and mechanical activation, respectively (O'Connor et al., 2005). Carbonation has been successfully performed after such pretreatment, but it is so expensive and energy-intensive that its feasibility is questionable (see Box 7.1 and O'Connor et al., 2005). Dissolution catalysts that can be added to the aqueous solution include strong and weak acids (Pundsack, 1967; Lackner et al., 1995; Fouda et al., 1996; Park et al., 2003; Maroto-Valer et al., 2005), bases (Blencoe et al., 2003) and chelating agents to extract SiO_2 or MgO groups from the mineral (Park et al., 2003). All three approaches have been studied and at least partially experimentally tested, but in all cases catalyst recovery represents the key hurdle. It is worth noting that the carbonation of metal oxides from industrial wastes can be faster than that of natural silicates (Johnson, 2000; Fernández Bertos et al., 2004; Huijgen et al., 2004; Iizuka et al., 2004; Stolaroff et al., 2005).

Hydrochloric acid (HCl) dissolution of serpentine or olivine was proposed first (Houston, 1945; Barnes et al., 1950; Wendt et al., 1998a). The process requires a number of steps to precipitate magnesium hydroxide ($Mg(OH)_2$), which can then directly react with gaseous CO_2, and to recover HCl. Exothermic and endothermic steps alternate and heat recovery is not always possible, thus making the overall process very energy-intensive and not viable (Wendt et al., 1998a; Newall et al., 2000; Lackner, 2002). Likewise, strong alkaline solutions (with NaOH) will dissolve the silica from the magnesium

Box 7.1 Wet mineral carbonation process.

A comprehensive energy and economic evaluation of the single-step wet carbonation process has been reported (O'Connor et al., 2005). Though limited to the specific carbonation process illustrated in Figure 7.3, this study is based on about 600 experimental tests and looks not only at the fundamental and technical aspects of the process, but also at the matching of carbon dioxide sources and potential sinks that in this case are natural silicate deposits. In particular, seven large ultramafic ores in the USA have been considered (two olivines, four serpentines (three lizardites and one antigorite) and one wollastonite). Three are located on the west coast, three on the east coast and one in Texas. The selection of the seven ores has also been based on considerations of regional coal consumption and potential CO_2 availability.

The three different minerals exhibit different reactivity, measured as the extent of the carbonation reaction after one hour under specified operating conditions. A trade-off has been observed between the extent of reaction and mineral pretreatment, thus higher reactivity is obtained for more intense pretreatment, which represents an energy cost. Mechanical activation is effective for the olivine and the wollastonite and involves the use of both conventional rod and ball milling techniques with an energy consumption of up to about 100 kWh t^{-1} mineral (standard pretreatment) and ultra-fine grinding for up to more than 200 kWh t^{-1} mineral (activated process). Conversion is no more than 60% in the former case and up to above 80% in the latter. In the case of the serpentine, after milling (standard pretreatment), thermal activation at 630°C is effective for the antigorite (up to 92% conversion) but only partially for the lizardite (maximum conversion not larger than 40%) and requires an energy consumption of about 350 kWh t^{-1} mineral. Optimal operating conditions for this wet process are mineral dependent and correspond to 185°C and 15 MPa for the olivine, 155°C and 11.5 MPa for the heat treated serpentine, and 100°C and 4 MPa for the wollastonite. In the first two cases, the carbonation reaction takes place in the presence of 0.64 mol L^{-1} sodium bicarbonate and 1 mol L^{-1} sodium chloride.

Table 7.1 Mineral carbonation storage costs for CO_2.

Ore (type of pre-treatment)	Conversion after 1 hour (%)	Cost (US$/t ore)	Energy input[a] (kWh/tCO$_2$ stored)	Cost (US$/tCO$_2$ stored)
Olivine (standard)	61	19	310	55
Olivine (activated)	81	27	640	59
Lizardite (standard)	9	15	180	430
Lizardite (activated)	40	44	180+2120=2300	210
Antigorite (standard)	62	15	180	250
Antigorite (activated)	92	48	180+830=1010	78
Wollastonite (standard)	43	15	190	91
Wollastonite (activated)	82	19	430	64

[a] The study assumes a coal fired power plant with 35% efficiency, corresponding to one tonne of CO_2 released per 1000 kWh electricity. The equivalent heat value for the same coal input is then 2,850 kWh. The two items in the sum break the total energy input into electrical + thermal; in all other cases it is pure electrical energy.

Process costs have been calculated for these seven ores in the case of both standard mineral pretreatment and activated process. Costs include only storage, thus neither CO_2 capture nor CO_2 transport and are based on the assumption that CO_2 is received pure at 15 MPa at the plant. Investment costs are calculated accounting for the different reactor costs depending on the different operating conditions corresponding to the different mineral ores. Storage costs are calculated per tonne of silicate ore and per tonne of CO_2 stored and are complemented by the energy consumption per tonne of CO_2 stored in the above Table. The table highlights a trade-off between energy input associated with the pretreatment procedure and cost per unit carbon dioxide stored. Assuming that the cheapest technology is used for each mineral, costs range from 55 US$/tCO$_2$ stored for olivine (standard pretreatment), to 64 US$/tCO$_2$ stored for wollastonite (activated), to 78 US$/tCO$_2$ stored for antigorite (activated), to 210 US$/tCO$_2$ stored for lizardite (activated). Since the last case requires too large an energy input, the cost of the most realistic technologies falls into a range from 50 to 100 US$/tCO$_2$ stored.

silicate, thus allowing for further digestion of the remaining $(Mg(OH)_2)$; however, also in this case the recovery of the NaOH catalyst seems to be very difficult (Blencoe *et al.*, 2003). To overcome the substantial energy penalty of water evaporation in the hydrochloric acid process, it was proposed to dissolve the silicate minerals in a magnesium chloride melt in order either to precipitate $Mg(OH)_2$ as before or to allow for direct carbonation in the melt (Wendt *et al.*, 1998a; 1998b; 1998c; 1998d). No experimental demonstration of this process has been provided, possibly also because of the corrosive conditions of the reaction; energy and material balances indicate that either version of the process will hardly be viable (Newall *et al.*, 2000; Haywood *et al.*, 2001).

Weaker acids that might reduce the energy requirements for recovery include acetic acid (Kakizawa *et al.*, 2001), oxalic acid (Park *et al.*, 2003), orthophosphoric acid (Park *et al.*, 2003) and ammonium bisulphate (Pundsack 1967). Among the possible chelating agents that keep either silicates or magnesium ions in solution by forming water-soluble complexes, is EDTA – ethylene-diamine-tetra-acetic acid (Carey *et al.*, 2003; Park *et al.*, 2003; Park and Fan, 2004). Citric acid is also effective because it combines its acidic properties with strong chelating properties (Carey *et al.*, 2003). All these additives have been proven to enhance the dissolution of silicate minerals, but only in the acetic acid case has a complete process scheme, including acid recovery, been described and evaluated (Kakizawa *et al.*, 2001). This is based on two steps, whereby the metal ions are extracted first using acetic acid and then the carbonate is

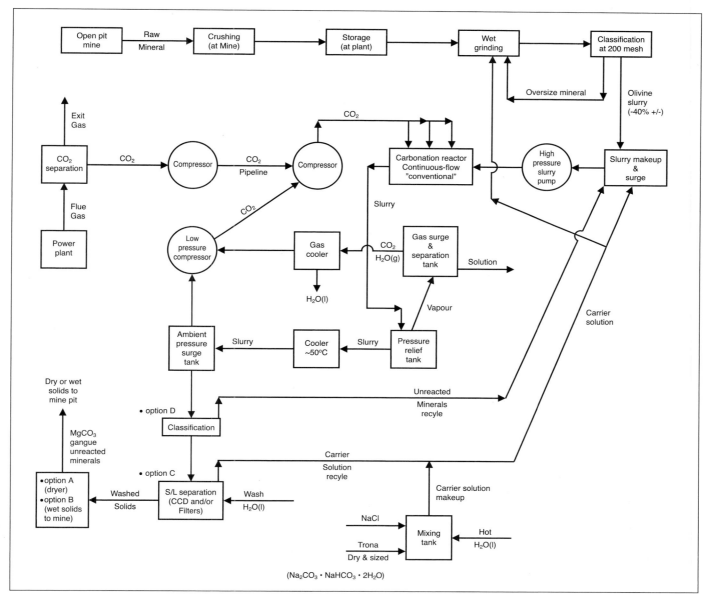

Figure 7.3 Process scheme of the single-step mineral carbonation of olivine in aqueous solution (Courtesy Albany Research Centre). 'Single-step' indicates that mineral dissolution and carbonate precipitation take place simultaneously in the same carbonation reactor, whereas more steps are of course needed for the whole process, including preparation of the reactants and separation of the products.

precipitated upon CO_2 addition. Acetic acid remains in solution as either calcium or magnesium acetate or free acid and can be recycled. The process has only been demonstrated for wollastonite. Experimental conversion levels of the wollastonite have not exceeded 20% (Kakizawa *et al.*, 2001).

7.2.4.5   A worked out example: single-step carbonation

Figure 7.3 illustrates the single step wet mineral carbonation process that can be applied to natural silicates as well as to industrial residues, for example steel slag (Huijgen *et al.*, 2004). The figure refers to the carbonation of olivine, whereby the mineral is ground first. Subsequently it is dissolved in an aqueous solution of sodium chloride (NaCl, 1 mol L^{-1}) and sodium bicarbonate ($NaHCO_3$, 0.64 mol L^{-1}) in contact with high pressure CO_2 and carbonated therein (O'Connor *et al.*, 2002; O'Connor *et al.*, 2005). The additives are easily recovered upon filtration of the solid particles, since the sodium and chloride ions do not participate in the reaction and remain in solution, whereas the bicarbonate ion is replenished by contacting the solution in the carbonation reactor with the CO_2 atmosphere. A maximum conversion of 81% in one hour was obtained with an olivine of 37 μm particle size, at a temperature of 185°C and a CO_2 partial pressure of 15 MPa. An important element of the process scheme in Figure 7.3 is the classification (sieving) that allows separating the carbonate and silica products from the olivine that has to be recycled. This is possible since non-reacted olivine minerals are coarse, whereas the carbonate and silica consist of finer particles (O'Connor *et al.*, 2002). An additional difficulty of single-step carbonation is when, upon extraction of the metal oxide from the solid particles, a silica layer forms or a carbonate layer precipitates on the particles themselves, thus hindering further dissolution. Experimental evidence indicates that this does not occur in the case of olivine (O'Connor *et al.*, 2002), whereas it does occur in the case of steel slag (Huijgen *et al.*, 2004).

Using the process scheme illustrated in Figure 7.3, it is possible to calculate the material balances by considering that the molecular mass of carbon dioxide is 44.0 g mol^{-1}, of magnesium carbonate is 84.3 g mol^{-1}, of silica is 60.1 g mol^{-1} and of olivine is 140.7 g mol^{-1}. For the sake of simplicity only two assumptions are made, namely the degree of conversion in the carbonation reactor – the fraction of olivine fed to the reactor that is converted to carbonate in a single pass – and the fraction of non-reacted mineral in the classifier that is not recycled, but ends up with the material for disposal. Based on the stoichiometry of the carbonation reaction, 1.6 tonnes of olivine would be needed to fix one tonne of CO_2, thus producing 2.6 tonnes of solid material for disposal. Assuming 90% carbonation conversion and 10% losses in the classifier, 1.62 tonnes of olivine would be needed and 2.62 tonnes of solids per tonne of CO_2 mineralized would be for disposal. Assuming only 50% conversion and 20% losses, for one tonne of CO_2 stored, 1.87 tonnes of olivine would be needed and 2.87 tonnes would be disposed of. In the latter case however the carbonation reactor would be twice as big as in the former case.

Olivine has the highest concentration of reactive magnesium

oxide among the natural minerals (57% by weight). Other minerals in general contain a lower concentration. For pure serpentine the magnesium oxide concentration is about 44% and for typical ores about 50% of that of the pure mineral. Therefore, the mineral feedstock required to fix 1 tonne of CO_2 in carbonates is between 1.6 and 3.7 tonnes and the process yields between 2.6 and 4.7 tonnes of products to be handled. The carbonation process consumes energy and thus causes CO_2 emissions that reduce the net storage of CO_2 accordingly. For the olivine carbonation process, having the lowest unit cost among those described in Box 7.1, the energy requirement is 1.1 GJ/tCO_2. If this is provided by the same coal derived electricity it would cause CO_2 emissions equal to 30% of the fixed CO_2.

7.2.5   Product handling and disposal

Disposal options for mineral carbonates are determined by the mass of the resulting material (see Figure 7.2). It is not cost-effective to ship the bulk of these materials over long distances. As a result the obvious disposal location is at the mine site. As in any large-scale mining operation, the logistics of mining a site and reclaiming it after refilling it with the tailings is substantial, but it does not pose novel problems (Newall *et al.*, 2000). The amount of material to be disposed of is between 50 and 100% by volume more than that originally mined. These volumes are comparable to volumes commonly handled in mining operations and are subject to standard mine reclamation practice (Lackner *et al.*, 1997; Newall *et al.*, 2000).

The fine grinding of the mineral ore might allow for the extraction of valuable mineral constituents. Serpentine and olivine mines could provide iron ore that either would be removed as magnetite by magnetic separation or result from chemical precipitation during magnesium extraction, yielding concentrated iron oxide or hydroxide (Park and Fan, 2004). Peridotite rocks may contain chromite, elements like nickel and manganese and also elements in the platinum group, but how these can be recovered has still to be studied (Goff and Lackner, 1998). It has been suggested, that magnesium carbonate and silica may find uses as soil enhancers, roadfill or filler for mining operations. Eventually mineral carbonation would have to operate at scales that would saturate any product or byproduct market, but products and byproducts, when usable, could help make a demonstration of the process more viable (Lackner *et al.*, 1997; Goff and Lackner, 1998).

7.2.6   Environmental impact

The central environmental issue of mineral carbonation is the associated large-scale mining, ore preparation and waste-product disposal (Goff and Lackner, 1998). It can directly lead to land clearing and to the potential pollution of soil, water and air in surrounding areas. It may also indirectly result in habitat degradation. An environmental impact assessment would be required to identify and prevent or minimize air emissions, solid waste disposal, wastewater discharges, water use, as well as social disturbances. As for many other mining activities, the preventing and mitigating practices are relatively basic and well developed.

Land clearing: The amount of material required to store CO_2 involves extensive land clearing and the subsequent displacement of millions of tonnes of earth, rock and soil, increasing the potential for erosion, sedimentation and habitat loss in the area. Access roads would also lead to clearing of vegetation and soil. Standard practices recommended to minimize these impacts include storage of topsoil removed for use in future reclamation activities, use of existing tracks when constructing access roads and pipelines and use of drainage and sediment collection systems to catch runoff or divert surface water, minimizing erosion.

Air quality: Mining activities like blasting, drilling, earth moving and grading can generate dust and fine particulate matter that affect visibility and respiration and pollute local streams and vegetation. Dust prevention measures are widely applied at mining operations today, but if not properly controlled, dust can threaten human respiratory health. This is particularly important in serpentine mining because serpentine often contains chrysotile, a natural form of asbestos. Even though chrysotile is not as hazardous as amphibole asbestos (tremolite, actinolite) (Hume and Rimstidt, 1992), the presence of chrysotile requires covering of exposed veins and monitoring of air quality (Nichols, 2000). On the other hand, mineral carbonation products are asbestos free, as the reaction destroys chrysotile, which reacts faster than other serpentines, even if conversion of the starting material is not complete. This makes mineral carbonation a potentially effective method for the remediation of asbestos in serpentine tailing (O'Connor *et al.*, 2000). The resulting mineral carbonates are inert, but large volumes of powders would also have to be controlled, for example by cementing them together to avoid contamination of soil and vegetation, as well as habitat destruction.

Tailings: Tailings consist of finely ground particles, including ground-up ore and process byproducts. Tailings management systems should be designed and implemented from the earliest stages of the project. Usually tailings are stored in tailings impoundments designed to hold tailings behind earth-fill dams (Newall *et al.*, 2000). Other control measures depend on whether tailings are dry or wet, on particle size and chemical reactivity.

Leaching of metals: Although the low acidity of the resulting byproducts reduces the possibility of leaching, certainty about leaching can only be obtained by conducting tests. If necessary, a lining system would prevent ground water contamination. Leaching containment is also possible without lining where underlying rock has been shown to be impermeable.

Reclamation: To minimize water contamination, restore wildlife habitat and ecosystem health and improve the aesthetics of the landscape, a comprehensive reclamation programme has to be designed during the planning phase of the mining project and be implemented concurrently throughout operations. Concurrent incorporation of reclamation with the mining of the site reduces waste early, prevents clean-up costs and decreases potential liabilities. Land rehabilitation will involve the re-shaping of landform, because the volume of tailings will be larger than the mined rock. The main environmental concern regarding reclamation is major soil movements by erosion or landslides. This can be controlled by adequate vegetation cover and by covering the soil with protective mulch, by maintaining moisture in the soil, or by constructing windbreaks to protect the landform from exposure to high winds.

7.2.7   *Life Cycle Assessment and costs*

At the current stage of development, mineral carbonation consumes additional energy and produces additional CO_2 compared to other storage options. This is shown in Figure 7.1 and is why a Life Cycle Assessment of the specific process routes is particularly important. The potential of mineral carbonation depends on the trade-off between costs associated with the energy consuming steps (mining, pre-processing of the mineral ore, its subsequent disposal and mine reclamation) and benefits (the large potential capacity due to the vast availability of natural metal oxide bearing silicates and the permanence of CO_2 storage).

A life cycle analysis of the mining, size reduction process, waste disposal and site restoration calculated additional annual CO_2 emissions of 0.05 tCO_2/tCO_2 stored (Newall *et al.*, 2000). This included grinding of the rock to particle sizes less than 100 microns; a ratio of 2.6 tonnes of serpentine per tonne of CO_2 was assumed. The cost was assessed to be about 14 US$/t$CO_2$ stored; the capital cost being about 20% of the total. All cost estimates were based on OECD Western labour costs and regulations. The conversion factor from electrical energy to CO_2 emissions was 0.83 tCO_2/MWh electricity. Costs were calculated on the basis of an electricity price of US$ 0.05 kWh^{-1} electricity. Results from other studies were converted using these values (Newall *et al.*, 2000). Other estimates of these costs are between 6 and 10 US$/t$CO_2$ stored, with 2% additional emissions (Lackner *et al.*, 1997).

As far as the scale of mining and disposal is concerned – about 1.6 to 3.7 tonnes of silicate and 2.6 to 4.7 tonnes of disposable materials per tonne of CO_2 fixed in carbonates, as reported in Section 7.2.4 – this is of course a major operation. When considering that one tonne of carbon dioxide corresponds to 0.27 tonnes of carbon only in theory, but in practice to about 2 tonnes of raw mineral due to the overburden, it follows that mineral carbonation to store the CO_2 produced by burning coal would require the installation of a mining industry of a scale comparable to the coal industry itself. Such large mining operations are significant, but placing them in the context of the operations needed for the use of fossil fuels and geological or ocean storage, the volumes are comparable.

The energy requirements and the costs of the carbonation reaction are very much process dependent and more difficult to estimate, due to scarcity of data. The most detailed study has been carried out for the process where the silicates are dissolved in a magnesium chloride melt (Newall *et al.*, 2000). An overall cost (including the operations mentioned in the previous paragraph) of 80 US$/tCO$_2$ stored was obtained, with 27.5% additional CO$_2$ emissions, thus leading to 110 US$/tCO$_2$ avoided. In the case of the two-step acetic acid process, an overall cost of 27 US$/tCO$_2$ avoided has been reported, but the assumptions are based on a rather limited set of experimental data (Kakizawa *et al.*, 2001). A comprehensive energy and economic evaluation of the single step wet carbonation process illustrated in Figure 7.3 has been recently reported (O'Connor *et al.*, 2005) and is discussed in detail in Box 7.1. This study calculates storage costs between 50 and 100 US$/tCO$_2$ stored, with between 30% and 50% of the energy produced needed as input to the mineral carbonation step, i.e. a corresponding reduction of power plant efficiency from 35% for instance to 25% and 18%, respectively. This implies that a full CCS system with mineral carbonation would need 60-180% more energy than a power plant with equivalent output without CCS, when the 10-40% energy penalty in the capture plant is accounted too. No similar economic evaluation is available for either dry mineral carbonation or carbonation using industrial residues. However, it is worth pointing out that the carbonation of toxic wastes may lead to stabilized materials with reduced leaching of heavy metals. Therefore these materials might be disposed of more easily or even used for applications such as in construction work (see Figure 7.2) (Venhuis and Reardon, 2001; Meima *et al.*, 2002).

Once the carbon has been stored through mineral carbonation, there are virtually no emissions of CO$_2$ due to leakage. To the extent that weathering at the disposal site occurs and leaches out magnesium carbonate from the carbonation products, additional CO$_2$ would be bound in the transformation of solid magnesium carbonate to dissolved magnesium bicarbonate (Lackner, 2002). It can therefore be concluded that the fraction of carbon dioxide stored through mineral carbonation that is retained after 1000 years is virtually certain to be 100%. As a consequence, the need for monitoring the disposal sites will be limited in the case of mineral carbonation.

7.2.8   *Future scope*

7.2.8.1   *Public acceptance*
Public acceptance of mineral carbonation is contingent on the broader acceptance of CCS. Acceptance might be enhanced by the fact that this method of storage is highly verifiable and unquestionably permanent. On the downside, mineral carbonation involves large-scale mining and associated environmental concerns: terrain changes, dust pollution exacerbated by potential asbestos contamination and potential trace element mobilization. Generally, public acceptance will require a demonstration that everything possible is done to minimize secondary impacts on the environment.

7.2.8.2   *Gap analysis*
Mineral carbonation technology must reduce costs and reduce the energy requirements associated with mineral pretreatment by exploiting the exothermic nature of the reaction. Mineral carbonation will always be more expensive than most

applications of geological storage, but in contrast has a virtually unlimited permanence and minimal monitoring requirements. Research towards reducing costs for the application of mineral carbonation to both natural silicates and industrial wastes, where the kinetics of the reaction is believed to be more favourable, is ongoing. Moreover, an evaluation is needed to determine the fraction of the natural reserves of silicates, which greatly exceed the needs, that can be effectively exploited for mineral carbonation. This will require thorough study, mapping the resources and matching sources and sinks, as in O'Connor *et al.* (2005). The actual size of the resource base will be significantly influenced by the legal and societal constraints at a specific location. Integrating power generation, mining, carbonation reaction, carbonates' disposal and the associated transport of materials and energy needs to be optimized in a site-specific manner. A final important gap in mineral carbonation is the lack of a demonstration plant.

7.3   Industrial uses of carbon dioxide and its emission reduction potential

7.3.1   *Introduction*

As an alternative to storing captured CO$_2$ in geological formations (see Chapter 5), in the oceans (see Chapter 6), or in mineral form as carbonates (see Section 7.2), this section of the report assesses the potential for reducing net CO$_2$ emissions to the atmosphere by using CO$_2$ either directly or as a feedstock in chemical processes that produce valuable carbon containing products. The utilization of CO$_2$ establishes an inventory of stored CO$_2$, the so-called carbon chemical pool, primarily in the form of carbon-containing fuels, chemicals and other products (Xiaoding and Moulijn, 1996). The production and use of these products involve a variety of different 'life cycles' (i.e., the chain of processes required to manufacture a product from raw materials, to use the product for its intended purpose and ultimately to dispose of it or to reuse it in some fashion). Depending on the product life-cycle, CO$_2$ is stored for varying periods of time and in varying amounts. As long as the recycled carbon remains in use, this carbon pool successfully stores carbon. Withdrawal from this pool, by decay or by disposal typically re-injects this carbon into the atmospheric pool.

CO$_2$ that has been captured using one of the options described in Chapter 3 could reduce net CO$_2$ emissions to the atmosphere if used in industrial processes as a source of carbon, only if the following criteria are met:

1. The use of captured CO$_2$ must not simply replace a source of CO$_2$ that would then be vented to the atmosphere. Replacement of CO$_2$ derived from a lime kiln or a fermentation process would not lead to a net reduction in CO$_2$ emissions, while on the other hand replacement of CO$_2$ derived from natural geological deposits, which would thus be left undisturbed, would lead to a net reduction of CO$_2$ emissions. This would apply to the majority of the CO$_2$ used for enhanced oil recovery in the USA (see Section 5.3.2)

that is currently provided from natural geological deposits (Audus et Oonk, 1997).

2. The compounds produced using captured CO_2 must have a long lifetime before the CO_2 is liberated by combustion or other degradation processes.

3. When considering the use of captured CO_2 in an industrial process, the overall system boundary must be carefully defined to include all materials, fossil fuels, energy flows, emissions and products in the full chain of processes used to produce a unit of product in order to correctly determine the overall (net) CO_2 avoided.

CO_2 reductions solely due to energy efficiency improvements are not within the scope of this report, which is focused on capture and storage rather than efficiency improvements. Similarly while environmental benefits like those obtained in replacing organic solvents with supercritical CO_2 may slightly increase the carbon chemical pool, these primary drivers are not discussed in this report. Similarly, this report specifically excludes all uses of captured CO_2 to replace other chemicals that are released into the atmosphere and that have high greenhouse-gas potential, fluorocarbons for example. This area is covered by the IPCC/TEAP Special Report on Safeguarding the Ozone Layer and the Global Climate System: issues related to Hydrofluorocarbons and Perfluorocarbons (IPCC/TEAP, 2005).

The third point is especially important in any effort to estimate the potential for net CO_2 reductions from the substitution of a CO_2-utilizing process for alternative routes to manufacturing a desired product. In particular, it is essential that the system boundary encompasses all 'upstream' processes in the overall life cycle and does not focus solely on the final production process of interest. The appropriate system boundary is shown schematically in Figure. 7.4 This is an extension of the system boundary diagrams shown earlier in Section 7.2 (Figure 7.1) and in Chapter 1 (Figure 1.4) in the context of a CO_2 capture and storage system. The inputs include all fossil fuels together with all other materials used within the system. The fossil fuel input provides energy to the power or industrial plant, including the CO_2 capture system, as well as the elemental carbon used as building blocks for the new chemical compound. Flows of CO_2, energy and materials pass from the primary fuel-consuming processes to the industrial process that utilizes the captured CO_2. This produces a desired product (containing carbon derived from captured CO_2) together with other products (such as useful energy from the power plant) and environmental emissions that may include CO_2 plus other gaseous, liquid or solid residuals.

Once the overall system has been defined and analyzed in this way, it can also be compared to an alternative system that does not involve the use of captured CO_2. Using basic mass and energy balances, the overall avoided CO_2 can then be assessed as the difference in net emissions associated with the production of a desired product. In general, the difference could be either positive or negative, thus meaning that utilization of CO_2 could result in either a decrease or increase in net CO_2 emissions, depending on the details of the processes being compared. Note that only fossil fuels as a primary energy source are considered in this framework. Renewable energy sources and nuclear power are specifically excluded, as their availability would have implications well beyond the analysis of CO_2 utilization options (see Chapter 8 for further discussion). Note too that other emissions from the process may include toxic or harmful materials, whose flows also could be either reduced or increased by the adoption of a CO_2-based process.

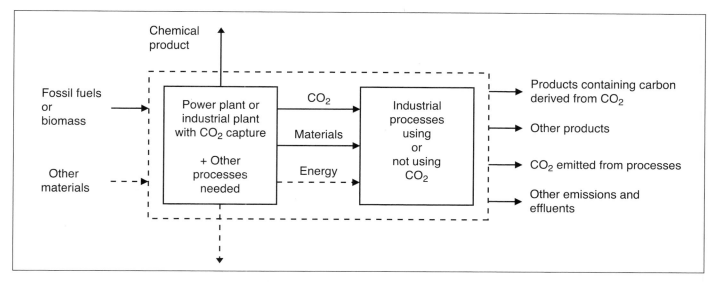

Figure 7.4 Material and energy balances through the system boundaries for a power plant or an industrial plant with CO_2 capture, followed by an industrial process using CO_2. The inputs include all fossil fuels together with all other materials used within the system. The fossil fuel input provides energy to the power or industrial plant, including the CO_2 capture system, as well as the elemental carbon used as building blocks for the new chemical compound. From the primary fuel-consuming processes, flows of CO_2, energy and materials pass to the industrial process, which utilizes the captured CO_2. This produces a desired product (containing carbon, derived from captured CO_2) together with other products (such as useful energy from the power plant) and environmental emissions that may include CO_2 plus other gaseous, liquid or solid residuals.

The application of this framework to the assessment of CO_2 utilization processes is discussed in more detail later in this chapter. First, however, we will examine current uses of CO_2 in industrial processes and their potential for long-term CO_2 storage.

7.3.2 Present industrial uses of carbon dioxide

Carbon dioxide is a valuable industrial gas with a large number of uses that include production of chemicals, for example urea, refrigeration systems, inert agent for food packaging, beverages, welding systems, fire extinguishers, water treatment processes, horticulture, precipitated calcium carbonate for the paper industry and many other smaller-scale applications. Large quantities of carbon dioxide are also used for enhanced oil recovery, particularly in the United States (see Section 5.3.2). Accordingly, there is extensive technical literature dealing with CO_2 uses in industry and active research groups are exploring new or improved CO_2 utilization processes.

Much of the carbon dioxide used commercially is recovered from synthetic fertilizer and hydrogen plants, using either a chemical or physical solvent scrubbing system (see Section 3.5.2). Other industrial sources of CO_2 include the fermentation of sugar (dextrose) used to produce ethyl alcohol:

$$C_6H_{12}O_6 \rightarrow 2C_2H_5OH + 2CO_2 \qquad (3)$$

Industrial CO_2 is also produced from limekilns, such as those used in the production of sodium carbonate and in the Kraft wood pulping process. This involves the heating (calcining) of a raw material such as limestone:

$$CaCO_3 \rightarrow CaO + CO_2 \qquad (4)$$

In some parts of the world, such as the United States, Italy, Norway and Japan, some CO_2 is extracted from natural CO_2 wells. It is also recovered during the production and treatment of raw natural gas that often contains CO_2 as an impurity (see Chapter 2 for more details about CO_2 sources).

A large proportion of all CO_2 recovered is used at the point of production to make further chemicals of commercial importance, chiefly urea and methanol. The CO_2 recovered for other commercial uses is purified, liquefied, delivered and stored mostly as a liquid, typically at 20 bar and $-18°C$ (Pierantozzi, 2003).

Table 7.2 shows the worldwide production and CO_2 usage rates for the major chemical or industrial applications currently using CO_2 (excluding enhanced oil recovery, which is dealt with in Chapter 5). The approximate lifetime of stored carbon before it is degraded to CO_2 that is emitted to the atmosphere is also shown. Such values mean that the fraction of the CO_2 used to produce the compounds in the different chemical classes or for the different applications, which is still stored after the period of time indicated in the last column of Table 7.2 drops to zero.

7.3.3 New processes for CO_2 abatement

7.3.3.1 Organic chemicals and polymers
A number of possible new process routes for the production of chemicals and polymers have been considered in which CO_2 is used as a substitute for other C_1 building blocks, such as carbon monoxide, methane and methanol. The use of CO_2, an inert gas whose carbon is in a highly oxidized state, requires development of efficient catalytic systems and, in general, the use of additional energy for CO_2 reduction. Chemicals that have been considered include polyurethanes and polycarbonates, where the motivation has primarily been to avoid the use of phosgene because of its extreme toxicity, rather than to find a sink for CO_2. The proposed processes can have a lower overall energy consumption than the current phosgene-based routes leading to further CO_2 emission reductions. Current world consumption of polycarbonates is about 2.7 Mt yr^{-1}. If all polycarbonate production was converted to CO_2-based processes the direct consumption of CO_2 would be about 0.6 MtCO$_2$yr^{-1}. Some CO_2

Table 7.2 Industrial applications of CO_2 (only products or applications at the Mtonne-scale): yearly market, amount of CO_2 used, its source, and product lifetime (Aresta and Tommasi, 1997; Hallman and Steinberg, 1999; Pelc et al., 2005). The figures in the table are associated with a large uncertainty.

Chemical product class or application	Yearly market (Mt yr^{-1})	Amount of CO_2 used per Mt product (MtCO$_2$)	Source of CO_2	Lifetime[b]
Urea	90	65	Industrial	Six months
Methanol (additive to CO)	24	<8	Industrial	Six months
Inorganic carbonates	8	3	Industrial, Natural[a]	Decades to centuries
Organic carbonates	2.6	0.2	Industrial, Natural[a]	Decades to centuries
Polyurethanes	10	<10	Industrial, Natural[a]	Decades to centuries
Technological	10	10	Industrial, Natural[a]	Days to years
Food	8	8	Industrial, Natural[a]	Months to years

[a] Natural sources include both geological wells and fermentation.
[b] The fraction of used CO_2 that is still stored after the indicated period of time drops to zero.

savings that are difficult to quantify from current published data are claimed for energy/materials changes in the process.

Similarly, if all world polyurethane production was converted, then direct CO_2 consumption would be about 2.7 $MtCO_2$/yr. However, little progress in commercial application of CO_2-based production has been reported. And as indicated earlier, these possible CO_2 applications directly affect only a very small fraction of the anthropogenic CO_2 emitted to the atmosphere. The net savings in CO_2 would be even smaller or could be negative, as the energy that was available in the hydrocarbon resource is missing in the CO_2 feedstock and unless compensated for by improved process efficiency it would have to be made up by additional energy supplies and their associated CO_2 emissions.

7.3.3.2   Fuel production using carbon dioxide

Liquid carbon-based fuels, gasoline and methanol for example, are attractive because of their high energy density and convenience of use, which is founded in part on a well-established infrastructure. Carbon dioxide could become the raw material for producing carbon-based fuels with the help of additional energy. Since energy is conserved, this cannot provide a net reduction in carbon dioxide emissions as long as the underlying energy source is fossil carbon. If a unit of energy from a primary resource produces a certain amount of CO_2, then producing a fuel from CO_2 will recycle CO_2 but release an equivalent amount of CO_2 to provide the necessary energy for the conversion. Since all these conversion processes involve energy losses, the total CO_2 generated during fuel synthesis tends to exceed the CO_2 converted, which once used up, is also emitted.

Production of liquid carbon-based fuels from CO_2 only reduces CO_2 emissions if the underlying energy infrastructure is not based on fossil energy. For example, one could still use gasoline or methanol rather than converting the transport sector to hydrogen, by using hydrogen and CO_2 as feedstocks for producing gasoline or methanol. The hydrogen would be produced from water, using hydropower, nuclear energy, solar energy or wind energy. As long as some power generation using fossil fuels remains, carbon dioxide for this conversion will be available (Eliasson, 1994). Alternatively, it might be possible to create a closed cycle with CO_2 being retrieved from the atmosphere by biological or chemical means. Such cycles would rely on the availability of cheap, clean and abundant non-fossil energy, as would the hydrogen economy, and as such they are beyond the scope of this report.

Methanol production is an example of the synthesis of liquid fuels from CO_2 and hydrogen. Today a mixture of CO, CO_2 and hydrogen is produced through reforming or partial oxidation or auto thermal reforming of fossil fuels, mainly natural gas. The methanol producing reactions, which are exothermic, take place over a copper/zinc/alumina catalyst at about 260°C (Inui, 1996; Arakawa, 1998; Ushikoshi *et al.*, 1998; Halmann and Steinberg, 1999):

$$CO + 2H_2 \rightarrow CH_3OH \qquad (5)$$

$$CO_2 + 3H_2 \rightarrow CH_3OH + H_2O \qquad (6)$$

Alternatively one could exploit only reaction (6), by using captured CO_2 and hydrogen from water hydrolysis powered for instance by solar energy (Sano *et al.*, 1998).

7.3.3.3   Capture of CO_2 in biomass

Biomass production of fuels also falls into the category of generating fuels from CO_2. With the help of photosynthesis, solar energy can convert water and CO_2 into energetic organic compounds like starch. These in turn can be converted into industrial fuels like methane, methanol, hydrogen or bio-diesel (Larson, 1993). Biomass can be produced in natural or agricultural settings, or in industrial settings, where elevated concentrations of CO_2 from the off-gas of a power plant would feed micro-algae designed to convert CO_2 into useful chemicals (Benemann, 1997). Since biological processes collect their own CO_2, they actually perform CO_2 capture (Dyson, 1976). If the biomass is put to good use, they also recycle carbon by returning it to its energetic state. Biomass production eliminates the need for fossil fuels, because it creates a new generation of biomass-based carbonaceous fuels. As a replacement for fossil energy it is outside the scope of this report. As a CO_2 capture technology, biomass production is ultimately limited by the efficiency of converting light into chemically stored energy. Currently solar energy conversion efficiencies in agricultural biomass production are typically below 1% (300 GJ ha^{-1} yr^{-1} or 1 W m^{-2} (Larson, 1993)). Micro-algae production is operating at slightly higher rates of 1 to 2% derived by converting photon utilization efficiency into a ratio of chemical energy per unit of solar energy (Melis *et al.*, 1998; Richmond and Zou, 1999). Hence the solar energy collection required for micro-algae to capture a power plant's CO_2 output is about one hundred times larger than the power plant's electricity output. At an average of 200 W m^{-2} solar irradiation, a 100 MW power plant would require a solar collection area in the order of 50 km^2.

7.3.4   Assessment of the mitigation potential of CO_2 utilization

This final section aims at clarifying the following points: (i) to what extent the carbon chemical pool stores CO_2; (ii) how long CO_2 is stored in the carbon chemical pool; (iii) how large the contribution of the carbon chemical pool is to emission mitigation.

To consider the first point, the extent of CO_2 storage provided by the carbon chemical pool, it is worth referring again to Table 7.2. As reported there, total industrial CO_2 use is approximately 115 $MtCO_2$ yr^{-1}. Production of urea is the largest consumer of CO_2, accounting for over 60% of that total. To put it in perspective, the total is only 0.5% of total anthropogenic CO_2 emissions – about 24 $GtCO_2$ yr^{-1}. However, it is essential to realize that these figures represent only the yearly CO_2 flux in and out of the carbon chemical pool, and not the actual size of the pool, which is controlled by marketing and product distribution considerations and might be rather smaller than the total yearly CO_2 consumption. Moreover, the contribution to the storage of carbon – on a yearly basis for instance – does not correspond to the size of the pool, but to its size variation on a yearly basis, or in general on its rate of change that might be positive (increase of carbon storage and reduction of CO_2 emissions) or negative (decrease of carbon storage and increase of CO_2 emissions) depending on the evolution of the markets and of the distribution systems (see also Box 7.2 for a quantitative example). Data on the amount of carbon stored as inventory of these materials in the supply chain and on the rate of change of this amount is not available, but the figures in Table 7.2 and the analysis above indicate that the quantity of captured carbon that could be stored is very small compared with total anthropogenic carbon emissions. Thus, the use of captured CO_2 in industrial processes could have only a minute (if any) effect on reduction of net CO_2 emissions.

As to the second point, the duration of CO_2 storage in the carbon chemical pool and typical lifetime of the CO_2 consuming chemicals when in use before being degraded to CO_2 that is emitted to the atmosphere, are given in the last column of Table 7.2 Rather broad ranges are associated with classes of compounds consisting of a variety of different chemicals. The lifetime of the materials produced that could use captured CO_2 could vary from a few hours for a fuel such as methanol, to a few months for urea fertilizer, to decades for materials such as plastics and laminates, particularly those materials used in the construction industry. This indicates that even when there is a net storage of CO_2 as discussed in the previous paragraph, the duration of such storage is limited.

As to the last point, the extent of emission mitigation provided by the use of captured CO_2 to produce the compounds in the carbon chemical pool. Replacing carbon derived from a fossil fuel in a chemical process, for example a hydrocarbon, with captured CO_2 is sometimes possible, but does not affect the overall carbon budget, thus CO_2 does not replace the fossil fuel feedstock. The hydrocarbon has in fact two functions – it provides energy and it provides carbon as a building block. The CO_2 fails to provide energy, since it is at a lower energy level than the hydrocarbon (see Box 7.3). The energy of the hydrocarbon is often needed in the chemical process and, as in the production of most plastics, it is embodied in the end product. Alternatively, the energy of the hydrocarbon is available and likely to be utilized in other parts of the process, purification, pretreatment for example, or in other processes within the same plant. If this energy is missing, since CO_2 is used as carbon source, it has to be replaced somehow to close the energy balance of the plant. As long as the replacement energy is provided from fossil fuels, net CO_2 emissions will remain unchanged. It is worth noting that an economy with large non-fossil energy resources could consider CO_2 feedstocks to replace hydrocarbons in chemical synthesis. Such approaches are not covered here, since they are specific examples of converting to non-fossil energy and as such are driven by the merits of the new energy source rather than by the need for capture and storage of CO_2.

7.3.5   *Future scope*

The scale of the use of captured CO_2 in industrial processes is too small, the storage times too short and the energy balance too unfavourable for industrial uses of CO_2 to become significant as a means of mitigating climate change. There is a lack of data available to adequately assess the possible overall CO_2 inventory of processes that involve CO_2 substitution with associated energy balances and the effects of changes in other feedstocks

Box 7.2 Carbon chemical pool.

The carbon chemical pool is the ensemble of anthropogenic carbon containing organic chemicals. This box aims to provide criteria for measuring the quantitative impact on carbon mitigation of such a pool. If this impact were significant, using carbon from CO_2 could be an attractive storage option for captured CO_2.

Considering a specific chemical A, whose present worldwide production is 12 Mt yr^{-1}, whose worldwide inventory is 1 Mt – the monthly production – and whose lifetime before degradation to CO_2 and release to the atmosphere is less than one year. If next year production and inventory of A do not change, the contribution to CO_2 storage of this member of the chemical pool will be null. If production increased by a factor ten to 120 Mt yr^{-1}, whereas inventory were still 1 Mt, again the contribution of A to CO_2 storage would be null.

If on the contrary next year production increases and inventory also increases, for example to 3 Mt, to cope with increased market demand, the contribution of A to CO_2 storage over the year will be equivalent to the amount of CO_2 stoichiometrically needed to produce 2 Mt of A. However, if due to better distribution policies and despite increased production, the worldwide inventory of A decreased to 0.7 Mt, then A would yield a negative contribution to CO_2 storage, thus over the year the amount of CO_2 stoichiometrically needed to produce 0.3 Mt of A would be additionally emitted to the atmosphere.

Therefore, the impact on carbon dioxide mitigation of the carbon chemical pool does not depend on the amounts of carbon containing chemical products produced; there is CO_2 emission reduction in a certain time only if the pool has grown during that time. With increasing production, such impact can be positive or negative, as shown above. It is clear that since this would be a second or third order effect with respect to the overall production of carbon containing chemicals – itself much smaller in terms of fossil fuel consumption than fossil fuel combustion – this impact will be insignificant compared with the scale of the challenge that carbon dioxide capture and storage technologies have to confront.

Box 7.3. Energy gain or penalty in using CO_2 as a feedstock instead of carbon.

CO_2 can be used as a provider of carbon atoms for chemical synthesis, as an alternative to standard processes where the carbon atom source is fossil carbon, as coal or methane or other. This includes processes where the carbon atom in the CO_2 molecule is either reduced by providing energy, for example methanol synthesis, or does not change its oxidation state and does not need energy, synthesis of polycarbonates for example.

For the sake of simplicity let us consider a reaction from carbon to an organic final product A (containing n carbon atoms) that takes place in a chemical plant (standard process):

$$nC \rightarrow A \qquad (7)$$

Let us also consider the alternative route whereby CO_2 captured from the power plant where carbon has been burnt is used in the chemical plant where the synthesis of A is carried out. In this case the sequence of reactions would be:

$$nC \rightarrow nCO_2 \rightarrow A \qquad (8)$$

The overall energy change upon transformation of C into A, ΔH, is the same in both cases. The difference between the two cases is that in case (8) this overall energy change is split into two parts – $\Delta H = \Delta H_{com} + \Delta H_{syn}$ – one for combustion in the power plant and the other for the synthesis of A from CO_2 in the chemical plant (ΔH_{com} will be –400 which means 400 are made available by the combustion of carbon). If ΔH is negative, that means an overall exothermic reaction (1), then ΔH_{syn} will be either negative or even positive. If ΔH is positive, that means an overall endothermic reaction (7), then ΔH_{syn} will be even more positive. In both cases, exothermic or endothermic reaction, the chemical plant will lack 400 kJ/molC energy in case (2) with respect to case (1). This energy has already been exploited in the power plant and is no longer available in the chemical plant. It is worth noting that large-scale chemical plants (these are those of interest for the purpose of carbon dioxide emission mitigation) make the best possible use of their energy by applying so-called heat integration, for example by optimizing energy use through the whole plant and not just for individual processes. In case (1) chemical plants make good use of the 400 kJ/molC that are made available by the reaction (7) in excess of the second step of reaction (8).

Therefore, in terms of energy there is no benefit in choosing path (8) rather than path (7). In terms of efficiency of the whole chemical process there might be a potential improvement, but there might also be a potential disadvantage, since route (7) integrates the heat generation associated with the oxidation of carbon and the conversion to product A. These effects are of second order importance and have to be evaluated on a case-by-case basis. Nevertheless, the scale of the reduction in CO_2 emissions would be rather small, since it would be even smaller than the scale of the production of the chemicals that might be impacted by the technology change, that is by the change from path (7) to path (8) (Audus and Oonk, 1997).

and emissions. However, the analysis above demonstrates that, although the precise figures are difficult to estimate and even their sign is questionable, the contribution of these technologies to CO_2 storage is negligible. Research is continuing on the use of CO_2 in organic chemical polymer and plastics production, but the drivers are generally cost, elimination of hazardous chemical intermediates and the elimination of toxic wastes, rather than the storage of CO_2.

References

Arakawa, H., 1998: Research and development on new synthetic routes for basic chemicals by catalytic hydrogenation of CO_2. In *Advances in Chemical Conversions for Mitigating Carbon Dioxide*, Elsevier Science B.V., p 19-30.

Aresta, M., I. Tommasi, 1997: Carbon dioxide utilization in the chemical industry. *Energy Convers. Mgmt* 38, S373-S378.

Audus, H. and Oonk, H., 1997, An assessment procedure for chemical utilization schemes intended to reduce CO_2 emission to atmosphere, *Energy Conversion and Management*, 38 (suppl,

Proceedings of the Third International Conference on Carbon Dioxide Removal, 1996), S 409- S 414

Barnes, V. E., D. A. Shock, and W. A. Cunningham, 1950: Utilization of Texas Serpentine, No. 5020. *Bureau of Economic Geology: The University of Texas.*

Bearat, H., M. J. McKelvy, A. V. G. Chizmeshya, R. Sharma, R. W. Carpenter, 2002: Magnesium Hydroxide Dehydroxylation/ Carbonation Reaction Processes: Implications for Carbon Dioxide Mineral Sequestration. *Journal of the American Ceramic Society,* **85** (4), 742-48.

Benemann, J. R., 1997: CO_2 Mitigation with Microalgae Systems. *Energy Conversion and Management* **38,** Supplement 1, S475-S79.

Blencoe, J.G., L.M. Anovitz, D.A. Palmer, J.S. Beard, 2003: Carbonation of metal silicates for long-term CO_2 sequestration, U.S. patent application.

Brownlow, A. H., 1979. *Geochemistry.* Englewood Cliffs, NJ: Prentice-Hall

Butt, D.P., Lackner, K.S., Wendt., C.H., Conzone, S.D., Kung, H., Lu., Y.-C., Bremser, J.K., 1996. Kinetics of thermal dehydroxilation and carbonation of magnesium hydroxide. *J. Am. Ceram. Soc.*

79(7), 1892-1988.

Butt, D. P., K. S. Lackner,1997 : A Method for Permanent Disposal of CO_2 in Solid Form. *World Resource Review* **9**(3), 324-336.

Carey, J. W., Lichtner, P. C., Rosen, E. P., Ziock, H.-J., and Guthrie, G. D., Jr. (2003) Geochemical mechanisms of serpentine and olivine carbonation. In Proceedings of the Second National Conference on Carbon Sequestration, Washington, DC, USA May 5-8, 2003.

Coleman, R.G., 1977: Ophiolites: Springer-Verlag, Berlin, 229 pp.

Drăgulescu, C., P. Tribunescu, and O. Gogu, 1972: Lösungsgleichgewicht von MgO aus Serpentinen durch Einwirkung von CO_2 und Wasser. *Revue Roumaine de Chimie,* **17** (9), 1517-24.

Dunsmore, H. E., 1992: A Geological Perspective on Global Warming and the Possibility of Carbon Dioxide Removal as Calcium Carbonate Mineral. *Energy Convers. Mgmgt.,* **33**, 5-8,565-72.

Dyson, F., 1976: Can We Control the Amount of Carbon Dioxide in the Atmosphere? *IEA Occasional Paper*, IEA (O)-76-4: Institute for Energy Analysis, Oak Ridge Associated Universities

Eliasson, B., 1994: CO_2 Chemistry: An Option for CO_2 Emission Control. In *Carbon Dioxide Chemistry: Environmental Issues*, J. Paul and C.-M. Pradier Eds., The Royal Society of Chemistry, Cambridge, p 5-15.

Fernández Bertos, M., Simons, S.J.R., Hills, C.D., Carey, P.J., 2004. A review of accelerated carbonation technology in the treatment of cement-based materials and sequestration of CO_2. *J. Hazard. Mater.* **B112**, 193-205.

Fouda M. F. R., R. E. Amin and M. Mohamed, 1996: Extraction of magnesia from Egyptian serpentine ore via reaction with different acids. 2. Reaction with nitric and acetic acids. *Bulletin of the chemical society of Japan*, **69** (7): 1913-1916.

Goff, F. and K. S. Lackner, 1998: Carbon Dioxide Sequestering Using Ultramafic Rocks. *Environmental Geoscience* **5**(3): 89-101.

Goff, F., G. Guthrie, *et al.*, 2000: Evaluation of Ultramafic Deposits in the Eastern United States and Puerto Rico as Sources of Magnesium for Carbon Dioxide Sequestration. *LA-13694-MS.* Los Alamos, New Mexico, USA - Los Alamos National Laboratory.

Halmann, M.M. and M. Steinberg (eds.), 1999: *Greenhouse Gas Carbon Dioxide Mitigation Science and Technology.* Lewis Publishers, USA, 568 pp.

Hartman, H.L. (ed.), 1992: *SME mining engineering handbook, 2nd ed.* Society for Mining, Metallurgy, and Exploration, Inc., USA.

Haywood, H. M., J. M. Eyre and H. Scholes, 2001: Carbon dioxide sequestration as stable carbonate minerals-environmental barriers. *Environ. Geol.* **41**, 11–16.

Houston, E. C., 1945: Magnesium from Olivine. Technical Publication No. 1828, American Institute of Mining and Metallurgical Engineers.

Huijgen, W., G.-J. Witkamp, R. Comans, 2004: Mineral CO_2 sequestration in alkaline solid residues. In Proceedings of the GHGT-7 Conference, Vancouver, Canada September 5-9, 2004.

Hume, L. A., and J. D. Rimstidt, 1992: The biodurability of chrysotile asbestos. *Am. Mineral*, **77**, 1125-1128.

Iizuka, A., Fujii, M., Yamasaki, A., Yanagisawa, Y., 2004. Development of a new CO_2 sequestration process utilizing the carbonation of waste cement. *Ind. Eng. Chem. Res.* **43**, 7880-7887.

IPCC/TEAP (Intergovernmental Panel on climate Change and Technology and Economic Assessment Panel), 2005: Special

Report on Safeguarding the Ozone Layer and the Global Climate System: issues related to Hydrofluorocarbons and Perfluorocarbons, Cambridge University Press, Cambridge, UK.

Inui, T., 1996: Highly effective conversion of carbon dioxide to valuable compounds on composite catalysts. *Catal. Today*, **29**(1-4), 329-337.

Johnson D.C. 2000: Accelerated carbonation of waste calcium silicate materials, SCI Lecture Papers Series 108/2000, 1-10.

Kakizawa, M., A. Yamasaki, Y. Yanagisawa, 2001: A new CO_2 disposal process via artificial weathering of calcium silicate accelerated by acetic acid. *Energy* **26**(4): 341-354.

Kim, D. J. and H. S. Chung, 2002: Effect of grinding on the structure and chemical extraction of metals from serpentine. *Particulate Science and Technology.* **20**(2), 159-168.

Lackner, K. S., 2002: Carbonate Chemistry for Sequestering Fossil Carbon. *Annu. Rev. Energy Environ.* **27**, (1), 193-232.

Lackner, K. S., C. H. Wendt, D. P. Butt, E. L. Joyce and D. H. Sharp, 1995: Carbon dioxide disposal in carbonate minerals. *Energy*, **20** 1153-1170.

Lackner, K. S., D. P. Butt, C. H. Wendt, F. Goff and G. Guthrie, 1997: Carbon Dioxide Disposal in Mineral Form: Keeping Coal Competitive *Tech. Report No. LA-UR-97-2094* (Los Alamos National Laboratory).

Larson, E. D., 1993: Technology for Electricity and Fuels from Biomass, *Annual Review of Energy and Environment* **18**, 567-630.

Lasaga, A. C. and R. A. Berner 1998: Fundamental aspects of quantitative models for geochemical cycles. *Chemical Geology* **145** (3-4), 161-175.

Maroto-Valer, M.M., Fauth, D.J., Kuchta, M.E., Zhang, Y., Andrésen, J.M.: 2005. Activation of magnesium rich minerals as carbonation feedstock materials for CO_2 sequestration. *Fuel Process. Technol.*, **86**, 1627-1645.

Meima, J.A., van der Weijden, R.D., Eighmy T.T., Comans, R.N.J, 2002: Carbonation processes in municipal solid waste incinerator bottom ash and their effect on the leaching of copper and molybdenum. *Applied Geochem.*, **17**, 1503-1513.

Melis, A., J. Neidhardt, and J. R. Benemann, 1998: Dunaliella Salina (Chlorophyta) with Small Chlorophyll Antenna Sizes Exhibit Higher Photosynthetic Productivities and Photon Use Efficiencies Than Normally Pigmented Cells. *Journal of Applied Phycology* **10** (6), 515-25.

Newall, P. S., Clarke, S.J., Haywood, H.M., Scholes, H., Clarke, N.R., King, P.A., Barley, R.W., 2000: *CO_2 storage as carbonate minerals*, report PH3/17 for IEA Greenhouse Gas R&D Programme, CSMA Consultants Ltd, Cornwall, UK

Nichols, M. D., 2000: A General Location Guide for Ultramafic Rocks in California - Areas More Likely to Contain Naturally Occurring Asbestos. Sacramento, CA: California Department of Conservation, Division of Mines and Geology.

O'Connor, W. K., D.C. Dahlin, D.N. Nilsen, G.E. Rush, R.P. Walters, P.C. Turner, 2000: CO_2 Storage in Solid Form: A Study of Direct Mineral Carbonation. In *Proceedings of the 5th International Conference on Greenhouse Gas Technologies.* Cairns, Australia.

O'Connor, W. K., D. C. Dahlin, G. E. Rush, C. L. Dahlin, W. K. Collins, 2002: Carbon dioxide sequestration by direct mineral carbonation: process mineralogy of feed and products. *Minerals & metallurgical processing* **19** (2): 95-101.

O'Connor, W.K., D.C. Dahlin, G.E. Rush, S.J. Gedermann, L.R. Penner, D.N. Nilsen, Aqueous mineral carbonation, Final Report, DOE/ARC-TR-04-002 (March 15, 2005).

O'Hanley, D. S., 1996: Serpentinites: records of tectonic and petrological history, Oxford University Press, New York

Park, A.-H., A., R. Jadhav, and L.-S. Fan, 2003: CO_2 mineral sequestration: chemical enhanced aqueous carbonation of serpentine, *Canadian J. Chem. Eng.*, **81**, 885-890.

Park, A.-H., A., L.-S. Fan, 2004: CO_2 mineral sequestration: physically activated dissolution of serpentine and pH swing process, *Chem. Eng. Sci.*, **59**, 5241-5247.

Pelc, H., B. Elvers, S. Hawkins, 2005: Ullmann's Encyclopedia of Industrial Chemistry, Wiley-VCH Verlag GmbH & Co. KGaA.

Pierantozzi, R., 2003: Carbon Dioxide, Kirk Othmer Encyclopaedia of Chemical Technology, John Wiley and Sons.

Pundsack, F. L. 1967: Recovery of Silica, Iron Oxide and Magnesium Carbonate from the Treament of Serpentine with Ammonium Bisulfate, *United States Patent* No. 3,338,667.

Richmond, A., and N. Zou, 1999: Efficient Utilisation of High Photon Irradiance for Mass Production of Photoautotrophic Micro-Organisms. *Journal of Applied Phycology* **11**.1, 123-27.

Robie, R. A., Hemingway, B. S., Fischer, J. R. 1978: Thermodynamic properties of minerals and related substances at 298.15 K and 1 bar (10^5 Pascal) pressure and at higher temperature*s*, *US Geological Bulletin* 1452, Washington DC

Sano, H., Tamaura, Y.Amano, H. and Tsuji, M. (1998): Global carbon recycling energy delivery system for CO_2 mitigation (1) Carbon one-time recycle system towards carbon multi-recycle system, *Advances in Chemical Conversions for Mitigating Carbon Dioxide*, Elsevier Science B.V., p 273-278.

Seifritz, W., 1990: CO_2 disposal by means of silicates. *Nature* **345**, 486

Stolaroff, J.K., G.V. Lowry, D.W. Keith, 2005: Using CaO- and MgO-rich industrial waste streams for carbon sequestration. *Energy Conversion and Management*. **46**, 687-699.

Ushikoshi, K., K. Mori, T. Watanabe, M. Takeuchi and M. Saito, 1998: A 50 kg/day class test plant for methanol synthesis from CO_2 and H_2, *Advances in Chemical Conversions for Mitigating Carbon Dioxide*, Elsevier Science B.V., p 357-362.

Venhuis, M.A., E.J. Reardon, 2001: Vacuum method for carbonation of cementitious wasteforms. *Environ. Sci. Technol.* **35**, 4120-4125.

Xiaoding, X., Moulijn, J.A., 1996: Mitigation of CO_2 by chemical conversion: plausible chemical reactions and promising products. *Energy and Fuels*, **10**, 305-325

Wendt, C. H., D. P. Butt, K. S. Lackner, H.-J. Ziock *et al.*, 1998a: Thermodynamic Considerations of Using Chlorides to Accelerate the Carbonate Formation from Magnesium Silicates. *Fourth International Conference on Greenhouse Gas Control Technologies*, 30 August - 2 September. Eds. B. Eliasson, P. W. F. Riemer and A. Wokaun. Interlaken Switzerland.

Wendt, C. H., K. S. Lackner, D. P. Butt, H.-J. Ziock, 1998b: Thermodynamic Calculations for Acid Decomposition of Serpentine and Olivine in $MgCl_2$ Melts, I. Description of Concentrated $MgCl_2$ Melts. *Tech. Report No. LA-UR-98-4528*

(Los Alamos National Laboratory).

Wendt, C. H., K. S. Lackner, D. P. Butt and H.-J. Ziock, 1998c: Thermodynamic Calculations for Acid Decomposition of Serpentine and Olivine in $MgCl_2$ Melts, II. Reaction Equilibria in $MgCl_2$ Melts. *Tech. Report No. LA-UR-98-4529* (Los Alamos National Laboratory)

Wendt, C. H., K. S. Lackner, D. P. Butt and H.-J. Ziock, 1998d: Thermodynamic Calculations for Acid Decomposition of Serpentine and Olivine in $MgCl_2$ Melts, III. Heat Consumption in Process Design, *Tech. Report No. LA-UR-98-4529* (Los Alamos National Laboratory).

Zevenhoven, R., Kavaliauskaite, I. 2004: Mineral carbonation for long-term CO_2 storage: an exergy analysis. *Int. J. Thermodynamics*, **7**(1) 23-31

Ziock, H.-J., K. S. Lackner, 2000: Zero Emission Coal. *Contribution to the 5th International Conference on Greenhouse Gas Technologies*, Cairns, Australia, August 14-18, *Tech. Report No. LAUR-00-3573* (Los Alamos National Laboratory.

8

Cost and economic potential

Coordinating Lead Authors
Howard Herzog (United States), Koen Smekens (Belgium)

Lead Authors
Pradeep Dadhich (India), James Dooley (United States), Yasumasa Fujii (Japan), Olav Hohmeyer (Germany), Keywan Riahi (Austria)

Contributing Authors
Makoto Akai (Japan), Chris Hendriks (Netherlands), Klaus Lackner (United States), Ashish Rana (India), Edward Rubin (United States), Leo Schrattenholzer (Austria), Bill Senior (United Kingdom)

Review Editors
John Christensen (Denmark), Greg Tosen (South Africa)

Contents

EXECUTIVE SUMMARY

The major components of a carbon dioxide capture and storage (CCS) system include capture (separation plus compression), transport, and storage (including measurement, monitoring and verification). In one form or another, these components are commercially available. However, there is relatively little commercial experience with configuring all of these components into fully integrated CCS systems at the kinds of scales which would likely characterize their future deployment. The literature reports a fairly wide range of costs for employing CCS systems with fossil-fired power production and various industrial processes. The range spanned by these cost estimates is driven primarily by site-specific considerations such as the technology characteristics of the power plant or industrial facility, the specific characteristics of the storage site, and the required transportation distance of carbon dioxide (CO_2). In addition, estimates of the future performance of components of the capture, transport, storage, measurement and monitoring systems are uncertain. The literature reflects a widely held belief that the cost of building and operating CO_2 capture systems will fall over time as a result of technological advances.

The cost of employing a full CCS system for electricity generation from a fossil-fired power plant is dominated by the cost of capture. The application of capture technology would add about 1.8 to 3.4 US$ct kWh^{-1} to the cost of electricity from a pulverized coal power plant, 0.9 to 2.2 US$ct kWh^{-1} to the cost for electricity from an integrated gasification combined cycle coal power plant, and 1.2 to 2.4 US$ct kWh^{-1} from a natural-gas combined-cycle power plant. Transport and storage costs would add between –1 and 1 US$ct kWh^{-1} to this range for coal plants, and about half as much for gas plants. The negative costs are associated with assumed offsetting revenues from CO_2 storage in enhanced oil recovery (EOR) or enhanced coal bed methane (ECBM) projects. Typical costs for transportation and geological storage from coal plants would range from 0.05–0.6 US$ct kWh^{-1}. CCS technologies can also be applied to other industrial processes, such as hydrogen (H_2) production. In some of these non-power applications, the cost of capture is lower than for capture from fossil-fired power plants, but the concentrations and partial pressures of CO_2 in the flue gases from these sources vary widely, as do the costs. In addition to fossil-based energy conversion processes, CCS may be applied to biomass-fed energy systems to create useful energy (electricity or transportation fuels). The product cost of these systems is very sensitive to the potential price of the carbon permit and the associated credits obtained with systems resulting in negative emissions. These systems can be fuelled solely by biomass, or biomass can be co-fired in conventional coal-burning plants, in which case the quantity is normally limited to about 10–15% of the energy input.

Energy and economic models are used to study future scenarios for CCS deployment and costs. These models indicate that CCS systems are unlikely to be deployed on a large scale in the absence of an explicit policy that substantially limits greenhouse gas emissions to the atmosphere. The literature and current industrial experience indicate that, in the absence of measures to limit CO_2 emissions, there are only small, niche opportunities for the deployment of CCS technologies. These early opportunities for CCS deployment – that are likely to involve CO_2 captured from high-purity, low-cost sources and used for a value-added application such as EOR or ECBM production – could provide valuable early experience with CCS deployment, and create parts of the infrastructure and knowledge base needed for the future large-scale deployment of CCS systems.

With greenhouse gas emission limits imposed, many integrated assessment analyses indicate that CCS systems will be competitive with other large-scale mitigation options, such as nuclear power and renewable energy technologies. Most energy and economic modelling done to date suggests that the deployment of CCS systems starts to be significant when carbon prices begin to reach approximately 25–30 US$/t$CO_2$ (90–110 US$/tC). They foresee the large-scale deployment of CCS systems within a few decades from the start of any significant regime for mitigating global warming. The literature indicates that deployment of CCS systems will increase in line with the stringency of the modelled emission reduction regime. Least-cost CO_2 concentration stabilization scenarios, that also take into account the economic efficiency of the system, indicate that emissions mitigation becomes progressively more stringent over time. Most analyses indicate that, notwithstanding significant penetration of CCS systems by 2050, the majority of CCS deployment will occur in the second half of this century. They also indicate that early CCS deployment will be in the industrialized nations, with deployment eventually spreading worldwide. While different scenarios vary the quantitative mix of technologies needed to meet the modelled emissions constraint, the literature consensus is that CCS could be an important component of a broad portfolio of energy technologies and emission reduction approaches. In addition, CCS technologies are compatible with the deployment of other potentially important long-term greenhouse gas mitigation technologies such as H_2 production from biomass and fossil fuels.

Published estimates (for CO_2 stabilization scenarios between 450–750 ppmv) of the global cumulative amount of CO_2 that might be stored over the course of this century in the ocean and various geological formations span a wide range: from very small contributions to thousands of gigatonnes of CO_2. This wide range can largely be explained by the uncertainty of long-term, socio-economic, demographic and technological change, the main drivers of future CO_2 emissions. However, it is important to note that the majority of stabilization scenarios from 450–750 ppmv tend to cluster in the range of 220–2200 GtCO_2 (60–600 GtC). This demand for CO_2 storage appears to be within global estimates of total CO_2 storage capacity. The actual use of CCS is likely to be lower than the estimates for economic potential indicated by these energy and economic models, as there are other barriers to technology development not adequately accounted for in these modelling frameworks. Examples include concerns about environmental impact, the lack of a clear legal framework and uncertainty about how quickly learning-by-doing will lower costs. This chapter concludes with a review of knowledge gaps that affect the reliability of these model results.

Given the potential for hundreds to thousands of gigatonnes of CO_2 to be stored in various geological formations and the ocean, questions have been raised about the implications of gradual leakage from these reservoirs. From an economic perspective, such leakage – if it were to occur – can be thought of as another potential source of future CO_2 emissions, with the cost of offsetting this leaked CO_2 being equal to the cost of emission offsets when the stored CO_2 leaks to the atmosphere. Within this purely economic framework, the few studies that have looked at this topic indicate that some CO_2 leakage can be accommodated while progressing towards the goal of stabilizing atmospheric concentrations of CO_2.

8.1   Introduction

In this chapter, we address two of the key questions about any CO_2 mitigation technology: 'How much will it cost?' and 'How do CCS technologies fit into a portfolio of greenhouse gas mitigation options?' There are no simple answers to these questions. Costs for CCS technologies depend on many factors: fuel prices, the cost of capital, and costs for meeting potential regulatory requirements like monitoring, to just name a few. Add to this the uncertainties associated with technology development, the resource base for storage potential, the regulatory environment, etc., and it becomes obvious why there are many answers to what appear to be simple questions.

This chapter starts (in Section 8.2) by looking at the costs of the system components, namely capture and compression, transport, and storage (including monitoring costs and by-product credits from operations such as EOR). The commercial operations associated with each of these components provide a basis for the assessment of current costs. Although it involves greater uncertainty, an assessment is also included of how these costs will change in the future. The chapter then reviews the findings from economic modelling (Section 8.3). These models take component costs at various levels of aggregation and then model how the costs change with time and how CCS technologies compete with other CO_2 mitigation options given a variety of economic and policy assumptions. The chapter concludes with an examination of the economic implications of different storage times (Section 8.4) and a summary of the known knowledge gaps (Section 8.5).

8.2   Component costs

This section presents cost summaries for the three key components of a CCS system, namely capture (including compression), transport, and storage. Sections 8.2.1–8.2.3 summarize the results from Chapters 3–7. Readers are referred to those chapters for more details of component costs. Results are presented here in the form most convenient for each section. Transport costs are given in US\$/tCO_2 per kilometre, while storage costs are stated in US\$/tCO_2 stored. Capture costs for different types of power plants are represented as an increase in the electricity generation cost (US\$ MWh^{-1}). A discussion of how one integrates the costs of capture, transport and storage for a particular system into a single value is presented in Section 8.2.4.

8.2.1   *Capture and compression*[1]

For most large sources of CO_2 (e.g., power plants), the cost of capturing CO_2 is the largest component of overall CCS costs. In this report, capture costs include the cost of compressing the CO_2 to a pressure suitable for pipeline transport (typically about 14 MPa). However, the cost of any additional booster compressors that may be needed is included in the cost of transport and/or storage.

The total cost of CO_2 capture includes the additional capital requirements, plus added operating and maintenance costs incurred for any particular application. For current technologies, a substantial portion of the overall cost is due to the energy requirements for capture and compression. As elaborated in Chapter 3, a large number of technical and economic factors related to the design and operation of both the CO_2 capture system, and the power plant or industrial process to which it is applied, influence the overall cost of capture. For this reason, the reported costs of CO_2 capture vary widely, even for similar applications.

Table 8.1 summarizes the CO_2 capture costs reported in Chapter 3 for baseload operations of new fossil fuel power plants (in the size range of 300–800 MW) employing current commercial technology. The most widely studied systems are new power plants based on coal combustion or gasification. For costs associated with retrofitting existing power plants, see Table 3.8. For a modern (high-efficiency) coal-burning power plant, CO_2 capture using an amine-based scrubber increases the cost of electricity generation (COE) by approximately 40 to 70 per cent while reducing CO_2 emissions per kilowatt-hour (kWh) by about 85%. The same CO_2 capture technology applied to a new natural gas combined cycle (NGCC) plant increases the COE by approximately 40 to 70 per cent. For a new coal-based plant employing an integrated gasification combined cycle (IGCC) system, a similar reduction in CO_2 using current technology (in this case, a water gas shift reactor followed by a physical absorption system) increases the COE by 20 to 55%. The lower incremental cost for IGCC systems is due in large part to the lower gas volumes and lower energy requirements for CO_2 capture relative to combustion-based systems. It should be noted that the absence of industrial experience with large-scale capture of CO_2 in the electricity sector means that these numbers are subject to uncertainties, as is explained in Section 3.7.

---

[1] This section is based on material presented in Section 3.7. The reader is referred to that section for a more detailed analysis and literature references.

Table 8.1 Summary of new plant performance and CO_2 capture cost based on current technology.

Performance and Cost Measures	New NGCC Plant Range low	high	Rep. Value	New PC Plant Range low	high	Rep. Value	New IGCC Plant Range low	high	Rep. Value	New Hydrogen Plant Range low	high	Rep. Value	(Units for H_2 Plant)
Emission rate without capture (kg CO_2 MWh^{-1})	344	379	367	736	811	762	682	846	773	78	174	137	kg CO_2 GJ^{-1} (without capture)
Emission rate with capture (kg CO_2 MWh^{-1})	40	66	52	92	145	112	65	152	108	7	28	17	kg CO_2 GJ^{-1} (with capture)
Percent CO_2 reduction per kWh (%)	83	88	86	81	88	85	81	91	86	72	96	86	% reduction/unit of product
Plant efficiency with capture, LHV basis (%)	47	50	48	30	35	33	31	40	35	52	68	60	Capture plant efficiency (% LHV)
Capture energy requirement (% more input MWh^{-1})	11	22	16	24	40	31	14	25	19	4	22	8	% more energy input per GJ product
Total capital requirement without capture (US$ kW^{-1})	515	724	568	1161	1486	1286	1169	1565	1326	[No unique normalization for multi-product plants]			Capital requirement without capture
Total capital requirement with capture (US$ kW^{-1})	909	1261	998	1894	2578	2096	1414	2270	1825				Capital requirement with capture
Percent increase in capital cost with capture (%)	64	100	76	44	74	63	19	66	37	-2	54	18	% increase in capital cost
COE without capture (US$ MWh^{-1})	31	50	37	43	52	46	41	61	47	6.5	10.0	7.8	H_2 cost without capture (US$ GJ^{-1})
COE with capture only (US$ MWh^{-1})	43	72	54	62	86	73	54	79	62	7.5	13.3	9.1	H_2 cost with capture (US$ GJ^{-1})
Increase in COE with capture (US$ MWh^{-1})	12	24	17	18	34	27	9	22	16	0.3	3.3	1.3	Increase in H_2 cost (US$ GJ^{-1})
Percent increase in COE with capture (%)	37	69	46	42	66	57	20	55	33	5	33	15	% increase in H_2 cost
Cost of CO_2 captured (US$/t$CO_2$)	33	57	44	23	35	29	11	32	20	2	39	12	US$/t$CO_2$ captured
Cost of CO_2 avoided (US$/t$CO_2$)	37	74	53	29	51	41	13	37	23	2	56	15	US$/t$CO_2$ avoided
Capture cost confidence Level (see Table 3.7)	moderate			moderate			moderate			moderate to high			Confidence Level (see Table 3.7)

COE = Cost of electricity

Notes: [a] Ranges and representative values are based on data from Tables 3.7, 3.9, 3.10 and 3.11. All costs in this table are for capture only and do not include the costs of CO_2 transport and storage; see Chapter 8 for total CCS costs. [b] All PC and IGCC data are for bituminous coals only at costs of 1.0-1.5 US$ GJ^{-1} (LHV); all PC plants are supercritical units. [c] NGCC data based on natural gas prices of 2.8-4.4 US$ GJ^{-1} (LHV basis). [d] Costs are in constant US$ (approx. year 2002 basis). [e] Power plant sizes range from approximately 400-800 MW without capture and 300-700 MW with capture. [f] Capacity factors vary from 65-85% for coal plants and 50-95% for gas plants (average for each=80%). [g] Hydrogen plant feedstocks are natural gas (4.7-5.3 US$ GJ^{-1}) or coal (0.9-1.3 US$ GJ^{-1}); some plants in dataset produce electricity in addition to hydrogen. [h] Fixed charge factors vary from 11-16% for power plants and 13-20% for hydrogen plants. [i] All costs include CO_2 compression but not additional CO_2 transport and storage costs.

Studies indicate that, in most cases, IGCC plants are slightly higher in cost without capture and slightly lower in cost with capture than similarly sized PC plants fitted with a CCS system. On average, NGCC systems have a lower COE than both types of new coal-based plants with or without capture for baseload operation. However, the COE for each of these systems can vary markedly due to regional variations in fuel cost, plant utilization, and a host of other parameters. NGCC costs are especially sensitive to the price of natural gas, which has risen significantly in recent years. So comparisons of alternative power system costs require a particular context to be meaningful.

For existing, combustion-based, power plants, CO_2 capture can be accomplished by retrofitting an amine scrubber to the existing plant. However, a limited number of studies indicate that the post-combustion retrofit option is more cost-effective when accompanied by a major rebuild of the boiler and turbine to increase the efficiency and output of the existing plant by converting it to a supercritical unit. For some plants, similar benefits can be achieved by repowering with an IGCC system that includes CO_2 capture technology. The feasibility and cost of any of these options is highly dependent on site-specific circumstances, including the size, age and type of unit, and the availability of space for accommodating a CO_2 capture system. There has not yet been any systematic comparison of the feasibility and cost of alternative retrofit and repowering options for existing plants, as well as the potential for more cost-effective options employing advanced technology such as oxyfuel combustion.

Table 8.1 also illustrates the cost of CO_2 capture in the production of H_2, a commodity used extensively today for fuels and chemical production, but also widely viewed as a potential energy carrier for future energy systems. Here, the cost of CO_2 capture is mainly due to the cost of CO_2 compression, since separation of CO_2 is already carried out as part of the H_2 production process. Recent studies indicate that the cost of CO_2 capture for current processes adds approximately 5 to 30 per cent to the cost of the H_2 product.

In addition to fossil-based energy conversion processes, CO_2 could also be captured in power plants fuelled with biomass. At present, biomass plants are small in scale (<100 MW_e). Hence, the resulting costs of capturing CO_2 are relatively high compared to fossil alternatives. For example, the capturing of 0.19 $MtCO_2$ yr^{-1} in a 24 MW_e biomass IGCC plant is estimated to be about 82 US\$/tCO_2 (300 US\$/tC), corresponding to an increase of the electricity costs due to capture of about 80 US\$ MWh^{-1} (Audus and Freund, 2004). Similarly, CO_2 could be captured in biomass-fuelled H_2 plants. The cost is reported to be between 22 and 25 US\$/tCO_2 avoided (80–92 US\$/tC) in a plant producing 1 million Nm^3 d^{-1} of H_2 (Makihira *et al.*, 2003). This corresponds to an increase in the H_2 product costs of about 2.7 US\$ GJ^{-1} (i.e., 20% of the H_2 costs without CCS). The competitiveness of biomass CCS systems is very sensitive to the value of CO_2 emission reductions, and the associated credits obtained with systems resulting in negative emissions. Moreover, significantly larger biomass plants could benefit from

economies of scale, bringing down costs of the CCS systems to broadly similar levels as those in coal plants. However, there is too little experience with large-scale biomass plants as yet, so that their feasibility has still not been proven and their costs are difficult to estimate.

CCS technologies can also be applied to other industrial processes. Since these other industrial processes produce off-gases that are very diverse in terms of pressure and CO_2 concentration, the costs range very widely. In some of these non-power applications where a relatively pure CO_2 stream is produced as a by-product of the process (e.g., natural gas processing, ammonia production), the cost of capture is significantly lower than capture from fossil-fuel-fired power plants. In other processes like cement or steel production, capture costs are similar to, or even higher than, capture from fossil-fuel-fired power plants.

New or improved technologies for CO_2 capture, combined with advanced power systems and industrial process designs, can significantly reduce the cost of CO_2 capture in the future. While there is considerable uncertainty about the magnitude and timing of future cost reductions, studies suggest that improvements to current commercial technologies could lower CO_2 capture costs by at least 20–30%, while new technologies currently under development may allow for more substantial cost reductions in the future. Previous experience indicates that the realization of cost reductions in the future requires sustained R&D in conjunction with the deployment and adoption of commercial technologies.

8.2.2 Transport[2]

The most common and usually the most economical method to transport large amounts of CO_2 is through pipelines. A cost-competitive transport option for longer distances at sea might be the use of large tankers.

The three major cost elements for pipelines are construction costs (e.g., material, labour, possible booster station), operation and maintenance costs (e.g., monitoring, maintenance, possible energy costs) and other costs (e.g., design, insurance, fees, right-of-way). Special land conditions, like heavily populated areas, protected areas such as national parks, or crossing major waterways, may have significant cost impacts. Offshore pipelines are about 40% to 70% more costly than onshore pipes of the same size. Pipeline construction is considered to be a mature technology and the literature does not foresee many cost reductions.

Figure 8.1 shows the transport costs for 'normal' terrain conditions. Note that economies of scale dramatically reduce the cost, but that transportation in mountainous or densely populated areas could increase cost.

Tankers could also be used for transport. Here, the main cost elements are the tankers themselves (or charter costs), loading and unloading facilities, intermediate storage facilities, harbour

---

[2] This section is based on material presented in Section 4.6. The reader is referred to that section for a more detailed analysis and literature references.

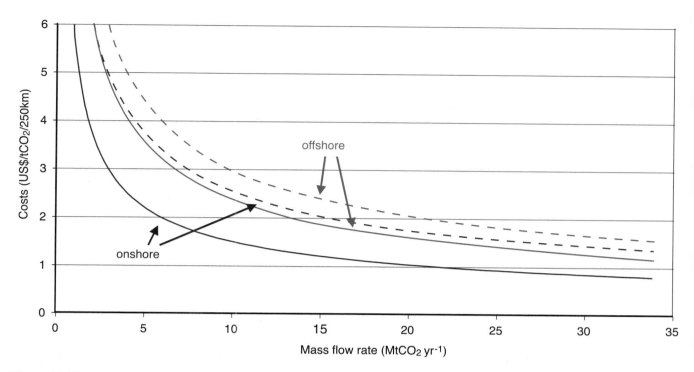

Figure 8.1 CO_2 transport costs range for onshore and offshore pipelines per 250 km, 'normal' terrain conditions. The figure shows low (solid lines) and high ranges (dotted lines). Data based on various sources (for details see Chapter 4).

fees, and bunker fuel. The construction costs for large special-purpose CO_2 tankers are not accurately known since none have been built to date. On the basis of preliminary designs, the costs of CO_2 tankers are estimated at US\$ 34 million for ships of 10,000 tonnes, US\$ 58 million for 30,000-tonne vessels, and US\$ 82 million for ships with a capacity of 50,000 tonnes.

To transport 6 $MtCO_2$ per year a distance of 500 km by ship would cost about 10 US\$/tCO_2 (37 US\$/tC) or 5 US\$/tCO_2/250km (18 US\$/tC/250km). However, since the cost is relatively insensitive to distance, transporting the same 6 $MtCO_2$ a distance of 1250 km would cost about 15 US\$/tCO_2 (55 US\$/tC) or 3 US\$/tCO_2/250km (11 US\$/tC/250km). This is close to the cost of pipeline transport, illustrating the point that ship transport becomes cost-competitive with pipeline transport if CO_2 needs to be transported over larger distances. However, the break-even point beyond which ship transportation becomes cheaper than pipeline transportation is not simply a matter of distance; it involves many other aspects.

8.2.3　Storage

8.2.3.1　Geological storage[3]
Because the technologies and equipment used for geological storage are widely used in the oil and gas industries, the cost estimates can be made with confidence. However, there will be a significant range and variability of costs due to site-specific factors: onshore versus offshore, the reservoir depth

and the geological characteristics of the storage formation (e.g., permeability, thickness, etc.). Representative estimates of the cost for storage in saline formations and disused oil and gas fields (see Table 8.2) are typically between 0.5–8.0 US\$/tCO_2 stored (2–29 US\$/tC), as explained in Section 5.9.3. The lowest storage costs will be associated with onshore, shallow, high permeability reservoirs and/or the reuse of wells and infrastructure in disused oil and gas fields.

The full range of cost estimates for individual options is very large. Cost information for storage monitoring is currently limited, but monitoring is estimated to add 0.1–0.3 US\$ per tonne of CO_2 stored (0.4–1.1 US\$/tC). These estimates do not include any well remediation or long-term liabilities. The costs of storage monitoring will depend on which technologies are used for how long, regulatory requirements and how long-term monitoring strategies evolve.

When storage is combined with EOR, enhanced gas recovery (EGR) or ECBM, the benefits of enhanced production can offset some of the capture and storage costs. Onshore EOR operations have paid in the range of 10–16 US\$ per tonne of CO_2 (37–59 US\$/tC). The economic benefit of enhanced production depends very much on oil and gas prices. It should be noted that most of the literature used as the basis for this report did not take into account the rise in oil and gas prices that started in 2003. For example, oil at 50 US\$/barrel could justify a credit of 30 US\$/tCO_2 (110 US\$/tC). The economic benefits from enhanced production make EOR and ECBM potential early cost-effective options for geological storage.

---

[3] This section is based on material presented in Section 5.9. The reader is referred to that section for a more detailed analysis and literature references.

Table 8.2 Estimates of CO_2 storage costs.

Option	Representative Cost Range (US$/tonne CO_2 stored)	Representative Cost Range (US$/tonne C stored)
Geological - Storage[a]	0.5-8.0	2-29
Geological - Monitoring	0.1-0.3	0.4-1.1
Ocean[b]		
Pipeline	6-31	22-114
Ship (Platform or Moving Ship Injection)	12-16	44-59
Mineral Carbonation[c]	50-100	180-370

[a] Does not include monitoring costs.
[b] Includes offshore transportation costs; range represents 100-500 km distance offshore and 3000 m depth.
[c] Unlike geological and ocean storage, mineral carbonation requires significant energy inputs equivalent to approximately 40% of the power plant output.

8.2.3.2   Ocean storage[4]

The cost of ocean storage is a function of the distance offshore and injection depth. Cost components include offshore transportation and injection of the CO_2. Various schemes for ocean storage have been considered. They include:

- tankers to transport low temperature (–55 to –50°C), high pressure (0.6–0.7 MPa) liquid CO_2 to a platform, from where it could be released through a vertical pipe to a depth of 3000 m;
- carrier ships to transport liquid CO_2, with injection through a towed pipe from a moving dispenser ship;
- undersea pipelines to transport CO_2 to an injection site.

Table 8.2 provides a summary of costs for transport distances of 100–500 km offshore and an injection depth of 3000 m.

Chapter 6 also discusses the option of carbonate neutralization, where flue-gas CO_2 is reacted with seawater and crushed limestone. The resulting mixture is then released into the upper ocean. The cost of this process has not been adequately addressed in the literature and therefore the possible cost of employing this process is not addressed here.

8.2.3.3   Storage via mineral carbonation[5]

Mineral carbonation is still in its R&D phase, so costs are uncertain. They include conventional mining and chemical processing. Mining costs include ore extraction, crushing and grinding, mine reclamation and the disposal of tailings and carbonates. These are conventional mining operations and several studies have produced cost estimates of 10 US$/tCO$_2$ (36 US$/tC) or less. Since these estimates are based on similar mature and efficient operations, this implies that there is a strong lower limit on the cost of mineral storage. Carbonation costs include chemical activation and carbonation. Translating today's laboratory implementations into industrial practice yields rough cost estimates of about 50–100 US$/tCO$_2$ stored

(180–370 US$/tC). Costs and energy penalties (30–50% of the power plant output) are dominated by the activation of the ore necessary to accelerate the carbonation reaction. For mineral storage to become practical, additional research must reduce the cost of the carbonation step by a factor of three to four and eliminate a significant portion of the energy penalty by, for example, harnessing as much as possible the heat of carbonation.

8.2.4   Integrated systems

The component costs given in this section provide a basis for the calculation of integrated system costs. However, the cost of mitigating CO_2 emissions cannot be calculated simply by summing up the component costs for capture, transport and storage in units of 'US$/tCO$_2$'. This is because the amount of

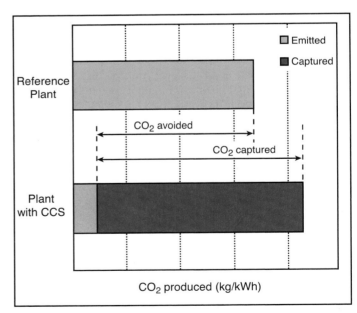

Figure 8.2 CO_2 capture and storage from power plants. The increased CO_2 production resulting from loss in overall efficiency of power plants due to the additional energy required for capture, transport and storage, and any leakage from transport result in a larger amount of 'CO_2 produced per unit of product' (lower bar) relative to the reference plant (upper bar) without capture

---

[4] This section is based on material presented in Section 6.9. The reader is referred to that section for a more detailed analysis and literature references.
[5] This section is based on material presented in Section 7.2. The reader is referred to that section for a more detailed analysis and literature references.

Box 8.1 Defining avoided costs for a fossil fuel power plant

In general, the capture, transport, and storage of CO_2 require energy inputs. For a power plant, this means that amount of fuel input (and therefore CO_2 emissions) increases per unit of net power output. As a result, the amount of CO_2 produced per unit of product (e.g., a kWh of electricity) is greater for the power plant with CCS than the reference plant, as shown in Figure 8.2 To determine the CO_2 reductions one can attribute to CCS, one needs to compare CO_2 emissions of the plant with capture to those of the reference plant without capture. These are the avoided emissions. Unless the energy requirements for capture and storage are zero, the amount of CO_2 avoided is always less than the amount of CO_2 captured. The cost in US$/tonne avoided is therefore greater than the cost in US$/tonne captured.

CO_2 captured will be different from the amount of atmospheric CO_2 emissions 'avoided' during the production of a given amount of a useful product (e.g., a kilowatt-hour of electricity or a kilogram of H_2). So any cost expressed per tonne of CO_2 should be clearly defined in terms of its basis, e.g., either a *captured* basis or an *avoided* basis (see Box 8.1). Mitigation cost is best represented as avoided cost. Table 8.3 presents ranges for total avoided costs for CO_2 capture, transport, and storage from four types of sources.

The mitigation costs (US$/tCO$_2$ avoided) reported in Table 8.3 are context-specific and depend very much on what is chosen as a reference plant. In Table 8.3, the reference plant is a power plant of the same type as the power plant with CCS. The mitigation costs here therefore represent the incremental cost of capturing and storing CO_2 from a particular type of plant.

In some situations, it can be useful to calculate a cost of CO_2 avoided based on a reference plant that is different from the CCS plant (e.g., a PC or IGCC plant with CCS using an NGCC reference plant). In Table 8.4, the reference plant represents the least-cost plant that would 'normally' be built at a particular location in the absence of a carbon constraint. In many regions today, this would be either a PC plant or an NGCC plant.

A CO_2 mitigation cost also can be defined for a collection of plants, such as a national energy system, subject to a given level of CO_2 abatement. In this case the plant-level product costs presented in this section would be used as the basic inputs to energy-economic models that are widely used for policy analysis and for the quantification of overall mitigation strategies and costs for CO_2 abatement. Section 8.3 discusses the nature of these models and presents illustrative model results, including the cost of CCS, its economic potential, and its relationship to other mitigation options.

Table 8.3a Range of total costs for CO_2 capture, transport, and geological storage based on current technology for new power plants.

	Pulverized Coal Power Plant	Natural Gas Combined Cycle Power Plant	Integrated Coal Gasification Combined Cycle Power Plant
Cost of electricity without CCS (US$ MWh^{-1})	43-52	31-50	41-61
Power plant with capture			
Increased Fuel Requirement (%)	24-40	11-22	14-25
CO_2 captured (kg MWh^{-1})	820-970	360-410	670-940
CO_2 avoided (kg MWh^{-1})	620-700	300-320	590-730
% CO_2 avoided	81-88	83-88	81-91
Power plant with capture and geological storage[6]			
Cost of electricity (US$ MWh^{-1})	63-99	43-77	55-91
Electricity cost increase (US$ MWh^{-1})	19-47	12-29	10-32
% increase	43-91	37-85	21-78
Mitigation cost (US$/tCO$_2$ avoided)	30-71	38-91	14-53
Mitigation cost (US$/tC avoided)	110-260	140-330	51-200
Power plant with capture and enhanced oil recovery[7]			
Cost of electricity (US$ MWh^{-1})	49-81	37-70	40-75
Electricity cost increase (US$ MWh^{-1})	5-29	6-22	(-5)-19
% increase	12-57	19-63	(-10)-46
Mitigation cost (US$/tCO$_2$ avoided)	9-44	19-68	(-7)-31
Mitigation cost (US$/tC avoided)	31-160	71-250	(-25)-120

[6] Capture costs represent range from Tables 3.7, 3.9 and 3.10. Transport costs range from 0–5 US$/tCO$_2$. Geological storage cost (including monitoring) range from 0.6–8.3 US$/tCO$_2$.

[7] Capture costs represent range from Tables 3.7, 3.9 and 3.10. Transport costs range from 0–5 US$/tCO$_2$ stored. Costs for geological storage including EOR range from −10 to −16 US$/tCO$_2$ stored.

Table 8.3b Range of total costs for CO$_2$ capture, transport, and geological storage based on current technology for a new hydrogen production plant.

	Hydrogen Production Plant
Cost of H$_2$ without CCS (US$ GJ^{-1})	6.5-10.0
Hydrogen plant with capture	
Increased fuel requirement (%)	4-22
CO$_2$ captured (kg GJ^{-1})	75-160
CO$_2$ avoided (kg GJ^{-1})	60-150
% CO$_2$ avoided	73-96
Hydrogen plant with capture and geological storage[8]	
Cost of H$_2$ (US$ GJ^{-1})	7.6-14.4
H$_2$ cost increase (US$ GJ^{-1})	0.4-4.4
% increase	6-54
Mitigation cost (US$/tCO$_2$ avoided)	3-75
Mitigation cost (US$ tC avoided)	10-280
Hydrogen plant with capture and enhanced oil recovery[9]	
Cost of H$_2$ (US$ GJ^{-1})	5.2-12.9
H$_2$ cost increase (US$ GJ^{-1})	(-2.0)-2.8
% increase	(-28)-28
Mitigation cost (US$/tCO$_2$ avoided)	(-14)-49
Mitigation cost (US$/tC avoided)	(-53)-180

8.3   CCS deployment scenarios

Energy-economic models seek the mathematical representation of key features of the energy system in order to represent the evolution of the system under alternative assumptions, such as population growth, economic development, technological change, and environmental sensitivity. These models have been employed increasingly to examine how CCS technologies would deploy in a greenhouse gas constrained environment. In this section we first provide a brief introduction to the types of energy and economic models and the main assumptions driving future greenhouse gas emissions and the corresponding measures to reduce them. We then turn to the principal focus of this section: an examination of the literature based on studies using these energy and economic models, with an emphasis on what they say about the potential use of CCS technologies.

8.3.1   Model approaches and baseline assumptions

The modelling of climate change abatement or mitigation scenarios is complex and a number of modelling techniques have been applied, including input-output models, macroeconomic (top-down) models, computable general equilibrium (CGE) models and energy-sector-based engineering models (bottom-up).

Table 8.4 Mitigation cost for different combinations of reference and CCS plants based on current technology and new power plants.

	NGCC Reference Plant		PC Reference Plant	
	US$/tCO$_2$ avoided	US$/tC avoided	US$/tCO$_2$ avoided	US$/tC avoided
Power plant with capture and geological storage				
NGCC	40-90	140-330	20-60	80-220
PC	70-270	260-980	30-70	110-260
IGCC	40-220	150-790	20-70	80-260
Power plant with capture and EOR				
NGCC	20-70	70-250	1-30	4-130
PC	50-240	180-890	10-40	30-160
IGCC	20 – 190	80 – 710	1 – 40	4 – 160

[8] Capture costs represent range from Table 3.11. Transport costs range from 0–5 US$/tCO$_2$. Geological storage costs (including monitoring) range from 0.6–8.3 US$/tCO$_2$.

[9] Capture costs represent range from Table 3.11. Transport costs range from 0–5 US$/tCO$_2$. EOR credits range from 10–16 US$/tCO$_2$.

8.3.1.1 Description of bottom-up and top-down models

The component and systems level costs provided in Section 8.2 are based on technology-based bottom-up models. These models can range from technology-specific, engineering-economic calculations embodied in a spreadsheet to broader, multi-technology, integrated, partial-equilibrium models. This may lead to two contrasting approaches: an engineering-economic approach and a least-cost equilibrium one. In the first approach, each technology is assessed independently, taking into account all its parameters; partial-equilibrium least-cost models consider all technologies simultaneously and at a higher level of aggregation before selecting the optimal mix of technologies in all sectors and for all time periods.

Top-down models evaluate the system using aggregate economic variables. Econometric relationships between aggregated variables are generally more reliable than those between disaggregated variables, and the behaviour of the models tends to be more stable. It is therefore common to adopt high levels of aggregation for top-down models; especially when they are applied to longer-term analyses. Technology diffusion is often described in these top-down models in a more stylized way, for example using aggregate production functions with price-demand or substitution elasticities.

Both types of models have their strengths and weaknesses. Top-down models are useful for, among other things, calculating gross economic cost estimates for emissions mitigation. Most of these top-down macro-economic models tend to overstate costs of meeting climate change targets because, among other reasons, they do not take adequate account of the potential for no-regret measures and they are not particularly adept at estimating the benefits of climate change mitigation. On the other hand, many of these models – and this also applies to bottom-up models – are not adept at representing economic and institutional inefficiencies, which would lead to an underestimation of emissions mitigation costs.

Technologically disaggregated bottom-up models can take some of these benefits into account but may understate the costs of overcoming economic barriers associated with their deployment in the market. Recent modelling efforts have focused on the coupling of top-down and bottom-up models in order to develop scenarios that are consistent from both the macroeconomic and systems engineering perspectives. Readers interested in a more detailed discussion of these modelling frameworks and their application to understanding future energy, economic and emission scenarios are encouraged to consult the IPCC's Working Group III's assessment of the international work on both bottom-up and top-down analytical approaches (Third Assessment Report; IPCC, 2001).

8.3.1.2 Assumptions embodied in emissions baselines

Integrated Assessment Models (IAMs) constitute a particular category of energy and economic models and will be used here to describe the importance of emissions baselines before examining model projections of potential future CCS use. IAMs integrate the simulation of climate change dynamics with the modelling of the energy and economic systems. A common and illuminating type of analysis conducted with IAMs, and with other energy and economic models, involves the calculation of the cost differential or the examination of changes in the portfolio of energy technologies used when moving from a baseline (i.e., no climate policy) scenario to a control scenario (i.e., a case where a specific set of measures designed to constrain GHG emissions is modelled). It is therefore important to understand what influences the nature of these baseline scenarios. A number of parameters spanning economic, technological, natural and demographic resources shape the energy use and resulting emissions trajectories of these baseline cases. How these parameters change over time is another important aspect driving the baseline scenarios. A partial list of some of the major parameters that influence baseline scenarios include, for example, modelling assumptions centring on:

- global and regional economic and demographic developments;
- costs and availability of
  1) global and regional fossil fuel resources;
  2) fossil-based energy conversion technologies (power generation, H_2 production, etc.), including technology-specific parameters such as efficiencies, capacity factors, operation and maintenance costs as well as fuel costs;
  3) zero-carbon energy systems (renewables and nuclear), which might still be non-competitive in the baseline but may play a major role competing for market shares with CCS if climate policies are introduced;
- rates of technological change in the baseline and the specific way in which technological change is represented in the model;
- the relative contribution of CO_2 emissions from different economic sectors.

Modelling all of these parameters as well as alternative assumptions for them yields a large number of 'possible futures'. In other words, they yield a number of possible baseline scenarios. This is best exemplified by the Special Report on Emission Scenarios (SRES, 2000): it included four different narrative storylines and associated scenario families, and identified six 'illustrative' scenario groups – labelled A1FI, A1B, A1T, A2, B1, B2 – each representing different plausible combinations of socio-economic and technological developments in the absence of any climate policy (for a detailed discussion of these cases, see SRES, 2000). The six scenario groups depict alternative developments of the energy system based on different assumptions about economic and demographic change, hydrocarbon resource availability, energy demand and prices, and technology costs and their performance. They lead to a wide range of possible future worlds and CO_2 emissions consistent with the full uncertainty range of the underlying literature (Morita and Lee, 1998). The cumulative emissions from 1990 to 2100 in the scenarios range from less than 2930 to 9170 $GtCO_2$ (800 to 2500 GtC). This range is divided into four intervals, distinguishing between scenarios

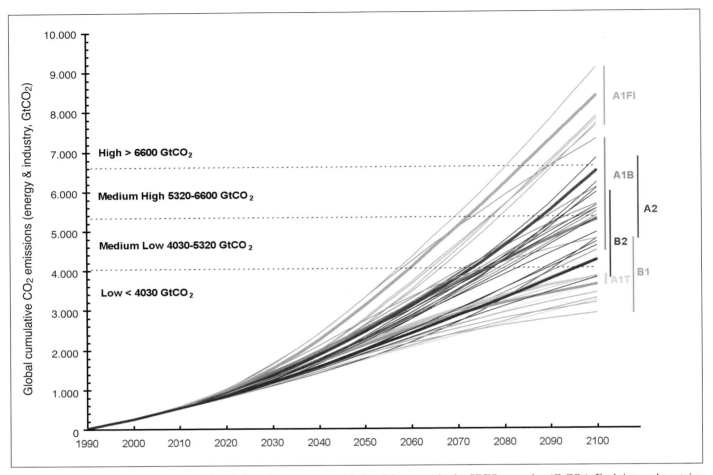

Figure 8.3 Annual and cumulative global emissions from energy and industrial sources in the SRES scenarios ($GtCO_2$). Each interval contains alternative scenarios from the six SRES scenario groups that lead to comparable cumulative emissions. The vertical bars on the right-hand side indicate the ranges of cumulative emissions (1990–2100) of the six SRES scenario groups.

with high, medium-high, medium-low, and low emissions:
- high ($\geq$6600 $GtCO_2$ or $\geq$1800 GtC);
- medium-high (5320–6600 $GtCO_2$ or 1450–1800 GtC);
- medium-low (4030–5320 $GtCO_2$ or 1100–1450 GtC);
- low ($\leq$4030 $GtCO_2$ or $\leq$1100 GtC).

As illustrated in Figure 8.3, each of the intervals contains multiple scenarios from more than one of the six SRES scenario groups (see the vertical bars on the right side of Figure 8.3, which show the ranges for cumulative emissions of the respective SRES scenario group). Other scenario studies, such as the earlier set of IPCC scenarios developed in 1992 (Pepper *et al.*, 1992) project similar levels of cumulative emissions over the period 1990 to 2100, ranging from 2930 to 7850 $GtCO_2$ (800 to 2,140 GtC). For the same time horizon, the IIASA-WEC scenarios (Nakicenovic *et al.*, 1998) report 2,270–5,870 $GtCO_2$ (620–1,600 GtC), and the Morita and Lee (1998) database – which includes more than 400 emissions scenarios – report cumulative emissions up to 12,280 $GtCO_2$ (3,350 GtC).

The SRES scenarios illustrate that similar future emissions can result from very different socio-economic developments, and that similar developments in driving forces can nonetheless result in wide variations in future emissions. The scenarios also indicate that the future development of energy systems will play a central role in determining future emissions and suggests that technological developments are at least as important a driving force as demographic change and economic development. These findings have major implications for CCS, indicating that the pace at which these technologies will be deployed in the future – and therefore their long-term potential – is affected not so much by economic or demographic change but rather by the choice of the technology path of the energy system, the major driver of future emissions. For a detailed estimation of the technical potential of CCS by sector for some selected SRES baseline scenarios, see Section 2.3.2. In the next section we shall discuss the economic potential of CCS in climate control scenarios.

8.3.2   CCS economic potential and implications

As shown by the SRES scenarios, uncertainties associated with alternative combinations of socio-economic and technological developments may lead to a wide range of possible future emissions. Each of the different baseline emissions scenarios has different implications for the potential use of CCS technologies in emissions control cases.[10] Generally, the size of the future market for CCS depends mostly on the carbon intensity of the baseline scenario and the stringency of the assumed climate stabilization target. The higher the CO_2 emissions in the baseline, the more emissions reductions are required to achieve a given level of allowable emissions, and the larger the markets for CCS. Likewise, the tighter the modelled constraint on CO_2 emissions, the more CCS deployment there is likely to be. This section will examine what the literature says about possible CCS deployment rates, the timing of CCS deployment, the total deployment of these systems under various scenarios, the economic impact of CCS systems and how CCS systems interact with other emissions mitigation technologies.

8.3.2.1   Key drivers for the deployment of CCS

Energy and economic models are increasingly being employed to examine how CCS technologies would deploy in environments where CO_2 emissions are constrained (i.e., in control cases). A number of factors have been identified that drive the rate of CCS deployment and the scale of its ultimate deployment in modelled control cases:[11]

1. *The policy regime*; the interaction between CCS deployment and the policy regime in which energy is produced and consumed cannot be overemphasized; the magnitude and timing of early deployment depends very much on the policy environment; in particular, the cumulative extent of deployment over the long term depends strongly on the stringency of the emissions mitigation regime being modelled; comparatively low stabilization targets (e.g., 450 ppmv) foster the relatively faster penetration of CCS and the more intensive use of CCS (where 'intensity of use' is measured both in terms of the percentage of the emissions reduction burden shouldered by CCS as well as in terms of how many cumulative gigatonnes of CO_2 is to be stored) (Dooley *et al.*, 2004b; Gielen and Podanski, 2004; Riahi and Roehrl, 2000);

2. *The reference case (baseline)*; storage requirements for stabilizing CO_2 concentrations at a given level are very sensitive to the choice of the baseline scenario. In other words, the assumed socio-economic and demographic trends, and particularly the assumed rate of technological change, have a significant impact on CCS use (see Section 8.3.1, Riahi and Roehrl, 2000; Riahi *et al.*, 2003);

3. *The nature, abundance and carbon intensity of the energy resources / fuels* assumed to exist in the future (e.g., a future world where coal is abundant and easily recoverable would use CCS technologies more intensively than a world in which natural gas or other less carbon-intensive technologies are inexpensive and widely available). See Edmonds and Wise (1998) and Riahi and Roehrl (2000) for a comparison of two alternative regimes of fossil fuel availability and their interaction with CCS;

4. *The introduction of flexible mechanisms such as emissions trading* can significantly influence the extent of CCS deployment. For example, an emissions regime with few, or significantly constrained, emissions trading between nations entails the use of CCS technologies sooner and more extensively than a world in which there is efficient global emissions trading and therefore lower carbon permit prices (e.g., Dooley *et al.*, 2000 and Scott *et al.*, 2004). Certain regulatory regimes that explicitly emphasize CCS usage can also accelerate its deployment (e.g., Edmonds and Wise, 1998).

5. *The rate of technological change (induced through learning or other mechanisms)* assumed to take place with CCS and other salient mitigation technologies (e.g., Edmonds *et al.*, 2003, or Riahi *et al.*, 2003). For example, Riahi *et al.* (2003) indicate that the long-term economic potential of CCS systems would increase by a factor of 1.5 if it assumed that technological learning for CCS systems would take place at rates similar to those observed historically for sulphur removal technologies when compared to the situation where no technological change is specified.[12]

The marginal value of CO_2 emission reduction permits is one of the most important mechanisms through which these factors impact CCS deployment. CCS systems tend to deploy quicker and more extensively in cases with higher marginal carbon values. Most energy and economic modelling done to date suggests that CCS systems begin to deploy at a significant level when carbon dioxide prices begin to reach approximately 25–30 US\$/tCO_2 (90–110 US\$/tC) (IEA, 2004; Johnson and Keith, 2004; Wise and Dooley, 2004; McFarland *et al.*, 2004). The only caveat to this carbon price as a lower limit for the deployment of these systems is the 'early opportunities' literature discussed below.

Before turning to a specific focus on the possible contribution of CCS in various emissions mitigation scenarios, it is worth reinforcing the point that there is a broad consensus in the

---

[10] As no climate policy is assumed in SRES, there is also no economic value associated with carbon. The potential for CCS in SRES is therefore limited to applications where the supplementary benefit of injecting CO_2 into the ground exceeds its costs (e.g., EOR or ECBM). The potential for these options is relatively small as compared to the long-term potential of CCS in stabilization scenarios. Virtually none of the global modelling exercises in the literature that incorporate SRES include these options and so there is also no CCS system deployment assumed in the baseline scenarios.

[11] Integrated assessment models represent the world in an idealized way, employing different methodologies for the mathematical representation of socio-economic and technological developments in the real world. The representation of some real world factors, such as institutional barriers, inefficient legal frameworks, transaction costs of carbon permit trading, potential free-rider behaviour of geopolitical agents and the implications of public acceptance has traditionally been a challenge in modelling. These factors are represented to various degrees (often generically) in these models

---

[12] The factor increase of 1.5 corresponds to about 250 to 360 GtCO_2 of additional capture and storage over the course of the century.

technical literature that no single mitigation measure will be adequate to achieve a stable concentration of CO_2. This means that the CO_2 emissions will most likely be reduced from baseline scenarios by a portfolio of technologies in addition to other social, behavioural and structural changes (Edmonds *et al.*, 2003; Riahi and Roehrl, 2000). In addition, the choice of a particular stabilization level from any given baseline significantly affects the technologies needed for achieving the necessary emissions reduction (Edmonds *et al.*, 2000; Roehrl and Riahi, 2000). For example, a wider range of technological measures and their widespread diffusion, as well as more intensive use, are required for stabilizing at 450 ppmv compared with stabilization at higher levels (Nakicenovic and Riahi, 2001). These and other studies

(e.g., IPCC, 2001) have identified several classes of robust mitigation measures: reductions in demand and/or efficiency improvements; substitution among fossil fuels; deployment of non-carbon energy sources (i.e., renewables and nuclear); CO_2 capture and storage; and afforestation and reforestation.

8.3.3   *The share of CCS in total emissions mitigation*

When used to model energy and carbon markets, the aim of integrated assessment models is to capture the heterogeneity that characterizes energy demand, energy use and the varying states of development of energy technologies that are in use at any given point in time, as well as over time. These integrated

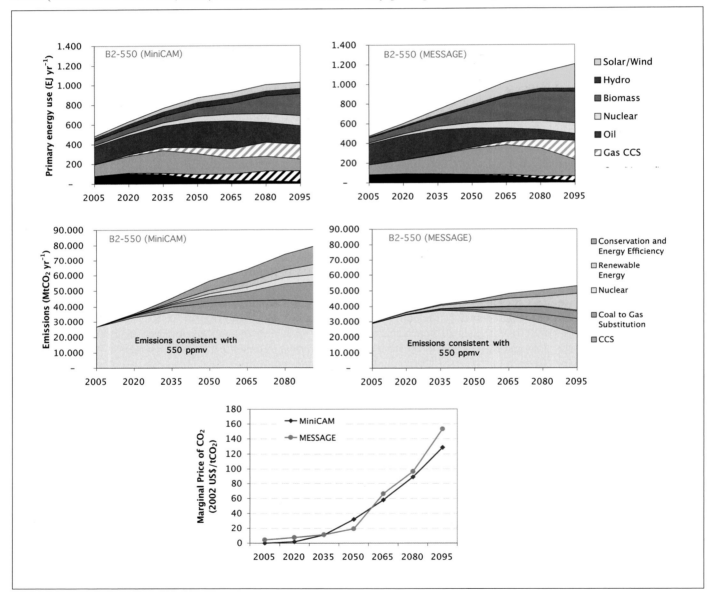

Figure 8.4 The set of graphs shows how two different integrated assessment models (MiniCAM and MESSAGE) project the development of global primary energy (upper panels) and the corresponding contribution of major mitigation measures (middle panels). The lower panel depicts the marginal carbon permit price in response to a modelled mitigation regime that seeks to stabilize atmospheric concentrations of CO_2 at 550 ppmv. Both scenarios adopt harmonized assumptions with respect to the main greenhouse gas emissions drivers in accordance with the IPCC-SRES B2 scenario (Source: Dooley et al., 2004b; Riahi and Roehrl, 2000).

Box 8.2 Two illustrative 550 ppmv stabilization scenarios based on IPCC SRES B2

The MESSAGE and MiniCAM scenarios illustrated in Figure 8.4 represent two alternative quantifications of the B2 scenario family of the IPCC SRES. They are used for subsequent CO_2 mitigation analysis and explore the main measures that would lead to the stabilization of atmospheric concentrations at 550 ppmv.

The scenarios are based on the B2 storyline, a narrative description of how the world will evolve during the twenty-first century, and share harmonized assumptions concerning salient drivers of CO_2 emissions, such as economic development, demographic change, and final energy demand.

In accordance with the B2 storyline, gross world product is assumed to grow from US\$ 20 trillion in 1990 to about US\$ 235 trillion in 2100 in both scenarios, corresponding to a long-term average growth rate of 2.2%. Most of this growth takes place in today's developing countries. The scenarios adopt the UN median 1998 population projection (UN, 1998), which assumes a continuation of historical trends, including recent faster-than-expected fertility declines, towards a completion of the demographic transition within the next century. Global population increases to about 10 billion by 2100. Final energy intensity of the economy declines at about the long-run historical rate of about one per cent per year through 2100. On aggregate, these trends constitute 'dynamics-as-usual' developments, corresponding to middle-of-the-road assumptions compared to the scenario uncertainty range from the literature (Morita and Lee, 1999).

In addition to the similarities mentioned above, the MiniCAM and MESSAGE scenarios are based on alternative interpretations of the B2 storyline with respect to a number of other important assumptions that affect the potential future deployment of CCS. These assumptions relate to fossil resource availability, long-term potentials for renewable energy, the development of fuel prices, the structure of the energy system and the sectoral breakdown of energy demand, technology costs, and in particular technological change (future prospects for costs and performance improvements for specific technologies and technology clusters).

The two scenarios therefore portray alternative but internally consistent developments of the energy technology portfolio, associated CO_2 emissions, and the deployment of CCS and other mitigation technologies in response to the stabilization target of 550 ppmv CO_2, adopting the same assumptions for economic, population, and aggregated demand growth. Comparing the scenarios' portfolio of mitigation options (Figure 8.4) illustrates the importance of CCS as part of the mitigation portfolio. For more details, see Dooley *et al.* (2004b) and Riahi and Roehrl (2000).

assessment tools are also used to model changes in market conditions that would alter the relative cost-competitiveness of various energy technologies. For example, the choice of energy technologies would vary as carbon prices rise, as the population grows or as a stable population increases its standard of living.

The graphs in Figure 8.4 show how two different integrated assessment models (MiniCAM and MESSAGE) project the development of global primary energy (upper panels), the contribution of major mitigation measures (middle panels), and the marginal carbon permit price in response to a modelled policy that seeks to stabilize atmospheric concentrations of CO_2 at 550 ppmv in accordance with the main greenhouse gas emissions drivers of the IPCC-SRES B2 scenario (see Box 8.2). As can be seen from Figure 8.4, CCS coupled with coal and natural-gas-fired electricity generation are key technologies in the mitigation portfolio in both scenarios and particularly in the later half of the century under this particular stabilization scenario. However, solar/wind, biomass, nuclear power, etc. still meet a sizeable portion of the global demand for electricity. This demonstrates that the world is projected to continue to use a multiplicity of energy technologies to meet its energy demands and that, over space and time, a large portfolio of these technologies will be used at any one time.

When assessing how various technologies will contribute to the goal of addressing climate change, these technologies are modelled in such a way that they all compete for market share to provide the energy services and emissions reduction required by society, as this is what would happen in reality. There are major uncertainties associated with the potential and costs of these options, and so the absolute deployment of CCS depends on various scenario-specific assumptions consistent with the underlying storyline and the way they are interpreted in the different models. In the light of this competition and the wide variety of possible emissions futures, the contribution of CCS to total emissions reduction can only be assessed within relatively wide margins.

The uncertainty with respect to the future deployment of CCS and its contribution to total emissions reductions for achieving stabilization of CO_2 concentrations between 450 and 750 ppmv is illustrated by the IPCC TAR mitigation scenarios (Morita *et al.*, 2000; 2001). The TAR mitigation scenarios are based upon SRES baseline scenarios and were developed by nine different modelling teams. In total, 76 mitigation scenarios were developed for TAR, and about half of them (36 scenarios from three alternative models: DNE21, MARIA, and MESSAGE) consider CO_2 capture and storage explicitly as a mitigation option. An overview of the TAR scenarios is presented in Morita *et al.* (2000). It includes eleven publications from individual modelling teams about their scenario assumptions and results.

As illustrated in Figure 8.5, which is based upon the TAR mitigation scenarios, the average share of CCS in total emissions reductions may range from 15% for scenarios aiming

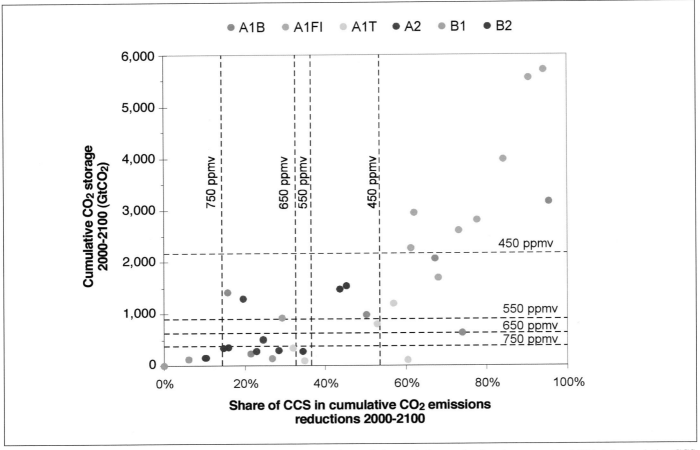

Figure 8.5 Relationship between (1) the imputed share of CCS in total cumulative emissions reductions in per cent and (2) total cumulative CCS deployment in GtCO$_2$ (2000–2100). The scatter plots depict values for individual TAR mitigation scenarios for the six SRES scenario groups. The vertical dashed lines show the average share of CCS in total emissions mitigation across the 450 to 750 ppmv stabilization scenarios, and the dashed horizontal lines illustrate the scenarios' average cumulative storage requirements across 450 to 750 ppmv stabilization.

at the stabilization of CO$_2$ concentrations at 750 ppmv to 54% for 450 ppmv scenarios.[13] However, the full uncertainty range of the set of TAR mitigation scenarios includes extremes on both the high and low sides, ranging from scenarios with zero CCS contributions to scenarios with CCS shares of more than 90% in total emissions abatement.

8.3.3.1 Cumulative CCS deployment

Top-down and bottom-up energy-economic models have been used to examine the likely total deployment of CCS technologies (expressed in GtC). These analyses reflect the fact that the future usage of CCS technologies is associated with large uncertainties. As illustrated by the IPCC-TAR mitigation scenarios, global cumulative CCS during the 21st century could range – depending on the future characteristics of the reference world (i.e., baselines) and the employed stabilization target

(450 to 750 ppmv) – from zero to more than 5500 GtCO$_2$ (1500 GtC) (see Figure 8.6). The average cumulative CO$_2$ storage (2000–2100) across the six scenario groups shown in Figure 8.6 ranges from 380 GtCO$_2$ (103 GtC) in the 750 ppmv stabilization scenarios to 2160 GtCO$_2$ (590 GtC) in the 450 ppmv scenarios (Table 8.5).[14] However, it is important to note that the majority of the six individual TAR scenarios (from the 20th to the 80th percentile) tend to cluster in the range of 220–2200 GtCO$_2$ (60–600 GtC) for the four stabilization targets (450–750 ppmv).

The deployment of CCS in the TAR mitigation scenarios is comparable to results from similar scenario studies projecting storage of 576–1370 GtCO$_2$ (157–374 GtC) for stabilization scenarios that span 450 to 750 ppmv (Edmonds *et al.*, 2000) and storage of 370 to 1250 GtCO$_2$ (100 to 340 GtC) for stabilization scenarios that span 450 to 650 ppmv (Dooley and Wise, 2003). Riahi *et al.* (2003) project 330–890 GtCO$_2$ (90–243 GtC) of stored CO$_2$ over the course of the current century for various

---

[13] The range for CCS mitigation in the TAR mitigation scenarios is calculated on the basis of the cumulative emissions reductions from 1990 to 2100, and represents the average contribution for 450 and 750 ppmv scenarios across alternative modelling frameworks and SRES baseline scenarios. The full range across all scenarios for 450 ppmv is 20 to 95% and 0 to 68% for 750 ppmv scenarios respectively.

[14] Note that Table 8.5 and Figure 8.6 show average values of CCS across alternative modelling frameworks used for the development of the TAR mitigation scenarios. The deployment of CCS over time, as well as cumulative CO$_2$ storage in individual TAR mitigation scenarios, are illustrated in Figures 8.5 and 8.7.

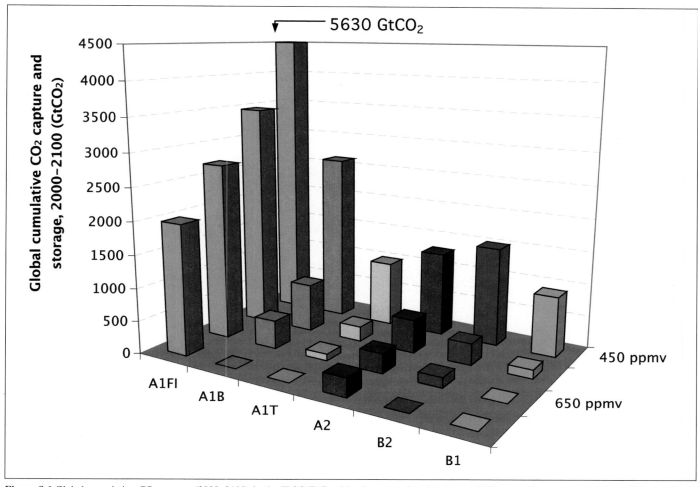

Figure 8.6 Global cumulative CO_2 storage (2000–2100) in the IPCC TAR mitigation scenarios for the six SRES scenario groups and CO_2 stabilization levels between 450 and 750 ppmv. Values refer to averages across scenario results from different modelling teams. The contribution of CCS increases with the stringency of the stabilization target and differs considerably across the SRES scenario groups.

550 ppmv stabilization cases. Fujii and Yamaji (1998) have also included ocean storage as an option. They calculate that, for a stabilization level of 550 ppmv, 920 GtCO$_2$ (250 GtC) of the emissions reductions could be provided by the use of CCS technologies and that approximately one-third of this could be stored in the ocean. This demand for CO_2 storage appears to be within global estimates of total CO_2 storage capacity presented in Chapters 5 and 6.

8.3.3.2 Timing and deployment rate

Recently, two detailed studies of the cost of CO_2 transport and storage costs have been completed for North America (Dooley *et al.*, 2004a) and Western Europe (Wildenborg *et al.*, 2004). These studies concur about the large potential of CO_2 storage capacity in both regions. Well over 80% of the emissions from current CO_2 point sources could be transported and stored in candidate geologic formations for less than 12–15 US$/tCO$_2$ in North America and 25 US$/tCO$_2$ in Western Europe. These studies are the first to define at a continental scale a 'CO_2 storage supply curve', conducting a spatially detailed analysis in order to explore the relationship between the price of CO_2

transport and storage and the cumulative amount of CO_2 stored. Both studies conclude that, at least for these two regions, the CO_2 storage supply curves are dominated by a very large single plateau (hundreds to thousands of gigatonnes of CO_2), implying roughly constant costs for a wide range of storage capacity[15]. In other words, at a practical level, the cost of CO_2 transport and storage in these regions will have a cap. These studies and a handful of others (see, for example, IEA GHG, 2002) have also shown that early (i.e., low cost) opportunities for CO_2 capture and storage hinge upon a number of factors: an inexpensive (e.g., high-purity) source of CO_2; a (potentially) active area of advanced hydrocarbon recovery (either EOR or ECBM); and the relatively close proximity of the CO_2 point source to the candidate storage reservoir in order to minimize transportation costs. These bottom-up studies provide some of the most detailed insights into the graded CCS resources presently available, showing that the set of CCS opportunities likely to be encountered in the real world will be very heterogeneous. These

---

[15] See Chapter 5 for a full assessment of the estimates of geological storage capacity.

Table 8.5 Cumulative CO_2 storage (2000 to 2100) in the IPCC TAR mitigation scenarios in $GtCO_2$. CCS contributions for the world and for the four SRES regions are shown for four alternative stabilization targets (450, 550, 650, and 750 ppmv) and six SRES scenario groups. Values refer to averages across scenario results from different modelling teams.

	All scenarios (average)	A1			A2	B2	B1
		A1FI	**A1B**	**A1T**			
WORLD							
450 ppmv	2162	5628	2614	1003	1298	1512	918
550 ppmv	898	3462	740	225	505	324	133
650 ppmv	614	2709	430	99	299	149	0
750 ppmv	377	1986	0	0	277	0	0
OECD90*							
450 ppmv	551	1060	637	270	256	603	483
550 ppmv	242	800	202	82	174	115	80
650 ppmv	172	654	166	54	103	55	0
750 ppmv	100	497	0	0	104	0	0
REF*							
450 ppmv	319	536	257	152	512	345	110
550 ppmv	87	233	99	42	55	79	16
650 ppmv	55	208	56	0	31	37	0
750 ppmv	36	187	0	0	28	0	0
ASIA*							
450 ppmv	638	2207	765	292	156	264	146
550 ppmv	296	1262	226	47	153	67	20
650 ppmv	223	1056	162	20	67	33	0
750 ppmv	111	609	0	0	57	0	0
ROW*							
450 ppmv	652	1825	955	289	366	300	179
550 ppmv	273	1167	214	54	124	63	17
650 ppmv	164	791	45	24	99	25	0
750 ppmv	130	693	0	0	89	0	0

* The OECD90 region includes the countries belonging to the OECD in 1990. The REF ('reforming economies') region aggregates the countries of the Former Soviet Union and Eastern Europe. The ASIA region represents the developing countries on the Asian continent. The ROW region covers the rest of the world, aggregating countries in sub-Saharan Africa, Latin America and the Middle East. For more details see SRES, 2000.

studies, as well as those based upon more top-down modelling approaches, also indicate that, once the full cost of the complete CCS system has been accounted for, CCS systems are unlikely to deploy on a large scale in the absence of an explicit policy or regulatory regime that substantially limits greenhouse gas emissions to the atmosphere. The literature and current industrial experience indicate that, in the absence of measures to limit CO_2 emissions, there are only small, niche opportunities for the deployment of CCS technologies. These early opportunities could provide experience with CCS deployment, including the creation of parts of the infrastructure and the knowledge base needed for the future large-scale deployment of CCS systems.

Most analyses of least-cost CO_2 stabilization scenarios indicate that, while there is significant penetration of CCS systems over the decades to come, the majority of CCS deployment will occur in the second half of this century

(Edmonds *et al.*, 2000, 2003; Edmonds and Wise, 1998; Riahi *et al.*, 2003). One of the main reasons for this trend is that the stabilization of CO_2 concentrations at relatively low levels (<650 ppmv) generally leads to progressively more constraining mitigation regimes over time, resulting in carbon permit prices that start out quite low and steadily rise over the course of this century. The TAR mitigation scenarios (Morita *et al.*, 2000) based upon the SRES baselines report cumulative CO_2 storage due to CCS ranging from zero to 1100 $GtCO_2$ (300 GtC) for the first half of the century, with the majority of the scenarios clustering below 185 $GtCO_2$ (50 GtC). By comparison, the cumulative contributions of CCS range from zero to 4770 $GtCO_2$ (1300 GtC) in the second half of the century, with the majority of the scenarios stating figures below 1470 $GtCO_2$ (400 GtC). The deployment of CCS over time in the TAR mitigation scenarios is illustrated in Figure 8.7. As can be seen, the use

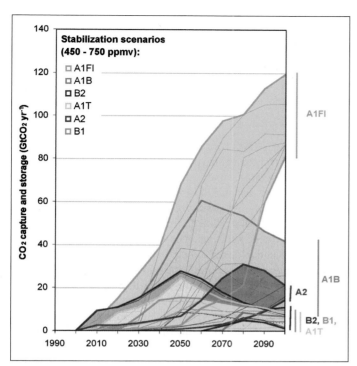

Figure 8.7 Deployment of CCS systems as a function of time from 1990 to 2100 in the IPCC TAR mitigation scenarios where atmospheric CO_2 concentrations stabilize at between 450 to 750 ppmv. Coloured thick lines show the minimum and maximum contribution of CCS for each SRES scenario group, and thin lines depict the contributions in individual scenarios. Vertical axes on the right-hand side illustrate the range of CCS deployment across the stabilization levels for each SRES scenario group in the year 2100.

of CCS is highly dependent upon the underlying base case. For example, in the high economic growth and carbon-intensive baseline scenarios (A1FI), the development path of CCS is characterized by steadily increasing contributions, driven by the rapidly growing use of hydrocarbon resources. By contrast, other scenarios (e.g., A1B and B2) depict CCS deployment to peak during the second half of the century. In a number of these scenarios, the contribution of CCS declines to less than 11 $GtCO_2$ per year (3 GtC per year) until the end of the century. These scenarios reflect the fact that CCS could be viewed as a transitional mitigation option (bridging the transition from today's fossil-intensive energy system to a post-fossil system with sizable contributions from renewables).

Given these models' relatively coarse top-down view of the world, there is less agreement about when the first commercial CCS units will become operational. This is – at least in part – attributable to the importance of policy in creating the context in which initial units will deploy. For example, McFarland *et al.* (2003) foresee CCS deployment beginning around 2035. Other modelling exercises have shown CCS systems beginning to deploy – at a lower level of less than 370 $MtCO_2$ a year (100 MtC a year) – in the period 2005–2020 (see, for example, Dooley *et al.*, 2000). Moreover, in an examination of CCS deployment in Japan, Akimoto *et al.* (2003) show CCS deployment beginning in 2010–2020. In a large body of literature (Edmonds *et al.*

2003; Dooley and Wise, 2003; Riahi *et al.* 2003; IEA, 2004), there is agreement that, in a CO_2-constrained world, CCS systems might begin to deploy in the next few decades and that this deployment will expand significantly after the middle of the century. The variation in the estimates of the timing of CCS-system deployment is attributable to the different ways energy and economic models parameterize CCS systems and to the extent to which the potential for early opportunities – such as EOR or ECBM – is taken into account. Other factors that influence the timing of CCS diffusion are the rate of increase and absolute level of the carbon price.

8.3.3.3 Geographic distribution

McFarland *et al.* (2003) foresee the eventual deployment of CCS technologies throughout the world but note that the timing of the entry of CCS technologies into a particular region is influenced by local conditions such as the relative price of coal and natural gas in a region. Dooley *et al.* (2002) show that the policy regime, and in particular the extent of emissions trading, can influence where CCS technologies are deployed. In the specific case examined by this paper, it was demonstrated that, where emissions trading was severely constrained (and where the cost of abatement was therefore higher), CCS technologies tended to deploy more quickly and more extensively in the US and the EU. On the other hand, in the absence of an efficient emissions-trading system spanning all of the Annex B nations, CCS was used less intensively and CCS utilization was spread more evenly across these nations as the EU and US found it cheaper to buy CCS-derived emission allowances from regions like the former Soviet Union.

Table 8.5 gives the corresponding deployment of CCS in the IPCC TAR mitigation scenarios for four world regions. All values are given as averages across scenario results from different modelling teams. The data in this table (in particular the far left-hand column which summarizes average CO_2 storage across all scenarios) help to demonstrate a common and consistent finding of the literature: over the course of this century, CCS will deploy throughout the world, most extensively in the developing nations of today (tomorrow's largest emitters of CO_2). These nations will therefore be likely candidates for adopting CCS to control their growing emissions.[16]

Fujii *et al.* (2002) note that the actual deployment of CCS technologies in any given region will depend upon a host of geological and geographical conditions that are, at present, poorly represented in top-down energy and economic models. In an attempt to address the shortcomings noted by Fujii *et al.* (2002) and others, especially in the way in which the cost of CO_2 transport and storage are parameterized in top-down models, Dooley *et al.* (2004b) employed graded CO_2 storage supply curves for all regions of the world based upon a preliminary assessment of the literature's estimate of regional CO_2 storage

---

[16] This trend can be seen particularly clearly in the far left-hand column of Table 8.5, which gives the average CCS deployment across all scenarios from the various models. Note, nevertheless, a few scenarios belonging to the B1 and B2 scenario family, which suggest larger levels of deployment for CCS in the developed world.

capacity. In this framework, where the cost of CO_2 storage varies across the globe depending upon the quantity, quality (including proximity) and type of CO_2 storage reservoirs present in the region, as well as upon the demand for CO_2 storage (driven by factors such as the size of the regional economy, the stringency of the modelled emissions reduction regime), the authors show that the use of CCS across the globe can be grouped into three broad categories: (1) countries in which the use of CCS does not appear to face either an economic or physical constraint on CCS deployment given the large potential CO_2 storage resource compared to projected demand (e.g., Australia, Canada, and the United States) and where CCS should therefore deploy to the extent that it makes economic sense to do so; (2) countries in which the supply of potential geological storage reservoirs (the authors did not consider ocean storage) is small in comparison to potential demand (e.g., Japan and South Korea) and where other abatement options must therefore be pressed into service to meet the modelled emissions reduction levels; and (3) the rest of the world in which the degree to which CCS deployment is constrained is contingent upon the stringency of the emission constraint and the useable CO_2 storage resource. The authors note that discovering the true CO_2 storage potential in regions of the world is a pressing issue; knowing whether a country or a region has 'sufficient' CO_2 storage capacity is a critical variable in these modelling analyses because it can fundamentally alter the way in which a country's energy infrastructure evolves in response to various modelled emissions constraints.

8.3.3.4 Long-term economic impact
An increasing body of literature has been analyzing short- and long-term financial requirements for CCS. The World Energy Investment Outlook 2003 (IEA, 2003) estimates an upper limit for investment in CCS technologies for the OECD of about US$ 350 to 440 billion over the next 30 years, assuming that all new power plant installations will be equipped with CCS. Similarly, Riahi *et al.* (2004) estimate that up-front investments for initial niche market applications and demonstration plants could amount to about US$ 70 billion or 0.2% of the total global energy systems costs over the next 20 years. This would correspond to a market share of CCS of about 3.5% of total installed fossil-power generation capacities in the OECD countries by 2020, where most of the initial CCS capacities are expected to be installed.

Long-term investment requirements for the full integration of CCS in the electricity sector as a whole are subject to major uncertainties. Analyses with integrated assessment models indicate that the costs of decarbonizing the electricity sector via CCS might be about three to four per cent of total energy-related systems costs over the course of the century (Riahi *et al.*, 2004). Most importantly, these models also point out that the opportunity costs of CCS not being part of the CO_2 mitigation portfolio would be significant. Edmonds *et al.* (2000) indicate that savings over the course of this century associated with the wide-scale deployment of CCS technologies when compared to a scenario in which these technologies do not exist could be in the range of tens of billions of 1990 US dollars for high

CO_2 concentrations limits such as 750 ppmv, to trillions of dollars for more stringent CO_2 concentrations such as 450 ppm [17]. Dooley *et al.* (2002) estimate cost savings in excess of 36% and McFarland *et al.* (2004) a reduction in the carbon permit price by 110 US$/t$CO_2$ in scenarios where CCS technologies are allowed to deploy when compared to scenarios in which they are not.

8.3.3.5 Interaction with other technologies
As noted above, the future deployment of CCS will depend on a number of factors, many of which interact with each other. The deployment of CCS will be impacted by factors such as the development and deployment of renewable energy and nuclear power (Mori, 2000). Edmonds *et al.* (2003) report that CCS technologies can synergistically interact with other technologies and in doing so help to lower the cost and therefore increase the overall economic potential of less carbon-intensive technologies. The same authors note that these synergies are perhaps particularly important for the combination of CCS, H_2 production technologies and H_2 end-use systems (e.g., fuel cells). On the other hand, the widespread availability of CCS technologies implies an ability to meet a given emissions reduction at a lower marginal cost, reducing demand for substitute technologies at the margin. In other words, CCS is competing with some technologies, such as energy-intensity improvements, nuclear, fusion, solar power options, and wind. The nature of that interaction depends strongly on the climate policy environment and the costs and potential of alternative mitigation options, which are subject to large variations depending on site-specific, local conditions (IPCC, 2001). At the global level, which is spatially more aggregated, this variation translates into the parallel deployment of alternative options, taking into account the importance of a diversified technology portfolio for addressing emissions mitigation in a cost-effective way.

An increasing body of literature (Willams, 1998; Obersteiner *et al.*, 2001; Rhodes and Keith, 2003; Makihira *et al.*, 2003; Edmonds *et al.*, 2003, Möllersten *et al.*, 2003) has begun to examine the use of CCS systems with biomass-fed energy systems to create useful energy (electricity or transportation fuels) as well as excess emissions credits generated by the system's resulting 'negative emissions'. These systems can be fuelled solely by biomass, or biomass can be co-fired in conventional coal-burning plants, in which case the quantity is normally limited to about 10–15% of the energy input. Obersteiner *et al.* (2001) performed an analysis based on the SRES scenarios, estimating that 880 to 1650 GtCO_2 (240 to 450 GtC) of the scenario's cumulative emissions that are vented during biomass-based energy-conversion processes could potentially be available for capture and storage over the course of the century. Rhodes and Keith (2003) note that, while this coupled bio-energy CCS system would generate expensive

---

[17] Savings are measured as imputed gains of GDP due to CCS deployment, in contrast to a world where CCS is not considered to be part of the mitigation portfolio.

electricity in a world of low carbon prices, this system could produce competitively priced electricity in a world with carbon prices in excess of 54.5 US$/tCO$_2$ (200 US$/tC). Similarly, Makihira *et al.* (2003) estimate that CO$_2$ capture during hydrogen production from biomass could become competitive at carbon prices above 54.5 to 109 US$/tCO$_2$ (200 to 400 US$/tC).

8.4   Economic impacts of different storage times

As discussed in the relevant chapters, geological and ocean storage might not provide permanent storage for all of the CO$_2$ injected. The question arises of how the possibility of leakage from reservoirs can be taken into account in the evaluation of different storage options and in the comparison of CO$_2$ storage with mitigation options in which CO$_2$ emissions are avoided.

Chapters 5 and 6 discuss the expected fractions of CO$_2$ retained in storage for geological and ocean reservoirs respectively. For example, Box 6.7 suggests four types of measures for ocean storage: storage efficiency, airborne fraction, net present value, and global warming potential. Chapter 9 discusses accounting issues relating to the possible impermanence of stored CO$_2$. Chapter 9 also contains a review of the broader literature on the value of delayed emissions, primarily focusing on sequestration in the terrestrial biosphere. In this section, we focus specifically on the economic impacts of differing storage times in geological and ocean reservoirs.

Herzog *et al.* (2003) suggest that CO$_2$ storage and leakage can be looked upon as two separate, discrete events. They represent the value of temporary storage as a familiar economic problem, with explicitly stated assumptions about the discount rate and carbon prices. If someone stores a tonne of CO$_2$ today, they will be credited with today's carbon price. Any future leakage will have to be compensated by paying the carbon price in effect at that time. Whether non-permanent storage options will be economically attractive depends on assumptions about the leakage rate, discount rate and relative carbon permit prices. In practice, this may turn out to be a difficult issue since the commercial entity that undertakes the storage may no longer exist when leakage rates have been clarified (as Baer (2003) points out), and hence governments or society at large might need to cover the leakage risk of many storage sites rather than the entity that undertakes the storage.

Ha-Duong and Keith (2003) explore the trade-offs between discounting, leakage, the cost of CO$_2$ storage and the energy penalty. They use both an analytical approach and an integrated assessment numerical model in their assessment. In the latter case, with CCS modelled as a backstop technology, they find that, for an optimal mix of CO$_2$ abatement and CCS technologies, 'an (annual) leakage rate of 0.1% is nearly the same as perfect storage while a leakage rate of 0.5% renders storage unattractive'.

Some fundamental points about the limitations of the economic valuation approaches presented in the literature have been raised by Baer (2003). He argues that financial efficiency, which is at the heart of the economic approaches to the valuation of, and decisions about, non-permanent storage is only one of a number of important criteria to be considered. Baer points out that at least three risk categories should to be taken into account as well:

- ecological risk: the possibility that 'optimal' leakage may preclude future climate stabilization;
- financial risk: the possibility that future conditions will cause carbon prices to greatly exceed current expectations, with consequences for the maintenance of liability and distribution of costs; and
- political risk: the possibility that institutions with an interest in CO$_2$ storage may manipulate the regulatory environment in their favour.

As these points have not been extensively discussed in the literature so far, the further development of the scientific debate on these issues must be followed closely.

In summary, within this purely economic framework, the few studies that have looked at this topic indicate that some CO$_2$ leakage can be accommodated while still making progress towards the goal of stabilizing atmospheric concentrations of CO$_2$. However, due to the uncertainties of the assumptions, the impact of different leakage rates and therefore the impact of different storage times are hard to quantify.

8.5   Gaps in knowledge

Cost developments for CCS technologies are now estimated based on literature, expert views and a few recent CCS deployments. Costs of large-scale integrated CCS applications are still uncertain and their variability depends among other things on many site-specific conditions. Especially in the case of large-scale CCS biomass based applications, there is a lack of experience and therefore little information in the literature about the costs of these systems.

There is little empirical evidence about possible cost decreases related to 'learning by doing' for integrated CCS systems since the demonstration and commercial deployment of these systems has only recently begun. Furthermore, the impact of targeted research, development and deployment (RD&D) of CCS investments on the level and rate of CCS deployment is poorly understood at this time. This lack of knowledge about how technologies will deploy in the future and the impact of RD&D on the technology's deployment is a generic issue and is not specific to CCS deployment.

In addition to current and future CCS technological costs, there are other possible issues that are not well known at this point and that would affect the future deployment of CCS systems: for example, costs related to the monitoring and regulatory framework, possible environmental damage costs, costs associated with liability and possible public-acceptance issues.

There are at present no known, full assessments of life-cycle costs for deployed CCS systems, and in particular the economic impact of the capture, transport and storage of non-pure CO$_2$ streams.

The development of bottom-up CCS deployment cost

curves that take into account the interplay between large CO_2 point sources and available storage capacity in various regions of the world should continue; these cost curves would help to show how CCS technologies will deploy in practice and would also help improve the economic modelling of CCS deployment in response to various modelled scenarios.

Recent changes in energy prices and changes in policy regimes related to climate change are not fully reflected in the literature available as this chapter was being written. This suggests a need for a continuous effort to update analyses and perhaps draft a range of scenarios with a wider range of assumptions (e.g., fuel prices, climate policies) in order to understand better the robustness and sensitivity of the current outcomes.

References

Akimoto, K., Kotsubo, H., Asami, T., Li, X., Uno, M., Tomoda, T., and T. Ohsumi, 2003: Evaluation of carbon sequestrations in Japan with a mathematical model. Greenhouse Gas Control Technologies: Proceedings of the Sixth International Conference on Greenhouse Gas Control Technologies, J. Gale and Y. Kaya (eds.), Kyoto, Japan, Elsevier Science, Oxford, UK.

Audus, H. and P. Freund, 2004: Climate change mitigation by biomass gasification combined with CO_2 capture and storage. In, E.S. Rubin, D.W. Keith, and C.F. Gilboy (eds.), Proceedings of 7th International Conference on Greenhouse Gas Control Technologies. Volume 1: Peer-Reviewed Papers and Plenary Presentations, IEA Greenhouse Gas Programme, Cheltenham, UK, 2004.

Baer, P., 2003: An Issue of Scenarios: Carbon Sequestration as Investment and the Distribution of Risk. An Editorial Comment. *Climate Change,* **59,** 283–291.

Dooley, J.J., R.T. Dahowski, C.L. Davidson, S. Bachu, N. Gupta, and H. Gale, 2004a: A CO_2 storage supply curve for North America and its implications for the deployment of carbon dioxide capture and storage systems. In, E.S. Rubin, D.W. Keith, and C.F. Gilboy (eds.), Proceedings of 7th International Conference on Greenhouse Gas Control Technologies. Volume 1: Peer-Reviewed Papers and Plenary Presentations, IEA Greenhouse Gas Programme, Cheltenham, UK, 2004.

Dooley, J.J., S.K. Kim, J.A. Edmonds, S.J. Friedman, and M.A. Wise, 2004b: A First Order Global Geologic CO_2 Storage Potential Supply Curve and Its Application in a Global Integrated Assessment Model. In, E.S. Rubin, D.W. Keith, and C.F. Gilboy (eds.), Proceedings of 7th International Conference on Greenhouse Gas Control Technologies. Volume 1: Peer-Reviewed Papers and Plenary Presentations, IEA Greenhouse Gas Programme, Cheltenham, UK, 2004.

Dooley, J.J., C.L. Davidson, M.A. Wise, R.T. Dahowski, 2004: Accelerated Adoption of Carbon Dioxide Capture and Storage within the United States Electric Utility Industry: the Impact of Stabilizing at 450 ppmv and 550 ppmv. In, E.S. Rubin, D.W. Keith and C.F. Gilboy (eds.), Proceedings of 7th International Conference on Greenhouse Gas Control Technologies. Volume 1: Peer-Reviewed Papers and Plenary Presentations, IEA Greenhouse Gas Programme, Cheltenham, UK, 2004.

Dooley, J.J. and M.A. Wise, 2003: Potential leakage from geologic sequestration formations: Allowable levels, economic considerations, and the implications for sequestration R&D. In: J. Gale and Y. Kaya (eds.), Greenhouse Gas Control Technologies: Proceedings of the Sixth International Conference on Greenhouse Gas Control Technologies, Kyoto, Japan, Elsevier Science, Oxford, UK, ISBN 0080442765.

Dooley, J.J., S.H. Kim, and P.J. Runci, 2000: The role of carbon capture, sequestration and emissions trading in achieving short-term carbon emissions reductions. Proceedings of the Fifth International Conference on Greenhouse Gas Control Technologies. Sponsored by the IEA Greenhouse Gas R&D Programme.

Edmonds, J., and M. Wise, 1998: The economics of climate change: Building backstop technologies and policies to implement the Framework Convention on Climate Change. *Energy & Environment,* **9**(4), 383–397.

Edmonds, J.A., J. Clarke, J.J. Dooley, S.H. Kim, R. Izaurralde, N. Rosenberg, G.M. Stokes, 2003: The potential role of biotechnology in addressing the long-term problem of climate change in the context of global energy and economic systems. In: J. Gale and Y. Kaya (eds.), Greenhouse Gas Control Technologies: Proceedings of the Sixth International Conference on Greenhouse Gas Control Technologies, Kyoto, Japan, Elsevier Science, Oxford, UK, pp. 1427–1433, ISBN 0080442765.

Edmonds, J., J. Clarke, J.J. Dooley, S.H. Kim, S.J. Smith, 2004: Stabilization of CO_2 in a B2 world: insights on the roles of carbon capture and disposal, hydrogen, and transportation technologies. *Energy Economics,* **26**(4), 501–755.

Edmonds, J.A., P. Freund, and J.J. Dooley, 2000: The role of carbon management technologies in addressing atmospheric stabilization of greenhouse gases. Published in the proceedings of the Fifth International Conference on Greenhouse Gas Control Technologies. Sponsored by the IEA Greenhouse Gas R&D Programme.

Fujii, Y. and K. Yamaji, 1998: Assessment of technological options in the global energy system for limiting the atmospheric CO_2 concentration, *Environmental Economics and Policy Studies,* **1** pp.113–139.

Fujii, Y., R. Fukushima, and K. Yamaji, 2002: Analysis of the optimal configuration of energy transportation infrastructure in Asia with a linear programming energy system model, *Int. Journal Global Energy Issues,* **18**, No.1, pp.23–43.

Gielen, D. and J. Podkanski. 2004: The Future Role of CO_2 Capture in the Electricity Sector. In, E.S. Rubin, D.W. Keith and C.F. Gilboy (eds.), Proceedings of 7th International Conference on Greenhouse Gas Control Technologies. Volume 1: Peer-Reviewed Papers and Plenary Presentations, IEA Greenhouse Gas Programme, Cheltenham, UK, 2004.

Ha-Duong, M. and D.W. Keith, 2003: CO_2 sequestration: the economics of leakage. *Clean Technology and Environmental Policy,* 5, 181–189.

Herzog, H., K. Caldeira, and J. Reilly, 2003: An Issue of Permanence: Assessing the Effectiveness of Temporary Carbon Storage, *Climatic Change,* **59.**

IEA, 2002: Greenhouse gas R&D programme. Opportunities for the early application of CO_2 sequestration technology. Report Number PH4/10, IEA, Paris, France.

IEA, 2003: World Energy Investment Outlook 2003. OECD/IEA, 75775 Paris Cedex 16, France, ISBN: 92-64-01906-5.

IEA, 2004: The Prospects for CO_2 Capture and Storage, OECD/IEA, 75775 Paris Cedex 16, France, ISBN 92-64-10881-5.

IPCC, 2001: Climate Change 2001: Mitigation, Contribution of Working Group III to the Third Assessment Report of the Intergovernmental Panel on Climate Change, Cambridge University Press, Cambridge, UK. 752 pp, ISBN: 0521015022.

Johnson, T.L. and D.W. Keith (2004). Fossil Electricity and CO_2 Sequestration: How Natural Gas Prices, Initial Conditions and Retrofits Determine the Cost of Controlling CO_2 Emissions. *Energy Policy,* **32,** p. 367–382.

Makihira, A., Barreto, L., Riahi, K., 2003: Assessment of alternative hydrogen pathways: Natural gas and biomass. IIASA Interim Report, IR-03-037, Laxenburg, Austria.

McFarland, J.R., Herzog, H.J., Reilly, J.M. 2003: Economic modeling of the global adoption of carbon capture and sequestration technologies, In: J. Gale and Y. Kaya (eds.), Greenhouse Gas Control Technologies: Proceedings of the Sixth International Conference on Greenhouse Gas Control Technologies, Kyoto, Japan, Elsevier Science, Oxford, UK.

McFarland, J.R., J.M. Reilly, and H.J. Herzog, 2004: Representing energy technologies in top-down economic models using bottom-up information, *Energy Economics,* **26,** 685–707.

Möllersten, K., J. Yan, and J. Moreira, 2003: Potential market niches for biomass energy with CO_2 capture and storage - opportunities for energy supply with negative CO_2 emissions, *Biomass and Bioenergy,* **25,** 273–285.

Mori, S., 2000: Effects of carbon emission mitigation options under carbon concentration stabilization scenarios, *Environmental Economics and Policy Studies,* **3,** pp.125–142.

Morita, T. and H.-C. Lee, 1998: Appendix to Emissions Scenarios Database and Review of Scenarios. *Mitigation and Adaptation Strategies for Global Change,* 3(2–4), 121–131.

Morita, T., N. Nakicenovic and J. Robinson, 2000: Overview of mitigation scenarios for global climate stabilization based on new IPCC emissions scenarios, *Environmental Economics and Policy Studies,* 3(2), 65–88.

Morita, T., J. Robinson, A. Adegbulugbe, J. Alcamo, D. Herbert, E.L. La Rovere, N. Nakicenovic, H. Pitcher, P. Raskin, K. Riahi, A. Sankovski, V. Sokolov, H.J.M. Vries, Z. Dadi, 2001: Greenhouse Gas Emission Mitigation Scenarios and Implications. In: Metz, B., O. Davidson, R. Swart, and J. Pan (eds.), 2001, Climate Change 2001: Mitigation, Contribution of Working Group III to the Third Assessment Report of the Intergovernmental Panel on Climate Change, Cambridge University Press, Cambridge, UK. 700 pp, ISBN: 0521015022.

Nakicenovic, N. and Riahi, K., 2001: An assessment of technological change across selected energy scenarios. In: Energy Technologies for the Twenty-First Century, World Energy Council (WEC), London, UK.

Nakicenovic, N., Grübler, A., and McDonald, A., eds., 1998: Global Energy Perspectives. Cambridge University Press, Cambridge, UK.

Obersteiner, M., Ch. Azar, P. Kauppi, K. Möllersten, J. Moreira, S. Nilsson, P. Read, K. Riahi, B. Schlamadinger, Y. Yamagata, J. Yan, and J.-P. van Ypersele, 2001: Managing climate risk, *Science* **294,** 786–787.

Pepper, W.J., J. Leggett, R. Swart, R.T. Watson, J. Edmonds, and I. Mintzer, 1992: Emissions scenarios for the IPCC. An update: Assumptions, methodology, and results. Support document for Chapter A3. In Climate Change 1992: Supplementary Report to the IPCC Scientific Assessment. J.T. Houghton, B.A. Callandar, and S.K. Varney (eds.), Cambridge University Press, Cambridge, UK.

Rhodes, J.S. and Keith, D.W., 2003: Biomass Energy with Geological Sequestration of CO_2: Two for the Price of One? In: J. Gale and Y. Kaya (eds.), Greenhouse Gas Control Technologies: Proceedings of the Sixth International Conference on Greenhouse Gas Control Technologies, Kyoto, Japan, Elsevier Science, Oxford, UK, pp. 1371–1377, ISBN 0080442765.

Riahi, K. and Roehrl, R.A., 2000: Energy technology strategies for carbon dioxide mitigation and sustainable development. *Environmental Economics and Policy Studies,* **63,** 89–123.

Riahi, K., E.S. Rubin, and L. Schrattenholzer, 2003: Prospects for carbon capture and sequestration technologies assuming their technological learning. In: J. Gale and Y. Kaya (eds.), Greenhouse Gas Control Technologies: Proceedings of the Sixth International Conference on Greenhouse Gas Control Technologies, Kyoto, Japan, Elsevier Science, Oxford, UK, pp. 1095–1100, ISBN 0080442765.

Riahi, K., L. Barreto, S. Rao, E.S. Rubin, 2004: Towards fossil-based electricity systems with integrated CO_2 capture: Implications of an illustrative long-term technology policy. In: E.S. Rubin, D.W. Keith and C.F. Gilboy (eds.), Proceedings of the 7th International Conference on Greenhouse Gas Control Technologies. Volume 1: Peer-Reviewed Papers and Plenary Presentations, IEA Greenhouse Gas Programme, Cheltenham, UK, 2004.

Roehrl, R.A. and K. Riahi, 2000: Technology dynamics and greenhouse gas emissions mitigation: A cost assessment, *Technological Forecasting & Social Change,* **63,** 231–261.

Scott, M.J., J.A. Edmonds, N. Mahasenan, J.M. Roop, A.L. Brunello, E.F. Haites, 2004: International emission trading and the cost of greenhouse gas emissions mitigation and sequestration. *Climatic Change,* **63,** 257–287.

SRES, 2000: Special Report on Emissions Scenarios (SRES) for the Intergovernmental Panel on Climate Change. Nakićenović *et al.,* Working Group III, Intergovernmental Panel on Climate Change (IPCC), Cambridge University Press, Cambridge, UK, ISBN: 0-521-80493-0.

UN (United Nations), 1998: World Population Projections to 2150. United Nations Department of Economic and Social Affairs Population Division, New York, NY, U.S.A.

Wildenborg, T., J. Gale, C. Hendriks, S. Holloway, R. Brandsma, E. Kreft, A. Lokhorst, 2004: Cost curves for CO_2 Storage: European Sector. In, E.S. Rubin, D.W. Keith and C.F. Gilboy (eds.), Proceedings of 7th International Conference on Greenhouse Gas Control Technologies. Volume 1: Peer-Reviewed Papers and Plenary Presentations, IEA Greenhouse Gas Programme, Cheltenham, UK, 2004.

Williams, R.H., 1998: Fuel decarbonisation for fuel cell applications and sequestration of the separated CO_2 in Eco-Restructuring: Implications for Sustainable Development, R.W. Ayres (ed.), United Nations University Press, Tokyo, pp. 180–222.

Wise, M.A. and J.J. Dooley. Baseload and Peaking Economics and the Resulting Adoption of a Carbon Dioxide Capture and Storage System for Electric Power Plants. In, E.S. Rubin, D.W. Keith and C.F. Gilboy (eds.), Proceedings of 7th International Conference on Greenhouse Gas Control Technologies. Volume 1: Peer-Reviewed Papers and Plenary Presentations, IEA Greenhouse Gas Programme, Cheltenham, UK, 2004.

9

Implications of carbon dioxide capture and storage for greenhouse gas inventories and accounting

Coordinating Lead Authors
Balgis Osman-Elasha (Sudan), Riitta Pipatti (Finland)

Lead Authors
William Kojo Agyemang-Bonsu (Ghana), A.M. Al-Ibrahim (Saudi Arabia), Carlos Lopez (Cuba), Gregg Marland (United States), Huang Shenchu (China), Oleg Tailakov (Russian Federation)

Review Editors
Takahiko Hiraishi (Japan), José Domingos Miguez (Brazil)

Contents

EXECUTIVE SUMMARY

This chapter addresses how methodologies to estimate and report reduced or avoided greenhouse gas emissions from the main options for CO_2 capture and storage (CCS) systems could be included in national greenhouse gas inventories, and in accounting schemes such as the Kyoto Protocol.

The *IPCC Guidelines* and Good Practice Guidance reports (*GPG2000* and *GPG-LULUCF*)[1] are used in preparing national inventories under the UNFCCC. These guidelines do not specifically address CO_2 capture and storage, but the general framework and concepts could be applied for this purpose. The IPCC guidelines give guidance for reporting on annual emissions by gas and by sector. The amount of CO_2 captured and stored can be measured, and could be reflected in the relevant sectors and categories producing the emissions, or in new categories created specifically for CO_2 capture, transportation and storage in the reporting framework. In the first option, CCS would be treated as a mitigation measure and, for example, power plants with CO_2 capture or use of decarbonized fuels would have lower emissions factors ($kgCO_2$/kg fuel used) than conventional systems. In the second option, the captured and stored amounts would be reported as removals (sinks) for CO_2. In both options, emissions from fossil fuel use due to the additional energy requirements in the capture, transportation and injection processes would be covered by current methodologies. But under the current framework, they would not be allocated to the CCS system.

Methodologies to estimate, monitor and report physical leakage from storage options would need to be developed. Some additional guidance specific to the systems would need to be given for fugitive emissions from capture, transportation and injection processes. Conceptually, a similar scheme could be used for mineral carbonation and industrial use of CO_2. However, detailed methodologies would need to be developed for the specific processes.

Quantified commitments, emission trading or other similar mechanisms need clear rules and methodologies for accounting for emissions and removals. There are several challenges for the accounting frameworks. Firstly, there is a lack of knowledge about the rate of physical leakage from different storage options including possibilities for accidental releases over a very long time period (issues of permanence and liability). Secondly, there are the implications of the additional energy requirements of the options; and the issues of liability and economic leakage where CO_2 capture and storage crosses the traditional accounting boundaries.

The literature on accounting for the potential impermanence of stored CO_2 focuses on sequestration in the terrestrial biosphere. Although notably different from CCS in oceans or in geological reservoirs (with respect to ownership, the role of management, measurement and monitoring, expected rate of physical leakage; modes of potential physical leakage; and assignment of liability), there are similarities. Accounting approaches, such as discounting, the ton-year approach, and rented or temporary credits, are discussed. Ultimately, political processes will decide the value of temporary storage and allocation of responsibility for stored carbon. Precedents set by international agreements on sequestration in the terrestrial biosphere provide some guidance, but there are important differences that will have to be considered.

9.1 Introduction

CO_2 capture and storage (CCS) can take a variety of forms. This chapter discusses how the main CCS systems as well as mineral carbonation and industrial uses of CO_2, described in the previous chapters could be incorporated into national greenhouse gas inventories and accounting schemes. However, inventory or accounting issues specific to enhanced oil recovery or enhanced coal bed methane are not addressed here.

The inclusion of CCS systems in national greenhouse gas inventories is discussed in Section 9.2 (Greenhouse gas inventories). The section gives an overview of the existing framework, the main concepts and methodologies used in preparing and reporting national greenhouse gas emissions and removals with the aim of identifying inventory categories for reporting CCS systems. In addition, areas are identified where existing methodologies could be used to include these systems in the inventories, and areas where new methodologies (including emission/removal factors and uncertainty estimates) would need to be developed. Treatment of CCS in corporate or company reporting is beyond the scope of the chapter.

Issues related to accounting[2] under the Kyoto Protocol; or under other similar accounting schemes that would limit emissions, provide credits for emission reductions, or encourage emissions trading; are addressed in Section 9.3 (Accounting issues). The section addresses issues that could warrant special rules and modalities in accounting schemes because of specific features of CCS systems, such as permanence of CO_2 storage and liability issues related to transportation and storage in international territories and across national borders. Specific consideration is also given to CCS systems in relation to the mechanisms of the Kyoto Protocol (Emission Trading, Joint Implementation and the Clean Development Mechanism).

---

[1] Revised 1996 IPCC Guidelines for National Greenhouse Gas Inventories (IPCC 1997) – abbreviated as IPCC Guidelines in this chapter; IPCC Good Practice Guidance and Uncertainty Management in National Greenhouse Gas Inventories (IPCC 2000) – abbreviated as GPG2000; and IPCC Good Practice Guidance for Land Use, Land-Use Change and Forestry (IPCC 2003) – abbreviated as GPG-LULUCF.

[2] 'Accounting' refers to the rules for comparing emissions and removals as reported with commitments. In this context, 'estimation' is the process of calculating greenhouse gas emissions and removals, and 'reporting' is the process of providing the estimates to the UNFCCC (IPCC 2003).

9.2   National greenhouse gas inventories

Information on pollutant emissions is usually compiled in 'emission inventories'. Emissions are listed according to categories such as pollutants, sectors, and source and compiled per geographic area and time interval. Many different emission inventories have been prepared for different purposes. Among the commitments in the United Nations Framework Convention on Climate Change (UNFCCC, 1992) all Parties, taking into account their common but differentiated responsibilities, and their specific national and regional development priorities, objectives and circumstances, shall: 'Develop, periodically update, publish and make available to the Conference of the Parties, national inventories of anthropogenic emissions by sources and removals by sinks of all greenhouse gases not controlled by the Montreal Protocol, using comparable methodologies to be agreed upon by the Conference of the Parties'.[3]

Industrialized countries (Annex I Parties) are required to report annually and developing countries (non-Annex I Parties) to report on greenhouse gas emissions and removals to the Convention periodically, as part of their National Communications to the UNFCCC. National greenhouse gas inventories are prepared using the methodologies in the *IPCC Guidelines* as complemented by the *GPG2000* and *GPGLULUCF*, or methodologies consistent with these. These inventories should include all anthropogenic greenhouse gas emissions by sources and removals by sinks not covered by the Montreal Protocol. To ensure high quality and accuracy, inventories by Annex I Parties are reviewed by expert review teams coordinated by the UNFCCC Secretariat. The review reports are published on the UNFCCC website[4].

The rules and modalities for accounting are elaborated under the Kyoto Protocol (UNFCCC, 1997) and the Marrakech Accords[5] (UNFCCC, 2002). The Kyoto Protocol specifies emission limitation or reduction commitments by the Annex I Parties for six gases/gas groups: carbon dioxide (CO_2), methane (CH_4), nitrous oxide (N_2O), hydrofluorocarbons (HFCs), perfluorocarbons (PFCs) and sulphur hexafluoride (SF_6).

At present, CCS is practiced on a very small scale. CCS projects have not generally been described in the national inventory reports of the countries where they take place. An exception is the Sleipner CCS project, which is included in Norway's inventory report.[6] Norway provides information on the annual captured and stored amounts, as well as on the amounts of CO_2 that escape to the atmosphere during the injection process (amounts have varied from negligible to about 0.8% of the captured amount). The escaping CO_2 emissions are

included in the total emissions of Norway. The spread of the CO_2 in the storage reservoir has been monitored by seismic methods. No physical leakage has been detected. An uncertainty estimate has not been performed but it is expected to be done when more information is available from the project's monitoring programme.

The scarce reporting of current CCS projects is due largely to the small number and size of industrial CCS projects in operation, as well as to the lack of clarity in the reporting methodologies.

9.2.1   Revised 1996 IPCC Guidelines and IPCC Good Practice Guidance

The reporting guidelines under the UNFCCC[7], and under the Kyoto Protocol as specified in the Marrakech Accords require Annex I Parties to use the *IPCC Guidelines[1]*, as elaborated by the *GPG2000[1]*, in estimating and reporting national greenhouse gas inventories. The use of the *GPG-LULUCF[1]* will start in 2005 with a one-year trial period[8]. Non-Annex I Parties also use the *IPCC Guidelines* in their reporting, and use of *GPG2000* and *GPG-LULUCF* reports is encouraged.[9] The main reporting framework (temporal, spatial and sectoral) and the guiding principles of the *IPCC Guidelines* and good practice guidance reports are given in Box 9.1.

The IPCC Guidelines will be revised and updated by early 2006[10]. In the draft outline for the 2006 IPCC Guidelines for National Greenhouse Gas Inventories, CCS is mentioned in a footnote in the Energy Sector: 'It is recognized that CO_2 capture and storage is an important emerging issue in inventory development. The coverage of CO_2 storage in this report will be closely coordinated with progress on IPCC SR on CO_2 capture and storage. CO_2 capture activities will be integrated as appropriate into the methods presented for source categories where it may occur.'

9.2.2   Methodological framework for CO_2 capture and storage systems in national greenhouse gas inventories

The two main options for including CCS in national greenhouse gas inventories have been identified and analysed using the current methodological framework for total chain from capture to storage (geological and ocean storage). These options are:
- Source reduction: To evaluate the CCS systems as mitigation options to reduce emissions to the atmosphere;

---

[3] Commitment related to the Articles 4.1 (a) and 12.1 (a) of the United Nations Framework Convention of Climate Change (UNFCCC).

[4] http://unfccc.int

[5] The Marrakech Accords refer to the Report of the Conference of the Parties of the UNFCCC on its seventh session (COP7), held in Marrakech 29 October to 10 November 2001.

[6] Norway's inventory report can be found at http://cdr.eionet. eu.int/no/un/UNFCCC/envqh6rog.

[7] FCCC/CP2002/7/Add.2: Annexes to Decision 17/CP.8 Guidelines for the preparation of national communications from Parties not included in Annex I to the Convention and 18/CP.8 Guidelines for the preparation of national communications by Parties included in Annex I to the Convention, part I: UNFCCC reporting guidelines on annual inventories.

[8] FCCC/SBSTA/2003/L.22 and FCCC/SBSTA/2003/L.22/Add.1.

[9] FCCC/CP/2002/7/Add.2.

[10] http://www.ipcc.ch/meet/session21.htm: IPCC XXI/Doc.10.

Box 9.1 Main reporting framework (temporal, spatial and sectoral) and guiding principles of the IPCC Guidelines and good practice guidance reports.

The IPCC methodologies for estimating and reporting **national** greenhouse gas inventories are based on **sectoral** guidance for reporting of actual emissions and removals of greenhouse gases by gas and **by year**. The *IPCC Guidelines* give the framework for the reporting (sectors, categories and sub-categories), default methodologies and default emission/removal factors (the so called Tier 1 methodologies) for the estimation. Higher tier methodologies are based on more sophisticated methods for estimating emissions/removals and on the use of national or regional parameters that accommodate the specific national circumstances. These methodologies are not always described in detail in the IPCC Guidelines. Use of transparent and well-documented national methodologies consistent with those in the *IPCC Guidelines* is encouraged.

The Good Practice Guidance (GPG) reports facilitate the development of inventories in which the emissions/removals are not over- or under-estimated, so far as can be judged, and in which the uncertainties are reduced as far as practicable. Further aims are to produce transparent, documented, consistent, complete, comparable inventories, which are i) assessed for uncertainties, ii) subject to quality assurance and quality control, and iii) efficient in the use of resources. The GPG reports give guidance on how to choose the appropriate methodologies for specific categories in a country, depending on the importance of the category (key category analysis is used to determine the importance) and on availability of data and resources for the estimation. Decision trees guide the choice of estimation method most suited to the national circumstances. The category-specific guidance linked to the decision trees also provides information on the choice of emission factors and activity data. The GPG reports give guidance on how to meet the requirements of transparency, consistency, completeness, comparability, and accuracy required by the national greenhouse gas inventories.

The **Sectors** covered in the *IPCC Guidelines* are: (i) Energy, (ii) Industrial Processes, (iii) Solvent and Other Product Use, (iv) Agriculture, (v) Land Use Change and Forestry, (vi) Waste and (vii) Other. The use of the seventh sector 'Other' is discouraged: 'Efforts should be made to fit all emission sources/sinks into the six categories described above. If it is impossible to do so, however, this category can be used, accompanied by a detailed explanation of the source/sink activity'' (IPCC 1997).

- Sink enhancement: To evaluate the CCS systems using an analogy with the treatment made to CO_2 removals by sinks in the sector Land Use, Land-Use Change and Forestry. A balance is made of the CO_2 emissions and removals to obtain the net emission or removal. In this option, removals by sinks are related to CO_2 storage.

In both options, estimation methodologies could be developed to cover most of the emissions in the CCS system (see Figure 9.1), and reporting could use the current framework for preparation of national greenhouse gas inventories.

In the first option, reduced emissions could be reported in the category where capture takes place. For instance, capture in power plants could be reported using lower emission factors than for plants without CCS. But this could reduce transparency of reporting and make review of the overall impact on emissions more difficult, especially if the capture process and emissions from transportation and storage are not linked. This would be emphasized where transportation and storage includes captured CO_2 from many sources, or when these take place across national borders. An alternative would be to track CO_2 flows through the entire capture and storage system making transparent how much CO_2 was produced, how much was emitted to the atmosphere at each process stage, and how much CO_2 was transferred to storage. This latter approach, which appears fully transparent and consistent with earlier UNFCCC agreements, is described in this chapter.

The second option is to report the impact of the CCS system as a sink. For instance, reporting of capture in power plants

would not alter the emissions from the combustion process but the stored amount of CO_2 would be reported as a removal in the inventory. Application of the second option would require adoption of new definitions not available in the UNFCCC or in the current methodological framework for the preparation of inventories. UNFCCC (1992) defines a sink as 'any process, activity or mechanism which removes a greenhouse gas, an aerosol, or a precursor of a greenhouse gas *from the atmosphere*'. Although 'removal' was not included explicitly in the UNFCCC definitions, it appears associated with the 'sink' concept. CCS[11] systems do not meet the UNFCCC definition for a sink, but given that the definition was agreed without having CCS systems in mind, it is likely that this obstacle could be solved (Torvanger *et al.*, 2005).

General issues of relevance to CCS systems include system boundaries (sectoral, spatial and temporal) and these will vary in importance with the specific system and phases of the system. The basic methodological approaches for system components, together with the status of the methods and availability of data for these are discussed below. Mineral carbonation and industrial use of CO_2 are addressed separately.

- Sectoral boundaries: The draft outline for the *2006 IPCC Guidelines* (see Section 9.2.1) states that: 'CO_2 capture activities will be integrated as appropriate into the methods presented for source/sink categories where they may

---

[11] Few cases are nearer to the 'sink' definition. For example, mineralization can also include fixation from the atmosphere.

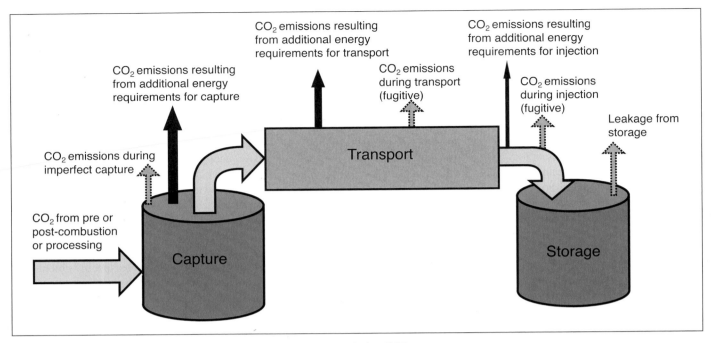

Figure 9.1 Simplified flow diagram of possible CO_2 emission sources during CCS

occur'. This approach is followed here when addressing the sectors under which the specific phases of the CCS systems could be reported. The reporting of emissions/removals associated with CO_2 capture, transportation, injection and storage processes should be described clearly to fulfil the requirement of transparent reporting.

- Spatial boundaries: National inventories include greenhouse gas emissions and removals taking place within national (including administered) territories and offshore areas over which that country has jurisdiction. Some of the emissions and removals of CCS systems could occur outside the areas under the jurisdiction of the reporting country, an aspect that requires additional consideration and is addressed mainly in Section 9.3.
- Temporal boundaries: Inventories are prepared on a calendar year basis. Some aspects of CCS systems (such as the amount of CO_2 captured or fugitive emissions from transportation) could easily be incorporated into an annual reporting system (yearly estimates would be required). However, other emissions (for example, physical leakage of CO_2 from geological storage) can occur over a very long period after the injection has been completed - time frames range from hundreds to even millions of years (see further discussion in Section 9.3).

Table 9.1 lists potential sources and emissions of greenhouse gases in the different phases of a CCS system and their relationship with the framework for the reporting (sectors, categories and sub-categories) of the *IPCC Guidelines*. The relative importance of these potential sources for the national greenhouse inventory can vary from one CCS project to another, depending on factors such as capture technologies

and storage site characteristics. Emissions from some of these sources are probably very small, sometimes even insignificant, but to guarantee an appropriate completeness[12] of the national inventory, it is necessary to evaluate their contribution.

Some important considerations relative to the source categories and emissions included in Table 9.1 are the following:

- Capture, transportation and injection of CO_2 into storage requires energy (the additional energy requirements have been addressed in previous chapters). Greenhouse gas emissions from this energy use are covered by the methodologies and reporting framework in the *IPCC Guidelines* and *GPG2000*. Additional methodologies and emission factors can be found in other extensive literature, such as EEA (2001) and US EPA (1995, 2000). Where capture processes take place at the fuel production site, the emissions from the fuel used in the capture process may not be included in the national statistics. Additional methods to cover emissions from this source may be needed. In the current reporting framework, emissions from the additional energy requirements would not be linked to the CCS system.
- Fugitive emissions from CCS systems can occur during capture, compression, liquefaction, transportation and injection of CO_2 to the storage reservoir. A general framework for estimation of fugitive emissions is included in the *IPCC Guidelines* in the Energy sector. The estimation and reporting of fugitive emissions from CCS need further

---

[12] Completeness means that an inventory covers all sources and sinks, as well as all gases included in the IPCC Guidelines and also other existing relevant source/sink categories specific to individual Parties, and therefore may not be included in the IPCC Guidelines. Completeness also means full geographic coverage of sources and sinks of a Party (FCCC/CP/1999/7).

Table 9.1 Potential sources and emissions of greenhouse gases (GHG) in the general phases of a CCS system.

IPCC guidelines		Emissions	Capture	Transportation [b]	Injection	Storage [c]
Sector [a]	Source category [a]					
1 Energy	GHG emissions from stationary combustion 1A1; 1A2	CO_2, CH_4, N_2O, NO_x, CO, NMVOCs, SO_2	•		•	
1 Energy	GHG emissions from mobile combustion — Water-borne navigation 1A3di [d] 1A3dii [e]	CO_2, CH_4, N_2O, NO_x, CO, NMVOCs, SO_2		•		
	Other transportation (pipeline transportation) 1A3ei	CO_2, CH_4, N_2O, NO_x, CO, NMVOCs, SO_2		•		
1 Energy	Fugitive emissions from fuels 1B — Oil and natural gas 1B2 [f]	CO_2; CH_4; N_2O NMVOCs	•		•	
2 Industrial processes (excluding emissions from fuel combustion)	Mineral products 2A (e.g., cement)	CO_2, SO_2	•		•	
	Chemical industry 2B (e.g., ammonia)	CO_2, NMVOCs, CO, SO_2	•		•	
	Metal production 2C (e.g., iron and steel)	CO_2, NO_x, NMVOCs, CO, SO_2	•		•	
	Other production 2D (e.g. food and drink)	CO_2, NMVOCs	•		•	
6 Waste	Industrial wastewater handling 6B1	CH_4	•			
	Fugitive CO_2 emissions from capture, transportation and injection processes [g] — Normal operations	CO_2	•	•	•	
	Repair and maintenance	CO_2	•	•	•	
	Systems upsets and accidental discharges	CO_2	•	•	•	

a) IPCC source/sink category numbering (see also IPCC (1997), Vol.1, Common Reporting Framework).

b) Emissions from transportation include both GHG emissions from fossil fuel use and fugitive emissions of CO_2 from pipelines and other equipment/processes. Besides ships and pipelines, limited quantities of CO_2 could be transported by railway or by trucks, source categories identified in the IPCC Guidelines/ GPG2000.

c) Long-term physical leakage of stored CO_2 is not covered by the existing framework for reporting of emissions in the *IPCC Guidelines*. Different potential options exist to report these emissions in the inventories (for example, in the relevant sectors/categories producing the emissions, creating a separate and new category for the capture, transportation and/or storage industry). No conclusion can yet be made on the most appropriate reporting option taking into account the different variants adopted by the CCS systems.

d) International Marine (Bunkers). Emissions based on fuel sold to ships engaged in international transport should not be included in national totals but reported separately under Memo Items.

e) National Navigation.

f) Emissions related to the capture (removal) of CO_2 in natural gas processing installations to improve the heating valued of the gas or to meet pipeline specifications.

g) A general framework for estimation of fugitive emissions is included in the *IPCC Guidelines* in the Energy sector. However, estimation and reporting of fugitive emissions from CCS needs further elaboration of the methodologies.

elaboration in methodologies.
- The long-term physical leakage of stored CO_2 (escape of CO_2 from a storage reservoir) is not covered by the existing framework for reporting emissions in the *IPCC Guidelines*. Different options exist to report these emissions in the inventories (for example, in the relevant sectors/categories producing the emissions initially, by creating a separate and new category under fugitive emissions, or by creating a new category for the capture, transportation and/or storage industry).
- Application of CCS to CO_2 emissions from biomass combustion, and to other CO_2 emissions of biological origin (for example, fermentation processes in the production of food and drinks) would require specific treatment in inventories. It is generally assumed that combustion of biomass fuels results in zero net CO_2 emissions if the biomass fuels are produced sustainably. In this case, the CO_2 released by combustion is balanced by CO_2 taken up during photosynthesis. In greenhouse gas inventories, CO_2 emissions from biomass combustion are, therefore, not reported under Energy. Any unsustainable production should be evident in the calculation of CO_2 emissions and removals in Land Use, Land-Use Change and Forestry Sector. Thus, CCS from biomass sources would be reported as negative CO_2 emissions.

9.2.2.1   Capture

The capture processes are well defined in space and time, and their emissions (from additional energy use, fugitives, etc.) could be covered by current national and annual inventory systems. The capture processes would result in reduced emissions from industrial plants, power plants and other sites of fuel combustion. For estimation purposes, the reduced CO_2 emissions could be determined by measuring the amount of CO_2 captured and deducting this from the total amount of CO_2 produced (see Figure 8.2 in Chapter 8).

The total amount of CO_2, including emissions from the additional energy consumption necessary to operate the capture process, could be estimated using the methods and guidance in the *IPCC Guidelines* and *GPG2000*. The capture process could produce emissions of other greenhouse gases, such as CH_4 from treatment of effluents (for example, from amine decomposition). These emissions are not included explicitly in the *IPCC Guidelines* and *GPG2000*. Estimates on the significance of these emissions are not available, but are likely to be small or negligible compared to the amount of captured CO_2.

Although not all possible CCS systems can be considered here, it is clear that some cases would require different approaches. For example, pre-combustion decarbonization in fuel production units presents some important differences compared to the post-combustion methods, and the simple estimation process described above might not be applicable. For example, the capture of CO_2 may take place in a different country than the one in which the decarbonized fuel is used. This would mean that emissions associated with the capture process

(possible fugitive CO_2 emissions) would need to be estimated and reported separately to those resulting from the combustion process (see also Section 9.3 on issues relating to accounting and allocation of the emissions and emissions reductions).

9.2.2.2   Transportation

Most research on CCS systems focuses on the capture and storage processes and fugitive emissions from CO_2 transportation are often overlooked (Gale and Davison, 2002). CO_2 transportation in pipelines and ships is discussed in Chapter 4. Limited quantities of CO_2 could also be transported via railway or by trucks (Davison *et al.*, 2001). The additional energy required for pipeline transport is mostly covered by compression at the capture site. Additional compression may be required when CO_2 is transported very long distances. The emissions from fossil fuel in transportation by ships, rail or trucks would be covered under the category on mobile combustion and other subcategories in the Energy sector. However, according to the current IPCC guidelines, emissions from fuels sold to any means of international transport should be excluded from the national total emissions and be reported separately as emissions from international bunkers. These emissions are not included in national commitments under the Kyoto Protocol (e.g., IPCC 1997 and 2000, see also Section 9.3).

Any fugitive emissions or accidental releases from transportation modes could be covered in the Energy sector under the category 'Fugitive Emissions'. CO_2 emissions from a pipeline can occur at the intake side during pumping and compression, at the pipeline joints, or at the storage site. Emission rates can differ from surface, underground and sub-sea pipelines. Explicit guidance for CO_2 transportation in pipelines is not given in the current IPCC methodologies, but a methodology for natural gas pipelines is included. A distinction is to be made between leakage during normal operation and CO_2 losses during accidents or other physical disruptions. As described in Chapter 4, statistics on the incident rate in pipelines for natural gas and CO_2 varied from 0.00011 to 0.00032 incidents km^{-1} $year^{-1}$ (Gale and Davison, 2002). However, as an analogy of CO_2 transportation to natural gas transportation, Gielen (2003) reported that natural gas losses during transportation can be substantial.

Total emissions from pipelines could be calculated on the basis of the net difference between the intake and discharge flow rates of the pipelines. Because CO_2 is transported in pipelines as a supercritical or dense phase fluid, the effect of the surrounding temperature on the estimated flow rate would need to be taken into account. Volumetric values would need to be corrected accordingly when CO_2 is transmitted from a cooler climate to a moderate or hot climate, and vice versa. In some cases, fugitive losses could be lower than metering accuracy tolerances. Hence, all metering devices measuring CO_2 export and injection should be to a given standard and with appropriate tolerances applied. But metering uncertainties may prohibit measurement of small quantities of losses during transportation. For transportation by CO_2 pipeline across the borders of several countries, emissions would need to be allocated to the countries where they occur.

No methodologies for estimation of fugitive emission from ship, rail or road transportation are included in the IPCC Guidelines.

9.2.2.3 Storage

Some estimates of CO_2 emissions (physical leakage rates) from geological and ocean storage are given in Chapters 5 and 6. Physical leakage rates are estimated to be very small for geological formations chosen with care. In oil reservoirs and coal seams, storage times could be significantly altered if exploitation or mining activities in these fields are undertaken after CO_2 storage. Some of the CO_2 injected into oceans would be released to the atmosphere over a period of hundreds to thousands of years, depending on the depth and location of injection.

The amount of CO_2 injected or stored could be easily measured in many CCS systems. Estimation of physical leakage rates would require the development of new methodologies. Very limited data are available in relation to the physical leakage of CO_2.

Despite the essential differences in the nature of the physical processes of CO_2 retention in oceans, geological formations, saline aquifers and mineralized solids, the mass of CO_2 stored over a given time interval can be defined by the Equation 1.

$$CO_2 \ stored = \int_0^T (CO_2 \ injected(t) - CO_2 \ emitted(t)dt \qquad (1)$$

where t is time and T is the length of the assessment time period.

Use of this simple equation requires estimates or measurements of the injected CO_2 mass and either default values of the amount of CO_2 emitted from the different storage types, or rigorous source-specific evaluation of mass escaped CO_2. This approach would be possible when accurate measurements of mass of injected and escaped CO_2 are applied on site. Thus, for monitoring possible physical leakage of CO_2 from geological formations, direct measurement methods for CO_2 detection, geochemical methods and tracers, or indirect measurement methods for CO_2 plume detection could be applied (see Section 5.6, Monitoring and verification technology).

Physical leakage of CO_2 from storage could be defined as follows (Equation 2):

$$Emissions \ of \ CO_2 \ from \ storage = \int_0^T m(t)dt \qquad (2)$$

where m(t) is the mass of CO_2 emitted to the atmosphere per unit of time and T is the assessment time period.

This addresses physical leakage that might occur in a specific timeframe after the injection, perhaps far into the future. The issue is discussed further in Section 9.3.

9.2.2.4 Mineral carbonation

Mineral carbonation of CO_2 captured from power plants and industrial processes is discussed in Chapter 7. These processes are still under development and aim at permanent fixation of the CO_2 in a solid mineral phase. There is no discussion in the literature about possible modes and rates of physical leakage of CO_2 from mineral carbonation, probably because investigations in this field have been largely theoretical character (for example, Goldberg *et al.*, 2000). However, the carbonate produced would be unlikely to release CO_2. Before and during the carbonation process, some amount of gas could escape into the atmosphere.

The net benefits of mineral carbonation processes would depend on the total energy use in the chain from capture to storage. The general framework discussed above for CCS systems can also be applied in preparing inventories of emissions from these processes. The emissions from the additional energy requirements would be seen in the energy sector under the current reporting framework. The amount of CO_2 captured and mineralized could be reported in the category where the capture takes place, or as a specific category addressing mineral carbonation, or in the sector 'Other'.

9.2.2.5 Industrial uses

Most industrial uses of CO_2 result in release of the gas to the atmosphere, often after a very short time period. Because of the short 'storage times', no change may be required in the inventory systems provided they are robust enough to avoid possible double counting or omission of emissions. The benefits of these systems are related to the systems they substitute for, and the relative net efficiencies of the alternate systems. Comparison of the systems would need to take into account the whole cycle from capture to use of CO_2. As an example, methanol production by CO_2 hydrogenation could be a substitute for methanol production from fossil fuels, mainly natural gas. The impacts of the systems are in general covered by current inventory systems, although they are not addressed explicitly, because the emissions and emission reductions are related to relative energy use (reduction or increase depending on the process alternatives).

In cases where industrial use of CO_2 would lead to more long-term carbon storage in products, inventory methodologies would need to be tailored case by case.

9.2.3 Monitoring, verification and uncertainties

The IPCC Guidelines and good practice reports give guidance on monitoring, verification and estimation of uncertainties, as well as on quality assurance and quality control measures. General guidance is given on how to plan monitoring, what to monitor and how to report on results. The purpose of verifying national inventories is to establish their reliability and to check the accuracy of the reported numbers by independent means.

Section 5.6, on monitoring and verification technology, assesses the current status of monitoring and verification techniques for CCS systems. The applicability of monitoring techniques as well as associated detection limits and uncertainties vary greatly depending on the type and specific characteristics of the CCS projects. There is insufficient experience in monitoring CCS projects to allow conclusions to be drawn on physical leakage rates.

Reporting of uncertainties in emission and removal estimates, and how they have been derived, is an essential part of national greenhouse gas inventories. Uncertainty estimates can be based on statistical methods where measured data are available, or on expert judgement. No information on uncertainties related to emissions from different phases of CCS systems was available. In Section 5.7.3, the probability of release from geological storage is assessed based on data from analogous natural or engineered systems, fundamental physical and chemical processes, as well as from experience with current geological storage projects. The probabilities of physical leakage are estimated to be small and the risks are mainly associated with leakage from well casings of abandoned wells.

9.3   Accounting issues

One of the goals of an accounting system is to ensure that CCS projects produce real and quantifiable environmental benefits. One ton of CO_2 permanently stored has the same benefit in terms of atmospheric CO_2 concentrations as one ton of CO_2 emissions avoided. But one ton of CO_2 temporarily stored has less value than one ton of CO_2 emissions avoided. This difference can be reflected in the accounting system. Accounting for CCS may have to go beyond measuring the amount of CO_2 stored in order to ensure the credibility of storage credits and that credits claimed are commensurate with benefits gained. CO_2 storage should not avoid properly accounting for emissions that have been moved to other times, other places, or other sectors. Yet, Kennett (2003) notes that if there is benefit to potentially permanent or even to known temporary storage, accounting systems should contribute to their credibility and transparency while minimizing transaction costs.

In a political environment where only some parties have commitments to limit greenhouse gas emissions and where emissions from all sources are not treated the same, the amount by which emissions are reduced may not be equal to the amount of CO_2 stored. Differences can occur because CO_2 can be captured in one country but released in another country or at a later time. Also, CCs requires energy and likely additional emissions of CO_2 to produce this additional energy. Yoshigahara *et al.* (2004) note that emission reduction through CCS technology differs from many other modes of emission reduction. Although the former avoids CO_2 release to the atmosphere, it creates the long-term possibility that stored CO_2 could eventually flow to the atmosphere through physical leakage.

In this Chapter, the general term 'leakage' is used in the economist's sense, to describe displacement of greenhouse gas emissions beyond the boundaries of the system under discussion. The term 'physical leakage' refers to escape of CO_2 from a storage reservoir. As discussed above, some physical leakage effects and the additional energy requirements will be reported within standard, national reporting procedures for greenhouse gas emissions. Additional complexities arise when new or unexpected sources of emissions occur, for example, if CO_2 injected into an uneconomic coal seam forces the release of methane from that seam. Complexities also arise when new

or unexpected sources of emissions occur in different countries, for example, if CO_2 is captured in one country but released in another, or at later times, for example, if CO_2 is captured during one time period and physically leaked to the atmosphere at a later time.

The problems of economic leakage are not unique to CCS systems, but the problems of physical leakage are unique to CCS. In particular, when emission inventories are done by country and year they may fail to report emissions that are delayed in time, displaced to other countries or to international waters, or that stimulate emissions of other greenhouse gases not identified as sources or for which methodologies have not been developed.

In this section, ideas on the issues involved in accounting are summarized for the stored CO_2 of CCS systems. The consequences for mitigating greenhouse gas emissions are discussed, and ideas on alternative accounting strategies to address them are presented. Figure 9.2 provides a simple flow diagram of how CCS emissions can create flows of greenhouse gases that transcend traditional accounting boundaries. The diagram also shows how emissions might escape reporting because they occur outside normal system boundaries (sectoral, national, or temporal) of reporting entities.

Concern about displacement of emissions across national boundaries is a consequence of the political and economic constructs being developed to limit greenhouse gas emissions. Most notably, the Kyoto Protocol imposes limits on greenhouse gas emissions from developed countries and from countries with economies in transition, but no such limits on emissions from developing countries or international transport.

Concern about displacement of emissions across temporal boundaries is essentially the widely posed question: 'if we store carbon away from the atmosphere, how long must it be stored?' The same question is phrased by Herzog *et al.* (2003) as 'What is the value of temporary storage?'

Concern about leakage among countries, sectors, or gases; or physical leakage from reservoirs is largely about the completeness and accuracy of emissions accounting. Kennett (2003), for example, emphasizes the importance of 'establishing general rules and procedures to simplify transactions, and increasing certainty by defining legal rights and by providing dispute resolution and enforcement procedures' and of ensuring the credibility of sinks-based emissions offsets or storage-based emissions reductions. The operation of a market requires clearly defined rights (i.e. who has the rights to the carbon stored), what those rights entail, how those rights can be transferred, and liability and remedies in the event of unanticipated release (Kennett, 2003). The core of establishing rights, liabilities, and markets will be the accounting and certification systems. Yet, a well-designed accounting system should not lead to transaction costs that unnecessarily discourage meritorious activities.

9.3.1 Uncertainty, non-permanence and discounting methodology

9.3.1.1 Dealing with the impermanence of carbon dioxide storage

CO_2 storage is not necessarily permanent. Physical leakage from storage reservoirs is possible via (1) gradual and long-term release or (2) sudden release of CO_2 caused by disruption of the reservoir. There is very little literature on accounting for the potential impermanence of CCS. There are, however, a significant number of publications on accounting for the impermanence of CO_2 sequestration in the terrestrial biosphere. Although sequestration in the terrestrial biosphere is notably different from CO_2 storage in the ocean or in geological reservoirs, there are also similarities. $^{13}CO_2$ stored in the terrestrial biosphere is subject to potential future release if, for example, there is a wildfire, change in land management practices, or climate change renders the vegetative cover unsustainable. Although the risks of CO_2 loss from well-chosen geological reservoirs are very different, such risks do exist. The literature suggests various accounting strategies so that sequestration in the biosphere could be treated as the negative equivalent of emissions. Sequestration could be shown in national emission accounts and trading of emissions credits, and debits between parties could occur for sequestration activities in the terrestrial biosphere. Whether CCS is treated as a CO_2 sink or as a reduction in emissions, the issues of accounting for physical leakage from storage are similar.

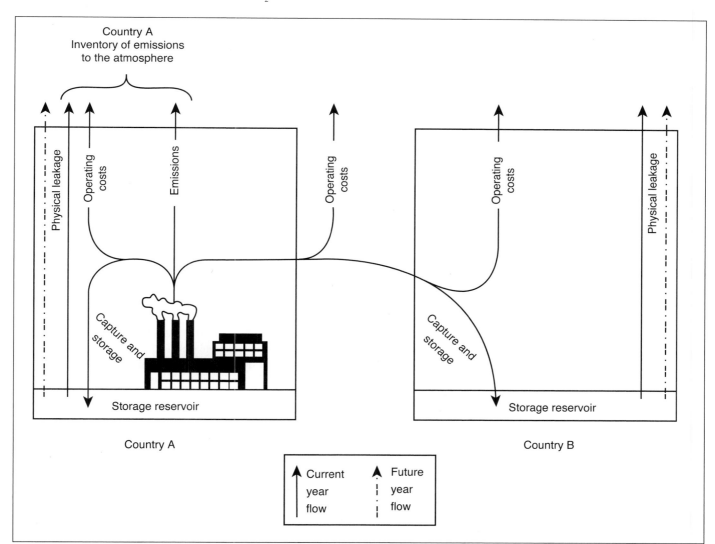

Figure 9.2 Simplified flow diagram showing how CCS could transcend traditional accounting boundaries[13]

---

[13] The operating cost shown are the CO_2 emitted as a result of the additional energy required to operate the system, plus fugitive emissions from separation, transport and injection.

Chomitz (2000) suggests two primary approaches to accounting for stored CO_2: (1) acknowledge that CO_2 storage is likely not permanent, assess the environmental and economic benefits of limited-term storage, and allot credits in proportion to the time period over which CO_2 is stored, and (2) provide reasonable assurance of indefinite storage. Examples discussed for sequestration in the terrestrial biosphere include (under the first approach) ton-year accounting (described below); and (under the second approach) various combinations of reserve credits and insurance replacing lost CO_2 by sequestration reserves or other permanent emissions reductions. For further discussion on these issues, see Watson *et al.*, 2000; Marland *et al.*, 2001; Subak, 2003; Aukland *et al.*, 2003; Wong and Dutschke, 2003; and Herzog *et al.*, 2003. There are also proposals to discount credits so that there is a margin of conservativeness in the number of credits acknowledged. With this kind of discussion and uncertainty, negotiations toward the Kyoto Protocol have chosen to place limits on the number of credits that can be claimed for some categories of terrestrial CO_2 sequestration during the Protocol's first commitment period (UNFCCC, 2002).

To illustrate the concept of allotting credits in proportion to storage time, one alternative, the ton-year approach is described. The ton-year alternative for accounting defines an artificial equivalence so that capture and storage for a given time interval (for example, t years) are equated with permanent storage. Availability of credits can be defined in different ways but typically capture and storage for one year would result in a number of credits equal to 1/t, and thus storage for t years would result in one full credit (Watson *et al.*, 2000). A variety of constructs have been proposed for defining the number of storage years that would be equated with permanent storage (see, for example, Marland *et al.*, 2001). But as Chomitz (2000) points out, despite being based on scientific and technical considerations, this equivalence is basically a political decision. Although ton-year accounting typifies the first approach, it has been subject to considerable discussion. Another derivative of Chomitz's first approach that has been further developed within negotiations on the Kyoto Protocol (Columbia, 2000; UNFCCC, 2002; UNFCCC, 2004) is the idea of expiring credits or rented temporary credits (Marland et el., 2001; Subak, 2003). Temporary or rented credits would have full value over a time period defined by rule or by contract, but would result in debits or have to be replaced by permanent credits at expiration. In essence, credit for stored CO_2 would create liability for the possible subsequent CO_2 release or commitment to storage was ended.

UNFCCC (2002), Marland et al. (2001), Herzog et al. (2003), and others agree that the primary issue for stored CO_2 is liability. They argue that if credit is given for CO_2 stored, there should be debits if the CO_2 is subsequently released. Physical leakage from storage and current emissions produce the same result for the atmosphere. Accounting problems arise if ownership is transferred or stored CO_2 is transferred to a place or party that does not accept liability (for example, if CO_2 is stored in a developing country without commitments

under the Kyoto protocol). Accounting problems also arise if potential debits are transferred sufficiently far into the future with little assurance that the systems and institutions of liability will still be in place if and when CO_2 is released. The system of expiring credits in the Marrakech Accords for sequestration in the terrestrial biosphere fulfils the requirement of continuing liability. Limiting these credits to five years provides reasonable assurance that the liable institutions will still be responsible. This arrangement also addresses an important concern of those who might host CO_2 storage projects, that they might be liable in perpetuity for stored CO_2. Under most proposals, the hosts for CO_2 storage would be liable for losses until credits expire and then liability would return to the purchaser/renter of the expiring credits. Kennett (2003) suggests that long-term responsibility for regulating, monitoring, certifying, and supporting credits will ultimately fall to governments (see also section 5.8.4). With this kind of ultimate responsibility, governments may wish to establish minimum requirements for CCS reservoirs and projects (see Torvanger *et al.*, 2005).

The published discussions on 'permanence' have largely been in the context of sequestration in the terrestrial biosphere. It is not clear whether the evolving conclusions are equally appropriate for CCS in the ocean or in geological reservoirs. Important differences between modes of CCS may influence the accounting scheme chosen (see Table 9.2). An apparent distinction is that sequestration in the terrestrial biosphere involves initial release of CO_2 to the atmosphere and subsequent removal by growing plants. But as storage in geological reservoirs does not generally involve release to the atmosphere, it might be envisioned as a decrease in emissions rather than as balancing source with sink. In either case, a mass of CO_2 must be managed and isolated from the atmosphere. Storage in the terrestrial biosphere leaves open the possibility that sequestration will be reversed because of decisions on maintenance or priorities for resource management. Ocean and geological storage have very different implications for the time scale of commitments and for the role of physical processes versus decisions in potential physical releases.

An important question for crediting CCS is whether future emissions have the same value as current emissions. Herzog *et al.* (2003) define 'sequestration effectiveness' as the net benefit from temporary storage compared to the net benefit of permanent storage, but this value cannot be known in advance. They go one step further and argue that while CO_2 storage is not permanent, reducing emissions may not be permanent either, unless some backstop energy technology assures all fossil fuel resources are not eventually consumed. According to Herzog *et al.* (2003), stored CO_2 emissions are little different, to fossil fuel resources left in the ground. Most analysts, however, assume that all fossil fuels will never be consumed so that refraining from emitting fossil-fuel CO_2 does not, like CO_2 storage, give rise directly to a risk of future emissions. Wigley *et al.* (1996) and Marland *et al.* (2001) argue that there is value in delaying emissions. If storage for 100 years were to be defined as permanent, then virtually all carbon injected below 1500 m in the oceans would be considered to be permanent storage (Herzog *et al.*, 2003).

Table 9.2 Differences between forms of carbon storage with potential to influence accounting method.

Property	Terrestrial biosphere	Deep ocean	Geological reservoirs
CO_2 sequestered or stored	Stock changes can be monitored over time.	Injected carbon can be measured	Injected carbon can be measured
Ownership	Stocks will have a discrete location and can be associated with an identifiable owner.	Stocks will be mobile and may reside in international waters.	Stocks may reside in reservoirs that cross national or property boundaries and differ from surface boundaries.
Management decisions	Storage will be subject to continuing decisions about land-use priorities.	Once injected, no further human decisions on maintenance.	Once injected, human decisions to influence continued storage involve monitoring and perhaps maintenance, unless storage interferes with resource recovery.
Monitoring	Changes in stocks can be monitored.	Changes in stocks will be modelled.	Release of CO_2 might be detected by physical monitoring but because of difficulty in monitoring large areas may also require modelling.
Time scale with expected high values for fraction CO_2 retained	Decades, depending on management decisions.	Centuries, depending on depth and location of injection.	Very small physical leakage from well-designed systems expected, barring physical disruption of the reservoir.
Physical leakage	Losses might occur due to disturbance, climate change, or land-use decisions.	Losses will assuredly occur as an eventual consequence of marine circulation and equilibration with the atmosphere.	Losses are likely to be small for well-designed systems except where reservoir is physically disrupted.
Liability	A discrete land-owner can be identified with the stock of sequestered carbon.	Multiple parties may contribute to the same stock of stored carbon and the carbon may reside in international waters.	Multiple parties may contribute to the same stock of stored carbon lying under several countries.

At the other temporal extreme, Kheshgi *et al.* (1994) point out that over the very long term of equilibration between the ocean and atmosphere (over 1000 years), capture and storage in the ocean will lead to higher CO_2 levels in the atmosphere than without emissions controls, because of the additional energy requirements for operating the system. It is also true that chronic physical leakage over long time periods could increase the difficulty of meeting targets for net emissions at some time in the future (see Hawkins, 2003; Hepple and Benson, 2003; and Pacala, 2003).

The fundamental question is then, how to deal with impermanent storage of CO_2. Although Findsen *et al.* (2003) detail many circumstances where accounting for CCS is beginning or underway, and although the rates of physical leakage for well-designed systems may sometimes be in the range of the uncertainty of other components of emissions, the risks of physical leakage need to be acknowledged. A number of questions remains to be answered: how to deal with liability and continuity of institutions in perpetuity, how to quantify the benefits of temporary storage; the needs in terms of monitoring and verification, whether or not there is a need for a reserve of credits or other ways to assure that losses will be replaced, whether or not there is need for a system of discounting to

consider expected or modelled duration of storage, the utility of expiring, temporary, or rented credits over very long time periods, whether there is a need to consider different accounting practices as a function of expected duration of storage or mode of storage. The implications if storage in the terrestrial biosphere and in geological formations are sufficiently different that the former might be considered carbon management and the latter CO_2 waste disposal.

Ultimately, the political process will decide the value of temporary storage and the allocation of responsibility for stored CO_2. Some guidance is provided by precedents set by international agreements on sequestration in the terrestrial biosphere. But there are important differences to be considered. The reason for rules and policies is presumably to influence behaviour. Accounting rules for CO_2 storage can best influence permanence if they are aimed accordingly: at liability for CO_2 stored in the terrestrial biosphere but at the initial design and implementation requirements for CCS in the oceans or geological reservoirs.

9.3.1.2   Attribution of physical leakage from storage in international/regional territories or shared facilities and the use of engineering standards to limit physical leakage

The previous section deals largely with the possibility that CO_2 emissions stored now will be released at a later time. It also introduces the possibility that emissions stored now will result in additional, current emissions in different countries or in different sectors. CO_2 injected into the ocean could leak physically from international waters. Accounting for stored CO_2 raises questions such as responsibility for the emissions from energy used in CO_2 transport and injection, especially if transport and/or storage is in a developing country or in international waters. Similarly, questions about physical leakage of stored CO_2 will need to address liability for current year physical leakage that occurs in developing countries or from international waters. These questions may be especially complex when multiple countries have injected CO_2 into a common reservoir such as the deep Atlantic Ocean, or into a deep aquifer under multiple countries, or if multiple countries share a common pipeline for CO_2 transport.

There may also be a need for international agreement on certification of CCS credits or performance standards for CCS projects. Standards would minimize the risk of leakage and maximize the time for CO_2 storage. Performance standards could minimize the possibility of parties looking for the least cost, lowest quality storage opportunities - opportunities most susceptible to physical leakage - when liability for spatial or temporal leakage is not clear. Performance standards could be used to limit the choice of technologies, quality of operations, or levels of measurement and monitoring.

9.3.2   Accounting issues related to Kyoto mechanisms (JI[14], CDM[15], and ET[16])

CCS is not currently addressed in the decisions of the COP to the UNFCCC in relation to the Kyoto mechanisms. Little guidance has been provided so far by international negotiations regarding the methodologies to calculate and account for project-related CO_2 reductions from CCS systems under the various project-based schemes in place or in development. The only explicit reference to CCS in the Kyoto Protocol states that Annex I countries need to "research, promote, develop and increasingly use CO_2 sequestration technologies"[17]. The Marrakech Accords further clarify the Protocol regarding technology cooperation, stating that Annex I countries should indicate how they give priority to cooperation in the development and transfer of technologies relating to fossil fuel that capture and store greenhouse gases (Paragraph 26, Decision 5/CP.7). No text referring explicitly to CCS project-based activities can be found in the CDM and JI-related decisions (Haefeli *et al.*, 2004).

Further, Haefeli *et al.* (2004) note that CCS is not explicitly addressed in any form in CO_2 reporting schemes that include projects (i.e., the Chicago Climate Exchange and the EU Directive for Establishing a Greenhouse Gas Emissions Trading Scheme (implemented in 2005) along with the EU Linking Directive (linking the EU Emissions Trading Scheme with JI and the CDM). At present, it is unclear how CCS will be dealt with in practice. According to Haines *et al.* (2004), the eligibility of CCS under CDM could be resolved in a specific agreement similar to that for land use, land-use change and forestry (LULUCF) activities. As with biological sinks, there will be legal issues as well as concerns about permanence and economic leakage, or emissions outside a system boundary. At the same time, CCS could involve a rather less complex debate because of the geological time scales involved. Moreover, Haefeli *et al.* (2004) noted that guidelines on how to account for CO_2 transfers between countries would need to be agreed either under the UNFCCC or the Kyoto Protocol. Special attention would need to be given to CO_2 exchange between an Annex I country and a non-Annex I country, and between an Annex I country party to the Kyoto Protocol and an Annex I country that has not ratified the Kyoto Protocol.

9.3.2.1   Emission baselines

The term 'baseline', used mostly in the context of project-based accounting, is a hypothetical scenario for greenhouse gas emissions in the absence of a greenhouse gas reduction project or activity (WRI, 2004). Emission baselines are the basis for calculation of net reductions (for example, storage) of emissions from any project-based activity. Baselines need to be established to show the net benefits of emissions reductions. The important issue is to determine which factors need to be taken into account when developing an emissions baseline. At present, there is little guidance on how to calculate net reductions in CO_2 emissions through CCS project-based activities. An appropriate baseline scenario could minimize the risk that a project receives credits for avoiding emissions that would have been avoided in the absence of the project (Haefeli *et al.*, 2004).

9.3.2.2   Leakage in the context of the Kyoto mechanisms

The term 'Leakage' is defined according to Marrakech Accords as 'the net change of anthropogenic emissions by sources and/or removals by sinks of greenhouse gases which occurs outside

---

[14] Kyoto Protocol Article 6.1 'For the purpose of meeting its commitments under Article 3, any Party included in Annex I may transfer to, or acquire from, any other such Party emission reduction units resulting from projects aimed at reducing anthropogenic emissions by sources or enhancing anthropogenic removals by sinks of greenhouse gases in any sector of the economy...'

[15] Kyoto Protocol Article 12.2 'The purpose of the clean development mechanism shall be to assist Parties not included in Annex I in achieving sustainable development and in contributing to the ultimate objective of the Convention, and to assist Parties included in Annex I in achieving compliance with their quantified emission limitation and reduction commitments under Article 3.'

[16] Kyoto Protocol Article 17 'The Conference of the Parties shall define the relevant principles, modalities, rules and guidelines, in particular for verification, reporting and accountability for emissions trading. The Parties included in Annex B may participate in emissions trading for the purpose of fulfilling their commitments under Article 3. Any such trading shall be supplemental to domestic actions for the purpose of meeting quantified emission limitation and reduction commitments under that Article.'

[17] Article 2, 1(a) (iv) of the Kyoto Protocol.

Table 9.3 Accounting issues related to Kyoto Mechanisms.

Mechanism	Article in the Kyoto Protocol	Principle	Requirements in relation to CCS	Basic considerations
Joint Implementation (JI)	Article 6.1	As a general principle, any Annex I party may transfer to or obtain from another Annex I party Emission Reduction Units (ERUs) that shall result from projects that seek to reduce GHG emissions by sources and/or enhance removals by sinks.	• Set modalities and procedures to set the project in a transparent manner • Procedures for verification and certification of ERU.	Important to ensure that credits received from projects in Annex I countries result from emission reductions that are real and additional to what would have happened in the absence of the project i.e. are measured against baselines.
Clean Development Mechanism (CDM)	Article 12.2	• Intended to promote sustainable develop-ment in developing countries through the allowance of trade between developed and developing countries. • Refers to the establishment of a CDM with the objective of assisting Annex I parties to achieve part of their Article 3 KP emission reduction commitments through the implementation of project-based activities generating emission cut-backs and/or enhanced sink removals.	Highly detailed set of modalities and procedures regarding issues such as: • project level versus national level obligations • modelled versus actual amounts of credits • timing of storage and liabilities in the long term.	• Overall baseline methodology • Annex I parties shall be able to acquire Certified Emission Reductions (CERs) from projects implemented in non Annex I countries. • Should provide real, measurable and long-term benefits related to the mitigation of climate change, i.e. will be measured against baselines.
Emission Trading (ET)	Article 17	Allows for trading between developed countries that have targets and assigned amount units (AAUs) allocated to them through the KP, it endorses the basic principle of the use of ET as a mean available to Annex I parties to achieve their emission commitment.	• Cap (emission trading) i.e. the maximum amount of allowable emission offsets between Annex I countries; • Net versus gross accounting (measures in non-Annex I).	• Trade is based on national Assigned Amounts (AAUs) to individual countries. • The proposed guidelines for ET contain provisions on the amount of AAUs that may be traded between Annex I parties so as to avoid overselling of quotas. It also contains several options that would impose a quantified upper limit on the amount of AAUs that a transferring party could trade. • A successful carbon trading system must accurately measure the offsets and credits to assure companies that they will receive the reductions.

the project boundary, and that is measurable and attributable to the Article 6 project'. The term has been proposed for leakage of emissions resulting from capture, transport and injection, which should not be confused with releases of CO_2 from a geological reservoir (escaped CO_2). According to Haefeli *et al.* (2004), current legislation does not deal with cross-border CCS projects and would need further clarification. Guidance would be especially needed to deal with cross-border projects involving CO_2 capture in an Annex I country that is party to the Kyoto Protocol and storage in a country not party to the Kyoto Protocol or in an Annex I country not bound by the Kyoto Protocol.

Table 9.3 provides an overview of the Kyoto mechanisms and the general principles and requirements of each (practical indices and specific accounting rules and procedures) for developing CCS accounting systems that can be employed for emissions control and reduction within these mechanisms. Although the political process has not yet decided how CCS systems will be accepted under the Kyoto mechanisms, these general procedures could be applicable to them as well as to other similar schemes on emission trading and projects.

9.4   Gaps in knowledge

Methodologies for incorporating CCS into national inventories and accounting schemes are under development. CCS (see Sections 9.2 and 9.3) can be incorporated in different ways and data requirements may differ depending on the choices made. The following gaps in knowledge and need for decisions by the political process have been identified:

- Methodologies to estimate physical leakage from storage, and emission factors (fugitive emissions) for estimating emissions from capture systems and from transportation and injection processes are not available.
- Geological and ocean storage open new challenges regarding a) uncertainty on the permanence of the stored emissions, b) the need for protocols on transboundary transport and storage, c) accounting rules for CCS and, d) insight on issues such as emission measurement, long term monitoring, timely detection and liability/responsibility.
- Methodologies for reporting and verification of reduced emission under the Kyoto Mechanisms have not been agreed upon.
- Methodologies for estimating and dealing with potential emissions resulting from system failures, such as sudden geological faults and seismic activities or pipeline disruptions have not been developed.

References

Aukland, L., P. Moura Costa, and S. Brown, 2003: A conceptual framework and its application for addressing leakage: the case of avoided deforestation. *Climate Policy*, **3**, 123-136.

Chomitz, K.M., 2000: Evaluating carbon offsets for forestry and energy projects: how do they compare? World Bank Policy Research Working Paper 2357, New York, p. 25, see http://wbln0018.worldbank.org/research/workpapers.nsf.

Columbia Ministry of the Environment, 2000: Expiring CERs, A proposal to addressing the permanence issue, pp. 23-26 in United Nations Framework Convention on Climate Change, UN-FCCC/SBSTA/2000/MISC.8, available at www.unfccc.de.

Davison, J.E., P. Freund, A. Smith, 2001: Putting carbon back in the ground, published by IEA Greenhouse Gas R&D Programme, Cheltenham, U.K. ISBN 1 898373 28 0.

EEA, 2001: Joint EMEP/CORINAIR Atmospheric Emission Inventory Guidebook - 3rd Edition, Copenhagen: European Environment Agency, 2001.

Findsen, J., C. Davies, and S. Forbes, 2003: Estimating and reporting GHG emission reductions from CO_2 capture and storage activities, paper presented at the second annual conference on carbon sequestration, Alexandria, Virginia, USA, May 5-8, 2003, US Department of Energy, 14 pp.

Gale, J., and J. Davison, 2002: Transmission of CO_2: Safety and Economic Considerations, Proceedings of the 6th International Conference on Greenhouse Gas Control Technologies, 1-4 October, 2002, Kyoto, Japan. pp. 517-522.

Gielen, D.J., 2003: Uncertainties in Relation to CO_2 capture and sequestration. Preliminary Results. IEA/EET working Paper, March.

Goldberg, P., R. Romanosky, Z.-Y. Chen, 2002: CO_2 Mineral Sequestration Studies in US. Proceedings of the fifth international conference on greenhouse gas control technologies, 13-16 August 2000, Australia.

Haefeli, S., M. Bosi, and C. Philibert, 2004: Carbon dioxide capture and storage issues - accounting and baselines under the United Nations Framework Convention on Climate Change. IEA Information Paper. IEA, Paris, 36 p.

Haines, M. *et al.,* 2004: Leakage under CDM/Use of the Clean Development Mechanism for CO_2 Capture and Storage.Based on a study commisioned by the IEA GHG R&D Programme.

Hawkins, D.G., 2003: Passing gas: policy implications of leakage from geologic carbon storage sites, pp. 249-254 in J. Gale and Y. Kaya (eds.) Greenhouse gas control technologies, proceedings of the 6th international conference on greenhouse gas control technologies, Pergamon Press, Amsterdam.

Hepple, R.P. and S. M. Benson, 2003: Implications of surface seepage on the effectiveness of geologic storage of carbon dioxide as a climate change mitigation strategy, pp. 261-266 in J. Gale and Y. Kaya (eds.) Greenhouse gas control technologies, Proceedings of the 6th International Conference on Greenhouse Gas Control Technologies, Pergamon Press, Amsterdam.

Herzog, H., K. Caldeira, and J. Reilly, 2003: An issue of permanence: assessing the effectiveness of temporary carbon storage, Climatic Change, **59** (3), 293-310.

IPCC, 2003: Good Practice Guidance for Land Use, Land-Use Change and Forestry. Penman, J. *et al.* (eds), IPCC/IGES, Japan.

IPCC, 2000: Good Practice Guidance and Uncertainty Management in National Greenhouse Gas Inventories, J. Perman *et al.* (eds), IPCC/IEA/OECD/IGES, Japan.

IPCC, 1997: Revised 1996 IPCC Guidelines for National Greenhose Gas Inventories, J. T. Houghton *et al.* (eds), IPCC/OECD/IEA, Paris, France.

Kennett, S.A., 2003: Carbon sinks and the Kyoto Protocol: Legal and Policy Mechanisms for domestic implementation, *Journal of Energy and Natural Resources Law*, **21**, 252-276.

Kheshgi, H.S., B.P. Flannery, M.I. Hoffert, and A.G. Lapenis, 1994: The effectiveness of marine CO_2 disposal, *Energy,* **19**, 967-974.

Marland, G., K. Fruit, and R. Sedjo, 2001: Accounting for sequestered carbon: the question of permanence. *Environmental Science and Policy*, **4**, 259-268.

Pacala, S.W., 2003: Global Constraints on Reservoir Leakage, pp. 267-272 in J. Gale and Y. Kaya (eds.). Greenhouse gas control technologies, proceedings of the 6th international conference on greenhouse gas control technologies, Pergamon Press, Amsterdam.

Subak, S., 2003: Replacing carbon lost from forests; an assessment of insurance, reserves, and expiring credits. *Climate Policy*, **3**, 107-122.

Torvanger, A., K. Rypdal, and S. Kallbekken, 2005: Geological CO_2 storage as a climate change mitigation option, Mitigation and Adaptation Strategies for Global Change, in press.

UNFCCC, 2004: Report of the conference of the parties on is ninth session, held at Milan from 1 to 12 December, 2003. United Nations Framework Convention on Climate Change FCCC/CP/2003/6/Add.2, 30 March 2004. Decision 19/CP.9. www.unfccc.int.

UNFCCC, 2002: Report of the conference of the parties on is seventh session, held at Marrakesh from 29 October to 10 November, 2001. United Nations Framework Convention on Climate Change FCCC/CP/2001/13/Add.1 - Add.3, 21 January 2002. www.unfccc.int.

UNFCCC, 1997: The Kyoto Protocol to the United Nations Framework Convention on Climate Change. UNEP-IU, France, 34 p.

UNFCCC, 1992: United Nations Framework Convention on Climate Change. UNEP/IUC. Switzerland. 30 p.

US EPA, 1995: Compilation of Air Pollutant Emisión Factors AP-42, Fifth Edition, Volume 1: Stationary Point and Area Sources. U.S. Environment Protection Agency, Research Triangle Park, NC, January 1995.

US EPA, 2000: Supplements to the Compilation of air Pollutant Emission Factors AP-42, Fifth Edition, Volume I; Stationary Point and Area Sources, U.S. Environment Protection Agency, January 1995-September 2000.

Watson, R.T., I.R. Noble, B. Bolin, N.H. Ravindranath, D.J. Verardo, and D. J. Dokken (eds.), 2000: Land use, land-use change, and forestry, A special report of the Intergovernmental Panel on Climate Change, Cambridge University Press, Cambridge, UK.

Wigley, T.M.L., R. Richels, and J.A. Edmonds, 1996: Economic and environmental choices in the stabilization of CO_2 concentrations, *Nature*, **379**, 240-243.

Wong, J. and M. Dutschke, 2003: Can permanence be insured? Consideration of some technical and practical issues of insuring carbon credits for afforestation and reforestation. HWMA discussion paper 235, Hamburgisches Welt-Wirtschafts-Archiv, Hamburg Institute of International Economics, Hamburg, Germany.

WRI, 2004: The Greenhouse Gas Protocol/ A Corporate Accounting and Reporting Standard. (Revised edition) ISBN 1-56973-568-9 (112 pages)

Yoshigahara, *et al.*, 2004: Draft Accounting Rules For Carbon Capture And Storage Technology. " Proceedings of 7th International Conference on Greenhouse Gas Control Technologies. E.S. Rubin, D.W. Keith, and C.F. Gilboy (eds.), Volume II, Pergamon Press, Amsterdam, 2005.

Annexes

Annex I

Properties of CO$_2$ and carbon-based fuels

Coordinating Lead Author
Paul Freund (United Kingdom)

Lead Authors
Stefan Bachu (Canada), Dale Simbeck (United States), Kelly (Kailai) Thambimuthu (Australia and Canada)

Contributing Authors
Murlidhar Gupta (Canada and India)

Contents

AI.1   Introduction

This Annex presents data about the relevant physical and chemical properties of CO$_2$ together with an outline of the effects of CO$_2$ on human health and a summary of some of the typical recommendations for avoiding harm to humans. Established uses for CO$_2$ are listed and some common conversion factors relevant to this report are presented. An introduction is also provided to the main types of fossil fuels and other carbon-containing fuels, as background to considering how their use produces CO$_2$.

AI.2   Carbon dioxide

Carbon dioxide is a chemical compound of two elements, carbon and oxygen, in the ratio of one to two; its molecular formula is CO$_2$. It is present in the atmosphere in small quantities (370 ppmv) and plays a vital role in the Earth's environment as a necessary ingredient in the life cycle of plants and animals. During photosynthesis plants assimilate CO$_2$ and release oxygen. Anthropogenic activities which cause the emission of CO$_2$ include the combustion of fossil fuels and other carbon-containing materials, the fermentation of organic compounds such as sugar and the breathing of humans. Natural sources of CO$_2$, including volcanic activity, dominate the Earth's carbon cycle.

CO$_2$ gas has a slightly irritating odour, is colourless and is denser than air. Although it is a normal, if minor, constituent of air, high concentrations of CO$_2$ can be dangerous.

AI.2.1   *Physical properties of CO$_2$*

AI.2.1.1   General

At normal temperature and pressure, carbon dioxide is a gas. The physical state of CO$_2$ varies with temperature and pressure as shown in Figure AI.1 – at low temperatures CO$_2$ is a solid; on warming, if the pressure is below 5.1 bar, the solid will sublime directly into the vapour state. At intermediate temperatures (between −56.5°C, the temperature of the triple point, and 31.1°C, the critical point), CO$_2$ may be turned from a vapour into a liquid by compressing it to the corresponding liquefaction pressure (and removing the heat produced).

At temperatures higher than 31.1°C (if the pressure is greater than 73.9 bar, the pressure at the critical point), CO$_2$ is said

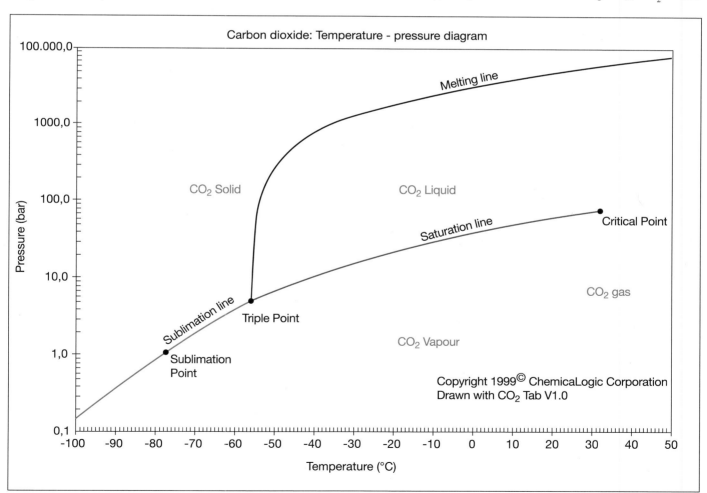

Figure AI.1 Phase diagram for CO$_2$. Copyright © 1999 ChemicaLogic Corporation, 99 South Bedford Street, Suite 207, Burlington, MA 01803 USA. All rights reserved.

to be in a supercritical state where it behaves as a gas; indeed under high pressure, the density of the gas can be very large, approaching or even exceeding the density of liquid water (also see Figure AI.2). This is an important aspect of CO_2's behaviour and is particularly relevant for its storage.

Heat is released or absorbed in each of the phase changes across the solid-gas, solid-liquid and liquid-gas boundaries (see Figure AI.1). However, the phase changes from the supercritical condition to liquid or from supercritical to gas do not require or release heat. This property is useful for the design of CO_2 compression facilities since, if this can be exploited, it avoids the need to handle the heat associated with the liquid-gas phase change.

AI.2.1.2   Specific physical properties

There is a substantial body of scientific information available on the physical properties of CO_2. Selected physical properties of CO_2 are given in Table AI.1 The phase diagram for CO_2 is shown in Figure AI.1 Many authors have investigated the

equation of state for CO_2 (e.g., Span and Wagner, 1996). The variation of the density of CO_2 as a function of temperature and pressure is shown in Figure AI.2, the variation of vapour pressure of CO_2 with temperature in Figure AI.3, and the variation of viscosity with temperature and pressure in Figure AI.4 Further information on viscosity can be found in Fenghour *et al.* (1998). The pressure-enthalpy chart for CO_2 is shown in Figure AI.5. The solubility of CO_2 in water is described in Figure AI.6.

AI.2.2   Chemical properties of CO_2

AI.2.2.1   General
Some thermodynamic data for CO_2 and a few related compounds are given in Table AI.2.

In an aqueous solution CO_2 forms carbonic acid, which is too unstable to be easily isolated. The solubility of CO_2 in water (Figure AI.6) decreases with increasing temperature and increases with increasing pressure. The solubility of CO_2 in

Table AI.1 Physical properties of CO_2.

Property	Value
Molecular weight	44.01
Critical temperature	31.1°C
Critical pressure	73.9 bar
Critical density	467 kg m^{-3}
Triple point temperature	-56.5 °C
Triple point pressure	5.18 bar
Boiling (sublimation) point (1.013 bar)	-78.5 °C
Gas Phase	
Gas density (1.013 bar at boiling point)	2.814 kg m^{-3}
Gas density (@ STP)	1.976 kg m^{-3}
Specific volume (@ STP)	0.506 m^3 kg^{-1}
Cp (@ STP)	0.0364 kJ (mol^{-1} K^{-1})
Cv (@ STP)	0.0278 kJ (mol^{-1} K^{-1})
Cp/Cv (@ STP)	1.308
Viscosity (@ STP)	13.72 μN.s m^{-2} (or μPa.s)
Thermal conductivity (@ STP)	14.65 mW (m K^{-1})
Solubility in water (@ STP)	1.716 vol vol^{-1}
Enthalpy (@ STP)	21.34 kJ mol^{-1}
Entropy (@ STP)	117.2 J mol K^{-1}
Entropy of formation	213.8 J mol K^{-1}
Liquid Phase	
Vapour pressure (at 20 °C)	58.5 bar
Liquid density (at -20 °C and 19.7 bar)	1032 kg m^{-3}
Viscosity (@ STP)	99 μN.s m^{-2} (or μPa.s)
Solid Phase	
Density of carbon dioxide snow at freezing point	1562 kg m^{-3}
Latent heat of vaporisation (1.013 bar at sublimation point)	571.1 kJ kg^{-1}

Where STP stands for Standard Temperature and Pressure, which is 0°C and 1.013 bar.
Sources: Air Liquide gas data table; Kirk-Othmer (1985); NIST (2003).

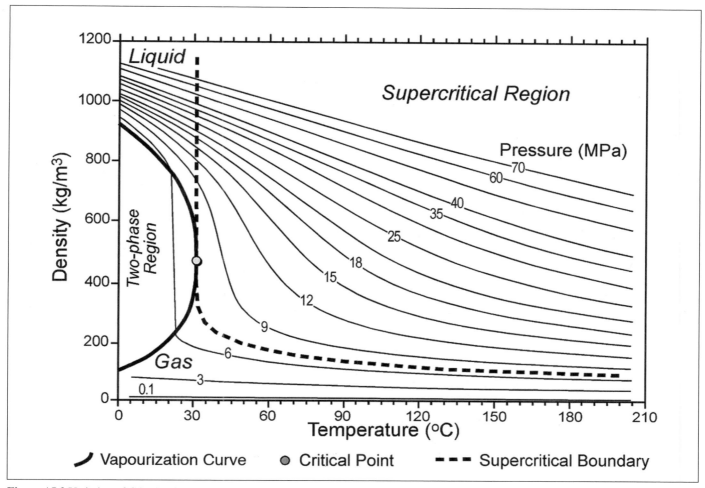

Figure AI.2 Variation of CO_2 density as a function of temperature and pressure (Bachu, 2003).

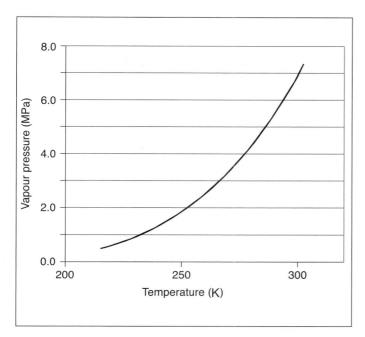

Figure AI.3 Vapour pressure of CO_2 as a function of temperature (Span and Wagner, 1996).

water also decreases with increasing water salinity by as much as one order of magnitude (Figure AI.7). The following empirical relation (Enick and Klara, 1990) can be used to estimate CO_2 solubility in brackish water and brine:

$$w_{CO2,\,b} = w_{CO2,\,w} \cdot (1.0 - 4.893414 \cdot 10^{-2} \cdot S +$$

$$0.1302838 \cdot 10^{-2} \cdot S^2 - 0.1871199 \cdot 10^{-4} \cdot S^3) \qquad (1)$$

where w_{CO2} is CO_2 solubility, S is water salinity (expressed as total dissolved solids in % by weight) and the subscripts w and b stand for pure water and brine, respectively. A solid hydrate separates from aqueous solutions of CO_2 that are chilled (below about 11°C) at elevated pressures. A hydrate is a crystalline compound consisting of the host (water) plus guest molecules. The host is formed from a tetrahedral hydrogen-bonding network of water molecules; this network is sufficiently open to create pores (or cavities) that are large enough to contain a variety of other small molecules (the guests). Guest molecules can include CH_4 and CO_2. CO_2 hydrates have similar (but not identical) properties to methane hydrates, which have been extensively studied due to their effects on natural gas production and their potential as future sources of hydrocarbons.

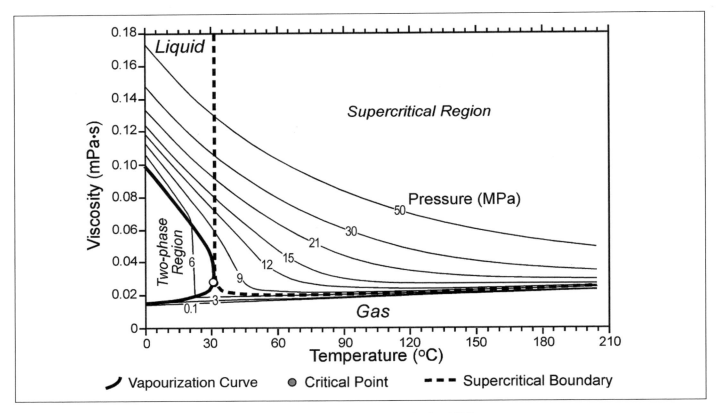

Figure AI.4 Variation of CO_2 viscosity as a function of temperature and pressure (Bachu, 2003).

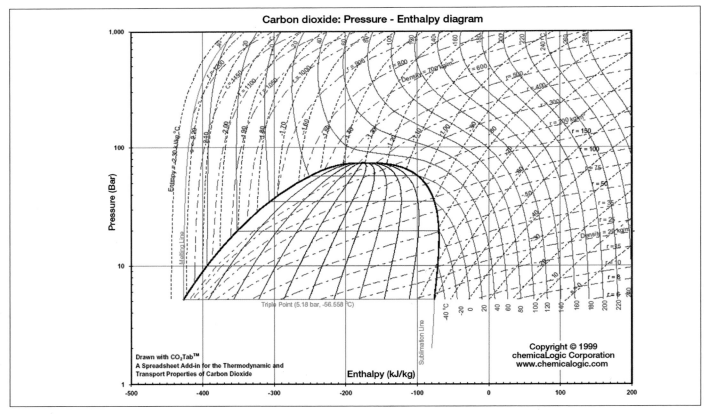

Figure AI.5 Pressure-Enthalpy chart for CO_2. Copyright © 1995-2003 ChemicaLogic Corporation, 99 South Bedford Street, Suite 207, Burlington, MA 01803 USA. All rights reserved.

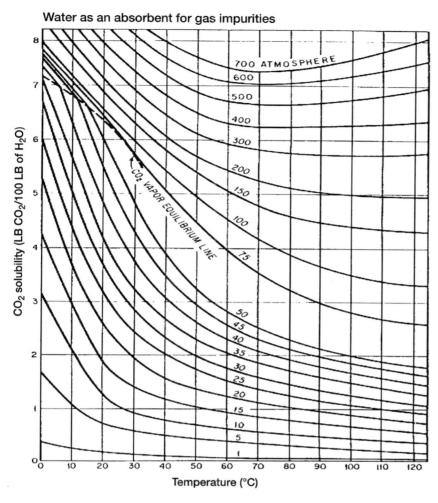

Water as an absorbent for gas impurities

Figure AI.6 Solubility of CO$_2$ in water (Kohl and Nielsen, 1997).

Table AI.2 Thermodynamic data for selected carbon-containing compounds (ref. Cox *et al.*, 1989 and other sources).

Compound	Heat of Formation ΔH_f° (kJ mol^{-1})	Gibbs free energy of formation ΔG_f° (kJ mol^{-1})	Standard molar entropy S_f° (J mol^{-1} K^{-1})
CO (g)	−110.53	−137.2	197.66
CO$_2$ (g)	−393.51	−394.4	213.78
CO$_2$ (l)		−386	
CO$_2$ (aq)	−413.26		119.36
CO$_3^{2-}$ (aq)	−675.23		−50.0
CaO (s)	−634.92		38.1
HCO$_3^-$ (aq)	−689.93	−603.3	98.4
H$_2$O (l)	−285.83		69.95
H$_2$O (g)	−241.83		188.84
CaCO$_3$ (s)	−1207.6 (calcite) −1207.8 (aragonite)	−1129.1 −1128.2	91.7 88
MgCO$_3$ (s)	−1113.28 (magnesite)	−1029.48	65.09
CH$_4$ (g)	−74.4	−50.3	186.3
CH$_3$OH (l)	−239.1	−166.6	126.8
(g)	−201.5	−162.6	239.8

CO_2 hydrates have not been studied as extensively.

AI.2.2.2 Impact of CO_2 on pH of water

The dissolution of CO_2 in water (this may be sea water, or the saline water in geological formations) involves a number of chemical reactions between gaseous and dissolved carbon dioxide (CO_2), carbonic acid (H_2CO_3), bicarbonate ions (HCO_3^-) and carbonate ions (CO_3^{2-}) which can be represented as follows:

$$CO_{2\,(g)} \leftrightarrow CO_{2\,(aq)} \qquad (2)$$

$$CO_{2\,(aq)} + H_2O \leftrightarrow H_2CO_{3\,(aq)} \qquad (3)$$

$$H_2CO_{3\,(aq)} \leftrightarrow H^+_{(aq)} + HCO_3^-{}_{(aq)} \qquad (4)$$

$$HCO_3^-{}_{(aq)} \leftrightarrow H^+_{(aq)} + CO_3^{2-}{}_{(aq)} \qquad (5)$$

Addition of CO_2 to water initially leads to an increase in the amount of dissolved CO_2. The dissolved CO_2 reacts with water to form carbonic acid. Carbonic acid dissociates to form bicarbonate ions, which can further dissociate into carbonate ions. The net effect of dissolving *anthropogenic* CO_2 in water is the removal of carbonate ions and production of bicarbonate ions, with a lowering in *pH*.

Figure AI.8 shows the dependence of *pH* on the extent to which CO_2 dissolves in sea water at temperatures of 0°C and 25°C based on theoretical calculations (IEA Greenhouse Gas R&D Programme, 2000) by iterative solution of the relationships (Horne, 1969) for the carbonic acid/bicarbonate/carbonate equilibria combined with activity coefficients for the bicarbonate and carbonate ions in sea water. The temperature dependence of the ionization of water and the bicarbonate equilibria were also included in this calculation. This gives values for the *pH* of typical sea water of 7.8–8.1 at 25°C and 8.1–8.4 at 0°C. These values, which are strongly dependent on carbonate/bicarbonate buffering, are in line with typical data for sea water (Figure AI.8 shows 2 experimental data points reported by Nishikawa *et al.*, 1992).

Figure AI.8 also shows that there is a small effect of temperature on the reduction in *pH* that results from dissolution of CO_2. A minor pressure dependence of water ionization is also reported (Handbook of Chemistry and Physics, 2000). The effect on water ionization of an increase in pressure from atmospheric to 250 bar (equivalent to 2500 m depth) is minor and about the same as would result from increasing temperature by about 2°C. The effect of pressure can therefore be ignored.

AI.2.3 Health and safety aspects of exposure to CO_2

As a normal constituent of the atmosphere, where it is present in low concentrations (currently 370 ppmv), CO_2 is considered harmless. CO_2 is non-flammable.

As it is 1.5 times denser than air at normal temperature and pressure, there will be a tendency for any CO_2 leaking from pipework or storage to collect in hollows and other low-lying confined spaces which could create hazardous situations. The hazardous nature of the release of CO_2 is enhanced because the gas is colourless, tasteless and is generally considered odourless

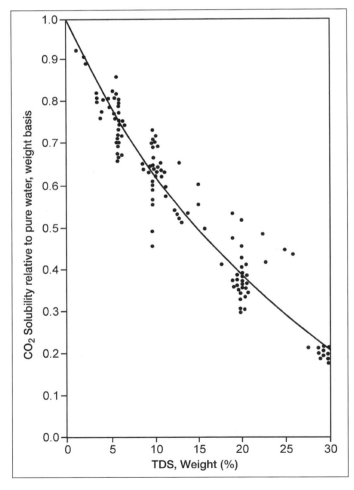

Figure AI.7 Solubility of CO_2 in brine relative to that in pure water, showing experimental points reported by Enick and Klara (1990) and correlation developed by those authors (TDS stands for total dissolved solids).

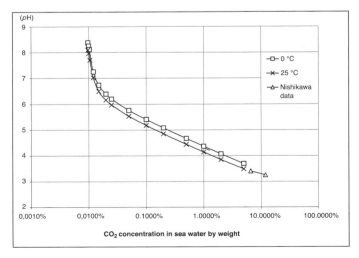

Figure AI.8 Dependence of *pH* on CO_2 concentration in sea water.

unless present in high concentrations.

When contained under pressure, escape of CO_2 can present serious hazards, for example asphyxiation, noise level (during pressure relief), frostbite, hydrates/ice plugs and high pressures (Jarrell *et al.*, 2002). The handling and processing of CO_2 must be taken into account during the preparation of a health, safety and environment plan for any facility handling CO_2.

AI.2.3.1 Effects of exposure to CO_2

At normal conditions, the atmospheric concentration of CO_2 is 0.037%, a non-toxic amount. Most people with normal cardiovascular, pulmonary-respiratory and neurological functions can tolerate exposure of up to 0.5–1.5% CO_2 for one to several hours without harm.

Higher concentrations or exposures of longer duration are hazardous – either by reducing the concentration of oxygen in the air to below the 16% level required to sustain human life[1], or by entering the body, especially the bloodstream, and/or altering the amount of air taken in during breathing; such physiological effects can occur faster than the effects resulting from the displacement of oxygen, depending on the concentration of CO_2. This is reflected in, for example, the current US occupational exposure standard of 0.5% for the maximum allowable concentration of CO_2 in air for eight hours continuous exposure; the maximum concentration to which operating personnel may be exposed for a short period of time is 3.0%.

The impact of elevated CO_2 concentrations on humans depends on the concentration and duration of exposure. At concentrations up to 1.5%, there are no noticeable physical consequences for healthy adults at rest from exposure for an hour or more (Figure AI.9); indeed, exposure to slightly elevated concentrations of CO_2, such as in re-breathing masks on aeroplanes at high altitude, may produce beneficial effects (Benson *et al.*, 2002). Increased activity or temperature may affect how the exposure is perceived. Longer exposure, even to less than 1% concentration, may significantly affect health. Noticeable effects occur above this level, particularly changes in respiration and blood pH level that can lead to increased heart rate, discomfort, nausea and unconsciousness.

It is noted (Rice, 2004) that most studies of the effects of CO_2 have involved healthy young male subjects, especially in controlled atmospheres such as submarines. Carbon dioxide tolerance in susceptible subgroups, such as children, the elderly, or people with respiratory deficiency, has not been studied to such an extent.

Acute exposure to CO_2 concentrations at or above 3% may significantly affect the health of the general population. Hearing loss and visual disturbances occur above 3% CO_2. Healthy young adults exposed to more than 3% CO_2 during exercise experience adverse symptoms, including laboured

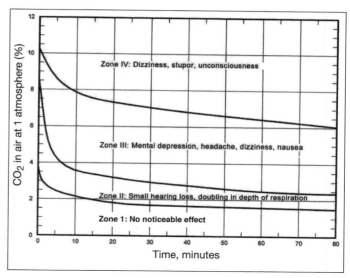

Figure AI.9 Effects of CO_2 exposure on humans (Fleming *et al.*, 1992).

breathing, headache, impaired vision and mental confusion. CO_2 acts as an asphyxiant in the range 7–10% and can be fatal at this concentration; at concentrations above 20%, death can occur in 20 to 30 minutes (Fleming *et al.*, 1992). The effects of CO_2 exposure are summarized in Table AI.3, which shows the consequences at different concentrations.

Health risks to the population could therefore occur if a release of CO_2 were to produce:

*   relatively low ambient concentrations of CO_2 for prolonged periods;
*   or intermediate concentrations of CO_2 in relatively anoxic environments;
*   or high concentrations of CO_2.

CO_2 intoxication is identified by excluding other causes, as exposure to CO_2 does not produce unique symptoms.

AI.2.3.2 Occupational standards

Protective standards have been developed for workers who may be exposed to CO_2 (Table AI.4 shows US standards but similar standards are understood to apply in other countries). These standards may or may not be relevant for protection of the general population against exposure to CO_2. Nevertheless, the occupational standards exist and provide a measure of the recommended exposure levels for this class of individual.

Site-specific risk assessments using these and other health data are necessary to determine potential health risks for the general population or for more sensitive subjects.

AI.2.3.3 Sensitive populations

Rice (2004) has indicated that there may be certain specific groups in the population which are more sensitive to elevated CO_2 levels than the general population. Such groups include those suffering from certain medical conditions including cerebral disease as well as patients in trauma medicated patients and those experiencing panic disorder, as well as individuals

---

[1] Signs of asphyxia will be noted when atmospheric oxygen concentration falls below 16%. Unconsciousness, leading to death, will occur when the atmospheric oxygen concentration is reduced to ≤ 8% although, if strenuous exertion is being undertaken, this can occur at higher oxygen concentrations (Rice, 2004).

Table AI.3 Some reports of reactions to exposure to elevated concentrations of CO_2.

CO_2 Concentration	Exposure reactions Air Products (2004)	Rice (2004)
1%	Slight increase in breathing rate.	Respiratory rate increased by about 37%.
2%	Breathing rate increases to 50% above normal level. Prolonged exposure can cause headache, tiredness.	Ventilation rate raised by about 100%. Respiratory rate raised by about 50%; increased brain blood flow.
3%	Breathing increases to twice normal rate and becomes laboured. Weak narcotic effect. Impaired hearing, headache, increase in blood pressure and pulse rate.	Exercise tolerance reduced in workers when breathing against inspiratory and expiratory resistance.
4-5%	Breathing increases to approximately four times normal rate; symptoms of intoxication become evident and slight choking may be felt.	Increase in ventilation rate by ~200%; Respiratory rate doubled, dizziness, headache, confusion, dyspnoea.
5-10%	Characteristic sharp odour noticeable. Very laboured breathing, headache, visual impairment and ringing in the ears. Judgment may be impaired, followed within minutes by loss of consciousness.	At 8-10%, severe headache, dizziness, confusion, dyspnoea, sweating, dim vision. At 10%, unbearable dyspnoea, followed by vomiting, disorientation, hypertension, and loss of consciousness.
50-100%	Unconsciousness occurs more rapidly above 10% level. Prolonged exposure to high concentrations may eventually result in death from asphyxiation.	

with pulmonary disease resulting in acidosis, children and people engaged in complex tasks.

CO_2 is a potent cerebrovascular dilator and significantly increases the cerebral blood flow. CO_2 exposure can seriously compromise patients in a coma or with a head injury, with increased intra-cranial pressure or bleeding, or with expanding lesions. An elevated partial pressure of CO_2 in arterial blood can further dilate cerebral vessels already dilated by anoxia.

Anoxia and various drugs (Osol and Pratt, 1973) can depress the stimulation of the respiratory centre by CO_2. In such patients, as well as patients with trauma to the head, the normal compensatory mechanisms will not be effective against exposure to CO_2 and the symptoms experienced will not necessarily alert the individuals or their carers to the presence of high CO_2 levels.

Patients susceptible to panic disorder may experience an increased frequency of panic attacks at 5% CO_2 (Woods *et al.*, 1988). Panic attack and significant anxiety can affect the ability of the individual to exercise appropriate judgment in dangerous situations.

CO_2 exposure can increase pulmonary pressure as well as systemic blood pressure and should be avoided in individuals with systemic or pulmonary hypertension. The rise in cardiac work during CO_2 inhalation could put patients with coronary artery disease or heart failure in jeopardy (Cooper *et al.*, 1970).

Infants and children breathe more air than adults relative to their body size and they therefore tend to be more susceptible to respiratory exposures (Snodgrass, 1992). At moderate to high CO_2 concentrations, the relaxation of blood vessels and enhanced ventilation could contribute to rapid loss of body heat in humans of any age. Carbon dioxide can significantly diminish an individual's performance in carrying out complex tasks.

AI.2.3.4 CO_2 control and response procedures

Suitable control procedures have been developed by industries which use CO_2, for example, minimizing any venting of CO_2 unless this cannot be avoided for safety or other operational reasons. Adequate ventilation must be provided when CO_2 is discharged into the air to ensure rapid dispersion.

Due its high density, released CO_2 will flow to low-levels and collect there, especially under stagnant conditions. High concentrations can persist in open pits, tanks and buildings. For this reason, monitors should be installed in areas where CO_2 might concentrate, supplemented by portable monitors. If CO_2 escapes from a vessel, the consequent pressure drop can cause a hazardous cold condition with danger of frostbite from contact with cold surfaces, with solid CO_2 (dry ice) or with escaping liquid CO_2. Personnel should avoid entering a CO_2 vapour

Table AI.4 Occupational exposure standards.

	Time-weighted average (8 hour day/40 hour week)	Short-term exposure limit (15 minute)	Immediately dangerous to life and health
OSHA permissible exposure limit[a]	5000 ppm (0.5%)		
NIOSH recommended exposure limit[b]	5000 ppm (0.5%)	30,000 ppm (3%)	40,000 ppm (5%)
ACGIH threshold limit value[c]	5000 ppm (0.5%)		

[a] OSHA - US Occupational Safety and Health Administration (1986).
[b] NIOSH - US National Institute of Occupational Safety and Health (1997).
[c] ACGIH - American Conference of Governmental Industrial Hygienists.

cloud not only because of the high concentration of CO_2 but also because of the danger of frostbite.

Hydrates, or ice plugs, can form in the piping of CO_2 facilities and flowlines, especially at pipe bends, depressions and locations downstream of restriction devices. Temperatures do not have to fall below 0°C for hydrates to form; under elevated pressures this can occur up to a temperature of 11°C.

AI.2.4   Established uses for CO₂

A long-established part of the industrial gases market involves the supply of CO_2 to a range of industrial users (source: Air Liquide). In several major industrial processes, CO_2 is manufactured on site as an intermediate material in the production of chemicals. Large quantities of CO_2 are used for enhanced oil recovery. Other uses of CO_2 include:

- Chemicals
  - Carbon dioxide is used in synthesis chemistry and to control reactor temperatures. CO_2 is also employed to neutralize alkaline effluents.
  - The main industrial use of CO_2 is in the manufacture of urea, as a fertilizer.
  - Large amounts of CO_2 are also used in the manufacture of inorganic carbonates and a lesser amount is used in the production of organic monomers and polycarbonates.
  - Methanol is manufactured using a chemical process which makes use of CO_2 in combination with other feedstocks.
  - CO_2 is also used in the manufacture of polyurethanes.
- Pharmaceuticals
  - CO_2 is used to provide an inert atmosphere, for chemical synthesis, supercritical fluid extraction and for acidification of waste water and for product transportation at low temperature (−78°C).
- Food and Beverage
  - CO_2 is used in the food business in three main areas: Carbonation of beverages; packaging of foodstuffs and as cryogenic fluid in chilling or freezing operations or as dry ice for temperature control during the distribution of foodstuffs.
- Health care
  - Intra-abdominal insufflation during medical procedures to expand the space around organs or tissues for better visualization.
- Metals industry
  - CO_2 is typically used for environmental protection; for example for red fume suppression during scrap and carbon charging of furnaces, for nitrogen pick-up reduction during tapping of electric arc furnaces and for bottom stirring.
  - In non-ferrous metallurgy, carbon dioxide is used for fume suppression during ladle transfer of matte (Cu/Ni production) or bullion (Zn/Pb production).

- A small amount of liquid CO_2 is used in recycling waters from acid mine drainage.
- Pulp and paper
  - CO_2 enables fine-tuning of the pH of recycled mechanical or chemical pulps after an alkaline bleaching. CO_2 can be used for increasing the performance of paper production machines.
- Electronics
  - CO_2 is used in waste water treatment and as a cooling medium in environmental testing of electronic devices. CO_2 can also be used to add conductivity to ultra-pure water and, as CO_2 snow, for abrasive cleaning of parts or residues on wafers; CO_2 can also be used as a supercritical fluid for removing photoresist from wafers, thus avoiding use of organic solvents.
- Waste treatment
  - Injection of CO_2 helps control the pH of liquid effluents.
- Other applications
  - CO_2 snow is used for fire extinguishers, for pH control and for regulation of waste waters in swimming pools.

AI.3   Conversion factors

Some conversion factors relevant to CO_2 capture and storage are given in Table AI.5 Other, less precise conversions and some approximate equivalents are given in Table AI.6.

AI.4   Fuels and emissions

AI.4.1   Carbonaceous fuels

Carbonaceous fuels can be defined as materials rich in carbon and capable of producing energy on oxidation. From a historical perspective, most of these fuels can be viewed as carriers of solar energy, having been derived from plants which depended on solar energy for growth. Thus, these fuels can be distinguished by the time taken for their formation, which is millions of years for fossil fuels, hundreds of years for peat and months-to-years for biofuels. On the scale of the human lifespan, fossil fuels are regarded as non-renewable carbonaceous fuels while biofuels are regarded as renewable. Coal, oil and natural gas are the major fossil fuels. Wood, agro-wastes, etcetera are the main biofuels for stationary uses but, in some parts of the world, crops such as soya, sugar cane and oil-seed plants are grown specifically to produce biofuels, especially transport fuels such as bioethanol and biodiesel. Peat is close to being a biofuel in terms of its relatively short formation time compared with fossil fuels.

AI.4.1.1 Coal
Coal is the most abundant fossil fuel present on Earth. Coal originated from the arrested decay of the remains of plant life which flourished in swamps and bogs many millions of years ago in a humid, tropical climate with abundant rainfall.

Table AI.5 Some conversion factors.

To convert:	Into the following units:	Multiply by:
US gallon	litre	3.78541
barrels (bbl)	m^3	0.158987
ton (Imperial)	tonne	1.01605
short ton (US)	tonne	0.907185
lbf	N	4.44822
kgf	N	9.80665
lbf in^{-2}	Bar	0.0689476
Bar	MPa	0.1
Btu	MJ	0.00105506
Btu	kWh	0.000293071
kWh	MJ	3.60000
Btu lb^{-1}	MJ kg^{-1}	0.00232600
Btu ft^{-3}	MJ m^{-3}	0.0372589
Btu/h	kW	0.000293071
Btu (lb.°F) $^{-1}$	kJ (kg.°C) $^{-1}$	4.18680
Btu (ft^2.h) $^{-1}$	kW m^{-2}	0.00315459
Btu (ft^3.h) $^{-1}$	kW m^{-3}	0.0103497
Btu (ft^2.h.°F) $^{-1}$	W (m^2.°C) $^{-1}$	5.67826
1 MMT[a]	million tonnes	0.907185
°F	°C	$°C = \dfrac{(°F - 32)}{1.8}$

[a] The abbreviation MMT is used in the literature to denote both Millions of short tons and Millions of metric tonnes. The conversion given here is for the former.

Table AI.6 Approximate equivalents and other definitions.

To convert	Into the following units	Multiply by
1 tC	tCO_2	3.667
1 tCO_2	$m^3 CO_2$ (at 1.013 bar and 15 °C)	534
1 t crude oil	Bbl	7.33
1 t crude oil	m^3	1.165

Fractions retained			
Release rate (fraction of stored amount released per year)	Fraction retained over 100 years	Fraction retained over 500 years	Fraction retained over 5000 years
0.001	90%	61%	1%
0.0001	99%	95%	61%
0.00001	100%	100%	95%

Other definitions	
Standard Temperature and Pressure	0 °C and 1.013 bar

Subsequent action of heat and pressure and other physical phenomena metamorphosed it into coal. Because of various degrees of metamorphic change during the process, coal is not a uniform substance; no two coals are the same in every respect. The composition of coal is reported in two different ways: The proximate analysis and the ultimate analysis, both expressed in % by weight. In a proximate analysis, moisture, volatile matter, fixed carbon and ash are measured using prescribed methods, which enable the equipment designer to determine how much air is to be supplied for efficient combustion, amongst other things. An ultimate analysis determines the composition in terms of the elements that contribute to the heating value, such as carbon, hydrogen, nitrogen, sulphur, the oxygen content (by difference), as well as ash. Along with these analyses, the heating value (expressed as kJ kg^{-1}) is also determined.

Carpenter (1988) describes the various coal classification systems in use today. In general, these systems are based on hierarchy and rank. The rank of a coal is the stage the coal has reached during the coalification[2] process – that is its degree of metamorphism or maturity. Table AI.7 shows the classification system adopted by the American Society for Testing Materials (ASTM), D388-92A (Carpenter, 1988; Perry and Green, 1997). This rank-based system is extensively used in North America and many other parts of the world. This system uses two parameters to classify coals by rank, fixed carbon (dry, mineral-matter-free) for the higher rank coals and gross calorific value (moist, mineral-matter-free) for the lower rank coals. The agglomerating character of the coals is used to differentiate between adjacent coal groups.

AI.4.1.2   Oil and petroleum fuels

During the past 600 million years, the remains of incompletely decayed plant have become buried under thick layers of rock and, under high pressure and temperature, have been converted to petroleum which may occur in gaseous, liquid or solid form. The fluid produced from petroleum reservoirs may be crude oil (a mixture of light and heavy hydrocarbons and bitumen) or natural gas liquids. Hydrocarbons can also be extracted from tar sands or oil shales; this takes place in several parts of the world.

Fuels are extracted from crude oil through fractional distillation, with subsequent conversion and upgrading. Such fuels are used for vehicles (gasoline, jet fuel, diesel fuel and liquefied petroleum gases (LPG)), heating oils, lighting oils, solvents, lubricants and building materials such as asphalts, plus a variety of other products. The compositions of heating fuels may differ in their composition, density, etcetera but general categories are recognized worldwide: kerosene-type vaporizing fuel, distillate (or 'gas oil') and more viscous blends and residuals. Tables AI.8 and AI.9 provide typical specifications of some common fuels (Perry and Green, 1997; Kaantee *et al.*, 2003).

AI.4.1.3   Natural gas

Natural gas is combustible gas that occurs in porous rock of the Earth's crust; it is often found with or near accumulations of crude oil. It may occur in separate reservoirs but, more commonly, it forms a gas cap entrapped between the petroleum and an impervious, capping rock layer in a petroleum reservoir. Under high-pressure conditions, it becomes partially mixed with or dissolved in the crude oil. Methane (CH$_4$) is the main component of natural gas, usually making up more than 80% of the constituents by volume. The remaining constituents are ethane (C$_2$H$_6$), propane (C$_3$H$_8$), butane (C$_4$H$_{10}$), hydrogen sulphide (H$_2$S) and inerts (N$_2$, CO$_2$ and He). The amounts of

these compounds can vary greatly depending on location. Natural gas is always treated prior to use, mainly by drying, and by removing H$_2$S and, depending on the amount present, CO$_2$. There are no universally accepted specification systems for marketed natural gas; however a typical composition of natural gas is given in Table AI.10 (Spath and Mann, 2000).

AI.4.1.4   Biofuels

Biofuels may be defined as fuels produced from organic matter or combustible oils produced by plants (IPCC, 2001). Dedicated energy crops, including short-rotation woody crops such as hardwood trees and herbaceous crops such as switch grass, are agricultural crops that are solely grown for use as biofuels. These crops have very fast growth rates and can therefore provide a regular supply of fuel. The category of biofuels also includes wood from trees and wood waste products (e.g., sawdust, wood chips, etc.), crop residues (e.g., rice husks, bagasse, corn husks, wheat chaff, etc.). This category of fuel is often taken to include some types of municipal, animal and industrial wastes (e.g., sewage sludge, manure, etc.). These would be combusted in stationary plants. Chemical properties of typical biofuels, including peat, are given in Table AI.11 (Sami *et al.*, 2001; Hower, 2003).

Biomass-derived fuels can also be manufactured for use as transport fuels, for example ethanol from fermentation of plant material or biodiesel produced by transesterification of vegetable oils. The energy efficiency of fermentation systems can be improved by combustion of the solid residues to produce electricity.

AI.4.2   Examples of emissions from carbonaceous fuels

Depending on the fuel type and application, the utilization of carbonaceous fuels causes direct and indirect emissions of one or more of the following: SO$_x$, NO$_x$, particulate matter, trace metals and elements, volatile organic carbons and greenhouse gases (e.g., CO$_2$, CH$_4$, N$_2$O). Direct emissions are usually confined to the point of combustion of the fuel. Indirect emissions include those that arise from the upstream recovery, processing and distribution of the fuel. Life cycle analysis (LCA) can be used to account for all emissions (direct as well as indirect) arising from the recovery, processing, distribution and end-use of a fuel. Table AI.12 (Cameron, 2002) and Table AI.13 (EPA, 2004) give an idea of some direct and indirect emissions anticipated, but these should only be viewed as examples due to the considerable variation there can be in many of these values.

---

[2] Coalification refers to the progressive transformation of peat through lignite/brown coal, to sub-bituminous, bituminous and anthracite coals.

Table AI.7 Characterization of coals by rank (according to ASTM D388-92A).

Class Group	Fixed Carbon Limits (dmmf basis)[a] %		Volatile Matter Limits (dmmf basis)[a] %		Gross Calorific Value Limits (mmmf basis)[b] MJ kg^{-1}		Agglomerating Character
	Equal to or greater than	Less than	Greater than	Equal to or less than	Equal to or greater than	Less than	
Anthracite							Non-agglomerating
Meta-anthracite	98	-	-	2	-	-	
Anthracite	92	98	2	8	-	-	
Semi-anthracite [c]	86	92	8	14	-	-	
Bituminous coal							Commonly agglomerating
Low volatile	78	86	14	22	-	-	
Medium volatile	69	78	22	31	-	-	
High volatile A	-	69	31	-	32.6 [d]	-	
High volatile B	-	-	-	-	30.2 [d]	32.6	
High volatile C	-	-	-	-	26.7	30.2	
					24.4	26.7	Agglomerating
Sub-bituminous coal							Non-agglomerating
A	-	-	-	-	24.4	26.7	
B	-	-	-	-	22.1	24.4	
C	-	-	-	-	19.3	22.1	
Lignite							
A	-	-	-	-	14.7	19.3	
B	-	-	-	-	-	14.7	

[a] Indicates dry-mineral-matter-free basis (dmmf).
[b] mmmf indicates moist mineral-matter-free basis; moist refers to coal containing its natural inherent moisture but not including visible water on the surface of the coal.
[c] If agglomerating, classified in the low volatile group of the bituminous class.
[d] Coals having 69% or more fixed carbon (dmmf) are classified according to fixed carbon, regardless of gross calorific value.

Table AI.8 Typical specifications of petroleum-based heating fuels.

Specifier	Number	Category
Canadian Government Specification Board, Department of Defense Production, Canada	3-GP-2	Fuel oil, heating
Deutsches Institut fur Normung e.V., Germany	DIN 51603	Heating (fuel) oils
British Standards Institution, UK	B.S. 2869	Petroleum fuels for oil engines and burners
Japan	JIS K2203	Kerosene
	JIS K2204	Gas oil
	JIS K2205	Fuel oil
Federal Specifications, United States	ASTM D 396	Fuel oil, burner

Table AI.9 Typical ultimate analysis of petroleum-based heating fuels.

Composition %	No. 1 fuel oil (41.5°API[a])	No. 2 fuel oil (33°API[a])	No. 4 fuel oil (23.2°API[a])	Low sulphur, No. 6 fuel oil (33°API[a])	High sulphur, No. 6 fuel oil (15.5o APIa)	Petroleum coke[b]
Carbon	86.4	87.3	86.47	87.26	84.67	89.5
Hydrogen	13.6	12.6	11.65	10.49	11.02	3.08
Oxygen	0.01	0.04	0.27	0.64	0.38	1.11
Nitrogen	0.003	0.006	0.24	0.28	0.18	1.71
Sulphur	0.09	0.22	1.35	0.84	3.97	4.00
Ash	<0.01	<0.01	0.02	0.04	0.02	0.50
C/H Ratio	6.35	6.93	7.42	8.31	7.62	29.05

[a]   Degree API = (141.5/s) -131.5; where s is the specific density at 15°C.
[b]   Reference: Kaantee *et al.* (2003).

Table AI.10 Typical natural gas composition.

Component	Pipeline composition used in analysis	Typical range of wellhead components (mol%)	
	Mol% (dry)	Low value	High value
Carbon dioxide CO_2	0.5	0	10
Nitrogen N_2	1.1	0	15
Methane CH_4	94.4	75	99
Ethane C_2H_6	3.1	1	15
Propane C_3H_8	0.5	1	10
Isobutane $C4H_{10}$	0.1	0	1
N-butane C_4H_{10}	0.1	0	2
Pentanes + (C_5+)	0.2	0	1
Hydrogen sulphide (H_2S)	0.0004	0	30
Helium (He)	0.0	0	5
Heat of combustion (LHV)	48.252 MJ kg^{-1}	-	-
Heat of combustion (HHV)	53.463 MJ kg^{-1}	-	-

Table AI.11 Chemical analysis and properties of some biomass fuels (Sami *et al.*, 2001; Hower, 2003).

	Peat	Wood (saw dust)	Crop residues (sugar cane bagasse)	Municipal solid waste	Energy crops (Eucalyptus)
Proximate Analysis					
Moisture	70–90	7.3	-	16–38	-
Ash	-	2.6	11.3	11–20	0.52
Volatile matter	45–75	76.2	-	67–78	-
Fixed carbon	-	13.9	14.9	6–12	16.9
Ultimate Analysis					
C	45–60	46.9	44.8	-	48.3
H	3.5–6.8	5.2	5.4	-	5.9
O	20–45	37.8	39.5	-	45.1
N	0.75–3	0.1	0.4	-	0.2
S	-	0.04	0.01	-	0.01
Heating Value, MJ kg$^{-1,}$ (HHV)	**17–22**	**18.1**	**17.3**	**15.9–17.5**	**19.3**

Table AI.12 Direct emissions of non-greenhouse gases from two examples of coal and natural gas plants based on best available control technology, burning specific fuels (Cameron, 2002).

Emissions	Coal (supercritical PC with best available emission controls)	Natural gas (NGCC with SCR)
NO_x, g GJ^{-1}	4-5	5
SO_x, g GJ^{-1}	4.5-5	0.7
Particulates, g GJ^{-1}	2.4-2.8	2
Mercury, mg GJ^{-1}	0.3-0.5	N/A

Table AI.13 Direct CO_2 emission factors for some examples of carbonaceous fuels.

Carbonaceous Fuel	Heat Content (HHV) MJ $kg^{-1 a}$	Emission Factor gCO_2 $MJ^{-1 a}$
Coal		
Anthracite	26.2	96.8
Bituminous	27.8	87.3
Sub-bituminous	19.9	90.3
Lignite	14.9	91.6
Biofuel		
Wood (dry)	20.0	78.4
Natural Gas	kJ m^{-3}	
	37.3	50
Petroleum Fuel	MJ m^{-3}	
Distillate Fuel Oil (#1, 2 & 4)	38,650	68.6
Residual Fuel Oil (#5 & 6)	41,716	73.9
Kerosene	37,622	67.8
LPG (average for fuel use	25,220	59.1
Motor Gasoline	-	69.3

[a] Reported values converted to SI units (NIES, 2003).

References

Air Liquide: http://www.airliquide.com/en/business/products/gases/gasdata/index.asp.

Air Products, 2004: Safetygram-18, Carbon Dioxide. http://www.airproducts.com/Responsibility/EHS/ProductSafety/ProductSafetyInformation/safetygrams.htm.

Bachu, S. 2003: Screening and ranking sedimentary basins for sequestration of CO_2 in geological media in response to climate change. Environmental Geology, **44**, pp 277–289.

Benson, S.M., R. Hepple, J. Apps, C.F. Tsang, and M. Lippmann, 2002: Lessons Learned from Natural and Industrial Analogues for Storage of Carbon Dioxide in Deep Geological Formations. Lawrence Berkeley National Laboratory, USA, LBNL-51170.

Cameron, D.H., 2002: Evaluation of Retrofit Emission Control Options: Final Report. A report prepared by Neill and Gunter Limited, ADA Environmental Solutions, LLC, for Canadian Clean Coal Power Coalition (CCPC), Project No. 40727, Canada, 127 pp.

Carpenter, A.M., 1988: Coal Classification. Report prepared for IEA Coal Research, London, UK, IEACR/12, 104 pp.

Cooper, E.S., J.W. West, M.E. Jaffe, H.I. Goldberg, J. Kawamura, L.C. McHenry Jr., 1970: The relation between cardiac function and cerebral blood flow in stroke patients. 1. Effect of CO_2 Inhalation. Stroke, **1**, pp 330–347.

Cox, J.D., D.D. Wagman, and V.A. Medvedev, 1989: CODATA Key Values for Thermodynamics, Hemisphere Publishing Corp., New York.

Enick, R.M. and S.M. Klara, 1990: CO_2 solubility in water and brine under reservoir conditions. Chem. Eng. Comm., **90**, pp 23–33.

EPA, 2004: Direct Emissions from Stationary Combustion. Core Module Guidance in Climate Leaders Greenhouse Gas Inventory Protocol. US Environmental Protection Agency. Available at http://www.epa.gov/climateleaders/pdf/stationarycombustionguidance.pdf.

Fenghour, A., W.A. Wakeham, and V. Vesovic, 1998: The Viscosity of Carbon Dioxide. J. Phys. Chem. Ref. Data, **27**, 1, pp 31–44.

Fleming, E.A., L.M. Brown, and R.I. Cook, 1992: Overview of Production Engineering Aspects of Operating the Denver Unit CO_2 Flood, paper SPE/DOE 24157 presented at the 1992 SPE/DOE Enhanced Oil Recovery Symposium, Tulsa, 22–24 April. Society of Petroleum Engineers Inc., Richardson, TX, USA.

Handbook of Chemistry and Physics, 2000: Lide, D.R. (ed). The Chemical Rubber Company, CRC Press LLC, Boca Raton, FL, USA.

Horne, R.A., 1969: Marine Chemistry; the structure of water and the chemistry of the hydrosphere. Wiley.

Hower, J., 2003: *Coal*, in Kirk-Othmer Encyclopedia of Chemical Technology, John Wiley & Sons, New York.

IEA Greenhouse Gas R&D Programme, 2000: Capture of CO_2 using water scrubbing. Report Ph3/26, IEA Greenhouse Gas R&D Programme, Cheltenham, UK

IPCC, 2001: Climate Change 2001: Mitigation. Contribution of Working Group III to the Third Assessment Report of the Intergovernmental Panel on Climate Change. Metz, B., O.R. Davidson, R. Swart, J. Pan (eds.). Cambridge University Press, Cambridge, UK

Jarrell, P.M., C.E. Fox, M.H. Stein, S.L. Webb, 2002: CO_2 flood environmental, health and safety planning, chapter 9 of Practical Aspects of CO_2 flooding. Monograph 22. Society of Petroleum Engineers, Richardson, TX, USA.

Kaantee, U., R. Zevenhoven, R. Backman, and M. Hupa, 2003: Cement manufacturing using alternative fuels and the advantages of process modelling. Fuel Processing Technology, 85 pp. 293–301.

Kirk-Othmer, 1985: Concise Encyclopaedia of Chemical Technology, 3ʳᵈ Edition. Wiley, New York, USA.

Kohl, A. L. and R.B. Nielsen, 1997: Gas Purification. Gulf Publishing Company, Houston, TX, USA.

NIOSH, 1997: National Institute for Occupational Safety and Health (NIOSH) Pocket Guide to Chemical Hazards. DHHS publication no. 97-140. US Government Printing Office, Washington, DC, USA.

Nishikawa, N., M. Morishita, M. Uchiyama, F. Yamaguchi, K. Ohtsubo, H. Kimuro, and R. Hiraoka, 1992: CO_2 clathrate formation and its properties in the simulated deep ocean. Proceedings of the first international conference on carbon dioxide removal. *Energy Conversion and Management,* **33,** pp. 651–658.

NIST, 2003: National Institute of Standards and Technology Standard Reference Database Number 69, March 2003, P.J. Linstrom and W.G. Mallard (eds.).

OSHA, 1986: Occupational Safety & Health Administration (OSHA) Occupational Health and Safety Standards Number 1910.1000 Table Z-1: Limits for Air Contaminants. US Department of Labor, Washington, DC, USA.

Osol, A. and R. Pratt, (eds.), 1973: The United States Dispensatory, 27ᵗʰ edition. J. B. Lippincott, Philadelphia, PA, USA.

Perry, R.H. and D.W. Green, 1997: Perry's Chemical Engineer's Handbook. D.W. Green, (ed.), McGraw-Hill, Montreal, pp. 27.4–27.24.

Rice, S.A., 2004: Human health risk assessment of CO_2: survivors of acute high-level exposure and populations sensitive to prolonged low level exposure. Poster 11-01 presented at 3ʳᵈ Annual conference on carbon sequestration, 3-6 May 2004, Alexandria, VA, USA.

Sami, M., K. Annamalai, and M. Worldridge, 2001: Cofiring of coal and biomass fuel blend. *Progress in Energy and Combustion Science,* **27**, pp. 171–214.

Snodgrass, W. R., 1992: Physiological and biochemical differences between children and adults as determinants of toxic exposure to environmental pollutants. In Similarities and differences between children and adults: Implications for risk assessment. Guzelain, P.S., C.J. Henry, S.S. Olin (eds.), ILSI Press, Washington, DC, USA.

Span, R and W. Wagner, 1996: A new equation of state for carbon dioxide covering the fluid region from the triple-point temperature to 1100K at pressures up to 800 MPa. Journal of Phys. Chem. Data, 25(6), pp. 1509–1596.

Spath, P.L. and M.K. Mann, 2000: Life Cycle Assessment of a Natural Gas Combined Cycle Power Generation System. Report no. NREL/TP-570-27715 prepared for National Renewable Energy Laboratory (NREL), US, 32 pp.

Woods, S.W., D.S. Charney, W.K. Goodman, G.R. Heninger, 1988: Carbon dioxide-induced anxiety. Behavioral, physiologic, and biochemical effects of carbon dioxide in patients with panic disorders and healthy subjects. *Arch. Gen Psychiatry,* **45**, pp 43–52.

Annex II

Glossary, acronyms and abbreviations

Coordinating Lead Author
Philip Lloyd (South Africa)

Lead Authors
Peter Brewer (United States), Chris Hendriks (Netherlands), Yasumasa Fujii (Japan), John Gale (United Kingdom), Balgis Osman Elasha (Sudan), Jose Moreira (Brazil), Juan Carlos Sanchez (Venezuela), Mohammad Soltanieh (Iran), Tore Torp (Norway), Ton Wildenborg (Netherlands)

Contributing Authors
Jason Anderson (United States), Stefan Bachu (Canada), Sally Benson (United States), Ken Caldeira (United States), Peter Cook (United States), Richard Doctor (United States), Paul Freund (United Kingdom), Gabriela von Goerne (Germany)

Note: the definitions in this Annex refer to the use of the terms in the context of this report. It provides an explanation of specific terms as the authors intend them to be interpreted in this report.

Abatement
Reduction in the degree or intensity of emissions or other pollutants.

Absorption
Chemical or physical take-up of molecules into the bulk of a solid or liquid, forming either a solution or compound.

Acid gas
Any gas mixture that turns to an acid when dissolved in water (normally refers to $H_2S + CO_2$ from sour gas (q.v.)).

Adiabatic
A process in which no heat is gained or lost by the system.

Adsorption
The uptake of molecules on the surface of a solid or a liquid.

Afforestation
Planting of new forests on lands that historically have not contained forests.

Aluminium silicate mineral
Natural mineral – such as feldspar, clays, micas, amphiboles – composed of Al_2O_3 and SiO_2 plus other cations.

Amine
Organic chemical compound containing one or more nitrogens in $-NH_2$, $-NH$ or $-N$ groups.

Anaerobic condition
Reducing condition that only supports life which does not require free oxygen.

Anhydrite
Calcium sulphate: the common hydrous form is called gypsum.

Antarctic Treaty
Applies to the area south of 60 degrees South, and declares that Antarctica shall be used for peaceful purposes only.

Anthracite
Coal with the highest carbon content and therefore the highest rank (q.v.).

Anthropogenic source
Source which is man-made as opposed to natural.

Anticline
Folded geological strata that is convex upwards.

API
American Petroleum Institute; degree API is a measure of oil density given by (141.5/specific gravity) -131.5.

Aquifer
Geological structure containing water and with significant permeability to allow flow; it is bound by seals.

Assessment unit
A geological province with high petroleum potential.

Assigned amount
The amount by which a Party listed in Annex B of the Kyoto Protocol agrees to reduce its anthropogenic emissions.

ATR
Auto thermal reforming: a process in which the heat for the reaction of CH_4 with steam is generated by partial oxidation of CH_4.

Autoproduction
The production of electricity for own use.

Basalt
A type of basic igneous rock which is typically erupted from a volcano.

Basel Convention
UN Convention on the Control of Transboundary Movements of Hazardous Wastes and their Disposal, which was adopted at Basel on 22 March 1989.

Baseline
The datum against which change is measured.

Basin
A geological region with strata dipping towards a common axis or centre.

Bathymetric
Pertaining to the depth of water.

Benthic
Pertaining to conditions at depth in bodies of water.

Bicarbonate ion
The anion formed by dissolving carbon dioxide in water, HCO_3^-.

Biomass
Matter derived recently from the biosphere.

Biomass-based CCS
Carbon capture and storage in which the feedstock (q.v.) is biomass

Bituminous coal
An intermediate rank of coal falling between the extremes of peat and anthracite, and closer to anthracite.

Blow-out
Refers to catastrophic failure of a well when the petroleum fluids or water flow unrestricted to the surface.

Bohr effect
The *p*H-dependent change in the oxygen affinity of blood.

Bottom-up model
A model that includes technological and engineering details in the analysis.

Boundary
In GHG accounting, the separation between accounting units, be they national, organizational, operational, business units or sectors.

Break-even price
The price necessary at a given level of production to cover all costs.

Buoyancy
Tendency of a fluid or solid to rise through a fluid of higher density.

Cap rock
Rock of very low permeability that acts as an upper seal to prevent fluid flow out of a reservoir.

Capillary entry pressure
Additional pressure needed for a liquid or gas to enter a pore and overcome surface tension.

Capture efficiency
The fraction of CO_2 separated from the gas stream of a source

Carbon credit
A convertible and transferable instrument that allows an organization to benefit financially from an emission reduction.

Carbon trading
A market-based approach that allows those with excess emissions to trade that excess for reduced emissions elsewhere.

Carbonate
Natural minerals composed of various anions bonded to a CO_3^{2-} cation (e.g. calcite, dolomite, siderite, limestone).

Carbonate neutralization
A method for storing carbon in the ocean based upon the reaction of CO_2 with a mineral carbonate such as limestone to produce bicarbonate anions and soluble cations.

Casing
A pipe which is inserted to stabilize the borehole of a well after it is drilled.

CBM
Coal bed methane

CCS
Carbon dioxide capture and storage

CDM
Clean development mechanism: a Kyoto Protocol mechanism to assist non-Annex 1 countries to contribute to the objectives of the Protocol and help Annex I countries to meet their commitments.

Certification
In the context of carbon trading, certifying that a project achieves a quantified reduction in emissions over a given period.

Chemical looping combustion
A process in which combustion of a hydrocarbon fuel is split into separate oxidation and reduction reactions by using a metal oxide as an oxygen carrier between the two reactors.

Chlorite
A magnesium-iron aluminosilicate sheet silicate clay mineral.

Class "x" well
A regulatory classification for wells used for the injection of fluids into the ground.

Claus plant
A plant that transforms H_2S into elemental sulphur.

Cleats
The system of joints, cleavage planes, or planes of weakness found in coal seams along which the coal fractures.

CO_2 avoided
The difference between CO_2 captured, transmitted and/or stored, and the amount of CO_2 generated by a system without capture, net of the emissions not captured by a system with CO_2 capture.

CO_2 equivalent
A measure used to compare emissions of different greenhouse gases based on their global warming potential.

Co-benefit
The additional benefits generated by policies that are implemented for a specific reason.

COE
Cost of electricity, value as calculated by Equation 1 in Section 3.7.

Co-firing
The simultaneous use of more than one fuel in a power plant or industrial process.

Completion of a well
Refers to the cementing and perforating of casing and stimulation to connect a well bore to reservoir.

Congruence
The quality of agreement between two entities.

Conservative values
Parameter values selected so that a parameter, such as CO_2 leakage, is over-estimated.

Containment
Restriction of movement of a fluid to a designated volume (e.g. reservoir).

Continental shelf
The extension of the continental mass beneath the ocean.

COREX
A process for producing iron.

Cryogenic
Pertaining to low temperatures, usually under about -100°C.

D, Darcy
A non-SI unit of permeability, abbreviated D, and approximately = $1\mu m^2$.

Dawsonite
A mineral: dihydroxide sodium aluminium carbonate.

Deep saline aquifer
A deep underground rock formation composed of permeable materials and containing highly saline fluids.

Deep sea
The sea below 1000m depth.

Default emissions factor
An approximate emission factor that may be used in the absence of precise or measured values of an Emissions Factor.

Demonstration phase
Demonstration phase means that the technology is implemented in a pilot project or on a small scale, but not yet economically feasible at full scale.

Dense phase
A gas compressed to a density approaching that of the liquid.

Dense fluid
A gas compressed to a density approaching that of the liquid.

Depleted
Of a reservoir: one where production is significantly reduced.

Diagenesis
Processes that cause changes in sediment after it has been deposited and buried under another layer.

DIC
Dissolved Inorganic Carbon.

Dip
In geology, the angle below the horizontal taken by rock strata.

Discharge
The amount of water issuing from a spring or in a stream that passes a specific point in a given period of time.

Discordant sequence
In geology, sequence of rock strata that is markedly different from strata above or below.

Dolomite
A magnesium-rich carbonate sedimentary rock. Also, a magnesium-rich carbonate mineral ($CaMgCO_3$).

Double-grip packer
A device used to seal a drill string equipped with two gripping mechanisms.

Down-hole log
Record of conditions in a borehole.

Drill cuttings
The solid particles recovered during the drilling of a well.

Drill string
The assembly of drilling rods that leads from the surface to the drilling tool.

Drive
Fluid flow created in formations by pressure differences arising from borehole operations.

Dry ice
Solid carbon dioxide

Dynamic miscibility
The attainment of mixing following the prolonged injection of gas into an oilfield.

ECBM

Enhanced coal bed methane recovery; the use of CO_2 to enhance the recovery of the methane present in unminable coal beds through the preferential adsorption of CO_2 on coal.

Economic potential

The amount of greenhouse gas emissions reductions from a specific option that could be achieved cost-effectively, given prevailing circumstances (i.e. a market value of CO_2 reductions and costs of other options).

Economically feasible under specific conditions

A technology that is well understood and used in selected commercial applications, such as in a favourable tax regime or a niche market, processing at least 0.1 $MtCO_2$/yr, with a few (less than 5) replications of the technology.

EGR

Enhanced gas recovery: the recovery of gas additional to that produced naturally by fluid injection or other means.

Emission factor

A normalized measure of GHG emissions in terms of activity, e.g., tonnes of GHG emitted per tonne of fuel consumed.

Emissions credit

A commodity giving its holder the right to emit a certain quantity of GHGs (q.v.).

Emissions trading

A trading scheme that allows permits for the release of a specified number of tonnes of a pollutant to be sold and bought.

Endothermic

Concerning a chemical reaction that absorbs heat, or requires heat to drive it.

Enhanced gas recovery

See EGR.

Enhanced oil recovery

See EOR

Entrained flow

Flow in which a solid or liquid, in the form of fine particles, is transported in diluted form by high velocity gas.

Entrainment gas

The gas employed in entrained flow (q.v.).

EOR

Enhanced oil recovery: the recovery of oil additional to that produced naturally by fluid injection or other means.

Euphotic zone

The zone of the ocean reached by sunlight.

Evaporite

A rock formed by evaporation.

Exothermic

Concerning a chemical reaction that releases heat, such as combustion.

Ex-situ mineralization

A process where minerals are mined, transferred to an industrial facility, reacted with carbon dioxide and processed.

Exsolution

The formation of different phases during the cooling of a homogeneous fluid.

Extended reach well

Borehole that is diverted into a more horizontal direction to extend its reach.

Extremophile

Microbe living in environments where life was previously considered impossible.

Far field

A region remote from a signal source.

Fault

In geology, a surface at which strata are no longer continuous, but displaced.

Fault reactivation

The tendency for a fault to become active, i.e. for movement to occur.

Fault slip

The extent to which a fault has slipped in past times.

FBC

Fluidized bed combustion: – combustion in a fluidized bed (q.v.).

Feldspar

A group of alumino-silicate minerals that makes up much of the Earth's crust.

Feedstock

The material that is fed to a process

FGD

Flue gas desulphurization.

Fischer-Tropsch

A process that transforms a gas mixture of CO and H_2 into liquid hydrocarbons and water.

Fixation
The immobilization of CO_2 by its reaction with another material to produce a stable compound

Fixed bed
A gas-solid contactor or reactor formed by a bed of stationary solid particles that allows the passage of gas between the particles.

Flood
The injection of a fluid into an underground reservoir.

Flue gas
Gases produced by combustion of a fuel that are normally emitted to the atmosphere.

Fluidized bed
A gas-solid contactor or reactor comprising a bed of fine solid particles suspended by passing a gas through the bed at sufficiently high velocity.

Folding
In geology, the bending of rock strata from the plane in which they were formed.

Formation
A body of rock of considerable extent with distinctive characteristics that allow geologists to map, describe, and name it.

Formation water
Water that occurs naturally within the pores of rock formations.

Fouling
Deposition of a solid on the surface of heat or mass transfer equipment that has the effect of reducing the heat or mass transfer.

Fracture
Any break in rock along which no significant movement has occurred.

Fuel cell
Electrochemical device in which a fuel is oxidized in a controlled manner to produce an electric current and heat directly.

Fugitive emission
Any releases of gases or vapours from anthropogenic activities such as the processing or transportation of gas or petroleum.

FutureGen Project
US Government initiative for a new power station with low CO_2 emissions.

Gas turbine
A machine in which a fuel is burned with compressed air or oxygen and mechanical work is recovered by the expansion of the hot products.

Gasification
Process by which a carbon-containing solid fuel is transformed into a carbon- and hydrogen-containing gaseous fuel by reaction with air or oxygen and steam.

Geochemical trapping
The retention of injected CO_2 by geochemical reactions.

Geological setting
The geological environment of various locations.

Geological time
The time over which geological processes have taken place.

Geomechanics
The science of the movement of the Earth's crust.

Geosphere
The earth, its rocks and minerals, and its waters.

Geothermal
Concerning heat flowing from deep in the earth.

GHG
Greenhouse gases: carbon dioxide (CO_2), methane (CH_4), nitrous oxide (N_2O), hydroflurocarbons (HFCs), perfluorocarbons (PFCs), and sulphur hexafluoride (SF_6).

Hazardous and non-hazardous waste
Potentially harmful and non-harmful substances that have been released or discarded into the environment.

Hazardous waste directive
European directive in force to regulate definitions of waste classes and to regulate the handling of the waste classes.

HAZOP
HAZard and OPerability, a process used to assess the risks of operating potentially hazardous equipment.

Helsinki Convention
International legal convention protecting the Baltic water against pollution.

Henry's Law
States that the solubility of a gas in a liquid is proportional to the partial pressure of the gas in contact with the liquid.

HHV
Higher heating value: the energy released from the combustion of a fuel that includes the latent heat of water.

Host rock
In geology, the rock formation that contains a foreign material.

Hybrid vehicle
Vehicle that combines a fossil fuel internal combustion engine and an alternative energy source, typically batteries.

Hydrate
An ice-like compound formed by the reaction of water and CO_2, CH_4 or similar gases.

Hydrodynamic trap
A geological structure in which fluids are retained by low levels of porosity in the surrounding rocks.

Hydrogeological
Concerning water in the geological environment.

Hydrostatic
Pertaining to the properties of a stationary body of water.

Hypercapnia
Excessively high CO_2 levels in the blood.

Hypoxia
Having low rates of oxygen transfer in living tissue.

Hysteresis
The phenomenon of a lagging recovery from deformation or other disturbance.

IEA GHG
International Energy Agency – Greenhouse Gas R&D Programme.

IGCC
Integrated gasification combined cycle: power generation in which hydrocarbons or coal are gasified (q.v.) and the gas is used as a fuel to drive both a gas and a steam turbine.

Igneous
Rock formed when molten rock (magma) has cooled and solidified (crystallized).

Immature basin
A basin in which the processes leading to oil or gas formation have started but are incomplete.

Infrared spectroscopy
Chemical analysis using infrared spectroscope method.

Injection
The process of using pressure to force fluids down wells.

Injection well
A well in which fluids are injected rather than produced.

Injectivity
A measure of the rate at which a quantity of fluid can be injected into a well.

In-situ mineralization
A process where minerals are not mined: carbon dioxide is injected in the silicate formation where it reacts with the minerals, forming carbonates and silica.

International Seabed Authority
An organization established under the 1982 UN Convention on the Law of the Sea, headquartered in Kingston, Jamaica.

Ion
An atom or molecule that has acquired a charge by either gaining or losing electrons.

IPCC
Intergovernmental Panel on Climate Change

JI
Joint Implementation: under the Kyoto Protocol, it allows a Party with a GHG emission target to receive credits from other Annex 1 Parties.

Kyoto Protocol
Protocol to the United Nations Framework Convention on Climate Change, which was adopted at Kyoto on 11 December 1997.

Leach
To dissolve a substance from a solid.

Leakage
In respect of carbon trading, the change of anthropogenic emissions by sources or removals by sinks which occurs outside the project boundary.

Leakage
In respect of carbon storage, the escape of injected fluid from storage.

Levellized cost
The future values of an input or product that would make the NPV (q.v.) of a project equal to zero.

LHV
Lower heating value: energy released from the combustion of a fuel that excludes the latent heat of water.

Lignite/sub-bituminous coal
Relatively young coal of low rank with a relatively high hydrogen and oxygen content.

Limestone
A sedimentary rock made mostly of the mineral calcite (calcium carbonate), usually formed from shells of dead organisms.

LNG
Liquefied natural gas

Lithology
Science of the nature and composition of rocks

Lithosphere
The outer layer of the Earth, made of solid rock, which includes the crust and uppermost mantle up to 100 km thick.

Log
Records taken during or after the drilling of a well.

London Convention
On the Prevention of Marine Pollution by Dumping of Wastes and Other Matter, which was adopted at London, Mexico City, Moscow and Washington on 29 December 1972.

London Protocol
Protocol to the Convention adopted in London on 2 November 1996 but which had not entered into force at the time of writing.

Low-carbon energy carrier
Fuel that provides low fuel-cycle-wide emissions of CO_2, such as methanol.

Macro-invertebrate
Small creature living in the seabed and subsoil, like earthworms, snails and beetles.

Madrid Protocol
A protocol to the 11th Antarctic Treaty to provide for Antarctica's environmental protection.

Mafic
Term used for silicate minerals, magmas, and rocks, which are relatively high in the heavier elements.

Magmatic activity
The flow of magma (lava).

Marginal cost
Additional cost that arises from the expansion of activity. For example, emission reduction by one additional unit.

Maturation
The geological process of changing with time. For example, the alteration of peat into lignite, then into sub-bituminous and bituminous coal, and then into anthracite.

Mature sedimentary basins
Geological provinces formed by the deposition of particulate matter under water when the deposits have matured into hydrocarbon reserves.

MEA
Mono-ethanolamine

Medium-gravity oil
Oil with a density of between about 850 and 925kg/m^3 (between 20 and 30 API).

Membrane
A sheet or block of material that selectively separates the components of a fluid mixture.

Metamorphic
Of rocks that have been altered by heat or pressure.

Mica
Class of silicate minerals with internal plate structure.

Microseismicity
Small-scale seismic tremors.

Migration
The movement of fluids in reservoir rocks.

Mineral trap
A geological structure in which fluids are retained by the reaction of the fluid to form a stable mineral.

Miscible displacement
Injection process that introduces miscible gases into the reservoir, thereby maintaining reservoir pressure and improving oil displacement.

Mitigation
The process of reducing the impact of any failure.

Monitoring
The process of measuring the quantity of carbon dioxide stored and its location.

Monte Carlo
A modelling technique in which the statistical properties of outcomes are tested by random inputs.

Mudstone
A very fine-grained sedimentary rock formed from mud.

MWh
Megawatt-hour

National Greenhouse Gas Inventory
An inventory of anthropogenic emissions by sources and removals by sinks of greenhouse gases prepared by Parties to the UNFCCC.

Natural analogue
A natural occurrence that mirrors in most essential elements an intended or actual human activity.

Natural underground trap
A geological structure in which fluids are retained by natural processes.

Navier-Stokes equations
The general equations describing the flow of fluids.

Near-field
The region close to a signal source.

NGCC
Natural gas combined cycle: natural-gas-fired power plant with gas and steam turbines.

Non-hazardous waste
Non-harmful substances that have been released or discarded into the environment.

NPV
Net present value: the value of future cash flows discounted to the present at a defined rate of interest.

Numerical approximation
Representation of physico-mathematical laws through linear approximations.

Observation well
A well installed to permit the observation of subsurface conditions.

OECD
Organization for Economic Co-operation and Development

OSPAR
Convention for the Protection of the Marine Environment of the North-East Atlantic, which was adopted at Paris on 22 September 1992.

Outcrop
The point at which a particular stratum reaches the earth's surface.

Overburden
Rocks and sediments above any particular stratum.

Overpressure
Pressure created in a reservoir that exceeds the pressure inherent at the reservoir's depth.

Oxidation
The loss of one or more electrons by an atom, molecule, or ion.

Oxyfuel combustion
Combustion of a fuel with pure oxygen or a mixture of oxygen, water and carbon dioxide.

Packer
A device for sealing off a section of a borehole or part of a borehole.

Partial oxidation
The oxidation of a carbon-containing fuel under conditions that produce a large fraction of CO and hydrogen.

Partial pressure
The pressure that would be exerted by a particular gas in a mixture of gases if the other gases were not present.

pCO_2
The partial pressure (q.v.) of CO_2.

PC
Pulverized coal: usually used in connection with boilers fed with finely ground coal.

Pejus level
The level in the ocean below which the functioning of animals deteriorates significantly.

Pelagic
Relating to, or occurring, or living in, or frequenting, the open ocean.

Perfluorocarbon
Synthetically produced halocarbons containing only carbon and fluorine atoms. They are characterized by extreme stability, non-flammability, low toxicity and high global warming potential.

Permeability
Ability to flow or transmit fluids through a porous solid such as rock.

Permian
A geological age between 290 and 248 million years ago.

Phytotoxic
Poisonous to plants.

Piezo-electric transducer
Crystals or films that are able to convert mechanical energy in electrical energy or vice-versa.

Pig
A device that is driven down pipelines to inspect and/or clean them.

Point source
An emission source that is confined to a single small location

Polygeneration
Production of more than one form of energy, for example synthetic liquid fuels plus electricity.

Pore space
Space between rock or sediment grains that can contain fluids.

Poroelastic
Elastic behaviour of porous media.

Porosity
Measure for the amount of pore space in a rock.

Post-combustion capture
The capture of carbon dioxide after combustion.

POX
Partial oxidation (q.v.)

Pre-combustion capture
The capture of carbon dioxide following the processing of the fuel before combustion.

Primary legal source
Legal source not depending on authority given by others.

Probability density function
Function that describes the probability for a series of parameter values.

Prospectivity
A qualitative assessment of the likelihood that a suitable storage location is present in a given area based on the available information

Proven reserve
For oil declared by operator to be economical; for gas about which a decision has been taken to proceed with development and production; see Resource.

Province
An area with separate but similar geological formations.

PSA
Pressure swing adsorption: a method of separating gases using the physical adsorption of one gas at high pressure and releasing it at low pressure.

Rank
Quality criterion for coal.

Reduction
The gain of one or more electrons by an atom, molecule, or ion

Reduction commitment
A commitment by a Party to the Kyoto Protocol to meet its quantified emission limit.

Reforestation
Planting of forests on lands that have previously contained forests but that have been converted to some other use.

Regional scale
A geological feature that crosses an entire basin.

Remediation
The process of correcting any source of failure.

Renewables
Energy sources that are inherently renewable such as solar energy, hydropower, wind, and biomass.

Rep. Value
Representative value

Reproductive dysfunction
Inability to reproduce.

Reserve
A resource (q.v.) from which it is generally economic to produce valuable minerals or hydrocarbons.

Reservoir
A subsurface body of rock with sufficient porosity and permeability to store and transmit fluids.

Residual saturation
The fraction of the injected CO_2 that is trapped in pores by capillary forces.

Resource
A body of a potentially valuable mineral or hydrocarbon.

Retrofit
A modification of the existing equipment to upgrade and incorporate changes after installation.

Risk assessment
Part of a risk-management system.

Root anoxia
Lack, or deficiency, of oxygen in root zone.

Root zone
Part of the soil in which plants have their roots.

Safe Drinking Water Act
An Act of the US Congress originally passed in 1974. It regulates, among other things, the possible contamination of underground water.

Saline formation
Sediment or rock body containing brackish water or brine.

Saline groundwater
Groundwater in which salts are dissolved.

Sandstone
Sand that has turned into a rock due to geological processes.

Saturated zone
Part of the subsurface that is totally saturated with groundwater.

Scenario
A plausible description of the future based on an internally consistent set of assumptions about key relationships and driving forces. Note that scenarios are neither predictions nor forecasts.

SCR
Selective catalytic reduction

Scrubber
A gas-liquid contacting device for the purification of gases or capture of a gaseous component.

Seabed
Borderline between the free water and the top of the bottom sediment.

Seal
An impermeable rock that forms a barrier above and around a reservoir such that fluids are held in the reservoir.

Secondary recovery
Recovery of oil by artificial means, after natural production mechanisms like overpressure have ceased.

Sedimentary basin
Natural large-scale depression in the earth's surface that is filled with sediments.

Seismic profile
A two-dimensional seismic image of the subsurface.

Seismic technique
Measurement of the properties of rocks by the speed of sound waves generated artificially or naturally.

Seismicity
The episodic occurrence of natural or man-induced earthquakes.

Selexol
A commercial physical absorption process to remove CO_2 using glycol dimethylethers.

Shale
Clay that has changed into a rock due to geological processes.

Shift convertor
A reactor in which the water-gas shift reaction, $CO + H_2O = CO_2 + H_2$ takes place.

Simplex orifice fitting
An apparatus for measuring the flow rate of gases or liquids.

Sink
The natural uptake of CO_2 from the atmosphere, typically in soils, forests or the oceans.

SMR
Steam methane reforming: a catalytic process in which methane reacts with steam to produce a mixture of H_2, CO and CO_2.

SNG
Synthetic natural gas: fuel gas with a high concentration of methane produced from coal or heavy hydrocarbons.

SOFC
Solid oxide fuel cell: a fuel cell (q.v.) in which the electrolyte is a solid ceramic composed of calcium- or yttrium-stabilized zirconium oxides.

Soil gas
Gas contained in the space between soil grains

Solubility trapping
A process in which fluids are retained by dissolution in liquids naturally present.

Sour gas
Natural gas containing significant quantities of acid gases like H_2S and CO_2.

Source
Any process, activity or mechanism that releases a greenhouse gas, an aerosol, or a precursor thereof into the atmosphere.

Speciation
The determination of the number of species into which a single species will divide over time.

Spill point
The structurally lowest point in a structural trap (q.v.) that can retain fluids lighter than background fluids.

Spoil pile
Heap of waste material derived from mining or processing operations.

SRES
Special Report on Emissions Scenarios; used as a basis for the climate projections in the TAR (q.v.).

Stabilization
Relating to the stabilization atmospheric concentrations of greenhouse gases.

Stable geological formation
A formation (q.v.) that has not recently been disturbed by tectonic movement.

Steam reforming
A catalytic process in which a hydrocarbon is reacted with steam to produce a mixture of H_2, CO and CO_2.

Storage
A process for retaining captured CO_2 so that it does not reach the atmosphere.

Strain gauge
Gauge to determine the deformation of an object subjected to stress.

Stratigraphic
The order and relative position of strata.

Stratigraphic column
A column showing the sequence of different strata.

Stratigraphic trap
A sealed geological container capable of retaining fluids, formed by changes in rock type, structure or facies.

Stimulation
The enhancement of the ability to inject fluids into, or recover fluids from, a well.

Stripper
A gas-liquid contacting device, in which a component is transferred from liquid phase to the gas phase.

Structural trap
Geological structure capable of retaining hydrocarbons, sealed structurally by a fault or fold.

Structure
Geological feature produced by the deformation of the Earth's crust, such as a fold or a fault; a feature within a rock such as a fracture; or, more generally, the spatial arrangement of rocks.

Structure contour map
Map showing the contours of geological structures.

Subsoil
Term used in London and OSPAR conventions, meaning the sediments below the seabed.

Sub-bituminous coal
Coal of a rank between lignite (q.v.) and bituminous (q.v.) coal.

Sustainable
Of development, that which is sustainable in ecological, social and economic areas.

Supercritical
At a temperature and pressure above the critical temperature and pressure of the substance concerned. The critical point represents the highest temperature and pressure at which the substance can exist as a vapour and liquid in equilibrium

Syngas
Synthesis gas (q.v.)

Synthesis gas
A gas mixture containing a suitable proportion of CO and H_2 for the synthesis of organic compounds or combustion.

Synfuel
Fuel, typically liquid fuel, produced by processing fossil fuel.

Tail gas
Effluent gas at the end of a process.

Tailing
The waste resulting from the extraction of value from ore.

TAR
Third Assessment Report of the Intergovernmental Panel on Climate Change

TCR
Total capital requirement

Technical Potential
The amount by which it is possible to reduce greenhouse gas emissions by implementing a technology or practice that has reached the demonstration phase.

Tectonically active area
Area of the Earth where deformation is presently causing structural changes.

Tertiary
Geological age about 65 to 2 million years ago.

Tertiary recovery
Oil generated by a third method; the first is by pressure release or depletion, and the second by oil driven out by the injection of water.

Thermocline
The ocean phenomenon characterized by a sharp change in temperature with depth.

Thermohaline
The vertical overturning of water masses due to seasonal heating, evaporation, and cooling.

Top-down model
A model based on applying macro-economic theory and econometric techniques to historical data about consumption, prices, etc.

Toxemia
Poisoning, usually of the blood.

Toxicology
Scientific study of poisons and their effects.

Tracer
A chemical compound or isotope added in small quantities to trace flow patterns.

Transaction cost
The full cost of transferring property or rights between parties.

Trap
A geological structure that physically retains fluids that are lighter than the background fluids, e.g. an inverted cup.

Ultramafic rocks
An igneous rock consisting almost entirely of iron- and magnesium-rich minerals with a silica content typically less than 45%.

UNCLOS
United Nations Convention on the Law of the Sea, which was adopted at Montego Bay on 10 December 1982.

Unconformity
A geological surface separating older from younger rocks and representing a gap in the geological record.

Under-saturated
A solution that could contain more solute than is presently dissolved in it.

UNFCCC
United Nations Framework Convention on Climate Change, which was adopted at New York on 9 May 1992.

Unminable
Extremely unlikely to be mined under current or foreseeable economic conditions

Updip
Inclining upwards following a structural contour of strata.

Upper ocean
The ocean above 1000m depth.

Vacuum residue
The heavy hydrocarbon mixture that is produced at the bottom of vacuum distillation columns in oil refineries.

Vadose zone
Region from the water table to the ground surface, also called the unsaturated zone because it is partially water-saturated.

Validation
In the context of CDM (q.v.), the process of the independent evaluation of a project by a designated operational entity on the basis of set requirements.

Ventilation
The exchange of gases dissolved in sea-water with the atmosphere, or gas exchange between an animal and the environment.

Verification
The proving, to a standard still to be decided, of the results of monitoring (q.v.). In the context of CDM, the independent review by a designated operational entity of monitored reductions in anthropogenic emissions.

Viscous fingering
Flow phenomenon arising from the flow of two largely immiscible fluids through a porous medium.

Well
Manmade hole drilled into the earth to produce liquids or gases, or to allow the injection of fluids.

Well with multiple completions
Well drilled with multiple branching holes and more than one hole being made ready for use.

Well-bore annulus
The annulus between the rock and the well casing.

Wellhead pressure
Pressure developed on surface at the top of the well.

Wettability
Surface with properties allowing water to contact the surface intimately.

Zero-carbon energy carrier
Carbon-free energy carrier, typically electricity or hydrogen.

Annex III

Units

Table AIII.1 Basic SI units

Physical Quantity	Unit	
	Name	**Symbol**
Length	meter	m
Mass	kilogram	kg
Time	second	s
Thermodynamic temperature	kelvin	K
Amount of substance	mole	mol

Table AIII.2 Multiplication factors

Multiple	Prefix	Symbol	Multiple	Prefix	Symbol
10^{-1}	deci	d	10	deca	da
10^{-2}	centi	c	10^2	hecto	h
10^{-3}	milli	m	10^3	kilo	k
10^{-6}	micro	μ	10^6	mega	M
10^{-9}	nano	n	10^9	giga	G
10^{-12}	pico	p	10^{12}	tera	T
10^{-15}	femto	f	10^{15}	peta	P

Table AIII.3 Special names and symbols for certain SI-derived units

Physical Quantity	Unit		
	Name	**Symbol**	**Definition**
Force	newton	N	$kg\ m\ s^{-2}$
Pressure	pascal	Pa	$kg\ m^{-1}\ s^{-2}\ (= N\ m^{-2})$
Energy	joule	J	$kg\ m^2\ s^{-2}$
Power	watt	W	$kg\ m^2\ s^{-3}\ (= J\ s^{-1})$
Frequency	hertz	Hz	s^{-1} (cycles per second)

Table AIII.4 Decimal fractions and multiples of SI units having special names

Physical quantity	Unit		
	Name	**Symbol**	**Definition**
Length	micron	μm	10^{-6} m
Area	hectare	ha	10^4 m^2
Volume	litre	L	10^{-3} m^3
Pressure	bar	bar	10^5 N m^{-2} = 10^5 Pa
Pressure	millibar	mb	10^2 N m^{-2} = 1 hPa
Mass	tonne	t	10^3 kg
Mass	gram	g	10^{-3} kg

Table AIII.5 Other units

Symbol	Description
°C	Degree Celsius (0°C = 273 K approximately) Temperature differences are also given in °C (= K) rather than the more correct form of 'Celsius degrees'
D	Darcy, unit for permeability, 10^{-12} m^2
ppm	Parts per million (10^6), mixing ratio (μmol mol^{-1})
ppb	Parts per billion (10^9), mixing ratio (nmol mol^{-1})
h	Hour
yr	Year
kWh	Kilowatt hour
MWh	Megawatt hour
MtCO$_2$	Megatonnes (1 Mt = 10^9 kg = 1 Tg) CO$_2$
GtCO$_2$	Gigatonnes (1 Gt = 10^{12} kg = 1 Pg) CO$_2$
tCO$_2$ MWh^{-1}	tonne CO$_2$ per megawatt hour
US\$ kWh^{-1}	US dollar per kilowatt hour

Annex IV

Authors and reviewers

AIV.1 Authors and Review Editors

Technical Summary

Co-ordinating Lead Authors

Edward Rubin	Carnegie Mellon University, United States
Leo Meyer	TSU IPCC Working Group III, Netherlands Environment Assessment Agency (MNP), Netherlands
Heleen de Coninck	TSU IPCC Working Group III, Energy research Centre of the Netherlands (ECN), Netherlands

Lead Authors

Juan Carlos Abanades	Instituto Nacional del Carbon (CSIC), Spain
Makoto Akai	National Institute of Advanced Industrial Science and Technology, Japan
Sally Benson	Lawrence Berkeley National Laboratory, United States
Ken Caldeira	Carnegie Institution of Washington, United States
Peter Cook	Cooperative Research Centre for Greenhouse Gas Technologies (CO2CRC), Australia
Ogunlade Davidson	Co-chair IPCC Working Group III, Faculty of Engineering, University of Sierra Leone, Sierra Leone
Richard Doctor	Argonne National Laboratory, United States
James Dooley	Battelle, United States
Paul Freund	United Kingdom
John Gale	IEA Greenhouse Gas R&D Programme, United Kingdom
Wolfgang Heidug	Shell International Exploration and Production B.V., Netherlands (Germany)
Howard Herzog	MIT, United States
David Keith	University of Calgary, Canada
Marco Mazzotti	ETH Swiss Federal Institute of Technology Zurich, Switzerland (Italy)
Bert Metz	Co-chair IPCC Working Group III, Netherlands Environment Assessment Agency (MNP), Netherlands
Balgis Osman-Elasha	Higher Council for Environment and Natural Resources, Sudan
Andrew Palmer	University of Cambridge, United Kingdom
Riitta Pipatti	Statistics Finland, Finland
Koen Smekens	Energy research Centre of the Netherlands (ECN), Netherlands (Belgium)
Mohammad Soltanieh	Environmental Research Centre, Dept. of Environment, Climate Change Office, Iran
Kelly (Kailai) Thambimuthu	Centre for Low Emission Technology, CSIRO, Australia (Australia and Canada)
Bob van der Zwaan	Energy research Centre of the Netherlands (ECN), The Netherlands

Review Editor
Ismail El Gizouli   IPCC WGIII vice-chair, Higher Council for Environment & Natural Resources, Sudan

Chapter 1: Introduction

Co-ordinating Lead Author
Paul Freund   United Kingdom

Lead Authors
Anthony Adegbulugbe   Centre of Energy Research and Development, Nigeria
Øyvind Christophersen   Norwegian Pollution Control Authority, Norway
Hisashi Ishitani   School of Media and Governance, Keio University, Japan
William Moomaw   Tufts University, The Fletcher School of Law and Diplomacy, United States
Jose Moreira   University of Sao Paulo, National Reference Center on Biomass (CENBIO), Brazil

Review Editors
Eduardo Calvo   IPCC vice-chair WGIII, Peru
Eberhard Jochem   Fraunhofer Institut/ETH Zürich, Germany/Switzerland (Germany)

Chapter 2: Sources of CO_2

Co-ordinating Lead Author
John Gale   IEA Greenhouse Gas R&D Programme, United Kingdom

Lead Authors
John Bradshaw   Geoscience Australia, Australia
Zhenlin Chen   China Meteorological Administration, China
Amit Garg   Ministry of Railways, India
Dario Gomez   Comision Nacional de Energia Atomica (CNEA), Argentina
Hans-Holger Rogner   International Atomic Energy Agency (IAEA), Austria (Germany)
Dale Simbeck   SFA Pacific Inc., United States
Robert Williams   Center for Energy & Environmental Studies, Princeton University, United States

Contributing Authors
Ferenc Toth   International Atomic Energy Agency (IAEA), Austria
Detlef van Vuuren   Netherlands Environment Assessment Agency (MNP), Netherlands

Review Editors
Ismail El Gizouli   IPCC WGIII vice-chair, Sudan
Jürgen Friedrich Hake   Forschungszentrum Jülich, Germany

Chapter 3: Capture

Co-ordinating Lead Authors
Juan Carlos Abanades   Instituto Nacional del Carbon (CSIC), Spain
Mohammad Soltanieh   Environmental Research Centre, Dept. of Environment, Climate Change Office, Iran
Kelly (Kailai) Thambimuthu   Centre for Low Emission Technology, CSIRO, Australia (Australia and Canada)

Lead Authors
Rodney Allam   Air Products PLC, United Kingdom
Olav Bolland   Norwegian University of Science and Technology, Norway
John Davison   IEA Greenhouse Gas R&D Programme, United Kingdom
Paul Feron   TNO Science and Industry, Netherlands
Fred Goede   SHE Centre, Sasol Ltd, South Africa
Alice Herrera   Industrial Technology Development Institute, Department of Science and Technology,
                                                   Philippines

Masaki Iijima	Mitsubishi Heavy Industries, Japan
Daniël Jansen	Energy research Centre of the Netherlands (ECN), Netherlands
Iosif Leites	State Institute for Nitrogen Industry, Russian Federation
Philippe Mathieu	University of Liege, Belgium
Edward Rubin	Carnegie Mellon University, United States
	(Crosscutting Chair Energy Requirements)
Dale Simbeck	SFA Pacific Inc., United States
Krzysztof Warmuzinski	Polish Academy of Sciences, Institute of Chemical Engineering, Poland
Michael Wilkinson	BP Exploration, United Kingdom
Robert Williams	Center for Energy & Environmental Studies, Princeton University, United States

Contributing Authors

Manfred Jaschik	Polish Academy of Sciences, Institute of Chemical Engineering, Poland
Anders Lyngfelt	Chalmers University of Technology, Sweden
Roland Span	Institute for Thermodynamics & Energy Technologies, Germany
Marek Tanczyk	Polish Academy of Sciences, Institute of Chemical Engineering, Poland

Review Editors

Ziad Abu-Ghararah	IPCC WGIII vice-chair, Saudi Arabia
Tatsuaki Yashima	Nihon University, Advanced Research Institute for the Sciences and Humanities, Japan

Chapter 4: Transport of CO_2

Co-ordinating Lead Authors

Richard Doctor	Argonne National Laboratory, Hydrogen and Greenhouse Gas Engineering, United States
Andrew Palmer	University of Cambridge, United Kingdom

Lead Authors

David Coleman	Kinder Morgan, United Kingdom
John Davison	IEA Greenhouse Gas R&D Programme, United Kingdom
Chris Hendriks	Ecofys, Netherlands
Olav Kaarstad	Statoil ASA, Industry and Commercialisation, Norway
Masahiko Ozaki	Nagasaki R & D Centre, Mitsubishi Heavy Industries, Ltd., Japan

Contributing Author

Michael Austell	Kinder Morgan, United Kingdom

Review Editors

Ramon Pichs-Madruga	Centro de Investigaciones de Economia Mundial (CIEM), IPCC WGIII vice-chair, Cuba
Svyatoslav Timashev	Science and Engineering Center, Ural Branch, Russian Academy of Sciences, Russian Federation

Chapter 5: Underground geological storage

Co-ordinating Lead Authors

Sally Benson	Lawrence Berkeley National Laboratory, United States
Peter Cook	Cooperative Research Centre for Greenhouse Gas Technologies (CO2CRC), Australia

Lead Authors

Jason Anderson	Institute for European Environmental Policy (IEEP), Belgium (United States)
Stefan Bachu	Alberta Energy and Utilities Board, Canada
Hassan Bashir Nimir	University of Khartoum, Sudan
Biswajit Basu	Oil and Natural Gas Corporation Ltd., India
John Bradshaw	Geoscience Australia, Australia
Gota Deguchi	Japan Coal Energy Centre, Japan

John Gale	IEA Greenhouse Gas R&D Programme, United Kingdom
Gabriela von Goerne	Greenpeace, Germany
Bill Gunter	Alberta Research Council, Canada
Wolfgang Heidug	Shell International Exploration and Production B.V., Netherlands (Germany)
Sam Holloway	British Geological Survey, United Kingdom
Rami Kamal	Saudi Aramco, Saudi Arabia
David Keith	University of Calgary, Canada
Philip Lloyd	Energy Research Centre, University of Cape Town, South Africa
Paulo Rocha	Petrobras - Petroleo Brasiliero S.A., Brazil
Bill Senior	DEFRA, United Kingdom
Jolyon Thomson	Defra Legal Services, International Environmental Law, United Kingdom
Tore Torp	Statoil R&D Centre, Norway
Ton Wildenborg	TNO Built Environment and Geosciences, Netherlands
Malcolm Wilson	University of Regina, Canada
Francesco Zarlenga	ENEA-Cr. Casaccia PROT-PREV, Italy
Di Zhou	South China Sea Institute of Oceanology, Chinese Academy of Sciences, China

Contributing Authors

Michael Celia	Princeton University, United States
Jonathan Ennis King	Commonwealth Scientific and Industrial Research Organisation (CSIRO), Australia
Erik Lindeberg	SINTEF Petroleum Research, Norway
Salvatore Lombardi	University of Rome "La Sapienza", Laboratory of Fluid Chemistry, Italy
Curt Oldenburg	Lawrence Berkeley National Laboratory, United States
Karsten Pruess	Lawrence Berkeley National Laboratory, United States
Andy Rigg	CSIRO Petroleum Resources, Australia
Scott Stevens	Advanced Resources International, United States
Elizabeth Wilson	Office of Research and Development / US EPA, United States
Steve Whittaker	Saskatchewan Industry & Resources, Canada

Review Editors

Günther Borm	GeoForschungsZentrum Potsdam, Germany
David G. Hawkins	Natural Resources Defense Council, United States
Arthur Lee	Chevron Corporation, United States

Chapter 6: Ocean storage

Co-ordinating Lead Authors

Ken Caldeira	Carnegie Institution of Washington, United States
Makoto Akai	National Institute of Advanced Industrial Science and Technology, Japan

Lead Authors

Peter Brewer	Monterey Bay Aquarium Research Institute, United States
Baixin Chen	National Institute of Advanced Industrial Science and Technology (AIST), Japan (China)
Peter Haugan	Geophysical Institute, University of Bergen, Norway
Toru Iwama	Seinan Gakuin University, Faculty of Law, Japan
Paul Johnston	Greenpeace Research Laboratories, United Kingdom
Haroon Kheshgi	ExxonMobil Research & Engineering Company, United States
Qingquan Li	National Climate Centre, China Meteorological Administration, China
Takashi Ohsumi	Research Institute of Innovative Technology for the Earth, Japan
Hans Pörtner	Alfred-Wegener-Institute for Polar and Marine Research, Marine Animal Ecophysiology, Germany
Christopher Sabine	Global Carbon Programme; NOAA/PMEL, United States
Yoshihisa Shirayama	Seto Marine Biological Laboratory, Kyoto University, Japan
Jolyon Thomson	Defra Legal Services, International Environmental Law, United Kingdom

Contributing Authors

Jim Barry	Monterey Bay Aquarium Research Institute, United States
Lara Hansen	WWF, United States

Review Editors

Brad De Young	Memorial University of Newfoundland, Canada
Fortunat Joos	University of Bern, Switzerland

Chapter 7: Mineral carbonation and industrial uses of carbon dioxide

Co-ordinating Lead Author

Marco Mazzotti	ETH Swiss Federal Institute of Technology Zurich, Switzerland (Italy)

Lead Authors

Juan Carlos Abanades	Instituto Nacional del Carbon (CSIC), Spain
Rodney Allam	Air Products PLC, United Kingdom
Klaus S. Lackner	School of Engineering and Applied Sciences, Columbia University, United States
Francis Meunier	CNAM-IFFI, France
Edward Rubin	Carnegie Mellon University, United States
Juan Carlos Sánchez M.	Environmental consultant, Venezuela
Katsunori Yogo	Research Institute of Innovative Technology for the Earth (RITE), Japan
Ron Zevenhoven	Helsinki University of Technology, Finland (The Netherlands)

Review Editors

Baldur Eliasson	Eliasson & Associates, Switzerland
R.T.M. Sutamihardja	The Office of the State Minister for Environment Republic of Indonesia, Indonesia

Chapter 8: Costs and economic potential

Co-ordinating Lead Authors

Howard Herzog	MIT, United States
Koen Smekens	Energy research Centre of the Netherlands (ECN), Netherlands (Belgium)

Lead Authors

Pradeep Dadhich	The Energy Research Institute, India
James Dooley	Battelle, United States
Yasumasa Fujii	School of Frontier Sciences, University of Tokyo, Japan
Olav Hohmeyer	University of Flensburg, Germany
Keywan Riahi	International Institute for Applied Systems Analysis (IIASA), Austria

Contributing Authors

Makoto Akai	National Institute of Advanced Industrial Science and Technology, Japan
Chris Hendriks	Ecofys, Netherlands
Klaus Lackner	School of Engineering and Applied Sciences, Columbia University, United States
Ashish Rana	National Institute for Environmental Studies, India
Edward Rubin	Carnegie Mellon University, United States
Leo Schrattenholzer	International Institute for Applied Systems Analysis (IIASA), Austria
Bill Senior	DEFRA, United Kingdom

Review Editors

John Christensen	UNEP Collaborating Centre on Energy and Environment (UCCEE), Denmark
Greg Tosen	Eskom Resources and Strategy, South Africa

Chapter 9: Implications of carbon dioxide capture and storage for greenhouse gas inventories and accounting

Co-ordinating Lead Authors
Balgis Osman-Elasha   Higher Council for Environment and Natural Resources, Sudan
Riitta Pipatti   Statistics Finland, Finland

Lead Authors
William Kojo Agyemang-Bonsu   Environmental Protection Agency, Ghana
A.M. Al-Ibrahim   King Abdulaziz City for Science & Technology (KACST), Saudi Arabia
Carlos López   Institute of Meteorology, Cuba
Gregg Marland   Oak Ridge National Laboratory, United States
Huang Shenchu   China Coal Information Institute, China
Oleg Tailakov   International Coal and Methane Research Centre - UGLEMETAN, Russian Federation

Review Editors
Takahiko Hiraishi   Institute for Global Environmental Strategies, Japan
José Domingos Miguez   Ministry of Science and Technology, Brazil

Annex I: Properties of CO_2 and carbon-based fuels

Co-ordinating Lead Author
Paul Freund   United Kingdom

Lead Authors
Stefan Bachu   Alberta Energy and Utilities Board, Canada
Dale Simbeck   SFA Pacific Inc., United States
Kelly (Kailai) Thambimuthu   Centre for Low Emission Technology, CSIRO, Australia (Australia and Canada)

Contributing Author
Murlidhar Gupta   CANMET Energy Technology Centre, Natural Resources Canada (India)

Annex II: Glossary, acronyms and abbreviations

Co-ordinating Lead Author
Philip Lloyd   Energy Research Institute, University of Capetown, South Africa

Lead Authors
Peter Brewer   Monterey Bay Aquarium Research Institute, United States
Chris Hendriks   Ecofys, Netherlands
Yasumasa Fujii   School of Frontier Sciences, University of Tokyo, Japan
John Gale   IEA Greenhouse Gas R&D Programme, United Kingdom
Balgis Osman Elasha   Higher Council for Environment and Natural Resources, Sudan
Jose Moreira   University of Sao Paulo, Biomass Users Network (BUN), Brazil
Juan Carlos Sánchez M.   Environmental consultant, Venezuela
Mohammad Soltanieh   Environmental Research Centre, Dept. of Environment, Climate Change Office, Iran
Tore Torp   Statoil R&D Centre, Corporate Strategic Technology, Norway
Ton Wildenborg   TNO Built Environment and Geosciences, Netherlands

Contributing Authors
Jason Anderson   Institute for European Environmental Policy (IEEP), Belgium (United States)
Stefan Bachu   Alberta Energy and Utilities Board, Canada
Sally Benson   Ernest Orlando Lawrence Berkeley National Laboratory, United States
Ken Caldeira   Carnegie Institution of Washington, United States
Peter Cook   Cooperative Research Centre for Greenhouse Gas Technologies (CO2CRC), Australia
Richard Doctor   Argonne National Laboratory, Hydrogen and Greenhouse Gas Engineering, United States

Paul Freund   United Kingdom
Gabriela von Goerne   Greenpeace, Germany

AIV.2 Crosscutting Chairs

Crosscutting group	*Chair*
Costs	Howard Herzog, MIT, United States
Energy requirements	Ed Rubin, Carnegie Mellon University, United States
Legal issues and environmental impacts	Wolfgang Heidug, Shell International Exploration and Production B.V., Netherlands (Germany)
Public perception and risks	David Keith, University of Calgary, Canada
Technical and economic potential	James Dooley, Battelle, United States

AIV.3 Expert Reviewers

Argentina
Charles Balnaves   BHP Petroleum Pty Ltd
Gustavo Galliano   Repsol YPF, Argentina Technology Centre
Héctor Ginzo   Ministry of Foreign Affairs, International Trade and Worship
Martiros Tsarukyan   Department of Atmosphere protection, Ministry of Nature Protection

Australia
Barry Hooper   CO2CRC, Department of Chemical and Biomolecular Engineering
Bill Koppe   Anglo Coal Australia
Brian Evans   Curtin University
Iain MacGill   School of Engineering & Telecommunications
John Torkington   Chevron Australia Pty Ltd
Jonathan Ennis-King   CSIRO Petroleum
Lincoln Paterson   CSIRO
Peter McNally   Greenhouse & Climate Change Co-ordinator
Robert Durie   CSIRO, Division of Energy Technology
Tristy Fairfield   Conservation Council of Western Australia

Austria
Klaus Radunsky   Umweltbundesamt
Margit Kapfer   Denkstatt
Torsten Clemens   OMV E&P

Belgium
Aviel Verbruggen   Universiteit Antwerpen
Ben Laenen   VITO
Dolf Gielen   International Energy Agency
Kris Piessens   Geological Survey of Belgium

Benin
Sabin Guendéhou   Benin Centre of Scientific and Technical Research

Brazil
Paulo Antônio De Souza   Companhia Vale do Rio Doce, Department of Environmental and Territorial Management
Paulo Cunha   Petrobras

Bulgaria
Teodor Ivanov   Ministry of Environment and Water

Canada
Bob Stobbs   Canadian Clean Power Coalition
C.S. Wong   OSAP, Institute of Ocean Sciences
Carolyn Preston   Canmet Energy Technology Centre, Natural Resources
Chris Hawkes   University of Saskatchewan
Don Lawton   University of Calgary, Department of Geology and Geophysics
Steve Whittaker   Saskatchewan Industry and Resources

China
Chen-Tung Arthur Chen   National Sun Yat-Sen University
Xiaochun Li   RITE
Xu Huaqing   ERI
Zhang ChengYi   National Climate Center

Denmark
Flemming Ole Rasmussen   Danish Energy Authority, Ministry of Economic and Business Affairs
Kim Nissen   Elsam Kraft A/S
Kristian Thor Jakobsen   University of Copenhagen

Dominican Republic
Rene Ledesma   Ministry of Environment and Natural Resources

Egypt
Mohamed Ahmed Darwish   The Egyptian Meteorological Authority

Finland
Allan Johansson   Technical Research Centre of Finland
Ilkka Savolainen   VTT Processes, Emission Control - Greenhouse Gases

France
Isabelle Czernichowski   BRGM (French Geological Survey)
Jean-Xavier Morin   Alstom
Marc Gillet   Observatoire National sur les Effets du Réchauffement Climatique
Martin Pêcheux   Institut des Foraminifères Symbiotiques
Paul Broutin   IFP-Lyon
Rene Ducroux   Centre of Initiative and Research on Energy and the Environment
Yann Le Gallo   Institut Français du Pétrole

Germany
Axel Michaelowa   Hamburg Institute of International Economics
Franz May   Federal Institute for Geosciences and Natural Resources
Gert Müller-Syring   DBI Gas- und Umwelttechnik GmbH
Jochen Harnisch   Ecofys Gmbh
Jürgen Engelhardt   Forschung und Entwicklung Rheinbraun AG
Martina Jung   Hamburg Institute of International Economics
Matthias Duwe   Climate Action Network Europe
Peter Markewitz   Forschungszentrum Jülich GmbH
Sven Bode   Hamburg Institute of International Economics
Ulf Riebesell   AWI
Wilhelm Kuckshinrichs   Forschungszentrum Jülich GmbH

India
Auro Ashish Saha   Department of Mechanical Engineering, Pondicherry Engineering College
M.M. Kapshe   MANIT

Ireland

Pat Finnegan   Greenhouse Ireland Action Network

Israel

Martin Halmann   Weizmann Institute of Science

Italy

Fedora Quattrocchi   Institute of Geophysics and Volcanology
Umberto Desideri   University of Perugia

Japan

Atsushi Ishimatsu   Nagasaki University
Hitoshi Koide   Waseda University
Imai Nobuo   Mitsubishi Heavy Industries, Ltd.
Koh Harada   Institute for Environmental Management, Institute of Advanced Industrial, Science and
                                               Technology
Kozo Sato   Geosystem Engineering, The University of Tokyo
Mikiko Kainuma   NIES, National Institute for Environmental Studies
Shigeo Murai   Research Institute of Innovative Technology for the Earth
Takahisa Yokoyama   CS Promotion Office
Yoichi Kaya   The Japan Committee for IIASA, Research Institute of Innovative Technology for the
                                               Earth (RITE)

Mexico

Antonio Juarez Alvarado   Directore Energy development, Min of Energy
Aquileo Guzman   National Institute of Ecology
Jose Antonio Benjamin Ordóňez-Díaz   National Autonomous University of Mexico, PhD. Student
Julia Martinez   Instituto National Ecologica
Lourdes Villers-Ruiz   Andador Epigmenio Ibarra

The Netherlands

Annemarie van der Rest   Shell Nederland BV
Bert van der Meer   TNO Built Environment and Geosciences
Bob van der Zwaan   ECN Policy Studies
Dorothee Bakker   University of East Anglia, United Kingdom
Evert Wesker   Shell Global Solutions International, Dept. OGIR
Ipo Ritsema   TNO Built Environment and Geosciences - National Geological Survey
Jos Cozijnsen   Consulting Attorney Energy & Environment
Jos Keurentjes   Eindhoven University of Technology
Karl-Heinz Wolf   Delft University of Technology
Nick ten Asbroek   TNO Science and Industry
Rob Arts   TNO Built Environment and Geosciences
Suzanne Hurter   Shell International Exploration and Production
Wouter Huijgen   ECN Clean Fossil Fuels

New Zealand

David Darby   Institute of Geological and Nuclear Sciences
Peter Read   Massey University
Wayne Hennessy   CRL Energy Ltd

Nigeria

Christopher Ugwu   Nigeria Society for the Improvement of Rural People (NSIRP), University of Nigeria

Norway

Asbjørn Torvanger   Centre for International Climate and Environmental Research
Bjorn Kvamme   Institute of Physics, University of Bergen
Gelein de Koeijer   Statoil ASA

Guttorm Alendal   Bergen Centre for Computational Science
Hallvard Fjøsne Svendsen   NTNU, Chemical Engineering Department
Hans Aksel Haugen   Norsk Hydro ASA, Environment and Energy Consulting
Kristin Rypdal   CICERO
Lars Eide   Norsk Hydro
Lars Golmen   Norwegian Institute for Water Research (NIVA)
Martin Hovland   Statoil ASA
Todd Allyn Flach   DNV Research, Energy and Resources, Det Norske Veritas

Philippines
Flaviana Hilario   Climatology and Agrometeorology Branch

Russian Federation
Michael Gytarsky   Institute of Global Climate and Ecology

Saudi Arabia
Ahmad Al-Hazmi   Sabic

Spain
Pilar Coca   ELCOGAS
Angel Maria Gutierrez   NaturCorp Redes, s.a.

Sweden
Christian Bernstone   Vattenfall Utveckling AB
Erlström Anders   Geological Survey of Sweden
Jerry Jinyue Yan   Luleå University of Technology
Kenneth Möllersten   Swedish Energy Agency Climate Change Division
Marie Anheden   Vattenfall Utveckling AB

Switzerland
Daniel Spreng   ETH Swiss Federal Institute of Technology Zürich
Jose Romero   Swiss Agency for the Environment, Forests and Landscape

United Kingdom
Andy Chadwick   British Geological Survey, Geophysics & Marine Geoscience
David Reiner   University of Cambridge, Judge Institute of Management
John Shepherd   Southhampton Oceanography Centre
Jon Gibbins   Energy Technology for Sustainable Development Group
Nick Riley   Reservoir Geoscience, British Geological Survey
Paul Zakkour   ERM Energy & Climate Change
Raymond Purdy   Centre for Law and the Environment, Faculty of Laws, University College London
Sevket Durucan   Imperial College of Science Technology and Medicine, Royal School of Mines
Simon Eggleston   IPCC NGGIP Technical Support Unit
Stef Simons   University College London, Centre for CO_2 Technology
Stefano Brandani   Centre for CO_2 Technology
Stuart Haszeldine   University of Edinburgh, Grant Institute
William Wilson   Cambrensis Ltd.

United States
Anand Gnanadesikan   GFDL
Craig Smith   University of Hawaii at Monoa
David Archer   University of Chicago
Don Seeburger   IPIECA
Eric Adams   MIT
Franklin Orr   Stanford University
Granger Morgan   Carnegie Mellon University

Grant Bromhal	U.S. Department of Energy, National Energy Technology Laboratory (NETL)
Greg Rau	Institute of Marine Sciences
Hamid Sarv	Babcock & Wilcox Research Center
Jette Findsen	Science Applications International Corporation
Julio Friedman	University of Maryland
Karl Turekian	Yale University
Larry Myer	Lawrence Berkeley National Laboratory
Leonard Bernstein	L.S. Bernstein & Associates, L.L.C.
Neeraj Gupta	Battelle
Neville Holt	EPRI
R.J. Batterham	Rio Tinto
Richard Rhudy	EPRI
Robert Buruss	US Geological Survey
Robert Finley	Illinois State Geological Survey
Robert Fledderman	MeadWestvaco Forestry Division
Robert Socolow	Princeton Environmental Institute, Princeton University
Scott Imbus	IPIECA
Seymour Alpert	Retired
Steve Crookshank	Policy Analysis & Statistics, American Petroleum Institute
Steven Kleespie	Rio Tinto
Susan Rice	Susan A. Rice and Associates, Inc
Tom Marrero	University of Missouri
Vello Kuuskra	Advanced Resources International, Inc.
Veronica Brieno Rankin	Michigan Technological University

Annex V

List of major IPCC reports

LIST OF MAJOR IPCC REPORTS

Climate Change - The IPCC Scientific Assessment
The 1990 report of the IPCC Scientific Assessment Working Group

Climate Change - The IPCC Impacts Assessment
The 1990 report of the IPCC Impacts Assessment Working Group

Climate Change - The IPCC Response Strategies
The 1990 report of the IPCC Response Strategies Working Group

Emissions Scenarios
Prepared by the IPCC Response Strategies Working Group, 1990

Assessment of the Vulnerability of Coastal Areas to Sea Level Rise - A Common Methodology, 1991

Climate Change 1992 - The Supplementary Report to the IPCC Scientific Assessment
The 1992 report of the IPCC Scientific Assessment Working Group

Climate Change 1992 - The Supplementary Report to the IPCC Impacts Assessment
The 1992 report of the IPCC Impacts Assessment Working Group

Climate Change: The IPCC 1990 and 1992 Assessments
IPCC First Assessment Report Overview and Policymaker Summaries, and 1992 IPCC Supplement

Global Climate Change and the Rising Challenge of the Sea
Coastal Zone Management Subgroup of the IPCC Response Strategies Working Group, 1992

Report of the IPCC Country Study Workshop, 1992

Preliminary Guidelines for Assessing Impacts of Climate Change, 1992

IPCC Guidelines for National Greenhouse Gas Inventories (3 volumes), 1994

Climate Change 1994 - Radiative Forcing of Climate Change and An Evaluation of the IPCC IS92 Emission Scenarios

IPCC Technical Guidelines for Assessing Climate Change Impacts and Adaptations
1995

Climate Change 1995 - The Science of Climate Change
– Contribution of Working Group I to the Second Assessment Report

Climate Change 1995 - Scientific-Technical Analyses of Impacts, Adaptations and Mitigation of Climate Change
- Contribution of Working Group II to the Second Assessment Report

Climate Change 1995 - The Economic and Social Dimensions of Climate Change - Contribution of Working Group III to the Second Assessment Report

The IPCC Second Assessment Synthesis of Scientific-Technical Information Relevant to Interpreting Article 2 of the UN Framework Convention on Climate Change, 1995

Revised 1996 IPCC Guidelines for National Greenhouse Gas Inventories (3 volumes), 1996

Technologies, Policies and Measuares for Mitigating Climate Change - IPCC Technical Paper 1, 1996

An Introduction to Simple Climate Models Used in the IPCC Second Assessment Report - IPCC Technical Paper 2, 1997

Stabilisation of Atmospheric Greenhouse Gases: Physical, Biological and Socio-Economic Implications - IPCC Technical Paper 3, 1997

Implications of Proposed Co2 Emissions Limitations IPCC Technical Paper 4, 1997

The Regional Impacts of Climate Change: An Assessment of Vulnerability
IPCC Special Report, 1997

Aviation and the Global Atmosphere
IPCC Special Report, 1999

Methodological and Technological Issues in Technology Transfer
IPCC Special Report, 2000

Emissions Scenarios
IPCC Special Report, 2000

Land Use, Land Use Change and Forestry
IPCC Special Report, 2000

Good Practice Guidance and Uncertainty Management in National Greenhouse Gas Inventories
IPCC National Greenhouse Gas Inventories Programme, 2000

Climate Change and Biodiversity - IPCC Technical Paper V, 2002

Climate Change 2001: The Scientific Basis - Contribution of Working Group I to the Third Assessment Report

Climate Change 2001: Impacts, Adaptation & Vulnerability - Contribution of Working Group II to the Third Assessment Report

Climate Change 2001: Mitigation - Contribution of Working Group III to the Third Assessment Report

Climate Change 2001: Synthesis Report

Good Practice Guidance for Land Use, Land-use Change and Forestry
IPCC National Greenhouse Gas Inventories Programme, 2003

Safeguarding the Ozone Layer and the Global Climate System: Issues Related to Hydrofluorocarbons and Perfluorocarbons
IPCC/TEAP Special Report, 2005